Lecture Notes in Computer Scienc

Commenced Publication in 1973
Founding and Former Series Editors:
Gerhard Goos, Juris Hartmanis, and Jan van Leeuwen

Kazue Sako   Palash Sarkar (Eds.)

Advances in Cryptology – ASIACRYPT 2013

19th International Conference on the Theory
and Application of Cryptology and Information Security
Bengaluru, India, December 1-5, 2013
Proceedings, Part I

Springer

Volume Editors

Kazue Sako
NEC Corporation
Kawasaki, Japan
E-mail: k-sako@ab.jp.nec.com

Palash Sarkar
Indian Statistical Institute
Kolkata, India
E-mail: palash@isical.ac.in

ISSN 0302-9743 e-ISSN 1611-3349
ISBN 978-3-642-42032-0 e-ISBN 978-3-642-42033-7
DOI 10.1007/978-3-642-42033-7
Springer Heidelberg New York Dordrecht London

CR Subject Classification (1998): E.3, D.4.6, F.2, K.6.5, G.2, I.1, J.1

LNCS Sublibrary: SL 4 – Security and Cryptology

Typesetting: Camera-ready by author, data conversion by Scientific Publishing Services, Chennai, India

Printed on acid-free paper

Springer is part of Springer Science+Business Media (www.springer.com)

Preface

It is our great pleasure to present the proceedings of Asiacrypt 2013 in two volumes of *Lecture Notes in Computer Science* published by Springer. This was the 19th edition of the International Conference on Theory and Application of Cryptology and Information Security held annually in Asia by the International Association for Cryptologic Research (IACR). The conference was organized by IACR in cooperation with the Cryptology Research Society of India and was held in the city of Bengaluru in India during December 1–5, 2013.

About one year prior to the conference, an international Program Committee (PC) of 46 scientists assumed the responsibility of determining the scientific content of the conference. The conference evoked an enthusiastic response from researchers and scientists. A total of 269 papers were submitted for possible presentations approximately six months before the conference. Authors of the submitted papers are spread all over the world. PC members were allowed to submit papers, but each PC member could submit at most two co-authored papers or at most one single-authored paper. The PC co-chairs did not submit any paper. All the submissions were screened by the PC and 54 papers were finally selected for presentations at the conference. These proceedings contain the revised versions of the papers that were selected. The revisions were not checked and the responsibility of the papers rests with the authors and not the PC members.

Selection of papers for presentation was made through a double-blind review process. Each paper was assigned three reviewers and submissions by PC members were assigned six reviewers. Apart from the PC members, 291 external reviewers were involved. The total number of reviews for all the papers was more than 900. In addition to the reviews, the selection process involved an extensive discussion phase. This phase allowed PC members to express opinion on all the submissions. The final selection of 54 papers was the result of this extensive and rigorous selection procedure. One of the final papers resulted from the merging of two submissions.

The best paper award was conferred upon the paper "Shorter Quasi-Adaptive NIZK Proofs for Linear Subspaces" authored by Charanjit Jutla and Arnab Roy. The decision was based on a vote among the PC members. In addition to the best paper, the authors of two other papers, namely, "Families of Fast Elliptic Curves from Q-Curves" authored by Benjamin Smith and "Key Recovery Attacks on 3-Round Even-Mansour, 8-Step LED-128, and Full AES^2" authored by Itai Dinur, Orr Dunkelman, Nathan Keller and Adi Shamir, were recommended by the Editor-in-Chief of the *Journal of Cryptology* to submit expanded versions to the journal.

A highlight of the conference was the invited talks. An extensive multi-round discussion was carried out by the PC to decide on the invited speakers. This resulted in very interesting talks on two different aspects of the subject. Lars Ramkilde Knudsen spoke on "Block Ciphers — Past and Present" a topic of classical and continuing importance, while George Danezis spoke on "Engineering Privacy-Friendly Computations," which is an important and a more modern theme.

Apart from the regular presentations and the invited talks, a rump session was organized on one of the evenings. This consisted of very short presentations on upcoming research results, announcements of future events, and other topics of interest to the audience.

We would like to thank the authors of all papers for submitting their research works to the conference. Such interest over the years has ensured that the Asiacrypt conference series remains a cherished venue of publication by scientists. Thanks are due to the PC members for their enthusiastic and continued participation for over a year in different aspects of selecting the technical program. External reviewers contributed by providing timely reviews and thanks are due to them. A list of external reviewers is provided in these proceedings. We have tried to ensure that the list is complete. Any omission is inadvertent and if there is an omission, we apologize to the person concerned.

Special thanks are due to Satyanarayana V. Lokam, the general chair of the conference. His message to the PC was to select the best possible scientific program without any other considerations. Further, he ensured that the PC co-chairs were insulated from the organizational work. This work was done by the Organizing Committee and they deserve thanks from all the participants for the wonderful experience. We thank Daniel J. Bernstein and Tanja Lange for expertly organizing and conducting the rump session.

The reviews and discussions were entirely carried out online using a software developed by Shai Halevi. At several times, we had to ask Shai for his help with some feature or the other of the software. Every time, we received immediate and helpful responses. We thank him for his support and also for developing the software. We also thank Josh Benaloh, who was our IACR liaison, for guidance on several issues. Springer published the volumes and made these available before the conference. We thank Alfred Hofmann and Anna Kramer and their team for their professional and efficient handling of the production process.

Last, but, not the least, we thank Microsoft Research; Google; Indian Statistical Institute, Kolkata; and National Mathematics Initiative, Indian Institute of Science, Bengaluru; for being generous sponsors of the conference.

December 2013

Kazue Sako
Palash Sarkar

Asiacrypt 2013

The 19th Annual International Conference on Theory and Application of Cryptology and Information Security

Sponsored by the *International Association for Cryptologic Research (IACR)*

December 1–5, 2013, Bengaluru, India

General Chair

Satyanarayana V. Lokam	Microsoft Research, India

Program Co-chairs

Kazue Sako	NEC Corporation, Japan
Palash Sarkar	Indian Statistical Institute, India

Program Committee

Michel Abdalla	École Normale Supérieure, France
Colin Boyd	Queensland University of Technology, Australia
Anne Canteaut	Inria Paris-Rocquencourt, France
Sanjit Chatterjee	Indian Institute of Science, India
Jung Hee Cheon	Seoul National University, Korea
Sherman S.M. Chow	Chinese University of Hong Kong, SAR China
Orr Dunkelmann	University of Haifa, Israel
Pierrick Gaudry	CNRS Nancy, France
Rosario Gennaro	City College of New York, USA
Guang Gong	University of Waterloo, Canada
Vipul Goyal	Microsoft Research, India
Eike Kiltz	University of Bochum, Germany
Tetsu Iwata	Nagoya University, Japan
Tanja Lange	Technische Universiteit Eindhoven, The Netherlands
Dong Hoon Lee	Korea University, Korea
Allison Lewko	Columbia University, USA
Benoit Libert	Technicolor, France
Dongdai Lin	Chinese Academy of Sciences, China
Anna Lysyanskaya	Brown University, USA
Subhamoy Maitra	Indian Statistical Institute, India

Willi Meier	University of Applied Sciences, Switzerland
Phong Nguyen	Inria, France and Tsinghua University, China
Kaisa Nyberg	Aalto University, Finland
Satoshi Obana	Hosei University, Japan
Kenny Paterson	Royal Holloway, University of London, UK
Krzysztof Pietrzak	Institute of Science and Technology, Austria
David Pointcheval	École Normale Supérieure, France
Manoj Prabhakaran	University of Illinois at Urbana-Champaign, USA
Vincent Rijmen	KU Leuven, Belgium
Rei Safavi-Naini	University of Calgary, Canada
Yu Sasaki	NTT, Japan
Nicolas Sendrier	Inria Paris-Rocquencourt, France
Peter Schwabe	Radboud University Nijmegen, The Netherlands
Thomas Shrimpton	Portland State University, USA
Nigel Smart	University of Bristol, UK
Francois-Xavier Standaert	Université Catholique de Louvain, Belgium
Damien Stehlé	École Normale Supérieure de Lyon, France
Willy Susilo	University of Wollongong, Australia
Tsuyoshi Takagi	Kyushu University, Japan
Vinod Vaikuntanathan	University of Toronto, Canada
Frederik Vercauteren	KU Leuven, Belgium
Xiaoyun Wang	Tsinghua University, China
Hoeteck Wee	George Washington University, USA and École Normale Supérieure, France
Hongjun Wu	Nanyang Technological University, Singapore

External Reviewers

Carlos Aguilar-Melchor
Masayuki Abe
Gergely Acs
Shashank Agrawal
Ahmad Ahmadi
Hadi Ahmadi
Mohsen Alimomeni
Joel Alwen
Prabhanjan Ananth
Gilad Asharov
Tomer Ashur
Giuseppe Ateniese
Man Ho Au
Jean-Philippe Aumasson
Pablo Azar
Foteini Baldimtsi
Subhadeep Banik
Paulo Barreto
Rishiraj Batacharrya
Lejla Batina
Anja Becker
Mihir Bellare
Fabrice Benhamouda
Debajyoti Bera
Daniel J. Bernstein
Rishiraj Bhattacharyya
Gaetan Bisson
Olivier Blazy
Céline Blondeau
Andrey Bogdanov

Alexandra Boldyreva
Joppe W. Bos
Charles Bouillaguet
Christina Boura
Elette Boyle
Fabian van den Broek
Billy Bob Brumley
Christina Brzuska
Angelo De Caro
Dario Catalano
André Chailloux
Melissa Chase
Anupam Chattopadhyay
Chi Chen
Jie Chen
Jing Chen
Yu Chen
Céline Chevalier
Ashish Choudhary
HeeWon Chung
Kai-Min Chung
Deepak Kumar Dalai
M. Prem Laxman Das
Gareth Davies
Yi Deng
Maria Dubovitskaya
François Durvaux
Barış Ege
Nicolas Estibals
Xinxin Fan
Pooya Farshim
Sebastian Faust
Nelly Fazio
Serge Fehr
Dario Fiore
Marc Fischlin
Georg Fuchsbauer
Eichiro Fujisaki
Jun Furukawa
Philippe Gaborit
Tommaso Gagliardoni
Martin Gagne
Steven Galbraith
David Galindo
Nicolas Gama
Sanjam Garg
Lubos Gaspar
Peter Gazi
Ran Gelles
Essam Ghadafi
Choudary Gorantla
Sergey Gorbunov
Dov S. Gordon
Louis Goubin
Matthew Green
Vincent Grosso
Jens Groth
Tim Güneysu
Fuchun Guo
Jian Guo
Divya Gupta
Sourav Sen Gupta
Benoît Gérard
Dong-Guk Han
Jinguang Han
Carmit Hazay
Nadia Heninger
Jens Hermans
Florian Hess
Shoichi Hirose
Viet Tung Hoang
Jaap-Henk Hoepmann
Dennis Hofheinz
Hyunsook Hong
Jin Hong
Qiong Huang
Tao Huang
Yan Huang
Fei Huo
Michael Hutter
Jung Yeon Hwang
Takanori Isobe
Mitsugu Iwamoto
Abhishek Jain
Stanislaw Jarecki
Mahavir Jhawar
Shoaquan Jiang
Ari Juels
Marc Kaplan
Koray Karabina

Aniket Kate
Jonathan Katz
Liam Keliher
Stéphanie Kerckhof
Hyoseung Kim
Kitak Kim
Minkyu Kim
Sungwook Kim
Taechan Kim
Yuichi Komano
Takeshi Koshiba
Anna Krasnova
Fabien Laguillaumie
Russell W.F. Lai
Adeline Langlois
Jooyoung Lee
Kwangsu Lee
Moon Sung Lee
Younho Lee
Tancrède Lepoint
Gaëtan Leurent
Anthony Leverrier
Huijia Rachel Lin
Feng-Hao Liu
Zhenhua Liu
Zongbin Liu
Adriana López-Alt
Atul Luykx
Vadim Lyubashevsky
Arpita Maitra
Hemanta Maji
Cuauhtemoc Mancillas-López
Kalikinkar Mandal
Takahiro Matsuda
Alexander May
Sarah Meiklejohn
Florian Mendel
Alfred Menezes
Kazuhiko Minematsu
Marine Minier
Rafael Misoczki
Amir Moradi
Tal Moran
Kirill Morozov
Pratyay Mukherjee
Yusuke Naito
María Naya-Plasencia
Gregory Neven
Khoa Nguyen
Antonio Nicolosi
Ivica Nikolić
Ryo Nishimaki
Ryo Nojima
Adam O'Neill
Cristina Onete
Elisabeth Oswald
Ilya Ozerov
Omkant Pandey
Tapas Pandit
Jong Hwan Park
Seunghwan Park
Michal Parusinski
Valerio Pastro
Arpita Patra
Goutam Paul
Roel Peeters
Christopher Peikert
Milinda Perera
Ludovic Perret
Thomas Peters
Christophe Petit
Duong Hieu Phan
Bertram Poettering
Joop van de Pol
Gordon Proctor
Emmanuel Prouff
Elizabeth Quaglia
Somindu C Ramanna
Mariana Raykova
Christian Rechberger
Francesco Regazzoni
Oscar Reparaz
Reza Reyhanitabar
Thomas Ristenpart
Damien Robert
Thomas Roche
Mike Rosulek
Sujoy Sinha Roy
Sushmita Ruj
Carla Ràfols

Santanu Sarkar
Michael Schneider
Dominique Schröder
Jacob Schuldt
Jae Hong Seo
Minjae Seo
Yannick Seurin
Hakan Seyalioglu
Setareh Sharifian
Abhi Shelat
Dale Sibborn
Dimitris E. Simos
Dave Singelee
William E. Skeith III
Boris Skoric
Adam Smith
Ben Smith
Hadi Soleimany
Katherine Stange
Douglas Stebila
John Steinberger
Ron Steinfeld
Mario Strefler
Donald Sun
Koutarou Suzuki
Yin Tan
Ying-Kai Tang
Sidharth Telang
Isamu Teranishi
R. Seth Terashima
Stefano Tessaro
Susan Thomson
Emmanuel Thomé
Gilles Van Assche
Konstantinos Vamvourellis
Alex Vardy
K. Venkata
Damien Vergnaud
Nicolas Veyrat-Charvillon
Gilles Villard
Ivan Visconti
Huaxiong Wang
Lei Wang
Meiqin Wang
Peng Wang
Pengwei Wang
Wenhao Wang
Gaven Watson
Carolyn Whitnall
Daniel Wichs
Michael J. Wiener
Shuang Wu
Teng Wu
Keita Xagawa
Haixia Xu
Rui Xue
Bohan Yang
Guomin Yang
Kan Yasuda
Takanori Yasuda
Kazuki Yoneyama
Hongbo Yu
Tsz Hon Yuen
Dae Hyun Yum
Aaram Yun
Hui Zhang
Liang Feng Zhang
Liting Zhang
Mingwu Zhang
Rui Zhang
Tao Zhang
Wentao Zhang
Zongyang Zhang
Colin Jia Zheng
Xifan Zheng
Hong-Sheng Zhou
Yongbin Zhou
Bo Zhu
Youwen Zhu
Vassilis Zikas
Paul Zimmermann

Organizing Committee

Raghav Bhaskar	Microsoft Research India, Bengaluru
Vipul Goyal	Microsoft Research India, Bengaluru
Neeraj Kayal	Microsoft Research India, Bengaluru
Satyanarayana V. Lokam	Microsoft Research India, Bengaluru
C. Pandurangan	Indian Institute of Technology, Chennai
Govindan Rangarajan	Indian Institute of Science, Bengaluru

Sponsors

Microsoft Research
Google
Indian Statistical Institute, Kolkata
National Mathematics Initiative, Indian Institute of Science, Bengaluru

Invited Talks

Block Ciphers – Past and Present

Lars Ramkilde Knudsen

DTU Compute, Denmark
lrkn@dtu.dk

Abstract. In the 1980s researchers were trying to understand the design of the DES, and breaking it seemed impossible. Other block ciphers were proposed, and cryptanalysis of block ciphers got interesting. The area took off in the 1990s where it exploded with the appearance of differential and linear cryptanalysis and the many variants thereof which appeared in the time after. In the 2000s AES became a standard and it was constructed specifically to resist the general attacks and the area of (traditional) block cipher cryptanalysis seemed saturated.... Much of the progress in cryptanalysis of the AES since then has come from side-channel attacks and related-key attacks.

Still today, for most block cipher applications the AES is a good and popular choice. However, the AES is perhaps not particularly well suited for extremely constrained environments such as RFID tags. Therefore, one trend in block cipher design has been to come up with ultra-lightweight block ciphers with good security and hardware efficiency. I was involved in the design of the ciphers Present (from CHES 2007), PrintCipher (presented at CHES 2010) and PRINCE (from Asiacrypt 2012). Another trend in block cipher design has been try to increase the efficiency by making certain components part of the secret key, e.g., to be able to reduce the number of rounds of a cipher.

In this talk, I will review these results.

Engineering Privacy-Friendly Computations

George Danezis [1,2]

[1] University College London
[2] Microsoft Research, Cambridge

Abstract. In the past few years tremendous cryptographic progress has been made in relation to primitives for privacy friendly-computations. These include celebrated results around fully homomorphic encryption, faster somehow homomorphic encryption, and ways to leverage them to support more efficient secret-sharing based secure multi-party computations. Similar break-through in verifiable computation, and succinct arguments of knowledge, make it practical to verify complex computations, as part of privacy-preserving client side program execution. Besides computations themselves, notions like differential privacy attempt to capture the essence of what it means for computations to leak little personal information, and have been mapped to existing data query languages.

So, is the problem of computation on private data solved, or just about to be solved? In this talk, I argue that the models of generic computation supported by cryptographic primitives are complete, but rather removed from what a typical engineer or data analyst expects. Furthermore, the use of these cryptographic technologies impose constrains that require fundamental changes in the engineering of computing systems. While those challenges are not obviously cryptographic in nature, they are nevertheless hard to overcome, have serious performance implications, and errors open avenues for attack.

Throughout the talk I use examples from our own work relating to privacy-friendly computations within smart grid and smart metering deployments for private billing, privacy-friendly aggregation, statistics and fraud detection. These experiences have guided the design of ZQL, a cryptographic language and compiler for zero-knowledge proofs, as well as more recent tools that compile using secret-sharing based primitives.

Table of Contents – Part I

Zero-Knowledge

Algebraic Cryptography

Theoretical Cryptography-I

Protocols

Theoretical Cryptography-II

Symmetric Key Cryptanalysis

Symmetric Key Cryptology: Schemes and Analysis

Side-Channel Cryptanalysis

Table of Contents – Part II

Message Authentication Codes

Signatures

Cryptography Based Upon Physical Assumptions

Multi-Party Computation

Cryptographic Primitives

Analysis, Cryptanalysis and Passwords

Leakage-Resilient Cryptography

Two-Party Computation

Hash Functions

Shorter Quasi-Adaptive NIZK Proofs for Linear Subspaces

Charanjit S. Jutla[1] and Arnab Roy[2]

[1] IBM T. J. Watson Research Center
Yorktown Heights, NY 10598, USA
csjutla@us.ibm.com

[2] Fujitsu Laboratories of America
Sunnyvale, CA 94085, USA
arnab@cs.stanford.edu

Abstract. We define a novel notion of quasi-adaptive non-interactive zero knowledge (NIZK) proofs for probability distributions on parametrized languages. It is quasi-adaptive in the sense that the common reference string (CRS) generator can generate the CRS depending on the language parameters. However, the simulation is required to be uniform, i.e., a single efficient simulator should work for the whole class of parametrized languages. For distributions on languages that are linear subspaces of vector spaces over bilinear groups, we give quasi-adaptive computationally sound NIZKs that are shorter and more efficient than Groth-Sahai NIZKs. For many cryptographic applications quasi-adaptive NIZKs suffice, and our constructions can lead to significant improvements in the standard model. Our construction can be based on any k-linear assumption, and in particular under the eXternal Diffie Hellman (XDH) assumption our proofs are even competitive with Random-Oracle based Σ-protocol NIZK proofs.

We also show that our system can be extended to include integer tags in the defining equations, where the tags are provided adaptively by the adversary. This leads to applicability of our system to many applications that use tags, e.g. applications using Cramer-Shoup projective hash proofs. Our techniques also lead to the shortest known (ciphertext) fully secure identity based encryption (IBE) scheme under standard static assumptions (SXDH). Further, we also get a short publicly-verifiable CCA2-secure IBE scheme.

Keywords: NIZK, Groth-Sahai, bilinear pairings, signatures, dual-system IBE, DLIN, SXDH.

1 Introduction

In [13] a remarkably efficient non-interactive zero-knowledge (NIZK) proof system [3] was given for groups with a bilinear map, which has found many applications in design of cryptographic protocols in the standard model. All earlier NIZK proof systems (except [12], which was not very efficient) were constructed

K. Sako and P. Sarkar (Eds.) ASIACRYPT 2013 Part I, LNCS 8269, pp. 1–20, 2013.

by reduction to Circuit Satisfiability. Underlying this system, now commonly known as Groth-Sahai NIZKs, is a homomorphic commitment scheme. Each variable in the system of algebraic equations to be proven is committed to using this scheme. Since the commitment scheme is homomorphic, group operations in the equations are translated to corresponding operations on the commitments and new terms are constructed involving the constants in the equations and the randomness used in the commitments. It was shown that these new terms along with the commitments to variables constitute a zero-knowledge proof [13].

While the Groth-Sahai system is quite efficient, it still falls short in comparison to Schnorr-based Σ-protocols [8] turned into NIZK proofs in the Random Oracle model [2] using the Fiat-Shamir paradigm [10]. Thus, the quest remains to obtain even more efficient NIZK Proofs. In particular, in a linear system of rank t, some t of the equations already serve as commitments to t variables. Thus, the question arises if, at the very least, fresh commitments to these variables as done in Groth-Sahai NIZKs can be avoided.

Our Contributions. In this paper, we show that for languages that are linear subspaces of vector spaces of the bilinear groups, one can indeed obtain more efficient computationally-sound NIZK proofs in a slightly different *quasi-adaptive* setting, which suffices for many cryptographic applications. In the quasi-adaptive setting, we consider a class of parametrized languages $\{L_\rho\}$, parametrized by ρ, and we allow the CRS generator to generate the CRS based on the language parameter ρ. However, the CRS simulator in the zero-knowledge setting is required to be a single efficient algorithm that works for the whole parametrized class or probability distributions of languages, by taking the parameter as input. We will refer to this property as *uniform simulation.*

Many hard languages that are commonly used in cryptography are distributions on class of parametrized languages, e.g. the DDH language based on the decisional Diffie-Hellman (DDH) assumption is hard only when in the tuple $\langle \mathbf{g}, \mathbf{f}, x \cdot \mathbf{g}, x \cdot \mathbf{f} \rangle$, even $\mathbf{f}$ is chosen at random (in addition to $x \cdot \mathbf{g}$ being chosen randomly). However, applications (or trusted parties) usually set $\mathbf{f}$, once and for all, by choosing it at random, and then all parties in the application can use *multiple* instances of the above language with the same fixed $\mathbf{f}$. Thus, we can consider $\mathbf{f}$ as a parameter for a class of languages that only specify the last two components above. If NIZK proofs are required in the application for this parametrized language, then the NIZK CRS can be generated by the trusted party that chooses the language parameter $\mathbf{f}$. Hence, it can base the CRS on the language parameter[1].

We remark that adaptive NIZK proofs [3] also allow the CRS to depend on the language, but without requiring uniform simulation. Such NIZK proofs that allow different efficient simulators for each particular language (from a parametrized class) are unlikely to be useful in applications. Thus, most NIZK proofs, including Groth-Sahai NIZKs, actually show that the same efficient

[1] However, in the security definition, the efficient CRS simulator does not itself generate $\mathbf{f}$, but is given $\mathbf{f}$ as input chosen randomly.

simulator works for the whole class, i.e. they show uniform simulation. The Groth-Sahai system achieves uniform simulation without making any distinction between different classes of parametrized languages, i.e. it shows a single efficient CRS simulator that works for *all* algebraic languages without taking any language parameters as input. Thus, there is potential to gain efficiency by considering quasi-adaptive NIZK proofs, i.e. by allowing the (uniform) simulator to take language parameters as input[2].

Our approach to building more efficient NIZK proofs for linear subspaces is quite different from the Groth-Sahai techniques. In fact, our system does not require any commitments to the witnesses at all. If there are t free variables in defining a subspace of the n-dimensional vector-space and assuming the subspace is full-ranked (i.e. has rank t), then t components of the vector already serve as commitment to the variables. As an example, consider the language L (over a cyclic group $\mathbb{G}$ of order q, in additive notation) to be

$$L = \{\langle \boldsymbol{l}_1, \boldsymbol{l}_2, \boldsymbol{l}_3\rangle \in \mathbb{G}^3 \mid \exists x_1, x_2 \in \mathbb{Z}_q : \boldsymbol{l}_1 = x_1 \cdot \mathbf{g},\ \boldsymbol{l}_2 = x_2 \cdot \mathbf{f},\ \boldsymbol{l}_3 = (x_1 + x_2) \cdot \mathbf{h}\}$$

where $\mathbf{g}, \mathbf{f}, \mathbf{h}$ are parameters defining the language. Then, $\boldsymbol{l}_1$ and $\boldsymbol{l}_2$ are already binding commitments to x_1 and x_2. Thus, we only need to show that the last component $\boldsymbol{l}_3$ is consistent.

The main idea underlying our construction can be summarized as follows. Suppose the CRS can be set to be a basis for the null-space $L_\rho^\perp$ of the language L_ρ. Then, just pairing a potential language candidate with $L_\rho^\perp$ and testing for all-zero suffices to prove that the candidate is in L_ρ, as the null-space of $L_\rho^\perp$ is just L_ρ. However, efficiently computing null-spaces in hard bilinear groups is itself hard. Thus, an efficient CRS simulator cannot generate $L_\rho^\perp$, but can give a (hiding) commitment that is computationally indistinguishable from a binding commitment to $L_\rho^\perp$. To achieve this we use a homomorphic commitment just as in the Groth-Sahai system, but we can use the simpler El-Gamal encryption style commitment as opposed to the more involved Groth-Sahai commitments, and this allows for a more efficient verifier[3]. As we will see later in Section 5, a more efficient verifier is critical for obtaining short identity based encryption schemes (IBE).

In fact, the idea of using the null-space of the language is reminiscent of Waters' dual-system IBE construction [24], and indeed our system is inspired by that construction[4], although the idea of using it for NIZK proofs, and in particular the proof of soundness is novel. Another contribution of the paper is in the definition of quasi-adaptive NIZK proofs.

---

[2] It is important to specify the information about the parameter which is supplied as input to the CRS simulator. We defer this important issue to Section 2 where we formally define quasi-adaptive NIZK proofs.

[3] Our quasi-adaptive NIZK proofs are already shorter than Groth-Sahai as they require no commitments to variables, and have to prove lesser number of equations, as mentioned earlier.

[4] In Section 5 and in the Appendix, we show that the design of our system leads to a shorter SXDH assumption based dual-system IBE.

For n equations in t variables, our quasi-adaptive computationally-sound NIZK proofs for linear subspaces require only $k(n-t)$ group elements, under the k-linear decisional assumption [23,5]. Thus, under the XDH assumption for bilinear groups, our proofs require only $(n-t)$ group elements. In contrast, the Groth-Sahai system requires $(n+2t)$ group elements. Similarly, under the decisional linear assumption (DLIN), our proofs require only $2(n-t)$ group elements, whereas the Groth-Sahai system requires $(2n+3t)$ group elements. These parameters are summarized in Table 1. While our CRS size grows proportional to $t(n-t)$, more importantly there is a significant comparative improvement in the number of pairings required for verification. Specifically, under XDH we require at most half the number of pairings, and under DLIN we require at most 2/3 the number of pairings. The Σ-protocol NIZK proofs based on the Random Oracle model require n group elements, t elements of $\mathbb{Z}_q$ and 1 hash value. Although our XDH based proofs require less number of group elements, the Σ-protocol proofs do not require bilinear groups and have the advantage of being proofs of knowledge (PoK). We remark that the Groth-Sahai system is also not a PoK for witnesses that are $\mathbb{Z}_q$ elements. A recent paper by Escala et al [9] has also optimized proofs of linear subspaces in a language dependent CRS setting. Their system also removes the need for commitment to witnesses but still implicitly uses Groth Sahai proofs. In comparison, our proofs are still much shorter.

Table 1. Comparison with Groth-Sahai NIZKs for Linear Subspaces. Parameter t is the number of unknowns or witnesses and n is the dimension of the vector space, or in other words, the number of equations.

	XDH			DLIN		
	Proof	CRS	#Pairings	Proof	CRS	#Pairings
Groth-Sahai	$n+2t$	4	$2n(t+2)$	$2n+3t$	9	$3n(t+3)$
This paper	$n-t$	$2t(n-t)+2$	$(n-t)(t+2)$	$2n-2t$	$4t(n-t)+3$	$2(n-t)(t+2)$

Thus, for the language L above, which is just a DLIN tuple used ubiquitously for encryption, our system only requires two group elements under the DLIN assumption, whereas the Groth-Sahai system requires twelve group elements (note, $t=2$, $n=3$ in L above). For the Diffie-Hellman analogue of this language $\langle x\cdot \mathbf{g}, x\cdot \mathbf{f}\rangle$, our system produces a *single* element proof under the XDH assumption, which we demonstrate in Section 3 (whereas the Groth-Sahai system requires $(n+2t=)$ 4 elements for the proof with $t=1$ and $n=2$).

Our NIZK proofs also satisfy some interesting new properties. Firstly, the proofs in our system are unique for each language member. This has interesting applications as we will see later in a CCA2-IBE construction. Secondly, the CRS in our system, though dependent on the language parameters, can be split into two parts. The first part is required only by the prover, and the second part is required only by the verifier, and the latter can be generated independent of the language. This is surprising since our verifier does not even take the language (parameters) as input. Only the randomization used in the verifier CRS generation is used in the prover CRS to link the two CRSes. This is in sharp contrast to Groth-Sahai NIZKs, where the verifier needs the language as input. This split-CRS property has interesting applications as we will see later.

Extension to Linear Systems with Tags. Our system does not yet extend naturally to quadratic or multi-linear equations, whereas the Groth-Sahai system does[5]. However, we can extend our system to include tags, and allow the defining equations to be polynomially dependent on tags. For example, our system can prove the following language:

$$L' = \left\{ \begin{array}{c} \langle \mathbf{l}_1, \mathbf{l}_2, \mathbf{l}_3, \textsc{tag} \rangle \in \mathbb{G}^3 \times \mathbb{Z}_q \mid \exists x_1, x_2 \in \mathbb{Z}_q : \\ \mathbf{l}_1 = x_1 \cdot \mathbf{f},\ \mathbf{l}_2 = x_2 \cdot \mathbf{g},\ \mathbf{l}_3 = (x_1 + \textsc{tag} \cdot x_2) \cdot \mathbf{h} \end{array} \right\}.$$

Note that this is a non-trivial extension since the TAG is adaptively provided by the adversary after the CRS has been set.

The extension to tags is very important, as we now discuss. Many applications require that the NIZK proof also be simulation-sound. However, extending NIZK proofs for bilinear groups to be unbounded simulation-sound requires handling quadratic equations (see [5] for a generic construction). On the other hand, many applications just require one-time simulation soundness, and as has been shown in [14], this can be achieved for linear subspaces by projective hash proofs [7]. Projective hash proofs can be defined by linear extensions, but require use of tags. Thus, our system can handle such equations. Many applications, such as signatures, can also achieve implicit unbounded simulation soundness using projective hash proofs, and such applications can utilize our system (see Section 5).

Applications. While the cryptographic literature is replete with NIZK proofs, we will demonstrate the applicability of quasi-adaptive NIZKs, and in particular our efficient system for linear subspaces, to a few recent applications such as signature schemes [5], UC commitments [11], password-based key exchange [16,14], key-dependent encryption [5]. For starters, based on [11], our system yields an adaptive UC-secure commitment scheme (in the erasure model) that has only four group elements as commitment, and another four as opening (under the DLIN assumption; and 3 + 2 under SXDH assumption), whereas the original scheme using Groth-Sahai NIZKs required 5 + 16 group elements.

We also obtain one of the shortest signature schemes under a static standard assumption, i.e. SXDH, that only requires five group elements. We also show how this signature scheme can be extended to a short fully secure (and perfectly complete) dual-system IBE scheme, and indeed a scheme with ciphertexts that are only four group elements plus a tag (under the SXDH assumption). This is the shortest IBE scheme under the SXDH assumption, and is technically even shorter than a recent and independently obtained scheme of [6] which requires five group elements as ciphertext. Table 2 depicts numerical differences between the parameter sizes of the two schemes. The SXDH-IBE scheme of [6] uses the concept of dual pairing vector spaces (due to Okamoto and Takashima [19,20],

[5] However, since commitments in Groth-Sahai NIZKs are linear, there is scope for mixing the two systems to gain efficiency.

and synthesized from Waters' dual system IBE). However, the dual vector space and its generalizations due to others [17] do not capture the idea of proof verification. Thus, one of our main contributions can be viewed as showing that the dual system not only does zero-knowledge simulation but also extends to provide a computationally sound verifier for general linear systems.

Table 2. Comparison with the SXDH-based IBE of Chen et al. [6]. The notation $|\cdot|$ denotes the bit length of an element of the given group.

	Public Key	Secret Key	Ciphertext	#Pairings	Anonymity
CLLWW12 [6]	$8\lvert\mathbb{G}_1\rvert + \lvert\mathbb{G}_T\rvert$	$4\lvert\mathbb{G}_2\rvert$	$4\lvert\mathbb{G}_1\rvert + \lvert\mathbb{G}_T\rvert$	4	yes
This paper	$5\lvert\mathbb{G}_1\rvert + \lvert\mathbb{G}_T\rvert$	$5\lvert\mathbb{G}_2\rvert$	$3\lvert\mathbb{G}_1\rvert + \lvert\mathbb{G}_T\rvert + \lvert\mathbb{Z}_q\rvert$	3	yes

Finally, using our QA-NIZKs we show a short *publicly-verifiable* CCA2-secure IBE scheme. Public verifiability is an informal but practically important notion which implies that one can publicly verify if the decryption will yield "invalid ciphertext". Thus, this can allow a network gateway to act as a filter. Our scheme only requires two additional group elements over the basic IBE scheme.

Organization of the Paper. We begin the rest of the paper with the definition of quasi-adaptive NIZKs in Section 2. In Section 3 we develop quasi-adaptive NIZKs for linear subspaces under the XDH assumption. In Section 4, we extend our system to include tags, define a notion called split-CRS QA-NIZKs. and extend our system to construct split-CRS NIZKs for affine spaces. Finally, we demonstrate applications of our system in Section 5. We defer detailed proofs and descriptions to the full paper [15]. We also describe our system based on the k-linear assumption in [15].

Notations. We will be dealing with witness-relations R that are binary relations on pairs (x, w), and where w is commonly referred to as the witness. Each witness-relation defines a language $L = \{x \mid \exists w : R(x, w)\}$. For every witness-relation R_ρ we will use L_ρ to denote the language it defines. Thus, a NIZK proof for a witness-relation R_ρ can also be seen as a NIZK proof for its language L_ρ.

Vectors will always be row-vectors and will always be denoted by an arrow over the letter, e.g. $\vec{\mathbf{r}}$ for (row) vector of $\mathbb{Z}_q$ elements, and $\vec{\mathbf{d}}$ as (row) vector of group elements.

2 Quasi-Adaptive NIZK Proofs

Instead of considering NIZK proofs for a (witness-) relation R, we will consider Quasi-Adaptive NIZK proofs for a probability distribution $\mathcal{D}$ on a collection of (witness-) relations $\mathcal{R} = \{R_\rho\}$. The quasi-adaptiveness allows for the common reference string (CRS) to be set based on R_ρ after the latter has been chosen according to $\mathcal{D}$. We will however require, as we will see later, that the simulator generating the CRS (in the simulation world) is a single probabilistic polynomial time algorithm that works for the whole collection of relations $\mathcal{R}$.

To be more precise, we will consider ensemble of distributions on witness-relations, each distribution in the ensemble itself parametrized by a security parameter. Thus, we will consider ensemble $\{\mathcal{D}_\lambda\}$ of distributions on collection of relations $\mathcal{R}_\lambda$, where each $\mathcal{D}_\lambda$ specifies a probability distribution on $\mathcal{R}_\lambda = \{R_{\lambda,\rho}\}$. When λ is clear from context, we will just refer to a particular relation as R_ρ, and write $\mathcal{R}_\lambda = \{R_\rho\}$.

Since in the quasi-adaptive setting the CRS could depend on the relation, we must specify what information about the relation is given to the CRS generator. Thus, we will consider an associated *parameter language* such that a member of this language is enough to characterize a particular relation, and this language member is provided to the CRS generator. For example, consider the class of parametrized relations $\mathcal{R} = \{R_\rho\}$, where parameter ρ is a tuple $\mathbf{g}, \mathbf{f}, \mathbf{h}$ of three group elements. Suppose, R_ρ (on $\langle \boldsymbol{l}_1, \boldsymbol{l}_2, \boldsymbol{l}_3 \rangle, \langle x_1, x_2 \rangle$) is defined as

$$R_{\langle \mathbf{g}, \mathbf{f}, \mathbf{h} \rangle}(\langle \boldsymbol{l}_1, \boldsymbol{l}_2, \boldsymbol{l}_3 \rangle, \langle x_1, x_2 \rangle) \stackrel{\text{def}}{=} \left(\begin{array}{c} x_1, x_2 \in \mathbb{Z}_q, \boldsymbol{l}_1, \boldsymbol{l}_2, \boldsymbol{l}_3 \in \mathbb{G} \textbf{ and} \\ \boldsymbol{l}_1 = x_1 \cdot \mathbf{g}, \boldsymbol{l}_2 = x_2 \cdot \mathbf{f}, \boldsymbol{l}_3 = (x_1 + x_2) \cdot \mathbf{h} \end{array} \right).$$

For this class of relations, one could seek a quasi-adaptive NIZK where the CRS generator is just given ρ as input. Thus in this case, the associated parameter language $\mathcal{L}_{\text{par}}$ will just be triples of group elements[6]. Moreover, the distribution $\mathcal{D}$ can just be on the parameter language $\mathcal{L}_{\text{par}}$, i.e. $\mathcal{D}$ just specifies a $\rho \in \mathcal{L}_{\text{par}}$. Again, $\mathcal{L}_{\text{par}}$ is technically an ensemble.

We call $(\mathsf{K}_0, \mathsf{K}_1, \mathsf{P}, \mathsf{V})$ a *QA-NIZK* proof system for witness-relations $\mathcal{R}_\lambda = \{R_\rho\}$ with parameters sampled from a distribution $\mathcal{D}$ over associated parameter language $\mathcal{L}_{\text{par}}$, if there exists a probabilistic polynomial time simulator $(\mathsf{S}_1, \mathsf{S}_2)$, such that for all non-uniform PPT adversaries $\mathcal{A}_1, \mathcal{A}_2, \mathcal{A}_3$ we have:

Quasi-Adaptive Completeness:

$$\begin{array}{l} \Pr[\lambda \leftarrow \mathsf{K}_0(1^m); \rho \leftarrow \mathcal{D}_\lambda; \psi \leftarrow \mathsf{K}_1(\lambda, \rho); (x, w) \leftarrow \mathcal{A}_1(\lambda, \psi, \rho); \\ \quad \pi \leftarrow \mathsf{P}(\psi, x, w) : \ \mathsf{V}(\psi, x, \pi) = 1 \text{ if } R_\rho(x, w)] = 1 \end{array}$$

Quasi-Adaptive Soundness:

$$\begin{array}{l} \Pr[\lambda \leftarrow \mathsf{K}_0(1^m); \rho \leftarrow \mathcal{D}_\lambda; \psi \leftarrow \mathsf{K}_1(\lambda, \rho); \\ \quad (x, \pi) \leftarrow \mathcal{A}_2(\lambda, \psi, \rho) : \ \mathsf{V}(\psi, x, \pi) = 1 \text{ and } \neg(\exists w : R_\rho(x, w))] \approx 0 \end{array}$$

Quasi-Adaptive Zero-Knowledge:

$$\Pr[\lambda \leftarrow \mathsf{K}_0(1^m); \rho \leftarrow \mathcal{D}_\lambda; \psi \leftarrow \mathsf{K}_1(\lambda, \rho) : \mathcal{A}_3^{\mathsf{P}(\psi, \cdot, \cdot)}(\lambda, \psi, \rho) = 1] \approx$$
$$\Pr[\lambda \leftarrow \mathsf{K}_0(1^m); \rho \leftarrow \mathcal{D}_\lambda; (\psi, \tau) \leftarrow \mathsf{S}_1(\lambda, \rho) : \mathcal{A}_3^{\mathsf{S}(\psi, \tau, \cdot, \cdot)}(\lambda, \psi, \rho) = 1],$$

6 It is worth remarking that alternatively the parameter language could also be discrete logarithms of these group elements (w.r.t. to some base), but a NIZK proof under this associated language may not be very useful. Thus, it is critical to define the proper associated parameter language.

where $\mathsf{S}(\psi, \tau, x, w) = \mathsf{S}_2(\psi, \tau, x)$ for $(x, w) \in R_\rho$ and both oracles (i.e. P and S) output failure if $(x, w) \notin R_\rho$.

Note that ψ is the CRS in the above definitions.

3 QA-NIZK for Linear Subspaces under the XDH Assumption

Setup. Let $\mathbb{G}_1, \mathbb{G}_2$ and $\mathbb{G}_T$ be cyclic groups of prime order q with a bilinear map $e : \mathbb{G}_1 \times \mathbb{G}_2 \to \mathbb{G}_T$ chosen by a group generation algorithm. Let $\mathbf{g}_1$ and $\mathbf{g}_2$ be generators of the group $\mathbb{G}_1$ and $\mathbb{G}_2$ respectively. Let $\mathbf{0}_1$, $\mathbf{0}_2$ and $\mathbf{0}_T$ be the identity elements in the three groups $\mathbb{G}_1, \mathbb{G}_2$ and $\mathbb{G}_T$ respectively. We use additive notation for the group operations in all the groups.

The bilinear pairing e naturally extends to $\mathbb{Z}_q$-vector spaces of $\mathbb{G}_1$ and $\mathbb{G}_2$ of the same dimension n as follows: $e(\vec{\mathbf{a}}, \vec{\mathbf{b}}^\top) = \sum_{i=1}^{n} e(\mathbf{a}_i, \mathbf{b}_i)$. Thus, if $\vec{\mathbf{a}} = \vec{\mathrm{x}} \cdot \mathbf{g}_1$ and $\vec{\mathbf{b}} = \vec{\mathrm{y}} \cdot \mathbf{g}_2$, where $\vec{\mathrm{x}}$ and $\vec{\mathrm{y}}$ are now vectors over $\mathbb{Z}_q$, then $e(\vec{\mathbf{a}}, \vec{\mathbf{b}}^\top) = (\vec{\mathrm{x}} \cdot \vec{\mathrm{y}}^\top) \cdot e(\mathbf{g}_1, \mathbf{g}_2)$. The operator "$\top$" indicates taking the transpose.

Linear Subspace Languages. To start off with an example, a set of equations $l_1 = x_1 \cdot \mathbf{g}$, $l_2 = x_2 \cdot \mathbf{f}$, $l_3 = (x_1 + x_2) \cdot \mathbf{h}$ will be expressed in the form $\vec{l} = \vec{\mathrm{x}} \cdot \mathbf{A}$ as follows:

$$\vec{l} = \begin{bmatrix} l_1 \ l_2 \ l_3 \end{bmatrix} = \begin{bmatrix} x_1 \ x_2 \end{bmatrix} \cdot \begin{bmatrix} \mathbf{g} & \mathbf{0}_1 & \mathbf{h} \\ \mathbf{0}_1 & \mathbf{f} & \mathbf{h} \end{bmatrix}$$

where $\vec{\mathrm{x}}$ is a vector of unknowns and $\mathbf{A}$ is a matrix specifying the group constants $\mathbf{g}, \mathbf{f}, \mathbf{h}$.

The scalars in this system of equations are from the field $\mathbb{Z}_q$. In general, we consider languages that are linear subspaces of vectors of $\mathbb{G}_1$ elements. These are just $\mathbb{Z}_q$-modules, and since $\mathbb{Z}_q$ is a field, they are vector spaces. In other words, the languages we are interested in can be characterized as languages parameterized by $\mathbf{A}$ as below:

$$L_{\mathbf{A}} = \{\vec{\mathrm{x}} \cdot \mathbf{A} \in \mathbb{G}_1^n \mid \vec{\mathrm{x}} \in \mathbb{Z}_q^t\}, \text{ where } \mathbf{A} \text{ is a } t \times n \text{ matrix of } \mathbb{G}_1 \text{ elements.}$$

Here $\mathbf{A}$ is an element of the associated *parameter language* $\mathcal{L}_{\text{par}}$, which is all $t \times n$ matrices of $\mathbb{G}_1$ elements. The parameter language $\mathcal{L}_{\text{par}}$ also has a corresponding witness relation $\mathcal{R}_{\text{par}}$, where the witness is a matrix of $\mathbb{Z}_q$ elements : $\mathcal{R}_{\text{par}}(\mathbf{A}, \mathsf{A})$ iff $\mathbf{A} = \mathsf{A} \cdot \mathbf{g}_1$.

Robust and Efficiently Witness-Samplable Distributions. Let the $t \times n$ dimensional matrix $\mathbf{A}$ be chosen according to a distribution $\mathcal{D}$ on $\mathcal{L}_{\text{par}}$. We will call the distribution $\mathcal{D}$ *robust* if with probability close to one the left-most t columns of $\mathbf{A}$ are full-ranked. We will call a distribution $\mathcal{D}$ on $\mathcal{L}_{\text{par}}$ *efficiently witness-samplable* if there is a probabilistic polynomial time algorithm such that it outputs a pair of matrices $(\mathbf{A}, \mathsf{A})$ that satisfy the relation $\mathcal{R}_{\text{par}}$ (i.e., $\mathcal{R}_{\text{par}}(\mathbf{A}, \mathsf{A})$ holds), and further the resulting distribution of the output $\mathbf{A}$ is same as $\mathcal{D}$. For example, the uniform distribution on $\mathcal{L}_{\text{par}}$ is efficiently witness-samplable, by first picking A at random, and then computing $\mathbf{A}$. As an example of a robust distribution, consider a distribution $\mathcal{D}$ on (2×3)-dimensional matrices $\begin{bmatrix} \mathbf{g} & \mathbf{0}_1 & \mathbf{h} \\ \mathbf{0}_1 & \mathbf{f} & \mathbf{h} \end{bmatrix}$ with $\mathbf{g}, \mathbf{f}$ and $\mathbf{h}$ chosen randomly from $\mathbb{G}_1$. It is easy to see that the first two columns are full-ranked if $\mathbf{g} \neq \mathbf{0}_1$ and $\mathbf{f} \neq \mathbf{0}_1$, which holds with probability $(1 - 1/q)^2$.

QA-NIZK Construction. We now describe a computationally sound quasi-adaptive NIZK $(\mathsf{K}_0, \mathsf{K}_1, \mathsf{P}, \mathsf{V})$ for linear subspace languages $\{L_{\mathbf{A}}\}$ with parameters sampled from a robust and efficiently witness-samplable distribution $\mathcal{D}$ over the associated parameter language $\mathcal{L}_{\text{par}}$.

Algorithm K_0**.** K_0 is same as the group generation algorithm for which the XDH assumption holds. $\lambda \stackrel{\text{def}}{=} (q, \mathbb{G}_1, \mathbb{G}_2, \mathbb{G}_T, e, \mathbf{g}_1, \mathbf{g}_2) \leftarrow \mathsf{K}_0(1^m)$, with $(q, \mathbb{G}_1, \mathbb{G}_2, \mathbb{G}_T, e, \mathbf{g}_1, \mathbf{g}_2)$ as described above.

We will assume that the size $t \times n$ of the matrix $\mathbf{A}$ is either fixed or determined by the security parameter m. In general, t and n could also be part of the parameter language, and hence t, n could be given as part of the input to CRS generator K_1.

Algorithm K_1**.** The algorithm K_1 generates the CRS as follows. Let $\mathbf{A}^{t \times n}$ be the parameter supplied to K_1. Let $s \stackrel{\text{def}}{=} n - t$: this is the number of equations in excess of the unknowns. It generates a matrix $\mathsf{D}^{t \times s}$ with all elements chosen randomly from $\mathbb{Z}_q$ and a single element b chosen randomly from $\mathbb{Z}_q$. The common reference string (CRS) has two parts $\mathbf{CRS}_p$ and $\mathbf{CRS}_v$ which are to be used by the prover and the verifier respectively.

$$\mathbf{CRS}_p^{t \times s} := \mathbf{A} \cdot \begin{bmatrix} \mathsf{D}^{t \times s} \\ b^{-1} \cdot \mathrm{I}^{s \times s} \end{bmatrix} \qquad \mathbf{CRS}_v^{(n+s) \times s} := \begin{bmatrix} b \cdot \mathsf{D} \\ \mathrm{I}^{s \times s} \\ -b \cdot \mathrm{I}^{s \times s} \end{bmatrix} \cdot \mathbf{g}_2$$

Here, I denotes the identity matrix. Note that $\mathbf{CRS}_v$ is independent of the parameter.

Prover P**.** Given candidate $\vec{l} = \vec{\mathrm{x}} \cdot \mathbf{A}$ with witness vector $\vec{\mathrm{x}}$, the prover generates the following proof consisting of s elements in $\mathbb{G}_1$:

$$\vec{\mathbf{p}} := \vec{\mathrm{x}} \cdot \mathbf{CRS}_p$$

Verifier V**.** Given candidate $\vec{l}$, and a proof $\vec{\mathbf{p}}$, the verifier checks the following:

$$e\left(\left[\vec{l} \mid \vec{\mathbf{p}}\right], \mathbf{CRS}_v\right) \stackrel{?}{=} \mathbf{0}_T^{1 \times s}$$

The security of the above system depends on the DDH assumption in group $\mathbb{G}_2$. Since $\mathbb{G}_2$ is a bilinear group, this assumption is known as the XDH assumption. These assumptions are standard and are formally described in [15].

Theorem 1. *The above algorithms* $(\mathsf{K}_0, \mathsf{K}_1, \mathsf{P}, \mathsf{V})$ *constitute a computationally sound quasi-adaptive NIZK proof system for linear subspace languages* $\{L_{\mathbf{A}}\}$ *with* *parameters* $\mathbf{A}$ *sampled from a robust and efficiently witness-samplable distribution* $\mathcal{D}$ *over the associated parameter language* $\mathcal{L}_{par}$*, given any group generation algorithm for which the DDH assumption holds for group* $\mathbb{G}_2$.

Remark. For language members, the proofs are unique as the bottom s rows of $\mathbf{CRS}_v$ are invertible.

Proof Intuition. A detailed proof of the theorem can be found in [15]. Here we give the main idea behind the working of the above quasi-adaptive NIZK, and in particular the soundness requirement which is the difficult part here. We first observe that completeness follows by straightforward bilinear manipulation. Zero Knowledge also follows easily: the simulator generates the same CRS as above but retains D and b as trapdoors. Now, given a language candidate $\vec{l}$, the proof is simply $\vec{\mathbf{p}} := \vec{l} \cdot \begin{bmatrix} \mathsf{D} \\ b^{-1} \cdot \mathrm{I}^{s\times s} \end{bmatrix}$. If $\vec{l}$ is in the language, i.e., it is $\vec{\mathrm{x}} \cdot \mathbf{A}$ for some $\vec{\mathrm{x}}$, then the distribution of the simulated proof is identical to the real world proof.

We now focus on the soundness proof which we establish by transforming the system over two games. Let Game $\mathbf{G}_0$ be the original system. Since $\mathcal{D}$ is efficiently witness samplable, in Game $\mathbf{G}_1$ the challenger generates both A and $\mathbf{A} = \mathsf{A} \cdot \mathbf{g}_1$. Then it computes a rank s matrix $\begin{bmatrix} \mathsf{W}^{t\times s} \\ \mathrm{I}^{s\times s} \end{bmatrix}$ of dimension $(t+s)\times s$ whose columns form a complete basis for the null-space of A, which means $\mathsf{A} \cdot \begin{bmatrix} \mathsf{W}^{t\times s} \\ \mathrm{I}^{s\times s} \end{bmatrix} = 0^{t\times s}$. Now statistically, the CRS in Game $\mathbf{G}_0$ is indistinguishable from the one where we substitute $\mathsf{D}' + b^{-1} \cdot \mathsf{W}$ for D, where D' itself is an independent random matrix. With this substitution, the $\mathbf{CRS}_p$ and $\mathbf{CRS}_v$ can be represented as

$$\mathbf{CRS}_p^{t\times s} = \mathbf{A} \cdot \begin{bmatrix} \mathsf{D}' \\ 0^{s\times s} \end{bmatrix}, \quad \mathbf{CRS}_v^{(n+s)\times s} = \begin{bmatrix} b \cdot \begin{bmatrix} \mathsf{D}' \\ 0^{s\times s} \end{bmatrix} + \begin{bmatrix} \mathsf{W} \\ \mathrm{I}^{s\times s} \end{bmatrix} \\ -b \cdot \mathrm{I}^{s\times s} \end{bmatrix} \cdot \mathbf{g}_2$$

Now we show that if an efficient adversary can produce a "proof" $\vec{\mathbf{p}}$ for which the above pairing test holds and yet the candidate $\vec{l}$ is not in $L_{\mathbf{A}}$, then it implies an efficient adversary that can break DDH in group $\mathbb{G}_2$. So consider a DDH game, where a challenger either provides a real DDH-tuple $\langle \mathbf{g}_2, b \cdot \mathbf{g}_2, r \cdot \mathbf{g}_2, \chi = br \cdot \mathbf{g}_2 \rangle$ or a fake DDH tuple $\langle \mathbf{g}_2, b \cdot \mathbf{g}_2, r \cdot \mathbf{g}_2, \chi = br' \cdot \mathbf{g}_2 \rangle$. Let us partition the $\mathbb{Z}_q$ matrix A as $\left[\mathsf{A}_0^{t\times t} | \mathsf{A}_1^{t\times s} \right]$ and the candidate vector $\vec{l}$ as $\left[\vec{l}_0^{\,1\times t} | \vec{l}_1^{\,1\times s} \right]$. Note that, since A_0 has rank t, the elements of $\vec{l}_0$ are 'free' elements and $\vec{l}_0$ can be extended to a unique n element vector $\vec{l}\,'$, which is a member of $L_{\mathbf{A}}$. This member vector $\vec{l}\,'$ can be computed as $\vec{l}\,' := \left[\vec{l}_0 \mid -\vec{l}_0 \cdot \mathsf{W} \right]$, nothing $\mathsf{W} = -\mathsf{A}_0^{-1}\mathsf{A}_1$. The proof of $\vec{l}\,'$ is computed as $\vec{\mathbf{p}}\,' := \vec{l}_0 \cdot \mathsf{D}'$. Since both $(\vec{l}, \vec{\mathbf{p}})$ and $(\vec{l}\,', \vec{\mathbf{p}}\,')$ pass the verification equation, we obtain: $\vec{l}_1' - \vec{l}_1 = b(\vec{\mathbf{p}}\,' - \vec{\mathbf{p}})$, where $\vec{l}_1' = -\vec{l}_0 \cdot \mathsf{W}$. In particular there exists $i \in [1, s]$, such that, $l_{1i}' - l_{1i} = b(\mathbf{p}_i' - \mathbf{p}_i) \neq \mathbf{0}_1$. This gives us a straightforward test for the DDH challenge: $e(l_{1i}' - l_{1i}, r \cdot \mathbf{g}_2) \stackrel{?}{=} e(\mathbf{p}_i' - \mathbf{p}_i, \chi)$. This leads to a proof of soundness of the QA-NIZK.

Remark. Observe from the proof above that the soundness can be based on the following *computational* assumption which is implied by XDH, which is a *decisional* assumption:

Definition 1. *Consider a generation algorithm $\mathcal{G}$ taking the security parameter as input, that outputs a tuple $(q, \mathbb{G}_1, \mathbb{G}_2, \mathbb{G}_T, e, \mathbf{g}_1, \mathbf{g}_2)$, where $\mathbb{G}_1, \mathbb{G}_2$ and $\mathbb{G}_T$ are groups of prime order q with generators $\mathbf{g}_1, \mathbf{g}_2$ and $e(\mathbf{g}_1, \mathbf{g}_2)$ respectively and which allow an efficiently computable $\mathbb{Z}_q$-bilinear pairing map $e : \mathbb{G}_1 \times \mathbb{G}_2 \to \mathbb{G}_T$. The assumption asserts that the following problem is hard: Given $\mathbf{f}, \mathbf{f}^b \xleftarrow{\$} \mathbb{G}_2$, output $\mathbf{h}, \mathbf{h}' \in \mathbb{G}_1$, such that $\mathbf{h}' = \mathbf{h}^b \neq \mathbf{0}_1$.*

Example: QA-NIZK for a DH tuple. In this example, we instantiate our general system to provide a NIZK for a DH tuple, that is a tuple of the form $(x \cdot \mathbf{g}, x \cdot \mathbf{f})$ for an a priori fixed base $(\mathbf{g}, \mathbf{f}) \in \mathbb{G}_1^2$. We assume DDH for the group $\mathbb{G}_2$.

As in the setup described before, we have $\mathbf{A} = \left[\mathbf{g}\ \mathbf{f}\right]$. The language is: $L = \{[x] \cdot \mathbf{A} \mid x \in \mathbb{Z}_q\}$.

Now proceeding with the framework, we generate D as $[d]$ and the element b where d and b are random elements of $\mathbb{Z}_q$. With this setting, the NIZK CRS is:

$$\mathbf{CRS}_p := \mathbf{A} \cdot \begin{bmatrix} \mathsf{D} \\ b^{-1} \cdot \mathrm{I}^{1\times 1} \end{bmatrix} = \left[d \cdot \mathbf{g} + b^{-1} \cdot \mathbf{f}\right], \ \mathbf{CRS}_v := \begin{bmatrix} b \cdot \mathsf{D} \\ \mathrm{I}^{1\times 1} \\ -b \cdot \mathrm{I}^{1\times 1} \end{bmatrix} \cdot \mathbf{g}_2 = \begin{bmatrix} bd \cdot \mathbf{g}_2 \\ \mathbf{g}_2 \\ -b \cdot \mathbf{g}_2 \end{bmatrix}$$

The proof of a tuple $(\mathbf{r}, \hat{\mathbf{r}})$ with witness r, is just the *single* element $r \cdot (d \cdot \mathbf{g} + b^{-1} \cdot \mathbf{f})$. In the proof of zero knowledge, the simulator trapdoor is (d, b) and the simulated proof of $(\mathbf{r}, \hat{\mathbf{r}})$ is just $(d \cdot \mathbf{r} + b^{-1} \cdot \hat{\mathbf{r}})$.

4 Extensions

In this section we consider some useful extensions of the concepts and constructions of QA-NIZK systems. We show how the previous system can be extended to include tags. The tags are elements of $\mathbb{Z}_q$, are included as part of the proof and are used as part of the defining equations of the language. We define a notion called split-CRS QA-NIZK system, where the prover and verifier use distinct parts of a CRS and we construct a split-CRS system for affine systems.

Tags. While our system works for any number of components in the tuple (except the first t) being dependent on any number of tags, to simplify the presentation we will focus on only one dependent element and only one tag. Also for simplicity, we will assume that this element is an affine function of the tag (the function being defined by parameters). We can handle arbitrary polynomial functions of the tags as well, but we will focus on affine functions here as most applications seem to need just affine functions. Then, the languages we handle can be characterized as

$$L_{\mathbf{A}, \vec{\mathbf{a}}_1, \vec{\mathbf{a}}_2} = \left\{\left\langle \vec{\mathrm{x}} \cdot \left[\mathbf{A} \mid (\vec{\mathbf{a}}_1^\top + \textsc{tag} \cdot \vec{\mathbf{a}}_2^\top)\right], \textsc{tag}\right\rangle \mid \vec{\mathrm{x}} \in \mathbb{Z}_q^t, \textsc{tag} \in \mathbb{Z}_q\right\}$$

where $\mathbf{A}^{t\times(n-1)}$, $\vec{a}_1^{1\times t}$ and $\vec{a}_2^{1\times t}$ are parameters of the language. A distribution is still called robust (as in Section 3) if with overwhelming probability the first t columns of $\mathbf{A}$ are full-ranked. Write $\mathbf{A}$ as $[\mathbf{A}_l^{t\times t} \mid \mathbf{A}_r^{t\times(n-1-t)}]$, where without loss of generality, $\mathbf{A}_l$ is non-singular. While the first $n-1-t$ components in excess of the unknowns, corresponding to $\mathbf{A}_r$, can be verified just as in Section 3, for the last component we proceed as follows.

Algorithm K_1**.** The CRS is generated as:

$$\mathbf{CRS}_{p,0}^{t\times 1} := \left[\mathbf{A}_l \mid \vec{a}_1^\top\right] \cdot \begin{bmatrix} \mathsf{D}_1 \\ b^{-1} \end{bmatrix} \qquad \mathbf{CRS}_{p,1}^{t\times 1} := \left[\mathbf{A}_l \mid \vec{a}_2^\top\right] \cdot \begin{bmatrix} \mathsf{D}_2 \\ b^{-1} \end{bmatrix}$$

$$\mathbf{CRS}_{v,0}^{(t+2)\times 1} := \begin{bmatrix} b \cdot \mathsf{D}_1 \\ 1 \\ -b \end{bmatrix} \cdot \mathsf{g}_2 \qquad \mathbf{CRS}_{v,1}^{(t+2)\times 1} := \begin{bmatrix} b \cdot \mathsf{D}_2 \\ 0 \\ 0 \end{bmatrix} \cdot \mathsf{g}_2$$

where D_1 and D_2 are random matrices of order $t \times 1$ independent of the matrix D chosen for proving the other components. The $\mathbb{Z}_q$ element b can be re-used from the other components.

Prover. Let $\vec{l}' \stackrel{\text{def}}{=} \vec{x} \cdot \left[\mathbf{A}_l \mid (\vec{a}_1^\top + \text{TAG} \cdot \vec{a}_2^\top)\right]$. The prover generates the following proof for the last component:

$$\vec{p} := \vec{x} \cdot (\mathbf{CRS}_{p,0} + \text{TAG} \cdot \mathbf{CRS}_{p,1})$$

Verifier. Given a proof $\vec{p}$ for candidate $\vec{l}'$ the verifier checks the following:

$$e\left(\left[\vec{l}' \mid \vec{p}\right], \mathbf{CRS}_{v,0} + \text{TAG} \cdot \mathbf{CRS}_{v,1}\right) \stackrel{?}{=} \mathbf{0}_T$$

The size of the proof is 1 element in the group $\mathbb{G}_1$. The proof of completeness, soundness and zero-knowledge for this quasi-adaptive system is similar to proof in Section 3 and a proof sketch can be found in [15].

Split-CRS QA-NIZK Proofs. We note that the QA-NIZK described in Section 3 has an interesting split-CRS property. In a **split-CRS QA-NIZK** for a distribution of relations, the CRS generator K_1 generates two CRS-es ψ_p and ψ_v, such that the prover P *only* needs ψ_p, and the verifier V *only* needs ψ_v. In addition, the CRS ψ_v is *independent* of the particular relation R_ρ. In other words the CRS generator K_1 can be split into two PPTs K_{11} and K_{12}, such that K_{11} generates ψ_v using just λ, and K_{12} generates ψ_p using ρ and a state output by K_{11}. The key generation simulator S_1 is also split similarly. The formal definition is given in [15].

In many applications, split-CRS QA-NIZKs can lead to simpler constructions (and their proofs) and possibly shorter proofs.

Split-CRS QA-NIZK for Affine Spaces. Consider languages that are affine spaces

$$L_{\mathbf{A},\vec{a}} = \{(\vec{x} \cdot \mathbf{A} + \vec{a}) \in \mathbb{G}_1^n \mid \vec{x} \in \mathbb{Z}_q^t\}$$

The parameter language $\mathcal{L}_{\text{par}}$ just specifies $\mathbf{A}$ and $\vec{a}$. A distribution over $\mathcal{L}_{\text{par}}$ is called robust if with overwhelming probability the left most $t \times t$ sub-matrix of $\mathbf{A}$

is non-singular (full-ranked). If $\vec{a}$ is given as part of the verifier CRS, then a QA-NIZK for distributions over this class follows directly from the construction in Section 3. However, that would make the QA-NIZK non split-CRS. We now show that the techniques of Section 3 can be extended to give a split-CRS QA-NIZK for (robust and witness-samplable) distributions over affine spaces.

The common reference string (CRS) has two parts ψ_p and ψ_v which are to be used by the prover and the verifier respectively. The split-CRS generator K_{11} and K_{12} work as follows. Let $s \stackrel{\text{def}}{=} n - t$: this is the number of equations in excess of the unknowns.

Algorithm K_{11}. The verifier CRS generator first generates a matrix $\mathsf{D}^{t \times s}$ with all elements chosen randomly from $\mathbb{Z}_q$ and a single element b chosen randomly from $\mathbb{Z}_q$. It also generates a row vector $\vec{\mathrm{d}}^{1 \times s}$ at random from $\mathbb{Z}_q$. Next, it computes

$$\mathbf{CRS}_v^{(n+s)\times s} := \begin{bmatrix} b \cdot \mathsf{D} \\ \mathrm{I}^{s\times s} \\ -b \cdot \mathrm{I}^{s \times s} \end{bmatrix} \cdot \mathbf{g}_2 \qquad \vec{\mathbf{f}}^{1\times s} := e(\mathbf{g}_1, b \cdot \vec{\mathrm{d}} \cdot \mathbf{g}_2)$$

The verifier CRS ψ_v is the matrix $\mathbf{CRS}_v$ and $\vec{\mathbf{f}}$.

Algorithm K_{12}. The prover CRS generator K_{12} generates

$$\mathbf{CRS}_p^{t\times s} = \begin{bmatrix} \mathbf{A}^{t\times n} \\ \vec{\mathrm{a}}^{1\times n} \end{bmatrix} \cdot \begin{bmatrix} \mathsf{D} \\ b^{-1} \cdot \mathsf{I}^{s\times s} \end{bmatrix} - \begin{bmatrix} 0^{t\times s} \\ \vec{\mathrm{d}}^{1\times s} \end{bmatrix} \cdot \mathbf{g}_1$$

The (prover) CRS ψ_p is just the matrix $\mathbf{CRS}_p$.

Prover. Given candidate $(\vec{\mathrm{x}} \cdot \mathbf{A} + \vec{\mathrm{a}})$ with witness vector $\vec{\mathrm{x}}$, the prover generates the following proof:

$$\vec{\mathbf{p}} := [\vec{\mathrm{x}} \mid 1] \cdot \mathbf{CRS}_p$$

Verifier. Given a proof $\vec{\mathbf{p}}$ of candidate $\vec{l}$, the verifier checks the following:

$$e\left(\left[\vec{l} \mid \vec{\mathbf{p}}\right], \mathbf{CRS}_v\right) \stackrel{?}{=} \vec{\mathbf{f}}$$

We provide a proof sketch in [15]. The split-CRS QA-NIZK for affine spaces also naturally extends to include tags as described before in this section.

5 Applications

In this section we mention several important applications of quasi-adaptive NIZK proofs. Before we go into the details of these applications, we discuss the general applicability of quasi-adaptive NIZKs. Recall in quasi-adaptive NIZKs, the CRS is set based on the language for which proofs are required. In many applications the language is set by a trusted party, and the most obvious example of this is the trusted party that sets the CRS in some UC applications, many of which have UC realizations only with a CRS. Another obvious example is the (H)IBE trusted party that issues secret keys to various identities. In many public key applications, the party issuing the public key is also considered trusted, i.e. incorruptible, as security is defined with respect to the public key issuing party (acting as challenger). Thus, in all these settings if the language for which proofs are required is determined by a incorruptible party, then that party can also issue the QA-NIZK CRS based on that language. It stands to reason that most languages for which proofs are required are ultimately set by an incorruptible party (at least as far as the security definitions are concerned), although they may not be linear subspaces, and can indeed be multi-linear or even quadratic. For example, suppose a potentially corruptible party P wants to (NIZK) prove that $x \in L_\rho$, where L_ρ is a language that it generated. However, this proof is unlikely to be of any use unless it also proves something about L_ρ, e.g., that ρ itself is in another language, say L'. In some applications, potentially corruptible parties generate new linear languages using random tags. However, the underlying basis for these languages is set by a trusted party, and hence the NIZK CRS for the whole range of tag based languages can be generated by that trusted party based on the original basis for these languages.

Adaptive UC Commitments in the Erasure Model. The SXDH-based commitment scheme from [11] requires the following quasi-adaptive NIZK proof (see [15] for details)

$$\{\langle R, S, T\rangle \mid \exists r : R = r \cdot \mathbf{g}, S = r \cdot \mathbf{h}, T = r \cdot (\mathbf{d}_1 + \text{TAG} \cdot \mathbf{e}_1)\}$$

with parameters $\mathbf{h}, \mathbf{d}_1, \mathbf{e}_1$ (chosen randomly), which leads to a UC commitment scheme with commitment consisting of 3 $\mathbb{G}_1$ elements, and a proof consisting of two $\mathbb{G}_2$ elements. Under DLIN, a similar scheme leads to a commitment consisting of 4 elements and an opening of another 4 elements, whereas [11] stated a scheme using Groth-Sahai NIZK proofs requiring $(5+16)$ elements. More details can be found in [15].

One-time (Relatively) Simulation-Sound NIZK for DDH and Others. In [14] it was shown that for linear subspace languages, such as the DDH or DLIN language, or the language showing that two El-Gamal encryptions are of the same message [18,22], the NIZK proof can be made one-time simulation sound using a projective hash proof [7] and proving in addition that the hash proof is correct. For the DLIN language, this one-time simulation sound proof (in Groth-Sahai system) required 15 group elements, whereas the quasi-adaptive proof in this paper leads to a proof of size only 5 group elements.

Signatures. We will now show a generic construction of existentially unforgeable signature scheme (against adaptive adversaries) from labeled CCA2-encryption schemes and split-CRS QA-NIZK proof system (as defined in Section 4) for a related language distribution. This construction is a generalization of a signature scheme from [5] which used (fully) adaptive NIZK proofs and *required* constructions based on groups in which the CDH assumption holds.

Let $\mathcal{E} = (\mathsf{KeyGen}, \mathsf{Enc}, \mathsf{Dec})$ be a labeled CCA-encryption scheme on messages. Let X_m be any subset of the message space of $\mathcal{E}$ such that $1/|X_m|$ is negligible in the security parameter m. Consider the following class of (parametrized) languages $\{L_\rho\}$:

$$L_\rho = \{(c, M) \mid \exists r : \; c = \mathsf{Enc}_{\mathsf{pk}}(\mathbf{u}; r; M)\}$$

with parameter $\rho = (\mathbf{u}, \mathsf{pk})$. The notation $\mathsf{Enc}_{\mathsf{pk}}(\mathbf{u}; r; M)$ means that $\mathbf{u}$ is encrypted under public key pk with randomness r and label M. Consider the following distribution $\mathcal{D}$ on the parameters: $\mathbf{u}$ is chosen uniformly at random from X_m and pk is generated using the probabilistic algorithm KeyGen of $\mathcal{E}$ on 1^m (the secret key is discarded). Note we have an ensemble of distributions, one for each value of the security parameter, but we will suppress these details.

Let $\mathcal{Q} = (\mathsf{K}_0, \langle \mathsf{K}_{11}, \mathsf{K}_{12} \rangle, \mathsf{P}, \mathsf{V})$ be a split-CRS QA-NIZK for distribution $\mathcal{D}$ on $\{L_\rho\}$. Note that the associated parameter language $\mathcal{L}_{\mathrm{par}}$ is just the set of pairs $(\mathbf{u}, \mathsf{pk})$, and $\mathcal{D}$ specifies a distribution on $\mathcal{L}_{\mathrm{par}}$.

Now, consider the following signature scheme $\mathcal{S}$.

Key Generation. On input a security parameter m, run $\mathsf{K}_0(1^m)$ to get λ. Let $\mathcal{E}.\mathsf{pk}$ be generated using KeyGen of $\mathcal{E}$ on 1^m (the secret key sk is discarded). Choose $\mathbf{u}$ at random from X_m. Let $\rho = (\mathbf{u}, \mathcal{E}.\mathsf{pk})$. Generate ψ_v by running K_{11} on λ (it also generates a state s). Generate ψ_p by running K_{12} on (λ, ρ) and state s. The public key $\mathcal{S}.\mathsf{pk}$ of the signature scheme is then ψ_v. The secret key $\mathcal{S}.\mathsf{sk}$ consists of $(\mathbf{u}, \mathcal{E}.\mathsf{pk}, \psi_p)$.

Sign. The signature on M just consists of a pair $\langle c, \pi \rangle$, where c is an $\mathcal{E}$-encryption of $\mathbf{u}$ with label M (using public key $\mathcal{E}.\mathsf{pk}$ and randomness r), and π is the QA-NIZK proof generated using prover P of $\mathcal{Q}$ on input $(\psi_p, (c, M), r)$. Recall r is the witness to the language member (c, M) of L_ρ (and $\rho = (\mathbf{u}, \mathcal{E}.\mathsf{pk})$).

Verify. Given the public key $\mathcal{S}.\mathsf{pk}\,(= \psi_v)$, and a signature $\langle c, \pi \rangle$ on message M, the verifier uses the verifier V of $\mathcal{Q}$ and outputs $\mathsf{V}(\psi_v, (c, M), \pi)$.

Theorem 2. *If $\mathcal{E}$ is a labeled CCA2-encryption scheme and $\mathcal{Q}$ is a split-CRS quasi-adaptive NIZK system for distribution $\mathcal{D}$ on class of languages $\{L_\rho\}$ described above, then the signature scheme described above is existentially unforgeable under adaptive chosen message attacks.*

The theorem is proved in [15]. It is worth remarking here that the reason one can use a quasi-adaptive NIZK here is because the language L_ρ for which (multiple) NIZK proof(s) is required is set (or chosen) by the (signature scheme) key generator, and hence the key generator can generate the CRS for the NIZK after it sets the language. The proof of the above theorem can be understood in terms of simulation-soundness. Suppose the above split-CRS QA-NIZK was also unbounded simulation-sound. Then, one can replace the CCA2 encryption scheme with just a CPA-encryption scheme, and still get a secure signature scheme. A proof sketch of this is as follows: an Adversary $\mathcal{B}$ is only given ψ_v (which is independent of parameters, including $\mathbf{u}$). Further, the simulator for the QA-NIZK can replace all proofs by simulated proofs (that do not use witness r used for encryption). Next, one can employ CPA-security to replace encryptions of $\mathbf{u}$ by encryptions of 1. By unbounded simulation soundness of the QA-NIZK it follows that if $\mathcal{B}$ produces a verifying signature then it must have produced an encryption of $\mathbf{u}$. However, the view of $\mathcal{B}$ is independent of $\mathbf{u}$, and hence its probability of forging a signature is negligible.

However, the best known technique for obtaining efficient unbounded simulation soundness itself requires CCA2 encryption (see [5]), and in addition NIZK proofs for quadratic equations. On the other hand, if we instantiate the above theorem with Cramer-Shoup encryption scheme, we get remarkably short signatures (in fact the shortest signatures under any static and standard assumption). The Cramer-Shoup encryption scheme PK consists of $\mathbf{g}, \mathbf{f}, \mathbf{k}, \mathbf{d}, \mathbf{e}$ chosen randomly from $\mathbb{G}_1$, along with a target collision-resistant hash function $\mathcal{H}$ (with a public random key). The set X from which $\mathbf{u}$ is chosen is just the whole group $\mathbb{G}_1$. Then an encryption of $\mathbf{u}$ is obtained by picking r at random, and obtaining the tuple

$$\langle R = r \cdot \mathbf{g},\ S = r \cdot \mathbf{f},\ T = \mathbf{u} + r \cdot \mathbf{k},\ H = r \cdot (\mathbf{d} + \text{TAG} \cdot \mathbf{e}) \rangle$$

where $\text{TAG} = \mathcal{H}(R, S, T, M)$. It can be shown that it suffices to hide $\mathbf{u}$ with the hash proof H (although one has to go into the internals of the hash-proof based CCA2 encryption; see Appendix in [14]). Thus, we just need a (split-CRS) QA-NIZK for the tag-based *affine* system (it is affine because of the additive constant $\mathbf{u}$). There is one variable r, and three equations (four if we consider the original CCA-2 encryption) Thus, we just need $(3-1)*1 (= 2)$ proof elements, leading to a total signature size of 5 elements (i.e. $R, S, \mathbf{u}+H$, and the two proof elements) under the SXDH assumption.

Dual-System Fully Secure IBE. It is well-known that Identity Based Encryption (IBE) implies signature schemes (due to Naor), but the question arises whether the above signature scheme using Cramer-Shoup CCA2-encryption and the related QA-NIZK can be converted into an IBE scheme. To achieve this, we take a hint from Naor's IBE to Signature Scheme conversion, and let the signatures (on identities) be private keys of the various identities. The verification of the QA-NIZK from Section 3 works by checking $e\left(\left[\vec{l} \mid \vec{\mathbf{p}}\right], \mathsf{CRS}_v\right) \stackrel{?}{=} \mathbf{0}_T^{1\times s}$ (or more precisely, $e\left(\left[\vec{l} \mid \vec{\mathbf{p}}\right], \mathsf{CRS}_v\right) \stackrel{?}{=} \vec{\mathbf{f}}$ for the affine language). However, there are two issues: (1) CRS_v needs to be randomized, (2) there are two equations to be verified (which correspond to the alternate decryption of Cramer-Shoup encryption, providing implicit simulation-soundness). Both these problems are resolved by first scaling CRS_v by a random value s, and then taking a linear combination of the two equations using a public random tag. The right hand side $s \cdot \vec{\mathbf{f}}$ can then serve as secret one-time pad for encryption. Rather than being a provable generic construction, this is more a hint to get to a really short IBE. We give the construction in Appendix A and a complete proof in [15]. It shows an IBE scheme under the SXDH assumption where the ciphertext has only four group

($\mathbb{G}_1$) elements plus a $\mathbb{Z}_q$-tag, which is the shortest IBE known under standard static assumptions[7].

Publicly-Verifiable CCA2 Fully-Secure IBE. We can also extend our IBE scheme above to be publicly-verifiable CCA2-secure [21,1]. Public verifiability is an informal but practical notion: most CCA2-secure schemes have a test of well-formedness of ciphertext, and on passing the test a CPA-secure scheme style decryption suffices. However, if this test can be performed publicly, i.e. without access to the secret key, then we call the scheme publicly-verifiable. While there is a well known reduction from hierarchical IBE to make an IBE scheme CCA2-secure [4], that reduction does not make the scheme publicly-verifiable CCA2 in a useful manner. In the IBE setting, publicly-verifiable *also* requires that it be verifiable if the ciphertext is *valid for the claimed identity.* This can have interesting applications where the network can act as a filter. We show that our scheme above can be extended to be publicly-verifiable CCA2-fully-secure IBE with *only* two additional group elements in the ciphertext (and two additional group elements in the keys). We give the construction in Appendix B and a complete proof in [15]. The IBE scheme above has four group elements (and a tag), where one group element serves as one-time pad for encrypting the plaintext. The remaining three group elements form a linear subspace with one variable as witness and three integer tags corresponding to: (a) the identity, (b) the tag needed in the IBE scheme, and (c) a 1-1 (or universal one-way) hash of some of the elements. We show that if these three group elements can be QA-NIZK proven to be consistent, and given the unique proof property of our QA-NIZKs, then the above IBE scheme can be made CCA2-secure - the dual-system already has implicit simulation-soundness as explained in the signature scheme above, and we show that this QA-NIZK need not be simulation-sound. Since, there are three components, and one variable (see the appendix for details), the QA-NIZK requires only two group elements under SXDH.

References

1. Bellare, M., Desai, A., Pointcheval, D., Rogaway, P.: Relations among notions of security for public-key encryption schemes. In: Krawczyk, H. (ed.) CRYPTO 1998. LNCS, vol. 1462, pp. 26–45. Springer, Heidelberg (1998)
2. Bellare, M., Rogaway, P.: Random oracles are practical: A paradigm for designing efficient protocols. In: Ashby, V. (ed.) ACM CCS 1993, pp. 62–73. ACM Press (November 1993)
3. Blum, M., Feldman, P., Micali, S.: Non-interactive zero-knowledge and its applications (extended abstract). In: STOC, pp. 103–112 (1988)
4. Boneh, D., Canetti, R., Halevi, S., Katz, J.: Chosen-ciphertext security from identity-based encryption. SIAM J. Comput. 36(5), 1301–1328 (2007)

[7] [6] have recently and independently obtained a short IBE under SXDH, but our IBE ciphertexts are even shorter. See Table 2 in the Introduction for detailed comparison.

5. Camenisch, J., Chandran, N., Shoup, V.: A public key encryption scheme secure against key dependent chosen plaintext and adaptive chosen ciphertext attacks. In: Joux, A. (ed.) EUROCRYPT 2009. LNCS, vol. 5479, pp. 351–368. Springer, Heidelberg (2009)
6. Chen, J., Lim, H.W., Ling, S., Wang, H., Wee, H.: Shorter IBE and signatures via asymmetric pairings. In: Abdalla, M., Lange, T. (eds.) Pairing 2012. LNCS, vol. 7708, pp. 122–140. Springer, Heidelberg (2013)
7. Cramer, R., Shoup, V.: Universal hash proofs and a paradigm for adaptive chosen ciphertext secure public-key encryption. In: Knudsen, L.R. (ed.) EUROCRYPT 2002. LNCS, vol. 2332, pp. 45–64. Springer, Heidelberg (2002)
8. Damgård, I.: On Σ protocols, `http://www.daimi.au.dk/~ivan/Sigma.pdf`
9. Escala, A., Herold, G., Kiltz, E., Ràfols, C., Villar, J.: An algebraic framework for diffie-hellman assumptions. In: Canetti, R., Garay, J.A. (eds.) CRYPTO 2013, Part II. LNCS, vol. 8043, pp. 129–147. Springer, Heidelberg (2013)
10. Fiat, A., Shamir, A.: How to prove yourself: Practical solutions to identification and signature problems. In: Odlyzko, A.M. (ed.) CRYPTO 1986. LNCS, vol. 263, pp. 186–194. Springer, Heidelberg (1987)
11. Fischlin, M., Libert, B., Manulis, M.: Non-interactive and re-usable universally composable string commitments with adaptive security. In: Lee, D.H., Wang, X. (eds.) ASIACRYPT 2011. LNCS, vol. 7073, pp. 468–485. Springer, Heidelberg (2011)
12. Groth, J.: Simulation-sound NIZK proofs for a practical language and constant size group signatures. In: Lai, X., Chen, K. (eds.) ASIACRYPT 2006. LNCS, vol. 4284, pp. 444–459. Springer, Heidelberg (2006)
13. Groth, J., Sahai, A.: Efficient non-interactive proof systems for bilinear groups. In: Smart, N.P. (ed.) EUROCRYPT 2008. LNCS, vol. 4965, pp. 415–432. Springer, Heidelberg (2008)
14. Jutla, C., Roy, A.: Relatively-sound NIZKs and password-based key-exchange. In: Fischlin, M., Buchmann, J., Manulis, M. (eds.) PKC 2012. LNCS, vol. 7293, pp. 485–503. Springer, Heidelberg (2012)
15. Jutla, C.S., Roy, A.: Shorter quasi-adaptive NIZK proofs for linear subspaces. Cryptology ePrint Archive, Report 2013/109 (2013), `http://eprint.iacr.org/`
16. Katz, J., Vaikuntanathan, V.: Round-optimal password-based authenticated key exchange. In: Ishai, Y. (ed.) TCC 2011. LNCS, vol. 6597, pp. 293–310. Springer, Heidelberg (2011)
17. Lewko, A.: Tools for simulating features of composite order bilinear groups in the prime order setting. In: Pointcheval, D., Johansson, T. (eds.) EUROCRYPT 2012. LNCS, vol. 7237, pp. 318–335. Springer, Heidelberg (2012)
18. Naor, M., Yung, M.: Public-key cryptosystems provably secure against chosen ciphertext attacks. In: 22nd ACM STOC Annual ACM Symposium on Theory of Computing. ACM Press (May 1990)
19. Okamoto, T., Takashima, K.: Homomorphic encryption and signatures from vector decomposition. In: Galbraith, S.D., Paterson, K.G. (eds.) Pairing 2008. LNCS, vol. 5209, pp. 57–74. Springer, Heidelberg (2008)
20. Okamoto, T., Takashima, K.: Hierarchical predicate encryption for inner-products. In: Matsui, M. (ed.) ASIACRYPT 2009. LNCS, vol. 5912, pp. 214–231. Springer, Heidelberg (2009)
21. Rackoff, C., Simon, D.R.: Non-interactive zero-knowledge proof of knowledge and chosen ciphertext attack. In: Feigenbaum, J. (ed.) CRYPTO 1991. LNCS, vol. 576, pp. 433–444. Springer, Heidelberg (1992)

22. Sahai, A.: Non-malleable non-interactive zero knowledge and adaptive chosen-ciphertext security. In: 40th FOCS Annual Symposium on Foundations of Computer Science, pp. 543–553. IEEE Computer Society Press (October 1999)
23. Shacham, H.: A Cramer-Shoup encryption scheme from the linear assumption and from progressively weaker linear variants. Cryptology ePrint Archive, Report 2007/074 (2007), http://eprint.iacr.org/
24. Waters, B.: Dual system encryption: Realizing fully secure IBE and HIBE under simple assumptions. In: Halevi, S. (ed.) CRYPTO 2009. LNCS, vol. 5677, pp. 619–636. Springer, Heidelberg (2009)

A Dual System IBE under SXDH Assumption

For ease of reading, we switch to multiplicative group notation in the following.
Setup: The authority uses a group generation algorithm for which the SXDH assumption holds to generate a bilinear group $(\mathbb{G}_1, \mathbb{G}_2, \mathbb{G}_T)$ with $\mathbf{g}_1$ and $\mathbf{g}_2$ as generators of $\mathbb{G}_1$ and $\mathbb{G}_2$ respectively. Assume that $\mathbb{G}_1$ and $\mathbb{G}_2$ are of order q, and let e be a bilinear pairing on $\mathbb{G}_1 \times \mathbb{G}_2$. Then it picks c at random from $\mathbb{Z}_q$, and sets $\mathbf{f} = \mathbf{g}_2^c$. It further picks Δ_1, Δ_2, Δ_3, Δ_4, b, d, e, u from $\mathbb{Z}_q$, and publishes the following public key **PK**:
$\mathbf{g}_1, \mathbf{g}_1^b, \mathbf{v}_1 = \mathbf{g}_1^{-\Delta_1 \cdot b+d}, \mathbf{v}_2 = \mathbf{g}_1^{-\Delta_2 \cdot b+e}, \mathbf{v}_3 = \mathbf{g}_1^{-\Delta_3 \cdot b+c}$, and $\mathbf{k} = e(\mathbf{g}_1, \mathbf{g}_2)^{-\Delta_4 \cdot b+u}$.
The authority retains the following master secret key **MSK**: $\mathbf{g}_2$, $\mathbf{f} = (\mathbf{g}_2^c)$, and Δ_1, Δ_2, Δ_3, Δ_4, d, e, u.

Encrypt(PK, i, M). The encryption algorithm chooses s and TAG at random from $\mathbb{Z}_q$. It then blinds M as $C_0 = M \cdot \mathbf{k}^s$, and also creates

$$C_1 = \mathbf{g}_1^s, C_2 = \mathbf{g}_1^{bs}, C_3 = \mathbf{v}_1^s \cdot \mathbf{v}_2^{i \cdot s} \cdot \mathbf{v}_3^{\text{TAG} \cdot s}$$

and the ciphertext is $C = \langle C_0, C_1, C_2, C_3, \text{TAG} \rangle$.

KeyGen(MSK, i). The authority chooses r at random from $\mathbb{Z}_q$ and creates

$$R = \mathbf{g}_2^r, S = \mathbf{g}_2^{r \cdot c}, T = \mathbf{g}_2^{u+r\cdot(d+i\cdot e)}, W_1 = \mathbf{g}_2^{-\Delta_4 - r\cdot(\Delta_1 + i \cdot \Delta_2)}, W_2 = \mathbf{g}_2^{-r\cdot\Delta_3}$$

as the secret key K_i for identity i.

Decrypt(K_i, C). Let TAG be the tag in C. Obtain

$$\kappa = \frac{e(C_1, S^{\text{TAG}} \cdot T) \cdot e(C_2, W_1 \cdot W_2^{\text{TAG}})}{e(C_3, R)}$$

and output C_0/κ.

Theorem 3. *Under the SXDH Assumption, the above scheme is a fully-secure IBE scheme.*

B Publicly Verifiable CCA2-IBE under SXDH Assumption

Setup. The authority uses a group generation algorithm for which the SXDH assumption holds to generate a bilinear group $(\mathbb{G}_1, \mathbb{G}_2, \mathbb{G}_T)$ with $\mathbf{g}_2$ and $\mathbf{g}_1$ as generators of $\mathbb{G}_1$ and $\mathbb{G}_2$ respectively. Assume that $\mathbb{G}_1$ and $\mathbb{G}_2$ are of order q, and let e be a bilinear pairing on $\mathbb{G}_1 \times \mathbb{G}_2$. Then it picks c at random from $\mathbb{Z}_q$, and sets $\mathbf{f} = \mathbf{g}_2^c$. It further picks Δ_1, Δ_2, Δ_3, Δ_4, Δ_5, b, d, e, u, z from $\mathbb{Z}_q$, and publishes the following public key **PK**:
$\mathbf{g}_1$, $\mathbf{g}_1^b$, $\mathbf{v}_1 = \mathbf{g}_1^{-\Delta_1 \cdot b+d}$, $\mathbf{v}_2 = \mathbf{g}_1^{-\Delta_2 \cdot b+e}$, $\mathbf{v}_3 = \mathbf{g}_1^{-\Delta_3 \cdot b+c}$, $\mathbf{v}_4 = \mathbf{g}_1^{-\Delta_4 \cdot b+z}$, and $\mathbf{k} = e(\mathbf{g}_1, \mathbf{g}_2)^{-\Delta_5 \cdot b+u}$.

Consider the language:

$$L = \{\langle C_1, C_2, C_3, i, \text{TAG}, h\rangle \mid \exists s : C_1 = \mathbf{g}_1^s, C_2 = \mathbf{g}_1^{bs}, C_3 = \mathbf{v}_1^s \cdot \mathbf{v}_2^{i \cdot s} \cdot \mathbf{v}_3^{\text{TAG} \cdot s} \cdot \mathbf{v}_4^{h \cdot s}\}$$

It also publishes the QA-NIZK CRS for the language L (which uses tags i, TAG and h). It also publishes a 1-1, or Universal One-Way Hash function (UOWHF) $\mathcal{H}$. The authority retains the following master secret key **MSK**: $\mathbf{g}_2$, $\mathbf{f}$ $(= \mathbf{g}_2^c)$, and Δ_1, Δ_2, Δ_3, Δ_4, Δ_5, d, e, u, z.

Encrypt(PK, i, M). The encryption algorithm chooses s and TAG at random from $\mathbb{Z}_q$. It then blinds M as $C_0 = M \cdot \mathbf{k}^s$, and also creates

$$C_1 = \mathbf{g}_1^s, C_2 = \mathbf{g}_1^{b \cdot s}, C_3 = \mathbf{v}_1^s \cdot \mathbf{v}_2^{i \cdot s} \cdot \mathbf{v}_3^{\text{TAG} \cdot s} \cdot \mathbf{v}_4^{h \cdot s},$$

where $h = \mathcal{H}(C_0, C_1, C_2, \text{TAG}, i)$. The ciphertext is then $C = \langle C_0, C_1, C_2, C_3, \text{TAG}, \mathbf{p}_1, \mathbf{p}_2\rangle$, where $\langle \mathbf{p}_1, \mathbf{p}_2\rangle$ is a QA-NIZK proof that $\langle C_0, C_1, C_2, C_3, i, \text{TAG}, h\rangle \in L$.

KeyGen(MSK, i). The authority chooses r at random from $\mathbb{Z}_q$ and creates

$$R = \mathbf{g}_2^r, S_1 = \mathbf{g}_2^{r \cdot c}, S_2 = \mathbf{g}_2^{r \cdot z}, T = \mathbf{g}_2^{u + r \cdot (d + i \cdot e)},$$
$$W_1 = \mathbf{g}_2^{-\Delta_5 - r \cdot (\Delta_1 + i \cdot \Delta_2)}, W_2 = \mathbf{g}_2^{-r \cdot \Delta_3}, W_3 = \mathbf{g}_2^{-r \cdot \Delta_4}$$

as the secret key K_i for identity i.

Decrypt(K_i, C). Let TAG be the tag in C. Let $h = \mathcal{H}(C_0, C_1, C_2, \text{TAG}, i)$. First (publicly) verify that the ciphertext satisfies the QA-NIZK for the language above. Then, obtain

$$\kappa = \frac{e(C_1, S_1^{\text{TAG}} \cdot S_2^h \cdot T) \cdot e(C_2, W_1 \cdot W_2^{\text{TAG}} \cdot W_3^h)}{e(C_3, R)}$$

and output C_0/κ. If the QA-NIZK does not verify, output $\perp$.

This public-verifiability of the consistency test is informally called the publicly-verifiable CCA2 security.

Theorem 4. *Under the SXDH Assumption, the above scheme is a CCA2 fully-secure IBE scheme.*

Constant-Round Concurrent Zero Knowledge in the Bounded Player Model

Vipul Goyal[1], Abhishek Jain[2], Rafail Ostrovsky[3], Silas Richelson[4], and Ivan Visconti[5]

[1] Microsoft Research, India
vipul@microsoft.com
[2] MIT and Boston University, USA
abhishek@csail.mit.edu
[3] UCLA, USA
rafail@cs.ucla.edu
[4] UCLA, USA
sirichel@math.ucla.edu
[5] University of Salerno, Italy
visconti@dia.unisa.it

Abstract. In [18] Goyal et al. introduced the bounded player model for secure computation. In the bounded player model, there are an a priori bounded number of players in the system, however, each player may execute any unbounded (polynomial) number of sessions. They showed that even though the model consists of a relatively mild relaxation of the standard model, it allows for round-efficient concurrent zero knowledge. Their protocol requires a super-constant number of rounds. In this work we show, constructively, that there exists a *constant-round* concurrent zero-knowledge argument in the bounded player model. Our result relies on a new technique where the simulator obtains a trapdoor corresponding to a player identity by putting together information obtained in multiple sessions. Our protocol is only based on the existence of a collision-resistance hash-function family and comes with a "straight-line" simulator.

We note that this constitutes the strongest result known on constant-round concurrent zero knowledge in the plain model (under well accepted relaxations) and subsumes Barak's constant-round bounded concurrent zero-knowledge result. We view this as a positive step towards getting constant round fully concurrent zero-knowledge in the plain model, without relaxations.

Keywords: concurrent zero knowledge, straight-line simulation, bounded player model.

1 Introduction

The notion of a zero-knowledge proof [17] is central in cryptography, both for its conceptual importance and for its wide ranging applications to the design of

K. Sako and P. Sarkar (Eds.) ASIACRYPT 2013 Part I, LNCS 8269, pp. 21–40, 2013.

secure cryptography protocols. Initial results for zero-knowledge were in the so called stand-alone setting where there is a single protocol execution happening in isolation.

The fact that on the Internet an adversary can control several players motivated the notion of concurrent zero knowledge [15] (cZK). Here the prover is simultaneously involved in several sessions and the scheduling of the messages is coordinated by the adversary who also keeps control of all verifiers. Concurrent zero knowledge is much harder to achieve than zero knowledge. Indeed, while we know how to achieve zero-knowledge in 4 rounds, a sequence of results [21,33,6] increased the lower bound on the round complexity of concurrent zero-knowledge with black-box simulation to almost logarithmic in the security parameter. In the meanwhile, the upper bound has been improved and now almost matches the logarithmic lower bound [31,20,30]. After almost a decade of research on this topic, the super-logarithmic round concurrent zero-knowledge protocol of Prabhakaran et al. [30] remains the best known in terms of round complexity.

Some hope for a better round complexity started from the breakthrough result of Barak [1] where non-black-box simulation under standard assumptions was proposed. His results showed how to obtain bounded-concurrent zero knowledge in constant rounds. This refers to the setting where there is an a priori fixed bound on the total number of concurrent executions (and the protocol may become completely insecure if the actual number of sessions exceed this bound). Unfortunately, since then, the question of achieving sub-logarithmic round complexity with unbounded concurrency using non-black-box techniques has remained open, and represents one of the most challenging open questions in the study of zero-knowledge protocols.[1]

Bounded player model. Recently, Goyal, Jain, Ostrovsky, Richelson and Visconti [18] introduced the so called bounded player model. In this model, it is only assumed that there is an a-priori (polynomial) upper-bound on the total number of players that may ever participate in protocol executions. There is no setup stage, or, trusted party, and the simulation must be performed in polynomial time. While there is a bound on the number of players, any player may join in at any time and may be subsequently involved in any unbounded (polynomial) number of concurrent sessions. Since there is no a priori bound on the number of sessions, it is a strengthening of the bounded-concurrency model used in Barak's result. The bounded player model also has some superficial similarities to the bare-public-key model of [5] which is discussed later in this section.

As an example, if we consider even a restriction to a single verifier that runs an unbounded number of sessions, the simulation strategy of [1] breaks down completely. Goyal et al. [18] gave a $\omega(1)$-round concurrent zero knowledge protocol in the bounded player model. The technique they proposed relies on the

[1] In this paper, we limit our discussion to results which are based on standard complexity-theoretic and number-theoretic assumptions. We note that constant round concurrent zero-knowledge is known to exist under non-standard assumptions such as a variation of the (non-falsifiable) knowledge of exponent assumption [19] or the existence of P-certificates [8].

fact that the simulator has several choices in every sessions on where to spend computation trying to extract a trapdoor, and, its running time is guaranteed to be polynomial as long as the number of such choices is super-constant. Their technique fails inherently if constant round-complexity is desired.

We believe the eventual goal of achieving round efficient concurrent zero-knowledge (under accepted assumptions) is an ambitious one. Progress towards this goal would not only impact how efficiently one can implement zero-knowledge (in the network setting), but also, will improve various secure computation protocol constructions in this setting (as several secure computation protocols use, e.g., PRS preamble [30] for concurrent input extraction). Bounded player model is somewhere between the standard model (where the best known protocols require super-logarithmic number of rounds), and, the bounded concurrency model (where constant round protocols are known). We believe the study of round complexity of concurrent zero-knowledge in the bounded player model might shed light on how to construct such protocols in the standard model as well.

Our Results. In this work, we give a constant-round protocol in the bounded player (BP) model. Our constructions inherently relies on non-black-box simulation. The simulator for our protocol does not rely on rewinding techniques and instead works in a "straight-line" manner (as in Barak [1]). Our construction is only based on the existence of a collision-resistant hash-function family.

Theorem 1. *Assuming the existence of a collision-resistance hash-function family, there exists a constant round concurrent zero-knowledge argument system with concurrent soundness in the bounded player model.*

We note that this constitutes the strongest result known on constant-round zero-knowledge in the concurrent setting (in the plain model). It subsumes Barak's result: now the total number of sessions no longer needs to be bounded; only the number of new players starting the interaction with the prover is bounded. A player might join in at anytime and may subsequently be involved in any unbounded (polynomial) number of sessions.

We further note that, as proved by Goyal et al. [18], unlike previously studied relaxations of the standard model (e.g., bounded number of sessions, timing assumptions, super- polynomial simulation), concurrent-secure computation is still impossible to achieve in the bounded player model. This gives evidence that the BP model is "closer" to the standard model than previously studied models, and study of this model might shed light on constructing constant-round concurrent zero-knowledge in the standard model as well. Moreover, despite the impossibility of concurrent-secure computation, techniques developed in the concurrent zero-knowledge literature have found applications in other areas in cryptography, including resettable security [5], non-malleability [14], and even in proving black-box lower bounds [27].

1.1 Technical Overview

In this section, first, we recall some observations by Goyal et al [18] regarding why simple approaches to extend the construction of Barak [1] to the bounded player model are bound to fail. We also recall the basic idea behind the protocol of [18]. Armed with this background, we then proceed to discuss the key technical ideas behind our constant round cZK protocol in the bounded player model. Initial parts of this section are borrowed verbatim from [18].

Why natural approaches fail. Recall that in the bounded player model, the only assumption is that the total number of players that will ever be present in the system is *a priori* bounded. Then, as observed by Goyal et al [18], the black-box lower-bound of Canetti et al. [6] is applicable to the bounded player model as well. Thus, it is clear that we must resort to non-black-box techniques. Now, a natural approach to leverage the bound on the number of players is to associate with each verifier V_i a public key pk_i and then design an FLS-style protocol [16] that allows the ZK simulator to extract, in a non-black-box manner, the secret key sk_i of the verifier and then use it as a "trapdoor" for "easy" simulation. The key intuition is that once the simulator extracts the secret key sk_i of a verifier V_i, it can perform easy simulation of *all* the sessions associated with V_i. Then, since the total number of verifiers is bounded, the simulator will need to perform non-black-box extraction only an *a priori* bounded number of times (once for each verifier), which can be handled in a manner similar to the setting of bounded-concurrency [1].

Unfortunately, as observed by Goyal et al. [18], the above intuition is misleading. In order to understand the problem with the above approach, let us first consider a candidate protocol more concretely. In fact, it suffices to focus on a preamble phase that enables non-black-box extraction (by the simulator) of a verifier's secret key since the remainder of the protocol can be constructed in a straightforward manner following the FLS approach. Now, consider the following candidate preamble phase (using the non-black-box extraction technique of [3]): first, the prover and verifier engage in a coin-tossing protocol where the prover proves "honest behavior" using a Barak-style non-black-box ZK protocol [1]. Then, the verifier sends an encryption of its secret key under the public key that is determined from the output of the coin-tossing protocol [18].

In order to analyze this protocol, we will restrict our discussion to the simplified case where only one verifier is present in the system (but the total number of concurrent sessions are unbounded). At this point, one may immediately object that in the case of a single verifier identity, the problem is not interesting since the bounded player model is identical to the bare-public key model, where one can construct four-round cZK protocols using rewinding based techniques. However, simulation techniques involving rewinding do not "scale" well to the case of polynomially many identities (unless we use a large number of rounds) and fail. In contrast, our simulation approach is "straight-line" for an unbounded number of sessions and scales well to a large bounded number of identities. Therefore, in the forthcoming discussion, we will restrict our discussion to straight-line simulation. In this case, we find it instructive to focus on the case of a single identity to explain the key issues and our ideas to resolve them.

We now turn to analyze the candidate protocol. Now, following the intuition described earlier, one may think that the simulator can simply cheat in the coin-tossing protocol in the "inner-most" session in order to extract the secret key, following which all the sessions can be simulated in a straight-line manner, without performing any additional non-black-box simulation. Consider, however, the following adversarial verifier strategy: the verifier schedules an unbounded number of sessions in such a manner that the coin-tossing protocols in all of these sessions are executed in a "nested" manner. Furthermore, the verifier sends the ciphertext (containing its secret key) in each session only *after* all the coin-tossing protocols across all sessions are completed. Note that in such a scenario, the simulator would be forced to perform non-black-box simulation in an unbounded number of sessions. Unfortunately, this is a non-trivial problem that we do not know how to solve.

The approach of Goyal et al. [18]. In an effort to bypass the above problem, Goyal et al. use multiple ($\omega(1)$, to be precise) preamble phases (instead of only one), such that the simulator is required to "cheat" in only one of these preambles. This, however, immediately raises a question: in which of the $\omega(1)$ preambles should the simulator cheat? This is a delicate question since if, for example, we let the simulator pick one of preambles uniformly at random, then with non-negligible probability, the simulator will end up choosing the first preamble phase. In this case, the adversary can simply perform the same attack as it did earlier playing only the first preamble phase, but for many different sessions so that the simulator will still have to cheat in many of them. Indeed, it would seem that any randomized oblivious simulation strategy can be attacked in a similar manner by simply identifying the first preamble phase where the simulator would cheat with a non-negligible probability.

The main idea in [18] is to use a specific probability distribution such that the simulator cheats in the first preamble phase with only negligible probability, while the probability of cheating in the later preambles increases gradually such that the "overall" probability of cheating is 1 (as required). Further, the distribution is such that the probability of cheating in the ith preamble is less than a fixed polynomial factor of the total probability of cheating in one of the previous $i-1$ blocks. This allows them (by a careful choice of parameters) to ensure that the probability of the simulator failing in more than a given polynomially bounded number of sessions w.r.t. any given verifier is negligible (and then rely on the techniques from the bounded-concurrency model [1] to handle the bounded number of non-black-box simulations).

Our Construction. The techniques used in our work are quite different and unrelated to the techniques in the work of Goyal et al. [18]. As illustrated in the discussion above, the key issue is the following. Say that a slot of the protocol completes. Then, the simulator starts the non-black-box simulation and computes the first "heavy" universal argument message, and, sends it across. However, before the simulator can finish this simulation successfully (and somehow learn a trapdoor from the verifier which can then be used to complete other sessions without non-black-box simulation), the verifier switches to another session. Then, in order to proceed, the simulator would have to perform non-black-box simulation and the heavy computation again (resulting in the number of sessions where non-black-box simulation is performed becoming unbounded). So overall, the problem is the "delay" between the heavy computation, and, the point at which the simulator extracts the verifier trapdoor (which can then be used to quickly pass through other sessions with this particular verifier without any heavy computation or non-black-box simulation).

Our basic approach is to "construct the trapdoor slowly as we go along": have any heavy computation done in any session (with this verifier) contribute to the construction of a trapdoor which can then be used to quickly pass through other sessions. To illustrate our idea, we shall focus on the case of a single verifier as before. The description below is slightly oversimplified for the sake of readability.

To start with, in the very first session, the verifier is supposed to choose a key pair of a signature scheme (this key pair remains the same across all sessions involving this verifier). As in Barak's protocol [1], we will just have a single slot followed by a universal argument (UA). However, now once a slot is complete, the verifier is required to immediately send a signature[2] on the transcript of the slot (i.e., on the prover commitment, and, the verifier random string) to the prover. This slot now constitutes a "hard statement" certified by the verifier: it could be used by the prover in any session (with this verifier). If the prover could prove that he has a signed slot such that the machine committed to in this slot could output the verifier random string in this slot, the verifier would be instructed to accept. Thus, the simulator would now simply take the first slot that completes (across all sessions), and, would prove the resulting "hard statement" in the universal arguments of all the sessions. This would allow him to presumably compute the required PCP only once and use it across all sessions. Are we done? Turns out that the answer is no.

Even if the prover is executing the UA corresponding to the same slot (on which he has obtained a signature) in every session, because of the interactive nature of UAs, the (heavy) computation the prover does in a session cannot be entirely used in another session. This is because the challenge of the verifier would be different in different sessions. To solve this problem and continue the construction of a single trapdoor (useful across all sessions), we apply our basic idea one more time. The prover computes and sends the first UA message. The verifier is required to respond with a random challenge and a signature on the UA transcript so far. The prover can compute the final UA message, and, the construction of the trapdoor is complete: the trapdoor constitutes of a signed slot, an accepting UA transcript (proving that the machine committed to in the slot indeed outputs the random string in that slot), and, a signature on the first two UA messages (proving that the challenge was indeed generated by the verifier after getting the first UA message). To summarize, the simulator would use the following two sessions for the construction of the trapdoor: the first session

[2] Signatures of committed messages computed by a verifier where previously used in [12] to allow the simulator to get through rewindings one more signature in order to cheat in the main thread. Here instead we insist with straight-line simulation.

where a slot completes, and, the first session where the verifier sends the UA random challenge.

The above idea indeed is oversimplified and ignores several problems. Firstly, since an honest prover executes each concurrent session oblivious of others, any correlations in the prover messages across different sessions (in particular, sending the same UA first message) would lead to the simulated transcript being distinguishable from the real one. Furthermore, the prover could be proving a different overall statement to the verifier in every session (and hence even a UA first message cannot be reused across different sessions). The detailed description of our construction is given in Section 3.

1.2 Related Work

Bare public key and other related models. The bare public key model was proposed in [5] where, before any interaction starts, every player is required to declare a public key and store it in a public file (which never changes once the sessions start). In this model it is known how to obtain constant-round concurrent zero knowledge with concurrent soundness under standard assumptions [13,35,36,34]. This model has also been used for constant-round concurrent non-malleable zero knowledge [25] and various constant-round resettable and simultaneously resettable protocols [22,39,11,9,10,38,37,7].

As discussed in [18], the crucial restriction of the BPK model is that all players who wish to ever participate in protocol executions must be fixed during the preprocessing phase, and new players cannot be added "on-the-fly" during the proof phase. We do *not* make such a restriction in our work and, despite superficial resemblance, the techniques useful in constructing secure protocols in the BPK model have limited relevance in our setting. In particular, constant round cZK is known to exist in the BPK model using only black-box simulation, while in our setting, non-black-box techniques are *necessary* to achieve sublogarithmic-round cZK.

In light of the above discussion, since the very premise of the BPK model (that all players are fixed ahead of time and declare a key) does not hold in the bounded player model, we believe that the bounded player model is much closer in spirit (as well as technically) to the bounded concurrency model of Barak. The bounded player model is a strict generalization of the bounded concurrency model. Thus, our constant-round construction is the first strict improvement to Barak's bounded concurrent ZK protocol. We stress that we improve the achieved security under concurrent composition, still under standard assumptions and without introducing any setup/weakness. Summing up, ours is a construction which is the closest known to achieving constant-round concurrent zero knowledge in the plain model.

Round efficient concurrent zero-knowledge is known in a number of other models as well (which do not seem to be directly relevant to our setting) such as the common-reference string model, the super-polynomial simulation model, etc. We refer the reader to [18] for a more detailed discussion.

2 Preliminaries and Definitions

Notation. We will use the symbol "||" to denote the concatenation of two strings appearing respectively before and after the symbol.

2.1 Bounded Player Model

We first recall the *bounded player model* for concurrent security, as introduced in [18]. In the bounded player model, there is an a-priori (polynomial) upper bound on the total number of player that will ever be present in the system. Specifically, let n denote the security parameter. Then, we consider an upper bound $N = \text{poly}(n)$ on the total number of players that can engage in concurrent executions of a protocol at any time. We assume that each player P_i ($i \in N$) has an associated unique identity id_i, and that there is an established mechanism to enforce that party P_i uses the same identity id_i in each protocol execution that it participates in. Note, however, that such identities do not have to be established in advance. In particular, new players can join the system with their own (new) identities, as long as the number of players does not exceed N. We stress that there is not bound on the number of protocol executions that can be started by each party.

The bounded player model is formalized by means of a functionality F_{bp}^N that registers the identities of the player in the system. Specifically, a player P_i that wishes to participate in protocol executions can, at any time, register an identity id_i with the functionality F_{bp}^N. The registration functionality does not perform any checks on the identities that are registered, except that each party P_i can register at most one identity id_i, and that the total number of identity registrations are bounded by N. In other words, F_{bp}^N refuses to register any new identities once N number of identities have already been registered. The functionality F_{bp}^N is formally defined in Figure 1.

Functionality F_{bp}^N

F_{bp}^N initializes a variable *count* to 0 and proceeds as follows.

- **Register commands:** Upon receiving a message (**register**, *sid*, id_i) from some party P_i, the functionality checks that no pair (P_i, id'_i) is already recorded and that $count < N$. If this is the case, it records the pair (P_i, id_i) and sets $count = count + 1$. Otherwise, it ignores the received message.
- **Retrieve commands:** Upon receiving a message (retrieve, *sid*, P_i) from some party P_j or the adversary A, the functionality checks if some pair (P_i, id_i) is recorded. If this the case, it sends $(sid, P_i, \mathsf{id}_i)$ to P_j (or A). Otherwise, it returns $(sid, P_i, \perp)$.

Fig. 1. The Bounded Player Functionality F_{bp}^N

In our constructions we will explicitly work in the setting where the identity of each party is a tuple (h, vk), where $h \leftarrow \mathcal{H}_n$ is a hash function chosen from a family $\mathcal{H}_n$ of collision resistant hash functions, and vk is a verification key for a signature scheme.

2.2 Concurrent Zero Knowledge in Bounded Player Model

In this section, we formally define concurrent zero knowledge in the bounded player model. The definition given below, is an adaptation of the one of [30] to the bounded player model, by also considering non-black-box simulation. Some of the text below is taken verbatim from [30].

Let PPT denote probabilistic-polynomial time. Let $\langle P, V \rangle$ be an interactive argument for a language L. Consider a concurrent adversarial verifier V^* that, given input $x \in L$, interacts with an unbounded number of independent copies of P (all on the same common input x and moreover equipped with a proper witness w), without any restriction over the scheduling of the messages in the different interactions with P. In particular, V^* has control over the scheduling of the messages in these interactions. Further, we say that V^* is an N-bounded concurrent adversary if it assumes at most N verifier identities during its (unbounded) interactions with P.[3]

The transcript of a concurrent interaction consists of the common input x, followed by the sequence of prover and verifier messages exchanged during the interaction. We denote by $\mathsf{view}^P_{V^*}(x, z, N)$ the random variable describing the content of the random tape of the N-bounded concurrent adversary V^* with auxiliary input z and the transcript of the concurrent interaction between P and V^* on common input x.

Definition 1 (Concurrent Zero Knowledge in Bounded Player Model). *Let $\langle P, V \rangle$ be an interactive argument system for a language L. We say that $\langle P, V \rangle$ is* concurrent zero-knowledge *in the bounded player model if for every N-bounded concurrent non-uniform* PPT *adversary V^*, there exists a* PPT *algorithm $\mathcal{S}$, such that the following ensembles are computationally indistinguishable, $\{\mathsf{view}^P_{V^*}(x, z, N)\}_{x \in L, z \in \{0,1\}^*}$ and $\{\mathcal{S}(x, z, N)\}_{x \in L, z \in \{0,1\}^*}$.*

As a final note, we remark that following previous work in the BPK model and in the BP model, we will consider the notion of concurrent soundness where the malicious prover is allowed to play any concurrent number of sessions with the same verifier. Indeed, this is notion is strictly stronger than sequential soundness.

2.3 Building Blocks

In this section, we discuss the main building blocks that we will use in our cZK construction.

[3] Thus, V^* can open multiple sessions with P for every unique verifier identity.

Statistically binding commitment schemes. In our constructions, we will make use of a statistically binding string commitment scheme, denoted **Com**. For simplicity of exposition, we will make the simplifying assumption that **Com** is a non-interactive perfectly binding commitment scheme. In reality, **Com** would be taken to be a standard 2-round commitment scheme, e.g. [24]. Unless stated otherwise, we will simply use the notation $\mathbf{Com}(x)$ to denote a commitment to a string x, and assume that the randomness (used to create the commitment) is implicit. We will denote by $\mathbf{Com}(x; r)$ a commitment to a string x with randomness r.

Witness indistinguishable arguments of knowledge. We will also make use of a witness-indistinguishable proof of knowledge (WIPOK) for all of $\mathcal{NP}$ in our construction. Such a scheme can be constructed, for example, by parallel repetition of the 3-round Blum's protocol for Graph Hamiltonicity [4]. We will denote such an argument system by $\langle P_{\mathsf{WI}}, V_{\mathsf{WI}} \rangle$.

The universal argument of [2]. In our construction, we will use the 4-round universal argument system (UA), denoted pUA presented in [2] and based on the existence of collision-resistant hash functions. We will assume without loss of generality that the initial commitment of the PCP sent by the prover in the second round also contains a commitment of the statement. We notice that such an argument system is still sound when the prover is required to open the commitment of the statement in the very last round.

Signature schemes. We will use a signature scheme (**KeyGen**, **Sign**, **Verify**) that is unforgeable against chosen message attacks. Note that such signatures schemes are known based on one way functions [32].

3 A Constant-Round Protocol

In this section, we describe our constant-round concurrent zero-knowledge protocol in the bounded player model.

Relation R_{sim}. We first recall a slight variant of Barak's [1] $\mathbf{NTIME}(T(n))$ relation R_{sim}, as used previously in [28]. Let $T : \mathbb{N} \rightarrow \mathbb{N}$ be a "nice" function that satisfies $T(n) = n^{\omega(1)}$. Let $\{\mathcal{H}_n\}_n$ be a family of collision-resistant hash functions where a function $h \in \mathcal{H}_n$ maps $\{0,1\}^*$ to $\{0,1\}^n$, and let **Com** be a perfectly binding commitment scheme for strings of length n, where for any $\alpha \in \{0,1\}^n$, the length of $\mathbf{Com}(\alpha)$ is upper bounded by $2n$. The relation R_{sim} is described in Figure 2.

Remark 1. The relation presented in Figure 2 is slightly oversimplified and will make Barak's protocol work only when $\{\mathcal{H}_n\}_n$ is collision-resistant against "slightly" super-polynomial sized circuits. For simplicity of exposition, in this manuscript, we will work with this assumption. We stress, however, that as discussed in prior works *[2,26,29,28,18]*, this assumption can be relaxed by using

Instance: A triplet $\langle h, c, r\rangle \in \mathcal{H}_n \times \{0,1\}^n \times \{0,1\}^{\text{poly}(n)}$.
Witness: A program $\Pi \in \{0,1\}^*$, a string $y \in \{0,1\}^*$ and a string $s \in \{0,1\}^{\text{poly}(n)}$.
Relation: $R_{\mathsf{sim}}(\langle h, c, r\rangle, \langle \Pi, y, s\rangle) = 1$ if and only if:

1. $|y| \leq |r| - n$.
2. $c = \mathbf{Com}(h(\Pi); s)$.
3. $\Pi(y) = r$ within $T(n)$ steps.

Fig. 2. R_{sim} - A variant of Barak's relation [28]

a "good" error-correcting code ECC (with constant distance and polynomial-time encoding and decoding procedures), and replacing the condition $c = \mathbf{Com}(h(\Pi); s)$ with $c = \mathbf{Com}(\mathsf{ECC}(h(\Pi)); s)$.

Our protocol. We are now ready to present our concurrent zero knowledge protocol, denoted $\langle P, V\rangle$. Let P and V denote the prover and verifier respectively. Let N denote the bound on the number of verifiers in the system. In our construction, the identity of a verifier V_i corresponds to a verification key vk_i of a secure signature scheme and a hash function $h_i \in \mathcal{H}_n$ from a family $\mathcal{H}_n$ of collision-resistant hash functions. Let (**KeyGen**, **Sign**, **Verify**) be a secure signature scheme. Let $\langle P_{\mathsf{WI}}, V_{\mathsf{WI}}\rangle$ be a witness-indistinguishable argument of knowledge system. Let pUA be the universal argument (UARG) system of [2] that we discussed previously; the transcript is composed by four messages $(h, \beta, \gamma, \delta)$ where h is a collision-resistant hash function.

The protocol $\langle P, V\rangle$ is described in Figure 3. For our purposes, we set the length parameter $\ell(N) = N \cdot P(n) + n$, where $P(n)$ is a polynomial upper bound on the total length of the prover messages in the UARG pUA plus the output length of a hash function $h \in \mathcal{H}_n$. For simplicity we omit some standard checks (e.g., the prover needs to check that vk and h are recorded, the prover needs to check that the signatures is valid).

The completeness property of $\langle P, V\rangle$ follows immediately from the construction. Next, in Section 3.2, we prove *concurrent soundness* of $\langle P, V\rangle$, i.e., we show that a computationally-bounded adversarial prover who engages in multiple concurrent executions of $\langle P, V\rangle$ (where the scheduling across the sessions is controlled by the adversary) can not prove a false statement in any of the executions, except with negligible probability. As observed in [18], "stand-alone" soundness does not imply concurrent soundness in the bounded player model. Informally speaking, this is because the standard approach of reducing concurrent soundness to stand-alone soundness by "internally" emulating all but one verifier does not work since the verifier's keys are private.[4]

[4] Indeed, Micali and Reyzin [23] gave concrete counter-examples to show that stand-alone soundness does not imply concurrent soundness in the bare public key model. It is not difficult to see that their results immediately extend to the bounded player model.

Parameters: Security parameter n, number of players $N = N(n)$, length parameter $\ell(N)$.
Common Input: $x \in \{0,1\}^{\mathrm{poly}(n)}$.
Private Input to P: A witness w s.t. $R_L(x, w) = 1$.
Private Input to V: A key pair $(sk, vk) \xleftarrow{R} \mathbf{KeyGen}(1^n)$, and a hash function $h \xleftarrow{R} \mathcal{H}_n$.

Stage 1 (Preamble Phase):
$V \to P$: Send vk, h.
$P \to V$: Send $c = \mathbf{Com}(0^n)$.
$V \to P$: Send $r \xleftarrow{R} \{0,1\}^{\ell(N)}$, and $\sigma = \mathbf{Sign}_{sk}(c \| r)$.
$P \to V$: Send $c' = \mathbf{Com}(0^n)$.
$V \to P$: Send $\gamma \xleftarrow{R} \{0,1\}^n$, and $\sigma' = \mathbf{Sign}_{sk}(c' \| \gamma)$.
Stage 2 (Proof Phase):
$P \leftrightarrow V$: An execution of WIPOK $\langle P_{\mathsf{WI}}, V_{\mathsf{WI}} \rangle$ to prove the OR of the following statements:
1. $\exists w \in \{0,1\}^{\mathrm{poly}(|x|)}$ s.t. $R_L(x, w) = 1$.
2. $\exists \langle c, r, \sigma \rangle$, and $\langle \beta, \gamma, \delta, c', t, \sigma' \rangle$ s.t.
   - $\mathbf{Verify}_{vk}(c \| r; \sigma) = 1$, and
   - $c' = \mathbf{Com}(\beta; t)$, and $\mathbf{Verify}_{vk}(c' \| \gamma; \sigma') = 1$, and
   - $(h, \beta, \gamma, \delta)$ is an accepting transcript for a UARG pUA proving the following statement: $\exists \langle \Pi, y, s \rangle$ s.t. $R_{\mathsf{sim}}(\langle h, c, r \rangle, \langle \Pi, y, s \rangle) = 1$.

Fig. 3. Protocol $\langle P, V \rangle$

We now turn to prove that protocol $\langle P, V \rangle$ is concurrent zero-knowledge in the bounded player model.

3.1 Proof of Concurrent Zero Knowledge

In this section, we prove that the protocol $\langle P, V \rangle$ described in Section 3 is concurrent zero-knowledge in the bounded player model. Towards this end, we will construct a non-black-box (polynomial-time) simulator and then prove that the concurrent adversary's view output by the simulator is indistinguishable from the real view. We start by giving an overview of the proof and then proceed to give details.

Overview. Recall that unlike the bounded concurrency model, the main challenge in the bounded player model is that the total number of sessions that a concurrent verifier may schedule is not a priori bounded. Thus, one can not directly employ Barak's simulation strategy of committing to a machine that takes only a bounded-length input y (smaller than the challenge string r) and outputs the next message of the verifier. Towards this end, the crucial observation in [18] is that in the bounded player model, once the simulator is able to "solve" the identity of a specific verifier, then it does not need to be perform any more "expensive" (Barak-style) non-black-box simulation for that identity. Then, the main challenge remaining is to ensure that the expensive non-black-box simulations that need to be performed *before* the simulator can solve a particular identity, can be a-priori bounded, regardless of the number of concurrent sessions opened by the verifier. Indeed, [18] use a randomized simulation strategy (that crucially relies on a super-constant number of rounds) to achieve this effect.

In our case, we also build on the same set of observations. However, we crucially follow a different strategy to a-priori bound the number of expensive non-black-box simulations that need to performed in order to solve a given identity. In particular, unlike [18], where the "trapdoor" for a given verifier simply corresponds to its secret key, in our case, the trapdoor consists of a signed statement and a corresponding universal argument proof transcript (where the signature is computed by the verifier using the signing key corresponding to its identity). Further, and more crucially, unlike [18], where the simulator makes a "disjoint" effort in each session corresponding to a verifier to extract the trapdoor, in our case, the simulator gradually builds the trapdoor by making "joint" effort across the sessions. In fact, our simulator only performs *one* expensive non-black-box simulation per identity; as such, the a-priori bound on the number of identities immediately yields us the desired effect. Indeed, this is why we can perform concurrent simulation in only a constant number of rounds.

The Simulator. We now proceed to describe our simulator $\mathcal{S}$. Let N denote the a priori bound on the number of verifiers in the system. Then, the simulator $\mathcal{S}$ interacts with an adversary $V^* = (V_1^*, \ldots, V_N^*)$ who controls verifiers $V_1, \ldots, V_N$. V^* interacts with $\mathcal{S}$ in m sessions, and controls the scheduling of the messages. $\mathcal{S}$ is given non-black-box access to V^*.

The simulator $\mathcal{S}$ consists of two main subroutines, namely, $\mathcal{S}_{\mathsf{easy}}$ and $\mathcal{S}_{\mathsf{heavy}}$. As the name suggests, the job of $\mathcal{S}_{\mathsf{heavy}}$ is to perform the "expensive" non-black-box simulation operations, namely, constructing the transcripts of universal arguments, which yield a trapdoor for every verifier V_i. On the other hand, $\mathcal{S}_{\mathsf{easy}}$ computes the actual (simulated) prover messages in both the preamble phase and the proof phase, by using the trapdoors. We now give more details.

SIMULATOR $\mathcal{S}$. Throughout the simulation, $\mathcal{S}$ maintains the following three data structures, each of which is initialized to $\perp$:

1. a list $\boldsymbol{\pi} = (\pi_1, \ldots, \pi_N)$, where each π_i is either $\perp$ or is computed to be $h_i(\Pi)$. Here, h_i is the hash function corresponding to V_i and Π is the augmented machine code that is used for non-black-box simulation. We defer the description of Π to below.
2. a list $\mathsf{trap}^{\mathsf{heavy}} = (\mathsf{trap}_1^{\mathsf{heavy}}, \ldots, \mathsf{trap}_N^{\mathsf{heavy}})$, where each $\mathsf{trap}_i^{\mathsf{heavy}}$ corresponds to a tuple $\langle h_i, c, r, \Pi, y, s\rangle$ s.t. $R_{\mathsf{sim}}(\langle h_i, c, r\rangle, \langle \Pi, y, s\rangle) = 1$.

3. a list $\mathsf{trap}^{\mathsf{easy}} = (\mathsf{trap}_1^{\mathsf{easy}}, \ldots, \mathsf{trap}_N^{\mathsf{easy}})$, where each $\mathsf{trap}_i^{\mathsf{easy}}$ corresponds to a tuple $\langle c, r, \sigma, \beta, \gamma, \delta, c', t, \sigma' \rangle$ s.t.
   - $\mathbf{Verify}_{vk_i}(c \| r; \sigma) = 1$, and
   - $c' = \mathbf{Com}(\beta; t)$, and $\mathbf{Verify}_{vk_i}(c' \| \gamma; \sigma') = 1$, and
   - $(h_i, \beta, \gamma, \delta)$ is an accepting transcript for a UARG pUA proving the following statement: $\exists \langle \Pi, y, s \rangle$ s.t. $R_{\mathsf{sim}}(\langle h_i, c, r \rangle, \langle \Pi, y, s \rangle) = 1$.

Augmented machine Π. The augmented machine code Π simply consists of the code of the adversarial verifier V^* and the code of the subroutine $\mathcal{S}_{\mathsf{easy}}$ (with a sufficiently long random tape hardwired, to compute the prover messages in each session) , i.e., $\Pi = (V^*, \mathcal{S}_{\mathsf{easy}})$. The input y to the machine Π consists of the lists $\boldsymbol{\pi}$ and $\mathsf{trap}^{\mathsf{easy}}$, i.e., $y = (\boldsymbol{\pi}, \mathsf{trap}^{\mathsf{easy}})$. Note that it follows from the description that $|y| \leq \ell(N) - n$.

We now describe the subroutines $\mathcal{S}_{\mathsf{easy}}$ and $\mathcal{S}_{\mathsf{heavy}}$, and then proceed to give a formal description of $\mathcal{S}$. For simplicity of exposition, in the discussion below, we assume that the verifier sends the first message in the WIPOK $\langle P_{\mathsf{WI}}, V_{\mathsf{WI}} \rangle$.

ALGORITHM $\mathcal{S}_{\mathsf{easy}}(i, \mathsf{msg}_j^V, \boldsymbol{\pi}, \mathsf{trap}^{\mathsf{easy}}; z)$. The algorithm $\mathcal{S}_{\mathsf{easy}}$ prepares the (simulated) messages of the prover P in the protocol. More specifically, when executed with input $(i, \mathsf{msg}_j^V, \boldsymbol{\pi}, \mathsf{trap}^{\mathsf{easy}}; z)$, $\mathcal{S}_{\mathsf{easy}}$ does the following:

1. If msg_j^V is the first verifier message of the preamble phase from V_i in a session, then $\mathcal{S}_{\mathsf{easy}}$ parses $\boldsymbol{\pi}$ as $\pi_1, \ldots, \pi_N$. It computes and outputs $c = \mathbf{Com}(\pi_i; z)$.
2. If msg_j^V is the second verifier message of the preamble phase from V_i in a session, then $\mathcal{S}_{\mathsf{easy}}$ computes and outputs $c = \mathbf{Com}(\beta; z)$, where β is the corresponding (i.e., fourth) entry in $\mathsf{trap}_i^{\mathsf{easy}} \in \mathsf{trap}^{\mathsf{easy}}$.
3. If msg_j^V is a verifier message of the WIPOK from V_i in the proof phase of a session, then if $\mathsf{trap}_i^{\mathsf{easy}} = \perp$, then $\mathcal{S}_{\mathsf{easy}}$ aborts and outputs $\perp$, otherwise $\mathcal{S}_{\mathsf{easy}}$ simply runs the code of the honest P_{WI} to compute the response using randomness z and the trapdoor witness $\mathsf{trap}_i^{\mathsf{easy}}$.

ALGORITHM $\mathcal{S}_{\mathsf{heavy}}(i, j, \gamma, \mathsf{trap}^{\mathsf{heavy}})$. The algorithm $\mathcal{S}_{\mathsf{heavy}}$ simply prepares *one* UARG transcript for every verifier V_i, which in turn is used as a trapdoor by the algorithm $\mathcal{S}_{\mathsf{easy}}$. More concretely, when executed with input $(i, j, \gamma, \mathsf{trap}^{\mathsf{heavy}})$, $\mathcal{S}_{\mathsf{heavy}}$ does the following:

1. If $j = 1$, then $\mathcal{S}_{\mathsf{heavy}}$ parses the ith entry $\mathsf{trap}_i^{\mathsf{heavy}}$ in $\mathsf{trap}^{\mathsf{heavy}}$ as (h_i, c, r, Π, y, s). It runs the honest prover algorithm P_{UA} and computes the first message β of a UARG for the statement: $\exists \langle \Pi, y, s \rangle$ s.t. $R_{\mathsf{sim}}(\langle h_i, c, r \rangle, \langle \Pi, y, s \rangle) = 1$. $\mathcal{S}_{\mathsf{heavy}}$ saves its internal state as state_i and outputs β.[5]
2. If $j = 2$, then $\mathcal{S}_{\mathsf{heavy}}$ uses state_i and γ to honestly compute the final prover message δ for the UARG with prefix (h_i, β, γ). It outputs δ.

ALGORITHM $\mathcal{S}$. Given the above subroutines, the simulator $\mathcal{S}$ works as follows. We assume that every time $\mathcal{S}$ updates the lists $\boldsymbol{\pi}$ and $\mathsf{trap}^{\mathsf{easy}}$, it also automatically updates the entry corresponding to y (i.e., the fifth entry) in each

[5] For simplicity of exposition, we describe $\mathcal{S}_{\mathsf{heavy}}$ as a stateful algorithm.

$\mathsf{trap}_i^{\mathsf{heavy}} \in \mathsf{trap}^{\mathsf{heavy}}$. For simplicity of exposition, we do not explicitly mention this below.

Preamble phase:

1. On receiving the first message $\mathsf{msg}_1^V = (vk_i, h_i)$ from V^* on behalf of V_i in the preamble phase of a session, $\mathcal{S}$ first checks whether $\pi_i = \bot$ (where π_i is the ith entry in the list $\boldsymbol{\pi}$); if the check succeeds, then $\mathcal{S}$ updates $\pi_i = h_i(\Pi)$. Next, $\mathcal{S}$ samples fresh randomness s from its random tape and runs $\mathcal{S}_{\mathsf{easy}}$ on input $(i, \mathsf{msg}_1^V, \boldsymbol{\pi}, \mathsf{trap}^{\mathsf{easy}}; s)$. $\mathcal{S}$ sends the output string c from $\mathcal{S}_{\mathsf{easy}}$ to V^*. Further, $\mathcal{S}$ adds $(h_i, c, \cdot, \Pi, y, s)$ to $\mathsf{trap}_i^{\mathsf{heavy}}$ and $(c, \cdot, \cdot, \cdot, \cdot, \cdot, \cdot, \cdot)$ to $\mathsf{trap}_i^{\mathsf{easy}}$.
2. On receiving the second message message $\mathsf{msg}_2^V = (r, \sigma)$ from V^* on behalf of V_i in the preamble phase of a session, $\mathcal{S}$ first verifies the validity of the signature σ w.r.t. vk_i. If the check fails, $\mathcal{S}$ considers this session aborted (as the prover would do) and ignores any additional message for this session. Otherwise, $\mathcal{S}$ checks whether the entries corresponding to r and σ (i.e., 2nd and 3rd entries) in $\mathsf{trap}_i^{\mathsf{easy}}$ are $\bot$. If the check succeeds, then:
   - $\mathcal{S}$ sets r as 3rd entry of $\mathsf{trap}_i^{\mathsf{heavy}}$ and r, σ as second and third entries of $\mathsf{trap}_i^{\mathsf{easy}}$.
   - Further, $\mathcal{S}$ runs $\mathcal{S}_{\mathsf{heavy}}$ on input[6] $(i, 1, \bot, \mathsf{trap}^{\mathsf{heavy}})$ to compute the message β of a UARG for the statement: $\exists \langle \Pi, y, s\rangle$ s.t. $R_{\mathsf{sim}}(\langle h_i, c, r\rangle, \langle \Pi, y, s\rangle) = 1$. Here h_i, c, r, Π, y, s are such that $\mathsf{trap}_i^{\mathsf{heavy}} = \langle h_i, c, r, \Pi, y, s\rangle$.
   - On receiving the output message β, $\mathcal{S}$ sets to β the fourth slot of $\mathsf{trap}_i^{\mathsf{easy}}$.

   Next, $\mathcal{S}$ samples fresh randomness t and runs $\mathcal{S}_{\mathsf{easy}}$ on input $(i, \mathsf{msg}_2^V, \boldsymbol{\pi}, \mathsf{trap}^{\mathsf{easy}}; t)$. On receiving the output string c' from $\mathcal{S}_{\mathsf{easy}}$, $\mathcal{S}$ forwards it to V^*. Further, $\mathcal{S}$ sets to (c', t) the 7th and 8th slot of $\mathsf{trap}_i^{\mathsf{easy}}$.
3. Finally, on receiving the last message $\mathsf{msg}_{\mathsf{fin}}^V = (\gamma, \sigma')$ from V^* on behalf of V_i in the preamble phase of a session, $\mathcal{S}$ first verifies the validity of the signature σ' w.r.t. vk_i. If the check fails, $\mathcal{S}$ considers this session aborted (as the prover would do) and ignores any additional message for this session. Otherwise, $\mathcal{S}$ checks whether the entries corresponding to γ and σ' in $\mathsf{trap}_i^{\mathsf{easy}}$ are $\bot$. If the check succeeds, then:
   - $\mathcal{S}$ sets to γ and σ' the 5th and 9th slot of $\mathsf{trap}_i^{\mathsf{easy}}$.
   - Further, $\mathcal{S}$ runs $\mathcal{S}_{\mathsf{heavy}}$ on input $(i, 2, \gamma, \mathsf{trap}^{\mathsf{heavy}})$ to compute the final prover message δ of the UARG with prefix (h_i, β, γ), where (β, γ) are the corresponding entries in $\mathsf{trap}_i^{\mathsf{easy}}$.
   - On receiving the output message δ, $\mathcal{S}$ sets to δ the 6th slot of $\mathsf{trap}_i^{\mathsf{easy}}$.

Proof phase: On receiving any message msg_j^V from V^* on behalf of V_i, $\mathcal{S}$ runs $\mathcal{S}_{\mathsf{easy}}$ on input $(i, \mathsf{msg}_j^V, \boldsymbol{\pi}, \mathsf{trap}^{\mathsf{easy}})$ and fresh randomness. $\mathcal{S}$ forwards the output message of $\mathcal{S}_{\mathsf{easy}}$ to V^*.

This completes the description of $\mathcal{S}$ and the subroutines $\mathcal{S}_{\mathsf{easy}}$, $\mathcal{S}_{\mathsf{heavy}}$. It follows immediately from the above description that $\mathcal{S}$ runs in polynomial time and outputs $\bot$ with probability negligibly close to an honest prover.

---

[6] For simplicity of exposition, we assume that randomness is hardwired in $\mathcal{S}^{\mathsf{heavy}}$ and do not mention it explicitly.

We now show through a series of hybrid experiments the simulator's output is computationally indistinguishable from the output of the adversary when interacting with honest provers. Our hybrid experiments will be H_i for $i = 0, \ldots, 3$. We write $H_i \approx H_j$ if no V^* can distinguish (except with negligible probability) between its interaction with H_i and H_j.

Hybrid H_0. Experiment H_0 corresponds to the honest prover. That is, in every session $j \in [m]$, H_0 sends c and c' as commitments to the all zeros string in the preamble phase. We provide H_0 with a witness that $x \in L$ which it uses to complete the both executions of the WIPOK $\langle P_{\mathsf{WI}}, V_{\mathsf{WI}} \rangle$ played in each session.

Hybrid H_1. Experiment H_1 is similar to H_0, except the following. For every $i \in [N]$, for every session corresponding to verifier V_i, the commitment c in the preamble phase is prepared as a commitment to $\pi_i = h_i(\Pi)$, where h_i is the hash function in the identity of V_i and Π is the augmented machine code as described above.

The computational hiding property of **Com** ensures that $H_1 \approx H_0$.

Hybrid H_2. Experiment H_2 is similar to H_1, except the following. For every $i \in [N]$, for every session corresponding to verifier V_i, the commitment c' in the preamble phase is prepared as a commitment to the string β with randomness t, where β is the first prover message of a UARG computed by $\mathcal{S}_{\mathsf{heavy}}$, in the manner as described above.

The computational hiding property of **Com** ensures that $H_2 \approx H_1$.

Hybrid H_3. Experiment H_3 is similar to H_2, except the following. For every $i \in [N]$, for every session corresponding to verifier V_i, the WIPOK $\langle P_{\mathsf{WI}}, V_{\mathsf{WI}} \rangle$ in the proof phase is executed using the trapdoor witness $\mathsf{trap}_i^{\mathsf{easy}}$, in the manner as described above. Note that this is our simulator $\mathcal{S}$.

The witness indistinguishability property of $\langle P_{\mathsf{WI}}, V_{\mathsf{WI}} \rangle$ ensures that $H_3 \approx H_2$.

3.2 Proof of Concurrent Soundness

Consider the interaction between a cheating P^* and an honest V. Suppose that P^* fools V into accepting a false proof in some session with non-negligible probability. We show how to reduce P^* to an adversary that breaks the security of one of the used ingredients. We will first consider P^* as a sequential malicious prover. We will discuss the issues deriving from a concurrent attack later.

First of all, notice that by the proof of knowledge property of the second WIPOK, we have that with non-negligible probability, an efficient adversary E can simply run as a honest verifier and extract a witness from that WIPOK of session l where the false statement is proved. Since the statement is false, the witness extracted will therefore be $(c, r, \sigma, \beta, \gamma, \delta, c', t, \sigma')$ such that $\mathbf{Verify}_{vk}(c \| r; \sigma) = 1$, $c' = \mathbf{Com}(\beta; t)$, $\mathbf{Verify}_{vk}(c' \| \gamma; \sigma') = 1$, and $(h, \beta, \gamma, \delta)$ is an accepting transcript for a UARG pUA proving the statement $\exists \langle \Pi, y, s \rangle$ s.t. $R_{\mathsf{sim}}(\langle h, c, r \rangle, \langle \Pi, y, s \rangle) = 1$, and h is the hash function corresponding to the verifier run by E in session l.

By the security of the signature scheme, it must be the case that signatures σ and σ' were generated and sent by E during the experiment (the reduction is standard and omitted).

Therefore we have that with non-negligible probability there is a session i where h and γ were played honestly by E, $(h, \beta, \gamma, \delta)$ is an accepting transcript for the UARG for $R_{\mathsf{sim}}(\langle h, c, r\rangle, \langle \Pi, y, s\rangle) = 1$, and a commitment to β was given before γ was sent. Moreover, there is a session j where c and r were played as commitment and challenge. Remember that the session l is the one where the false statement is proved.

We can now complete the proof by relying almost verbatim on the same analysis of [1,2]. Indeed, by rewinding the prover and changing the challenge r in session j, with another random string, we would have an execution identically distributed with respect to the previous one. Therefore it will happen with non-negligible probability that the prover succeeds in session l, still relying on the information obtained in sessions i and j. The analysis of [1,2] by relying on the weak proof of knowledge property of the UA, shows that this event can be reduced to finding a collision that contradicts the collision resistance of h.

We finally discuss the case of a concurrent adversarial prover. Such an attack is played by a prover aiming at obtaining from concurrent sessions some information to be used in the target session where the false theorem must be proved. In previous work in the BPK model and in the BP model this was a major problem because the verifier used to give a proof of knowledge of its secret key, and the malleability of such a proof of knowledge could be exploited by the malicious prover. Our protocol however bypasses this attack because our verifier does not give a proof of knowledge of the secret key of the signature scheme, but only gives signatures of specific messages. Indeed the only point in which the above proof of soundness needs to be upgraded is the claim that by the security of the signature scheme, it must be the case that signatures σ and σ' where generated and sent by E during the experiment. In case of sequential attack, this is true because running the extractor of the WIPOK in session l does not impact on other sessions since they were played in full either before or after session l. Instead, in case of a concurrent attack, while rewinding the adversarial prover, new sessions could be started and more signatures could be needed. As a result, it could happen that in such new sessions the prover would ask precisely the same signatures that are then extracted from the target session. We can conclude that this does not impact on the proof for the following two reasons. First, in the proof of soundness it does not matter if those signatures appear in the transcript of the attack, or just in the transcript of a rewinded execution. Second, the reduction on the security of the signature scheme works for any polynomial number of signatures asked to the oracle, therefore still holds in case of a concurrent attack. Indeed, the work of E is performed in polynomial time even when rewinding a concurrent malicious prover, therefore playing in total (i.e., summing sessions in the view of the prover and sessions played during rewinds) a polynomial number of sessions, and therefore asking a polynomial number of signatures only to the signature oracle.

Further details on the proof of soundness. Given a transcript $(h, UA1, UA2, UA3)$ for the universal argument of [2], we stress that soundness still works when the prover sends the statement to the verifier only at the 4th round, opening a commitment played in the second round. The proof of concurrent soundness of our protocol goes through a reduction to the soundness of the universal argument of [2] and goes as follows.

Let P^*_{ua} be the adversarial prover that we construct against the universal argument of [2], by making use of the adversary P^* of our protocol. Let V_{ua} be the honest verifier of the universal argument of [2]. P^*_{ua} gets "h" from V_{ua} and plays it in a random session s of the experiment (it could therefore be played in a rewinding thread) with P^*. Later on, since by contradiction P^* is successful, UA messages $(UA1, UA2, UA3)$ are extracted and with noticeable probability they correspond to session s. Therefore P^*_{ua} sends $UA1$ to V_{ua} and gets back $UA2'$. Then P^*_{ua} rewinds P* to the precise point where $UA2$ was played. Now P^*_{ua} plays $UA2'$. Again, later on, since by contradiction P^* is successful, P^*_{ua} will again extract from P^* and with noticeable probability (still because the number of sessions played in the experiment is polynomial), it will get an accepting transcript $(UA1, UA2', UA3^*)$ for the same statement (this is guaranteed by the security of the signature scheme and the binding of the commitment). Then P^*_{ua} can send $UA3^*$ to V_{ua} therefore proving a false statement.

Acknowledgments. Work supported in part by NSF grants 0830803, 09165174, 1065276, 1118126 and 1136174, US-Israel BSF grant 2008411, OKAWA Foundation Research Award, IBM Faculty Research Award, Xerox Faculty Research Award, B. John Garrick Foundation Award, Teradata Research Award, MIUR Project PRIN "GenData 2020" and Lockheed-Martin Corporation Research Award. This material is based upon work supported by the Defense Advanced Research Projects Agency through the U.S. Office of Naval Research under Contract N00014 – 11 – 1 – 0392. The views expressed are those of the author and do not reflect the official policy or position of the Department of Defense or the U.S. Government.

References

1. Barak, B.: How to go beyond the black-box simulation barrier. In: FOCS, pp. 106–115 (2001)
2. Barak, B., Goldreich, O.: Universal arguments and their applications. In: IEEE Conference on Computational Complexity, pp. 194–203 (2002)
3. Barak, B., Lindell, Y.: Strict polynomial-time in simulation and extraction. In: STOC, pp. 484–493 (2002)
4. Blum, M.: How to prove a theorem so no one else can claim it. In: Proceedings of the International Congress of Mathematicians, pp. 1444–1451 (1987)
5. Canetti, R., Goldreich, O., Goldwasser, S., Micali, S.: Resettable zero-knowledge (extended abstract). In: STOC, pp. 235–244 (2000)
6. Canetti, R., Kilian, J., Petrank, E., Rosen, A.: Black-box concurrent zero-knowledge requires $\tilde{\Omega}(\log n)$ rounds. In: STOC, pp. 570–579 (2001)

7. Cho, C., Ostrovsky, R., Scafuro, A., Visconti, I.: Simultaneously resettable arguments of knowledge. In: Cramer, R. (ed.) TCC 2012. LNCS, vol. 7194, pp. 530–547. Springer, Heidelberg (2012)
8. Chung, K.M., Lin, H., Pass, R.: Constant-round concurrent zero knowledge from p-certificates. In: FOCS. IEEE Computer Society (2013)
9. Deng, Y., Lin, D.: Instance-dependent verifiable random functions and their application to simultaneous resettability. In: Naor, M. (ed.) EUROCRYPT 2007. LNCS, vol. 4515, pp. 148–168. Springer, Heidelberg (2007)
10. Deng, Y., Lin, D.: Resettable zero knowledge with concurrent soundness in the bare public-key model under standard assumption. In: Pei, D., Yung, M., Lin, D., Wu, C. (eds.) Inscrypt 2007. LNCS, vol. 4990, pp. 123–137. Springer, Heidelberg (2008)
11. Di Crescenzo, G., Persiano, G., Visconti, I.: Constant-round resettable zero knowledge with concurrent soundness in the bare public-key model. In: Franklin, M. (ed.) CRYPTO 2004. LNCS, vol. 3152, pp. 237–253. Springer, Heidelberg (2004)
12. Di Crescenzo, G., Persiano, G., Visconti, I.: Improved setup assumptions for 3-round resettable zero knowledge. In: Lee, P.J. (ed.) ASIACRYPT 2004. LNCS, vol. 3329, pp. 530–544. Springer, Heidelberg (2004)
13. Di Crescenzo, G., Visconti, I.: Concurrent zero knowledge in the public-key model. In: Caires, L., Italiano, G.F., Monteiro, L., Palamidessi, C., Yung, M. (eds.) ICALP 2005. LNCS, vol. 3580, pp. 816–827. Springer, Heidelberg (2005)
14. Dolev, D., Dwork, C., Naor, M.: Nonmalleable cryptography. SIAM J. Comput. 30(2), 391–437 (2000)
15. Dwork, C., Naor, M., Sahai, A.: Concurrent zero-knowledge. In: STOC, pp. 409–418 (1998)
16. Feige, U., Lapidot, D., Shamir, A.: Multiple non-interactive zero knowledge proofs based on a single random string (extended abstract). In: FOCS, pp. 308–317 (1990)
17. Goldwasser, S., Micali, S., Rackoff, C.: The knowledge complexity of interactive proof-systems (extended abstract). In: STOC, pp. 291–304 (1985)
18. Goyal, V., Jain, A., Ostrovsky, R., Richelson, S., Visconti, I.: Concurrent zero knowledge in the bounded player model. In: Sahai, A. (ed.) TCC 2013. LNCS, vol. 7785, pp. 60–79. Springer, Heidelberg (2013)
19. Gupta, D., Sahai, A.: On constant-round concurrent zero-knowledge from a knowledge assumption. IACR Cryptology ePrint Archive 2012, 572 (2012)
20. Kilian, J., Petrank, E.: Concurrent and resettable zero-knowledge in polyloalgorithm rounds. In: STOC, pp. 560–569 (2001)
21. Kilian, J., Petrank, E., Rackoff, C.: Lower bounds for zero knowledge on the internet. In: FOCS, pp. 484–492 (1998)
22. Micali, S., Reyzin, L.: Soundness in the public-key model. In: Kilian, J. (ed.) CRYPTO 2001. LNCS, vol. 2139, pp. 542–565. Springer, Heidelberg (2001)
23. Micali, S., Reyzin, L.: Soundness in the public-key model. In: Kilian, J. (ed.) CRYPTO 2001. LNCS, vol. 2139, pp. 542–565. Springer, Heidelberg (2001)
24. Naor, M.: Bit commitment using pseudorandomness. J. Cryptology 4(2), 151–158 (1991)
25. Ostrovsky, R., Persiano, G., Visconti, I.: Constant-round concurrent non-malleable zero knowledge in the bare public-key model. In: Aceto, L., Damgård, I., Goldberg, L.A., Halldórsson, M.M., Ingólfsdóttir, A., Walukiewicz, I. (eds.) ICALP 2008, Part II. LNCS, vol. 5126, pp. 548–559. Springer, Heidelberg (2008)
26. Pass, R.: Bounded-concurrent secure multi-party computation with a dishonest majority. In: STOC, pp. 232–241 (2004)

27. Pass, R.: Limits of provable security from standard assumptions. In: STOC, pp. 109–118 (2011)
28. Pass, R., Rosen, A.: Concurrent non-malleable commitments. In: FOCS, pp. 563–572 (2005)
29. Pass, R., Rosen, A.: New and improved constructions of non-malleable cryptographic protocols. In: STOC, pp. 533–542 (2005)
30. Prabhakaran, M., Rosen, A., Sahai, A.: Concurrent zero knowledge with logarithmic round-complexity. In: FOCS, pp. 366–375 (2002)
31. Richardson, R., Kilian, J.: On the concurrent composition of zero-knowledge proofs. In: Stern, J. (ed.) EUROCRYPT 1999. LNCS, vol. 1592, pp. 415–431. Springer, Heidelberg (1999)
32. Rompel, J.: One-way functions are necessary and sufficient for secure signatures. In: STOC, pp. 387–394 (1990)
33. Rosen, A.: A note on the round-complexity of concurrent zero-knowledge. In: Bellare, M. (ed.) CRYPTO 2000. LNCS, vol. 1880, pp. 451–468. Springer, Heidelberg (2000)
34. Scafuro, A., Visconti, I.: On round-optimal zero knowledge in the bare public-key model. In: Pointcheval, D., Johansson, T. (eds.) EUROCRYPT 2012. LNCS, vol. 7237, pp. 153–171. Springer, Heidelberg (2012)
35. Visconti, I.: Efficient zero knowledge on the internet. In: Bugliesi, M., Preneel, B., Sassone, V., Wegener, I. (eds.) ICALP 2006. LNCS, vol. 4052, pp. 22–33. Springer, Heidelberg (2006)
36. Yao, A.C., Yung, M., Zhao, Y.: Concurrent knowledge extraction in the public-key model. In: Abramsky, S., Gavoille, C., Kirchner, C., Meyer auf der Heide, F., Spirakis, P.G. (eds.) ICALP 2010, Part I. LNCS, vol. 6198, pp. 702–714. Springer, Heidelberg (2010)
37. Deng, Y., Feng, D., Goyal, V., Lin, D., Sahai, A., Yung, M.: Resettable cryptography in constant rounds – the case of zero knowledge. In: Lee, D.H., Wang, X. (eds.) ASIACRYPT 2011. LNCS, vol. 7073, pp. 390–406. Springer, Heidelberg (2011)
38. Yung, M., Zhao, Y.: Generic and practical resettable zero-knowledge in the bare public-key model. In: Naor, M. (ed.) EUROCRYPT 2007. LNCS, vol. 4515, pp. 129–147. Springer, Heidelberg (2007)
39. Zhao, Y.: Concurrent/resettable zero-knowledge with concurrent soundness in the bare public-key model and its applications. Cryptology ePrint Archive, Report 2003/265 (2003), `http://eprint.iacr.org/`

Succinct Non-Interactive Zero Knowledge Arguments from Span Programs and Linear Error-Correcting Codes

Helger Lipmaa

Institute of Computer Science, University of Tartu, Estonia

Abstract. Gennaro, Gentry, Parno and Raykova proposed an efficient NIZK argument for CIRCUIT-SAT, based on non-standard tools like conscientious and quadratic span programs. We propose a new linear PCP for the CIRCUIT-SAT, based on a combination of *standard* span programs (that verify the correctness of every individual gate) and high-distance linear error-correcting codes (that check the consistency of wire assignments). This allows us to simplify all steps of the argument, which results in significantly improved efficiency. We then construct an NIZK CIRCUIT-SAT argument based on existing techniques.

Keywords: Circuit-SAT, linear error-correcting codes, linear PCP, non-interactive zero knowledge, polynomial algebra, quadratic span program, span program, verifiable computation.

1 Introduction

By using non-interactive zero knowledge (NIZK, [3]), the prover can create a proof π, s.t. any verifier can later, given access to a common reference string, the statement, and π, verify the truth of the intended statement without learning any side information. Since a single proof might get transferred and verified many times, one often requires sublinear communication and verifier's computation. (Unless stated explicitly, we measure the communication in group elements, and the computation in group operations.) While succinct NIZK proofs are important in many cryptographic applications, there are only a few different generic methodologies to construct them efficiently.

Groth [16] proposed the first sublinear-communication NIZK argument (computationally-sound proof, [4]) for an **NP**-complete language. His construction was improved by Lipmaa [19]. Their CIRCUIT-SAT argument consists of efficient arguments for more primitive tasks like Hadamard sum, Hadamard product and permutation. The CIRCUIT-SAT arguments of [16,19] have constant communication, quadratic prover's computation, and linear verifier's computation in s (the circuit size). In [16], the CRS length is $\Theta(s^2)$, and in [19], it is $\Theta(r_3^{-1}(s)) = o(s2^{2\sqrt{2\log_2 s}})$, where $r_3(N) = \Omega(N \log^{1/4} N / 2^{2\sqrt{2\log_2 N}})$ [9] is the cardinality of the largest progression-free subset of $[N]$. Because of the quadratic prover's computation, the arguments of Groth and Lipmaa are not applicable in

K. Sako and P. Sarkar (Eds.) ASIACRYPT 2013 Part I, LNCS 8269, pp. 41–60, 2013.

practice, unless s is really small. Very recently, Fauzi, Lipmaa and Zhang [10] constructed arguments for **NP**-complete languages Set Partition, Subset Sum and Decision Knapsack with the CRS length $\Theta(r_3^{-1}(s))$ and prover's computation $\Theta(r_3^{-1}(s) \log s)$. They did not propose a similar argument for the CIRCUIT-SAT.

Gennaro, Gentry, Parno and Raykova [15] constructed a CIRCUIT-SAT NIZK argument based on efficient (quadratic) span programs. Their argument consists of two steps. The first step is an information-theoretic reduction from the CIRCUIT-SAT to QSP-SAT [2], the satisfaction problem of quadratic span programs (QSPs, [15]). The second step consists of cryptographic tools that allow one to succinctly verify the satisfiability of a QSP.

Intuitively, a span program consists of vectors $\boldsymbol{u_i}$ for $i > 0$, a target vector $\boldsymbol{u_0}$, and a labelling of every vector $\boldsymbol{u_i}$ by a literal $x_\iota = x_\iota^1$ or $\bar{x}_\iota = x_\iota^0$ or by $\perp$. A span program accepts an input $\boldsymbol{w}$ iff $\boldsymbol{u_0}$ belongs to the span of the vectors $\boldsymbol{u_i}$ that are labelled by literals $x_\iota^{w_\iota}$ (or by $\perp$) that are consistent with the assignment $\boldsymbol{w} = (w_\iota)$ to the input $\boldsymbol{x} = (x_\iota)$. I.e., $\boldsymbol{u_0} = \sum_{i>0} a_i \boldsymbol{u_i}$, where $a_i \neq 0$ if the labelling of $\boldsymbol{u_i}$ is not consistent with $\boldsymbol{w}$. (See Sect. 3 for more background.)

Briefly, the first step constructs span programs (which satisfy a non-standard conscientiousness property) that verify the correct evaluation of every individual gate. Conscientiousness means that the span program accepts only if all inputs to the span program were actually used (in the case of CIRCUIT-SAT, this means that the prover has set some value to every input and output wire of the gate, and that exactly the same value can be uniquely extracted from the argument). The gate checkers are aggregated to obtain a single large conscientious span program that verifies the operation of every individual gate in parallel. They then construct a *weak wire checker* that verifies consistency, i.e., that all individual gate checkers work on an unequivocally defined set of wire values. The weak wire checker of [15] guarantees consistency only if all gate checkers are conscientious. They define quadratic span programs (QSPs, see [15]) and construct a QSP that implements both the aggregate gate checker and the weak wire checker.

In the second step, Gennaro et al. construct a non-adaptively sound NIZK argument that verifies the QSP, with a linear CRS length, $\Theta(s \log^2 s)$ prover's computation, and linear-in-input size verifier's computation. It can be made adaptively sound by using universal circuits [25], see [15] for more information.

The construction of [15] is quite monolithic and while containing many new ideas, they are not sufficiently clarified in [15]. Bitansky et al [2] simplified the second step of the construction from [15], by first constructing a linear PCP [2], then a linear interactive proof, and finally a NIZK argument for CIRCUIT-SAT. Their more modular approach makes the ideas behind the second step more accessible. Unfortunately, [2] is slightly less efficient than [15], and uses a (presumably) stronger security assumption.

We improve the construction of [15] in several aspects. Some improvements are conceptual (e.g., we provide cleaner definitions, that allow us to offer more efficient constructions) and some of the improvements are technical (with special emphasis on concrete efficiency). More precisely, we modularize — thus making its ideas more clear and accessible — the first step of [15] to construct a succinct non-adaptive 3-query linear PCP [2] for CIRCUIT-SAT. Then we use the techniques of [2], together with several new techniques, to modularize the second step of [15]. Importantly and contrarily to [2], by doing so we both improve on the efficiency of both steps and relax the security assumptions. We outline our construction below, and sketch the differences compared to [15].

The main body of the current work consists of a cleaner and more efficient reduction from CIRCUIT-SAT to QSP-SAT (another **NP**-complete language, defined later). Given a circuit C, we construct an efficient circuit checker, a QSP that is satisfiable iff C is satisfiable.

To verify whether circuit C accepts an input, we use a small *standard* (i.e., not necessarily conscientious) span program to verify an individual gate. For example, a NAND checker is a span program that accepts if the gate implements NAND correctly. We construct efficient span programs for gate checkers, needed for the CIRCUIT-SAT argument. E.g., we construct a size 6 and dimension 3 NAND checker; this can be compared to size 12 and dimension 9 conscientious NAND checker from [15]. By using the AND composition of span programs, we construct a single large span program that verifies every gate in parallel.

Unfortunately, simple AND composition of the gate checkers is not secure, because it allows "double-assignments". More precisely, some vectors of several adjacent gate checkers are labelled by the variable corresponding to the same wire. While every individual checker might be locally correct, one checker could work with value 0 while another checker could work with value 1 assigned to the same wire. Clearly, such bad cases should be detected. More precisely, it must be possible to verify efficiently that the coefficients a_i that were used in the gate checkers adjacent to some wire are consistent with a unique wire value.

We solve this issue as follows. Let Code be an efficient high-distance linear $[N, K, D]$ error-correcting code with $D > N/2$. For any wire η, consider all vectors from adjacent gate checkers that correspond to the claimed value x_η of this wire. Some of those vectors (say $\boldsymbol{u_i}$) are labelled by the positive literal x_η and some (say $\boldsymbol{v_i}$) by the negative literal $\bar{x}_\eta$. The individual gate checker's acceptance "fixes" certain coefficients a_i (that are used with $\boldsymbol{u_i}$) and b_i (that are used with $\boldsymbol{v_i}$) for all adjacent gate checkers. Roughly stating, for consistency of wire η one requires that either all a_i are zero (then unequivocally $x_\eta = 0$), or all b_i are zero (then unequivocally $x_\eta = 1$). We verify that this is the case by applying Code separately to the vectors $\boldsymbol{a}$ and $\boldsymbol{b}$. The high-distance property of Code guarantees that if $\boldsymbol{a}$ and $\boldsymbol{b}$ are not consistent, then there exists a coefficient i, s.t. $\mathsf{Code}(\boldsymbol{a})_i \cdot \mathsf{Code}(\boldsymbol{b})_i \neq 0$.

Motivated by this construction, we redefine QSPs [15] as follows. Let $\circ$ denote the pointwise product of two vectors. A QSP (that consists of two target vectors $\boldsymbol{u_0} = (u_{0j}) \in \mathbb{F}^d$ and $\boldsymbol{v_0} = (v_{0j}) \in \mathbb{F}^d$ and two $m \times d$ matrices $U = (u_{ij})$ and $V = (v_{ij})$ for $i \in [m]$ and $j \in [d]$) over some field $\mathbb{F}$ accepts an input iff for some vectors $\boldsymbol{a}$ and $\boldsymbol{b}$, consistent with this input,

$$(\boldsymbol{a}^\top \cdot U - \boldsymbol{u_0}) \circ (\boldsymbol{b}^\top \cdot V - \boldsymbol{v_0}) = \boldsymbol{0} \ . \tag{1}$$

Clearly, Eq. (1) is equivalent to the requirement that for all $j \in [d]$, $(\sum_{i=1}^m a_i u_{ij} - u_{0j}) \cdot (\sum_{i=1}^m b_i v_{ij} - v_{0j}) = 0$. Since $\mathbb{F}$ is an integral domain, the latter holds iff for all $j \in [d]$, either $\sum_{i=1}^{m} a_i u_{ij} = u_{0j}$ or $\sum_{i=1}^{m} b_i v_{ij} = v_{0j}$, which can be seen as an element-wise OR of two span programs. This can be compared to the element-wise AND of two span programs that accepts iff for all $j \in [d]$, both $\sum_{i=1}^{m} a_i u_{ij} = u_{0j}$ and $\sum_{i=1}^{m} b_i v_{ij} = v_{0j}$ iff two span programs accept simultaneously, i.e., $\sum a_i \boldsymbol{u_i} = \boldsymbol{u_0}$ and $\sum b_i \boldsymbol{v_i} = \boldsymbol{v_0}$. On the other hand, it is not known how to implement an element-wise OR composition of two span programs as a small span program. QSPs add an element-wise OR to an element-wise AND, and thus it is not surprising that they increase the expressiveness of span programs significantly.

The above linear error-correcting code based construction implements a QSP (a *wire checker*), with U and V being related to the generating matrices of the code. (See Def. 2.) Basically, the wire checker verifies the consistency of vectors $\boldsymbol{a}$ and $\boldsymbol{b}$ with the input.

We use the systematic Reed-Solomon code, since it is a maximum distance separable code with optimal support (i.e., it has the minimal possible number of non-zero elements in its generating matrix). It also results in the smallest degree of certain polynomials in the full NIZK argument. While no connection to error-correcting codes was made in [15], their wire checker can be seen as a suboptimal (overdefined) variant of the systematic Reed-Solomon code. Due to the better theoretical foundation, the new wire checker is more efficient, and optimal in its size and support. Moreover, one can use any efficient high-distance ($D > N/2$) linear error-correcting code, e.g., a near-MDS code [7]. Whether this would result in any improvement in the computational complexity of the final NIZK argument is an interesting open question.

Moreover, the wire checker of [15] is consistent (and thus their NIZK argument is sound) only if the gate checkers are conscientious. The new wire checker does not have this requirement. This not only enables one to use more efficient gate checkers but also potentially enables one to use known techniques (combinatorial characterization of span program size [11], semidefinite programming [24]) to construct more efficient checkers for larger unit computations.

We construct an aggregate wire checker by applying an AND composition to wire checkers, and then construct a single QSP (the *circuit checker*) that implements both the aggregate gate checker and the aggregate wire checker. At this point, the approach of the current paper pays off also conceptually: one can compare the description of the circuit checker (called a canonical QSP) in [15, Sect. 2.4], that takes about 3/4 of a page, with the description from the current paper (Def. 3) that takes only a couple of lines.

We prove that the circuit checker (the QSP) is satisfiable iff the original circuit is satisfiable. Since the efficiency of the new circuit checker depends on the fan-out of the circuit, we use the classical result from [17] about constructing low fan-out circuits that allows us to optimize the worst case size and other parameters, especially support, of the circuit checker.

To summarize, the new circuit checker consists of two elements. First, an aggregate gate checker (a span program) that verifies that every individual gate is executed correctly on their local variables. Second, an aggregate wire checker (a QSP, based on a high-distance linear error-correcting code) that verifies that individual gates are executed on the consistent assignments to the variables. Importantly (for the computational complexity of the NIZK argument), the circuit checker is a composition of small (quadratic) span programs, and has only a constant number of non-zero elements per vector.

This finishes the description of the CIRCUIT-SAT to QSP-SAT reduction. To construct an efficient NIZK argument for CIRCUIT-SAT, we need several extra steps. Based on the new circuit checker, we first construct a non-adaptive 2-query linear PCP([2], see Sect. 8 for a definition) for CIRCUIT-SAT with linear communication. This seems to be the first known non-trivial 2-query linear PCP. Moreover, we use a more elaborate extraction technique which, differently from the one from [15], also works with non-conscientious gate checkers. This improves the efficiency of the linear PCP. In particular, the computation of the decision functionality of the linear PCP is dominated by a small constant number of field operations. The same functionality required $\Theta(n)$ operations in [15,2]. Interestingly, this construction by itself is purely linear-algebraic, by using concepts like span programs, linear error-correcting codes, and linear PCPs.

To improve the communication of the linear PCP, as in [15], we define polynomial span programs and polynomial QSPs. Differently from [15] (that only gave the polynomial definition), our main definition of QSPs — as sketched above — is linear-algebraic, and we then use a transformation to get a QSP to a "polynomial" form. We feel the linear-algebraic definition is much more natural, and describes the essence of QSPs better. Based on the polynomial redefinition of QSPs and the Schwartz-Zippel lemma, we construct a succinct non-adaptive 3-query linear PCP for CIRCUIT-SAT. The prover's computation in this linear PCP is $\Theta(s \log s)$, where s is the size of the circuit, and the verifier's computation is again $\Theta(1)$. In [15], the corresponding parameters were $\Theta(s \log^2 s)$ and $\Theta(n)$. Thus, the new 3-query linear PCP is more efficient and conceptually simpler than the previously known 3-query linear PCPs [2].

By using techniques of [2], we convert the linear PCP to a succinct non-adaptive linear interactive proof, and then to a succinct non-adaptive NIZK argument. (See the full version, [20].) As in the case of the argument from [15], the latter can be made adaptive by using universal circuits [25].

Since the reduction from linear PCP to NIZK from [2] loses some efficiency and relies on a stronger security assumption than stated in [15], we also describe a direct NIZK argument with a (relatively complex) soundness proof that follows the outline of the soundness proof from [15]. The main difference in the proof is that we rephrase certain proof techniques from [15] in the language of multilinear universal hash functions. This might be an interesting contribution by itself. Apart from a more clear proof, this results in a slightly weaker security assumption. (See the full version [20] of this paper.)

The new non-adaptive CIRCUIT-SAT argument has CRS length $\Theta(s)$, prover's computation $\Theta(s \log s)$, verifier's computation $\Theta(1)$, and communication $\Theta(1)$. In all cases, the efficiency has been improved as compared to the (QSP-based) argument from [15]. Moreover, all additional optimization techniques applicable to the argument from [15] (e.g., the use of collision-resistant hash functions) are also applicable to the new argument.

We hope that by using our techniques, one can construct efficient NIZK arguments for other languages, like the techniques of [19] were used in [5] to construct an efficient range argument, and in [21] to construct an efficient shuffle. QSPs have more applications than just in the NIZK construction. We only mention that one can construct a related zap [8], and a related (public or designated-verifier) succinct non-interactive argument of knowledge (SNARK, see [22,6]) by using the techniques of [1,14].

It is also natural to apply our techniques to verifiable computation [13]: instead of gates, one can talk about small (but possibly much larger) computational units, and instead of wires, about the values transferred between the computational units. Since here one potentially deals with much larger span programs than in the case of the CIRCUIT-SAT argument, the use of standard (non-conscientious) span programs is especially beneficial. Since in the case of verifiable computation, the computed function F (and thus also the circuit C) is known while generating the CRS, one can use the non-adaptively sound version of the new argument [23].

Gennaro et al. [15] also proposed a NIZK argument that is based on quadratic arithmetic programs (QAP-s), a novel computational model for arithmetic circuits. QAP-based arguments are often significantly more efficient than QSP-based arguments, see [15,23]. We can use our techniques to improve on QAP-based arguments, but here the improvements are less significant and thus we have omitted full discussion. (See the full version.) Briefly, differently from [15], we give an (again, more clean) linear-algebraic definition of QAP-s. This enables us to present a short alternative proof of the result from [15] that any arithmetic circuit with n inputs and s multiplication gates can be computed by a QAP of size $n + s$ and dimension s. We remark that the QAP-based construction results in a 4-query linear PCP, while the QSP-based construction from the current paper results in a 3-query linear PCP.

Due to the lack of space, many proofs are given only in the full version [20].

2 Preliminaries: Circuits and Circuit-SAT

For a fixed circuit C, let $s = |C|$ be its size (the number of gates), s_{e} its number of wires, and n be its input size. Every gate ι computes some unary or binary function $f_\iota : \{0,1\}^{\leq 2} \to \{0,1\}$. We denote the set of gates of C by $[s]$ and the set of wires of C by $[s_{\mathsf{e}}]$. Assume that the first n wires, $\eta \in [n]$, start from n input gates $\iota \in [n]$. Every wire $\eta \in [s_{\mathsf{e}}]$ corresponds to a formal variable x_η in a natural way. This variable obtains an assignment w_η, $\eta \in [s_{\mathsf{e}}]$, computed by C from the input assignment $(w_i)_{i=1}^n$. Denote $\boldsymbol{w} := (w_\eta)_{\eta=1}^{s_{\mathsf{e}}}$. We write $C(\boldsymbol{w}) := C((w_i)_{i=1}^n)$. For a gate ι of C, let $\deg^+(\iota)$ be its fan-out, and let $\deg^-(\iota)$ be its fan-in. Let $\deg(\iota) = \deg^-(\iota) + \deg^+(\iota)$.

Let $\mathrm{poly}(x) := x^{O(1)}$. Let $\mathcal{R} = \{(C, \boldsymbol{w})\}$ be an efficiently computable binary relation with $|\boldsymbol{w}| = \mathrm{poly}(|C|)$ and $s := |C| = \mathrm{poly}(|\boldsymbol{w}|)$. Here, C is a statement, and $\boldsymbol{w}$ is a witness. Let $\mathcal{L} = \{C : \exists \boldsymbol{w}, (C, \boldsymbol{w}) \in \mathcal{R}\}$ be the related **NP**-language. For fixed s, we have a relation $\mathcal{R}_s$ and a language $\mathcal{L}_s$.

The *language* CIRCUIT-SAT consists of all (strings representing) circuits that produce a single bit of output and that have a satisfying assignment. That is, a string representing a circuit C is in CIRCUIT-SAT if there exists $\boldsymbol{w} \in \{0,1\}^{s_e}$ such that $C(\boldsymbol{w}) = 1$.

As before, we assume that $s = |C|$ is the number of gates, not the bitlength needed to represent C. Thus, $\mathcal{L}_s = \{C : |C| = s \wedge (\exists \boldsymbol{w} \in \{0,1\}^{s_e}, C(\boldsymbol{w}) = 1)\}$ and $\mathcal{R}_s = \{(C, \boldsymbol{w}) : |C| = s \wedge \boldsymbol{w} \in \{0,1\}^{s_e} \wedge C(\boldsymbol{w}) = 1\}$.

Let $G = (V, E)$ be the hypergraph of the circuit C. The vertices of G correspond to the gates of C. A hyperedge η connects the input gate of some wire to (potentially many) output gates of the same wire. In C, an edge η (except input edges, that have ϕ adjacent vertices) has $\phi + 1$ adjacent vertices, where ϕ is the fan-out of η's designated input gate. Every vertex of G can only be the starting gate of one hyperedge and the final gate of two hyperedges (since we only consider unary and binary gate operations). Thus, $|E(G)| \leq 2(|V(G)| - n)$.

3 Preliminaries: Span Programs

Let $\mathbb{F} = \mathbb{Z}_q$ be a finite field of size $q \gg 2$, where q is a prime. However, most of the results can be generalized to arbitrary fields. By default, vectors like $\boldsymbol{u}$ denote row vectors. For matrix U, let $\boldsymbol{u_i}$ be its ith row vector. For an $m \times d$ matrix U over $\mathbb{F}$, let $\mathsf{span}(U) := \{\sum_{i=1}^{m} a_i \boldsymbol{u_i} : \boldsymbol{a} \in \mathbb{F}^m\}$. Let x_ι, $\iota \in [n]$, be formal variables. Denote the positive literals x_ι by x_ι^1 and the negative literals $\bar{x}_\iota$ by x_ι^0.

A *span program* [18] $P = (\boldsymbol{u_0}, U, \varrho)$ over a field $\mathbb{F}$ is a linear-algebraic computation model. It consists of a non-zero target vector $\boldsymbol{u_0} \in \mathbb{F}^d$, an $m \times d$ matrix U over $\mathbb{F}$, and a labelling $\varrho : [m] \to \{x_\iota, \bar{x}_\iota : \iota \in [n]\} \cup \{\perp\}$ of U's rows by one of $2n$ literals or by $\perp$. Let $U_{\boldsymbol{w}}$ be the submatrix of U consisting of those rows whose labels are satisfied by the assignment $\boldsymbol{w} \in \{0,1\}^n$, that is, belong to $\{x_\iota^{w_\iota} : \iota \in [n]\} \cup \{\perp\}$. *$P$ computes a function f*, if for all $\boldsymbol{w} \in \{0,1\}^n$: $\boldsymbol{u_0} \in \mathsf{span}(U_{\boldsymbol{w}})$ if and only if $f(\boldsymbol{w}) = 1$.

Let $\varrho_{\boldsymbol{w}}^{-1} = \{i \in [m] : \varrho(i) \in \{x_\iota^{w_\iota} : \iota \in [n]\} \cup \{\perp\}\}$ be the set of rows whose labels are satisfied by the assignment $\boldsymbol{w}$. The *size*, $\mathsf{size}(P)$, of P is m. The dimension, $\mathsf{sdim}(P)$, is equal to d. P has *support* $\mathsf{supp}(P)$, if all vectors $\boldsymbol{u} \in U$ have altogether $\mathsf{supp}(P)$ non-zero elements. Clearly, $\boldsymbol{u_0}$ can be replaced by an arbitrary non-zero vector; one obtains the corresponding new span program (of the same size and dimension, but possibly different support) by applying a basis change matrix. Let $D(x_\iota) := \max_{j \in \{0,1\}} |\varrho^{-1}(x_\iota^j)|$, for each $\iota \in [n]$ and $j \in \{0,1\}$, be the maximum number of vectors that have the same label (ι, j); this parameter is needed when we construct wire checkers.

Complex span programs are constructed by using simple span programs and their composition rules. The Boolean function NAND $\bar{\wedge}$ is defined as $\bar{\wedge}(x, y) = x \bar{\wedge} y = \neg(x \wedge y)$. Span programs for AND, NAND, OR, XOR, and equality of two variables x and y are as in Fig. 1. Given span programs $P_0 = SP(f_0)$ an $P_1 = SP(f_1)$ for functions f_0 and f_1, one uses well-known AND and OR compositions to construct span programs for $f_0 \wedge f_1$ and $f_0 \vee f_1$.

$$\left(\begin{array}{c|cc} & 1 & 1 \\ \hline x & 1 & 0 \\ y & 0 & 1 \end{array}\right) \left(\begin{array}{c|c} & 1 \\ \hline \bar{x} & 1 \\ \bar{y} & 1 \end{array}\right) \left(\begin{array}{c|c} & 1 \\ \hline x & 1 \\ y & 1 \end{array}\right) \left(\begin{array}{c|rr} & 0 & 1 \\ \hline x & 1 & 1 \\ y & 1 & 1 \\ \bar{x} & -1 & 0 \\ \bar{y} & -1 & 0 \end{array}\right) \left(\begin{array}{c|rr} & 0 & 1 \\ \hline x & 1 & 1 \\ y & -1 & 0 \\ \bar{x} & -1 & 0 \\ \bar{y} & 1 & 1 \end{array}\right) \left(\begin{array}{c|ccc} & 1 & 1 & 1 \\ \hline x & 1 & 0 & 0 \\ y & 0 & 1 & 0 \\ z & 1 & 1 & 0 \\ \bar{x} & 0 & 0 & 1 \\ \bar{y} & 0 & 0 & 1 \\ \bar{z} & 0 & 0 & 1 \end{array}\right) \left(\begin{array}{c|ccc} & 1 & 1 & 1 \\ \hline x & 0 & 1 & 0 \\ y_1 & 0 & 0 & 1 \\ y_2 & 1 & 0 & 0 \\ \bar{x} & 1 & 0 & 0 \\ \bar{y}_1 & 0 & 1 & 0 \\ \bar{y}_2 & 0 & 0 & 1 \end{array}\right)$$

Fig. 1. From left to right: standard span programs $SP(\wedge)$, $SP(\bar{\wedge})$, $SP(\vee)$, $SP(\oplus)$, $SP(=)$ and new span programs $SP(c_{\bar{\wedge}})$ and $SP(c_{\mathsf{Y}})$

A span program $(\boldsymbol{u}_0, U, \varrho)$ is *conscientious* [15] if a linear combination associated to a satisfying assignment must use at least one vector associated to either x_ι or $\bar{x}_\iota$ for every $\iota \in [n]$. Clearly, $SP(\wedge)$, $SP(\oplus)$ and $SP(=)$ are conscientious, while $SP(\vee)$ is not.

4 Efficient Gate Checkers

A *gate checker* for a gate that implements $f : \{0,1\}^n \to \{0,1\}$ is a function $c_f : \{0,1\}^{n+1} \to \{0,1\}$, s.t. $c_f(\boldsymbol{x}, y) = 1$ iff $f(\boldsymbol{x}) = y$. The NAND-checker $c_{\bar{\wedge}} : \{0,1\}^3 \to \{0,1\}$ outputs 1 iff $z = x \bar{\wedge} y$.

Lemma 1. *$SP(c_{\bar{\wedge}})$ on Fig. 1 is a span program for $c_{\bar{\wedge}}$. It has size* 6, *dimension* 3, *and support* 7.

As seen from the proof , given an accepting assignment (x, y, z), one can efficiently find small values $a_i \in [-2, 1]$ such that $\sum_{i \geq 1} a_i \boldsymbol{u}_i = \boldsymbol{u}_0$. However, a satisfying input to $SP(c_{\bar{\wedge}})$ does not fix the values a_i unequivocally: if $(x, y, z) = (0, 0, 1)$ (that is, $a_1 = a_2 = a_6 = 0$), then one can choose an arbitrary a_4 and set $a_5 \leftarrow 1 - a_4$. Since one can set $a_4 = 0$, $SP(c_{\bar{\wedge}})$ is not conscientious.

Given $SP(c_{\bar{\wedge}})$, one can construct a size 6 and dimension 3 span program for the AND-checker $c_\wedge(x, y, z) := (x \wedge y) \oplus \bar{z}$ by interchanging in $SP(c_{\bar{\wedge}})$ the rows labelled by z and $\bar{z}$. Similarly, one can construct a size 6 and dimension 3 span program for the OR-checker $c_\vee(x, y, z) := (\bar{x} \wedge \bar{y}) \oplus z$ by interchanging in $SP(c_{\bar{\wedge}})$ the rows labelled by x and $\bar{x}$, and the rows labelled by y and $\bar{y}$. NOT-checker $[x \neq y] = x \oplus y$ is just the XOR function, and thus one can construct a size 4 and dimension 2 span program for the NOT-checker function.

We need the dummy gates $y \leftarrow x$, and corresponding dummy checkers $c_=(x, y) = [x = y]$. Clearly, the dummy checker function is just to the equality test, and thus has a conscientious span program of size 4 and dimension 2. Moreover, if $x = y \in \{0,1\}$, then $a_1 = a_2 = x$, while $a_3 = a_4 = 1 - x$.

We need the fork-checker $c_{\mathsf{Y}}(x, y_1, y_2)$ for the fork gate that computes $y_1 \leftarrow x$, $y_2 \leftarrow x$. In the CNF form, $c_{\mathsf{Y}}(x, y_1, y_2) = (\bar{x} \vee y_2) \wedge (x \vee \bar{y}_1) \wedge (y_1 \vee \bar{y}_2)$. Since every literal is mentioned once in the CNF, we can use AND and OR compositions to derive the span program on Fig. 1. It has size 6, dimension 3, and support 6.

We also need a 1-to-ϕ fork-checker that has 1 input x and ϕ outputs y_ι, with $y_\iota \leftarrow x$. The ϕ-fork checker is $c_{\mathsf{Y}}^{\phi}(x, \boldsymbol{y}) = (x \wedge y_1 \wedge \cdots \wedge y_\phi) \vee (\bar{x} \wedge \bar{y}_1 \wedge \cdots \wedge \bar{y}_\phi)$. Clearly, c_{Y}^{ϕ} has CNF $c_{\mathsf{Y}}^{\phi}(x, \boldsymbol{y}) = (x \vee \bar{y}_1) \wedge (y_1 \vee \bar{y}_2) \wedge \cdots \wedge (y_{\phi-1} \vee \bar{y}_\phi) \wedge (y_\phi \vee \bar{x})$. From this we construct a span program exactly as in the case $\phi = 2$, with size $2(\phi + 1)$ and dimension $\phi + 1$. It has only one vector labelled with every x_ι/y or its negation, thus $D(x) = D(y_\iota) = 1$ for all ι. To compute the support, note that $SP^*(c_{\mathsf{Y}}^{\phi})$ has two 1-entries in every column, and one in every row. Thus, $\mathsf{supp}(SP(c_{\mathsf{Y}}^{\phi})) = \sum_{i=1}^{\phi+1} 2 = 2\phi + 2$.

5 Aggregate Gate Checker

Given a circuit that consists of NAND, AND, OR, XOR, and NOT gates, we combine the individual gate checkers by using the AND composition rule. In addition, for the wire checker of Sect. 6.2 (and thus also the final NIZK argument) to be more efficient, all gates of the circuit C need to have a small fan-out. In [15], the authors designed a circuit of size $3 \cdot |C|$ that implements the functionality of C but only has fan-out 2 except for a specially introduced dummy input. Their aggregate gate checker (AGC) has size $36 \cdot |C|$ and dimension $27 \cdot |C|$. By using the techniques of [17] (that replaces every high fan-out gate with an inverse binary tree of fork gates, and then gives a more precise upper bound of the resulting circuit size), we prove a more precise result. We do not introduce the dummy input but we still add a dummy gate for every input. We then say that we deal with a circuit with dummy gates.

Since we are interested in circuit satisfiability, the X-checker (where say X = NAND) of the circuit's output gate simplifies to the X gate (e.g., NAND checker simplifies to NAND). Since X has a more efficient span program than X checker, then for the sake of simplicity, we will not mention this any more.

Let C be a circuit. The *AGC function* agc of a circuit C is a function $\mathsf{agc} : \{0,1\}^{\sum_{\iota=1}^{|C|} \deg(\iota)} \rightarrow \{0,1\}^{|C|}$. I If c_ι is the gate checker of the ιth gate and $\boldsymbol{x}_\iota$ has dimension $\deg(\iota)$, then $\mathsf{agc}(\boldsymbol{x}_1, \ldots, \boldsymbol{x}_{|C|}) = (c_1(\boldsymbol{x}_1), \ldots, c_{|C|}(\boldsymbol{x}_{|C|}))$.

As in [15], we construct the AGC by AND-composition of the gate checkers of the individual gate checkers. Since for an individual gate checker and a satisfying assignment, one can compute the corresponding coefficient vector $\boldsymbol{a}$ in constant time, the aggregate coefficient vector $\boldsymbol{a}$ can be computed from w in time $\Theta(s)$. Let $\boldsymbol{a} \leftarrow \mathsf{c2q}(w)$ be the corresponding algorithm.

Theorem 1. *Let $f : \{0,1\}^n \rightarrow \{0,1\}$ be the function computed by a fan-in ≤ 2 circuit C with $s = |C|$ NAND, AND, OR, XOR, and NOT gates. There exists a fan-in ≤ 2 and fan-out $\leq \phi$ circuit with dummy gates C_{bnd} for f, that has the same s gates as C, n additional dummy gates, and up to $(s - 2n)/(\phi - 1)$ additional ϕ-fork gates. Let $\phi^* := 1/(\phi - 1)$. The AGC $\mathsf{agc}(C_{\mathsf{bnd}})$ has a span program P with $\mathsf{size}(P) \leq (8 + 4\phi^*)\, s - (6 + 8\phi^*)\, n$, $\mathsf{sdim}(P) \leq (4 + 2\phi^*)\, s - (3 + 4\phi^*)\, n$, and $\mathsf{supp}(P) \leq (9 + 4\phi^*)\, s - (5 + 8\phi^*)\, n$. If $\phi = 3$, then $\mathsf{size}(P) \leq 10s - 10n$, $\mathsf{sdim}(P) \leq 5s - 5n$, and $\mathsf{supp}(P) \leq 11s - 9n$.*

The upper bounds of this theorem are worst-case, and often imprecise. The optimal choice of ϕ depends on the parameter that we are going to optimize. The AGC has optimal size, dimension and support if ϕ is large (preferably even if the fan-out bounding procedure of Thm. 1 is not applied at all). The support of the aggregate wire checker (see Sect. 6.3) is minimized if $\phi = 2$. To balance the parameters, we concentrate on the case $\phi = 3$.

6 Quadratic Span Programs and Wire Checker

6.1 Quadratic Span Programs

An intuitive definition of quadratic span programs (QSPs) was given in the introduction and will not be repeated here. We now give a formal (linear-algebraic) definition of QSPs. In Sect. 9, we will provide an equivalent polynomial redefinition of QSPs that is the same as the definition given in [15].

Definition 1. *A* quadratic span program (QSP) $Q = (\boldsymbol{u_0}, \boldsymbol{v_0}, U, V, \varrho)$ *over a field* $\mathbb{F}$ *consists of two target vectors* $\boldsymbol{u_0}, \boldsymbol{v_0} \in \mathbb{F}^d$*, two* $m \times d$ *matrices* U *and* V*, and a common labelling* $\varrho : [m] \to \{x_\iota, \bar{x}_\iota : \iota \in [n]\} \cup \{\perp\}$ *of the rows of* U *and* V*.* Q *accepts an input* $\boldsymbol{w} \in \{0,1\}^n$ *iff there exist* $(\boldsymbol{a}, \boldsymbol{b}) \in \mathbb{F}^m \times \mathbb{F}^m$*, with* $a_i = 0 = b_i$ *for all* $i \notin \varrho_{\boldsymbol{w}}^{-1}$*, such that* $(\boldsymbol{a}^\top \cdot \mathcal{V} - \boldsymbol{u_0}) \circ (\boldsymbol{b}^\top \cdot \mathcal{W} - \boldsymbol{v_0}) = \boldsymbol{0}$*, where* $\boldsymbol{x} \circ \boldsymbol{y}$ *denotes the pointwise (Hadamard) product of* $\boldsymbol{x}$ *and* $\boldsymbol{y}$*.* Q computes a function f *if for all* $\boldsymbol{w} \in \{0,1\}^n$*:* $f(\boldsymbol{w}) = 1$ *iff* Q *accepts* $\boldsymbol{w}$*.*

We remark that one can have $\boldsymbol{u_0} = \boldsymbol{v_0} = \boldsymbol{0}$. (See Def. 2, for example.)

The size, $\mathsf{size}(Q)$, of Q is m. The dimension, $\mathsf{sdim}(Q)$, of Q is d. The *support*, $\mathsf{supp}(Q)$, of Q is equal to the sum of the supports (that is, the number of non-zero elements) of all vectors $\boldsymbol{u_i}$ and $\boldsymbol{v_i}$. Clearly, one can compose QSPs by using the AND and OR composition rules of span programs, though one has to take care to apply the same transformation to both U and V simultaneously.

The language QSP-SAT consists of all (strings representing) QSPs that produce a single bit of output and that have a satisfying assignment. I.e., a string representing an n-input QSP Q is in QSP-SAT if there exists $\boldsymbol{w} \in \{0,1\}^n$, such that $Q(\boldsymbol{w}) = 1$. The witness of this fact is $(\boldsymbol{a}, \boldsymbol{b})$, and we write $Q(\boldsymbol{a}, \boldsymbol{b}) = Q(\boldsymbol{w})$.

6.2 Wire Checker

Gate checkers verify that every individual gate is followed correctly, i.e., that its output wire obtains a value which is consistent with its input wires. One also requires inter-gate (wire) consistency that ensures that adjacent gate checkers do not make double assignments to any of the wires. Here, we consider hyperwires that have one input gate and potentially many output gates. Following [15], for this purpose we construct a wire checker. We first construct a wire checker for every single wire (that verifies that the variables involved in the span programs of the vertices that are adjacent to this concrete wire do not get inconsistent assignments), and then aggregate them by using an AND composition.

For a (hyper)wire η, let $N(\eta)$ be the set of η's adjacent gates. For gate $\iota \in N(\eta)$, let $P_\iota = (\boldsymbol{u}_0^{(\iota)}, U^{(\iota)}, \varrho^{(\iota)})$ be its gate checker. For every $\iota \in N(\eta)$, one of the input or output variables of P_ι (that we denote by $x_{\iota:\eta}$) corresponds to x_η. Recall that for a local variable y of a span program P_ι, $D(y) = \max(|\varrho^{-1}(y)|, |\varrho^{-1}(\bar{y})|)$. We assume $|\varrho^{-1}(y)| = |\varrho^{-1}(y)|$, by adding zero vectors to the span programs if necessary. Let $D(\eta) := \sum_{\iota \in N(\eta)} D(x_{\iota:\eta})$ be the number of the times the rows of adjacent gate checkers have been labelled by a local copy of x_η^1.

We define the ηth wire checker between the rows of adjacent gates $i \in N(\eta)$ in the AGC that are labelled either by the local variable $x_{i:\eta}$ or its negation $\bar{x}_{i:\eta}$, i.e., between $2D(\eta)$ rows $\{i : \exists k \in N(\eta) \text{ s.t. } \varrho^{(k)}(i) = x_{k:\eta} \vee \varrho^{(k)}(i) = \bar{x}_{k:\eta}\}$. Let ψ be the natural labelling of the wire checkers, with $\psi(i) = x_\eta^j$ iff $\varrho^{(k)}(i) = x_{k:\eta}^j$ for some $k \in N(\eta)$.

Example 1. Consider a (hyper)wire η that has one input gate ι_1 and two output gates ι_2 and ι_3. Assume that all three gates implement NAND, and thus they have gate checkers $SP(c_{\bar{\wedge}})$ from Fig. 1. Assume that $x_\eta = z_{\iota_1} = x_{\iota_2} = y_{\iota_3}$. Thus, the ηth wire checker is defined between the rows 3 and 6 of the checker for ι_1, rows 1 and 4 of the checker for ι_2, and rows 2 and 5 of the checker for ι_3. Thus, $D(\eta) = D(z_{\iota_1}) + D(x_{\iota_2}) + D(y_{\iota_3}) = 6$. □

We first define the wire checker for a wire η and thus for one variable x_η. In Sect. 6.3, we will give a definition and a construction in the aggregate case.

For $\boldsymbol{y} = (y_1, \ldots, y_{2D})^\top$, let $\boldsymbol{y}^{(1)} := (y_1, \ldots, y_D)^\top$ and $\boldsymbol{y}^{(2)} := (y_{D+1}, \ldots, y_{2D})^\top$. Fix a wire η. Assume that $D = D(\eta)$. Let $Q = (\boldsymbol{u}_0, \boldsymbol{v}_0, U, V, \psi)$, with $m \times d$ matrices U and V, be a QSP. Q is a *wire checker*, if for any $\boldsymbol{a}, \boldsymbol{b} \in \mathbb{F}^{2D}$, Eq. (1) holds iff $\boldsymbol{a}$ and $\boldsymbol{b}$ are consistent bit assignments in the following sense: for both $k \in \{1, 2\}$, either $\boldsymbol{a}^{(k)} = \boldsymbol{0}$ or $\boldsymbol{b}^{(k)} = \boldsymbol{0}$.

We propose a new wire checker that is based on the properties of high-distance linear error-correcting codes, see the introduction for some intuition. To obtain optimal efficiency, we choose particular codes (namely, systematic Reed-Solomon codes).

Definition 2. *Let $D^* := 2D - 1$. Let RS_D be the $D \times D^*$ generator matrix of the $[D^*, D, D]_q$ systematic Reed-Solomon code. Let $m = 2D$ and $d = 2D^*$. Let* $U = U_D = \begin{pmatrix} RS_D & 0_{D\times D^*} \\ 0_{D\times D^*} & RS_D \end{pmatrix}$ *and* $V = V_D = \begin{pmatrix} 0_{D\times D^*} & RS_D \\ RS_D & 0_{D\times D^*} \end{pmatrix}$. *Let $Q_{\mathsf{wc}} := (\boldsymbol{0}, \boldsymbol{0}, U, V, \psi)$, where $\psi^{-1}(\bar{x}_\eta) = [1, D]$ and $\psi^{-1}(x_\eta) = [D+1, 2D]$.*

We informally define the degree $\mathsf{sdeg}(Q)$ of a (quadratic) span program Q as the degree of the interpolating polynomial that obtains the value u_{ij} at point j. See Sect. 9 for a formal definition.

Lemma 2. *Q_{wc} is a wire checker of size $2D$, degree $D + D^* = 3D - 1$, dimension $2D^* = 4D - 2$, and support $4D^2$.*

Proof. The claim about the parameters follows straightforwardly from the properties of the code. It is easy to see that if $\boldsymbol{a}$ and $\boldsymbol{b}$ are consistent bit assignments, then Q_{wc} accepts. For example, if $\boldsymbol{a}^{(1)} = \boldsymbol{b}^{(2)} = \boldsymbol{0}$, then clearly

$(\boldsymbol{a}^\top \cdot U)_j = \sum_{i=1}^m a_i u_{ij} = 0$ for $j \in [1, D^*]$ and $(\boldsymbol{b}^\top \cdot V)_j = \sum_{i=1}^m b_i v_{ij} = 0$ for $j \in [D^* + 1, 2D^*]$. Thus, $(\boldsymbol{a}^\top \cdot U)_j \cdot (\boldsymbol{b}^\top \cdot V)_j = 0$ for $j \in [1, 2D^*]$, and thus $(\boldsymbol{a}^\top \cdot U - \mathbf{0}) \circ (\boldsymbol{b}^\top \cdot V - \mathbf{0}) = \mathbf{0}$.

Now, assume that $\boldsymbol{a}$ and $\boldsymbol{b}$ are inconsistent bit assignments, i.e., $\boldsymbol{a}^{(k)} \neq \mathbf{0}$ and $\boldsymbol{b}^{(k)} \neq \mathbf{0}$ for $k \in \{1, 2\}$. W.l.o.g., let $k = 1$. Since RS_D is the generator matrix of the systematic Reed-Solomon code, the vectors $\boldsymbol{a}^\top \cdot RS_D$ and $\boldsymbol{b}^\top \cdot RS_D$ have at least $D > D^*/2$ non-zero coefficients among its first D^* coefficients. Thus, both $\sum_{i=1}^m a_i u_{ij}$ and $\sum_{i=1}^m b_i v_{ij}$ are non-zero for more than $D^*/2$ different values $j \in [D^*]$. Hence, there exists a coefficient $j \in [D^*]$, such that $(\sum_{i=1}^m a_i u_{ij})(\sum_{i=1}^m b_i v_{ij}) \neq 0$. Thus, Q_{wc} does not accept. □

We chose a Reed-Solomon code since it is a maximum distance separable (MDS) code and thus minimizes the number of columns in RS_D. It also naturally minimizes the degree of the wire checker. Moreover, RS_D has D^2 non-zero elements. Clearly (and this is the reason we use a systematic code), D^2 is also the smallest support a generator matrix G of an $[n = 2D - 1, k = D, d = D]_q$ code can have, since every row of G is a codeword and thus must have at least d non-zero entries. Thus, G must have at least $dD \geq D^2$ non-zero entries, where the last inequality is due to the singleton bound.

The (weak) wire checker of [15], while described by using a completely different terminology, can be seen as implementing an overdefined version (with $D^* = 3D - 2$) of the construction from Def. 2. The linear-algebraic reinterpretation of QSPs together with the introducing of coding-theoretic terminology allowed us to better exposit the essence of wire checkers. It also allowed us to improve on the efficiency, and prove the optimality of the new construction.

A wire checker with $U = V = RS_D$ satisfies the even stronger security requirement that Eq. (1) holds iff either $\boldsymbol{a} = 0$ or $\boldsymbol{b} = 0$. One may hope to pair up literals corresponding to x_η in the U part and literals corresponding to $\bar{x}_\eta$ in the V part. This is impossible in our application: when we aggregate the wire checkers, we must use vectors labelled with both negative and positive literals in the same part, U or V, and we cannot pair up columns from U and V that have different indices. (See Def. 3.) The construction of Def. 2 allows one to do it, though one has to use V that is a dual of U according to the following definition.

For a labelling ψ, we define the *dual labelling* ψ_{dual}, such that $\psi_{\mathsf{dual}}(i) = x_\eta^j$ iff $\psi(i) = x_\eta^{1-j}$. Let $V = U_{\mathsf{dual}}$ be the same matrix as U, except that it has rows from $\psi^{-1}(\bar{x}_\eta)$ and $\psi^{-1}(x_\eta)$ switched, for every η. To simplify the notation, we will not mention the dual labelling ψ_{dual} unless absolutely necessary, and we will assume implicitly that (as it was in Def. 2) always $V = U_{\mathsf{dual}}$. Now, [15] constructed a *weak* wire checker that guarantees consistency if all individual gate checkers are conscientious. The new wire checker is both more efficient and more secure.

6.3 Aggregate Wire Checker

Let $Q = (\mathbf{0}, \mathbf{0}, U, V, \psi)$, with two $m \times d$ matrices U and $V = U_{\mathsf{dual}}$, be a QSP. Q is *an aggregate wire checker (AWC)* for circuit C, if Eq. (1) holds iff $\boldsymbol{a}, \boldsymbol{b} \in \mathbb{F}^m$ are consistent bit assignments in the following sense: for each $\eta \in [s_{\mathsf{e}}]$ and $k \in \{0, 1\}$, either $a_i = 0$ for all $i \in \psi^{-1}(x_\eta^k)$ or $b_i = 0$ for all $i \in \psi^{-1}(x_\eta^k)$.

We construct the AWC by AND-composing wire checkers for the individual wires. The AWC first resets all vectors $\boldsymbol{u}_i$ and $\boldsymbol{v}_i$ to $\mathbf{0}$, and precomputes RS_{D_η} for all relevant values $D_\eta \leq 2(\phi + 1)$. After that, for every wire η, it sets the entries in rows, labelled by either x_η or $\bar{x}_\eta$, and columns corresponding to wire η, according to the ηth wire checker.

We recall from Sect. 6.2 that for the wire checker of some wire to work, the vectors in U and V of this wire checker must have dual orderings. To keep notation simple, we will not mention this in what follows.

Theorem 2. *Let $\phi \geq 2$. Assume that C_{bnd} is the circuit, obtained by the transformation described in Thm. 1 (including the added dummy gates). For $\eta \in E(C_{\mathsf{bnd}})$, denote $D^*_\eta = 2D_\eta - 1$. Let $d \leftarrow \sum D^*_\eta$. We obtain the AWC Q_{awc} by merging wire checkers for the individual wires $\eta \in E(C_{\mathsf{bnd}})$ as described above.*

Proof. Let m be the size of the AWC (see Thm. 3). If $\boldsymbol{a}, \boldsymbol{b}$ are consistent assignments, then their restrictions to $\psi^{-1}(\bar{x}_\eta) \cup \psi^{-1}(x_\eta)$ are consistent assignments of the ηth wire. For every $\eta \in E(C_{\mathsf{bnd}})$, the ηth wire checker guarantees that $(\sum_{i=1}^m a_i u_{ij})(\sum_{i=1}^m b_i v_{ij}) = 0$, for columns j corresponding to this wire, iff the bit assignments of the ηth wire are consistent. Thus, $(\sum_{i=1}^m a_i u_{ij})(\sum_{i=1}^m b_i v_{ij}) = 0$ for $j \in [1, d]$ iff the bit assignments of all wires are consistent. □

Theorem 3. *Let $\phi^* := 1/(\phi - 1)$. Assume C implements $f : \{0,1\}^n \to \{0,1\}$, and $s = |C|$. Then $\mathsf{size}(Q_{\mathsf{awc}}) \leq (6 + 4\phi^*)s - (2 + 8\phi^*)n - 4$, $\mathsf{sdim}(Q_{\mathsf{awc}}) \leq (12 + 8\phi^*)s - (6 + 16\phi^*)n - 8$, $\mathsf{sdeg}(Q_{\mathsf{awc}}) \leq (9 + 6\phi^*)s - (4 + 12\phi^*)n - 6$, $\mathsf{supp}(Q_{\mathsf{awc}}) \leq 4(\phi + 1)^2((1 + \phi^*)s + (4 - 2\phi^*)n - 1)$. If $\phi = 3$, then $\mathsf{size}(Q_{\mathsf{awc}}) \leq 8s - 6n - 4$, $\mathsf{sdim}(Q_{\mathsf{awc}}) \leq 16s - 14n - 8$, $\mathsf{sdeg}(Q_{\mathsf{awc}}) \leq 12s - 10n - 6$, and $\mathsf{supp}(Q_{\mathsf{awc}}) \leq 72s - 68n - 36$.*

Clearly, other parameters but support are minimized when ϕ is large. If support is not important, then one can dismiss the bounding fan-out step, and get size $2s$, dimension $12s$, and degree $9s$.

Like in the case of wire checkers, [15] constructed a *weak AWC* that guarantees the required "no double assignments" property only if the individual gate checkers are conscientious. The new AWC does not have this restriction. The size of the weak AWC from [15] is $24s$ and the degree of it is $76s$.

7 Circuit Checker

Next, we combine the aggregate gate and wire checkers into a circuit checker, that can be seen as a reduction from CIRCUIT-SAT to QSP-SAT. Circuit checker was called a canonical quadratic span program in [15]. Since [18] introduced canonical span programs in a completely different context, we changed the terminology.

Let C be a circuit, and let $P^{\mathsf{w}} = (\mathbf{0}, \mathbf{0}, U^{\mathsf{w}}, V^{\mathsf{w}}, \psi)$ be an AWC for C_{bnd}. Let $P^{\mathsf{g}} = (\boldsymbol{u}_0, U^{\mathsf{g}}, \varrho)$ be an AGC for C_{bnd}. Let $P^{\mathsf{g}}_{\mathsf{dual}} = (\boldsymbol{u}_0, V^{\mathsf{g}}, \varrho_{\mathsf{dual}})$ be the corresponding dual span program. As before, $V^{\mathsf{g}} = U^{\mathsf{g}}_{\mathsf{dual}}$ and $V^{\mathsf{w}} = U^{\mathsf{w}}_{\mathsf{dual}}$, and ϱ and ψ are related as in Sect. 6.3. Let $m_{\mathsf{g}} = \mathsf{size}(P^{\mathsf{w}}) = \mathsf{size}(P^{\mathsf{g}}) = \mathsf{size}(P^{\mathsf{g}}_{\mathsf{dual}})$. Assume that $U^{\mathsf{w}} = \{\boldsymbol{u}^{\mathsf{w}}_1, \ldots, \boldsymbol{u}^{\mathsf{w}}_{m_{\mathsf{g}}}\}$ and $U^{\mathsf{g}} = \{\boldsymbol{u}^{\mathsf{g}}_1, \ldots, \boldsymbol{u}^{\mathsf{g}}_{m_{\mathsf{g}}}\}$ (and similarly, $V^{\mathsf{w}} = \{\boldsymbol{v}^{\mathsf{w}}_1, \ldots, \boldsymbol{v}^{\mathsf{w}}_{m_{\mathsf{g}}}\}$ and V^{g}) are ordered consistently (see Sect. 6.3).

Definition 3. *For* $m_{\mathsf{g}} = \mathsf{size}(P^{\mathsf{g}})$, $d_{\mathsf{g}} = \mathsf{sdim}(P^{\mathsf{g}})$ *and* $d_{\mathsf{w}} = \mathsf{sdim}(P^{\mathsf{w}})$, *define the* circuit checker *to be the QSP* $c_\Lambda(C) = (\boldsymbol{u_0}, \boldsymbol{v_0}, U, V, \varrho)$, *where*

$$\begin{pmatrix} \boldsymbol{u_0} \\ U \\ \hline \boldsymbol{v_0} \\ V \end{pmatrix} = \begin{pmatrix} \boldsymbol{u_0} & \boldsymbol{1}_{d_{\mathsf{g}}} & \boldsymbol{0}_{d_{\mathsf{w}}} \\ U^{\mathsf{g}} & 0_{m_{\mathsf{g}} \times d_{\mathsf{g}}} & U^{\mathsf{w}} \\ \hline \boldsymbol{1}_{d_{\mathsf{g}}} & \boldsymbol{u_0} & \boldsymbol{0}_{d_{\mathsf{w}}} \\ 0_{m_{\mathsf{g}} \times d_{\mathsf{g}}} & V^{\mathsf{g}} & V^{\mathsf{w}} \end{pmatrix} . \tag{2}$$

Here, $U = (\boldsymbol{u_1}, \dots, \boldsymbol{u_m})^\top$, $V = U_{\mathsf{dual}} = (\boldsymbol{v_1}, \dots, \boldsymbol{v_m})^\top$.

Recall that we denoted by c2q that computed the witness $\boldsymbol{a}$ of the AGC from w. We also denote $(\boldsymbol{a}, \boldsymbol{b}) \leftarrow \mathsf{c2q}(w)$, given that $\boldsymbol{b}$ is the dual of $\boldsymbol{a}$.

Theorem 4. *Let* $\boldsymbol{w} \in \{0,1\}^{s_e}$. $C(\boldsymbol{w}) = 1$ *iff* $c_\Lambda(C)(\mathsf{c2q}(\boldsymbol{w})) = 1$.

Proof. Clearly, $c_\Lambda(C)(\boldsymbol{a}, \boldsymbol{b}) = 1$ iff P^{g}, $P^{\mathsf{g}}_{\mathsf{dual}}$ and P^{w} all accept with the same witness $(\boldsymbol{a}, \boldsymbol{b})$: (i) $(\sum_{i=1}^m a_i u^{\mathsf{g}}_{ij} - u_{0j})(0-1) = 0$ for $j \in [d_{\mathsf{g}}]$ iff $\sum_{i=1}^m a_i u^{\mathsf{g}}_{ij} = u_{0j}$ for $j \in [d_{\mathsf{g}}]$ iff $\sum_{i=1}^m a_i \boldsymbol{u}^{\mathsf{g}}_i = \boldsymbol{u_0}$, (ii) $(0-1)(\sum_{i=1}^m b_i v^{\mathsf{g}}_{ij} - u_{0j}) = 0$ for $j \in [d_{\mathsf{g}}]$ iff $\sum_{i=1}^m b_i v^{\mathsf{g}}_{ij} = u_{0j}$ for $j \in [d_{\mathsf{g}}]$ iff $\sum_{i=1}^m b_i \boldsymbol{v}^{\mathsf{g}}_i = \boldsymbol{u_0}$, (iii) $(\sum_{i=1}^m a_i u^{\mathsf{w}}_{ij}) \cdot (\sum_{i=1}^m b_i v^{\mathsf{w}}_{ij}) = 0$ for $j \in [d_{\mathsf{w}}]$.

Assume $C(\boldsymbol{w}) = 1$. By the construction of P^{g}, there exists $\boldsymbol{a} \in \mathbb{F}^m$, with $a_i = 0$ for $i \notin \psi_{\boldsymbol{w}}^{-1}$, s.t. $\boldsymbol{a}^\top \cdot U^{\mathsf{g}} = \boldsymbol{u_0}$. Let $\boldsymbol{b} \leftarrow \boldsymbol{a}$, then also $\boldsymbol{b}^\top \cdot V^{\mathsf{g}} = \boldsymbol{u_0}$. Since $\boldsymbol{a}$ and $\boldsymbol{b}$ are consistent bit assignments in the evaluation of $C(\boldsymbol{w})$, P^{w} accepts.

Second, assume that there exist $(\boldsymbol{a}, \boldsymbol{b})$, s.t. $c_\Lambda(C)(\boldsymbol{a}, \boldsymbol{b}) = 1$. Since P^{w} accepts, there are no double assignments. That means, that for each η, for some (possibly non-unique) bit $w_\eta \in \{0,1\}$ and all $i \in \psi^{-1}(x_\eta^{\bar{w}_\eta})$, $a_i = 0$. Dually, $b_i = 0$ for all $i \in \psi_{\mathsf{dual}}^{-1}(x_\eta^{\bar{w}_\eta})$ (w_η clearly has to be the same in both cases). Since this holds for every wire, there exists an assignment $\boldsymbol{w}$ of input values, s.t. for all $i \notin \psi_{\boldsymbol{w}}^{-1}$ and $j \notin (\psi_{\mathsf{dual}}^{-1})_{\boldsymbol{w}}$, $a_i = b_j = 0$. Moreover, $C(\boldsymbol{w}) = 1$. □

We will explain in the full version how the parameters of $Q := c_\Lambda(C)$ influence the efficiency of the CIRCUIT-SAT NIZK argument. For example, the support of Q affects the prover's computation, while its degree d affects the CRS length but also the prover's computation and the security assumption. More precisely, the prover's computation of the non-adaptive NIZK argument is $\Theta(\mathsf{supp}(Q) + d \cdot \log d)$ non-cryptographic operations and $\Theta(d)$ cryptographic operations. One should choose ϕ such that the prover's computation will be minimal. This value depends on the constants in Θ. For simplicity, we will consider the case $\phi = 3$.

Theorem 5. *Let* $s = |C|$ *and* $Q := c_\Lambda(C_{\mathsf{bnd}})$. *Let* ϕ *be the fanout of* C_{bnd}, *and* $\phi^* = 1/(\phi - 1)$. *Then* $\mathsf{sdeg}(Q) \leq (17 + 10\phi^*)s - (6 + 20\phi^*)n - 6$, $\mathsf{supp}(Q) \leq (50 + 8\phi(3+\phi) + 40\phi^*)s + 2(-13 + 8\phi(3+2\phi) - 40\phi^*)n - 8(1+\phi)^2$, *and* $\mathsf{size}(Q) \leq \mathsf{size}(P^{\mathsf{w}}) + \mathsf{size}(P^{\mathsf{g}}) \leq 2(7 + 4\phi^*)s - (8 + 16\phi^*)n - 4$. *If* $\phi = 3$, *then* $\mathsf{sdeg}(Q) \leq 22s - 16n - 6$, $\mathsf{size}(Q) \leq 18s - 16n - 4$, *and* $\mathsf{supp}(Q) \leq 214s + 366n - 128$.

The degree of the circuit checker from [15] is $130s$ and its size is $36s$. Thus, even when $\phi = 3$, we have improved on their construction about 6 times degree-wise and 2 times size-wise. The QSP-SAT witness $(\boldsymbol{a}, \boldsymbol{b})$ can be computed in linear time $\Theta(s)$ by using the algorithm c2q.

8 Two-Query Linear PCP for Circuit-SAT

In Thm. 4, we presented a reduction from CIRCUIT-SAT to QSP-SAT. That is, we showed that if for some $\boldsymbol{w}$, $C(\boldsymbol{w}) = 1$, then one can efficiently construct a witness $(\boldsymbol{a}, \boldsymbol{b}) = \mathsf{c2q}(\boldsymbol{w})$ such that $c_\Lambda(C)(\boldsymbol{a}, \boldsymbol{b}) = 1$. In this section, we construct a two-query non-adaptive linear PCP [2] for CIRCUIT-SAT. In the rest of the paper, we modify this to succinct three-query non-adaptive linear PCP, to a non-adaptive linear interactive proof and finally to a non-adaptive non-interactive zero knowledge argument. Here, non-adaptivity means that the query algorithm (in the linear PCP and linear interactive proof) or the CRS generation algorithm (in the NIZK argument) may depend on the statement C.

Let $\mathcal{R} = \{(C, \boldsymbol{w})\}$ be a binary relation, $\mathbb{F}$ be a finite field, $\mathcal{P}_{\mathsf{lpcp}}$ be a deterministic prover algorithm and $\mathcal{V}_{\mathsf{lpcp}} = (\mathcal{Q}_{\mathsf{lpcp}}, \mathcal{D}_{\mathsf{lpcp}})$, where $\mathcal{Q}_{\mathsf{lpcp}}$ is a probabilistic query algorithm and $\mathcal{D}_{\mathsf{lpcp}}$ is an oracle deterministic decision algorithm. The pair $(\mathcal{P}_{\mathsf{lpcp}}, \mathcal{V}_{\mathsf{lpcp}})$ is a *non-adaptive k-query linear PCP* [2] for $\mathcal{R}$ over $\mathbb{F}$ with query length m if it satisfies the following conditions.

Syntax: on any input C and oracle $\boldsymbol{\pi}$, the verifier $\mathcal{V}_{\mathsf{lpcp}}$ works as follows. $\mathcal{Q}_{\mathsf{lpcp}}(C)$ generates k queries $\boldsymbol{q_1}, \ldots, \boldsymbol{q_k} \in \mathbb{F}^m$ to $\boldsymbol{\pi}$, and a state information st. Given k oracle answers $z_1 \leftarrow \langle \boldsymbol{\pi}, \boldsymbol{q_1} \rangle$, ..., $z_k \leftarrow \langle \boldsymbol{\pi}, \boldsymbol{q_k} \rangle$, such that $\boldsymbol{z} = (z_1, \ldots, z_k)$, $\mathcal{D}^{\boldsymbol{\pi}}_{\mathsf{lpcp}}(\mathsf{st}; \boldsymbol{w}) = \mathcal{D}_{\mathsf{lpcp}}(\mathsf{st}, \boldsymbol{z}; \boldsymbol{w})$ accepts or rejects.

Completeness: for every $(C, \boldsymbol{w}) \in \mathcal{R}$, the output of $\mathcal{P}_{\mathsf{lpcp}}(C, \boldsymbol{w})$ is a description of a linear function $\boldsymbol{\pi} : \mathbb{F}^m \to \mathbb{F}$ such that $\mathcal{D}^{\boldsymbol{\pi}}_{\mathsf{lpcp}}(\mathsf{st}; \boldsymbol{w})$ accepts with probability 1.

Knowledge: there exists a knowledge extractor $\mathcal{X}_{\mathsf{lpcp}}$, such that for every linear function $\boldsymbol{\pi}^* : \mathbb{F}^m \to \mathbb{F}$: if the probability that $\mathcal{V}^{\boldsymbol{\pi}^*}_{\mathsf{lpcp}}(C)$ accepts is at least ε, then $\mathcal{X}^{\boldsymbol{\pi}^*}_{\mathsf{lpcp}}(C)$ outputs $\boldsymbol{w}$ such that $(C, \boldsymbol{w}) \in \mathcal{R}$.

$(\mathcal{P}_{\mathsf{lpcp}}, \mathcal{V}_{\mathsf{lpcp}})$ has *degree* $(d_\mathcal{Q}, d_\mathcal{D})$, if $\mathcal{Q}_{\mathsf{lpcp}}$ (resp., $\mathcal{D}_{\mathsf{lpcp}}$) can be computed by an arithmetic circuit of degree $d_\mathcal{Q}$ (resp., $d_\mathcal{D}$).

We remark that in the following non-adaptive linear PCP, $\mathcal{D}_{\mathsf{lpcp}}$ does not depend on w.

Theorem 6. *Let* $\mathbb{F}$ *be a field, and let* C *be a circuit with dummy gates. Let* $\mathcal{P}^{(2)}_{\mathsf{lpcp}}$ *and* $\mathcal{V}^{(2)}_{\mathsf{lpcp}} = (\mathcal{Q}^{(2)}_{\mathsf{lpcp}}, \mathcal{D}^{(2)}_{\mathsf{lpcp}})$ *be as follows:*

$\mathcal{Q}^{(2)}_{\mathsf{lpcp}}(C)$: $Q \leftarrow c_\Lambda(C)$; $m \leftarrow \mathsf{size}(Q)$; $\boldsymbol{q_u} \leftarrow (\boldsymbol{u_i}, \boldsymbol{0}_m)_{i=1}^m$; $\boldsymbol{q_v} \leftarrow (\boldsymbol{0}_m, \boldsymbol{v_i})_{i=1}^m$; $\boldsymbol{q} \leftarrow (\boldsymbol{q_u}, \boldsymbol{q_v})$; $\mathsf{st} \leftarrow (\boldsymbol{u_0}, \boldsymbol{v_0})$; *return* $(\boldsymbol{q}, \mathsf{st})$;

$\mathcal{P}^{(2)}_{\mathsf{lpcp}}(C, \boldsymbol{w})$: $Q \leftarrow c_\Lambda(C)$; $(\boldsymbol{\pi_u}, \boldsymbol{\pi_v}) = (\boldsymbol{a}, \boldsymbol{b}) \leftarrow \mathsf{c2q}(\boldsymbol{w})$; *return* $\boldsymbol{\pi} = (\boldsymbol{\pi_u}, \boldsymbol{\pi_v})$;

$\mathcal{D}^{(2)}_{\mathsf{lpcp}}(\mathsf{st}, (\boldsymbol{z_u}, \boldsymbol{z_v}); \boldsymbol{w})$: *if* $(\boldsymbol{z_u} - \boldsymbol{u_0}) \circ (\boldsymbol{z_v} - \boldsymbol{v_0}) = \boldsymbol{0}$ *then return* 1 *else return* 0;

$(\mathcal{P}^{(2)}_{\mathsf{lpcp}}, \mathcal{V}^{(2)}_{\mathsf{lpcp}})$ *is a non-adaptive* 2*-query linear PCP for* CIRCUIT-SAT *with query length* $2md$ *and knowledge error* 0.

Proof. COMPLETENESS: Clearly, $\boldsymbol{z_u} \leftarrow \langle \boldsymbol{\pi}, \boldsymbol{q_u} \rangle = \sum_{i=1}^m a_i \boldsymbol{u_i}$, $\boldsymbol{z_v} \leftarrow \langle \boldsymbol{\pi}, \boldsymbol{q_v} \rangle = \sum_{i=1}^m b_i \boldsymbol{v_i}$. Thus, $\boldsymbol{z_u} - \boldsymbol{u_0} = \boldsymbol{a}^\top \cdot U - \boldsymbol{u_0}$ and $\boldsymbol{z_v} - \boldsymbol{v_0} = \boldsymbol{b}^\top \cdot V - \boldsymbol{v_0}$, and the circuit checker accepts.

KNOWLEDGE PROPERTY: Due to the construction of $\mathcal{Q}^{(2)}_{\mathsf{lpcp}}$, $\boldsymbol{z_u} = \sum_{i=1}^m a_i \boldsymbol{u_i}$, and $\boldsymbol{z_v} = \sum_{i=1}^m b_i \boldsymbol{v_i}$. If $\mathcal{D}^{(2)}_{\mathsf{lpcp}}$ accepts, then by Thm. 4, the wire checker implies that no wire η gets a double assignment. However, it may be the case that some wire has no assignment. Nevertheless, on input (st, C) and access to the oracle $\boldsymbol{\pi}^*$, we will now extract a CIRCUIT-SAT witness $\boldsymbol{w} = (w_\eta)_{i=\eta}^{s_e}$ (i.e., the vector of wire values) such that $C(\boldsymbol{w}) = 1$.

First, the extractor obtains the whole linear function $\boldsymbol{\pi}^* = (\boldsymbol{a}, \boldsymbol{b})$, by querying the oracle $\boldsymbol{\pi}^*$ up to $2m$ times. We deduce $\boldsymbol{w}$ from $\boldsymbol{\pi}^*$ as follows.

Let η be any wire of the circuit C. Since the wire checker accepts, the gate checkers of its neighbouring gates do not assign multiple values to the wire η. There are two different cases.

If η is an input wire to the circuit, then its output gate ι is a conscientious dummy gate. Therefore, the value w_η can be extracted from the local values of a_i corresponding to the gate ι.

Assume that η is an internal wire. Since all gates implement functions with well-defined outputs, the gate checker of the input gate of η assigns some value w_η to this wire. Moreover, every output gate ι of η either assigns the same value w_η or does not assign any value. In the latter case, the output value of ι does not depend on w_η, and thus assigning w_η to η is consistent with the output value of ι. Therefore, also here the value w_η can be extracted, but this time from the local values of a_i and b_i corresponding to the input gate of η. □

A simple corollary of this theorem is that the algorithm $\mathsf{c2q}$ is efficiently invertible. Thus, the constructed **NP**-reduction from CIRCUIT-SAT to QSP-SAT preserves knowledge (i.e., it is a Levin reduction).

Note that the communication and computation can be optimized by defining $\boldsymbol{q_u} \leftarrow (\boldsymbol{u_i})_{i=1}^m$, $\boldsymbol{q_v} \leftarrow (\boldsymbol{v_i})_{i=1}^m$, and computing say $\boldsymbol{z_u} \leftarrow \langle \boldsymbol{\pi_u}, \boldsymbol{q_u} \rangle$.

9 Succinct 3-Query Linear PCP from Polynomial QSPs

Since we are interested in succinct arguments, we need to be able to compress the witness vectors $\boldsymbol{a}$ and $\boldsymbol{b}$. As in [15], we will do it by using polynomial interpolation to define polynomial QSPs. We employ the Schwartz-Zippel lemma to show that the resulting succinct 3-query linear PCP has the knowledge property.

9.1 Polynomial Span Programs and QSPs

Instead of considering the target and row vectors of a span program or a QSP as being members of the vector space $\mathbb{F}^d$, interpret them as degree-$(d-1)$ polynomials in $\mathbb{F}[X]$. The map $\boldsymbol{u} \to \hat{u}(X)$ is implemented by choosing d different field elements (that are the same for all vectors $\boldsymbol{u}$) $r_j \leftarrow \mathbb{F}$, and then defining a degree-$(\leq d-1)$ polynomial $\hat{u}(X)$ via polynomial interpolation, so that $\hat{u}(r_j) = u_j$ for all $j \in [d]$. This maps the vectors $\boldsymbol{u_i}$ of the original span program P to polynomials $\hat{u}_i(X)$, and the target vector $\boldsymbol{u_0}$ to the polynomial $\hat{u}_0(X)$. Finally, let $Z(X) := \prod_{j=1}^d (X - r_j)$; this polynomial can be thought of as a mapping of the all-zero vector $\mathbf{0} = (0, \ldots, 0)$.

The choice of r_j influences efficiency. If r_j are arbitrary, then multipoint evaluation and polynomial interpolation take time $O(d \log^2 d)$ [12]. If d is a power of

2 and $r_j = \omega_d^j$, where ω_d is the dth primitive root of unity, then both operations can be done in time $O(d \log d)$ by using Fast Fourier Transform [12]. In what follows, d and r_j are chosen as in the current paragraph.

Clearly, $\boldsymbol{u_0}$ is in the span of the vectors that belong to $\varrho_{\boldsymbol{w}}^{-1}$ iff $\boldsymbol{u_0} = \sum_{i\in\varrho_{\boldsymbol{w}}^{-1}} a_i \boldsymbol{u_i}$ for some $a_i \in \mathbb{F}$. The latter is equivalent to the requirement that $Z(X)$ divides $\hat{u}(X) := \sum_{i\in\varrho_{\boldsymbol{w}}^{-1}} a_i\hat{u}_i(X) - \hat{u}_0(X)$. Really, $\boldsymbol{u_0}$ is the vector of evaluations of $\hat{u}_0(X)$, and $\boldsymbol{u_i}$ is the vector of evaluations of $\hat{u}_i(X)$. Thus, $\sum a_i\boldsymbol{u_i} - \boldsymbol{u_0} = \mathbf{0}$ iff $\sum a_i\hat{u}_i(X) - \hat{u}_0(X)$ evaluates to 0 at all r_j, and hence is divisible by $Z(X)$.

A *polynomial span program* $P = (\hat{u}_0, U, \varrho)$ over a field $\mathbb{F}$ consists of a target polynomial $\hat{u}_0(X) \in \mathbb{F}[X]$, a tuple $U = (\hat{u}_i(X))_{i=1}^m$ of polynomials from $\mathbb{F}[X]$, and a labelling $\varrho : [m] \to \{x_\iota, \bar{x}_\iota : \iota \in [n]\} \cup \{\perp\}$ of the polynomials from U. Let $U_{\boldsymbol{w}}$ be the subset of U consisting of those polynomials whose labels are satisfied by the assignment $\boldsymbol{w} \in \{0,1\}^n$, that is, by $\{x_\iota^{w_\iota} : \iota \in [n]\} \cup \{\perp\}$. The span program P *computes a function* f, if for all $\boldsymbol{w} \in \{0,1\}^n$: there exists $\boldsymbol{a} \in \mathbb{F}^m$ such that $Z(X) \mid (\hat{u}_0(X) + \sum_{u\in U_{\boldsymbol{w}}} a_i\hat{u}(X))$ (P accepts) iff $f(\boldsymbol{w}) = 1$.

Alternatively, P accepts $\boldsymbol{w} \in \{0,1\}^n$ iff there exists a vector $\boldsymbol{a} \in \mathbb{F}^m$, with $a_i = 0$ for all $i \notin \varrho_{\boldsymbol{w}}^{-1}$, s.t. $Z(X) \mid \sum_{i=1}^m a_i\hat{u}_i(X) - \hat{u}_0(X)$. The size of P is m and the degree of P is $\deg Z(X)$.

Definition 4. *A* polynomial QSP $Q = (\hat{u}_0, \hat{v}_0, U, V, \varrho)$ *over a field* $\mathbb{F}$ *consists of target polynomials* $\hat{u}_0(X) \in \mathbb{F}[X]$ *and* $\hat{v}_0(X) \in \mathbb{F}[X]$, *two tuples* $U = (\hat{u}_i(X))_{i=1}^m$ *and* $V = (\hat{v}_i(X))_{i=1}^m$ *of polynomials from* $\mathbb{F}[X]$, *and a labelling* $\varrho : [m] \to \{x_\iota, \bar{x}_\iota : \iota \in [n]\} \cup \{\perp\}$. Q accepts an input $\boldsymbol{w} \in \{0,1\}^n$ *iff there exist two vectors* $\boldsymbol{a}$ *and* $\boldsymbol{b}$ *from* $\mathbb{F}^m$, *with* $a_i = 0 = b_i$ *for all* $i \notin \varrho_{\boldsymbol{w}}^{-1}$, *s.t.* $Z(X) \mid (\sum_{i=1}^m a_i\hat{u}_i(X) - \hat{u}_0(X))(\sum_{i=1}^m b_i\hat{v}_i(X) - \hat{v}_0(X))$. Q computes a Boolean function $f : \{0,1\}^n \to \{0,1\}$ *if* Q *accepts* $\boldsymbol{w}$ *iff* $f(\boldsymbol{w}) = 1$.

The size of Q is m and the degree of Q is $\deg Z(X)$. Keeping in mind the reinterpretation of span programs, Def. 4 is clearly equivalent to Def. 1. (Also here, $V = U_{\mathsf{dual}}$, with the dual operation defined appropriately.)

To get from the linear-algebraic interpretation to polynomial interpretation, one has to do the following. Assume that the dimension of the QSP is d and that the size is m. Let $r_j \leftarrow \omega_d^j$, $j \in [d]$. For $i \in [m]$, interpolate the polynomial $\hat{u}_i(X)$ (resp., $\hat{v}_i(X)$) from the values $\hat{u}_i(r_j) = u_{ij}$ (resp., $\hat{v}_i(r_j) = v_{ij}$) for $j \in [d]$. Set $Z(X) := \prod_{j=1}^d (X - r_j)$. The labelling ψ is left unchanged. It is clear that the resulting polynomial QSP $(\hat{u}_0, \hat{v}_0, U, V, \psi)$ computes the same Boolean function as the original QSP.

The *polynomial circuit checker* $c_\Lambda^{\mathsf{poly}}(C) = (\hat{u}_0, \hat{v}_0, U, V, \psi)$, with $U = (\hat{u}_0, \dots, \hat{u}_m)$ and $V = (\hat{v}_0, \dots, \hat{v}_m)$, is the polynomial version of $c_\Lambda(C)$.

Theorem 7. *Let* $\boldsymbol{w} \in \{0,1\}^n$. $C(\boldsymbol{w}) = 1$ *iff* $c_\Lambda^{\mathsf{poly}}(C)(\mathsf{c2q}(\boldsymbol{w})) = 1$.

Proof. Follows from Thm. 4 and the construction of polynomial QSPs. □

9.2 Succinct Three-Query Linear PCP

To achieve better efficiency, following [2], we define a 3-query linear PCP with $|z| = \Theta(1)$ that is based on the polynomial QSPs. For a set $\mathcal{P}$ of polynomials, let $\mathsf{span}(\mathcal{P})$ be their span (i.e., the set of $\mathbb{F}$-linear combinations). Then, $\boldsymbol{u}$ is in the span of vectors $\boldsymbol{u_i}$, $\boldsymbol{u} = \sum_{i=1}^m a_i \boldsymbol{u_i}$, iff the corresponding interpolated polynomial $\hat{u}(X)$ is in the span of polynomials $\hat{u}_i(X)$, i.e., $\hat{u}(X) = \sum_{i=1}^m a_i \hat{u}_i(X)$.

Let $\mathbb{F}$ be any field. We recall that according to the Schwartz-Zippel lemma, for any nonzero polynomial $f : \mathbb{F}^m \to \mathbb{F}$ of total degree d and any finite subset S of $\mathbb{F}$, $\Pr_{\boldsymbol{x} \leftarrow S^m}[f(\boldsymbol{x}) = 0] \le d/|S|$.

Theorem 8. *Let $\mathbb{F}$ be a field, and C a circuit with dummy gates. Let $\mathcal{P}_{\mathsf{lpcp}}^{(3)}$ and $\mathcal{V}_{\mathsf{lpcp}}^{(3)} = (\mathcal{Q}_{\mathsf{lpcp}}^{(3)}, \mathcal{D}_{\mathsf{lpcp}}^{(3)})$ be as follows. Here, $\mathsf{PolyInt}$ is polynomial interpolation.*

$\mathcal{Q}_{\mathsf{lpcp}}^{(3)}(C)$: *$Q \leftarrow c_\Lambda(C)$; $m \leftarrow \mathsf{size}(Q)$; $d \leftarrow \mathsf{sdeg}(Q)$; For $i \leftarrow 1$ to d do: $r_i \leftarrow \omega_d^i$; $\sigma \leftarrow_r \mathbb{F}$; Compute $(\sigma^i)_{i=0}^{d-1}$; $Z(\sigma) \leftarrow \prod_{j=1}^d (\sigma - r_j)$; Compute $(\hat{u}_i(\sigma))_{i=0}^m$, $(\hat{v}_i(\sigma))_{i=0}^m$; $\mathsf{st} \leftarrow (Z(\sigma), \hat{u}_0(\sigma), \hat{v}_0(\sigma))$; $\boldsymbol{q_u} \leftarrow (((\hat{u}_i(\sigma))_{i=1}^m, \boldsymbol{0}_m, \boldsymbol{0}_d)$; $\boldsymbol{q_v} \leftarrow (\boldsymbol{0}_m, (\hat{v}_i(\sigma))_{i=1}^m, \boldsymbol{0}_d)$ $\boldsymbol{q_h} \leftarrow (\boldsymbol{0}_m, \boldsymbol{0}_m, (\sigma^i)_{i=0}^{d-1})$; $\boldsymbol{q} \leftarrow (\boldsymbol{q_u}, \boldsymbol{q_v}, \boldsymbol{q_h})$; return $(\boldsymbol{q}, \mathsf{st})$;*

$\mathcal{P}_{\mathsf{lpcp}}^{(3)}(C, \boldsymbol{w})$: *Compute $(Q, m, (r_i)_{i=1}^d)$ as in $\mathcal{Q}_{\mathsf{lpcp}}^{(3)}(C)$; $(\boldsymbol{a}, \boldsymbol{b}) \leftarrow \mathsf{c2q}(\boldsymbol{w})$; $\boldsymbol{u}^\dagger \leftarrow \boldsymbol{u_0} + \sum_{i=1}^m a_i \boldsymbol{u_i}$; $\hat{u}^\dagger(X) \leftarrow \mathsf{PolyInt}((r_i, u_i^\dagger)_{i=1}^d)$; $\boldsymbol{v}^\dagger \leftarrow \boldsymbol{v_0} + \sum_{i=1}^m a_i \boldsymbol{v_i}$; $\hat{v}^\dagger(X) \leftarrow \mathsf{PolyInt}((r_i, v_i^\dagger)_{i=1}^d)$; $Z(X) \leftarrow \prod_{i=1}^d (X - r_i)$; $\hat{h}(X) = \sum_{i=0}^{d-1} h_i X^i \leftarrow \hat{u}^\dagger(X)\hat{v}^\dagger(X)/Z(X) \in \mathbb{F}^{d-2}$; return $\boldsymbol{\pi} = (\boldsymbol{\pi_u}, \boldsymbol{\pi_v}, \boldsymbol{\pi_h}) \leftarrow (\boldsymbol{a}, \boldsymbol{b}, \hat{\boldsymbol{h}}) \in \mathbb{F}^{2m+d}$;*

$\mathcal{D}_{\mathsf{lpcp}}^{(3)}(\mathsf{st}, (z_u, z_v, z_h); \boldsymbol{w})$: *if $(z_u - \hat{u}_0(\sigma)) \cdot (z_v - \hat{v}_0(\sigma)) = Z(\sigma) \cdot z_h$ then return 1 else return 0;*

$(\mathcal{P}_{\mathsf{lpcp}}^{(3)}, \mathcal{V}_{\mathsf{lpcp}}^{(3)})$ is a non-adaptive 3-query linear PCP over $\mathbb{F}$ for CIRCUIT-SAT *with query length $2m + d$ and knowledge error $2d/|\mathbb{F}|$.*

Proof. COMPLETENESS: again straightforward, since $z_u = \hat{u}_w(\sigma) \leftarrow \langle \boldsymbol{\pi}, \boldsymbol{q_u} \rangle = \sum_{i=1}^m a_i \hat{u}_i(\sigma)$, $z_v = \hat{v}(\sigma) \leftarrow \langle \boldsymbol{\pi}, \boldsymbol{q_v} \rangle = \sum_{i=1}^m b_i \hat{v}_i(\sigma)$, and $z_h = \hat{h}(\sigma) \leftarrow \langle \boldsymbol{\pi}, \boldsymbol{q_h} \rangle = \sum_{i=0}^{d-1} \hat{h}_i \sigma^i$. KNOWLEDGE: assume that the verifier accepts with probability $\varepsilon \ge 2d/|\mathbb{F}|$. That is, $\Pr_{\sigma \leftarrow \mathbb{F}}[(\sum_{i=1}^m a_i \hat{u}_i(\sigma) - \hat{u}_0(\sigma))(\sum_{i=1}^m a_i \hat{v}_i(\sigma) - \hat{v}_0(\sigma)) = Z(\sigma) \cdot (\sum_{i=0}^{d-1} h_i \sigma^i)] = \varepsilon$. Due to the Schwartz-Zippel lemma, since $\varepsilon \ge 2d/|\mathbb{F}|$, $(\sum_{i=1}^m a_i \hat{u}_i(X) - \hat{u}_0(X))(\sum_{i=1}^m a_i \hat{v}_i(X) - \hat{v}_0(X)) = Z(X) \cdot (\sum_{i=0}^{d-1} h_i X^i)$, and due to the equivalence between QSPs and polynomial QSPs, Eq. (1) holds. The claim now follows from Thm. 6. □

Theorem 9. *Assume d is a power of 2. $\mathcal{P}_{\mathsf{lpcp}}^{(3)}$ runs in time $\Theta(d \log d)$, $\mathcal{Q}_{\mathsf{lpcp}}^{(3)}$ runs in time $\Theta(d \log d)$, and the time of $\mathcal{D}_{\mathsf{lpcp}}^{(3)}$ is dominated by 2 $\mathbb{F}$-additions and by 2 $\mathbb{F}$-multiplications. $\mathcal{V}_{\mathsf{lpcp}}^{(3)}$ has degree $(d, 2)$.*

A similar result was proven in [15] (though without using the terminology of linear PCPs) in the case of conscientious gate checkers. We only require the dummy gates to be conscientious.

In [15], it was only shown that $\hat{h}(X)$ can be computed by using multipoint evaluation and polynomial interpolation in time $\Theta(d \log^2 d)$. Moreover, the computation of $\mathcal{D}$ was $\Theta(n)$ due to a different extraction technique.

10 From Non-Adaptive Linear PCP to Adaptive NIZK

Given the 3-query linear PCP of Thm. 8, one can use the transformation [2] to construct first a non-adaptive NIZK argument for CIRCUIT-SAT. See the full version. The The non-adaptive NIZK argument can be made adaptive by using universal circuits [25], see [15] for details.

We will provide more details in the full version [20]. There, we will also provide a direct construction of the non-adaptive NIZK argument. The latter has a (quite complex) soundness proof related to the soundness proof from [15] that results in the use of a weaker security assumption. Here, we state only the following straightforward corollary of Thm. 9 and the transformations from [2].

Theorem 10. *Assume d is a power of 2. There exists a non-adaptive NIZK* CIRCUIT-SAT *argument, s.t. the prover and the CRS generation take $\Theta(d \log d)$ cryptographic operations, the verification time is dominated by $\Theta(1)$ pairings, and the communication is a $\Theta(1)$ group elements.*

Acknowledgements. We thank Andris Ambainis, Aleksandrs Belovs, Vitaly Skachek, Hendri Tan, and anonymous reviewers for useful comments and discussions. The author was supported by the Estonian Research Council, and European Union through the European Regional Development Fund.

References

1. Bitansky, N., Canetti, R., Chiesa, A., Tromer, E.: From Extractable Collision Resistance to Succinct Non-Interactive Arguments of Knowledge, and Back Again. In: Goldwasser, S. (ed.) ITCS 2012, pp. 326–349. ACM Press (2012)
2. Bitansky, N., Chiesa, A., Ishai, Y., Paneth, O., Ostrovsky, R.: Succinct Non-interactive Arguments via Linear Interactive Proofs. In: Sahai, A. (ed.) TCC 2013. LNCS, vol. 7785, pp. 315–333. Springer, Heidelberg (2013)
3. Blum, M., Feldman, P., Micali, S.: Non-Interactive Zero-Knowledge and Its Applications. In: STOC 1988, pp. 103–112. ACM Press (1988)
4. Brassard, G., Chaum, D., Crépeau, C.: Minimum Disclosure Proofs of Knowledge. Journal of Computer and System Sciences 37(2), 156–189 (1988)
5. Chaabouni, R., Lipmaa, H., Zhang, B.: A Non-interactive Range Proof with Constant Communication. In: Keromytis, A.D. (ed.) FC 2012. LNCS, vol. 7397, pp. 179–199. Springer, Heidelberg (2012)
6. Di Crescenzo, G., Lipmaa, H.: Succinct NP Proofs from an Extractability Assumption. In: Beckmann, A., Dimitracopoulos, C., Löwe, B. (eds.) CiE 2008. LNCS, vol. 5028, pp. 175–185. Springer, Heidelberg (2008)
7. Dodunekov, S., Landgev, I.: On Near-MDS Codes. Journal of Geometry 54(1-2), 30–43 (1995)
8. Dwork, C., Naor, M.: Zaps and Their Applications. In: FOCS 2000, pp. 283–293. IEEE Computer Society Press (2000)
9. Elkin, M.: An Improved Construction of Progression-Free Sets. Israel J. of Math. 184, 93–128 (2011)

10. Fauzi, P., Lipmaa, H., Zhang, B.: Efficient Modular NIZK Arguments from Shift and Product. In: Abdalla, M. (ed.) CANS 2013. LNCS, vol. 8257, pp. 92–121. Springer, Heidelberg (2013)
11. Gál, A.: A Characterization of Span Program Size and Improved Lower Bounds for Monotone Span Programs. Computational Complexity 10(4), 277–296 (2001)
12. Gathen, J., Gerhard, J.: Modern Computer Algebra, 2nd edn. Cambridge University Press (2003)
13. Gennaro, R., Gentry, C., Parno, B.: Non-interactive Verifiable Computing: Outsourcing Computation to Untrusted Workers. In: Rabin, T. (ed.) CRYPTO 2010. LNCS, vol. 6223, pp. 465–482. Springer, Heidelberg (2010)
14. Gennaro, R., Gentry, C., Parno, B., Raykova, M.: Quadratic Span Programs and Succinct NIZKs without PCPs. Tech. Rep. 2012/215, IACR (April 19, 2012), `http://eprint.iacr.org/2012/215` (last retrieved version from June 18, 2012)
15. Gennaro, R., Gentry, C., Parno, B., Raykova, M.: Quadratic Span Programs and Succinct NIZKs without PCPs. In: Johansson, T., Nguyen, P.Q. (eds.) EUROCRYPT 2013. LNCS, vol. 7881, pp. 626–645. Springer, Heidelberg (2013)
16. Groth, J.: Short Pairing-Based Non-interactive Zero-Knowledge Arguments. In: Abe, M. (ed.) ASIACRYPT 2010. LNCS, vol. 6477, pp. 321–340. Springer, Heidelberg (2010)
17. Hoover, H.J., Klawe, M.M., Pippenger, N.: Bounding Fan-out in Logical Networks. Journal of the ACM 31(1), 13–18 (1984)
18. Karchmer, M., Wigderson, A.: On Span Programs. In: Structure in Complexity Theory Conference 1993, pp. 102–111. IEEE Computer Society Press (1993)
19. Lipmaa, H.: Progression-Free Sets and Sublinear Pairing-Based Non-Interactive Zero-Knowledge Arguments. In: Cramer, R. (ed.) TCC 2012. LNCS, vol. 7194, pp. 169–189. Springer, Heidelberg (2012)
20. Lipmaa, H.: Succinct Non-Interactive Zero Knowledge Arguments from Span Programs and Linear Error-Correcting Codes. Tech. Rep. 2013/121, IACR (February 28, 2013), `http://eprint.iacr.org/2013/121`
21. Lipmaa, H., Zhang, B.: A More Efficient Computationally Sound Non-Interactive Zero-Knowledge Shuffle Argument. In: Visconti, I., De Prisco, R. (eds.) SCN 2012. LNCS, vol. 7485, pp. 477–502. Springer, Heidelberg (2012)
22. Micali, S.: CS Proofs. In: Goldwasser, S. (ed.) FOCS 1994, pp. 436–453. IEEE, IEEE Computer Society Press (1994)
23. Parno, B., Gentry, C., Howell, J., Raykova, M.: Pinocchio: Nearly Practical Verifiable Computation. In: IEEE Symposium on Security and Privacy, pp. 238–252. IEEE Computer Society
24. Reichardt, B.: Reflections for Quantum Query Algorithms. In: Randall, D. (ed.) SODA 2011, pp. 560–569. SIAM (2011)
25. Valiant, L.G.: Universal Circuits (Preliminary Report). In: STOC 1976, pp. 196–203. ACM (1976)

Families of Fast Elliptic Curves from $\mathbb{Q}$-curves

Benjamin Smith

Team GRACE, INRIA Saclay–Île-de-France
and Laboratoire d'Informatique de l'École polytechnique (LIX)
Bâtiment Alan Turing, 1 rue Honoré d'Estienne d'Orves
Campus de l'École polytechnique, 91120 Palaiseau, France
smith@lix.polytechnique.fr

Abstract. We construct new families of elliptic curves over $\mathbb{F}_{p^2}$ with efficiently computable endomorphisms, which can be used to accelerate elliptic curve-based cryptosystems in the same way as Gallant–Lambert–Vanstone (GLV) and Galbraith–Lin–Scott (GLS) endomorphisms. Our construction is based on reducing quadratic $\mathbb{Q}$-curves (curves defined over quadratic number fields, without complex multiplication, but with isogenies to their Galois conjugates) modulo inert primes. As a first application of the general theory we construct, for every prime $p > 3$, two one-parameter families of elliptic curves over $\mathbb{F}_{p^2}$ equipped with endomorphisms that are faster than doubling. Like GLS (which appears as a degenerate case of our construction), we offer the advantage over GLV of selecting from a much wider range of curves, and thus finding secure group orders when p is fixed. Unlike GLS, we also offer the possibility of constructing twist-secure curves. Among our examples are prime-order curves over $\mathbb{F}_{p^2}$, equipped with fast endomorphisms, and with almost-prime-order twists, for the particularly efficient primes $p = 2^{127} - 1$ and $p = 2^{255} - 19$.

Keywords: Elliptic curve cryptography, endomorphisms, GLV, GLS, exponentiation, scalar multiplication, $\mathbb{Q}$-curves.

1 Introduction

Let $\mathcal{E}$ be an elliptic curve over a finite field $\mathbb{F}_q$, and let $\mathcal{G} \subset \mathcal{E}(\mathbb{F}_q)$ be a cyclic subgroup of prime order N. When implementing cryptographic protocols in $\mathcal{G}$, the fundamental operation is *scalar multiplication* (or *exponentiation*):

$$\text{Given } P \text{ in } \mathcal{G} \text{ and } m \text{ in } \mathbb{Z}, \text{ compute } [m]P := \underbrace{P \oplus \cdots \oplus P}_{m \text{ times}}.$$

The literature on general scalar multiplication algorithms is vast, and we will not explore it in detail here (see [10, §2.8,§11.2] and [5, Chapter 9] for introductions to exponentiation and multiexponentiation algorithms). For our purposes, it suffices to note that the dominant factor in scalar multiplication time using conventional algorithms is the bitlength of m. As a basic example, if $\mathcal{G}$ is a generic cyclic abelian group, then we may compute $[m]P$ using a variant

K. Sako and P. Sarkar (Eds.) ASIACRYPT 2013 Part I, LNCS 8269, pp. 61–78, 2013.

of the binary method, which requires at most $\lceil \log_2 m \rceil$ doublings and (in the worst case) about as many addings in $\mathcal{G}$.

But elliptic curves are not generic groups: they have a rich and concrete geometric structure, which should be exploited for fun and profit. For example, endomorphisms of elliptic curves may be used to accelerate generic scalar multiplication algorithms, and thus to accelerate basic operations in curve-based cryptosystems.

Suppose $\mathcal{E}$ is equipped with an efficient endomorphism ψ, defined over $\mathbb{F}_q$. By *efficient*, we mean that we can compute the image $\psi(P)$ of any point P in $\mathcal{E}(\mathbb{F}_q)$ for the cost of $O(1)$ operations in $\mathbb{F}_q$. In practice, we want this to cost no more than a few doublings in $\mathcal{E}(\mathbb{F}_q)$.

Assume $\psi(\mathcal{G}) \subseteq \mathcal{G}$, or equivalently, that ψ restricts to an endomorphism of $\mathcal{G}$.[1] Now $\mathcal{G}$ is a finite cyclic group, isomorphic to $\mathbb{Z}/N\mathbb{Z}$; and every endomorphism of $\mathbb{Z}/N\mathbb{Z}$ is just an integer multiplication modulo N. Hence, ψ acts on $\mathcal{G}$ as multiplication by some integer eigenvalue λ_ψ: that is,

$$\psi|_{\mathcal{G}} = [\lambda_\psi]_{\mathcal{G}} \ .$$

The eigenvalue λ_ψ is a root of the characteristic polynomial of ψ in $\mathbb{Z}/N\mathbb{Z}$.

Returning to the problem of scalar multiplication: we want to compute $[m]P$. Rewriting m as

$$m = a + b\lambda_\psi \pmod{N}$$

for some a and b, we can compute $[m]P$ using the relation

$$[m]P = [a]P + [b\lambda_\psi]P = [a]P + [b]\psi(P)$$

and a two-dimensional multiexponentation such as Straus's algorithm [28], which has a loop length of $\log_2 \|(a,b)\|_\infty$ (ie, $\log_2 \|(a,b)\|_\infty$ doubles and as many adds; recall that $\|(a,b)\|_\infty = \max(|a|,|b|)$). If λ_ψ is not too small, then we can easily find (a,b) such that $\log_2 \|(a,b)\|_\infty$ is roughly half of $\log_2 N$. (We remove the "If" and the "roughly" for our ψ in §4.) The endomorphism lets us replace conventional $\log_2 N$-bit scalar multiplications with $\frac{1}{2}\log_2 N$-bit multiexponentiations. In terms of basic binary methods, we are halving the loop length, cutting the number of doublings in half.

Of course, in practice we are not halving the execution time. The precise speedup ratio depends on a variety of factors, including the choice of exponentiation and multiexponentiation algorithms, the cost of computing ψ, the shortness of a and b on the average, and the cost of doublings and addings in terms of bit operations—to say nothing of the cryptographic protocol, which may prohibit some other conventional speedups. For example: in [11], Galbraith, Lin,

[1] This assumption is satisfied almost by default in the context of classical discrete log-based cryptosystems. If $\psi(\mathcal{G}) \not\subseteq \mathcal{G}$, then $\mathcal{E}[N](\mathbb{F}_q) = \mathcal{G} + \psi(\mathcal{G}) \cong (\mathbb{Z}/N\mathbb{Z})^2$, so $N^2 \mid \#\mathcal{E}(\mathbb{F}_q)$ and $N \mid q-1$; such $\mathcal{E}$ are cryptographically inefficient, and discrete logs in $\mathcal{G}$ are vulnerable to the Menezes–Okamoto–Vanstone reduction [21]. However, these $\mathcal{G}$ do arise naturally in pairing-based cryptography; in that context the assumption should be verified carefully.

and Scott report experiments where cryptographic operations on GLS curves required between 70% and 83% of the time required for the previous best practice curves—with the variation depending on the architecture, the underyling point arithmetic, and the protocol.

To put this technique into practice, we need a source of cryptographic elliptic curves equipped with efficient endomorphisms. To date, in the large characteristic case[2], there have been essentially only two constructions:

1. The classic *Gallant–Lambert–Vanstone* (GLV) construction [12]. Here, elliptic curves over number fields with explicit complex multiplication (CM) by CM-orders with small discriminants are reduced modulo suitable primes p; an explicit endomorphism on the CM curve reduces to an efficient endomorphism over the finite field.
2. The more recent *Galbraith–Lin–Scott* (GLS) construction [11]. Here, curves over $\mathbb{F}_p$ are viewed over $\mathbb{F}_{p^2}$; the p-power sub-Frobenius induces an extremely efficient endomorphism on the quadratic twist (which can have prime order).

These constructions have since been combined to give 3- and 4-dimensional variants [18,32], and extended to hyperelliptic curves in a variety of ways [3,17,26,29]. However, basic GLV and GLS remain the archetypal constructions.

Our contribution: new families of endomorphisms. In this work, we propose a new source of elliptic curves over $\mathbb{F}_{p^2}$ with efficient endomorphisms: quadratic $\mathbb{Q}$-curves.

Definition 1. *A quadratic* $\mathbb{Q}$*-curve of degree* d *is an elliptic curve* $\mathcal{E}$ *without CM, defined over a quadratic number field* K*, such that there exists an isogeny of degree* d *from* $\mathcal{E}$ *to its Galois conjugate* ${}^{\sigma}\mathcal{E}$*, where* $\langle\sigma\rangle = \mathrm{Gal}(K/\mathbb{Q})$.[3]

$\mathbb{Q}$-curves are well-established objects of interest in number theory, where they formed a natural setting for generalizations of the Modularity Theorem. Ellenberg's survey [8] gives an excellent introduction to this beautiful theory.

Our application of quadratic $\mathbb{Q}$-curves is rather more prosaic: given a d-isogeny $\widetilde{\mathcal{E}} \to {}^{\sigma}\widetilde{\mathcal{E}}$ over a quadratic field, we reduce modulo an inert prime p to obtain an isogeny $\mathcal{E} \to {}^{\sigma}\mathcal{E}$ over $\mathbb{F}_{p^2}$. We then exploit the fact that the p-power Frobenius isogeny maps ${}^{\sigma}\mathcal{E}$ back onto $\mathcal{E}$; composing with the reduced d-isogeny, we obtain an endomorphism of $\mathcal{E}$ of degree dp. For efficiency reasons, d must be small; it turns out that for small values of d, we can write down one-parameter families of $\mathbb{Q}$-curves (our approach below was inspired by the explicit techniques of Hasegawa [15]). We thus obtain one-parameter families of elliptic curves over $\mathbb{F}_{p^2}$ equipped with efficient non-integer endomorphisms. For these endomorphisms we can give convenient explicit formulæ for short scalar decompositions (see §4).

For concrete examples, we concentrate on the cases $d = 2$ and 3 (in §5 and §6, respectively), where the endomorphism is more efficient than a single doubling

[2] We are primarily interested in the large characteristic case, where $q = p$ or p^2, so we will not discuss τ-adic/Frobenius expansion-style techniques here.

[3] The Galois conjugate ${}^{\sigma}\mathcal{E}$ is the curve formed by applying σ to all of the coefficients of the defining equation of $\mathcal{E}$; see §2.

(we briefly discuss higher degrees in §11). For maximum generality and flexibility, we define our curves in short Weierstrass form; but we include transformations to Montgomery, twisted Edwards, and Doche–Icart–Kohel models where appropriate in §8.

Comparison with GLV. Like GLV, our method involves reducing curves defined over number fields to obtain curves over finite fields with explicit CM. However, we emphasise a profound difference: in our method, the curves over number fields generally *do not have CM themselves.*

GLV curves are necessarily isolated examples—and the really useful examples are extremely limited in number (see [18, App. A] for a list of curves). The scarcity of GLV curves[4] is their Achilles' heel: as noted in [11], if p is fixed then there is no guarantee that there will exist a GLV curve with prime (or almost-prime) order over $\mathbb{F}_p$. Consider the situation discussed in [11, §1]: the most efficient GLV curves have CM discriminants -3 and -4. If we are working at a 128-bit security level, then the choice $p = 2^{255} - 19$ allows particularly fast arithmetic in $\mathbb{F}_p$. But the largest prime factor of the order of a curve over $\mathbb{F}_p$ with CM discriminant -4 (resp. -3) has 239 (resp. 230) bits: using these curves wastes 9 (resp. 13) potential bits of security. In fact, we are lucky with $D = -3$ and -4: for all of the other discriminants offering endomorphisms of degree at most 3, we can do no better than a 95-bit prime factor, which represents a catastrophic 80-bit loss of relative security.

In contrast, our construction yields true families of curves, covering $\sim p$ isomorphism classes over $\mathbb{F}_{p^2}$. This gives us a vastly higher probability of finding prime (or almost-prime)-order curves over practically important fields.

Comparison with GLS. Like GLS, we construct curves over $\mathbb{F}_{p^2}$ equipped with an inseparable endomorphism. While these curves are not defined over the prime field, the fact that the extension degree is only 2 means that Weil descent attacks offer no advantage when solving DLP instances (see [11, §9]). And like GLS, our families offer around p distinct isomorphism classes of curves, making it easy to find secure group orders when p is fixed.

But unlike GLS, our curves have j-invariants in $\mathbb{F}_{p^2}$: they are not isomorphic to or twists of subfield curves. This allows us to find twist-secure curves, which are resistant to the Fouque–Lercier–Réal–Valette fault attack [9]. As we will see in §9, our construction reduces to GLS in the degenerate case $d = 1$ (that is, where

[4] The scarcity of useful GLV curves is easily explained: efficient *separable* endomorphisms have extremely small degree (so that the dense defining polynomials can be evaluated quickly). But the degree of the endomorphism is the norm of the corresponding element of the CM-order; and to have non-integers of very small norm, the CM-order must have a tiny discriminant. Up to twists, the number of elliptic curves with CM discriminant D is the Kronecker class number $h(D)$, which is in $O(\sqrt{D})$. Of course, for the tiny values of D in question, the asymptotics of $h(D)$ are irrelevant; for the six D corresponding to endomorphisms of degree at most 3, we have $h(D) = 1$, so there is only one j-invariant. For $D = -4$ (corresponding to $j = 1728$) there are two or four twists over $\mathbb{F}_p$; for $D = -3$ (corresponding to $j = 0$) we have two or six, and otherwise we have only two. In particular, there are at most 18 distinct curves over $\mathbb{F}_p$ with a non-integer endomorphism of degree at most 3.

$\widetilde{\phi}$ is an isomorphism). Our construction is therefore a sort of generalized GLS—though it is not the higher-degree generalization anticipated by Galbraith, Lin, and Scott themselves, which composes the sub-Frobenius with a non-rational separable isogeny and its dual isogeny (cf. [11, Theorem 1]).

In §4, we prove that we can immediately obtain scalar decompositions of the same bitlength as GLS for curves over the same fields: the decompositions produced by Proposition 2 are identical to the GLS decompositions of [11, Lemma 2] when $d = 1$, up to sign. For this reason, we do not provide extensive implementation details in this paper: while our endomorphisms cost a few more $\mathbb{F}_q$-operations to evaluate than the GLS endomorphism, this evaluation is typically carried out only once per scalar multiplication. This evaluation is the only difference between a GLS scalar multiplication and one of ours: the subsequent multiexponentiations have exactly the same length as in GLS, and the underlying curve and field arithmetic is the same, too.

2 Notation and Conventions

Throughout, we work over fields of characteristic not 2 or 3. Let

$$\mathcal{E} : y^2 = x^3 + a_4 x + a_6$$

be an elliptic curve over such a field K.

Galois conjugates. For every automorphism σ of K, we define the conjugate curve

$${}^{\sigma}\mathcal{E} : y^2 = x^3 + {}^{\sigma}a_4 x + {}^{\sigma}a_6 .$$

If $\phi : \mathcal{E} \to \mathcal{E}_1$ is an isogeny, then we obtain a conjugate isogeny ${}^{\sigma}\phi : {}^{\sigma}\mathcal{E} \to {}^{\sigma}\mathcal{E}_1$ by applying σ to the defining equations of ϕ, $\mathcal{E}$, and $\mathcal{E}_1$.

Quadratic twists. For every $\lambda \neq 0$ in $\overline{K}$, we define a twisting isomorphism

$$\delta(\lambda) : \mathcal{E} \longrightarrow \mathcal{E}^{\lambda} : y^2 = x^3 + \lambda^4 a_4 x + \lambda^6 a_6$$

by

$$\delta(\lambda) : (x, y) \longmapsto (\lambda^2 x, \lambda^3 y) .$$

The twist $\mathcal{E}^{\lambda}$ is defined over $K(\lambda^2)$, and $\delta(\lambda)$ is defined over $K(\lambda)$.[5]

For every K-endomorphism ψ of $\mathcal{E}$, there is a twisted $K(\lambda^2)$-endomorphism

$$\psi^{\lambda} := \delta(\lambda)\psi\delta(\lambda^{-1})$$

of $\mathcal{E}^{\lambda}$. Observe that $\delta(\lambda_1)\delta(\lambda_2) = \delta(\lambda_1\lambda_2)$ for any λ_1, λ_2 in K, and $\delta(-1) = [-1]$. Also, ${}^{\sigma}(\mathcal{E}^{\lambda}) = ({}^{\sigma}\mathcal{E})^{{}^{\sigma}\lambda}$ for all automorphisms σ of $\overline{K}$.

If μ is a nonsquare in K, then $\mathcal{E}^{\sqrt{\mu}}$ is a *quadratic twist* of $\mathcal{E}$. If $K = \mathbb{F}_q$, then $\mathcal{E}^{\sqrt{\mu_1}}$ and $\mathcal{E}^{\sqrt{\mu_2}}$ are $\mathbb{F}_q$-isomorphic for all nonsquares μ_1, μ_2 in $\mathbb{F}_q$ (the isomorphism $\delta(\sqrt{\mu_1/\mu_2})$ is defined over $\mathbb{F}_q$ because μ_1/μ_2 must be a square).

[5] Throughout, conjugates are marked by left-superscripts, twists by right-superscripts.

When the choice of nonsquare is not important, $\mathcal{E}'$ denotes the quadratic twist. Similarly, if ψ is an $\mathbb{F}_q$-endomorphism of $\mathcal{E}$, then ψ' denotes the corresponding twisted $\mathbb{F}_q$-endomorphism of $\mathcal{E}'$.

The trace. If $K = \mathbb{F}_q$, then $\pi_{\mathcal{E}}$ denotes the q-power Frobenius endomorphism of $\mathcal{E}$. Recall that the characteristic polynomial of $\pi_{\mathcal{E}}$ has the form

$$\chi_{\mathcal{E}}(T) = T^2 - \operatorname{tr}(\mathcal{E})T + q, \qquad \text{with} \qquad |\operatorname{tr}(\mathcal{E})| \leq 2\sqrt{q} \; .$$

The *trace* $\operatorname{tr}(\mathcal{E})$ of $\mathcal{E}$ satisfies $\#\mathcal{E}(\mathbb{F}_q) = q + 1 - \operatorname{tr}(\mathcal{E})$ and $\operatorname{tr}(\mathcal{E}') = -\operatorname{tr}(\mathcal{E})$.

p-th powering. We write (p) for the p-th powering automorphism of $\overline{\mathbb{F}}_p$. Note that (p) is almost trivial to compute on $\mathbb{F}_{p^2} = \mathbb{F}_p(\sqrt{\Delta})$, because ${}^{(p)}(a + b\sqrt{\Delta}) = a - b\sqrt{\Delta}$ for all a and b in $\mathbb{F}_p$.

3 Quadratic $\mathbb{Q}$-curves and Their Reductions

Suppose $\widetilde{\mathcal{E}}/\mathbb{Q}(\sqrt{\Delta})$ is a quadratic $\mathbb{Q}$-curve of prime degree d (as in Definition 1), where Δ is a discriminant prime to d, and let $\widetilde{\phi} : \widetilde{\mathcal{E}} \to {}^{\sigma}\widetilde{\mathcal{E}}$ be the corresponding d-isogeny. In general, $\widetilde{\phi}$ is only defined over a quadratic extension $\mathbb{Q}(\sqrt{\Delta}, \gamma)$ of $\mathbb{Q}(\sqrt{\Delta})$. We can compute γ from Δ and $\ker \widetilde{\phi}$ using [13, Proposition 3.1], but after a suitable twist we can always reduce to the case where $\gamma = \sqrt{\pm d}$ (see [13, remark after Lemma 3.2]). The families of explicit $\mathbb{Q}$-curves of degree d that we treat below have their isogenies defined over $\mathbb{Q}(\sqrt{\Delta}, \sqrt{-d})$; so to simplify matters, from now on we will

$$\text{Assume } \widetilde{\phi} \text{ is defined over } \mathbb{Q}(\sqrt{\Delta}, \sqrt{-d}).$$

Let p be a prime of good reduction for $\widetilde{\mathcal{E}}$ that is inert in $\mathbb{Q}(\sqrt{\Delta})$ and prime to d. If $\mathcal{O}_{\Delta}$ is the ring of integers of $\mathbb{Q}(\sqrt{\Delta})$, then

$$\mathbb{F}_{p^2} = \mathcal{O}_{\Delta}/(p) = \mathbb{F}_p(\sqrt{\Delta}) \; .$$

Looking at the Galois groups of our fields, we have a series of injections

$$\langle (p) \rangle = \operatorname{Gal}(\mathbb{F}_p(\sqrt{\Delta})/\mathbb{F}_p) \hookrightarrow \operatorname{Gal}(\mathbb{Q}(\sqrt{\Delta})/\mathbb{Q}) \hookrightarrow \operatorname{Gal}(\mathbb{Q}(\sqrt{\Delta}, \sqrt{-d})/\mathbb{Q}) \; .$$

The image of (p) in $\operatorname{Gal}(\mathbb{Q}(\sqrt{\Delta})/\mathbb{Q})$ is σ, because p is inert in $\mathbb{Q}(\sqrt{\Delta})$. When extending σ to an automorphism of $\mathbb{Q}(\sqrt{\Delta}, \sqrt{-d})$, we extend it to be the image of (p): that is,

$${}^{\sigma}\!\left(\alpha + \beta\sqrt{\Delta} + \gamma\sqrt{-d} + \delta\sqrt{-d\Delta}\right) = \alpha - \beta\sqrt{\Delta} + (-d/p)\left(\gamma\sqrt{-d} - \delta\sqrt{-d\Delta}\right) \quad (1)$$

for all α, β, γ, and $\delta \in \mathbb{Q}$. (Recall that the Legendre symbol (n/p) is 1 if n is a square mod p, -1 if n is not a square mod p, and 0 if p divides n.)

Now let $\mathcal{E}/\mathbb{F}_{p^2}$ be the reduction modulo p of $\widetilde{\mathcal{E}}$. The curve ${}^{\sigma}\widetilde{\mathcal{E}}$ reduces to ${}^{(p)}\mathcal{E}$, while the d-isogeny $\widetilde{\phi} : \widetilde{\mathcal{E}} \to {}^{\sigma}\widetilde{\mathcal{E}}$ reduces to a d-isogeny $\phi : \mathcal{E} \to {}^{(p)}\mathcal{E}$ over $\mathbb{F}_{p^2}$.

Applying σ to $\widetilde{\phi}$, we obtain a second d-isogeny ${}^\sigma\widetilde{\phi} : {}^\sigma\widetilde{\mathcal{E}} \to \widetilde{\mathcal{E}}$ travelling in the opposite direction, which reduces mod p to a conjugate isogeny ${}^{(p)}\phi : {}^{(p)}\mathcal{E} \to \mathcal{E}$ over $\mathbb{F}_{p^2}$. Composing ${}^\sigma\widetilde{\phi}$ with $\widetilde{\phi}$ yields endomorphisms ${}^\sigma\widetilde{\phi} \circ \widetilde{\phi}$ of $\widetilde{\mathcal{E}}$ and $\widetilde{\phi} \circ {}^\sigma\widetilde{\phi}$ of ${}^\sigma\widetilde{\mathcal{E}}$, each of degree d^2. But (by definition) $\widetilde{\mathcal{E}}$ and ${}^\sigma\widetilde{\mathcal{E}}$ do not have CM, so all of their endomorphisms are integer multiplications; and since the only integer multiplications of degree d^2 are $[d]$ and $[-d]$, we conclude that

$${}^\sigma\widetilde{\phi} \circ \widetilde{\phi} = [\epsilon_p d]_{\widetilde{\mathcal{E}}} \quad \text{and} \quad \widetilde{\phi} \circ {}^\sigma\widetilde{\phi} = [\epsilon_p d]_{{}^\sigma\widetilde{\mathcal{E}}} , \quad \text{where} \quad \epsilon_p \in \{\pm 1\} .$$

Technically, ${}^\sigma\widetilde{\phi}$ and ${}^{(p)}\phi$ are—*up to sign*—the dual isogenies of $\widetilde{\phi}$ and ϕ, respectively. The sign ϵ_p depends on p (as well as on $\widetilde{\phi}$): if τ is the extension of σ to $\mathbb{Q}(\sqrt{\Delta}, \sqrt{-d})$ that is *not* the image of (p), then ${}^\tau\widetilde{\phi} \circ \widetilde{\phi} = [-\epsilon_p d]_{\widetilde{\mathcal{E}}}$. Reducing modulo p, we see that

$${}^{(p)}\phi \circ \phi = [\epsilon_p d]_{\mathcal{E}} \quad \text{and} \quad \phi \circ {}^{(p)}\phi = [\epsilon_p d]_{{}^{(p)}\mathcal{E}} .$$

The map $(x, y) \mapsto (x^p, y^p)$ defines p-isogenies

$$\pi_0 : {}^{(p)}\mathcal{E} \longrightarrow \mathcal{E} \quad \text{and} \quad {}^{(p)}\pi_0 : \mathcal{E} \longrightarrow {}^{(p)}\mathcal{E} .$$

Clearly, ${}^{(p)}\pi_0 \circ \pi_0$ (resp. $\pi_0 \circ {}^{(p)}\pi_0$) is the p^2-power Frobenius endomorphism of $\mathcal{E}$ (resp. ${}^{(p)}\mathcal{E}$). Composing π_0 with ϕ yields a degree-pd endomorphism

$$\psi := \pi_0 \circ \phi \in \operatorname{End}(\mathcal{E}) .$$

If d is very small—say, less than 10—then ψ is efficient because ϕ is defined by polynomials of degree about d, and π_0 acts as a simple conjugation on coordinates in $\mathbb{F}_{p^2}$, as in Eq. (1). (The efficiency of ψ depends primarily on its separable degree, d, and not on the inseparable part p.)

We also obtain an endomorphism ψ' on the quadratic twist $\mathcal{E}'$ of $\mathcal{E}$. Indeed, if $\mathcal{E}' = \mathcal{E}^{\sqrt{\mu}}$, then $\psi' = \psi^{\sqrt{\mu}}$, and ψ' is defined over $\mathbb{F}_{p^2}$.

Proposition 1. *With the notation above:*

$$\psi^2 = [\epsilon_p d]\pi_{\mathcal{E}} \quad \textit{and} \quad (\psi')^2 = [-\epsilon_p d]\pi_{\mathcal{E}'}.$$

There exists an integer r satisfying $dr^2 = 2p + \epsilon_p \operatorname{tr}(\mathcal{E})$ such that

$$\psi = \tfrac{1}{r}\left(\pi_{\mathcal{E}} + \epsilon_p p\right) \quad \textit{and} \quad \psi' = \tfrac{-1}{r}\left(\pi_{\mathcal{E}'} - \epsilon_p p\right) .$$

The characteristic polynomial of both ψ and ψ' is

$$P_\psi(T) = P_{\psi'}(T) = T^2 - \epsilon_p r d T + dp .$$

Proof. Clearly $\pi_0 \circ \phi = {}^{(p)}\phi \circ {}^{(p)}\pi_0$, so

$$\psi^2 = \pi_0\phi\pi_0\phi = \pi_0\phi^{(p)}\phi^{(p)}\pi_0 = \pi_0[\epsilon_p d]^{(p)}\pi_0 = [\epsilon_p d]\pi_0{}^{(p)}\pi_0 = [\epsilon_p d]\pi_{\mathcal{E}} .$$

Choosing a nonsquare μ in $\mathbb{F}_{p^2}$, so $\mathcal{E}' = \mathcal{E}^{\sqrt{\mu}}$ and $\psi' = \psi^{\sqrt{\mu}}$, we find

$$(\psi')^2 = \delta(\mu^{\frac{1}{2}})\psi^2\delta(\mu^{-\frac{1}{2}}) = \delta(\mu^{\frac{1}{2}})[\epsilon_p d]\pi_{\mathcal{E}}\delta(\mu^{-\frac{1}{2}})$$
$$= \delta(\mu^{\frac{1}{2}(1-p^2)})[\epsilon_p d]\pi_{\mathcal{E}'} = \delta(-1)[\epsilon_p d]\pi_{\mathcal{E}'} = [-\epsilon_p d]\pi_{\mathcal{E}'} \ .$$

Using $\pi_{\mathcal{E}}^2 - \mathrm{tr}(\mathcal{E})\pi_{\mathcal{E}} + p^2 = 0$ and $\pi_{\mathcal{E}'}^2 + \mathrm{tr}(\mathcal{E})\pi_{\mathcal{E}'} + p^2 = 0$, we verify that the expressions for ψ and ψ' give the two square roots of $\epsilon_p d\pi_{\mathcal{E}}$ in $\mathbb{Q}(\pi_{\mathcal{E}})$, and $-\epsilon_p d\pi_{\mathcal{E}'}$ in $\mathbb{Q}(\pi'_{\mathcal{E}})$, and that the claimed characteristic polynomial is satisfied. □

Now we just need a source of quadratic $\mathbb{Q}$-curves of small degree. Elkies [7] shows that all $\mathbb{Q}$-curves correspond to rational points on certain modular curves: Let $X^*(d)$ be the quotient of the modular curve $X_0(d)$ by all of its Atkin–Lehner involutions, let K be a quadratic field, and let σ be the involution of K over $\mathbb{Q}$. If e is a point in $X^*(d)(\mathbb{Q})$ and E is a preimage of e in $X_0(d)(K) \setminus X_0(d)(\mathbb{Q})$, then E parametrizes (up to $\overline{\mathbb{Q}}$-isomorphism) a d-isogeny $\widetilde{\phi} : \widetilde{\mathcal{E}} \to {}^{\sigma}\widetilde{\mathcal{E}}$ over K.

Luckily enough, for very small d, the curves $X_0(d)$ and $X^*(d)$ have genus zero—so not only do we get plenty of rational points on $X^*(d)$, we get a whole one-parameter family of $\mathbb{Q}$-curves of degree d. Hasegawa gives explicit universal curves for $d = 2$, 3, and 7 in [15, Theorem 2.2]: for each squarefree integer $\Delta \neq 1$, every $\mathbb{Q}$-curve of degree $d = 2, 3, 7$ over $\mathbb{Q}(\sqrt{\Delta})$ is $\overline{\mathbb{Q}}$-isomorphic to a rational specialization of one of these families. Hasegawa's curves for $d = 2$ and 3 ($\widetilde{\mathcal{E}}_{2,\Delta,s}$ in §5 and $\widetilde{\mathcal{E}}_{3,\Delta,s}$ in §6) suffice not only to illustrate our ideas, but also to give useful practical examples.

4 Short Scalar Decompositions

Before moving on to concrete constructions, we will show that the endomorphisms developed in §3 yield short scalar decompositions. Proposition 2 below gives explicit formulæ for producing decompositions of at most $\lceil \log_2 p \rceil$ bits.

Suppose $\mathcal{G}$ is a cyclic subgroup of $\mathcal{E}(\mathbb{F}_{p^2})$ such that $\psi(\mathcal{G}) = \mathcal{G}$; let $N = \#\mathcal{G}$. Proposition 1 shows that ψ acts as a square root of $\epsilon_p d$ on $\mathcal{G}$: its eigenvalue is

$$\lambda_\psi \equiv (1 + \epsilon_p p)/r \pmod{N} \ . \tag{2}$$

We want to compute a decomposition

$$m = a + b\lambda_\psi \pmod{N}$$

so as to efficiently compute

$$[m]P = [a]P + [b\lambda_\psi]P = [a]P + [b]\psi(P) \ .$$

The decomposition of m is not unique: far from it. The set of all decompositions (a, b) of m is the coset $(m, 0) + \mathcal{L}$, where

$$\mathcal{L} := \langle (N, 0), (-\lambda_\psi, 1) \rangle \subset \mathbb{Z}^2$$

is the lattice of decompositions of 0 (that is, of (a, b) such that $a + b\lambda_\psi \equiv 0 \pmod{N}$).

We want to find a decomposition where a and b have minimal bitlength: that is, where $\lceil \log_2 \|(a, b)\|_\infty \rceil$ is as small as possible. The standard technique is to (pre)-compute a short basis of $\mathcal{L}$, then use Babai rounding [1] to transform each scalar m into a short decomposition (a, b). The following lemma outlines this process; for further detail and analysis, see [12, §4] and [10, §18.2].

Lemma 1. *Let $\mathbf{e}_1, \mathbf{e}_2$ be linearly independent vectors in $\mathcal{L}$. Let m be an integer, and set*

$$(a, b) := (m, 0) - \lfloor \alpha \rceil \mathbf{e}_1 - \lfloor \beta \rceil \mathbf{e}_2 ,$$

where (α, β) is the (unique) solution in $\mathbb{Q}^2$ to the linear system $(m, 0) = \alpha \mathbf{e}_1 + \beta \mathbf{e}_2$. Then

$$m \equiv a + \lambda_\psi b \pmod{N} \qquad \text{and} \qquad \|(a, b)\|_\infty \leq \max\left(\|\mathbf{e}_1\|_\infty, \|\mathbf{e}_2\|_\infty\right) .$$

Proof. This is just [12, Lemma 2] (under the infinity norm). □

We see that better decompositions of m correspond to shorter bases for $\mathcal{L}$. If $|\lambda_\psi|$ is not unusually small, then we can compute a basis for $\mathcal{L}$ of size $O(\sqrt{N})$ using the Gauss reduction or Euclidean algorithms (cf. [12, §4] and [10, §17.1.1]).[6] The basis depends only on N and λ_ψ, so it can be precomputed.

In our case, lattice reduction is unnecessary: we can immediately write down two linearly independent vectors in $\mathcal{L}$ that are "short enough", and thus give explicit formulae for (a, b) in terms of m. These decompositions have length $\lceil \log_2 p \rceil$, which is near-optimal in cryptographic contexts: if $N \sim \#\mathcal{E}(\mathbb{F}_{p^2}) \sim p^2$, then $\log_2 p \sim \frac{1}{2} \log_2 N$.

Proposition 2. *With the notation above: given an integer m, let*

$$a = m - \lfloor m(1 + \epsilon_p p)/\#\mathcal{E}(\mathbb{F}_{p^2}) \rceil (1 + \epsilon_p p) + \lfloor mr/\#\mathcal{E}(\mathbb{F}_{p^2}) \rceil \epsilon_p dr \qquad \text{and}$$
$$b = \lfloor m(1 + \epsilon_p p)/\#\mathcal{E}(\mathbb{F}_{p^2}) \rceil r - \lfloor mr/\#\mathcal{E}(\mathbb{F}_{p^2}) \rceil (1 + \epsilon_p p) .$$

Then, assuming $d \ll p$ and $m \not\equiv 0 \pmod{N}$, we have

$$m \equiv a + b\lambda_\psi \pmod{N} \qquad \text{and} \qquad \lceil \log_2 \|(a, b)\|_\infty \rceil \leq \lceil \log_2 p \rceil .$$

Proof. Eq. (2) yields $r\lambda_\psi \equiv 1 + \epsilon_p p \pmod{N}$ and $r\epsilon_p d \equiv (1 + \epsilon_p p)\lambda_\psi \pmod{N}$, so $\mathbf{e}_1 = (1 + \epsilon_p p, -r)$ and $\mathbf{e}_2 = (-\epsilon_p dr, 1 + \epsilon_p p)$ are in $\mathcal{L}$ (they generate a sublattice of determinant $\#\mathcal{E}(\mathbb{F}_{p^2})$). Applying Lemma 1 with $\alpha = m(1 + \epsilon_p p)/\#\mathcal{E}(\mathbb{F}_{p^2})$ and $\beta = mr/\#\mathcal{E}(\mathbb{F}_{p^2})$, we see that $m \equiv a + b\lambda_\psi \pmod{N}$ and $\|(a, b)\|_\infty \leq \|\mathbf{e}_2\|_\infty$. But $d|r| \leq 2\sqrt{dp}$ (since $|\mathrm{tr}(\mathcal{E})| \leq 2p$) and $d \ll p$, so $\|\mathbf{e}_2\|_\infty = p + \epsilon_p$. The result follows on taking logs, and noting that $\lceil \log_2(p \pm 1) \rceil \leq \lceil \log_2 p \rceil$ (since $p > 3$). □

[6] General bounds on the constant hidden by the $O(\cdot)$ are derived in [26], but they are suboptimal for our endomorphisms in cryptographic contexts, where Proposition 2 gives better results.

5 Endomorphisms from Quadratic $\mathbb{Q}$-curves of Degree 2

Let Δ be a squarefree integer. Hasegawa defines a one-parameter family of elliptic curves over $\mathbb{Q}(\sqrt{\Delta})$ by

$$\widetilde{\mathcal{E}}_{2,\Delta,s} : y^2 = x^3 - 6(5 - 3s\sqrt{\Delta})x + 8(7 - 9s\sqrt{\Delta}) , \tag{3}$$

where s is a free parameter taking values in $\mathbb{Q}$ [15, Theorem 2.2]. The discriminant of $\widetilde{\mathcal{E}}_{2,\Delta,s}$ is $2^9 \cdot 3^6(1 - s^2\Delta)(1 + s\sqrt{\Delta})$, so $\widetilde{\mathcal{E}}_{2,\Delta,s}$ has good reduction at every $p > 3$ with $(\Delta/p) = -1$, for every s in $\mathbb{Q}$.

The curve $\widetilde{\mathcal{E}}_{2,\Delta,s}$ has a rational 2-torsion point $(4,0)$, which generates the kernel of a 2-isogeny $\widetilde{\phi}_{2,\Delta,s} : \widetilde{\mathcal{E}}_{2,\Delta,s} \to {}^\sigma\widetilde{\mathcal{E}}_{2,\Delta,s}$ defined over $\mathbb{Q}(\sqrt{\Delta}, \sqrt{-2})$. We construct $\widetilde{\phi}_{2,\Delta,s}$ explicitly: Vélu's formulae [30] define the (normalized) quotient $\widetilde{\mathcal{E}}_{2,\Delta,s} \to \widetilde{\mathcal{E}}_{2,\Delta,s}/\langle(4,0)\rangle$, and then the isomorphism $\widetilde{\mathcal{E}}_{2,\Delta,s}/\langle(4,0)\rangle \to {}^\sigma\widetilde{\mathcal{E}}_{2,\Delta,s}$ is the quadratic twist $\delta(1/\sqrt{-2})$. Composing, we obtain an expression for the isogeny as a rational map:

$$\widetilde{\phi}_{2,\Delta,t} : (x,y) \longmapsto \left(\frac{-x}{2} - \frac{9(1+s\sqrt{\Delta})}{x-4}, \frac{y}{\sqrt{-2}} \left(\frac{-1}{2} + \frac{9(1+s\sqrt{\Delta})}{(x-4)^2} \right) \right) .$$

Conjugating and composing, we see that ${}^\sigma\widetilde{\phi}_{2,\Delta,t}\widetilde{\phi}_{2,\Delta,t} = [2]$ if $\sigma(\sqrt{-2}) = -\sqrt{-2}$, and $[-2]$ if $\sigma(\sqrt{-2}) = \sqrt{-2}$: that is, the sign function for $\widetilde{\phi}_{2,\Delta,t}$ is

$$\epsilon_p = -\left(-2/p\right) = \begin{cases} +1 & \text{if } p \equiv 5,7 \pmod 8 , \\ -1 & \text{if } p \equiv 1,3 \pmod 8 . \end{cases} \tag{4}$$

Theorem 1. *Let $p > 3$ be a prime, and define ϵ_p as in Eq. (4). Let Δ be a nonsquare*[7] *in $\mathbb{F}_p$, so $\mathbb{F}_{p^2} = \mathbb{F}_p(\sqrt{\Delta})$. For each s in $\mathbb{F}_p$, let*

$$C_{2,\Delta}(s) := 9(1 + s\sqrt{\Delta})$$

and let $\mathcal{E}_{2,\Delta,s}$ be the elliptic curve over $\mathbb{F}_{p^2}$ defined by

$$\mathcal{E}_{2,\Delta,s} : y^2 = x^3 + 2(C_{2,\Delta}(s) - 24)x - 8(C_{2,\Delta}(s) - 16) .$$

Then $\mathcal{E}_{2,\Delta,s}$ has an efficient $\mathbb{F}_{p^2}$-endomorphism of degree $2p$ defined by

$$\psi_{2,\Delta,s} : (x,y) \longmapsto \left(\frac{-x^p}{2} - \frac{C_{2,\Delta}(s)^p}{x^p - 4}, \frac{y^p}{\sqrt{-2}} \left(\frac{-1}{2} + \frac{C_{2,\Delta}(s)^p}{(x^p-4)^2} \right) \right) ,$$

and there exists an integer r satisfying $2r^2 = 2p + \epsilon_p \mathrm{tr}(\mathcal{E}_{2,\Delta,s})$ such that

$$\psi_{2,\Delta,s} = \frac{1}{r}\left(\pi_{\mathcal{E}_{2,\Delta,s}} + \epsilon_p p\right) \qquad \textit{and} \qquad \psi_{2,\Delta,s}^2 = [\epsilon_p 2]\pi_{\mathcal{E}_{2,\Delta,s}} .$$

[7] The choice of Δ is (theoretically) irrelevant, since all quadratic extensions of $\mathbb{F}_p$ are isomorphic. If Δ and Δ' are two nonsquares in $\mathbb{F}_p$, then $\Delta/\Delta' = a^2$ for some a in $\mathbb{F}_p$, so $\mathcal{E}_{2,\Delta,t}$ and $\mathcal{E}_{2,\Delta',at}$ are identical. We are therefore free to choose any practically convenient value for Δ, such as one permitting faster arithmetic in $\mathbb{F}_p(\sqrt{\Delta})$.

The twisted endomorphism $\psi'_{2,\Delta,s}$ *on* $\mathcal{E}'_{2,\Delta,s}$ *satisfies* $\psi'_{2,\Delta,s} = \frac{-1}{r}(\pi_{\mathcal{E}'_{2,\Delta,s}} - \epsilon_p p)$ *and* $(\psi'_{2,\Delta,s})^2 = [-\epsilon_p 2]\pi_{\mathcal{E}'_{2,\Delta,s}}$. *The characteristic polynomial of* $\psi_{2,\Delta,s}$ *and* $\psi'_{2,\Delta,s}$ *is* $P_{2,\Delta,s}(T) = T^2 - \epsilon_p rT + 2p$.

Proof. Reduce $\widetilde{\mathcal{E}}_{2,\Delta,s}$ and $\widetilde{\phi}_{2,\Delta,s}$ mod p and compose with π_0 as in §3, then apply Proposition 1 using Eq. (4). □

If $\mathcal{G} \subset \mathcal{E}_{2,\Delta,s}(\mathbb{F}_{p^2})$ is a cyclic subgroup of order N such that $\psi_{2,\Delta,s}(\mathcal{G}) = \mathcal{G}$, then the eigenvalue of $\psi_{2,\Delta,s}$ on $\mathcal{G}$ is

$$\lambda_{2,\Delta,s} = \frac{1}{r}(1 + \epsilon_p p) \equiv \pm\sqrt{\epsilon_p 2} \pmod{N} .$$

Applying Proposition 2, we can decompose scalar multiplications in $\mathcal{G}$ as $[m]P = [a]P + [b]\psi_{2,\Delta,s}(P)$ where a and b have at most $\lceil \log_2 p \rceil$ bits.

Proposition 3. *Theorem 1 yields at least* $p - 3$ *non-isomorphic curves over* $\mathbb{F}_{p^2}$ *(and at least* $2p - 6$ *non-*$\mathbb{F}_{p^2}$*-isomorphic curves, if we count the quadratic twists) equipped with efficient endomorphisms.*

Proof. It suffices to show that the j-invariant $j(\mathcal{E}_{2,\Delta,s}) = \frac{2^6(5-3s\sqrt{\Delta})^3}{(1-s^2\Delta)(1+s\sqrt{\Delta})}$ takes at least $p - 3$ distinct values in $\mathbb{F}_{p^2}$ as s ranges over $\mathbb{F}_p$. If $j(\mathcal{E}_{2,\Delta,s_1}) = j(\mathcal{E}_{2,\Delta,s_2})$ with $s_1 \neq s_2$, then s_1 and s_2 satisfy $F_0(s_1, s_2) - 2\sqrt{\Delta}F_1(s_1, s_2) = 0$, where $F_1(s_1, s_2) = (s_1 + s_2)(63\Delta s_1 s_2 - 65)$ and $F_0(s_1, s_2) = (\Delta s_1 s_2 + 1)(81\Delta s_1 s_2 - 175) + 49\Delta(s_1 + s_2)^2$ are polynomials over $\mathbb{F}_p$. If s_1 and s_2 are in $\mathbb{F}_p$, then we must have $F_0(s_1, s_2) = F_1(s_1, s_2) = 0$. Solving the simultaneous equations, discarding the solutions that can never be in $\mathbb{F}_p$, and dividing by two (since (s_1, s_2) and (s_2, s_1) represent the same collision) yields at most 3 collisions $j(\mathcal{E}_{2,\Delta,s_1}) = j(\mathcal{E}_{2,\Delta,s_2})$ with $s_1 \neq s_2$ in $\mathbb{F}_p$. □

We observe that ${}^\sigma\widetilde{\mathcal{E}}_{2,\Delta,s} = \widetilde{\mathcal{E}}_{2,\Delta,-s}$, so we do not gain any more isomorphism classes in Proposition 3 by including the codomain curves.

6 Endomorphisms from Quadratic $\mathbb{Q}$-curves of Degree 3

Let Δ be a squarefree discriminant; Hasegawa defines a one-parameter family of elliptic curves over $\mathbb{Q}(\sqrt{\Delta})$ by

$$\widetilde{\mathcal{E}}_{3,\Delta,s} : y^2 = x^3 - 3(5 + 4s\sqrt{\Delta})x + 2(2s^2\Delta + 14s\sqrt{\Delta} + 11) , \tag{5}$$

where s is a free parameter taking values in $\mathbb{Q}$. As for the curves in §5, the curve $\widetilde{\mathcal{E}}_{3,\Delta,s}$ has good reduction at every inert $p > 3$ for every s in $\mathbb{Q}$.

The curve $\widetilde{\mathcal{E}}_{3,\Delta,s}$ has a subgroup of order 3 defined by the polynomial $x - 3$, consisting of 0 and $(3, \pm 2(1 - s\sqrt{\Delta}))$. Exactly as in §5, taking the Vélu quotient and twisting by $1/\sqrt{-3}$ yields an explicit 3-isogeny $\widetilde{\phi}_{3,\Delta,s} : \widetilde{\mathcal{E}}_{3,\Delta,s} \to {}^\sigma\widetilde{\mathcal{E}}_{3,\Delta,s}$; its sign function is

$$\epsilon_p = -\left(-3/p\right) = \begin{cases} +1 & \text{if } p \equiv 2 \pmod{3} , \\ -1 & \text{if } p \equiv 1 \pmod{3} . \end{cases} \tag{6}$$

Theorem 2. *Let $p > 3$ be a prime, and define ϵ_p as in Eq. (6). Let Δ be a nonsquare*[8] *in $\mathbb{F}_p$, so $\mathbb{F}_{p^2} = \mathbb{F}_p(\sqrt{\Delta})$. For each s in $\mathbb{F}_p$, let*

$$C_{3,\Delta}(s) := 2(1 + s\sqrt{\Delta})$$

and let $\mathcal{E}_{3,\Delta,s}$ be the elliptic curve over $\mathbb{F}_{p^2}$ defined by

$$\mathcal{E}_{3,\Delta,s} : y^2 = x^3 - 3\big(2C_{3,\Delta}(s) + 1\big)x + \big(C_{3,\Delta}(s)^2 + 10C_{3,\Delta}(s) - 2\big) .$$

Then $\mathcal{E}_{3,\Delta,s}$ has an efficient $\mathbb{F}_{p^2}$-endomorphism $\psi_{3,\Delta,s}$ of degree $3p$, mapping (x, y) to

$$\left(-\frac{x^p}{3} - \frac{4C_{3,\Delta}(s)^p}{x^p - 3} - \frac{4C_{3,\Delta}(s)^{2p}}{3(x^p - 3)^2}, \frac{y^p}{\sqrt{-3}}\left(\frac{-1}{3} + \frac{4C_{3,\Delta}(s)^p}{(x^p - 3)^2} + \frac{8C_{3,\Delta}(s)^{2p}}{3(x^p - 3)^3}\right)\right),$$

and there exists an integer r satisfying $3r^2 = 2p + \epsilon_p \mathrm{tr}(\mathcal{E}_{3,\Delta,s})$ such that

$$\psi_{3,\Delta,s}^2 = [\epsilon_p 3]\pi_{\mathcal{E}_{3,\Delta,s}} \qquad \text{and} \qquad \psi_{3,\Delta,s} = \frac{1}{r}\left(\pi + \epsilon_p p\right) .$$

The twisted endomorphism $\psi'_{3,\Delta,s}$ on $\mathcal{E}'_{3,\Delta,s}$ satisfies $(\psi'_{3,\Delta,s})^2 = [-\epsilon_p 3]\pi_{\mathcal{E}'_{3,\Delta,s}}$ and $\psi'_{3,\Delta,s} = (-\pi_{\mathcal{E}'_{3,\Delta,s}} + \epsilon_p p)/r$. Both $\psi_{3,\Delta,s}$ and $\psi'_{3,\Delta,s}$ have characteristic polynomial $P_{3,\Delta,s}(T) = T^2 - \epsilon_p rT + 3p$.

Proof. Reduce $\widetilde{\mathcal{E}}_{3,\Delta,s}$ and $\widetilde{\phi}_{3,\Delta,s}$ mod p, compose with π_0 as in §3, and apply Proposition 1 using Eq. (6). □

Proposition 4. *Theorem 2 yields at least $p - 8$ non-isomorphic curves over $\mathbb{F}_{p^2}$ (and counting quadratic twists, at least $2p - 16$ non-$\mathbb{F}_{p^2}$-isomorphic curves) equipped with efficient endomorphisms.*

Proof. The proof is exactly as for Proposition 3. □

7 Cryptographic-Sized Curves

We will now exhibit some curves with cryptographic parameter sizes, and secure and twist-secure group orders. We computed the curve orders below using Magma's implementation of the Schoof–Elkies–Atkin algorithm [25,19,4].

First consider the degree-2 curves of §5. By definition, $\mathcal{E}_{2,\Delta,s}$ and its quadratic twist $\mathcal{E}'_{2,\Delta,s}$ have points of order 2 over $\mathbb{F}_{p^2}$: they generate the kernels of our endomorphisms. If $p \equiv 2 \pmod 3$, then $2r^2 = 2p + \epsilon_p \mathrm{tr}(\mathcal{E})$ implies $\mathrm{tr}(\mathcal{E}) \not\equiv 0 \pmod 3$, so when $p \equiv 2 \pmod 3$ either $p^2 - \mathrm{tr}(\mathcal{E}) + 1 = \#\mathcal{E}_{2,\Delta,s}(\mathbb{F}_{p^2})$ or $p^2 + \mathrm{tr}(\mathcal{E}) + 1 = \#\mathcal{E}'_{2,\Delta,s}(\mathbb{F}_{p^2})$ is divisible by 3. However, when $p \equiv 1 \pmod 3$ we can hope to find curves of order twice a prime whose twist also has order twice a prime.

[8] As in Theorem 1, the particular value of Δ is theoretically irrelevant.

Example 1. Let $p = 2^{80} - 93$ and $\Delta = 2$. For $s = 4556$, we find a twist-secure curve: $\#\mathcal{E}_{2,2,4556}(\mathbb{F}_{p^2}) = 2N$ and $\#\mathcal{E}'_{2,2,4556}(\mathbb{F}_{p^2}) = 2N'$ where

$$N = 730750818665451459101729015265709251634505119843 \quad \text{and}$$
$$N' = 730750818665451459101730957248125446994932083047$$

are 159-bit primes. Proposition 2 lets us replace 160-bit scalar multiplications in $\mathcal{E}_{2,2,4556}(\mathbb{F}_{p^2})$ and $\mathcal{E}'_{2,2,4556}(\mathbb{F}_{p^2})$ with 80-bit multiexponentiations.

Now consider the degree-3 curves of §6. The order of $\mathcal{E}_{3,\Delta,s}(\mathbb{F}_{p^2})$ is always divisible by 3: the kernel of $\psi_{3,\Delta,s}$ is generated by the rational point $(3, C_{3,\Delta}(s))$. However, on the quadratic twist, the nontrivial points in the kernel of $\psi'_{3,\Delta,s}$ are *not* defined over $\mathbb{F}_{p^2}$ (they are conjugates), so $\mathcal{E}'_{3,\Delta,s}(\mathbb{F}_{p^2})$ can have prime order.

Example 2. Let $p = 2^{127} - 1$; then $\Delta = -1$ is a nonsquare in $\mathbb{F}_p$. The parameter value $s = 122912611041315220011572494331480107107$ yields

$$\#\mathcal{E}_{3,-1,s}(\mathbb{F}_{p^2}) = 3 \cdot N \qquad \text{and} \qquad \#\mathcal{E}'_{3,-1,s}(\mathbb{F}_{p^2}) = N' ,$$

where N is a 253-bit prime and N' is a 254-bit prime. Using Proposition 2, any scalar multiplication in $\mathcal{E}_{3,-1,s}(\mathbb{F}_{p^2})$ or $\mathcal{E}'_{3,-1,s}(\mathbb{F}_{p^2})$ can be computed via a 127-bit multiexponentiation.

Example 3. Let $p = 2^{255} - 19$; then $\Delta = -2$ is a nonsquare in $\mathbb{F}_p$. Taking

$$s = 52960937784593362700485649923279446947410945689208862015782690291692803003486$$

yields $\#\mathcal{E}_{3,-2,s}(\mathbb{F}_{p^2}) = 3 \cdot N$ and $\#\mathcal{E}_{3,-2,s}(\mathbb{F}_{p^2}) = N'$, where N and N' are 509- and 510-bit primes, respectively. Proposition 2 transforms any 510-bit scalar multiplication in $\mathcal{E}_{3,-2,s}(\mathbb{F}_{p^2})$ or $\mathcal{E}'_{2,-2,s}(\mathbb{F}_{p^2})$ into a 255-bit multiexponentiation.

8 Alternative Models: Montgomery, Twisted Edwards, and Doche–Icart–Kohel

Montgomery models. The curve $\mathcal{E}_{2,\Delta,s}$ has a Montgomery model over $\mathbb{F}_{p^2}$ if and only if $2C_{2,\Delta}(s)$ is a square in $\mathbb{F}_{p^2}$ (by [22, Proposition 1]): in that case, setting

$$B_{2,\Delta}(s) := \sqrt{2C_{2,\Delta}(s)} \qquad \text{and} \qquad A_{2,\Delta}(s) = 12/B_{2,\Delta}(s) ,$$

the birational mapping $(x, y) \mapsto (X/Z, Y/Z) = \left((x-4)/B_{2,\Delta}(s), y/B_{2,\Delta}(s)^2\right)$ takes us from $\mathcal{E}_{2,\Delta,s}$ to the projective Montgomery model

$$\mathcal{E}^{\mathrm{M}}_{2,\Delta,s} : B_{2,\Delta}(s)Y^2Z = X\left(X^2 + A_{2,\Delta}(s)XZ + Z^2\right) . \tag{7}$$

(If $2C_{2,\Delta}(s)$ is not a square, then $\mathcal{E}^{\mathrm{M}}_{2,\Delta,s}$ is $\mathbb{F}_{p^2}$-isomorphic to the quadratic twist $\mathcal{E}'_{2,\Delta,s}$.) These models offer a particularly efficient arithmetic, where we use only the X and Z coordinates [20]. The endomorphism is defined (on the X and Z coordinates) by

$$\psi_{2,\Delta,s} : (X : Z) \longmapsto (X^{2p} + A_{2,\Delta}(s)^p X^p Z^p + Z^{2p} : -2B_{2,\Delta}(s)^{1-p} X^p Z^p) .$$

<u>*Twisted Edwards models.*</u> Every Montgomery model corresponds to a twisted Edwards model (and vice versa) [2,16]. Let

$$a_2(s) = (A_{2,\Delta}(s) + 2)/B_{2,\Delta}(s) \quad \text{and} \quad d_2(s) = (A_{2,\Delta}(s) - 2)/B_{2,\Delta}(s) ;$$

then with $u = X/Z$ and $v = Y/Z$, the birational maps

$$(u, v) \mapsto (x_1, x_2) = \left(\frac{u}{v}, \frac{u-1}{u+1}\right) , \quad (x_1, x_2) \mapsto (u, v) = \left(\frac{1+x_2}{1-x_2}, \frac{1+x_2}{x_1(1-x_2)}\right)$$

take us between the Montgomery model of Eq. (7) and the twisted Edwards model

$$\mathcal{E}^{\mathrm{TE}}_{2,\Delta,s} : a_2(s)x_1^2 + x_2^2 = 1 + d_2(s)x_1^2x_2^2 .$$

<u>*Doche–Icart–Kohel models.*</u> Doubling-oriented Doche–Icart–Kohel models of elliptic curves are defined by equations of the form

$$y^2 = x(x^2 + Dx + 16D) .$$

These curves have a rational 2-isogeny ϕ with kernel $\langle(0,0)\rangle$, and ϕ and its dual isogeny $\phi^\dagger$ are both in a special form that allows us to double more quickly by using the decomposition $[2] = \phi^\dagger\phi$ (see [6, §3.1] for details).

Our curves $\mathcal{E}_{2,\Delta,s}$ come equipped with a rational 2-isogeny, so it is natural to try putting them in Doche–Icart–Kohel form. The isomorphism

$$\alpha : (x, y) \longmapsto (u, v) = \left(\mu^2(x + 4), \mu^3 y\right) \quad \text{with} \quad \mu = 4\sqrt{6/C_{2,\Delta}(s)}$$

takes us from $\mathcal{E}_{2,\Delta,s}$ into a doubling-oriented Doche–Icart–Kohel model

$$\mathcal{E}^{\mathrm{DIK}}_{2,\Delta,s} : v^2 = u\left(u^2 + D_{2,\Delta}(s)u + 16D_{2,\Delta}(s)\right) ,$$

where $D_{2,\Delta}(s) = 2^7/(1+s\sqrt{\Delta})$. While $\mathcal{E}^{\mathrm{DIK}}_{2,\Delta,s}$ is defined over $\mathbb{F}_{p^2}$, the isomorphism is only defined over $\mathbb{F}_{p^2}(\sqrt{1+s\sqrt{\Delta}})$; so if $1+s\sqrt{\Delta}$ is not a square in $\mathbb{F}_{p^2}$ then $\mathcal{E}^{\mathrm{DIK}}_{2,\Delta,s}$ is $\mathbb{F}_{p^2}$-isomorphic to $\mathcal{E}'_{2,\Delta,s}$. The endomorphism $\psi^{\mathrm{DIK}}_{2,\Delta,s} := \alpha\psi_{2,\Delta,s}\alpha^{-1}$ is $\overline{\mathbb{F}}_p$-isomorphic to the Doche–Icart–Kohel isogeny (they have the same kernel).

Similarly, we can exploit the rational 3-isogeny on $\mathcal{E}_{3,\Delta,s}$ for Doche–Icart–Kohel tripling (see [6, §3.2]). Let $a_{3,\Delta}(s) = 9/C_{3,\Delta}(s)$ and $b_{3,\Delta}(s) = a_{3,\Delta}(s)^{-1/2}$; then the isomorphism $(x, y) \mapsto (u, v) = \left(a_{3,\Delta}(s)(x/3 - 1), b_{3,\Delta}(s)^3 y\right)$ takes us from $\mathcal{E}_{3,\Delta,s}$ to the tripling-oriented Doche–Icart-Kohel model

$$\mathcal{E}^{\mathrm{DIK}}_{3,\Delta,s} : v^2 = u^3 + 3a_{3,\Delta}(s)(u+1)^2 .$$

9 Degree One: GLS as a Degenerate Case

Returning to the framework of §3, suppose $\widetilde{\mathcal{E}}$ is a curve defined over $\mathbb{Q}$ and base-extended to $\mathbb{Q}(\sqrt{D})$: then $\widetilde{\mathcal{E}} = {}^{\sigma}\widetilde{\mathcal{E}}$, and we can apply the construction of §3 taking $\widetilde{\phi} : \widetilde{\mathcal{E}} \to {}^{\sigma}\widetilde{\mathcal{E}}$ to be the identity map. Reducing modulo an inert prime p, the endomorphism ψ is nothing but π_0 (which is an endomorphism, since $\mathcal{E}$ is a subfield curve). We have $\psi^2 = \pi_0^2 = \pi_{\mathcal{E}}$, so the eigenvalue of ψ is ± 1 on cryptographic subgroups of $\mathcal{E}(\mathbb{F}_{p^2})$. Clearly, this endomorphism is of no use to us for scalar decompositions.

However, looking at the quadratic twist $\mathcal{E}'$, the twisted endomorphism ψ' satisfies $(\psi')^2 = -\pi_{\mathcal{E}'}$; the eigenvalue of ψ' on cryptographic subgroups is a square root of -1. We have recovered the Galbraith–Lin–Scott endomorphism (cf. [11, Theorem 2]).

More generally, suppose $\widetilde{\phi} : \widetilde{\mathcal{E}} \to {}^{\sigma}\widetilde{\mathcal{E}}$ is a $\overline{\mathbb{Q}}$-isomorphism: that is, an isogeny of degree 1. If $\widetilde{\mathcal{E}}$ does not have CM, then ${}^{\sigma}\widetilde{\phi} = \epsilon_p \phi^{-1}$, so $\psi^2 = [\epsilon_p]\pi_{\mathcal{E}}$ with $\epsilon_p = \pm 1$. This situation is isomorphic to GLS. In fact, $\widetilde{\mathcal{E}} \cong {}^{\sigma}\widetilde{\mathcal{E}}$ implies $j(\widetilde{\mathcal{E}}) = j({}^{\sigma}\widetilde{\mathcal{E}}) = {}^{\sigma}j(\widetilde{\mathcal{E}})$; so $j(\widetilde{\mathcal{E}})$ is in $\mathbb{Q}$, and $\widetilde{\mathcal{E}}$ is isomorphic to (or a quadratic twist of) a curve defined over $\mathbb{Q}$. We note that in the case $d = 1$, we have $r = \pm t_0$ in Proposition 1 where t_0 is the trace of π_0, and the basis constructed in the proof of Proposition 2 is (up to sign) the same as the basis of [11, Lemma 3].

While $\mathcal{E}'(\mathbb{F}_{p^2})$ may have prime order, $\mathcal{E}(\mathbb{F}_{p^2})$ cannot: the points fixed by π_0 form a subgroup of order $p+1-t_0$, where $t_0^2-2p = \operatorname{tr}(\mathcal{E})$ (the complementary subgroup, where π_0 has eigenvalue -1, has order $p+1+t_0$). We see that the largest prime divisor of $\#\mathcal{E}(\mathbb{F}_{p^2})$ can be no larger than $O(p)$. If we are in a position to apply the Fouque–Lercier–Réal–Valette fault attack [9]—for example, if Montgomery ladders are used for scalar multiplication and multiexponentiation—then we can solve DLP instances in $\mathcal{E}'(\mathbb{F}_{p^2})$ in $O(p^{1/2})$ group operations (in the worst case!). While $O(p^{1/2})$ is still exponentially difficult, it falls far short of the ideal $O(p)$ for general curves over $\mathbb{F}_{p^2}$. GLS curves should therefore be avoided where the fault attack can be put into practice.

10 CM Specializations

By definition, $\mathbb{Q}$-curves do not have CM. However, some exceptional fibres of the families $\widetilde{\mathcal{E}}_{2,\Delta,s}$ and $\widetilde{\mathcal{E}}_{3,\Delta,s}$ do have CM. There are only finitely many such curves over any given $\mathbb{Q}(\sqrt{\Delta})$; following Quer ([23, §5] and [24, §6]), we give an exhaustive list of the corresponding parameter values in Tables 1 and 2. In each table, if Δ is a squarefree discriminant and there exists s in $\mathbb{Q}$ such that $1/(s^2\Delta - 1)$ takes the first value in a column, then the curve $\widetilde{\mathcal{E}}_{d,\Delta,s}/\mathbb{Q}(\sqrt{\Delta})$ has CM by the quadratic order of discriminant D specified by the second value.

Suppose we have chosen d, Δ, and s such that $\widetilde{\mathcal{E}}_{d,\Delta,s}$ is a CM-curve. If the discriminant of the associated CM order is small, then we can compute an explicit endomorphism of $\widetilde{\mathcal{E}}_{d,\Delta,s}$ of small degree, which then yields an efficient endomorphism ρ (say) on the reduction $\mathcal{E}_{d,\Delta,s}$ modulo p (as in the GLV construction). If p is inert, then we also have the degree-dp endomorphism ψ constructed above.

Table 1. CM specializations of $\widetilde{\mathcal{E}}_{2,\Delta,s}$ (cf. Quer [23, §5])

$1/(s^2\Delta-1)$	4	−9	48	−81	324	−2401	−9801	25920	777924	−96059601
D	−20	−24	−36	−40	−52	−72	−88	−100	−148	−232

Table 2. CM specializations of $\widetilde{\mathcal{E}}_{3,\Delta,s}$ (cf. Quer [24, §6])

$1/(s^2\Delta-1)$	1/4	−2	−27/2	16	−125/4	80	1024	3024	250000
D	−15	−24	−48	−51	−60	−75	−123	−147	−267

Combinations of ρ and ψ may be used for four-dimensional scalar decompositions; for example, the endomorphisms $[1], \rho, \psi, \rho\psi$ can be used as a basis for the 4-dimensional decomposition techniques elaborated by Longa and Sica in [18].

In fact, reducing these CM fibres modulo a well-chosen p turns out to form a simple alternative construction for some of the curves investigated by Guillevic and Ionica in [14]: the twisted curve $\mathcal{E}_{2,\Delta,s}^{\sqrt{3}}$ coincides with the curve $E_{1,c}$ of [14, §2] when $c = s\sqrt{\Delta}$, while $\mathcal{E}_{3,\Delta,s}$ is the curve $E_{2,c}$ of [14, §2] when $c = -2s\sqrt{\Delta}$. The almost-prime-order 254-bit curve of [14, Example 1] corresponds to the reduction modulo p of a twist of one of the curves in the column of Table 1 with $1/(s^2\Delta - 1) = 4$. This curve has an efficient CM endomorphism (a square root of $[-5]$) as well as an endomorphism of degree $2p$; these endomorphisms are combined to compute short 4-dimensional scalar decompositions.

From the point of view of scalar multiplication, using CM fibres of these families allows us to pass from 2-dimensional to 4-dimensional scalar decompositions, with a consequent speedup. However, in restricting to CM fibres we also re-impose the chief drawback of GLV on ourselves: that is, as explained in the introduction, we cannot hope to find secure (and twist-secure) curves over $\mathbb{F}_{p^2}$ when p is fixed. In practice, this means that the 4-dimensional scalar decomposition speedup comes at the cost of suboptimal field arithmetic; we pay for shorter loop lengths with comparatively slower group operations.

We must therefore make a choice between 4-dimensional decompositions and fast underlying field arithmetic. In this article we have chosen the latter option, so we will not treat CM curves in depth here (we refer the reader to [14] instead).

11 Higher Degrees

We conclude with some brief remarks on $\mathbb{Q}$-curves of other degrees. Hasegawa provides a universal curve for $d = 7$ (and any Δ) in [15, Theorem 2.2], and our results for $d = 2$ and $d = 3$ carry over to $d = 7$ in an identical fashion, though the endomorphism is slightly less efficient in this case (its defining polynomials are sextic).

For $d = 5$, Hasegawa notes that it is impossible to give a universal $\mathbb{Q}$-curve for every discriminant Δ: there exists a quadratic $\mathbb{Q}$-curve of degree 5 over $\mathbb{Q}(\sqrt{\Delta})$ if and only if $(5/p_i) = 1$ for every prime $p_i \neq 5$ dividing Δ [15, Proposition 2.3]. But this is no problem when reducing modulo p, if we are prepared to give up total freedom in choosing Δ: we can take $\Delta = -11$ for $p \equiv 1 \pmod 4$ and $\Delta = -1$ for $p \equiv 3 \pmod 4$, and then use the curves defined in [15, Table 6]. The generic curves here do not have rational torsion points; it is therefore possible for the reductions and their twists to have prime order.

Composite degree $\mathbb{Q}$-curves (such as $d = 6$ and 10) promise more interesting results. Degrees greater than 10 yield less efficient endomorphisms, and so are less interesting from a practical point of view.

Acknowledgements. The author thanks François Morain, David Gruenewald, and Craig Costello for their helpful comments. These results crystallized during a workshop on Number Theory at the American University of Beirut in April 2013; the author thanks Wissam Raji, Kamal Khuri-Makdisi, and Martin Bright for their generous hospitality, and Evis Ieronymou for asking the hard questions.

References

1. Babai, L.: On Lovasz' lattice reduction and the nearest lattice point problem. Combinatorica 6, 1–13 (1986)
2. Bernstein, D.J., Birkner, P., Joye, M., Lange, T., Peters, C.: Twisted Edwards Curves. In: Vaudenay, S. (ed.) AFRICACRYPT 2008. LNCS, vol. 5023, pp. 389–405. Springer, Heidelberg (2008)
3. Bos, J.W., Costello, C., Hisil, H., Lauter, K.: Fast cryptography in genus 2. In: Johansson, T., Nguyen, P.Q. (eds.) EUROCRYPT 2013. LNCS, vol. 7881, pp. 194–210. Springer, Heidelberg (2013)
4. Bosma, W., Cannon, J.J., Fieker, C., Steel (eds.): Handbook of Magma functions, 2.19 edn. (2013)
5. Cohen, H., Frey, G. (eds.): Handbook of elliptic and hyperelliptic curve cryptography. Chapman & Hall / CRC (2006)
6. Doche, C., Icart, T., Kohel, D.R.: Efficient scalar multiplication by isogeny decompositions. In: Yung, M., Dodis, Y., Kiayias, A., Malkin, T. (eds.) PKC 2006. LNCS, vol. 3958, pp. 191–206. Springer, Heidelberg (2006)
7. Elkies, N.D.: On elliptic k-curves. In: Cremona, J., Lario, J.-C., Quer, J., Ribet, K. (eds.) Modular Curves and Abelian Varieties, pp. 81–92. Birkhäuser, Basel (2004)
8. Ellenberg, J.S.: $\mathbb{Q}$-curves and Galois representations. In: Cremona, J., Lario, J.-C., Quer, J., Ribet, K. (eds.) Modular Curves and Abelian Varieties, pp. 93–103. Birkhäuser, Basel (2004)
9. Fouque, P.-A., Lercier, R., Réal, D., Valette, F.: Fault attack on elliptic curve with Montgomery ladder. In: FDTC 2008, pp. 92–98. IEEE-CS (2008)
10. Galbraith, S.D.: Mathematics of public key cryptography. Cambridge University Press (2012)
11. Galbraith, S.D., Lin, X., Scott, M.: Endomorphisms for faster elliptic curve cryptography on a large class of curves. J. Crypt. 24(3), 446–469 (2011)

12. Gallant, R.P., Lambert, R.J., Vanstone, S.A.: Faster point multiplication on elliptic curves with efficient endomorphisms. In: Kilian, J. (ed.) CRYPTO 2001. LNCS, vol. 2139, pp. 190–200. Springer, Heidelberg (2001)
13. González, J.: Isogenies of polyquadratic $\mathbb{Q}$-curves to their Galois conjugates. Arch. Math. 77, 383–390 (2001)
14. Guillevic, A., Ionica, S.: Four-dimensional GLV via the Weil restriction. In: Sako, K., Sarkar, P. (eds.) ASIACRYPT 2013 Part I. LNCS, vol. 8269, pp. 79–96. Springer, Heidelberg (2013)
15. Hasegawa, Y.: $\mathbb{Q}$-curves over quadratic fields. Manuscripta Math. 94(1), 347–364 (1997)
16. Hisil, H., Wong, K.K.-H., Carter, G., Dawson, E.: Twisted Edwards curves revisited. In: Pieprzyk, J. (ed.) ASIACRYPT 2008. LNCS, vol. 5350, pp. 326–343. Springer, Heidelberg (2008)
17. Kohel, D.R., Smith, B.A.: Efficiently computable endomorphisms for hyperelliptic curves. In: Hess, F., Pauli, S., Pohst, M. (eds.) ANTS 2006. LNCS, vol. 4076, pp. 495–509. Springer, Heidelberg (2006)
18. Longa, P., Sica, F.: Four-dimensional Gallant–Lambert–Vanstone scalar multiplication. In: Wang, X., Sako, K. (eds.) ASIACRYPT 2012. LNCS, vol. 7658, pp. 718–739. Springer, Heidelberg (2012), `http://eprint.iacr.org/2011/608`
19. The Magma computational algebra system, `http://magma.maths.usyd.edu.au`
20. Montgomery, P.L.: Speeding the Pollard and Elliptic Curve Methods of factorization. Math. Comp. 48(177), 243–264 (1987)
21. Menezes, A., Okamoto, T., Vanstone, S.A.: Reducing elliptic curve logarithms to logarithms in a finite field. IEEE Trans. Inform. Theory 39(5), 1639–1646 (1993)
22. Okeya, K., Kurumatani, H., Sakurai, K.: Elliptic curves with the Montgomery-form and their cryptographic applications. In: Imai, H., Zheng, Y. (eds.) PKC 2000. LNCS, vol. 1751, pp. 238–257. Springer, Heidelberg (2000)
23. Quer, J.: Fields of definition of $\mathbb{Q}$-curves. J. Théor. Nombres Bordeaux 13(1), 275–285 (2001)
24. Quer, J.: $\mathbb{Q}$-curves and abelian varieties of GL_2-type. Proc. London Math. 81(2), 285–317 (2000)
25. Schoof, R.: Elliptic curves over finite fields and the computation of square roots mod p. Math. Comp. 44, 735–763 (1985)
26. Sica, F., Ciet, M., Quisquater, J.J.: Analysis of the Gallant-Lambert-Vanstone Method Based on Efficient Endomorphisms: Elliptic and Hyperelliptic Curves. In: Nyberg, K., Heys, H.M. (eds.) SAC 2002. LNCS, vol. 2595, pp. 21–36. Springer, Heidelberg (2003)
27. Silverman, J.H.: The arithmetic of elliptic curves. Grad. Texts in Math. 106(2e) (2009)
28. Straus, E.G.: Addition chains of vectors. Amer. Math. Monthly 71(7), 806–808 (1964)
29. Takashima, K.: A new type of fast endomorphisms on Jacobians of hyperelliptic curves and their cryptographic application. IEICE Trans. Fundamentals E89-A(1), 124–133 (2006)
30. Vélu, J.: Isogénies entre courbes elliptiques. C. R. Math. Acad. Sci. Paris 273, 238–241 (1971)
31. Verheul, E.: Evidence that XTR is more secure than supersingular elliptic curve cryptosystems. J. Crypt. 17, 277–296 (2004)
32. Zhou, Z., Hu, Z., Xu, M., Song, W.: Efficient 3-dimensional GLV method for faster point multiplication on some GLS elliptic curves. Inf. Proc. Lett. 110(22), 1003–1006 (2010)

Four-Dimensional GLV via the Weil Restriction

Aurore Guillevic[1,2] and Sorina Ionica[1]

[1] Crypto Team – DI – École Normale Supérieure
45 rue d'Ulm – 75230 Paris Cedex 05 – France
[2] Laboratoire Chiffre – Thales Communications and Security
4 Avenue des Louvresses – 92622 Gennevilliers Cedex – France
aurore.guillevic@ens.fr, sorina.ionica@m4x.org

Abstract. The Gallant-Lambert-Vanstone (GLV) algorithm uses efficiently computable endomorphisms to accelerate the computation of scalar multiplication of points on an abelian variety. Freeman and Satoh proposed for cryptographic use two families of genus 2 curves defined over $\mathbb{F}_p$ which have the property that the corresponding Jacobians are $(2,2)$-isogenous over an extension field to a product of elliptic curves defined over $\mathbb{F}_{p^2}$. We exploit the relationship between the endomorphism rings of isogenous abelian varieties to exhibit efficiently computable endomorphisms on both the genus 2 Jacobian and the elliptic curve. This leads to a four-dimensional GLV method on Freeman and Satoh's Jacobians and on two new families of elliptic curves defined over $\mathbb{F}_{p^2}$.

Keywords: GLV method, elliptic curves, genus 2 curves, isogenies.

1 Introduction

The scalar multiplication of a point on a small dimension abelian variety is one of the most important operations used in curve-based cryptography. Various techniques were introduced to speed-up the scalar multiplication. Firstly there exist exponent-recoding techniques such as sliding window and Non-Adjacent-Form representation [7]. These techniques are valid for generic groups and improved for elliptic curves as the inversion (or negation in additive notation) is free.

Secondly, in 2001, Gallant, Lambert and Vanstone [11] introduced a method which uses endomorphisms on the elliptic curve to decompose the scalar multiplication in a 2-dimensional multi-multiplication. Given an elliptic curve E over a finite field $\mathbb{F}_p$ with a fast endomorphism ϕ and a point P of large prime order r such that $\phi(P) = [\lambda]P$, the computation of $[k]P$ is decomposed as

$$[k]P = [k_1]P + [k_2]\phi(P),$$

with $k = k_1 + \lambda k_2 \pmod r$ such that $|k_1|, |k_2| \simeq \sqrt{r}$. Gallant et *al.* provided examples of curves whose endomorphism ϕ is given by complex-multiplication by $\sqrt{-1}$ (j-invariant $j = 1728$), $\frac{-1+\sqrt{-3}}{2}$ ($j = 0$), $\sqrt{-2}$ ($j = 8000$) and $\frac{1+\sqrt{-7}}{2}$ ($j = -3375$). In 2009 Galbraith, Lin and Scott [10] presented a method to construct an efficient endomorphism on elliptic curves E defined over $\mathbb{F}_{p^2}$ which are

K. Sako and P. Sarkar (Eds.) ASIACRYPT 2013 Part I, LNCS 8269, pp. 79–96, 2013.

quadratic twists of elliptic curves defined over $\mathbb{F}_p$. In this case, a fast endomorphism ψ is obtained by carefully exploiting the Frobenius endomorphism. This endomorphism verifies the equation $\psi^2+1=0$ when restricted to points defined over $\mathbb{F}_{p^2}$. In 2012, Longa and Sica improved the GLS construction, by showing that a 4-dimensional decomposition of scalar multiplication is possible, on GLS curves allowing efficient complex multiplication ϕ. Let λ, μ denote the eigenvalues of the two endomorphisms ϕ, ψ. Then we can decompose the scalar k into $k = k_0 + k_1\lambda + k_2\mu + k_3\lambda\mu$ and compute

$$[k]P = [k_0]P + [k_1]\phi(P) + [k_2]\psi(P) + [k_3]\phi \circ \psi(P).$$

Moreover, Longa and Sica provided an efficient algorithm to compute decompositions of k such that $|k_i| < Cr^{1/4}$, $i = 1, \ldots, 4$. Note that most curves presented in the literature have particular j-invariants. GLV curves have j-invariant 0, 1728, 8000, or -3375, while GLS curves have j-invariant in $\mathbb{F}_p$, even though they are defined over $\mathbb{F}_{p^2}$.

In 2013, Bos, Costello, Hisil and Lauter proposed in [3] a 4-dimensional GLV technique to speed-up scalar multiplication in genus 2. They considered the Buhler-Koblitz genus 2 curves $y^2 = x^5 + b$ and the Furukawa-Kawazoe-Takahashi curves $y^2 = x^5 + ax$. These two curves have a very efficient dimension-4 GLV technique available.

In this paper we study GLV decompositions on two types of abelian varieties:

- Elliptic curves defined over $\mathbb{F}_{p^2}$, with j-invariant defined over $\mathbb{F}_{p^2}$.
- Jacobians of genus 2 curves defined over $\mathbb{F}_p$, which are isogenous over an extension field to a product of elliptic curves defined over $\mathbb{F}_{p^2}$.

First, we study a family of elliptic curves whose equation is of the form $E_{1,c}(\mathbb{F}_{p^2}) : y^2 = x^3 + 27(10-3c)x + 14 - 9c$ with $c \in \mathbb{F}_{p^2} \setminus \mathbb{F}_p$, $c^2 \in \mathbb{F}_p$. These curves have an endomorphism Φ satisfying $\Phi^2 \pm 2 = 0$ for points defined over $\mathbb{F}_{p^2}$. Nevertheless, the complex multiplication discriminant of the curve is not 2, but of the form $-D = -2D'$. The second family is given by elliptic curves with equation of the form $E_{2,c}(\mathbb{F}_{p^2}) : y^2 = x^3 + 3(2c-5)x + c^2 + 14c + 22$ with $c \in \mathbb{F}_{p^2} \setminus \mathbb{F}_p$, $c^2 \in \mathbb{F}_p$. We show that these curves have an endomorphism Φ such that $\Phi^2 + 3 = 0$ for points defined over $\mathbb{F}_{p^2}$. The complex multiplication discriminant of the curve $E_{2,c}$ is of the form $-D = -3D'$. Our construction is a simple and efficient way to exploit the existence of a p-power Frobenius endomorphism on the Weil restriction of these curves. If the discriminant D is small, we propose a 4-dimensional GLV algorithm for the $E_{1,c}$ and $E_{2,c}$ families of curves. We use Velu's formulas to compute explicitly the endomorphisms on $E_{1,c}$ and $E_{2,c}$.

At last, we study genus 2 curves whose equations are $\mathcal{C}_1 : Y^2 = X^5 + aX^3 + bX$ and $\mathcal{C}_2 : Y^2 = X^6 + aX^3 + b$, with $a, b \in \mathbb{F}_p$. The Jacobians of these curves split over an extension field in two isogenous elliptic curves. More precisely, the Jacobian of $\mathcal{C}_1$ is isogenous to $E_{1,c} \times E_{1,c}$ and the Jacobian of $\mathcal{C}_2$ is isogenous to $E_{2,c} \times E_{2,-c}$. These two Jacobians were proposed for use in cryptography by Satoh [18] and Freeman and Satoh [9], who showed that they are isogenous over $\mathbb{F}_p$ to the Weil restriction of a curve of the form $E_{1,c}$ or $E_{2,c}$. This property is exploited to derive fast point counting algorithms and pairing-friendly constructions. We investigate efficient scalar multiplication via the GLV technique on Satoh and Freeman's Jacobians. We give explicit formulae for the $(2,2)$-isogeny between the product of elliptic curves and the Jacobian of the genus 2 curve. As a consequence, we derive a method to efficiently compute endomorphisms on the Jacobians of $\mathcal{C}_1$ and $\mathcal{C}_2$.

This paper is organized as follows. In Section 2 we review the construction of $(2,2)$-isogenies between Jacobians of $\mathcal{C}_1$ and $\mathcal{C}_2$ and products of elliptic curves. In Section 3 and 4 we give our construction of efficient endomorphisms on $E_{1,c}$ and $E_{2,c}$ and derive a four-dimensional GLV algorithm on these curves. Section 5 explains how to obtain a four-dimensional GLV method on the Jacobians of $\mathcal{C}_1$ and $\mathcal{C}_2$. Finally, in Section 6, our operation count at the 128 bit security level is proof that both elliptic curves defined over $\mathbb{F}_{p^2}$ and Satoh and Freeman's Jacobians yield scalar multiplication algorithms competitive with those of Longa and Sica and Bos *et al.*

2 Elliptic Curves with a Genus 2 Cover

In this paper we will work with two examples of genus 2 curves whose Jacobians allow over an extension field a $(2,2)$-isogeny to a product of elliptic curves. We first study the genus 2 curve

$$\mathcal{C}_1(\mathbb{F}_p) : Y^2 = X^5 + aX^3 + bX, \text{ with } a, b \neq 0 \in \mathbb{F}_p . \tag{1}$$

It was shown [15,18,9, §2, §3, §4.1] that the Jacobian of $\mathcal{C}_1$ is isogenous to $E_{1,c} \times E_{1,c}$, where

$$E_{1,c}(\mathbb{F}_p[\sqrt{b}]) : y^2 = (c+2)x^3 - (3c-10)x^2 + (3c-10)x - (c+2) \tag{2}$$

with $c = a/\sqrt{b}$. We recall the formulae for the cover maps from $\mathcal{C}_1$ to $E_{1,c}$. The reader is referred to the proof of Prop. 4.1 in [9] for details of the computations.

$$\begin{array}{ll} \varphi_1 : \mathcal{C}_1(\mathbb{F}_p) \to E_{1,c}(\mathbb{F}_p[\sqrt[8]{b}]) & \varphi_2 : \mathcal{C}_1(\mathbb{F}_q) \to E_{1,c}(\mathbb{F}_p[\sqrt[8]{b}]) \\ (x,y) \mapsto \left(\left(\frac{x+\sqrt[4]{b}}{x-\sqrt[4]{b}} \right)^2, \frac{8y\sqrt[8]{b}}{(x-\sqrt[4]{b})^3} \right) & (x,y) \mapsto \left(\left(\frac{x-\sqrt[4]{b}}{x+\sqrt[4]{b}} \right)^2, \frac{8iy\sqrt[8]{b}}{(x+\sqrt[4]{b})^3} \right), \end{array} \tag{3}$$

where $i = \sqrt{-1} \in \mathbb{F}_p$ or $\mathbb{F}_{p^2}$. The $(2,2)$-isogeny is given by

$$\begin{array}{c} I : J_{\mathcal{C}_1} \to E_{1,c} \times E_{1,c} \\ P + Q - 2P_\infty \mapsto (\varphi_{1*}(P) + \varphi_{1*}(Q), \varphi_{2*}(P) + \varphi_{2*}(Q)) \end{array} \tag{4}$$

and its dual is

$$\begin{array}{c} \hat{I} : E_{1,c} \times E_{1,c} \to J_{\mathcal{C}_1} \\ (S_1, S_2) \mapsto \varphi_1^*(S_1) + \varphi_2^*(S_2) - 4P_\infty \end{array}$$

with $\varphi_1^*(S_1) = \left(\frac{\sqrt{x_1}+1}{\sqrt{x_1}-1}\sqrt[4]{b}, \frac{y_1\sqrt[8]{b^5}}{(\sqrt{x_1}-1)^3}\right) + \left(\frac{-\sqrt{x_1}+1}{-\sqrt{x_1}-1}\sqrt[4]{b}, \frac{y_1\sqrt[8]{b^5}}{(-\sqrt{x_1}-1)^3}\right)$
and $\varphi_2^*(S_2) = \left(\frac{1+\sqrt{x_2}}{1-\sqrt{x_2}}\sqrt[4]{b}, \frac{-iy_2\sqrt[8]{b^5}}{(1-\sqrt{x_2})^3}\right) + \left(\frac{1-\sqrt{x_2}}{1+\sqrt{x_2}}\sqrt[4]{b}, \frac{-iy_2\sqrt[8]{b^5}}{(1+\sqrt{x_2})^3}\right)$.

Note that I and its dual are defined over an extension field of $\mathbb{F}_p$ of degree 1, 2, 4 or 8. One may easily check that $I \circ \hat{I} = [2]$ and $\hat{I} \circ I = [2]$. Since I splits multiplication by 2, an argument similar to [14, Prop. 21] implies that $2\mathrm{End}(J_{\mathcal{C}_1}) \subseteq \mathrm{End}(E_{1,c} \times E_{1,c})$ and $2\mathrm{End}(E_{1,c} \times E_{1,c}) \subseteq \mathrm{End}(J_{\mathcal{C}_1})$. We will use these inclusions to exhibit efficiently computable endomorphisms on both $J_{\mathcal{C}_1}$ and $E_{1,c}$.

Secondly, we consider an analogous family of degree 6 curves. These curves were studied by Duursma and Kiyavash [8] and by Gaudry and Schost [12].

$$\mathcal{C}_2(\mathbb{F}_p) : Y^2 = X^6 + aX^3 + b \text{ with } a, b \neq 0 \in \mathbb{F}_p\ . \tag{5}$$

The Jacobian of the curve denoted $J_{\mathcal{C}_2}$ is isogenous to the product of elliptic curves $E_{2,c} \times E_{2,-c}$, where

$$E_{2,c}(\mathbb{F}_p[\sqrt{b}]) : y^2 = (c+2)x^3 + (-3c+30)x^2 + (3c+30)x + (-c+2) \tag{6}$$
$$E_{2,-c}(\mathbb{F}_p[\sqrt{b}]) : y^2 = (-c+2)x^3 + (3c+30)x^2 + (-3c+30)x + (c+2), \tag{7}$$

with $c = a/\sqrt{b}$. The construction of the isogeny is similar to the one for I. We recall the formulae for cover maps from $\mathcal{C}_2$ to $E_{2,c}$ and to $E_{2,-c}$. For detailed computations, the reader is referred to Freeman and Satoh [9, Prop. 4].

$$\begin{aligned}\varphi_2 : \mathcal{C}_2(\mathbb{F}_p) &\to E_{2,c} \times E_{2,-c}(\mathbb{F}_p[\sqrt[6]{b}])\\ (X, Y) &\mapsto \left\{\left(\left(\frac{X+\sqrt[6]{b}}{X-\sqrt[6]{b}}\right)^2, \frac{8Y}{(X-\sqrt[6]{b})^3}\right), \left(\left(\frac{X-\sqrt[6]{b}}{X+\sqrt[6]{b}}\right)^2, \frac{8Y}{(X+\sqrt[6]{b})^3}\right)\right\}\end{aligned} \tag{8}$$

Note that the isogeny constructed using these cover maps is defined over an extension field of degree 1,2,3 or 6.

3 Four-Dimensional GLV on $E_{1,c}$

In this section, we construct two endomorphisms which may be used to compute scalar multiplication on $E_{1,c}$ using a 4-dimensional GLV algorithm. We assume that $c \in \mathbb{F}_{p^2} \setminus \mathbb{F}_p$ and $c^2 \in \mathbb{F}_p$.

3.1 First Endomorphism on $E_{1,c}$ with Vélu's Formulas

We aim to compute a 2-isogeny on $E_{1,c}(\mathbb{F}_{p^2})$. First we reduce the equation (2) of $E_{1,c}$ to

$$E_{1,c}(\mathbb{F}_{p^2}) : y^2 = x^3 + 27(3c-10)x - 108(9c-14) \tag{9}$$

through the change of variables $(x, y) \mapsto (3(c+2)x - (3c-10), (c+2)y)$. Note that we can write

$$E_{1,c}(\mathbb{F}_{p^2}) : y^2 = (x-12)(x^2 + 12x + 81c - 126). \tag{10}$$

Hence there always exists a 2-torsion point $P_2 = (12, 0)$ on $E_{1,c}(\mathbb{F}_{p^2})$. We apply Velu's formulas [20,6,14] to compute the isogeny whose kernel is generated by P_2. We obtain an isogeny from $E_{1,c}$ into $E_b : y^2 = x^3 + b_4x + b_6$ with $b_4 = -2^2 \cdot 27(3c+10)$, $b_6 = -2^2 \cdot 108(14+9c)$. We observe that E_b is isomorphic to the curve whose equation is

$$E_{1,-c}(\mathbb{F}_{p^2}) : y^2 = x^3 + 27(-3c-10)x + 108(14+9c) \tag{11}$$

through $(x_b, y_b) \mapsto (x_b/(-2), y_b/(-2\sqrt{-2}))$. Note that $\sqrt{-2} \in \mathbb{F}_{p^2}$ and thus this isomorphism is defined over $\mathbb{F}_{p^2}$. We define the isogeny

$$\begin{aligned} \mathcal{I}_2 : E_{1,c}(\mathbb{F}_{p^2}) &\to E_{1,-c}(\mathbb{F}_{p^2}) \\ (x,y) &\mapsto \left(\tfrac{-x}{2} + \tfrac{162+81c}{-2(x-12)}, \tfrac{-y}{2\sqrt{-2}}\left(1 - \tfrac{162+81c}{(x-12)^2}\right)\right). \end{aligned} \tag{12}$$

We show that we can use this isogeny to get an efficiently computable endomorphism on $E_{1,c}$. Observe that since $c \in \mathbb{F}_{p^2} \setminus \mathbb{F}_p$ and $c^2 \in \mathbb{F}_p$, we have that

$$\pi_p(c) = c^p = -c, \ \pi_p(j(E_{1,c})) = j(E_{1,-c}) \tag{13}$$

hence the curves $E_{1,c}$ and $E_{1,-c}$ are *isogenous* over $\mathbb{F}_{p^2}$ via the Frobenius map π_p. They are not isomorphic, because they do not have the same j-invariant.

To sum up, by composing $\pi_p \circ \mathcal{I}_2$, we obtain an efficiently computable endomorphism Φ_2 as follows:

$$\begin{aligned} \Phi_2 : E_{1,c}(\mathbb{F}_{p^2}) &\to E_{1,c}(\mathbb{F}_{p^2}) \\ (x,y) &\mapsto \left(\frac{-x^p}{2} - \frac{162-81c}{2(x^p-12)}, \frac{-y^p}{2\sqrt{-2}^p}\left(1 - \frac{162-81c}{(x^p-12)^2}\right)\right) \\ &= \left(\frac{x^{2p} - 12x^p + 162 - 81c}{-2(x^p-12)}, y^p \frac{x^{2p} - 24x^p - 18 + 81c}{-2\sqrt{-2}^p (x^p-12)^2}\right). \end{aligned}$$

If we compute formally[1] Φ_2^2 then we obtain exactly the formulas to compute $\pi_{p^2} \circ [-2]$ on $E_{1,c}(\mathbb{F}_{p^2})$ if $\sqrt{-2} \in \mathbb{F}_p$ or $\pi_{p^2} \circ [2]$ if $\sqrt{-2} \notin \mathbb{F}_p$. This difference occurs because a term $\sqrt{-2}\sqrt{-2}^p$ appears in the formula. If $p \equiv 1, 3 \mod 8$, $\sqrt{-2}^p = \sqrt{-2}$ and if $p \equiv 5, 7 \mod 8$, $\sqrt{-2}^p = -\sqrt{-2}$. Hence Φ_2 restricted to points defined over $\mathbb{F}_{p^2}$ verifies the equation

$$\Phi_2^2 \pm 2 = 0. \tag{14}$$

We note that the above construction does not come as a surprise. Since $2\mathrm{End}(J_{\mathcal{C}_1}) \subseteq \mathrm{End}(E_{1,c} \times E_{1,c})$ and since the Jacobian $J_{\mathcal{C}_1}$ is equipped with a p-power Frobenius endomorphism, we deduce that there are endomorphisms with inseparability degree p on the elliptic curve $E_{1,c}$. Our construction is simply an explicit method to compute such an endomorphism.

[1] E.g. Verification code with Maple can be found at the address `http://www.di.ens.fr/~ionica/VerificationMaple-Isogeny-2p-E1.maple`

Two-Dimensional GLV. By using Id and Φ_2, we get a two-dimensional GLV algorithm on the curve $E_{1,c}$. Smith [19] constructs families of 2-dimensional GLV curves by reducing mod p $\mathbb{Q}$-curves defined over quadratic number fields. $\mathbb{Q}$-curves are curves without complex multiplication with isogenies towards all their Galois conjugates. Since we are interested into designing a fast higher dimensional algorithm, we will study curves with small complex multiplication discriminant. In this purpose, our curves are constructed using the complex multiplication method. For a discussion on the advantages of using dimension 2 curves, see [19].

3.2 Efficient Complex Multiplication on $E_{1,c}(\mathbb{F}_{p^2})$

We suppose that the complex multiplication discriminant D of the curve $E_{1,c}$ is small. A natural way to obtain an efficiently computable endomorphism is to take Φ_D the generator for the endomorphism ring (i.e. $\sqrt{-D}$). Guillevic and Vergnaud [13, proof of Th. 1 (4.) §2.2] showed that $D = 2D'$, for some integer D'. Let t_{p^2} be the trace of $E_{1,c}(\mathbb{F}_{p^2})$. The equation of the complex multiplication is then

$$(t_{p^2})^2 - 4p^2 = -2D'\gamma^2, \tag{15}$$

for some $\gamma \in \mathbb{Z}$. We prove that there is an endomorphism on $E_{1,c}$ whose degree of separability is D'. In order to do that, we will need to compute first the general equation of Φ_2.

Lemma 1. *There are integers m and n such that if $p \equiv 1, 3 \pmod 8$, then*

$$t_{p^2} + 2p = D'm^2 \text{ and } t_{p^2} - 2p = -2n^2. \tag{16}$$

and if $p \equiv 5, 7 \pmod 8$, then

$$t_{p^2} + 2p = 2n^2 \text{ and } t_{p^2} - 2p = -D'm^2. \tag{17}$$

Moreover, the characteristic equation of Φ_2 is

$$\Phi_2^2 - 2n\Phi_2 + 2p\mathrm{Id} = 0\ . \tag{18}$$

Proof. We have that $\mathrm{Tr}(\Phi_2^2) - \mathrm{Tr}^2(\Phi_2) + 2\deg(\Phi_2) = 0$. We know that $\deg(\Phi_2) = 2p$ because $\Phi_2 = \pi_p \circ \mathcal{I}_2$ and $\deg(\pi_p) = p, \deg(\mathcal{I}_2) = 2$, so $\mathrm{Tr}^2(\Phi_2) = \mathrm{Tr}(\Phi_2^2) + 4p$. Now, if $p \equiv 1, 3 \mod 8$, $\mathrm{Tr}(\Phi_2^2) = \mathrm{Tr}(\pi_{p^2} \circ [-2]) = -2t_{p^2}$ and we get $\mathrm{Tr}^2(\Phi_2) = -2t_{p^2} + 4p = -2(t_{p^2} - 2p)$. We may thus write $t_{p^2} - 2p = -2n^2$, for some integer n. If $p \equiv 5, 7 \mod 8$, $\mathrm{Tr}(\Phi_2^2) = \mathrm{Tr}(\pi_{p^2} \circ [2]) = 2t_{p^2}$ and we get $\mathrm{Tr}^2(\phi_2) = 2t_{p^2} + 4p = 2(t_{p^2} + 2p)$. Hence $t_{p^2} + 2p = 2n^2$ again. Using the complex multiplication equation (15), we have that there is an integer m such that $t_{p^2} + 2p = D'm^2$, if $p \equiv 1, 3 \pmod 8$ and $t_{p^2}^2 - 2p = -D'm^2$, if $p \equiv 5, 7 \pmod 8$. Using these notations, the characteristic equation of Φ_2 is

$$\Phi_2^2 - 2n\ \Phi_2 + 2p\ \mathrm{Id} = 0\ .$$

Theorem 1. *Let $E_{1,c}$ be an elliptic curve given by equation* (10)*, defined over $\mathbb{F}_{p^2}$. Let $-D$ be the complex multiplication discriminant and consider D' such that $D = 2D'$. There is an endomorphism $\Phi_{D'}$ of $E_{1,c}$ with degree of separability D'. The characteristic equation of this endomorphism is*

$$\Phi_{D'}^2 - D^{'} m\, \Phi_{D'} + D^{'} p \,\mathrm{Id} = 0\ . \tag{19}$$

Proof. Since $D = 2D'$, we have that Φ_D is the composition of a horizontal isogeny of degree 2 with a horizontal[2] isogeny of degree D'. We denote by $\mathcal{I}_2 : E_{1,c} \to E_{1,-c}$ the isogeny given by equation (12). Note that $\mathcal{I}_2$ is a horizontal isogeny of degree 2. Indeed, since $\pi_p : E_{1,-c} \to E_{1,c}$, it follows that $(\mathrm{End}(E_{1,c}))_2 \simeq (\mathrm{End}(E_{1,-c}))_2$. Since $2|D$, there is a unique horizontal isogeny of degree 2 starting from $E_{1,c}$. Hence the complex multiplication endomorphism on $E_{1,c}$ is $\Phi_D = \mathcal{I}_{D'} \circ \mathcal{I}_2$, with $\mathcal{I}_{D'} : E_{1,-c} \to E_{1,c}$ a horizontal isogeny of degree D'. We define $\Phi_{D'} = \mathcal{I}_{D'} \circ \pi'_p$, with $\pi'_p : E_{1,c} \to E_{1,-c}$. To compute the characteristic polynomial of $\Phi_{D'}$, we observe that

$$\Phi_{D'} \circ \Phi_2 = \Phi_D \circ \pi_{p^2}.$$

Hence, by using equation (18), we obtain that $\Phi_{D'}$ seen as algebraic integer in $\mathbb{Z}[\sqrt{-D}]$ is $\frac{-D'm \pm n\sqrt{-2D'}}{2}$. Hence we have $\Phi_{D'}^2 - D^{'} m\, \Phi_{D'} + D^{'} p \,\mathrm{Id} = 0$.

The endomorphism $\Phi_{D'}$ constructed in Theorem 1 is thus computed as the composition of a horizontal isogeny with the p-power of the Frobenius. Since computing the p-power Frobenius for extension fields of degree 2 costs one negation, we conclude that $\Phi_{D'}$ may be computed with Vélu's formulae with half the operations needed to compute Φ_D over $\mathbb{F}_{p^2}$.

Four-Dimensional GLV Algorithm. Assume that $E_{1,c}$ is such that $\#E_{1,c}(\mathbb{F}_{p^2})$ is divisible by a large prime of cryptographic size. Let $\Psi = \Phi_{D'}$ and $\Phi = \Phi_2$. We observe Φ and Ψ viewed as algebraic integers generate disjoint quadratic extensions of $\mathbb{Q}$. Consequently, one may use $1, \Phi, \Psi, \Phi\Psi$ to compute the scalar multiple $[k]P$ of a point $P \in E_{1,c}(\mathbb{F}_{p^2})$ using a four-dimensional GLV algorithm. We do not give here the details of the algorithm which computes decompositions

$$k = k_1 + k_2\lambda + k_3\mu + k_4\lambda\mu,$$

with λ and μ the eigenvalues of Φ and Ψ and $|k_i| < Cr^{1/4}$. Such an algorithm is obtained by working over $\mathbb{Z}[\Phi, \Psi]$, using a similar analysis to the one proposed by Longa and Sica [16].

Eigenvalue Computation. From equation (14), we deduce that the eigenvalue of Φ_2 is $p\sqrt{-2}$ if $p \equiv 1.3 \mod 8$ and $p\sqrt{2}$ if $p \equiv 5, 7 \mod 8$. We explain how to compute this eigenvalue mod $\#E_{1,c}(\mathbb{F}_{p^2})$. We will use the formulas (16) and (1).

[2] An isogeny $I : E \to E'$ of degree ℓ is called horizontal if $(\mathrm{End}(E))_\ell \simeq (\mathrm{End}(E'))_\ell$.

If $p \equiv 1, 3 \mod 8$, we obtain

$$\begin{aligned}
\#E_{1,c}(\mathbb{F}_{p^2}) &= (p+1)^2 - D' m^2 &&\rightarrow \quad \sqrt{D'} \equiv (p+1)/m \\
&= (p-1)^2 + 2n^2 &&\rightarrow \quad \sqrt{-2} \equiv (p-1)/n, \\
&= (1 - t_{p^2}/2)^2 + 2D'(nm/2)^2 &&\rightarrow \sqrt{-2D'} \equiv (2 - t_{p^2})/(nm) .
\end{aligned}$$

If $p \equiv 5, 7 \mod 8$, we obtain

$$\begin{aligned}
\#E_{1,c}(\mathbb{F}_{p^2}) &= (p-1)^2 + D' m^2 &&\rightarrow \quad \sqrt{-D'} \equiv (p-1)/m \\
&= (p+1)^2 - 2n^2 &&\rightarrow \quad \sqrt{2} \equiv (p+1)/n, \\
&= (1 - t_{p^2}/2)^2 + 2D'(nm/2)^2 &&\rightarrow \sqrt{-2D'} \equiv (2 - t_{p^2})/(nm) .
\end{aligned}$$

The eigenvalue of Φ_2 on $E_{1,c}(\mathbb{F}_{p^2})$ is $p\sqrt{-2} \equiv p(p-1)/n \mod \#E_{1,c}(\mathbb{F}_{p^2})$ if $p \equiv 1, 3 \mod 8$ or $p\sqrt{2} \equiv p(p+1)/n \mod \#E_{1,c}(\mathbb{F}_{p^2})$ if $p \equiv 5, 7 \mod 8$.

The eigenvalue of $\Phi_{D'}$ on $E_{1,c}(\mathbb{F}_{p^2})$ is $p\sqrt{D'} \equiv p(p+1)/m \mod \#E_{1,c}(\mathbb{F}_{p^2})$ if $p \equiv 1, 3 \mod 8$ or $p\sqrt{-D'} \equiv p(p-1)/m \mod \#E_{1,c}(\mathbb{F}_{p^2})$ if $p \equiv 5, 7 \mod 8$.

3.3 Curve Construction and Examples

We construct curves $E_{1,c}$ with good cryptographic properties (i.e. a large prime divides the number of points of $E_{1,c}$ over $\mathbb{F}_{p^2}$) by using the complex multiplication algorithm. More precisely, we look for prime numbers p such that the complex multiplication equation

$$4p = 2n^2 + D'm^2$$

is verified. Once p is found, we compute the roots of the Hilbert polynomial in $\mathbb{F}_{p^2}$ to get the j-invariant of the curve $j(E_{1,c})$. We finally get the value of c by solving $j(E_{1,c}) = 2^6 \frac{(3c-10)^3}{(c-2)(c+2)^2}$ in $\mathbb{F}_{p^2}$ and choosing a solution satisfying $c^2 \in \mathbb{F}_p$.

We note that for a bunch of discriminants (such as $-20, -24, -36$ etc.), Hilbert polynomial precomputation may be avoided by using parameterizations computed by Quer [17]:

$$C_t : y^2 = x^3 - 6(5 + 3\sqrt{t})x + 8(7 + 9\sqrt{t}), \tag{20}$$

for some $t \in \mathbb{Q}$. For instance $t = \frac{5}{4}$ for $D = -20$, $t = \frac{8}{9}$ for $D = -24$ etc. Once p is found, one may directly reduce mod p the curve given by equation 20. Curves given by equation (20) are $\mathbb{Q}$-curves and for these discriminants, we obtain the same curves as in [19].

Complex multiplication algorithms may not be avoided in certain cryptographic frames, such as pairing-friendly constructions. One advantage of the construction is that one has the liberty to choose the value r of the large prime number dividing the curve group order. This helps in preventing certain attacks, such as Cheon's attack [4] on the q-DH assumption. On the negative side, we cannot construct curves with fixed p (such as the attractive $2^{127} - 1$).

Using Magma, we computed an example with $p \equiv 5 \mod 8$, $D = 40$, $D' = 20$.

Example 1. We first search 63-bit numbers n, m such that $p = (2n^2 + 20m^2)/4$ is prime and $\#E_{1,c}(\mathbb{F}_{p^2})$ is almost prime. We can expect an order of the form $4r$, with r prime. In a few seconds, we find the following parameters.

n = 0x55d23edfa6a1f7e4
m = 0x549906b3eca27851
t_{p^2} = - 0xfaca844b264dfaa353355300f9ce9d3a
p = 0x9a2a8c914e2d05c3f2616cade9b911ad
r = 0x1735ce0c4fbac46c2245c3ce9d8da0244f9059ae9ae4784d6b2f65b29c444309
c^2 = 0x40b634aec52905949ea0fe36099cb21a

with r, p prime and $\#E_{1,c}(\mathbb{F}_{p^2}) = 4r$.

We use Vélu's formulas to compute a degree-5 isogeny from $E_{1,c}$ into $E_{b,5}$. We find a 5-torsion point $P_5(X_5, Y_5)$ on $E_{1,c}(\mathbb{F}_{p^8})$. The function `IsogenyFromKernel` in Magma evaluated at $(E_{1,c}(\mathbb{F}_{p^8}), (X - X_{P_5})(X - X_{2P_5}))$ outputs a curve $E_{b,5}$: $y_b^2 = x_b^3 - 25 \cdot 27(3c + 10)x_b + 125 \cdot 108(9c + 14)$. The curve E_b is isomorphic to $E_{1,-c}$ over $\mathbb{F}_{p^2}$ through $i_{\sqrt{5}} : (x_b, y_b) \mapsto (x_b/5, y_b/(5\sqrt{5}))$. The above function outputs also the desired isogeny with coefficients in $\mathbb{F}_{p^2}$:

$$\begin{aligned}
&\mathcal{I}_5 : \\
&E_{1,c}(\mathbb{F}_{p^2}) \to E_{b,5}(\mathbb{F}_{p^2}) \\
&\quad (x, y) \mapsto \Bigg(x + \frac{2 \cdot 3^3\left(\frac{3}{5}(13c+40)x + 4(27c+28)\right)}{x^2 + \frac{27}{2}cx - \frac{81}{10}c + 162} \\
&\qquad + \frac{-2^3 \cdot 3^4((9c+16)x^2 + \frac{2}{5}11(27c+64)x + \frac{2}{5}3^3(53c+80)}{(x^2 + \frac{27}{2}cx - \frac{81}{10}c + 162)^2}, \\
&\qquad y\Bigg(1 + \frac{-2^4 \cdot 3^4((9c+16)x^3 + \frac{3}{5}11(27c+64)x^2 + \frac{2}{5}3^4(53c+80)x + \frac{2}{5^2}3^2(4419c+13360))}{(x^2 + \frac{27}{2}cx - \frac{81}{10}c + 162)^3} \\
&\qquad + \frac{2 \cdot 3^3\left(\frac{3}{5}(13c+40)x^2 + 2^3(27c+28)x + 2\frac{3}{5}(369c+1768)\right)}{(x^2 + \frac{27}{2}cx - \frac{81}{10}c + 162)^2}\Bigg)\Bigg)
\end{aligned} \tag{21}$$

We finally obtain a second computable endomorphism on $E_{1,c}$ in this example by composing $\pi_p \circ i_{\sqrt{5}} \circ \mathcal{I}_5$.

4 Four-Dimensional GLV on $E_{2,c}(\mathbb{F}_{p^2})$

The construction of two efficiently computable endomorphisms on $E_{2,c}$, with degree of inseparability p, is similar to the one we gave for $E_{1,c}$.

We consider the elliptic curve given by eq. (6) in the reduced form:

$$E_{2,c}(\mathbb{F}_{p^2}) : y^2 = x^3 + 3(2c - 5)x + c^2 - 14c + 22 \ . \tag{22}$$

We assume that $c \in \mathbb{F}_{p^2} \setminus \mathbb{F}_p$, $c^2 \in \mathbb{F}_p$, c is not a cube in $\mathbb{F}_{p^2}$. In this case the isogeny (8) between J_{C_2} and $E_{2,c} \times E_{2,-c}$ is defined over $\mathbb{F}_{p^6}$. The 3-torsion subgroup $E_{2,c}(\mathbb{F}_{p^2})[3]$ contains the order 3 subgroup $\{\mathcal{O}, (3, c+2), (3, -c-2)\}$. We compute an isogeny whose kernel is this 3-torsion subgroup. With Vélu's formulas we obtain the curve $E_b : y^2 = x^3 - 27(2c+5)x - 27(c^2+14c+22)$. The curve E_b is isomorphic to $E_{2,-c} : (\mathbb{F}_{p^2}) : y^2 = x^3 - 3(2c+5)x + c^2 + 14c + 22$, via the isomorphism $(x, y) \mapsto (x/(-3), y/(-3\sqrt{-3}))$. We define the isogeny

$$\begin{aligned} \mathcal{I}_3 : E_{2,c} &\to E_{2,-c} \\ (x, y) &\mapsto \left(\tfrac{-1}{3}\left(x + \tfrac{12(c+2)}{x-3} + \tfrac{4(c+2)^2}{(x-3)^2}\right), \tfrac{-y}{3\sqrt{-3}}\left(1 - \tfrac{12(c+2)}{(x-3)^2} - \tfrac{8(c+2)^2}{(x-3)^3}\right)\right). \end{aligned}$$

Finally, we observe that $\pi_p(c) = -c$ and $\pi_p(j(E_{2,c})) = j(E_{2,-c})$. This implies that $E_{2,c}$ and $E_{2,-c}$ are isogenous through the Frobenius map π_p. We obtain the isogeny $\Phi_3 = \mathcal{I}_3 \circ \pi_p$ which is given by the following formula

$$\begin{aligned} &\Phi_3 : \\ &E_{2,c}(\mathbb{F}_{p^2}) \to E_{2,c}(\mathbb{F}_{p^2}) \\ &\quad (x, y) \mapsto \left(\tfrac{-1}{3}\left(x^p + \tfrac{12(2-c)}{x^p-3} + \tfrac{4(2-c)^2}{(x^p-3)^2}\right), \tfrac{y^p}{-3\sqrt{-3}^p}\left(1 - \tfrac{12(2-c)}{(x^p-3)^2} - \tfrac{8(2-c)^2}{(x^p-3)^3}\right)\right). \end{aligned}$$

We compute formally Φ_3^2 and obtain $\Phi_3^2 = \pi_{p^2} \circ [\pm 3]$. There is a term $\sqrt{-3}\sqrt{-3}^p$ in the y-side of Φ_3^2. We observe that if $p \equiv 1 \mod 3$, then $\left(\frac{-3}{p}\right) = 1$, $\sqrt{-3}\sqrt{-3}^p = -3$ and $\Phi_3^2 = \pi_{p^2} \circ [-3]$. Similarly, if $p \equiv 2 \mod 3$, then $\Phi_3^2 = \pi_{p^2} \circ [3]$. We conclude that for points defined over $\mathbb{F}_{p^2}$, we have

$$\Phi_3^2 \pm 3 = 0 .$$

Guillevic and Vergnaud [13, Theorem 2] showed that the complex multiplication discriminant is of the form $3D'$. With the same arguments as for $E_{1,c}$, we deduce that there are integers m and n such that if $p \equiv 1 \pmod 3$, then

$$t_{p^2} + 2p = D' m^2 \text{ and } t_{p^2} - 2p = -2n^2.$$

and if $p \equiv 2 \pmod 3$, then

$$t_{p^2} + 2p = 2n^2 \text{ and } t_{p^2} - 2p = -D' m^2.$$

As a consequence, we have the following theorem, whose proof is similar to the proof of 1.

Theorem 2. *Let $E_{2,c}$ be an elliptic curve given by equation (22), defined over $\mathbb{F}_{p^2}$. Let $-D$ be the complex multiplication discriminant and consider D' such that $D = 3D'$. There is an endomorphism $\Phi_{D'}$ of $E_{2,c}$ with degree of separability D'. The characteristic equation of this endomorphism is*

$$\Phi_{D'}^2 - D' m\, \Phi_{D'} + D' p \,\mathrm{Id} = 0 . \tag{23}$$

We have thus proven that $\Phi = \Phi_3$ and $\Psi = \Phi_{D'}$, viewed as algebraic integers, generate different quadratic extensions of $\mathbb{Q}$. As a consequence, we obtain a four-dimensional GLV algorithm on $E_{2,c}$.

5 Four-Dimensional GLV on J_{C_1} and J_{C_2}

The first endomorphism Ψ on J_{C_1} is induced by the curve automorphism $(x, y) \rightarrow (-x, iy)$, with i a square root of -1. The characteristic polynomial is $X^2 + 1 = 1$. On J_{C_2} we consider Ψ the endomorphism induced by the curve automorphism $(x, y) \rightarrow (\zeta_3 x, y)$. Its characteristic equation is $X^2 + X + 1$. The second endomorphism is constructed as $\Phi = \hat{I}(\Phi_{D'}, \Phi_{D'})I$, where $\Phi_{D'}$ is the elliptic curve endomorphism constructed in Theorem 1. In order to compute the characteristic equation for Φ, we follow the lines of the proof of Theorem 1 in [10]. We reproduce the computation for the Jacobian of C_1.

Theorem 3. *Let $C_1 : y^2 = x^5 + ax^3 + b$ be a hyperelliptic curve defined over $\mathbb{F}_p$ with ordinary Jacobian and let r a prime number such that $r || J_{C_1}(\mathbb{F}_p)$. Let $I : J_{C_1} \rightarrow E_{1,c} \times E_{1,c}$ the $(2, 2)$-isogeny defined by equation (4) and assume I is defined over an extension field of degree $k > 1$. We define $\Phi = \hat{I}(\Phi_{D'} \times \Phi_{D'})I$. where $\Phi_{D'}$ is the endomorphism defined in Theorem 1. Then*

1. *For $P \in J_{C_1}[r](\mathbb{F}_p)$, we have $\Phi(P) = [\lambda]P$, with $\lambda \in \mathbb{Z}$.*
2. *The characteristic equation of Φ is $\Phi^2 - 2D^{'}m\,\Phi + 4D^{'}p\,\mathrm{Id} = 0$.*

Proof. 1. Note that $\mathrm{End}(J_{C_1})$ is commutative, and Φ is defined over $\mathbb{F}_p$ (see [2, Prop. III.1.3]). Hence, for $\mathcal{D} \in J_{C_1}(\mathbb{F}_p)$, we have that $\pi(\Phi(\mathcal{D})) = \Phi(\pi(\mathcal{D})) = \Phi(\mathcal{D})$. Since there is only one subgroup of order r in $J_{C_1}(\mathbb{F}_p)$, we obtain that $\Phi(\mathcal{D}) = \lambda\mathcal{D}$.

2. Since $\hat{I}I = [2]$ then

$$\Phi^2 = \hat{I}(\Phi_{D'} \times \Phi_{D'})I\hat{I}(\Phi_{D'} \times \Phi_{D'})I = 2\hat{I}(\Phi_{D'}^2, \Phi_{D'}^2)I. \tag{24}$$

Since $\Phi_{D'}$ verifies the equation

$$\Phi_{D'}^2 - D^{'}m\,\Phi_{D'} + D^{'}p\,\mathrm{Id} = 0, \tag{25}$$

we have

$$[2]\hat{I}((\Phi_{D'}^2, \Phi_{D'}^2) - D^{'}m\,(\Phi_{D'}, \Phi_{D'}) + D^{'}p\,(\mathrm{Id}, \mathrm{Id}))I = \mathcal{O}_{J_{C_1}}$$

Using equation (24), we conclude that $\Phi^2 - 2D^{'}m\,\Phi + 4D^{'}p\,\mathrm{Id} = 0$.

5.1 Computing I on $J_{C_1}(\mathbb{F}_p)$

We show first how to compute stately the $(2, 2)$-isogeny on $J_{C_1}(\mathbb{F}_p)$ with only a small number of operations over extension fields of $\mathbb{F}_p$.

Let $\mathcal{D}$ be a divisor in $J_{C_1}(\mathbb{F}_p)$ given by its Mumford coordinates

$$\mathcal{D} = [U, V] = [T^2 + u_1 T + u_0, v_1 T + v_0],\ u_0, u_1, v_0, v_1 \in \mathbb{F}_p\ .$$

It corresponds to two points $P_1(X_1, Y_1), P_2(X_2, Y_2) \in C_1(\mathbb{F}_p)$ or $C_1(\mathbb{F}_{p^2})$. We have

$$u_1 = -(X_1 + X_2), u_0 = X_1 X_2, v_1 = \frac{Y_2 - Y_1}{X_2 - X_1}, v_0 = \frac{X_1 Y_2 - X_2 Y_1}{X_1 - X_2}.$$

Explicit formula to compute $\varphi_{1*}(P_1) + \varphi_{1*}(P_2)$. Let $\varphi_{1*}(P_1) = (x_{1,1}, y_{1,1})$ and $\varphi_{1*}(P_2) = (x_{2,1}, y_{2,1})$ In the following we give the formulas to compute $S_1(x_{3,1}, y_{3,1}) = \varphi_{1*}(P_1) + \varphi_{1*}(P_2)$.

$$x_{3,1} = \tfrac{\lambda_1^2}{c+2} - (x_{1,1} + x_{2,1}) + \tfrac{3c-10}{c+2} \text{ with}$$

$$\lambda_1 = \frac{2}{\sqrt[8]{b}} \frac{\left[(v_0u_1 - v_1u_0)u_1 - v_0u_0\right] + \left[3(v_0u_1 - v_1u_0)\right]\sqrt[4]{b} + \left[3v_0\right]\sqrt{b} + \left[v_1\right]\sqrt[4]{b^3}}{\left[u_0^2 - b\right] + \left[u_0u_1\right]\sqrt[4]{b} + \left[-u_1\right]\sqrt{b}}.$$

We denote $\lambda_1 = \Lambda_1/\sqrt[8]{b}$. The computation of the numerator of Λ_1 costs $4M_p$ and the denominator costs $S_p + M_p$. We will use the Jacobian coordinates for S_1: $x_{3,1} = X_{3,1}/Z_{3,1}^2$, $y_{3,1} = Y_{3,1}/Z_{3,1}^3$ to avoid inversion in $\mathbb{F}_{p^4}$. We continue with

$$x_{1,1} + x_{2,1} = 2\frac{\left(\left[u_0^2 + b\right] + \left[u_1^2 - 6u_0\right]\sqrt{b}\right)\left(\left[u_0^2 + b\right] + \left[-2u_0\right]\sqrt{b}\right)}{\left(\left[u_0^2 - b\right] + \left[u_0u_1\right]\sqrt[4]{b} + \left[-u_1\right]\sqrt{b}\right)^2}$$

As u_0^2 was already computed in Λ_1, this costs one square (u_1^2) and a multiplication in $\mathbb{F}_{p^2}$, hence $S_p + M_{p^2}$. The denominator is the same as the one of Λ_1^2, that is, Z_3^2.

Then

$$\begin{aligned} x_{3,1} &= \frac{\Lambda_1^2}{\sqrt[4]{b}(c+2)} - (x_{1,1} + x_{2,1}) + \frac{3c-10}{c+2} \\ &= \frac{\sqrt[4]{b}\Lambda_1^2}{(a+2\sqrt{b})} - (x_{1,1} + x_{2,1}) + \frac{3a - 10\sqrt{b}}{a+2\sqrt{b}}. \end{aligned}$$

To avoid tedious computations, it is preferable to precompute both $1/(a+2\sqrt{b})$ and $(3a - 10\sqrt{b})/(a + 2\sqrt{b})$ with one inversion in $\mathbb{F}_{p^2}$ and one multiplication in $\mathbb{F}_{p^2}$.

Computing $\sqrt[4]{b}\Lambda_1^2$ is done by shifting to the right coefficients and costs one multiplication by b (as $\Lambda_1^2 \in \mathbb{F}_{p^4}$). Then $\sqrt[4]{b}\Lambda_1^2 \cdot (a + 2\sqrt{b})^{-1}$ costs $2M_{p^2}$. Finally we need to compute $\frac{3a-10\sqrt{b}}{a+2\sqrt{b}} \cdot Z_3^2$ which costs $S_{p^4} + 2M_{p^2}$. The total cost of $X_{3,1}$, $Z_{3,1}$ and $Z_{3,1}^2$ is $6M_p + 2S_p + 5M_{p^2} + S_{p^4}$.

Computing $y_{3,1}$ is quite complicated because we deal with divisors so we do not have directly the coefficients of the two points. We use this trick:

$$\begin{aligned} y_{3,1} &= \lambda_1(x_{1,1} - x_{3,1}) - y_{1,1} \\ y_{3,1} &= \lambda_1(x_{2,1} - x_{3,1}) - y_{2,1} \\ 2y_{3,1} &= \lambda_1(x_{1,1} + x_{2,1} - 2x_{3,1}) - (y_{1,1} + y_{2,1}) \end{aligned}$$

Since $x_{1,1} + x_{2,1}$ was already computed for $x_{3,1}$, getting $(x_{1,1} + x_{2,1} - 2x_{3,1})$ costs only additions. We multiply the numerators of λ_1 and $(x_{1,1} + x_{2,1} - 2x_{3,1})$ which costs $1M_{p^4}$. The denominator is $Z_{3,1}^3$ and as $Z_{3,1}^2$ is already computed, this costs $1M_{p^4}$. The numerator of $(y_{1,1} + y_{2,1})$ contains products of u_0, u_1, v_0, v_1 previously computed and its denominator is simply Z_3^3. The total cost of $y_{3,1}$ is then $2M_{p^4}$. Finally, computing $(x_{3,1}, y_{3,1})$ costs

$$6M_p + 2S_p + 5M_{p^2} + S_{p^4} + 2M_{p^4}\ .$$

Now we show that computing $S_2(x_{3,2}, y_{3,2})$ is free of cost. We notice that

$$\varphi_1(X_j, Y_j) = \varphi_2(-X_j, iY_j)$$

with i such that $i^2 = -1$ and $j \in \{1, 2\}$. Rewriting this equation in terms of divisors, we derive that

$$S_2(x_{3,2}, y_{3,2}) = \varphi_{1*}([-u_1, u_0, -iv_1, iv_0]) .$$

We can simply compute S_2 with φ_{1*}:

$$\begin{aligned} x_{3,2} &= x_{3,1}([-u_1, u_0, -iv_1, iv_0]) \text{ with} \\ \lambda_2 &= \lambda_1([-u_1, u_0, -iv_1, iv_0]) \\ &= \tfrac{2i}{\sqrt[8]{b}} \tfrac{(v_0u_1 - v_1u_0)(u_1 - 3\sqrt[4]{b}) - v_0u_0 + 3\sqrt{b}v_0 - \sqrt[4]{b}^3 v_1}{(u_0 - \sqrt{b})(u_0 - \sqrt[4]{b}u_1 + \sqrt{b})} = \pi_{p^2}(\lambda_1) \end{aligned}$$

and

$$(x_{1,1} + x_{2,1})([-u_1, u_0, -iv_1, iv_0]) = 2\tfrac{u_0^2 + \sqrt{b}u_1^2 - 6\sqrt{b}u_0 + b}{(u_0 - \sqrt[4]{b}u_1 + \sqrt{b})^2} = \pi_{p^2}(x_{1,1} + x_{2,1}) .$$

We deduce that $x_{3,2} = \pi_{p^2}(x_{3,1}),\ y_{3,2} = \pi_{p^2}(y_{3,1})$ and

$$\varphi_{2*}(\mathcal{D}) = \varphi_{2*}(P_1) + \varphi_{2*}(P_2) = \pi_{p^2}(\varphi_{1*}(P_1) + \varphi_{1*}(P_2)) .$$

Computing $(x_{3,2}, y_{3,2})$ costs two Frobenius π_{p^2} which are performed with four negations on $\mathbb{F}_{p^2}$.

5.2 Computing Endomorphisms on $E_{1,c}$

Here we apply the endomorphism $\Phi_{D'}$ on $S_1(x_{3,1}, y_{3,1})$. As $\Phi_{D'}$ is defined over $\mathbb{F}_{p^2}$, it commutes with π_{p^2} hence $\Phi_{D'}(x_{3,2}) = \pi_{p^2}(\Phi_{D'}(x_{3,1}))$ is free. Unfortunately S_1 has coefficients in $\mathbb{F}_{p^4}$ hence we need to perform some multiplications in $\mathbb{F}_{p^4}$. More precisely, $y_{3,1}$ is of the form $\sqrt[8]{b}y'_{3,1}$ with $y'_{3,1} \in \mathbb{F}_{p^4}$. As the endomorphism is of the form $\Phi_{D'}(x, y) = (\Phi_{D',x}(x), y\Phi_{D',y}(x))$ the $\sqrt[8]{b}y'_{3,1}$ term is not involved in the endomorphism computation.

5.3 Computing $\hat{I}$ on $J_{\mathcal{C}_1}(\mathbb{F}_p)$.

Then we go back to $J_{\mathcal{C}_1}$. We compute the divisor of these two points (with $\pm\sqrt{x_{3,1}}$) on $J_{\mathcal{C}_1}$ and get

$$\varphi_1^*(x_{3,1}, y_{3,1}) = T^2 - 2\sqrt[4]{b}\tfrac{x_{3,1}+1}{x_{3,1}-1}T + \sqrt{b}, \tfrac{\sqrt{b}y_{3,1}}{2(x_{3,1}-1)}\left(\tfrac{x_{3,1}+3}{x_{3,1}-1}T - \sqrt[4]{b}\right) .$$

If $(x_{3,1}, y_{3,1})$ is in Jacobian coordinates $(X_{3,1}, Y_{3,1}, Z_{3,1})$ then we compute $\frac{x_{3,1}+1}{x_{3,1}-1} = \frac{X_{3,1}+Z_{3,1}^2}{X_{3,1}-Z_{3,1}^2}$.

A similar computation gives

$$\varphi_2^*(x_{3,2}, y_{3,2}) = T^2 + 2\sqrt[4]{b}\frac{x_{3,2}+1}{x_{3,2}-1}T + \sqrt{b}, \frac{\sqrt{b}y_{3,2}}{2(x_{3,2}-1)}\left(\frac{x_{3,2}+3}{x_{3,2}-1}T + \sqrt[4]{b}\right) .$$

Since $x_{3,2} = \pi_{p^2}(x_{3,1})$ and $y_{3,2} = \pi_{p^2}(y_{3,1})$, we have

$$\varphi_2^*(x_{3,2}, y_{3,2}) = T^2 + 2\sqrt[4]{b}\frac{\pi_{p^2}(x_{3,1})+1}{\pi_{p^2}(x_{3,1})-1}T + \sqrt{b}, \frac{\sqrt{b}\pi_{p^2}(y_{3,1})}{2(\pi_{p^2}(x_{3,1})-1)}\left(\frac{\pi_{p^2}(x_{3,1})+3}{\pi_{p^2}(x_{3,1})-1}T + \sqrt[4]{b}\right) .$$

Hence $\varphi_2^*(x_{3,2}, y_{3,2}) = \pi_{p^2}(\varphi_1^*(x_{3,1}, y_{3,1}))$.
Finally, we have

$$\varphi_2^*(\varphi_{2*}(P_1) + \varphi_{2*}(P_2)) = \pi_{p^2}(\varphi_1^*((\varphi_{1*}(P_1) + \varphi_{1*}(P_2)))) .$$

and, with similar arguments,

$$\varphi_2^*(\Phi_{D'}(\varphi_{2*}(P_1) + \varphi_{2*}(P_2))) = \pi_{p^2}(\varphi_1^*(\Phi_{D'}((\varphi_{1*}(P_1) + \varphi_{1*}(P_2))))) .$$

The computation of the sum $\varphi_1^*(\Phi_{D'}(\varphi_{1*}(\mathcal{D}))) + \pi_{p^2} \circ \varphi_1^*(\Phi_{D'}(\varphi_{1*}(\mathcal{D})))$ involves terms in $\mathbb{F}_{p^4}$ but thanks to its special form, we need to perform the operations in $\mathbb{F}_{p^2}$ only. We give the table of computations in Appendix A and show that most multiplications are performed over $\mathbb{F}_{p^2}$. We have followed computations for a multiplication in Mumford coordinates provided in [5].

We conclude that applying $\varphi_{1*}(P_1) + \varphi_{1*}(P_2)$ costs roughly as much as an addition on J_{C_1} over $\mathbb{F}_p$, $\varphi_{2*}(P_1) + \varphi_{2*}(P_2)$ is cost free. Computing $\Phi_{D'}$ depends on the size of D' and costs few multiplications over $\mathbb{F}_{p^4}$. Finally adding $\varphi_1^* + \varphi_2^*$ costs roughly an addition of divisors over $\mathbb{F}_{p^2}$.

6 Complexity Analysis and Comparison to GLS-GLV Curves

We explain that our construction is valid for GLS curves with discriminants -3 and -4. These curves are particularly interesting for cryptography, because their simple equation forms result into simple and efficient point additions. A four-dimensional GLV algorithm on these curves was proposed by Longa and Sica [16]. Although the endomorphisms we construct do not allow to derive a higher dimension algorithm, they offer an alternative to Longa and Sica's construction.

The Case $D = -4$. We consider a curve with CM discriminant $D = -4$, defined over $\mathbb{F}_{p^2}$, with $p \equiv 1 \mod 8$. Assume that the curve is of the form $E_\alpha(\mathbb{F}_{p^2}) : y^2 = x^3 + \alpha x$ with $\alpha \in \mathbb{F}_{p^2}$. A 2-torsion point is $P_2(0, 0)$. Using Vélu's formulas, we get the isogeny with kernel generated by P_2, whose equation is

$$(x, y) \mapsto \left(x + \frac{\alpha}{x}, y - y\frac{\alpha}{x^2}\right) .$$

This isogeny sends points on E_α on the curve $E_b : y^2 = x^3 - 4\alpha x$. We use the same trick as previously. If $\alpha \in \mathbb{F}_{p^2}$ is such that $\pi_p(\alpha) = \alpha^p = -\alpha$ (this is the case for example if $\alpha = \sqrt{a}$ with $a \in \mathbb{F}_p$ a non-square) then by composing with $(x_b, y_b) \mapsto \left(x_b^p/(-2), y_b^p/(-2\sqrt{-2})\right)$, we get an endomorphism Φ_2. Note that $\sqrt{-1} \in \mathbb{F}_p$ since $p \equiv 1 \mod 8$. We obtain

$$\Phi_2 : E_\alpha(\mathbb{F}_{p^2}) \to E_\alpha(\mathbb{F}_{p^2})$$
$$(x, y) \mapsto \begin{cases} \mathcal{O} & \text{if } (x, y) = (0, 0), \\ \left(\frac{(x^p)^2+\alpha}{2x^p}, \frac{y^p}{2\sqrt{2}}\left(1 - \frac{\alpha}{(x^p)^2}\right)\right) & \text{otherwise.} \end{cases}$$

We obtained an endomorphism Φ_2 such that $\Phi_2^2 - 2 = 0$, when restricted to points defined over $\mathbb{F}_{p^2}$. The complex multiplication endomorphism Φ on E_α is $(x, y) \to (-x, iy)$ and verifies the equation $\Phi^2 + 1 = 0$. The 4-dimensional GLV algorithm of Longa and Sica on this curve uses an endomorphism Ψ such that $\Psi^4 + 1 = 0$. With our method we obtain two distinct endomorphisms, but the three ones Ψ, Φ_2, Φ are not "independent" on the subgroup $E(\mathbb{F}_{p^2}) \setminus E[2]$. Indeed, we have $\Phi_2 + \Phi\Phi_2 = 2\Psi$.

Note that in this case the corresponding Jacobian splits into two isogenous elliptic curves over $\mathbb{F}_p$, namely the two quartic twists defined over $\mathbb{F}_p$ of $E_{1,c}$.

The Case $D = -3$. We consider the curve E_β whose Weierstrass equation is

$$y^2 = x^3 + \beta, \tag{26}$$

where $\beta^2 \in \mathbb{F}_p$. Our construction yields the following efficiently computable endomorphism

$$\Phi_3(x, y) = \left(\frac{1}{3}\left(x^p + \frac{4\beta^p}{x^{2p}}\right), \frac{y^p}{\sqrt{3}}\left(1 + \frac{8\beta^p}{x^{3p}}\right)\right).$$

When restricted to points defined over $\mathbb{F}_{p^2}$, this endomorphism verifies the equation $\Phi_3^2 - 3 = 0$, while the complex multiplication endomorphism Φ has characteristic equation $\Phi^2 + \Phi + 1 = 0$. Longa and Sica's algorithm uses the complex multiplication Φ and an endomorphism Ψ verifying $\Psi^2 + 1 = 0$ for points defined over $\mathbb{F}_{p^2}$. We observe that $2\Phi_3\Psi - 1 = 2\Phi$.

We give in Table 6 the operation count of a computation of one scalar multiplication using two-dimensional and four-dimensional GLV on E and E_β given by equation (26). We denote by m, s and by M, S the cost of multiplication and squaring over $\mathbb{F}_p$ and over $\mathbb{F}_{p^2}$, respectively. We denote by c the cost of multiplication by a constant in $\mathbb{F}_{p^2}$. In order to give global estimates, we will assume that $m \sim s$ and that $M \sim 3m$ and $S \sim 3s$. Additions in $\mathbb{F}_p$ are not completely negligible compared to multiplications, but we do not count additions here. We counted operations by using formulæ from Bernstein and Lange's database [1] for addition and doubling in projective coordinates. On the curve $E_{1,c}$ addition costs $12M + 2S$, while doubling costs $5S + 6M + 1c$. For E_β , addition costs $12M + 2S$, while doubling is

$3M+5S+1c$. Note that by using Montgomery's simultaneous inversion method, we could also obtain all points in the look-up table in affine coordinates and use mixed additions for the addition step of the scalar multiplication algorithm. This variant adds one inversion and $3(n-1)$ multiplications, where n is the length of the look-up table. We believe this is interesting for implementations of cryptographic applications which need to perform several scalar multiplications. For genus 2 arithmetic on curves of the form $y^2 = x^5 + ax^3 + bx$, we used formulæ given by Costello and Lauter [5] in projective coordinates. An addition costs $43M + 4S$ and a doubling costs $30M + 9S$.

Table 1. Total cost of scalar multiplication at a 128-bit security level

Curve	Method	Operation count	Global estimation
$E_{1,c}$	4-GLV, 16 pts.	$1168M + 440S$	$4797m$
E_β	4-GLV, 16 pts.	$976M + 440S$	$4248m$
$E_{1,c}$	2-GLV, 4 pts.	$2048M + 832S$	$8640m$
E_β	2-GLV, 4 pts.	$1664M + 832S$	$7488m$
J_{C_1}	4-GLV, 16 pts.	$4500m + 816s$	$5316m$
J_{C_1}	2-GLV, 4 pts.	$7968m + 1536s$	$9504m$
FKT [3]	4-GLV, 16 pts.	$4500m + 816s$	$5316m$
Kummer [3]	–	$3328m + 2304s$	$5632m$

The practical gain of the 4-dimensional GLV on $E_{1,c}$, when compared to the 2-dimensional GLV method, is of 44%. Curves with discriminant -3, defined over $\mathbb{F}_{p^2}$, which belong both to the family of curves we propose and to the one proposed by Longa and Sica, offer a 12% speed-up, thanks to their efficient arithmetic.

7 Conclusion

We have studied two families of elliptic curves defined over $\mathbb{F}_{p^2}$ which have the property that the Weil restriction is isogenous over $\mathbb{F}_p$ to the Jacobian of a genus 2 curve. We have proposed a four dimensional GLV algorithm on these families of elliptic curves and on the corresponding Jacobians of genus 2 curves. Our complexity estimates show that these abelian varieties offer efficient scalar multiplication, competitive to GLV algorithms on other families in the literature, having two efficiently computable and "independent" endomorphisms.

Acknowledgements. We are grateful to Damien Vergnaud and Léo Ducas for many helpful discussions on the GLV algorithm and lattice reduction. We thank the anonymous reviewers of the Asiacrypt conference for their remarks. This work was supported in part by the French ANR-09-VERS-016 BEST Project.

References

1. Bernstein, D., Lange, T.: Explicit-Formulas Database, `http://www.hyperelliptic.org/EFD/`
2. Bisson, G.: Endomorphism rings in cryptography. PhD thesis, Institut National Polytechnique de Lorraine (2011)
3. Bos, J.W., Costello, C., Hisil, H., Lauter, K.: Fast cryptography in genus 2. In: Johansson, T., Nguyen, P.Q. (eds.) EUROCRYPT 2013. LNCS, vol. 7881, pp. 194–210. Springer, Heidelberg (2013)
4. Cheon, J.H.: Security analysis of the strong diffie-hellman problem. In: Vaudenay, S. (ed.) EUROCRYPT 2006. LNCS, vol. 4004, pp. 1–11. Springer, Heidelberg (2006)
5. Costello, C., Lauter, K.: Group Law Computations on Jacobians of Hyperelliptic Curves. In: Miri, A., Vaudenay, S. (eds.) SAC 2011. LNCS, vol. 7118, pp. 92–117. Springer, Heidelberg (2012)
6. Dewaghe, L.: Un corollaire aux formules de Vélu. Draft (1995)
7. Doche, C.: Exponentiation. In: Handbook of Elliptic and Hyperelliptic Curve Cryptography, ch. 9, pp. 145–168. Chapman and Hall/CRC, Taylor and Francis Group (2006)
8. Duursma, I., Kiyavash, N.: The vector decomposition problem for elliptic and hyperelliptic curves. Journal of the Ramanujan Mathematical Society 20(1), 59–76 (2005)
9. Freeman, D.M., Satoh, T.: Constructing pairing-friendly hyperelliptic curves using Weil restriction. Journal of Number Theory 131(5), 959–983 (2011)
10. Galbraith, S.D., Lin, X., Scott, M.: Endomorphisms for faster elliptic curve cryptography on a large class of curves. In: Joux, A. (ed.) EUROCRYPT 2009. LNCS, vol. 5479, pp. 518–535. Springer, Heidelberg (2009)
11. Gallant, R.P., Lambert, R.J., Vanstone, S.A.: Faster point multiplication on elliptic curves with efficient endomorphisms. In: Kilian, J. (ed.) CRYPTO 2001. LNCS, vol. 2139, pp. 190–200. Springer, Heidelberg (2001)
12. Gaudry, P., Schost, É.: On the invariants of the quotients of the jacobian of a curve of genus 2. In: Bozta, S., Sphparlinski, I. (eds.) AAECC 2001. LNCS, vol. 2227, pp. 373–386. Springer, Heidelberg (2001)
13. Guillevic, A., Vergnaud, D.: Genus 2 Hyperelliptic Curve Families with Explicit Jacobian Order Evaluation and Pairing-Friendly Constructions. In: Abdalla, M., Lange, T. (eds.) Pairing 2012. LNCS, vol. 7708, pp. 234–253. Springer, Heidelberg (2013)
14. Kohel, D.: Endomorphism rings of elliptic curves over finite fields. Ph.D. thesis, University of California at Berkeley (1996)
15. Leprévost, F., Morain, F.: Revêtements de courbes elliptiques à multiplication complexe par des courbes hyperelliptiques et sommes de caractères. Journal of Number Theory 64, 165–182 (1997), `http://www.lix.polytechnique.fr/Labo/Francois.Morain/Articles/LIX-RR-94-07-revetement.ps.gz`
16. Longa, P., Sica, F.: Four dimensional Gallant-Lambert-Vanstone scalar multiplication. Journal of Cryptology, 1–36 (2013)
17. Quer, J.: Fields of definition of $\mathbb{Q}$-curves. Journal de Théorie des Nombres de Bordeaux 13(1), 275–285 (2001)
18. Satoh, T.: Generating genus two hyperelliptic curves over large characteristic finite fields. In: Joux, A. (ed.) EUROCRYPT 2009. LNCS, vol. 5479, pp. 536–553. Springer, Heidelberg (2009)

19. Smith, B.: Families of fast elliptic curves from Q-curves. In: Sako, K., Sarkar, P. (eds.) ASIACRYPT 2013, Part I. LNCS, vol. 8269, pp. 61–78. Springer, Heidelberg (2013), http://eprint.iacr.org/2013/312
20. Vélu, J.: Isogenies entre courbes elliptiques. Comptes Rendus De l'Académie Des Sciences Paris, Série I-Mathèmatique, Série A 273, 238–241 (1971)

A Appendix 1

Following [5], we explain here the step addition of two divisors in the isogeny computation in Section 5.3. We denote by m_n and s_n the cost of multiplication and squaring, respectively, in an extension field $\mathbb{F}_{p^n}$.

$\sigma_1 = u_1 + \pi_{p^2}(u_1)$, $\Delta_0 = v_0 - \pi_{p^2}(v_0)$, $\Delta_1 = v_1 - \pi_{p^2}(v_1)$, $U_1 = u_1^2$ $(1m_4)$
$M_1 = u_1^2 - \pi_{p^2}(u_1^2)$,$M_2 = \sqrt{b}(\pi_{p^2}(u_1) - u_1)$, $M_3 = u_1 - \pi_{p^2}(u_1)$;
$l_2 = 2(M_2 \cdot \Delta_1 + \Delta_0 \cdot M_1)$; $l_3 = \Delta_0 \cdot M_3$; $d = -2M_2 \cdot M_3$; $(4m_2)$
$A = 1/(d \cdot l3)$; $B = d \cdot A$; $C = d \cdot B$; $D = l_2 \cdot B$; $(3m_2{+}1m_4)$
$E = l_3^2 \cdot A$; $CC = C^2$; $u_1'' = 2 \cdot D - CC - \sigma_1$ $(1m_2{+}2s_2)$
$u_0'' = D^2 + C \cdot (v_1 + \pi_{p^2}(v_1)) - ((u_1'' - CC) \cdot \sigma_1 + (U_1 + \pi_{p^2}(U_1)))/2$ $(2m_2{+}1s_4)$
$U_0'' = \pi_{p^2}(u_1) \cdot u_0''$; $v_1'' = D \cdot (u_1 - u_1'') + u_1^{2''} - u_0'' - U_1$; $(2m_4{+}1s_1)$
$v_0'' = D \cdot (u_0 - u_0'') + U_0''$;$v_1'' = -(E \cdot v_1'' + v_1)$; $v_0'' = -(E \cdot v_0'' + v_0)$; $(3m_4)$

Discrete Gaussian Leftover Hash Lemma over Infinite Domains

Shweta Agrawal[1], Craig Gentry[2], Shai Halevi[2], and Amit Sahai[1]

[1] UCLA
[2] IBM Research

Abstract. The classic Leftover Hash Lemma (LHL) is often used to argue that certain distributions arising from modular subset-sums are close to uniform over their finite domain. Though very powerful, the applicability of the leftover hash lemma to lattice based cryptography is limited for two reasons. First, typically the distributions we care about in lattice-based cryptography are *discrete Gaussians*, not uniform. Second, the elements chosen from these discrete Gaussian distributions lie in an infinite domain: a lattice rather than a finite field.

In this work we prove a "lattice world" analog of LHL over infinite domains, proving that certain "generalized subset sum" distributions are statistically close to well behaved discrete Gaussian distributions, even without any modular reduction. Specifically, given many vectors $\{\boldsymbol{x}_i\}_{i=1}^m$ from some lattice $L \subset \mathbb{R}^n$, we analyze the probability distribution $\sum_{i=1}^m z_i \boldsymbol{x}_i$ where the integer vector $\boldsymbol{z} \in \mathbb{Z}^m$ is chosen from a discrete Gaussian distribution. We show that when the $\boldsymbol{x}_i$'s are "random enough" and the Gaussian from which the $\boldsymbol{z}$'s are chosen is "wide enough", then the resulting distribution is statistically close to a near-spherical discrete Gaussian over the lattice L. Beyond being interesting in its own right, this "lattice-world" analog of LHL has applications for the new construction of multilinear maps [5], where it is used to sample Discrete Gaussians obliviously. Specifically, given encoding of the $\boldsymbol{x}_i$'s, it is used to produce an encoding of a near-spherical Gaussian distribution over the lattice. We believe that our new lemma will have other applications, and sketch some plausible ones in this work.

1 Introduction

The Leftover Hash Lemma (LHL) is a central tool in computer science, stating that universal hash functions are good randomness extractors. In a characteristic application, the universal hash function may often be instantiated by a simple inner product function, where it is used to argue that a random linear combination of some elements (that are chosen at random and then fixed "once and for all") is statistically close to the uniform distribution over some finite domain. Though extremely useful and powerful in general, the applicability of the leftover hash lemma to lattice based cryptography is limited for two reasons. First, typically the distributions we care about in lattice-based cryptography are *discrete Gaussians*, not uniform. Second, the elements chosen from these discrete

K. Sako and P. Sarkar (Eds.) ASIACRYPT 2013 Part I, LNCS 8269, pp. 97–116, 2013.

Gaussian distributions lie in an infinite domain: a lattice rather than a finite field.

The study of discrete Gaussian distributions underlies much of the advances in lattice-based cryptography over the last decade. A discrete Gaussian distribution is a distribution over some fixed lattice, in which every lattice point is sampled with probability proportional to its probability mass under a standard (n-dimensional) Gaussian distribution. Micciancio and Regev have shown in [10] that these distributions share many of the nice properties of their continuous counterparts, and demonstrated their usefulness for lattice-based cryptography. Since then, discrete Gaussian distributions have been used extensively in all aspects of lattice-based cryptography (most notably in the famous "Learning with Errors" problem and its variants [14]). Despite their utility, we still do not understand discrete Gaussian distributions as well as we do their continuous counterparts.

A Gaussian Leftover Hash Lemma for Lattices?
The LHL has been applied often in lattice-based cryptography, but sometimes awkwardly. As an example, in the integer-based fully homomorphic encryption scheme of van Dijk et al. [18], ciphertexts live in the lattice $\mathbb{Z}$. Roughly speaking, the public key of that scheme contains many encryptions of zero, and encryption is done by adding the plaintext value to a subset-sum of these encryptions of zero. To prove security of this encryption method, van Dijk et al. apply the left-over hash lemma in this setting, but with the cost of complicating their encryption procedure by reducing the subset-sum of ciphertexts modulo a single large ciphertext, so as to bring the scheme back in to the realm of finite rings where the leftover hash lemma is naturally applied.[1] It is natural to ask whether that scheme remains secure also without this artificial modular reduction, and more generally whether there is a more direct way to apply the LHL in settings with infinite rings.

As another example, in the recent construction of multilinear maps [5], Garg et. al. require a procedure to randomize "encodings" to break simple algebraic relations that exist between them. One natural way to achieve this randomization is by adding many random encodings of zero to the public parameters, and adding a random linear combination of these to re-randomize a given encoding (without changing the encoded value). However, in their setting, there is no way to "reduce" the encodings so that the LHL can be applied. Can they argue that the new randomized encoding yields an element from some well behaved distribution?

In this work we prove an analog of the leftover hash lemma over lattices, yielding a positive answers to the questions above. We use discrete Gaussian distributions as our notion of "well behaved" distributions. Then, for m vectors $\{\boldsymbol{x}_i\}_{i\in[m]}$ chosen "once and for all" from an n dimensional lattice $L \subset \mathbb{R}^n$, and a coefficient vector $\boldsymbol{z}$ chosen from a discrete Gaussian distribution over the

---

[1] Once in the realms of finite rings, one can alternatively use the generic proof of Rothblum [15], which also uses the LHL.

integers, we give sufficient conditions under which the distribution $\sum_{i=1}^{m} z_i \boldsymbol{x}_i$ is "well behaved."

Oblivious Gaussian Sampler
Another application of our work is in the construction of an extremely simple *discrete Gaussian sampler* [6,13]. Such samplers, that sample from a spherical discrete Gaussian distribution over a lattice have been constructed by [6] (using an algorithm by Klein [7]) as well as Peikert [13]. Here we consider a much simpler discrete Gaussian sampler (albeit a somewhat imperfect one). Specifically, consider the following sampler. In an offline phase, for $m > n$, the sampler samples a set of short vectors $\boldsymbol{x}_1, \boldsymbol{x}_2, \ldots, \boldsymbol{x}_m$ from L – e.g., using GPV or Peikert's algorithm. Then, in the online phase, the sampler generates $\boldsymbol{z} \in \mathbb{Z}^m$ according to a discrete Gaussian and simply outputs $\sum_{i=1}^{m} z_i \boldsymbol{x}_i$. But does this simpler sampler work – i.e., can we say anything about its output distribution? Also, how small can we make the dimension m of $\boldsymbol{z}$ and how small can we make the entries of $\boldsymbol{z}$? Ideally m would be not much larger than the dimension of the lattice and the entries of $\boldsymbol{z}$ have small variance – e.g., $\tilde{O}(\sqrt{n})$.

A very useful property of such a sampler is that it can be made *oblivious* to an explicit representation of the underlying lattice, which makes it applicable easily within an additively homomorphic scheme. Namely, if you are given lattice points encrypted under an additively homomorphic encryption scheme, you can use them to generate an encrypted well behaved Gaussian on the underlying lattice. Previous samplers [6,13] are too complicated to use within an additively homomorphic encryption scheme [2].

Our Results
In this work, we obtain a discrete Gaussian version of the LHL over infinite rings. Formally, consider an n dimensional lattice L and (column) vectors $X = [\boldsymbol{x}_1|\boldsymbol{x}_2|\ldots|\boldsymbol{x}_m] \in L$. We choose $\boldsymbol{x}_i$ according to a discrete Gaussian distribution $\mathcal{D}_{L,S}$, where $\mathcal{D}_{L,S}$ is defined as $\mathcal{D}_{L,S,\boldsymbol{c}}(\boldsymbol{x}) = \frac{\rho_{S,\boldsymbol{c}}(\boldsymbol{x})}{\rho_{S,\boldsymbol{c}}(L)}$ with $\rho_{S,\boldsymbol{c}}(\boldsymbol{x}) \stackrel{\text{def}}{=} \exp(-\pi\|\boldsymbol{x} - \boldsymbol{c}\|^2/s^2)$ and $\rho_{S,\boldsymbol{c}}(A)$ for set A denotes $\sum_{\boldsymbol{x}\in A} \rho_{S,\boldsymbol{c}}(\boldsymbol{x})$.

Let $\boldsymbol{z} \leftarrow \mathcal{D}_{\mathbb{Z}^m,s'}$, we analyze the conditions under which the vector $X \cdot \boldsymbol{z}$ is statistically close to a "near-spherical" discrete Gaussian. Formally, consider:

$$\mathcal{E}_{X,s'} \stackrel{\text{def}}{=} \{X \cdot \boldsymbol{z} : \boldsymbol{z} \leftarrow \mathcal{D}_{\mathbb{Z}^m,s'}\}$$

Then, we prove that $\mathcal{E}_{X,s'}$ is close to a discrete Gaussian over L of moderate "width". Specifically, we show that for large enough s', with overwhelming probability over the choice of X:

1. $\mathcal{E}_{X,s'}$ is statistically close to the ellipsoid Gaussian $\mathcal{D}_{L,s'X^\top}$, over L.
2. The singular values of the matrix X are of size roughly $s\sqrt{m}$, hence the shape of $\mathcal{D}_{L,s'X^\top}$ is "roughly spherical". Moreover, the "width" of $\mathcal{D}_{L,s'X^\top}$ is roughly $s's\sqrt{m} = \text{poly}(n)$.

[2] As noted by Peikert [13], one can generate an ellipsoidal Gaussian distribution over the lattice given a basis B by just outputting $\boldsymbol{y} \leftarrow B \cdot \boldsymbol{z}$ where $\boldsymbol{z}$ is a discrete Gaussian, but this ellipsoidal Gaussian distribution would typically be very skewed.

We emphasize that it is straightforward to show that the covariance matrix of $\mathcal{E}_{X,s'}$ is exactly $s'^2 XX^\top$. However, the technical challenge lies in showing that $\mathcal{E}_{X,s'}$ is close to a discrete Gaussian for a non-square X. Also note that for a square X, the shape of the covariance matrix $XX^\top$ will typically be very "skewed" (i.e., the least singular value of $X^\top$ is typically much smaller than the largest singular value). We note that the "approximately spherical" nature of the output distribution is important for performance reasons in applications such as GGH: These applications must choose parameters so that the least singular value of X "drowns out" vectors of a certain size, and the resulting vectors that they draw from $\mathcal{E}_{X,s'}$ grow in size with the largest singular value of X, hence it is important that these two values be as close as possible.

Our Techniques

Our main result can be argued along the following broad outline. Our first theorem (Theorem 2) says that the distribution of $X \cdot \boldsymbol{z} \leftarrow \mathcal{E}_{X,s'}$ is indeed statistically close to a discrete Gaussian over L, as long as s' exceeds the smoothing parameter of a certain "orthogonal lattice" related to X (denoted A). Next, Theorem 3 clarifies that A will have a small smoothing parameter as long as $X^\top$ is "regularly shaped" in a certain sense. Finally, we argue in Lemma 8 that when the columns of X are chosen from a discrete Gaussian, $\boldsymbol{x}_i \leftarrow \mathcal{D}_{L,S}$, then $X^\top$ is "regularly shaped," i.e. has singular values all close to $\sigma_n(S)\sqrt{m}$.

The analysis of the smoothing parameter of the "orthogonal lattice" A is particularly challenging and requires careful analysis of a certain "dual lattice" related to A. Specifically, we proceed by first embedding A into a full rank lattice A_q and then move to study M_q – the (scaled) dual of A_q. Here we obtain a lower bound on $\lambda_{n+1}(M_q)$, i.e. the $n+1^{th}$ minima of M_q. Next, we use a theorem by Banasczcyk to convert the lower bound on $\lambda_{n+1}(M_q)$ to an upper bound on $\lambda_{m-n}(A_q)$, obtaining $m-n$ linearly independent, bounded vectors in A_q. We argue that these vectors belong to A, thus obtaining an upper bound on $\lambda_{m-n}(A)$. Relating $\lambda_{m-n}(A)$ to $\eta_\epsilon(A)$ using a lemma by Micciancio and Regev completes the analysis. (We note that probabilistic bounds on the minima and smoothing parameter A_q, M_q are well known in the case when the entries of matrix X are uniformly random mod q (e.g. [6]), but here we obtain bounds in the case when X has Gaussian entries significantly smaller than q.)

To argue that $X^\top$ is regularly shaped, we begin with the literature of random matrices which establishes that for a matrix $H \in \mathbb{R}^{m\times n}$, where each entry of H is distributed as $\mathcal{N}(0, s^2)$ and m is sufficiently greater than n, the singular values of H are all of size roughly $s\sqrt{m}$. We extend this result to discrete Gaussians – showing that as long as each vector $\boldsymbol{x}_i \leftarrow \mathcal{D}_{L,S}$ where S is "not too small" and "not too skewed", then with high probability the singular values of $X^\top$ are all of size roughly $s\sqrt{m}$.

Related Work

Properties of linear combinations of discrete Gaussians have been studied before in some cases by Peikert [13] as well as more recently by Boneh and Freeman [3]. Peikert's "convolution lemma" (Theorem 3.1 in [13]) analyzes certain cases in which a linear combination of discrete Gaussians yields a discrete Gaussian, in the one dimensional case. More recently, Boneh and Freeman [3] observed that under certain conditions, a linear combination of discrete Gaussians over a lattice is also a discrete Gaussian. However, the deviation of the Gaussian needed to achieve this are quite large. Related questions were considered by Lyubashevsky [9] where he computes the expectation of the inner product of discrete Gaussians.

Discrete Gaussian samplers have been studied by [6] (who use an algorithm by [7]) and [13]. These works describe a discrete Gaussian sampling algorithm that takes as input a 'high quality' basis B for an n dimensional lattice L and output a sample from $\mathcal{D}_{L,s,\boldsymbol{c}}$. In [6], $s \geq \|\tilde{B}\| \cdot \omega(\sqrt{\log n})$, and $\tilde{B} = \max_i \|\tilde{b}_i\|$ is the Gram Schmidt orthogonalization of B. In contrast, the algorithm of [13] requires $s \geq \sigma_1(B)$, i.e. the largest singular value of B, but is fully parallelizable. Both these samplers take as input an explicit description of a "high quality basis" of the relevant lattice, and the quality of their output distribution is related to the quality of the input basis.

Peikert's sampler [13] is elegant and its complexity is difficult to beat: the only online computation is to compute $\boldsymbol{c} - B_1\lfloor B_1^{-1}(\boldsymbol{c} - \boldsymbol{x}_2)\rceil$, where $\boldsymbol{c}$ is the center of the Gaussian, B_1 is the sampler's basis for its lattice L, and $\boldsymbol{x}_2$ is a vector that is generated in an offline phase (freshly for each sampling) in a way designed to "cancel" the covariance of B_1 so as to induce a purely spherical Gaussian. However, since our sampler just directly takes an integer linear combination of lattice vectors, and does not require extra precision for handling the inverse B_1^{-1}, it might outperform Peikert's in some situations, at least when $\boldsymbol{c} = 0$.

2 Preliminaries

We say that a function $f : \mathbb{R}^+ \to \mathbb{R}^+$ is negligible (and write $f(\lambda) < \mathsf{negl}(\lambda)$) if for every d we have $f(\lambda) < 1/\lambda^d$ for sufficiently large λ. For two distributions $\mathcal{D}_1$ and $\mathcal{D}_2$ over some set Ω the statistical distance $\mathrm{SD}(\mathcal{D}_1, \mathcal{D}_2)$ is

$$\mathrm{SD}(\mathcal{D}_1, \mathcal{D}_2) \stackrel{\text{def}}{=} \frac{1}{2} \sum_{x \in \Omega} \left| \Pr_{\mathcal{D}_1}[x] - \Pr_{\mathcal{D}_2}[x] \right|$$

Two distribution ensembles $\mathcal{D}_1(\lambda)$ and $\mathcal{D}_2(\lambda)$ are statistically close or statistically indistinguishable if $\mathrm{SD}\big(\mathcal{D}_1(\lambda), \mathcal{D}_2(\lambda)\big)$ is a negligible function of λ.

2.1 Gaussian Distributions

For any real $s > 0$ and vector $\boldsymbol{c} \in \mathbb{R}^n$, define the (spherical) Gaussian function on $\mathbb{R}^n$ centered at $\boldsymbol{c}$ with parameter s as $\rho_{s,\boldsymbol{c}}(\boldsymbol{x}) = \exp(-\pi\|\boldsymbol{x} - \boldsymbol{c}\|^2/s^2)$ for all $\boldsymbol{x} \in \mathbb{R}^n$. The *normal distribution* with mean μ and deviation σ, denoted $\mathcal{N}(\mu, \sigma^2)$, assigns to each real number $x \in \mathbb{R}$ the probability density $f(x) = \frac{1}{\sigma\sqrt{2\pi}} \cdot \rho_{\sigma\sqrt{2\pi},\mu}(x)$. The n-dimensional (spherical) continuous Gaussian distribution with center $\boldsymbol{c}$ and uniform deviation σ^2, denoted $\mathcal{N}^n(\boldsymbol{c}, \sigma^2)$, just chooses each entry of a dimension-n vector independently from $\mathcal{N}(c_i, \sigma^2)$.

The n-dimensional spherical Gaussian function generalizes naturally to ellipsoid Gaussians, where the different coordinates are jointly Gaussian but are neither identical nor independent. In this case we replace the single variance parameter $s^2 \in \mathbb{R}$ by the covariance matrix $\Sigma \in \mathbb{R}^{n \times n}$ (which must be positive-definite and symmetric). To maintain consistency of notations between the spherical and ellipsoid cases, below we let S be a matrix such that $S^\top \times S = \Sigma$. Such a matrix S always exists for a symmetric Σ, but it is not unique. (In fact there exist such S'es that are not even n-by-n matrices, below we often work with such rectangular S'es.)

For a rank-n matrix $S \in \mathbb{R}^{m \times n}$ and a vector $\mathbf{c} \in \mathbb{R}^n$, the ellipsoid Gaussian function on $\mathbb{R}^n$ centered at $\mathbf{c}$ with parameter S is defined by

$$\rho_{S,\mathbf{c}}(\boldsymbol{x}) = \exp\left(-\pi(\boldsymbol{x}-\mathbf{c})^\top (S^\top S)^{-1}(\boldsymbol{x}-\mathbf{c})\right) \ \forall \boldsymbol{x} \in \mathbb{R}^n.$$

Obviously this function only depends on $\Sigma = S^\top S$ and not on the particular choice of S. It is also clear that the spherical case can be obtained by setting $S = sI_n$, with I_n the n-by-n identity matrix. Below we use the shorthand $\rho_s(\cdot)$ (or $\rho_S(\cdot)$) when the center of the distribution is $\mathbf{0}$.

2.2 Matrices and Singular Values

In this note we often use properties of rectangular (non-square) matrices. For $m \geq n$ and a rank-n matrix[3] $X' \in \mathbb{R}^{m \times n}$, the pseudoinverse of X' is the (unique) m-by-n matrix Y' such that $X'^\top Y' = Y'^\top X' = I_n$ and the columns of Y' span the same linear space as those of X'. It is easy to see that Y' can be expressed as $Y' = X'(X'^\top X')^{-1}$ (note that $X'^\top X'$ is invertible since X' has rank n).

For a rank-n matrix $X' \in \mathbb{R}^{m \times n}$, denote $U_{X'} = \{\|X'\boldsymbol{u}\| : \boldsymbol{u} \in \mathbb{R}^n, \|\boldsymbol{u}\| = 1\}$. The *least singular value* of X' is then defined as $\sigma_n(X') = \inf(U'_X)$ and similarly the *largest singular value* of X' is $\sigma_1(X') = \sup(U'_X)$. Some properties of singular values that we use later in the text are stated in Fact 1.

Fact 1. *For rank-n matrices $X', Y' \in \mathbb{R}^{m \times n}$ with $m \geq n$, the following holds:*

1. *If $X'^\top X' = Y'^\top Y'$ then X', Y' have the same singular values.*
2. *If Y' is the (pseudo)inverse of X' then the singular values of X', Y' are reciprocals.*
3. *If X' is a square matrix (i.e., $m = n$) then X', $X'^\top$ have the same singular values.*
4. *If $\sigma_1(Y') \leq \delta\sigma_n(X')$ for some constant $\delta < 1$, then $\sigma_1(X'+Y') \in [1-\delta, 1+\delta]\sigma_1(X')$ and $\sigma_n(X'+Y') \in [1-\delta, 1+\delta]\sigma_n(X')$.* □

It is well known that when m is sufficiently larger than n, then the singular values of a "random matrix" $X' \in \mathbb{R}^{m \times n}$ are all of size roughly $\sqrt{m}$. For example, Lemma 1 below is a special case of [8, Thm 3.1], and Lemma 2 can be proved along the same lines of (but much simpler than) the proof of [17, Corollary 2.3.5].

[3] We use the notation X' instead of X to avoid confusion later in the text where we will instantiate $X' = X^\top$.

Lemma 1. *There exists a universal constant $C > 1$ such that for any $m > 2n$, if the entries of $X' \in \mathbb{R}^{m\times n}$ are drawn independently from $\mathcal{N}(0,1)$ then* $\Pr[\sigma_n(X') < \sqrt{m}/C] < \exp(-O(m))$. □

Lemma 2. *There exists a universal constant $C > 1$ such that for any $m > 2n$, if the entries of $X' \in \mathbb{R}^{m\times n}$ are drawn independently from $\mathcal{N}(0,1)$ then* $\Pr[\sigma_1(X') > C\sqrt{m}] < \exp(-O(m))$. □

Corollary 1. *There exists a universal constant $C > 1$ such that for any $m > 2n$ and $s > 0$, if the entries of $X' \in \mathbb{R}^{m\times n}$ are drawn independently from $\mathcal{N}(0,s^2)$ then*

$$\Pr\left[s\sqrt{m}/C \;<\; \sigma_n(X') \;\le\; \sigma_1(X') \;<\; sC\sqrt{m}\right] > 1 - \exp(-O(m)). \quad \square$$

Remark. The literature on random matrices is mostly focused on analyzing the "hard cases" of more general distributions and m which is very close to n (e.g., $m = (1+o(1))n$ or even $m = n$). For our purposes, however, we only need the "easy case" where all the distributions are Gaussian and $m \gg n$ (e.g., $m = n^2$), in which case all the proofs are much easier (and the universal constant from Corollary 1 gets closer to one).

2.3 Lattices and Their Dual

A lattice $L \subset \mathbb{R}^n$ is an additive discrete sub-group of $\mathbb{R}^n$. We denote by span(L) the linear subspace of $\mathbb{R}^n$, spanned by the points in L. The rank of $L \subset \mathbb{R}^n$ is the dimension of span(L), and we say that L has full rank if its rank is n. In this work we often consider lattices of less than full rank.

Every (nontrivial) lattice has bases: a basis for a rank-k lattice L is a set of k linearly independent points $\boldsymbol{b}_1, \ldots, \boldsymbol{b}_k \in L$ such that $L = \{\sum_{i=1}^k z_i \boldsymbol{b}_i : z_i \in \mathbb{Z}\, \forall i\}$. If we arrange the vectors $\boldsymbol{b}_i$ as the columns of a matrix $B \in \mathbb{R}^{n\times k}$ then we can write $L = \{B\boldsymbol{z} : \boldsymbol{z} \in \mathbb{Z}^k\}$. If B is a basis for L then we say that B spans L.

Definition 1 (Dual of a Lattice). *For a lattice $L \subset \mathbb{R}^n$, its* dual lattice *consists of all the points in* span(L) *that are orthogonal to L modulo one, namely:*

$$L^* = \{\boldsymbol{y} \in \mathrm{span}(L) : \forall \boldsymbol{x} \in L, \langle \boldsymbol{x}, \boldsymbol{y}\rangle \in \mathbb{Z}\}$$

Clearly, if L is spanned by the columns of some rank-k matrix $X \in \mathbb{R}^{n\times k}$ then L^* is spanned by the columns of the pseudoinverse of X. It follows from the definition that for two lattices $L \subseteq M$ we have $M^* \cap \mathrm{span}(L) \subseteq L^*$.

Banasczcyk provided strong transference theorems that relate the size of short vectors in L to the size of short vectors in L^*. Recall that $\lambda_i(L)$ denotes the i-th minimum of L (i.e., the smallest s such that L contains i linearly independent vectors of size at most s).

Theorem 1 (Banasczcyk [2]). *For any rank-n lattice $L \subset \mathbb{R}^m$, and for all $i \in [n]$,*

$$1 \le \lambda_i(L) \cdot \lambda_{n-i+1}(L^*) \le n.$$

2.4 Gaussian Distributions over Lattices

The *ellipsoid discrete Gaussian distribution* over lattice L with parameter S, centered around $\mathbf{c}$, is

$$\forall\, \boldsymbol{x} \in L, \mathcal{D}_{L,S,\mathbf{c}}(\boldsymbol{x}) = \frac{\rho_{S,\mathbf{c}}(\boldsymbol{x})}{\rho_{S,\mathbf{c}}(L)},$$

where $\rho_{S,\mathbf{c}}(A)$ for set A denotes $\sum_{\boldsymbol{x}\in A} \rho_{S,\mathbf{c}}(\boldsymbol{x})$. In other words, the probability $\mathcal{D}_{L,S,\mathbf{c}}(\boldsymbol{x})$ is simply proportional to $\rho_{S,\mathbf{c}}(\boldsymbol{x})$, the denominator being a normalization factor. The same definitions apply to the spherical case, which is denoted by $\mathcal{D}_{L,s,\mathbf{c}}(\cdot)$ (with lowercase s). As before, when $\mathbf{c} = \mathbf{0}$ we use the shorthand $\mathcal{D}_{L,S}$ (or $\mathcal{D}_{L,s}$). The following useful fact that follows directly from the definition, relates the ellipsoid Gaussian distributions over different lattices:

Fact 2. *Let $L \subset \mathbb{R}^n$ be a full-rank lattice, $\mathbf{c} \in R^n$ a vector, and $S \in \mathbb{R}^{m\times n}$, $B \in \mathbb{R}^{n\times n}$ two rank-n matrices, and denote $L' = \{B^{-1}\boldsymbol{v} : \boldsymbol{v} \in L\}$, $\mathbf{c}' = B^{-1}\mathbf{c}$, and $S' = S\times(B^\top)^{-1}$. Then the distribution $\mathcal{D}_{L,S,\mathbf{c}}$ is identical to the distribution induced by drawing a vector $\boldsymbol{v} \leftarrow \mathcal{D}_{L',S',\mathbf{c}'}$ and outputting $\boldsymbol{u} = B\boldsymbol{v}$.* □

A useful special case of Fact 2 is when L' is the integer lattice, $L' = \mathbb{Z}^n$, in which case L is just the lattice spanned by the basis B. In other words, the ellipsoid Gaussian distribution on $L(B)$, $\boldsymbol{v} \leftarrow \mathcal{D}_{L(B),S,\mathbf{c}}$, is induced by drawing an integer vector according to $\boldsymbol{z} \leftarrow \mathcal{D}_{\mathbb{Z}^n,S',\mathbf{c}'}$ and outputting $\boldsymbol{v} = B\boldsymbol{z}$, where $S' = S(B^{-1})^\top$ and $\mathbf{c}' = B^{-1}\mathbf{c}$.

Another useful special case is where $S = sB^\top$, so S is a square matrix and $S' = sI_n$. In this case the ellipsoid Gaussian distribution $\boldsymbol{v} \leftarrow \mathcal{D}_{L,S,\mathbf{c}}$ is induced by drawing a vector according to the *spherical Gaussian* $\boldsymbol{u} \leftarrow \mathcal{D}_{L',s,\mathbf{c}'}$ and outputting $\boldsymbol{v} = \frac{1}{s}S^\top\boldsymbol{u}$, where $\mathbf{c}' = s(S^\top)^{-1}\mathbf{c}$ and $L' = \{s(S^\top)^{-1}\boldsymbol{v} : \boldsymbol{v} \in L\}$.

Smoothing parameter. As in [10], for lattice L and real $\epsilon > 0$, the *smoothing parameter* of L, denoted $\eta_\epsilon(L)$, is defined as the smallest s such that $\rho_{1/s}(L^* \setminus \{\mathbf{0}\}) \le \epsilon$. Intuitively, for a small enough ϵ, the number $\eta_\epsilon(L)$ is sufficiently larger than L's fundamental parallelepiped so that sampling from the corresponding Gaussian "wipes out the internal structure" of L. Thus, the sparser the lattice, the larger its smoothing parameter.

It is well known that for a spherical Gaussian with parameter $s > \eta_\epsilon(L)$, the size of vectors drawn from $\mathcal{D}_{L,s}$ is bounded by $s\sqrt{n}$ whp (cf. [10, Lemma 4.4], [12, Corollary 5.3]). The following lemma (that follows easily from the spherical case and Fact 2) is a generalization to ellipsoid Gaussians.

Lemma 3. *For a rank-n lattice L, vector $\mathbf{c} \in \mathbb{R}^n$, constant $0 < \epsilon < 1$ and matrix S s.t. $\sigma_n(S) \ge \eta_\epsilon(L)$, we have that for $\boldsymbol{v} \leftarrow \mathcal{D}_{L,S,\mathbf{c}}$,*

$$\Pr_{\boldsymbol{v}\leftarrow\mathcal{D}_{L,S,\mathbf{c}}}\left(\|\boldsymbol{v} - \mathbf{c}\| \ge \sigma_1(S)\sqrt{n}\right) \le \frac{1+\epsilon}{1-\epsilon}\cdot 2^{-n}.$$

Moreover, for every $\boldsymbol{z} \in \mathbb{R}^n$ $r > 0$ *it holds that*

$$\Pr_{\boldsymbol{v} \leftarrow \mathcal{D}_{L,S,\boldsymbol{c}}} (|\langle \boldsymbol{v} - \boldsymbol{c}, \boldsymbol{z} \rangle| \geq r\sigma_1(S)\|\boldsymbol{z}\|) \leq 2en \cdot \exp(-\pi r^2).$$

The proof can be found in the long version [1].

The next lemma says that the Gaussian distribution with parameter $s \geq \eta_\epsilon(L)$ is so smooth and "spread out" that it covers the approximately the same number of L-points regardless of where the Gaussian is centered. This is again well known for spherical distributions (cf. [6, Lemma 2.7]) and the generalization to ellipsoid distributions is immediate using Fact 2.

Lemma 4. *For any rank-n lattice* L, *real* $\epsilon \in (0,1)$, *vector* $\boldsymbol{c} \in \mathbb{R}^n$, *and rank-n matrix* $S \in \mathbb{R}^{m \times n}$ *such that* $\sigma_n(S) \geq \eta_\epsilon(L)$, *we have* $\rho_{S,\boldsymbol{c}}(L) \in [\frac{1-\epsilon}{1+\epsilon}, 1] \cdot \rho_S(L)$. □

Regev also proved that drawing a point from L according to a spherical discrete Gaussian and adding to it a spherical continuous Gaussian, yields a probability distribution close to a continuous Gaussian (independent of the lattice), provided that both distributions have parameters sufficiently larger than the smoothing parameter of L.

Lemma 5 (Claim 3.9 of [14]). *Fix any n-dimensional lattice* $L \subset \mathbb{R}^n$, *real* $\epsilon \in (0, 1/2)$, *and two reals* s, r *such that* $\frac{rs}{\sqrt{r^2+s^2}} \geq \eta_\epsilon(L)$, *and denote* $t = \sqrt{r^2 + s^2}$.

Let $\mathcal{R}_{L,r,s}$ *be a distribution induced by choosing* $\boldsymbol{x} \leftarrow \mathcal{D}_{L,s}$ *from the spherical discrete Gaussian on* L *and* $\boldsymbol{y} \leftarrow \mathcal{N}^n(0, r^2/2\pi)$ *from a continuous Gaussian, and outputting* $\boldsymbol{z} = \boldsymbol{x} + \boldsymbol{y}$. *Then for any point* $\boldsymbol{u} \in \mathbb{R}^n$, *the probability density* $\mathcal{R}_{L,r,s}(\boldsymbol{u})$ *is close to the probability density under the spherical continuous Gaussian* $\mathcal{N}^n(0, t^2/2\pi)$ *upto a factor of* $\frac{1-\epsilon}{1+\epsilon}$:

$$\tfrac{1-\epsilon}{1+\epsilon}\mathcal{N}^n(0, t^2/2\pi)(\boldsymbol{u}) \;\leq\; \mathcal{R}_{L,r,s}(\boldsymbol{u}) \;\leq\; \tfrac{1+\epsilon}{1-\epsilon}\mathcal{N}^n(0, t^2/2\pi)(\boldsymbol{u})$$

In particular, the statistical distance between $\mathcal{R}_{L,r,s}$ *and* $\mathcal{N}^n(0, t^2/2\pi)$ *is at most* 4ϵ.

More broadly, Lemma 5 implies that for any event $E(\boldsymbol{u})$, we have

$$\Pr_{\boldsymbol{u} \leftarrow \mathcal{N}(0,t^2/2\pi)}[E(\boldsymbol{u})] \cdot \tfrac{1-\epsilon}{1+\epsilon} \;\leq\; \Pr_{\boldsymbol{u} \leftarrow \mathcal{R}_{L,r,s}}[E(\boldsymbol{u})] \;\leq\; \Pr_{\boldsymbol{u} \leftarrow \mathcal{N}(0,t^2/2\pi)}[E(\boldsymbol{u})] \cdot \tfrac{1+\epsilon}{1-\epsilon}$$

Another useful property of "wide" discrete Gaussian distributions is that they do not change much by short shifts. Specifically, if we have an arbitrary subset of the lattice, $T \subseteq L$, and an arbitrary *short vector* $\boldsymbol{v} \in L$, then the probability mass of T is not very different than the probability mass of $T - \boldsymbol{v} = \{\boldsymbol{u} - \boldsymbol{v} : \boldsymbol{u} \in T\}$. Below let $\mathsf{erf}(\cdot)$ denote the Gauss error function.

Lemma 6. *Fix a lattice* $L \subset \mathbb{R}^n$, *a positive real* $\epsilon > 0$, *and two parameters* s, c *such that* $c > 2$ *and* $s \geq (1 + c)\eta_\epsilon(L)$. *Then for any subset* $T \subset L$ *and any additional vector* $\boldsymbol{v} \in L$, *it holds that* $\mathcal{D}_{L,s}(T) - \mathcal{D}_{L,s}(T - \boldsymbol{v}) \leq \frac{\mathsf{erf}(q(1+4/c)/2)}{\mathsf{erf}(2q)} \cdot \frac{1+\epsilon}{1-\epsilon}$, *where* $q = \|v\| \sqrt{\pi}/s$.

We provide the proof in A.1.

One useful special case of Lemma 6 is when $c = 100$ (say) and $\|\boldsymbol{v}\| \approx s$, where we get a bound $\mathcal{D}_{L,s}(T) - \mathcal{D}_{L,s}(T - \boldsymbol{v}) \leq \frac{\mathsf{erf}(0.52\sqrt{\pi})}{\mathsf{erf}(2\sqrt{\pi})} \cdot \frac{1+\epsilon}{1-\epsilon} \approx 0.81$. We note that when $\frac{\|\boldsymbol{v}\|}{s} \to 0$, the bound from Lemma 6 tends to (just over) 1/4, but we note that we can make it tend to zero with a different choice of parameters in the proof (namely making $H'_{\boldsymbol{v}}$ and $H''_{\boldsymbol{v}}$ thicker, e.g. $H''_{\boldsymbol{v}} = H_{\boldsymbol{v}}$ and $H'_{\boldsymbol{v}} = 2H_{\boldsymbol{v}}$). Lemma 6 extends easily also to the ellipsoid Gaussian case, using Fact 2:

Corollary 2. *Fix a lattice $L \subset \mathbb{R}^n$, a positive real $\epsilon > 0$, a parameter $c > 2$ and a rank-n matrix S such that $s \stackrel{\text{def}}{=} \sigma_n(S) \geq (1+c)\eta_\epsilon(L)$. Then for any subset $T \subset L$ and any additional vector $\boldsymbol{v} \in L$, it holds that $\mathcal{D}_{L,S}(T) - \mathcal{D}_{L,S}(T - \boldsymbol{v}) \leq \frac{\mathsf{erf}(q(1+4/c)/2)}{\mathsf{erf}(2q)} \cdot \frac{1+\epsilon}{1-\epsilon}$, where $q = \|v\|\sqrt{\pi}/s$.*

Micciancio and Regev give the following bound on the smoothing parameter in terms of the primal lattice.

Lemma 7. *[Lemma 3.3 of [10]] For any n-dimensional lattice L and positive real $\epsilon > 0$,*

$$\eta_\epsilon(L) \leq \lambda_n(L) \cdot \sqrt{\frac{\ln(2n(1+1/\epsilon))}{\pi}}.$$

In particular, for any superlogarithmic function $\omega(\log n)$, there exists a negligible function $\epsilon(n)$ such that $\eta_\epsilon(L) \leq \sqrt{\omega(\log n)} \cdot \lambda_n(L)$.

3 Our Discrete Gaussian LHL

Consider a full rank lattice $L \subseteq \mathbb{Z}^n$, some negligible $\epsilon = \epsilon(n)$, the corresponding smoothing parameter $\eta = \eta_\epsilon(L)$ and parameters $s > \Omega(\eta)$, $m > \Omega(n \log n)$, and $s' > \Omega(\text{poly}(n) \log(1/\epsilon))$. The process that we analyze begins by choosing "once and for all" m points in L, drawn independently from a discrete Gaussian with parameter s, $\boldsymbol{x}_i \leftarrow \mathcal{D}_{L,s}$.[4]

Once the $\boldsymbol{x}_i$'s are fixed, we arrange them as the columns of an n-by-m matrix $X = (\boldsymbol{x}_1|\boldsymbol{x}_2|\ldots|\boldsymbol{x}_m)$, and consider the distribution $\mathcal{E}_{X,s'}$, induced by choosing an integer vector $\boldsymbol{v}$ from a discrete spherical Gaussian with parameter s' and outputting $\boldsymbol{y} = X \cdot \boldsymbol{v}$:

$$\mathcal{E}_{X,s'} \stackrel{\text{def}}{=} \{X \cdot \boldsymbol{v} : \boldsymbol{v} \leftarrow \mathcal{D}_{\mathbb{Z}^m,s'}\}. \tag{1}$$

Our goal is to prove that $\mathcal{E}_{X,s'}$ is close to the ellipsoid Gaussian $\mathcal{D}_{L,s'X^\top}$, over L. We begin by proving that the singular values of $X^\top$ are all roughly of the size $s\sqrt{m}$[5].

[4] More generally, we can consider drawing the vectors $\boldsymbol{x}_i$ from an ellipsoid discrete Gaussian, $\boldsymbol{x}_i \leftarrow \mathcal{D}_{L,S}$, so long as the least singular value of S is at least s.

[5] Since we eventually apply the following lemmas to $X^\top$, we will use $X^\top$ in the statement of the lemmas for consistency at the risk of notational clumsiness.

Lemma 8. *There exists a universal constant $K > 1$ such that for all $m \geq 2n$, $\epsilon > 0$ and every n-dimensional real lattice $L \subset \mathbb{R}^n$, the following holds: choosing the rows of an m-by-n matrix $X^\top$ independently at random from a spherical discrete Gaussian on L with parameter $s > 2K\eta_\epsilon(L)$, $X^\top \leftarrow (\mathcal{D}_{L,s})^m$, we have*

$$\Pr\left[s\sqrt{2\pi m}/K < \sigma_n(X^\top) \leq \sigma_1(X^\top) < sK\sqrt{2\pi m}\right] > 1-(4m\epsilon+O(\exp(-m/K))).$$

The proof can be found in the long version [1].

3.1 The Distribution $\mathcal{E}_{X,s'}$ Over $\mathbb{Z}^n$

We next move to show that with high probability over the choice of X, the distribution $\mathcal{E}_{X,s'}$ is statistically close to the ellipsoid discrete Gaussian $\mathcal{D}_{L,s'X^\top}$. We first prove this for the special case of the integer lattice, $L = \mathbb{Z}^n$, and then use that special case to prove the same statement for general lattices. In either case, we analyze the setting where the columns of X are chosen from an ellipsoid Gaussian which is "not too small" and "not too skewed."

Parameters. Below n is the security parameters and $\epsilon =$ negligible(n). Let S be an n-by-n matrix such that $\sigma_n(S) \geq 2K\eta_\epsilon(\mathbb{Z}^n)$, and denote $s_1 = \sigma_1(S)$, $s_n = \sigma_n(S)$, and $w = s_1/s_n$. (We consider w to be a measure for the "skewness" of S.) Also let m, q, s' be parameters satisfying $m \geq 10n \log q$, $q > 8m^{5/2}n^{1/2}s_1 w$, and $s' \geq 4wm^{3/2}n^{1/2}\ln(1/\epsilon)$. An example setting of parameters to keep in mind is $m = n^2$, $s_n = \sqrt{n}$ (which implies $\epsilon \approx 2^{-\sqrt{n}}$), $s_1 = n$ (so $w = \sqrt{n}$), $q = 8n^7$, and $s' = n^5$.

Theorem 2. *For ϵ negligible in n, let $S \in \mathbb{R}^{n\times n}$ be a matrix such that $s_n = \sigma_n(S) \geq 18K\eta_\epsilon(\mathbb{Z}^n)$, and denote $s_1 = \sigma_1(S)$ and $w = s_1/s_n$. Also let m, s' be parameters such that $m \geq 10n\log(8m^{5/2}n^{1/2}s_1 w)$ and $s' \geq 4wm^{3/2}n^{1/2}\ln(1/\epsilon)$.*

Then, when choosing the columns of an n-by-m matrix X from the ellipsoid Gaussian over $\mathbb{Z}^n$, $X \leftarrow (\mathcal{D}_{\mathbb{Z}^n,S})^m$, we have with all but probability $2^{-O(m)}$ over the choice of X, that the statistical distance between $\mathcal{E}_{X,s'}$ and the ellipsoid Gaussian $\mathcal{D}_{\mathbb{Z}^n,s'X^\top}$ is bounded by 2ϵ.

The rest of this subsection is devoted to proving Theorem 2. We begin by showing that with overwhelming probability, the columns of X span all of $\mathbb{Z}^n$, which means also that the support of $\mathcal{E}_{X,s'}$ includes all of $\mathbb{Z}^n$.

Lemma 9. *With parameters as above, when drawing the columns of an n-by-m matrix X independently at random from $\mathcal{D}_{\mathbb{Z}^n,S}$ we get $X \cdot \mathbb{Z}^m = \mathbb{Z}^n$ with all but probability $2^{-O(m)}$.*

The proof can be found in the long version [1].

From now on we assume that the columns of X indeed span all of $\mathbb{Z}^n$. Now let $A = A(X)$ be the $(m-n)$-dimensional lattice in $\mathbb{Z}^m$ orthogonal to all the rows of X, and for any $\boldsymbol{z} \in \mathbb{Z}^n$ we denote by $A_{\boldsymbol{z}} = A_{\boldsymbol{z}}(X)$ the $\boldsymbol{z}$ coset of A:

$$A = A(X) \stackrel{\text{def}}{=} \{\boldsymbol{v} \in \mathbb{Z}^m : X\cdot\boldsymbol{v} = \boldsymbol{0}\} \text{ and } A_{\boldsymbol{z}} = A_{\boldsymbol{z}}(X) \stackrel{\text{def}}{=} \{\boldsymbol{v} \in \mathbb{Z}^m : X\cdot\boldsymbol{v} = \boldsymbol{z}\}.$$

Since the columns of X span all of $\mathbb{Z}^n$ then $A_{\boldsymbol{z}}$ is nonempty for every $\boldsymbol{z} \in \mathbb{Z}^n$, and we have $A_{\boldsymbol{z}} = \boldsymbol{v}_{\boldsymbol{z}} + A$ for any arbitrary point $\boldsymbol{v}_{\boldsymbol{z}} \in A_{\boldsymbol{z}}$.

Below we prove that the smoothing parameter of A is small (whp), and use that to bound the distance between $\mathcal{E}_{X,s'}$ and $\mathcal{D}_{\mathbb{Z}^n,s'X^\top}$. First we show that if the smoothing parameter of A is indeed small (i.e., smaller than the parameter s' used to sample the coefficient vector $\boldsymbol{v}$), then $\mathcal{E}_{X,s'}$ and $\mathcal{D}_{\mathbb{Z}^n,s'X^\top}$ must be close.

Lemma 10. *Fix X and $A = A(X)$ as above. If $s' \geq \eta_\epsilon(A)$, then for any point $\boldsymbol{z} \in \mathbb{Z}^n$, the probability mass assigned to $\boldsymbol{z}$ by $\mathcal{E}_{X,s'}$ differs from that assigned by $\mathcal{D}_{\mathbb{Z}^n,s'X^\top}$ by at most a factor of $(1-\epsilon)/(1+\epsilon)$, namely*

$$\mathcal{E}_{X,s'}(\boldsymbol{z}) \in \left[\tfrac{1-\epsilon}{1+\epsilon}, 1\right] \cdot \mathcal{D}_{\mathbb{Z}^n,s'X^\top}(\boldsymbol{z}).$$

In particular, if $\epsilon < 1/3$ then the statistical distance between $\mathcal{E}_{X,s'}$ and $\mathcal{D}_{\mathbb{Z}^n,s'X}$ is at most 2ϵ.

The proof can be found in Appendix A.2.

The Smoothing Parameter of A. We now turn our attention to proving that A is "smooth enough". Specifically, for the parameters above we prove that with high probability over the choice of X, the smoothing parameter $\eta_\epsilon(A)$ is bounded below $s' = 4wm^{3/2}n^{1/2}\ln(1/\epsilon)$.

Recall again that $A = A(X)$ is the rank-$(m-n)$ lattice containing all the integer vectors in $\mathbb{Z}^m$ orthogonal to the rows of X. We extend A to a full-rank lattice as follows: First we extend the rows space of X, by throwing in also the scaled standard unit vectors $q\boldsymbol{e}_i$ for the integer parameter q mentioned above ($q \geq 8m^{5/2}n^{1/2}s_1 w$). That is, we let $M_q = M_q(X)$ be the full-rank m-dimensional lattice spanned by the rows of X and the vectors $q\boldsymbol{e}_i$,

$$M_q = \{X^\top \boldsymbol{z} + q\boldsymbol{y} : \boldsymbol{z} \in \mathbb{Z}^n, \boldsymbol{y} \in \mathbb{Z}^m\} = \{\boldsymbol{u} \in \mathbb{Z}^m : \exists \boldsymbol{z} \in \mathbb{Z}_q^n \text{ s.t. } \boldsymbol{u} \equiv X^\top \boldsymbol{z} \pmod q\}$$

(where we identity $\mathbb{Z}_q$ above with the set $[-q/2, q/2) \cap \mathbb{Z}$). Next, let A_q be the dual of M_q, scaled up by a factor of q, i.e.,

$$\begin{aligned} A_q &= qM_q^* = \{\boldsymbol{v} \in \mathbb{R}^m : \forall \boldsymbol{u} \in M_q, \langle \boldsymbol{v}, \boldsymbol{u}\rangle \in q\mathbb{Z}\} \\ &= \{\boldsymbol{v} \in \mathbb{R}^m : \forall \boldsymbol{z} \in \mathbb{Z}_q^n, \boldsymbol{y} \in \mathbb{Z}^m, \ \ \boldsymbol{z}^\top X \cdot \boldsymbol{v} + q\langle \boldsymbol{v}, \boldsymbol{y}\rangle \in q\mathbb{Z}\} \end{aligned}$$

It is easy to see that $A \subset A_q$, since any $\boldsymbol{v} \in A$ is an integer vector (so $q\langle \boldsymbol{v}, \boldsymbol{y}\rangle \in q\mathbb{Z}$ for all $\boldsymbol{y} \in \mathbb{Z}^m$) and orthogonal to the rows of X (so $\boldsymbol{z}^\top X \cdot \boldsymbol{v} = 0$ for all $\boldsymbol{z} \in \mathbb{Z}_q^n$).

Obviously all the rows of X belong to M_q, and whp they are linearly independent and relatively short (i.e., of size roughly $s_1\sqrt{m}$). In Lemma 11 below we show, however, that whp over the choice of X's, these are essentially the *only* short vectors in M_q.

Lemma 11. *Recall that we choose X as $X \leftarrow (\mathcal{D}_{\mathbb{Z}^n,S})^m$, and let $w = \sigma_1(S)/\sigma_n(S)$ be a measure of the "skewness" of S. The $n+1$'st minima of the lattice $M_q = M_q(X)$ is at least $q/(4w\sqrt{mn})$, except with negligible probability over the choice of X. Namely, $\Pr_{X \leftarrow (\mathcal{D}_{\mathbb{Z}^n,S})^m}[\lambda_{n+1}(M_q) < q/(4w\sqrt{mn})] < 2^{-O(m)}$.*

Proof. We prove that with high probability over the choice of X, every vector in M_q which is *not* in the linear span of the rows of X is of size at least $q/4nw$.

Recall that every vector in M_q is of the form $X^\top \boldsymbol{z} + q\boldsymbol{y}$ for some $\boldsymbol{z} \in \mathbb{Z}_q^n$ and $\boldsymbol{y} \in \mathbb{Z}^m$. Let us denote by $[\boldsymbol{v}]_q$ the modular reduction of all the entries in $\boldsymbol{v}$ into the interval $[-q/2, q/2)$, then clearly for every $\boldsymbol{z} \in \mathbb{Z}_q^n$

$$\|[X^\top \boldsymbol{z}]_q\| \;=\; \inf\{\|X^\top \boldsymbol{z} + q\boldsymbol{y}\| : \boldsymbol{y} \in \mathbb{Z}^m\}.$$

Moreover, for every $\boldsymbol{z} \in \mathbb{Z}_q^n, \boldsymbol{y} \in \mathbb{Z}^m$, if $X^\top \boldsymbol{z} + q\boldsymbol{y} \neq [X^\top \boldsymbol{z}]_q$ then $\|X\boldsymbol{z} + q\boldsymbol{y}\| \geq q/2$. Thus it suffices to show that every vector of the form $[X^\top \boldsymbol{z}]_q$ which is not in the linear span of the rows of X has size at least $q/4nw$ (whp over the choice of X).

Fix a particular vector $\boldsymbol{z} \in \mathbb{Z}_q^n$ (i.e. an integer vector with entries in $[-q/2, q/2)$). For this fixed vector $\boldsymbol{z}$, let $i_{\max}$ be the index of the largest entry in $\boldsymbol{z}$ (in absolute value), and let $z_{\max}$ be the value of that entry. Considering the vector $\boldsymbol{v} = [X^\top \boldsymbol{z}]_q$ for a random matrix X whose columns are drawn independently from the distribution $\mathcal{D}_{\mathbb{Z}^n,S}$, each entry of $\boldsymbol{v}$ is the inner product of the fixed vector $\boldsymbol{z}$ with a random vector $\boldsymbol{x}_i \leftarrow \mathcal{D}_{\mathbb{Z}^n,S}$, reduced modulo q into the interval $[-q/2, +q/2)$.

Denoting $s_1 = \sigma_1(S)$ and $s_n = \sigma_n(S)$, we now have two cases, either $\boldsymbol{z}$ is "small", i.e., $|z_{\max}| < q/(2s_1\sqrt{mn})$ or it is "large", $|z_{\max}| \geq q/(2s_1\sqrt{mn})$. By the "moreover" part in Lemma 3 (with $r = \sqrt{m}$), for each $\boldsymbol{x}_i$ we have $|\langle \boldsymbol{x}_i, \boldsymbol{z}\rangle\| \leq s_1\sqrt{m}\|\boldsymbol{z}\|$ except with probability bounded below 2^{-m}. If $\boldsymbol{z}$ is "small" then $\|\boldsymbol{z}\| \leq q/(2s_1\sqrt{m})$ and so we get

$$|\langle \boldsymbol{x}_i, \boldsymbol{z}\rangle| \;\leq\; \|\boldsymbol{z}\| \cdot s_1\sqrt{m} \;<\; q/2$$

except with probability $< 2^{-m}$. Hence except with probability $m2^{-m}$ all the entries of $X^\top \boldsymbol{z}$ are smaller than $q/2$ in magnitude, which means that $[X^\top \boldsymbol{z}]_q = X^\top \boldsymbol{z}$, and so $[X^\top \boldsymbol{z}]_q$ belongs to the row space of X. Using the union bound again, we get that with all but probability $q^n \cdot m2^{-m} < m2^{-9m/10}$, the vectors $[X^\top \boldsymbol{z}]_q$ for all the "small" $\boldsymbol{z}$'s belong to the row space of X.

We next turn to analyzing "large" $\boldsymbol{z}$'s. Fix one "large" vector $\boldsymbol{z}$, and for that vector define the set of "bad" vectors $\boldsymbol{x} \in \mathbb{Z}^n$, i.e. the ones for which $|[\langle \boldsymbol{z}, \boldsymbol{x}\rangle]_q| < q/4nw$ (and the other vectors $\boldsymbol{x} \in \mathbb{Z}^n$ are "good"). Observe that if $\boldsymbol{x}$ is "bad", then we can get a "good" vector by adding to it the $i_{\max}$'th standard unit vector, scaled up by a factor of $\mu = \min\left(\lceil s_n \rceil,\ \lfloor q/|2z_{\max}| \rfloor\right)$, since

$$|[\langle \boldsymbol{z}, \boldsymbol{x} + \mu \boldsymbol{e}_{i_{\max}}\rangle]_q| \;=\; |[\langle \boldsymbol{z}, \boldsymbol{x}\rangle + \mu z_{\max}]_q| \;\geq\; \mu|z_{\max}| - |[\langle \boldsymbol{z}, \boldsymbol{x}\rangle]_q| \;\geq\; q/4nw.$$

(The last two inequalities follow from $q/2nw < \mu|z_{\max}| \leq q/2$ and $|[\langle \boldsymbol{z}, \boldsymbol{x}\rangle]_q| < q/(4w\sqrt{mn})$.) Hence the injunction $\boldsymbol{x} \mapsto \boldsymbol{x} + \mu \boldsymbol{e}_{i_{\max}}$ maps "bad" $\boldsymbol{x}$'es to "good" $\boldsymbol{x}$'es. Moreover, since the $\boldsymbol{x}$'es are chosen according to the wide ellipsoid Gaussian $\mathcal{D}_{\mathbb{Z}^n,S}$ with $\sigma_n(S) = s_n \geq \eta_\epsilon(\mathbb{Z}^n)$, and since the scaled standard unit vectors are short, $\mu < s_n + 1$, then by Lemma 6 the total probability mass of the "bad" vectors $\boldsymbol{x}$ differs from the total mass of the "good" vectors $\boldsymbol{x} + \mu \boldsymbol{e}_{i_{\max}}$ by at most 0.81. It follows that when choosing $\boldsymbol{x} \leftarrow \mathcal{D}_{\mathbb{Z}^n,S}$, we have $\Pr_{\boldsymbol{x}}\left[|[\langle \boldsymbol{z}, \boldsymbol{x}\rangle]_q| < q/(4w\sqrt{mn})\right] \leq (1 + 0.81)/2 < 0.91$. Thus the probability that all the entries of $[X^\top z]_q$ are smaller than $q/(4w\sqrt{nm})$ in magnitude is bounded by $(0.91)^m = 2^{-0.14m}$. Since $m > 10n \log q$, we can use the union bound to conclude that the probability that there exists some "large" vector for which $\|[X^\top z]_q\| < q/(4w\sqrt{mn})$ is no more than $q^n \cdot 2^{-0.14m} < 2^{-O(m)}$.

Summing up the two cases, with all but probability $2^{-O(m)}$) over the choice of X, there does not exist any vector $z \in \mathbb{Z}_q^n$ for which $[X^\top z]_q$ is linearly independent of the rows of X and yet $|[X^\top z]_q| < q/(4w\sqrt{mn})$.

Corollary 3. *With the parameters as above, the smoothing parameter of $A = A(X)$ satisfies $\eta_\epsilon(A) \le s' = 4wm^{3/2}n^{1/2}\ln(1/\epsilon)$, except with probability $2^{-O(m)}$.*

The proof can be found in the long version [1].
Putting together Lemma 10 and Corollary 3 completes the proof of Theorem 2. □

3.2 The Distribution $\mathcal{E}_{X,s'}$ over General Lattices

Armed with Theorem 2, we turn to prove the same theorem also for general lattices.

Theorem 3. *Let L be a full-rank lattice $L \subset \mathbb{R}^n$ and B a matrix whose columns form a basis of L. Also let $M \in \mathbb{R}^{n\times n}$ be a full rank matrix, and denote $S = M(B^\top)^{-1}$, $s_1 = \sigma_1(S)$, $s_n = \sigma_n(S)$, and $w = s_1/s_n$. Finally let ϵ be negligible in n and m, s' be parameters such that $m \ge 10n\log(8m^{5/2}n^{1/2}s_1 w)$ and $s' \ge 4wm^{3/2}n^{1/2}\ln(1/\epsilon)$.*

If $s_n \ge \eta_\epsilon(\mathbb{Z}^n)$, then, when choosing the columns of an n-by-m matrix X from the ellipsoid Gaussian over L, $X \leftarrow (\mathcal{D}_{L,M})^m$, we have with all but probability $2^{-O(m)}$ over the choice of X, that the statistical distance between $\mathcal{E}_{X,s'}$ and the ellipsoid Gaussian $\mathcal{D}_{L,s'X^\top}$ is bounded by 2ϵ.

This theorem is an immediate corollary of Theorem 2 and Fact 2. The proof can be found in the long version [1].

4 Applications

In this section, we discuss the application of our discrete Gaussian LHL in the construction of multilinear maps from lattices [5]. This construction is illustrative of a "canonical setting" where our lemma should be useful.

Brief overview of the GGH Construction. To begin, we provide a very high level overview of the GGH construction, skipping most details. We refer the reader to [5] for a complete description. In [5], the mapping $a \to g^a$ from bilinear maps is viewed as a form of "encoding" $a \mapsto \mathit{Enc}(a)$ that satisfies some properties:

1. Encoding is easy to compute in the forward direction and hard to invert.
2. Encoding is additively homomorphic and also one-time multiplicatively homomorphic (via the pairing).

3. Given $Enc(a), Enc(b)$ it is easy to test whether $a = b$.
4. Given encodings, it is hard to test more complicated relations between the underlying scalars. For example, BDDH roughly means that given $Enc(a)$, $Enc(b), Enc(c), Enc(d)$ it is hard to test if $d = abc$.

In [5], the authors construct encodings from ideal lattices that approximately satisfy (and generalize) the above properties. Skipping most of the details, [5] roughly used a specific (NTRU-like) lattice-based homomorphic encryption scheme, where $Enc(a)$ is just an encryption of a. The ability to add and multiply then just follows from the homomorphism of the underlying cryptosystem, and GGH described how to add to this cryptosystem a "broken secret key" that cannot be used for decryption but is good enough for testing if two ciphertexts encrypt the same element. (In the terminology from [5], this broken key is called the *zero-test parameter.*)

In the specific cryptosystem used in the GGH construction, ciphertexts are elements in some polynomial ring (represented as vectors in $\mathbb{Z}^n$), and additive/multiplicative homomorphism is implemented simply by addition and multiplication in the ring. A natural way to enable encoding is to publish a single ciphertext that encrypts/encodes 1, $\boldsymbol{y}_1 = Enc(1)$. To encode any other plaintext element a, we can use the multiplicative homomorphism by setting $Enc(a) = a \cdot \boldsymbol{y}_1$ in the ring. However this simple encoding is certainly not hard to decode: just dividing by $\boldsymbol{y}_1$ in the ring suffices! For the same reason, it is also not hard to determine "complex relations" between encoding.

Randomizing the encodings. To break these simple algebraic relations, the authors include in the public parameters also "randomizers" $\boldsymbol{x}_i$ $(i = 1, \ldots, m)$, which are just random encryptions/encodings of zero, namely $\boldsymbol{x}_i \leftarrow Enc(0)$. Then to re-randomize the encoding $\boldsymbol{u}_a = a \cdot \boldsymbol{y}_1$, they add to it a "random linear combination" of the $\boldsymbol{x}_i$'s, and (by additive homomorphism) this is another encoding of the same element. This approach seems to be thwart the simple algebraic decoding from above, but what can be said about the resulting encodings? Here is where GGH use our results to analyze the probability distribution of these re-randomized encodings.

In a little more detail, an instance of the GGH encoding includes an ideal lattice L and a secret ring element $\boldsymbol{z}$, and an encoding of an element a has the form $\boldsymbol{e}_a/\boldsymbol{z}$ where $\boldsymbol{e}_a$ is a short element that belongs to the same coset of L as the "plaintext" a. The $\boldsymbol{x}_i$'s are therefore ring elements of the form $\boldsymbol{b}_i/\boldsymbol{z}$ where the $\boldsymbol{b}_i$'s are short vectors in L. Denoting by X the matrix with the $\boldsymbol{x}_i$ as columns and by B the matrix with the numerators $\boldsymbol{b}_i$ as columns, i.e., $X = (\boldsymbol{x}_1 | \ldots | \boldsymbol{x}_m)$ and $B = (\boldsymbol{b}_1 | \ldots | \boldsymbol{b}_m)$. Re-randomizing the encoding $\boldsymbol{u}_a = \boldsymbol{e}_a/\boldsymbol{z}$ is obtained by choosing a random coefficient vector $\boldsymbol{r} \leftarrow D_{\mathbb{Z}^m, \sigma^*}$ (for large enough σ^*), and setting

$$\boldsymbol{u}' := \boldsymbol{u}_a + X\boldsymbol{r} = \frac{\boldsymbol{e}_a + B\boldsymbol{r}}{\boldsymbol{z}}.$$

Since all the $\boldsymbol{b}_i$'s are in the lattice L, then obviously $\boldsymbol{e}_a + B\boldsymbol{r}$ is in the same coset of L as $\boldsymbol{e}_a$ itself. Moreover since the $\boldsymbol{b}_i$'s are short and so are the coefficients of $\boldsymbol{r}$, then also so is $\boldsymbol{e}_a + B\boldsymbol{r}$. Hence $\boldsymbol{u}'$ is a valid encoding of the same plaintext a that was encoded in $\boldsymbol{u}_a$.

Finally, using our Theorem 3 from this work, GGH can claim that the distribution of $\boldsymbol{u}$ is nearly independent of the original $\boldsymbol{u}_a$ (conditioned on its coset). If the $\boldsymbol{b}_i$'s are chosen from a wide enough spherical distribution, then our Gaussian LHL allows them to conclude that $B\boldsymbol{r}$ is close to a wide ellipsoid Gaussian. With appropriate choice of σ^* the "width" of that distribution is much larger than the original $\boldsymbol{e}_a$, hence the distribution of $\boldsymbol{e}_a + B\boldsymbol{r}$ is nearly independent of $\boldsymbol{e}_a$, conditioned on the coset it belongs to.

5 Discussion

Unlike the classic LHL, our lattice version of LHL is less than perfect – instead of yielding a perfectly spherical Gaussian, it only gives us an approximately spherical one, i.e. $\mathcal{D}_{L,s'X^\top}$. Here approximately spherical means that all the singular values of the matrix $X^\top$ are within a small, constant sized interval. It is therefore natural to ask: 1) Can we do better and obtain a perfectly spherical Gaussian? 2) Is an approximately spherical Gaussian sufficient for cryptographic applications?

First let us consider whether we can make the Gaussian perfectly spherical. Indeed, as the number of lattice vectors m grows larger, we expect the greatest and least singular value of the discrete Gaussian matrix X to converge – this would imply that as $m \to \infty$, the linear combination $\sum_{i=1}^{m} z_i \boldsymbol{x}_i$ does indeed behave like a spherical Gaussian. While we do not prove this, we refer the reader to [16] for intuitive evidence. However, the focus of this work is small m (e.g., $m = \tilde{O}(n)$) suitable for applications, in which case we do not know how to prove the same.

This leads to the second question: is approximately spherical good enough? This depends on the application. We have already seen that it is sufficient for GGH encodings [5], where a canonical, wide-enough, but non-spherical Gaussian is used to "drown out" an initial encoding, and send it to a canonical distribution of encodings that encode the same value. Our LHL shows that one can sample from such a canonical approximate Gaussian distribution without using the initial Gaussian samples "wastefully".

On the other hand, we caution the reader that if the application requires the basis vectors $\boldsymbol{x}_1, \ldots, \boldsymbol{x}_m$ to be kept secret (such as when the basis is a trapdoor), then one must carefully consider whether our Gaussian sampler can be used safely. This is because, as demonstrated by [11] and [4], lattice applications where the basis is desired to be secret can be broken completely even if partial information about the basis is leaked. In an application where the trapdoor is available explicitly and oblivious sampling is not needed, it is safer to use the samplers of [6] or [13] to sample a perfectly spherical Gaussian that is statistically independent of the trapdoor.

Acknowledgments. The first and fourth authors were supported in part from a DARPA/ONR PROCEED award, NSF grants 1228984, 1136174, 1118096, and 1065276, a Xerox Faculty Research Award, a Google Faculty Research Award, an equipment grant from Intel, and an Okawa Foundation Research Grant. This material is based upon work supported by the Defense Advanced Research Projects Agency through the U.S. Office of Naval Research under Contract N00014-11-1-0389. The views expressed are those of the author and do not reflect the official policy or position of the Department of Defense, the National Science Foundation, or the U.S. Government.

The second and third authors were supported by the Intelligence Advanced Research Projects Activity (IARPA) via Department of Interior National Business Center (DoI/NBC) contract number D11PC20202. The U.S. Government is authorized to reproduce and distribute reprints for Governmental purposes notwithstanding any copyright annotation thereon. Disclaimer: The views and conclusions contained herein are those of the authors and should not be interpreted as necessarily representing the official policies or endorsements, either expressed or implied, of IARPA, DoI/NBC, or the U.S. Government.

References

1. Agrawal, S., Gentry, C., Halevi, S., Sahai, A.: Discrete gaussian leftover hash lemma over infinite domains (2012), `http://eprint.iacr.org/2012/714`
2. Banaszczyk, W.: New bounds in some transference theorems in the geometry of numbers. Mathematische Annalen 296(4), 625–635 (1993)
3. Boneh, D., Freeman, D.M.: Homomorphic signatures for polynomial functions. In: Paterson, K.G. (ed.) EUROCRYPT 2011. LNCS, vol. 6632, pp. 149–168. Springer, Heidelberg (2011)
4. Ducas, L., Nguyen, P.Q.: Learning a zonotope and more: Cryptanalysis of ntrusign countermeasures. In: Wang, X., Sako, K. (eds.) ASIACRYPT 2012. LNCS, vol. 7658, pp. 433–450. Springer, Heidelberg (2012)
5. Garg, S., Gentry, C., Halevi, S.: Candidate multilinear maps from ideal lattices and applications. In: Johansson, T., Nguyen, P.Q. (eds.) EUROCRYPT 2013. LNCS, vol. 7881, pp. 1–17. Springer, Heidelberg (2013), `http://eprint.iacr.org/2013/610`
6. Gentry, C., Peikert, C., Vaikuntanathan, V.: Trapdoors for hard lattices and new cryptographic constructions. In: Dwork, C. (ed.) STOC, pp. 197–206. ACM (2008)
7. Klein, P.: Finding the closest lattice vector when it's unusually close. In: Proceedings of the Eleventh Annual ACM-SIAM Symposium on Discrete Algorithms, SODA 2000, pp. 937–941 (2000)
8. Litvak, A.E., Pajor, A., Rudelson, M., Tomczak-Jaegermann, N.: Smallest singular value of random matrices and geometry of random polytopes. Advances in Mathematics 195(2) (2005)
9. Lyubashevsky, V.: Lattice signatures without trapdoors. In: Pointcheval, D., Johansson, T. (eds.) EUROCRYPT 2012. LNCS, vol. 7237, pp. 738–755. Springer, Heidelberg (2012)
10. Micciancio, D., Regev, O.: Worst-case to average-case reductions based on gaussian measures. SIAM J. Computing 37(1), 267–302 (2007)

11. Nguyen, P.Q., Regev, O.: Learning a parallelepiped: Cryptanalysis of GGH and NTRU signatures. J. Cryptol. 22(2), 139–160 (2009)
12. Peikert, C.: Limits on the hardness of lattice problems in l_p norms. Computational Complexity 17(2), 300–351 (2008)
13. Peikert, C.: An efficient and parallel gaussian sampler for lattices. In: Rabin, T. (ed.) CRYPTO 2010. LNCS, vol. 6223, pp. 80–97. Springer, Heidelberg (2010)
14. Regev, O.: On lattices, learning with errors, random linear codes, and cryptography. JACM 56(6) (2009)
15. Rothblum, R.: Homomorphic encryption: From private-key to public-key. In: Ishai, Y. (ed.) TCC 2011. LNCS, vol. 6597, pp. 219–234. Springer, Heidelberg (2011)
16. Rudelson, M., Vershynin, R.: Non-asymptotic theory of random matrices: extreme singular values. In: International Congress of Mathematicans (2010)
17. Tao, T.: Topics in random matrix theory. Graduate Studies in Mathematics, vol. 132. American Mathematical Society (2012)
18. van Dijk, M., Gentry, C., Halevi, S., Vaikuntanathan, V.: Fully Homomorphic Encryption over the Integers. In: Gilbert, H. (ed.) EUROCRYPT 2010. LNCS, vol. 6110, pp. 24–43. Springer, Heidelberg (2010)

A More Proofs

A.1 Proof of Lemma 6

Proof. Clearly for any fixed $\boldsymbol{v}$, the set that maximizes $\mathcal{D}_{L,s}(T) - \mathcal{D}_{L,s}(T-\boldsymbol{v})$ is the set of all vectors $\boldsymbol{u} \in L$ for which $\mathcal{D}_{L,s}(\boldsymbol{u}) > \mathcal{D}_{L,s}(\boldsymbol{u}-\boldsymbol{v})$, which we denote by $T_{\boldsymbol{v}} \stackrel{\text{def}}{=} \{\boldsymbol{u} \in L : \mathcal{D}_{L,s}(\boldsymbol{u}) > \mathcal{D}_{L,s}(\boldsymbol{u}-\boldsymbol{v})\}$. Observe that for any $\boldsymbol{u} \in L$ we have $\mathcal{D}_{L,s}(\boldsymbol{u}) > \mathcal{D}_{L,s}(\boldsymbol{u}-\boldsymbol{v})$ iff $\rho_s(\boldsymbol{u}) > \rho_s(\boldsymbol{u}-\boldsymbol{v})$, which is equivalent to $\|\boldsymbol{u}\| < \|\boldsymbol{u}-\boldsymbol{v}\|$. That is, $\boldsymbol{u}$ must lie in the half-space whose projection on $\boldsymbol{v}$ is less than half of $\boldsymbol{v}$, namely $\langle \boldsymbol{u}, \boldsymbol{v}\rangle < \|\boldsymbol{v}\|^2/2$. In other words we have

$$T_{\boldsymbol{v}} = \{\boldsymbol{u} \in L : \langle \boldsymbol{u}, \boldsymbol{v}\rangle < \|\boldsymbol{v}\|^2/2\},$$

which also means that $T_{\boldsymbol{v}} - \boldsymbol{v} = \{\boldsymbol{u} \in L : \langle \boldsymbol{u}, \boldsymbol{v}\rangle < -\|\boldsymbol{v}\|^2/2\} \subseteq T_{\boldsymbol{v}}$. We can therefore express the difference in probability mass as $\mathcal{D}_{L,s}(T_{\boldsymbol{v}}) - \mathcal{D}_{L,s}(T_{\boldsymbol{v}}-\boldsymbol{v}) = \mathcal{D}_{L,s}(T_{\boldsymbol{v}} \setminus (T_{\boldsymbol{v}}-\boldsymbol{v}))$. Below we denote this set-difference by

$$H_{\boldsymbol{v}} \stackrel{\text{def}}{=} T_{\boldsymbol{v}} \setminus (T_{\boldsymbol{v}} - \boldsymbol{v}) = \left\{\boldsymbol{u} \in L : \langle \boldsymbol{u}, \boldsymbol{v}\rangle \in (-\tfrac{\|\boldsymbol{v}\|^2}{2}, \tfrac{\|\boldsymbol{v}\|^2}{2}]\right\}.$$

That is, $H_{\boldsymbol{v}}$ is the "slice" in space of width $\|\boldsymbol{v}\|$ in the direction of $\boldsymbol{v}$, which is symmetric around the origin. The arguments above imply that for any set T we have $\mathcal{D}_{L,s}(T) - \mathcal{D}_{L,s}(T-\boldsymbol{v}) \le \mathcal{D}_{L,s}(H_{\boldsymbol{v}})$. The rest of the proof is devoted to upper-bounding the probability mass of that slice, i.e., $\mathcal{D}_{L,s}(H_{\boldsymbol{v}}) = \Pr_{\boldsymbol{u} \leftarrow \mathcal{D}_{L,s}}[\boldsymbol{u} \in H_{\boldsymbol{v}}]$.

To this end we consider the slightly thicker slice, say $H'_{\boldsymbol{v}} = (1+\frac{4}{c})H_{\boldsymbol{v}}$, and the random variable $\boldsymbol{w}$, which is obtained by drawing $\boldsymbol{u} \leftarrow \mathcal{D}_{L,s}$ and adding to it a continuous Gaussian variable of "width" s/c. We argue that $\boldsymbol{w}$ is somewhat likely to fall outside of the thick slice $H'_{\boldsymbol{v}}$, but conditioning on $\boldsymbol{u} \in H_{\boldsymbol{v}}$ we have that $\boldsymbol{w}$ is very unlikely to fall outside of $H'_{\boldsymbol{v}}$. Putting these two arguments together, we get that $\boldsymbol{u}$ must have significant probability of falling outside $H_{\boldsymbol{v}}$, thereby getting our upper bound.

In more detail, denoting $r = s/c$ we consider drawing $\boldsymbol{u} \leftarrow \mathcal{D}_{L,s}$ and $\boldsymbol{z} \leftarrow \mathcal{N}^n(0, r^2/2\pi)$, and setting $\boldsymbol{w} = \boldsymbol{u} + \boldsymbol{z}$. Denoting $t = \sqrt{r^2 + s^2}$, we have that $s \le t \le s(1 + \frac{1}{c})$ and $rs/t \ge s/(c+1) \ge \eta_\epsilon(L)$. Thus the conditions of Lemma 5 are met, and we get that $\boldsymbol{w}$ is distributed close to a normal random variable $\mathcal{N}^n(0, t^2/2\pi)$, upto a factor of at most $\frac{1+\epsilon}{1-\epsilon}$.

Since the continuous Gaussian distribution is spherical, we can consider expressing it in an orthonormal basis with one vector in the direction of $\boldsymbol{v}$. When expressed in this basis, we get the event $\boldsymbol{z} \in H'_{\boldsymbol{v}}$ exactly when the coefficient in the direction of $\boldsymbol{v}$ (which is distributed close to the 1-dimensional Gaussian $\mathcal{N}(0, t^2/2\pi)$) exceeds $\|\boldsymbol{v}(1 + \frac{4}{c})/2\|$ in magnitude. Hence we have

$$\begin{aligned}\Pr[\boldsymbol{w} \in H'_{\boldsymbol{v}}] &\le \Pr_{\alpha \leftarrow \mathcal{N}(0,t^2/2\pi)}[|\alpha| \le \|\boldsymbol{v}\|] \cdot \frac{1+\epsilon}{1-\epsilon} \\ &= \mathsf{erf}\left(\frac{\|\boldsymbol{v}\|\sqrt{\pi}(1+\frac{4}{c})}{2t}\right) \cdot \frac{1+\epsilon}{1-\epsilon} \le \mathsf{erf}\left(\frac{\|\boldsymbol{v}\|\sqrt{\pi}(1+\frac{4}{c})}{2s}\right) \cdot \frac{1+\epsilon}{1-\epsilon}\end{aligned}$$

On the other hand, consider the conditional probability $\Pr[\boldsymbol{w} \in H'_{\boldsymbol{v}} | \boldsymbol{u} \in H_{\boldsymbol{v}}]$: Let $H''_{\boldsymbol{v}} = \frac{4}{c} H_{\boldsymbol{v}}$, then if $\boldsymbol{u} \in H_{\boldsymbol{v}}$ and $\boldsymbol{z} \in H''_{\boldsymbol{v}}$, then it must be the case that $\boldsymbol{w} = \boldsymbol{u} + \boldsymbol{z} \in H'_{\boldsymbol{v}}$. As before, we can consider the continuous Gaussian on $\boldsymbol{z}$ in an orthonormal basis with one vector in the direction of $\boldsymbol{v}$, and we get

$$\begin{aligned}\Pr[\boldsymbol{w} \in H'_{\boldsymbol{v}} | \boldsymbol{u} \in H_{\boldsymbol{v}}] &\ge \Pr[\boldsymbol{z} \in H''_{\boldsymbol{v}} | \boldsymbol{u} \in H_{\boldsymbol{v}}] = \Pr[\boldsymbol{z} \in H''_{\boldsymbol{v}}] \\ &= \Pr_{\beta \leftarrow \mathcal{N}(0,r^2/2\pi)}[|\beta| \le 2\|\boldsymbol{v}\|/c] = \mathsf{erf}(\|\boldsymbol{v}\| 2\sqrt{\pi}/cr) = \mathsf{erf}(2\|\boldsymbol{v}\|\sqrt{\pi}/s)\end{aligned}$$

Putting the last two bounds together, we get

$$\begin{aligned}\mathsf{erf}\left(\frac{\|\boldsymbol{v}\|\sqrt{\pi}(1+\frac{4}{c})}{2s}\right) \cdot \frac{1+\epsilon}{1-\epsilon} \ge \Pr[\boldsymbol{w} \in H'_{\boldsymbol{v}}] &\ge \Pr[\boldsymbol{u} \in H_{\boldsymbol{v}}] \cdot \Pr[\boldsymbol{w} \notin H'_{\boldsymbol{v}} | \boldsymbol{u} \in H_{\boldsymbol{v}}] \\ &\ge \Pr[\boldsymbol{u} \in H_{\boldsymbol{v}}] \cdot \mathsf{erf}\left(\frac{\|\boldsymbol{v}\| 2\sqrt{\pi}}{s}\right)\end{aligned}$$

from which we conclude that $\Pr[\boldsymbol{u} \in H_{\boldsymbol{v}}] \le \frac{\mathsf{erf}\left(\|\boldsymbol{v}\|\sqrt{\pi}(1+4/c)/2s\right)}{\mathsf{erf}\left(\|\boldsymbol{v}\| 2\sqrt{\pi}/s\right)} \cdot \frac{1+\epsilon}{1-\epsilon}$, as needed.

A.2 Proof of Lemma 10

Proof. Fix some $\boldsymbol{z} \in \mathbb{Z}^n$. The probability mass assigned to $\boldsymbol{z}$ by $\mathcal{E}_{X,s'}$ is the probability of drawing a random vector according to the discrete Gaussian $\mathcal{D}_{\mathbb{Z}^m,s'}$ and hitting some $\boldsymbol{v} \in \mathbb{Z}^m$ for which $X \cdot \boldsymbol{v} = \boldsymbol{z}$. In other words, this is exactly the probability mass assigned by $\mathcal{D}_{\mathbb{Z}^m,s'}$ to the coset $A_{\boldsymbol{z}}$. Below let $T = T(X) \subseteq \mathbb{R}^m$ be the linear subspace containing the lattice A, and $T_{\boldsymbol{z}} = T_{\boldsymbol{z}}(X) \subseteq \mathbb{R}^m$ be the affine subspace containing the coset $A_{\boldsymbol{z}}$:

$$T = T(X) = \{\boldsymbol{v} \in \mathbb{R}^m : X \cdot \boldsymbol{v} = 0\}, \text{ and } T_{\boldsymbol{z}} = T_{\boldsymbol{z}}(X) = \{\boldsymbol{v} \in \mathbb{R}^m : X \cdot \boldsymbol{v} = \boldsymbol{z}\}.$$

Let Y be the pseudoinverse of X (i.e. $XY^\top = I_n$ and the rows of Y span the same linear sub-space as the rows of X). Let $\boldsymbol{u_z} = Y^\top \boldsymbol{z}$, and we note that $\boldsymbol{u_z}$ is the point in the affine space $T_{\boldsymbol{z}}$ closest to the origin: To see this, note that $\boldsymbol{u_z} \in T_{\boldsymbol{z}}$ since $X \cdot \boldsymbol{u_z} = X \times Y^\top \boldsymbol{z} = \boldsymbol{z}$. In addition, $\boldsymbol{u_z}$ belongs to the row space of Y, so also to the row space of X, and hence it is orthogonal to T.

Since $\boldsymbol{u_z}$ is the point in the affine space $T_{\boldsymbol{z}}$ closest to the origin, it follows that for every point in the coset $\boldsymbol{v} \in A_{\boldsymbol{z}}$ we have $\|\boldsymbol{v}\|^2 = \|\boldsymbol{u_z}\|^2 + \|\boldsymbol{v} - \boldsymbol{u_z}\|^2$, and therefore

$$\rho_{s'}(\boldsymbol{v}) = e^{-\pi(\|\boldsymbol{v}\|/s')^2} = e^{-\pi(\|\boldsymbol{u_z}\|/s')^2} \cdot e^{-\pi(\|\boldsymbol{v}-\boldsymbol{u_z}\|/s')^2} = \rho_{s'}(\boldsymbol{u_z}) \cdot \rho_{s'}(\boldsymbol{v} - \boldsymbol{u_z}).$$

This, in turn, implies that the total mass assigned to $A_{\boldsymbol{z}}$ by $\rho_{s'}$ is

$$\rho_{s'}(A_{\boldsymbol{z}}) = \sum_{\boldsymbol{v} \in A_{\boldsymbol{z}}} \rho_{s'}(\boldsymbol{v}) = \rho_{s'}(\boldsymbol{u_z}) \cdot \sum_{\boldsymbol{v} \in A_{\boldsymbol{z}}} \rho_{s'}(\boldsymbol{v} - \boldsymbol{u_z}) = \rho_{s'}(\boldsymbol{u_z}) \cdot \rho_{s'}(A_{\boldsymbol{z}} - \boldsymbol{u_z}) \quad (2)$$

Fix one arbitrary point $\boldsymbol{w_z} \in A_{\boldsymbol{z}}$, and let $\boldsymbol{\delta_z}$ be the distance from $\boldsymbol{u_z}$ to that point, $\boldsymbol{\delta_z} = \boldsymbol{u_z} - \boldsymbol{w_z}$. Since $A_{\boldsymbol{z}} = \boldsymbol{w_z} + A$, we get $A_{\boldsymbol{z}} - \boldsymbol{u_z} = A - \boldsymbol{\delta_z}$, and together with the equation above we have:

$$\begin{aligned} \rho_{s'}(A_{\boldsymbol{z}}) = \rho_{s'}(\boldsymbol{u_z}) \cdot \rho_{s'}(A_{\boldsymbol{z}} - \boldsymbol{u_z}) &= \rho_{s'}(\boldsymbol{u_z}) \cdot \rho_{s'}(A - \boldsymbol{\delta_z}) \\ = \rho_{s'}(\boldsymbol{u_z}) \cdot \rho_{s',\boldsymbol{\delta_z}}(A) &\overset{\text{Lemma 4}}{=} \rho_{s'}(\boldsymbol{u_z}) \cdot \rho_{s'}(A) \cdot \left[\tfrac{1-\epsilon}{1+\epsilon}, 1\right]. \end{aligned} \quad (3)$$

As a last step, recall that $\boldsymbol{u_z} = Y^\top \boldsymbol{z}$ where $YY^\top = (XX^\top)^{-1}$. Thus $\rho_{s'}(\boldsymbol{u_z}) =$

$$\rho_{s'}(Y^\top \boldsymbol{z}) = \exp(-\pi|\boldsymbol{z}^\top YY^\top \boldsymbol{z}|/s'^2) = \exp\left(-\pi\left|\boldsymbol{z}^\top \left((s'X)(s'X)^\top\right)^{-1} \boldsymbol{z}\right|\right) = \rho_{(s'X)^\top}(\boldsymbol{z}).$$

Putting everything together we get

$$\mathcal{E}_{X,s'}(\boldsymbol{z}) = \mathcal{D}_{\mathbb{Z}^m,s'}(A_{\boldsymbol{z}}) = \frac{\rho_{s'}(A_{\boldsymbol{z}})}{\rho_{s'}(\mathbb{Z}^m)} \in \rho_{(s'X^\top)}(\boldsymbol{z}) \cdot \frac{\rho_{s'}(A)}{\rho_{s'}(\mathbb{Z}^m)} \cdot \left[\tfrac{1-\epsilon}{1+\epsilon}, 1\right]$$

The term $\frac{\rho_{s'}(A)}{\rho_{s'}(\mathbb{Z}^m)}$ is a normalization factor independent of $\boldsymbol{z}$, hence the probability mass $\mathcal{E}_{X,s'}(\boldsymbol{z})$ is proportional to $\rho_{(s'X^\top)}(\boldsymbol{z})$, upto some "deviation factor" in $[\frac{1-\epsilon}{1+\epsilon}, 1]$.

New Insight into the Isomorphism of Polynomial Problem IP1S and Its Use in Cryptography

Gilles Macario-Rat[1], Jérôme Plut[2], and Henri Gilbert[2]

[1] Orange Labs
38–40, rue du Général Leclerc, 92794 Issy-les-Moulineaux Cedex 9, France
gilles.macariorat@orange.com
[2] ANSSI,
51 Boulevard de la Tour-Maubourg, 75007 Paris, France
{henri.gilbert,jerome.plut}@ssi.gouv.fr

Abstract. This paper investigates the mathematical structure of the "Isomorphism of Polynomial with One Secret" problem (IP1S). Our purpose is to understand why for practical parameter values of IP1S most random instances are easily solvable (as first observed by Bouillaguet et al.). We show that the structure of the equations is directly linked to a matrix derived from the polar form of the polynomials. We prove that in the likely case where this matrix is cyclic, the problem can be solved in polynomial time – using an algorithm that unlike previous solving techniques is not based upon Gröbner basis computation.

1 Introduction

Multivariate cryptography is a sub area of cryptography the development of which was initiated in the late 80's [13] and was motivated by the search for alternatives to asymmetric cryptosystems based on algebraic number theory. RSA and more generally most existing asymmetric schemes based on algebraic number theory use the difficulty of solving one univariate equation over a large group (e.g. $x^e = y$ where e and y are known). Multivariate cryptography as for it, aims at using the difficulty of solving systems of multivariate equations over a small field.

A limited number of multivariate problems have emerged that can be reasonably conjectured to possess intractable instances of relatively small size. Two classes of multivariate problems are underlying most multivariate cryptosystems proposed so far, the MQ problem of solving a multivariate system of m quadratic equations in n variables over a finite field $\mathbb{F}_q$ - that was shown to be NP-complete even over $\mathbb{F}_2$ for $m \approx n$ [10]- and the broad family of the so-called isomorphism of polynomials (IP) problems.

Isomorphism of Polynomial problems can be roughly described as the equivalence of multivariate polynomial systems of equations up to linear (or affine) bijective changes of variables. Two separate subfamilies of IP problems can be distinguished: isomorphism of polynomials with two secrets (IP2S for short) and isomorphism of polynomials with one secret (IP1S for short). A little more in

K. Sako and P. Sarkar (Eds.) ASIACRYPT 2013 Part I, LNCS 8269, pp. 117–133, 2013.

detail, given two m-tuples $a = (a_1, \ldots, a_m)$ and $b = (b_1, \ldots, b_m)$ of polynomials in n variables over $\mathbb{K} = \mathbb{F}_q$, IP2S consists of finding two linear bijective transformations S of $\mathbb{K}^n$ and T of $\mathbb{K}^m$, such that $b = T \circ a \circ S$. Respectively, (computational) IP1S consists of finding one linear bijective transformations S of $\mathbb{K}^n$, such that $b = a \circ S$. Many variants of both problems can be defined depending on the value of the triplet (n, m, q), the degree d of the polynomial equations of a and b, whether these polynomials are homogeneous or not, whether S and T are affine or linear, etc. It turns out that there are considerable security and simplicity advantages in restricting oneself, for cryptographic applications, to instances involving only homogeneous polynomials of degree d and linear transformations S and T. For performance reasons, the quadratic case $d = 2$ is most frequently encountered in cryptography. Due to the existence of an efficient canonical reduction algorithm for quadratic forms, instances such that $m \geq 2$ must then be considered. The cubic case $d = 3$ is also sometimes considered, then instances such that $m = 1$ are generally encountered.

Many asymmetric cryptosystems whose security is related to the hardness of special trapdoor instances of IP2S were proposed in which all or part of the m-tuple of polynomials b plays the role of the public key and is related by secret linear bijections S and T to a specially crafted, easy to invert multivariate polynomial mapping a. Most of these systems, e.g. Matsumoto and Imai's seminal multivariate scheme C* [13], but also reinforced variants such as SFLASH and HFE [18,16] were shown to be weak because the use of trapdoor instances of IP2S with specific algebraic properties considerably weakens the general IP2S problem. A survey of the status of the IP2S problems and improved techniques for solving homogeneous instances are presented in [1] and [4].

The IP1S problem was introduced in [16] by Patarin, who proposed in the same paper a zero-knowledge asymmetric authentication scheme named the IP identification scheme with one secret (IP1S scheme for short). This authentication scheme is inspired by the well known zero-knowledge proof for Graph Isomorphism by Goldreich et al. [11]. It can be converted into a (less practical) asymmetric signature scheme using the Fiat-Shamir transformation. The IP1S problem and the related identification scheme were believed to possess several attractive features:

- The conjecture that the IP1S problem is not solvable in polynomial time was supported by the proof in [17] that the quadratic version of IP1S (QIP1S for short) is at least as hard as the Graph Isomorphism problem (GI) [1], one of the most extensively studied problems in complexity theory. While the GI problem is not believed to be NP-complete since it is NP and co-NP and hard instances of GI are difficult to construct for small parameter values, GI is generally believed not to be solvable in polynomial time.
- unlike the encryption or signature schemes based on IP2S mentioned above, the IP1S scheme does not use special trapdoor instances of the IP1S problem

[1] However as mentioned in the conclusion of this paper, if the flaw recently discovered by the authors in the corresponding proof in [17] is confirmed, this casts some doubts on the fact that Quadratic IP1S is indeed as hard as GI.

and therefore its security is directly related to the intractability of general IP1S instances.

The IP1S problem also has some loose connections with the multivariate signature scheme UOV [12], that has until now remarkably well survived all advances in the cryptanalysis of multivariate schemes. While in UOV the public quadratic function b is related to the secret quadratic function by the equation $b = a \circ S$, both a and S are unknown whereas only S is unknown in the IP1S problem.

Former Results. Initial assessments of the security of practical instances of the IP1S problem suggested that relatively small public key and secret sizes - typically about 256 bits - could suffice to ensure a security level of more than 2^{64}. The IP1S scheme therefore appeared to favorably compare with many other zero-knowledge authentication schemes, e.g [21,22,20]. Moreover, despite advances in solving some particular instances of the IP1S problem, in particular Perret's Jacobian algorithm[2] [19], the four challenge parameter values proposed in 1996 [16] (with $q = 2$ or 2^{16}, $d = 2$ and $m = 2$, or $d = 3$ and $m = 1$) remained unbroken until 2011.

Significant advances on solving IP1S instances that are practically relevant for cryptography were made quite recently [2,1]. Dubois in [7] and the authors of [2] were the first to notice that the IP1S problem induces numerous linear equations in the coefficients of the matrix of S and of the inverse mapping $T = S^{-1}$. When $m \geq 3$, the number mn^2 of obtained linear equations is substantially larger than the number $2n^2$ of variables. While the system cannot have full rank since the dimension of the vector space of solutions is at least 1, it can heuristically be expected to have a very small vector space of solutions that can be tried exhaustively. The authors of [2] even state that they "empirically find one solution (when the polynomials are randomly chosen)".

Therefore the most interesting remaining case appears to be $m = 2$. It is shown in [2] that the vector space of solutions of the linear equations is then isomorphic to the commutant of a non-singular $n \times n$ matrix M and that its dimension r is lower bounded by n in odd characteristic and $2n$ in even characteristic. The reported computer experiments indicate that r is extremely likely to be close to these lower bounds in practice. While for typical values of q^n the vector space of solutions is too large to be exhaustively searched, one can try to solve the equation $b = a \circ S$ over this vector space. This provides a system of quadratic equations in a restricted variable set of $r \approx n$ (resp. $r \approx 2n$) coordinates. The approach followed in [2] in order to solve this system consisted of applying Gröbner basis algorithms such as Faugère's $F4$ [8] and related computer algebra tools such as FGLM [9]. This method turned out to be quite successful: all the IP1S challenges proposed by Patarin were eventually broken in computing times ranging from less than 1 s to 1 month. This led the authors of [2] to conclude that "[the] IP1S-Based identification scheme is no longer competitive

[2] This algorithm recovers mn linear equations in the coefficients of S and is therefore suited for solving IP1S instances such that $m \approx n$.

with respect to other combinatorial-based identification schemes". However, the heuristic explanation suggested in [2], namely that the obtained system was so massively over defined that a random system with the same number of random quadratic equations would be efficiently solvable in time $O(n^9)$ with overwhelming probability, was later on shown to be false by one of the authors of [2], due to an overestimate of the number of linearly independent quadratic equations.

This is addressed in Bouillaguet's PhD dissertation [1] where the results of [2] are revisited. The main discrepancy with the findings of [2] is the observation that in all the reported experiments in odd and even characteristic, the number of linearly independent quadratic equations, that was supposed in [2] to be close to n^2, is actually bounded over by a small multiple of n and only marginally larger than r. The author writes "This means that we cannot argue that solving these equations is doable in polynomial time. An explanation of this phenomenon has eluded us so far." Despite of the surprisingly small number of linearly independent quadratic equations, nearly all instances are confirmed to be efficiently solvable for all practical values of n when the size q of the field is sufficiently small (q=2 or 3) and still solvable efficiently up to values of n of about 20. The author writes "For instance, when $q = 2$ and $n = 128$ we are solving a system of 256 quadratic equations in 256 variables over $\mathbb{F}_2$. When the equations are random this is completely infeasible. In our case, it just takes 3 minutes ! We have no clear explanation of this phenomenon."

Our Contribution. The lack of explanation for the success of the attack – more precisely the puzzling fact that the number of linearly independent quadratic equations is close to n in odd characteristic and to $2n$ in even characteristic and the even more puzzling fact that nearly all instances are nevertheless solvable – motivated our research on IP1S. We revisited the former analysis and eventually found an algebraic explanation of why most random instances of the quadratic IP1S problem are efficiently solvable that leads to a new method (not based on Gröbner basis computations) to directly solve these instances. Our analysis shows in particular that in the likely cases where the characteristic is odd and the matrix M is cyclic or the characteristic is even and M is similar to a block-wise diagonal matrix with two equal cyclic $\frac{n}{2} \times \frac{n}{2}$ diagonal blocks, the quadratic equations split up in an appropriate base in small triangular quadratic systems that can be solved efficiently in polynomial time. The highlighted structure of the quadratic equations seems to be the essential reason why Gröbner basis computations behave so well on most instances.

The rest of this paper is organized as follows. In Section 2, we present the problem IP1S, its background and some major mathematical results used in the following sections. We then discuss in Section 3 and 4 the resolution of the problem over finite fieds of odd, resp. even characteristic.

2 The Isomorphism of Polynomial Problem with One Secret

2.1 Notations and First Definitions

Let $\mathbb{K}$ be a field; for practical considerations, we shall assume that $\mathbb{K}$ is the finite field $\mathbb{F}_q$ with q elements, although most of the discussion is true in the general case.

A *(homogeneous) quadratic form in n variables* over $\mathbb{K}$ is a homogeneous polynomial of degree two, of the form $q = \sum_{i,j=1\dots n} \alpha_{i,j} x_i x_j$, where the coefficients $\alpha_{i,j}$ belong to $\mathbb{K}$. For simplicity, we write $x = (x_i)$ for the vector with coordinates x_i. The quadratic form q can be described by the matrix with general term $\alpha_{i,j}$. Note that the matrix representation of a quadratic form is not unique: two matrices represent the same linear form if, and only if, their difference is skew-symmetric.

The *polar form* associated to a quadratic form q is the bilinear form $b = \mathcal{P}(q)$ defined by $b(x, y) = q(x + y) - q(x) - q(y)$. This is a symmetric bilinear form. This can be used to give an intrinsic definition of bilinear forms (which is useful to abstract changes of bases from some proofs below): given a vector space V, a quadratic form over V is a function $q : V \to \mathbb{K}$ such that

(i) for all $x \in V$ and $\lambda \in \mathbb{K}$, $q(\lambda x) = \lambda^2 q(x)$;
(ii) the polar form $\mathcal{P}(q)$ is bilinear.

For any matrix A, let ${}^tA$ be the transpose matrix of A and $\mathcal{P}(A)$ be the symmetric matrix ${}^tA + A$. Then if q is a quadratic form with matrix A, its polar form has matrix $\mathcal{P}(A)$. The quadratic form q is *regular* if its polar form is not singular, *i.e.* if it defines a bijection from V to its dual. In general, we define the *kernel* of a quadratic form to be the kernel of its polar form.

From the definition of $b = \mathcal{P}(q)$ we derive the *polarity identity*

$$2q(x) = b(x, x). \tag{1}$$

This identity obviously behaves very differently when 2 is a unit in $\mathbb{K}$ and when $2 = 0$ in $\mathbb{K}$. This forces us to use some quite different methods in both cases.

If 2 is invertible in $\mathbb{K}$ then the polarity identity (1) allows recovery of a quadratic form from its polar bilinear form. In other words, quadratic forms in n variables correspond to symmetric matrices.

Conversely, if $2 = 0$, then the polarity identity reads as $b(x, x) = 0$; in other words, the polar form is an alternating bilinear form. In this case, equality of polar forms does not imply equality of quadratic forms. Define $\Delta(A)$ as the matrix of diagonal entries of the matrix A. Then quadratic forms A and B are equal if, and only if, $\mathcal{P}(A) = \mathcal{P}(B)$ and $\Delta(A) = \Delta(B)$.

2.2 The Quadratic IP1S Problem

We now state the quadratic IP1S problem and give an account of its current status after the recent work of [2] and [1].

Problem 1 (Quadratic IP1S). Given two m-tuples $a = (a_1, \ldots, a_m)$ and $b = (b_1, \ldots, b_m)$ of quadratic homogeneous forms in n variables over $\mathbb{K} = \mathbb{F}_q$, find a non-singular linear mapping $S \in GL_n(\mathbb{K})$ (if any) such that $b = a \circ S$, i.e. $b_i = a_i \circ S$ for $i = 1, \ldots, m$.

Remark 1. In order not to unnecessarily complicate the presentation, our definition of the IP1S problem slightly differs[3] from the initial statement of the problem introduced in [16]. Though the name "quadratic homogeneous IP1S" might be more accurate to refer to the exact class of instances we consider, we will name it quadratic IP1S or IP1S in the sequel.

If we denote by A_i, resp. B_i any $n \times n$ matrices representing the a_i, resp. the b_i and denote by X the matrix representation of S, the conditions for the equality of two quadratic forms given in Section 2.1. allow to immediately translate the quadratic IP1S problem into equivalent matrix equations.

- If the characteristic of $\mathbb{K}$ is odd: the problem is equivalent to finding an invertible matrix X that satisfies the m polar equations: $\mathcal{P}(B_i) = {}^tX\mathcal{P}(A_i)X$
- If the characteristic of $\mathbb{K}$ is even: the problem is equivalent to finding an invertible matrix X that satisfies the polar and the diagonal equations: $\mathcal{P}(B_i) = {}^tX\mathcal{P}(A_i)X$; $\Delta(B_i) = \Delta({}^tXA_iX)$.

In the following sections we will consider IP1S instances such that $m = 2$, that are believed to represent the most "interesting" instances of IP1S as reminded above. Matrix pencils, that can be viewed as $n \times n$ matrices whose coefficients are polynomials of degree 1 of $\mathbb{K}[\lambda]$ represent a convenient way to capture the above equations in a more compact way. If we denote by A and B the matrix pencils $\lambda A_0 + A_1$ and $\lambda B_0 + B_1$, and by extension $\mathcal{P}(A)$ and $\mathcal{P}(B)$ the symmetric matrix pencils $\lambda\mathcal{P}(A_0) + \mathcal{P}(A_1)$ and $\lambda\mathcal{P}(B_0) + \mathcal{P}(B_1)$, the two polar equations can be written in one equation: $\mathcal{P}(B) = {}^tX\mathcal{P}(A)X$. However, as detailed in the next section, the theory of pencils is far more powerful than just a convenient notation for pairs of matrices. See for instance [3].

2.3 Mathematical Background

In this Section we briefly outline a few known definitions and results related to the classification of matrices and matrix pencils and known methods for solving matrix equations that are relevant for the investigation the IP1S problem.

[3] While in [16] the isomorphism of two m-tuples quadratic polynomials comprising also linear and constant terms through a non-singular affine transformation was considered, we consider here the isomorphism of two m-tuples of quadratic forms through a non-singular linear transformation. This replacement of the original definition by a simplified definition is justified by the fact that all instances of the initial problem can be shown to be either easily solvable due to the lower degree homogeneous equations they induce or efficiently reducible to an homogeneous quadratic instance.

Basic Facts about Matrices. Two matrices A and B are *similar* if there exists an invertible matrix P such that $P^{-1}AP = B$ and *congruent* if there exists an invertible P such that ${}^tPAP = B$.

The matrix A is called *cyclic* if its minimal and characteristic polynomials are equal.

For any matrix A, the *commutant* of A is the algebra $\mathcal{C}_A$ of all matrices commuting with A. It contains the algebra $\mathbb{K}[A]$, and this inclusion is an equality if, and only if, A is cyclic.

For any matrix A, let $\prod p_i^{e_i}$ be the prime factorization of its minimal polynomial. Then $\mathbb{K}[A]$ is the direct product of the algebras $\mathbb{K}[x]/p_i(x)^{e_i}$; each of these factors is a local algebra with residual field equal to the extension field $\mathbb{K}[x]/p_i$.

Pencils of Bilinear and Quadratic Forms. Let V be a $\mathbb{K}$-vector space and $Q(V)$ be the vector space of all quadratic forms on V. A *projective pencil of quadratic forms* on V is a projective line in $\mathbb{P}Q(V)$, *i.e.* a two-dimensional subspace of $Q(V)$. As a projective pencil is the image of the projective line $\mathbb{P}^1$ in $Q(V)$, it is determined by the images of the points ∞ and 0 in $\mathbb{P}^1$, which we write A_0 and A_∞.

An *affine pencil of quadratic forms* is an affine line in $Q(V)$, or equivalently a pair of elements of $Q(V)$. The affine pencil with basis (A_∞, A_0) may also be written as a polynomial matrix $A_\lambda = A_0 + \lambda A_\infty$. Given a projective pencil A of $Q(V)$, the choice of any basis (A_∞, A_0) of A determines an affine pencil.

A projective pencil is *regular* if it contains at least one regular quadratic form. An affine pencil (A_∞, A_0) is *regular* if A_∞ is regular; it is *degenerate* if the intersection of the kernels of the quadratic forms A_λ is nontrivial.

If an affine pencil is non-degenerate, then the polynomial $\det A_\lambda$ is non-zero; choosing any λ which is not a root of this polynomial proves that the associated projective pencil is regular (over $\mathbb{K}$ itself if it is infinite, and over a finite extension of $\mathbb{K}$ if it is finite). This gives a basis of the projective pencil which turns the affine pencil into a regular one. We shall therefore assume all affine pencils to be regular.

Two pencils A, B of quadratic forms are *congruent* if there exists an invertible matrix X such that ${}^tXA_\lambda X = B_\lambda$. The case $m = 2$ of the quadratic IP1S problem reduces to the Pencil congruence problem: given two affine pencils A and B, known to be congruent, exhibit a suitable congruence matrix X.

We first note that the IP1S problem easily reduces to the case where both pencils are regular. Namely, if one (and therefore both) is degenerate, then we may quotient out both spaces by the (isomorphic) kernels of the pencils; this defines non-degenerate affine pencils on the quotient vector spaces, which are still congruent. Since the associated projective pencils are regular, a change of basis in the pencils (and maybe an extension of scalars) brings us to the case of two regular affine pencils.

We define pencils of bilinear forms in the same way as pencils of quadratic forms. The pencil $b_\lambda = b_0 + \lambda b_\infty$ *regular* if b_∞ is; in this case, the *characteristic endomorphism* of the pencil is the endomorphism $f = b_\infty^{-1} \circ b_0$.

The following lemma allows to decompose pencils as direct sums, with each factor having a power of an irreducible polynomial as its characteristic endomorphism.

Lemma 1. *Let b be a regular pencil of symmetric bilinear forms. Then all primary subspaces of the characteristic endomorphism f are orthogonal with respect to all forms of b.*

Proof. We have to prove the following: given any two mutually prime factors p, q of f and any $x, y \in V$ such that $p(f)(x) = 0$ and $q(f)(y) = 0$, then for all λ, we have $b_\lambda(x, y) = 0$. For this it is enough to show that $b_\infty(x, y) = 0$.

Since p, q are mutually prime, there exist u, v such that $up + vq = 1$. Note that, for all $x, y \in V$, we have $b_\infty(x, fy) = b_0(x, y) = b_0(y, x) = b_\infty(fx, y)$; therefore, all elements of $\mathbb{K}[f]$ are self-adjoint with respect to b_∞. From this we derive the following:

$$\begin{aligned} b_\infty(x, y) &= b_\infty(x, u(f)p(f)y + v(f)q(f)y) \\ &= b_\infty(u(f)p(f)x, y) + b_\infty(x, v(f)q(f)y) \\ &= 0. \end{aligned} \tag{2}$$

□

Explicit Similarity of a Matrix and Its Transposed. The next result is intensively used in the sequel to deal with symmetric pencils. Although this result is classic [23], we are interested with the explicit form given below.

Theorem 1. *For any matrix M, there exists a non-singular symmetric matrix T such that ${}^tMT = TM$.*

Proof. Using primary decomposition for M, we may assume that it is of the form

$$M = \begin{pmatrix} M_0 & 1 & & 0 \\ & \ddots & \ddots & \\ & & \ddots & 1 \\ 0 & & & M_0 \end{pmatrix}, \tag{3}$$

where M_0 is the companion matrix of a polynomial $p(\lambda) = \lambda^n + \sum_{i=0}^{n-1} p_i \lambda^i$. We then define matrices T_0 and T by

$$T_0 = \begin{pmatrix} p_1 & \cdots & p_{n-1} & 1 \\ \vdots & \cdot^{\cdot^{\cdot}} & \cdot^{\cdot^{\cdot}} & \\ p_{n-1} & \cdot^{\cdot^{\cdot}} & & \\ 1 & & & 0 \end{pmatrix}, \quad T = \begin{pmatrix} 0 & & T_0 \\ & \cdot^{\cdot^{\cdot}} & \\ T_0 & & 0 \end{pmatrix}. \tag{4}$$

One can easily verify that T_0 is invertible, symmetric and ${}^tM_0T_0 = T_0M_0$, and that the same is true for T and M. □

3 IP1S in Characteristic Different from Two

Let $\mathbb{K}$ be a field of characteristic different from two[4]. In this case, the polarity identity (1) identifies quadratic forms with symmetric bilinear forms, or again with symmetric matrices with entries in $\mathbb{K}$. We shall therefore write a quadratic pencil A as $A_\lambda = A_0 + \lambda A_\infty$, where A_0 and A_∞ are symmetric matrices.

Proposition 1. *Let $A_\lambda = A_0 + \lambda A_\infty$, $B_\lambda = B_0 + \lambda B_\infty$ be two regular affine pencils.*

(i) If A_λ is congruent to B_λ, then the characteristic matrices

$$M_A = A_\infty^{-1} A_0 \quad \text{and} \quad M_B = B_\infty^{-1} B_0$$

are similar.

(ii) Assume that M_A and M_B are similar and choose P such that $P^{-1} M_A P = M_B$. Then ${}^tPA_\lambda P = {}^tPA_\infty P(\lambda + M_B)$.

(iii) Assume that $A_\lambda = A_\infty(\lambda + M)$ and $B_\lambda = B_\infty(\lambda + M)$. Then the solutions of the pencil congruence problem are exactly the invertible X such that

$$XM = MX \quad \text{and} \quad {}^tXA_\infty X = B_\infty. \tag{5}$$

Proof. (i). Since A_λ is regular, A_∞ is invertible and we may write $A_\lambda = A_\infty(\lambda + A_\infty^{-1} A_0)$; likewise, $B_\lambda = B_\infty(\lambda + B_\infty^{-1} B_0)$. Choose P such that ${}^tPA_\lambda P = B_\lambda$, then

$$B_\infty(\lambda + M_B) = {}^tPA_\lambda P = {}^tPA_\infty P(\lambda + P^{-1} M_A P), \tag{6}$$

which implies $P^{-1} M_A P = M_B$ as required. The same computations prove (ii).

The equations (5) follows directly from the equality ${}^tXA_\infty(\lambda + M)X = {}^tXA_\infty X(\lambda + X^{-1} MX)$. □

We now restrict ourselves to the case where the characteristic endomorphism is cyclic.

Proposition 2. *Let $A_\lambda = A_\infty(\lambda + M)$ and $B_\lambda = B_\infty(\lambda + M)$ be two regular symmetric pencils such that the matrix M is cyclic, that is, its minimal and characteristic polynomials are equal.*

Then the solutions X of the pencil congruence problem are the square roots of $A_\infty^{-1} B_\infty$ in the algebra $\mathbb{K}[M]$.

Proof. Since M is cyclic, its commutant is reduced to the algebra $\mathbb{K}[M]$; therefore, all solutions of the congruence problem are polynomials in M.

Since A_λ is symmetric, both matrices A_∞ and $A_0 = A_\infty M$ are symmetric; therefore, ${}^tMA_\infty = A_\infty M$. Since X is a polynomial in M, we deduce that also ${}^tXA_\infty = A_\infty X$.

The relation ${}^tXA_\infty X = B_\infty$ may therefore be rewritten as $A_\infty X^2 = B_\infty$, or $X^2 = A_\infty^{-1} B_\infty$. □

[4] Although this is not used in cryptography, we mention that this section also applies verbatim to the case of characteristic zero.

Theorem 2. *Let $\mathbb{K}$ be a finite field of odd characteristic and A_λ, B_λ be two regular pencils of quadrics over $\mathbb{K}^n$, congruent to each other, such that at least one is cyclic (and therefore both are). Then the pencil congruence problem may be solved using no more than $\widetilde{O}(n^3)$ operations in the field $\mathbb{K}$.*

Proof. The first step is to reduce to the case of primary components of the characteristic endomorphism. This may be done, using for example Frobenius reduction of both matrices $A_\infty^{-1}A_0$ and $B_\infty^{-1}B_0$, with a complexity of $\widetilde{O}(n^3)$ operations. This also provides the change of basis making the characteristic endomorphism of both pencils to have the same matrix.

There remains to compute a square root of $C = A_\infty^{-1}B_\infty$ in $\mathbb{K}[M]$, where now the minimal polynomial of M is p^e, with p irreducible. For this we first write C as a polynomial $g(M)$; this again requires $\widetilde{O}(n^3)$ operations. To solve the equation $y^2 = g(M)$ in the ring $\mathbb{K}[M] = \mathbb{K}[x]/p(x)^e$, we first solve it in the (finite) residual field $\mathbb{K}[x]/p(x)$, with complexity $\widetilde{O}(n^3)$ again; lifting the solution to the ring $\mathbb{K}[M]$ requires only $\widetilde{O}(n^2)$ with Hensel lifting. □

Solutions of the IP1S problem are square roots of an element C of the algebra $\mathbb{K}[M]$; therefore, the number of solutions is 2^s, where s is the number of connected components of $\mathbb{K}[M]$, that is, the number of prime divisors of the minimal polynomial of M.

Summary and Computer Experiments. The case where all the elementary divisors of $\mathcal{P}(A)$ are pairwise co-prime – or equivalently where M is cyclic – represents in practice a quite large fraction of random cases (see for instance [15]). In this case, as shown above, the number of solutions is exactly 2^s where s is the numbers of elementary divisors and solutions can be efficiently computed (in polynomial time $\widetilde{O}(n^3)$) by our method. The highlighted structure of the equations also provides some likely explanations of why Gröbner basis computation methods such as those presented in [2] were successful in this case. We give in next table results (timings) of our MAGMA script SOLVECYCLICODDPC, t is the mean execution time when solving 100 random cyclic IP1S instances, τ is the observed fraction in percent of such "cyclic" instances over random instances.

q	n	t	τ
3	80	5.s.	87.
3	128	34.s.	88.
3^{10}	32	15.s.	100.

q	n	t	τ
5	20	0.07s.	95.
5	32	0.28s.	95.
5	80	7.s.	95.
5^7	32	8.s.	100.

q	n	t	τ
7^6	32	11.s.	100.
65537	8	0.04s.	100.
65537	20	1.s.	100.

4 IP1S in Characteristic Two

Let $\mathbb{K}$ be a perfect field of characteristic two. In this case, the polarity identity (1) shows that the polar form $b = \mathcal{P}(q)$ attached to a quadratic form q is an alternating bilinear form.

4.1 Pencils of Alternating Bilinear Forms

This paragraph is a reminder of classical results. We refer the reader to [14] for the proofs.

If b is alternating and nondegenerate, then the vector space V has a *symplectic basis*, i.e. a basis $(e_1, \ldots, e_n, f_1, \ldots, f_n)$ such that $b(e_i, f_i) = 1$ and all other pairings are zero. In particular, the dimension of V is even. The vector E space generated by the e_i is equal to its orthogonal space $E^\perp$; such a space is called a *Lagrangian* space for b.

We recall that two matrices A and B define the same quadratic form if and only if $\mathcal{P}(A) = \mathcal{P}(B)$ and $\Delta(A) = \Delta(B)$.

Although quadratic forms only produce alternating bilinear forms in characteristic two, the following lemma about alternating forms is true in all characteristics. It proves that there exists a basis of V in which the pencil has the block-matrix decomposition

$$A_\infty = \begin{pmatrix} 0 & 1 \\ 1 & 0 \end{pmatrix}, \quad A_0 = \begin{pmatrix} 0 & {}^tF \\ F & 0 \end{pmatrix}; \qquad A_\infty^{-1} A_0 = \begin{pmatrix} F & 0 \\ 0 & {}^tF \end{pmatrix}. \tag{7}$$

The matrix F is called the *Pfaffian* endomorphism of A.

Lemma 2. *Let $b = (b_\infty, b_0)$ be a regular pencil of alternating bilinear forms on V. Then there exists a symplectic basis for b_∞ whose Lagrangian is stable by the characteristic endomorphism of b.*

Proof. Let f be the characteristic endomorphism of b. By Lemma 1, we may replace V by one of the primary components of f and therefore assume that the minimal polynomial of f is p^n where p is a prime polynomial. By extending scalars to $\mathbb{K}[\lambda]/p(\lambda)$ and replacing b_0 by $\lambda b_\infty + b_0$ we may assume that $p(t) = t$. We now prove the lemma by induction on $\dim V$.

Since t^n is the minimal polynomial of f and b_∞ is non-degenerate, there exists $x, y \in V$ such that $b_\infty(x, f^{n-1}y) = 1$. Let $W = \mathbb{K}[f]x \oplus \mathbb{K}[f]y$. Then we may write $V = W \oplus W^\perp$ where both W and its b_∞-orthogonal $W^\perp$ are stable by f; since $W^\perp$ satisfies the lemma by the induction hypothesis, we only need to prove it for W.

Let $a(t) = 1 + a_1 t + \cdots + a_{n-1} t^{n-1}$ be a polynomial and $x' = a(f)x$. Then we still have $b_\infty(x', f^{n-1}y) = 1$, and moreover we can choose a so that $b_\infty(x', f^i y) = 0$ for all $i = 0, \ldots, n-2$. In other words, $(x', fx', \ldots, f^{n-1}x', f^{n-1}y, f^{n-2}y, \ldots, fy, y)$ is a symplectic basis for b_∞ on W. By construction, its Lagrangian is $\mathbb{K}[f]x$, which is obviously stable by the characteristic endomorphism f. $\square$

Proposition 3. *Let $\mathbb{K}$ be a binary field. Any regular pencil of alternating bilinear forms is congruent to a pencil of the form*

$$A_\infty = \begin{pmatrix} 0 & T \\ T & 0 \end{pmatrix}, \quad A_0 = \begin{pmatrix} 0 & TM \\ TM & 0 \end{pmatrix}, \tag{8}$$

where M is in rational (Frobenius) normal form and T is the symmetric matrix defined in Theorem 1.

Proof. From the equation 7, choose a matrix P such that $M = P^{-1}FP$ is in rational normal form and define T as in Theorem 1. Then the coordinate change $\begin{pmatrix} P & 0 \\ 0 & {}^tP^{-1}T \end{pmatrix}$ produces the required form. □

Let A be a pencil as in (8). The automorphism group $O(A)$ of A is the set of matrices $X = \begin{pmatrix} X_1 & X_2 \\ X_3 & X_4 \end{pmatrix}$ such that ${}^tXAX = A$, that is all X_i commute with M and ${}^tX_1TX_4 + {}^tX_3TX_2 = T$.

From now, we suppose that M is cyclic and for the sake of simplicity that its primary decomposition has only one component.

Since M is cyclic, all X_i belong to $\mathbb{K}[M]$. The group $O(A)$ is generated by the elementary transformations

$$G_1(X) = \begin{pmatrix} 1 & X \\ 0 & 1 \end{pmatrix}, \quad G_2(X) = \begin{pmatrix} 1 & 0 \\ X & 1 \end{pmatrix}, \quad G_3(X) = \begin{pmatrix} X & 0 \\ 0 & X^{-1} \end{pmatrix}, \quad G_4 = \begin{pmatrix} 0 & 1 \\ 1 & 0 \end{pmatrix}, \quad (9)$$

where $X \in \mathbb{K}[M]$, X invertible for $G_3(X)$. The first three transformations generate the subgroup of *positive* automorphisms of A. This is a subgroup of order two of the orthogonal group [6].

4.2 Pencils of Quadratic Forms

The following proposition deals with the diagonal terms of a quadratic form in the cyclic case. We recall that, using the notations of Theorem 1, $\mathbb{K}[M_0]$ is an extension field of $\mathbb{K}$, and $\mathbb{K}[M]$ is the (local) $\mathbb{K}[M_0]$-algebra generated by

$$H = \begin{pmatrix} 0 & 1 & & 0 \\ & \ddots & \ddots & \\ & & \ddots & 1 \\ 0 & & & 0 \end{pmatrix}. \quad (10)$$

We write $\varphi(X) = X^2$ for the Frobenius map of $\mathbb{K}[M_0]$. Since this is a finite field, the Frobenius map is bijective. It extends to $\mathbb{K}[M]$ as $\varphi(\sum x_i H^i) = \sum x_i^2 H^i$.

Proposition 4. *Define matrices M of size n, M_0, T_0 of size $e = n/d$ as in Theorem 1.*

(i) The $\mathbb{K}$-linear map $\mathbb{K}[M_0] \mapsto \mathbb{K}^e$, $X \mapsto \Delta(T_0X)$ is an isomorphism.
(ii) For any diagonal matrix D of size e, there exists a (unique) matrix $C = \psi_0(D) \in \mathbb{K}[M_0]$ such that, for all $X \in \mathbb{K}[M_0]$:

$$\Delta({}^tXDX) = \Delta(T_0CX^2). \quad (11)$$

(iii) Let D be a diagonal matrix of size n, written as blocks $D_0, \dots, D_{d-1}$, and write $X \in \mathbb{K}[M]$ as $X = \sum x_i H^i$ with $x_i \in \mathbb{K}[M_0]$. Also define $\psi(D) = \sum \psi_0(D_i)H^i \in \mathbb{K}[M]$. Then we have the relation in $\mathbb{K}[M]$

$$\psi(\Delta({}^tXDX)) = \varphi(X) \cdot \psi(D). \quad (12)$$

Proof. (i) Since $2 = 0$ in $\mathbb{K}$, for any symmetric matrix A and any X, we have

$$\Delta({}^t X \Delta(A) X) = \Delta({}^t X A X). \tag{13}$$

Since the space $\mathbb{K}[M_0]$ has dimension e over $\mathbb{K}$, we only have to check injectivity. Assume $\Delta(T_0 X) = 0$ with $X \neq 0$; since $\mathbb{K}[M_0]$ is a field, X is invertible. Let $Y = \varphi^{-1}(X^{-1})$. We then have

$$\Delta(T_0) = \Delta(T_0 X Y^2) = \Delta({}^t Y (T_0 X) Y) = \Delta({}^t Y \Delta(T_0 X) Y) = 0. \tag{14}$$

Let $p(x) = p_0 + \cdots + p_{e-1} x^{e-1} + x^e$ be the minimal polynomial of M_0. From $\Delta(T_0) = 0$ we deduce that $p_{e-1} = p_{e-3} = \cdots = 0$, which contradicts the irreducibility of p.

(ii) Let $C \in \mathbb{K}[M_0]$ such that $\Delta(C) = D$; applying (13) to the symmetric matrix $T_0 C$ and using the symmetry of $T_0 M_0$ yields

$$\Delta(T_0 C X^2) = \Delta({}^t X T_0 C X) = \Delta({}^t X \Delta(T_0 C) X) = \Delta({}^t X D X). \tag{15}$$

(iii) From direct computation we find that the diagonal blocks of ${}^t X D X$ are $B_m = \sum_{i+j=m} {}^t X_i D_j X_i$; hence $\Delta(B_m) = \sum \Delta(T_0 \psi_0(D_j) X_i^2)$ and $\psi_0(B_m) = \sum \psi_0(D_j)\,\varphi(X_i)$. □

For any binary field $\mathbb{K}$, we write $\wp(\mathbb{K})$ for the set of elements $x^2 + x \in \mathbb{K}$. This is an additive subgroup of $\mathbb{K}$, and the characteristic-two analogue of the set of squares. For any element α of $\mathbb{K}[M]$, we call *valuation of regularity* of α that we simply note $\mathrm{val}(\alpha)$ the smallest integer m such that there exists an invertible α' of $\mathbb{K}[M]$ such that $\alpha = H^m \alpha'$.

Proposition 5. *Any regular pencil of quadratic forms is congruent to a pencil of the form*

$$A_\infty = \begin{pmatrix} D_1 & T \\ 0 & D_2 \end{pmatrix}, \quad A_0 = \begin{pmatrix} D_3 & TM \\ 0 & D_4 \end{pmatrix}, \tag{16}$$

where M, T are as in Prop. 3 and D_i are diagonal matrices whose values $\alpha_i = \psi(D_i)$ satisfy either one or the other of the following two kinds of canonical forms:

(i) $\alpha_1 = H^m$, $\mathrm{val}(\alpha_1 + \alpha_3) > m$, $\alpha_2 = 0$ or $\alpha_2 = \delta H^{d-1-m}$, $\mathrm{val}(\alpha_4) \geq m$, for some $m \in \{0, \ldots, d\}$, and some fixed $\delta \in \mathbb{K}[M_0] \smallsetminus \wp(\mathbb{K}[M_0])$;
(ii) $\alpha_1 = H^m$ or $\alpha_3 = H^m$, $\mathrm{val}(\alpha_1 + \alpha_3) = m$, $\alpha_2 = \alpha_4$, $\mathrm{val}(\alpha_2) > m$ for some $m \in \{0, \ldots, d\}$, and some fixed $\delta \in \mathbb{K}[M_0] \smallsetminus \wp(\mathbb{K}[M_0])$.

Proof. By Prop. 3, we may compute bases in which the pencils of polar forms have the form (8). In the same bases the pencils have the form (16) with M, T, M_0, T_0 as in Theorem 1 and D_i are some diagonal matrices. We now perform elementary transformations of the orthogonal group of $\mathcal{P}(A)$ to simplify the diagonal part of the quadratic pencil. We use the transformations $G_i(X)$ from (9) for a matrix $X = x_0 + \cdots + x_{d-1}H^{d-1} \in \mathbb{K}[M]$. The effects of the elementary transformations $G_i(X)$ on the coefficients α_i are:

$$\begin{aligned}
G_1(X): \quad & \alpha_1 \leftarrow \alpha_1 + \varphi(X)\,\alpha_2 + \psi(\Delta(TX)),\\
& \alpha_3 \leftarrow \alpha_3 + \varphi(X)\,\alpha_4 + \psi(\Delta(TX)),\\
& \alpha_2 \leftarrow \alpha_2, \quad \alpha_4 \leftarrow \alpha_4;\\
G_2(X): \quad & \alpha_2 \leftarrow \alpha_2 + \varphi(X)\,\alpha_1 + \psi(\Delta(TX)),\\
& \alpha_4 \leftarrow \alpha_4 + \varphi(X)\,\alpha_3 + \psi(\Delta(TX)),\\
& \alpha_1 \leftarrow \alpha_1, \quad \alpha_3 \leftarrow \alpha_3;\\
G_3(X): \quad & \alpha_1 \leftarrow \varphi(X)\,\alpha_1, \quad \alpha_2 \leftarrow \varphi(X^{-1})\,\alpha_2,\\
& \alpha_3 \leftarrow \varphi(X)\,\alpha_3, \quad \alpha_4 \leftarrow \varphi(X^{-1})\,\alpha_4;\\
G_4: \quad & \alpha_1 \leftrightarrow \alpha_2, \quad \alpha_3 \leftrightarrow \alpha_4.
\end{aligned}$$

.A direct computation gives

$$\psi(\Delta(TMX)) = \sum_{i \geq \frac{d-1}{2}} x_{2i-(d-1)} H^i.$$

As in Prop. 4, we write D_i as d blocks $D_{i,j}$ and define $\alpha_{i,j} = \psi_0 D_{i,j}$. From what we get above we explicit the effects of the elementary transformation $G_1(X)$ on the coefficients $\alpha_{i,j}$:

$$\begin{aligned}
G_1(X): \quad & \alpha_{1,m} \leftarrow \alpha_{1,m} + \sum_{i+j=m} \alpha_{2,i} x_j^2 \quad \text{for}\, m < \frac{d-1}{2},\\
& \alpha_{1,m} \leftarrow \alpha_{1,m} + \sum_{i+j=m} \alpha_{2,i} x_j^2 + x_{2m-(d-1)} \quad \text{for}\, m \geq \frac{d-1}{2};
\end{aligned}$$

If all $\alpha_i = 0$, we are done: the pencil is canonical. If not, we search the value α_i with smallest valuation. Using G_4, we may assume it is α_1 or α_3. We first suppose that we have $\mathrm{val}(\alpha_1 + \alpha_3) > m$, that is α_1 and α_3 have the same trailing term. We call this the case (i). Using G_3, we may assume $\alpha_1 = H^m$, and therefore $\alpha_3 = H^m + \alpha$, with $\mathrm{val}(\alpha) > m$. We look then for X such that $G_2(X)(\alpha_2) = 0$. We note that the corresponding system is triangular and all equations can be solved except maybe for this one: $\alpha_{2,d-1-m} = x_{d-1-2m}^2 + x_{d-1-2m}$. Therefore we may assume that $\alpha_2 = 0$ or $\alpha_2 = \delta H^{d-1-m}$ for some fixed $\delta \in \mathbb{K}[M_0] \smallsetminus \wp(\mathbb{K}[M_0])$. We note also that $G_2(X)$ does not decrease the valuation of α_4. We have therefore by hypothesis $\mathrm{val}(\alpha_4) \geq m$.

We now examine the case (ii) where $\mathrm{val}(\alpha_1 + \alpha_3) = m$. Using again G_3, we may assume that $\alpha_1 = H^m$ or $\alpha_3 = H^m$. Let's note $\alpha_1 + \alpha_3 = H^m \alpha$ where α is invertible. We are looking for X such that $G_2(X)(\alpha_2) = G_2(X)(\alpha_4)$. By hypothesis on the valuation, we can write $\alpha_2 + \alpha_4 = H^m \alpha'$ for some α'. We naturally choose $X = \varphi^{-1}(\alpha' \alpha^{-1})$. At this stage, we can consider that $\alpha_2 = \alpha_4$.

However, the condition on the valuation may not hold. If by chance $\mathrm{val}(\alpha_2) > m$, then we are done. If on the contrary $\mathrm{val}(\alpha_2) \leq m$, then by using G_4, we search instead for a canonical form of the kind (i). □

Theorem 3. *Let $\mathbb{K}$ be a finite field with characteristic two. The cyclic case of the IP1S problem is solvable using $\widetilde{O}(n^3)$ operations in the field $\mathbb{K}$. Moreover, in the generic case, the IP1S problem has exactly 2^s solutions, where s is the number of components within the primary decomposition of M.*

Proof. To solve the IP1S problem for two pencils A and B, we may reduce them to the same canonical form using Prop. 5, using first the primary decomposition. Following along the proof of the proposition, we see that it is constructive and that all linear algebra algorithms used require at most $\widetilde{O}(n^3)$ field operations.

Solutions of the IP1S problem correspond bijectively to automorphisms of the canonical pencil. In the generic case, the ideal generated by the values (α_1, α_2) is the full algebra $\mathbb{K}[M]$; the canonical pencil is then such that that $\alpha_1 = 1$ and $\alpha_2 \in \{0, \delta H^{d-1}\}$.

For both values of α_2, since the equation $x_{d-1}^2 + x_{d-1} = 0$ has only the solutions 0 and 1 in each component $\mathbb{K}[M_0]$, the IP1S problem has in this case exactly 2^s solutions. □

IP1S Problem for *a* and *b*: Summary and Computer Experiments. Next table gives timings of our MAGMA script SolveCyclicEvenIP1S, with the same convention as for the odd case : τ represents the observed fraction of cyclic cases and t the average computing time over these cases.

q	n	t	τ
2	32	0.07s.	96.
2	128	2.s.	95.
2	256	33.s.	94.
2^4	32	0.3s.	100.
2^7	32	0.5s.	100.

q	n	t	τ
2^8	20	0.2s.	100.
2^8	32	0.6s.	100.
2^8	80	20.s.	100.
2^8	128	133.s	100.

5 Conclusion and Future Work

We have shown that special instances of the quadratic homogeneous IP1S problem with $m = 2$ equations can be solved in polynomial time. These instances are those where the characteristic endomorphism of the pencil (or its Pfaffian when the characteristic of the field is 2) is cyclic, and represent in practice a large fraction of generic instances. In a subsequent work, we studied the case where the characteristic endomorphism is no longer cyclic and found similar results to be published – at least for odd characteristic fields. In a work still in progress, we try to extend these results to QIP1S problem with more than 2 equations, and therefore expect to confirm that QIP1S is not as hard as GI.

References

1. Bouillaguet, C.: Études d'hypothèses algorithmiques et attaques de primitives cryptographiques. PhD thesis, Université Paris-Diderot – École Normale Supérieure (2011)
2. Bouillaguet, C., Faugère, J.-C., Fouque, P.-A., Perret, L.: Practical cryptanalysis of the identification scheme based on the isomorphism of polynomial with one secret problem. In: Catalano, et al. [5], pp. 473–493
3. Bouillaguet, C., Fouque, P.-A., Macario-Rat, G.: Practical key-recovery for all possible parameters of sflash. In: Lee, D.H., Wang, X. (eds.) ASIACRYPT 2011. LNCS, vol. 7073, pp. 667–685. Springer, Heidelberg (2011)
4. Bouillaguet, C., Fouque, P.-A., Véber, A.: Graph-theoretic algorithms for the "isomorphism of polynomials" problem. IACR Cryptology ePrint Archive 2012, 607 (2012)
5. Catalano, D., Fazio, N., Gennaro, R., Nicolosi, A. (eds.): PKC 2011. LNCS, vol. 6571. Springer, Heidelberg (2011)
6. Dieudonné, J.: Pseudo-discriminant and Dickson invariant. Pacific J. Math 5, 907–910 (1955)
7. Dubois, V., Kammerer, J.-G.: Cryptanalysis of cryptosystems based on noncommutative skew polynomials. In: Catalano, et al. [5], pp. 459–472
8. Faugère, J.-C.: A new efficient algorithm for computing Gröbner bases (F4). Journal of Pure and Applied Algebra 139(1-3), 61–88 (1999)
9. Faugère, J.-C., Gianni, P., Lazard, D., Mora, T.: Efficient Computation of Zero-dimensional Gröbner Bases by Change of Ordering. Journal of Symbolic Computation 16(4), 329–344 (1993)
10. Garey, M.R., Johnson, D.S.: Computers and Intractability: A Guide to the Theory of NP-Completeness. W. H. Freeman & Co. (1979); Ch. 7.2: Algebraic Equations over GF(2)
11. Goldreich, O.: The Foundations of Cryptography — Volume 1, Basic Techniques. Cambridge University Press (2001)
12. Kipnis, A., Patarin, J., Goubin, L.: Unbalanced oil and vinegar signature schemes. In: Stern, J. (ed.) EUROCRYPT 1999. LNCS, vol. 1592, pp. 206–222. Springer, Heidelberg (1999)
13. Matsumoto, T., Imai, H.: Public quadratic polynomial-tuples for efficient signature-verification and message-encryption. In: Günther, C.G. (ed.) EUROCRYPT 1988. LNCS, vol. 330, pp. 419–453. Springer, Heidelberg (1988)
14. Milnor, J.W., Husemoller, D.: Symmetric bilinear forms. Springer-Verlag (1973)
15. Neumann, P.M., Praeger, C.E.: Cyclic matrices over finite fields. J. London Math. Soc. (2) 52(2), 263–284 (1995)
16. Patarin, J.: Hidden Fields Equations (HFE) and Isomorphisms of Polynomials (IP): Two New Families of Asymmetric Algorithms. In: Maurer, U.M. (ed.) EUROCRYPT 1996. LNCS, vol. 1070, pp. 33–48. Springer, Heidelberg (1996)
17. Patarin, J., Goubin, L., Courtois, N.T.: Improved algorithms for isomorphisms of polynomials. In: Nyberg, K. (ed.) EUROCRYPT 1998. LNCS, vol. 1403, pp. 184–200. Springer, Heidelberg (1998)
18. Patarin, J., Goubin, L., Courtois, N.T.: C^*_{-+} and HM: Variations Around Two Schemes of T. Matsumoto and H. Imai. In: Ohta, K., Pei, D. (eds.) ASIACRYPT 1998. LNCS, vol. 1514, pp. 35–50. Springer, Heidelberg (1998)
19. Perret, L.: A fast cryptanalysis of the isomorphism of polynomials with one secret problem. In: Cramer, R. (ed.) EUROCRYPT 2005. LNCS, vol. 3494, pp. 354–370. Springer, Heidelberg (2005)

20. Pointcheval, D.: A new identification scheme based on the perceptrons problem. In: Guillou, L.C., Quisquater, J.-J. (eds.) EUROCRYPT 1995. LNCS, vol. 921, pp. 319–328. Springer, Heidelberg (1995)
21. Shamir, A.: An efficient identification scheme based on permuted kernels. In: Brassard, G. (ed.) CRYPTO 1989. LNCS, vol. 435, pp. 606–609. Springer, Heidelberg (1990)
22. Stern, J.: Designing identification schemes with keys of short size. In: Desmedt, Y.G. (ed.) CRYPTO 1994. LNCS, vol. 839, pp. 164–173. Springer, Heidelberg (1994)
23. Taussky, O., Zassenhaus, H.: On the similarity transformation between a matrix and its transpose. Pacific J. Math. 9, 893–896 (1959)

A Complexity, Timings, and Other Considerations

All the experimental results have been obtained with an Opteron 850 2.2GHz, with 32 GBytes of Ram. The systems associated with the instance of the problems and their solutions have been generated using the MAGMA software, version 2.13-15. MAGMA scripts cited in this paper can be obtained from the authors.

Constructing Confidential Channels from Authenticated Channels—Public-Key Encryption Revisited

Sandro Coretti, Ueli Maurer, and Björn Tackmann

Department of Computer Science, ETH Zürich, Switzerland
{corettis,maurer,bjoernt}@inf.ethz.ch

Abstract. The security of public-key encryption (PKE), a widely-used cryptographic primitive, has received much attention in the cryptology literature. Many security notions for PKE have been proposed, including several versions of CPA-security, CCA-security, and non-malleability. These security notions are usually defined via a game that no efficient adversary can win with non-negligible probability or advantage.

If a PKE scheme is used in a larger protocol, then the security of this protocol is proved by showing a reduction of breaking a certain security property of the PKE scheme to breaking the security of the protocol. A major problem is that each protocol requires in principle its own tailor-made security reduction. Moreover, which security notion of the PKE scheme should be used in a given context is a priori not evident; the employed games model the use of the scheme abstractly through oracle access to its algorithms, and the sufficiency for specific applications is neither explicitly stated nor proven.

In this paper we propose a new approach to investigating the application of PKE, based on the constructive cryptography framework [24,25]. The basic use of PKE is to enable confidential communication from a sender A to a receiver B, assuming A is in possession of B's public key. One can distinguish two relevant cases: The (non-confidential) communication channel from A to B can be authenticated (e.g., because messages are signed) or non-authenticated. The application of PKE is shown to provide the construction of a secure channel from A to B from two (assumed) authenticated channels, one in each direction, or, alternatively, if the channel from A to B is completely insecure, the construction of a confidential channel without authenticity. Composition then means that the assumed channels can either be physically realized or can themselves be constructed cryptographically, and also that the resulting channels can directly be used in any applications that require such a channel. The composition theorem of constructive cryptography guarantees the soundness of this approach, which eliminates the need for separate reduction proofs.

We also revisit several popular game-based security notions (and variants thereof) and give them a constructive semantics by demonstrating which type of construction is achieved by a PKE scheme satisfying which notion. In particular, the necessary and sufficient security notions for the above two constructions to work are CPA-security and a variant of CCA-security, respectively.

K. Sako and P. Sarkar (Eds.) ASIACRYPT 2013 Part I, LNCS 8269, pp. 134–153, 2013.

1 Introduction

Public-key encryption (PKE) is a cryptographic primitive devised to achieve confidential communication in a context where only authenticated (but not confidential) communication channels are available [11,34]. The cryptographic security of PKE is traditionally defined in terms of a certain distinguishing game in which no efficient adversary is supposed to achieve a non-negligible advantage. There exists quite a wide spectrum of security notions and variants thereof. These notions are motivated by clearly captured attacks (e.g., a chosen-ciphertext attack) that should be prevented, but in some cases they seem to have been proposed mainly because they are stronger than previous notions or can be shown to be incomparable.

This raises the question of which security notion for PKE is suitable or necessary for a certain higher-level protocol (using PKE) to be secure. The traditional answer to this question is that for each protocol one (actually, a cryptography expert) needs to identify the right security notion and provide a reduction proof to show that a PKE satisfying this notion yields a secure protocol.[1]

An alternative approach is to capture the semantics of a security notion by characterizing directly what it achieves, making explicit in which applications it can be used securely. The constructive cryptography paradigm [24,25] was proposed with this general goal in mind. Resources such as different types of communication channels are modeled explicitly, and the goal of a cryptographic protocol or scheme π is to *construct* a stronger or more useful resource S from an assumed resource R, denoted as $R \overset{\pi}{\Longmapsto} S$. Two such construction steps can then be composed, i.e., if we additionally consider a protocol ψ that assumes the resource S and constructs a resource T, the composition theorem states that

$$R \overset{\pi}{\Longmapsto} S \quad \wedge \quad S \overset{\psi}{\Longmapsto} T \qquad \Longrightarrow \qquad R \overset{\psi \circ \pi}{\Longmapsto} T,$$

where $\psi \circ \pi$ denotes the composed protocol.

Following the constructive paradigm, a protocol is built in a modular fashion from isolated construction steps. A security proof guarantees the soundness of one such step, and each proof is independent of the remaining steps. The composition theorem then guarantees that several such steps can be composed. While the general approach to protocol design based on reduction proofs is in principle sound, it is substantially more complex, more error-prone, and not suitable for re-use. This is part of the reason why it is generally not applied to the design of real-world protocols (e.g., TLS), which in turn is the main reason for the large number of protocol flaws discovered in the past. A major goal in cryptography must be to break the cycle of flaw discovery and fixes by providing solid proofs. Modularity appears to be the key in achieving this goal.

[1] Note that this work is orthogonal to the foundational problem of designing practical PKE schemes provably satisfying certain security notions, based on realistic hardness assumptions. The seminal CCA-secure PKE scheme based on the DDH-assumption by Cramer and Shoup [9,10] falls into this category, as do, e.g., [13,32,19,21,35].

In this spirit, we treat the use of PKE as such a construction step. The contributions of this paper are two-fold. First, we show how one can construct, using PKE, confidential channels from authenticated and insecure channels (cf. Section 1.1 and Section 3). Second, we revisit several known game-based security notions (and variants thereof) and give them a constructive semantics, providing an explicit understanding of the application contexts for which a given notion is suitable (cf. Section 1.2 and Section 4). In Section 1.3 we describe how our results, although stated in a simpler setting, capture settings with multiple senders and the notion of corruption that exists in other frameworks, and in Section 1.4 we contrast the constructive paradigm with the approach of idealizing the properties of cryptographic schemes. Related work is discussed in Section 1.5.

1.1 Constructing Confidential Channels Using PKE

From the perspective of constructive cryptography [24,25], the purpose of a public-key encryption scheme is to construct a confidential channel from non-confidential channels. Here, a channel is a resource (or functionality) that involves a sender, a receiver, and—to model channels with different levels of security—an attacker. A channel generally allows the sender to transmit a message to the receiver; the security properties of a particular channel are captured by the capabilities available to the attacker, which might, e.g., include reading or modifying the messages in transmission.

The parties access the channel through interfaces that the channel provides and that are specific for each party. For example, the sender's interface allows to input messages, and the receiver's interface allows to receive them. We refer to the interfaces by labels A, B, and E, where A and B are the sender's and the receiver's interfaces, respectively, and E is the adversary's interface. In this work, we consider the following four types of channels (from A to B; channels in the opposite direction are defined analogously), using the notation from [27]:[2]

- An *insecure channel*, denoted $-\!\!\twoheadrightarrow$, allows the adversary to read, deliver, and to delete all messages input at A, as well as to inject its own messages.
- An *authenticated channel*, denoted $\bullet\!\!-\!\!\diamond\!\!\twoheadrightarrow$, still allows to read all messages, but the adversary is limited to forwarding or deleting messages input at A.
- A *confidential channel*, denoted $-\!\!\diamond\!\!\twoheadrightarrow\!\!\bullet$, only leaks the length of the messages but does not necessarily prevent injections.
- A *secure channel*, denoted $\bullet\!\!-\!\!\diamond\!\!\twoheadrightarrow\!\!\bullet$, also only leaks the message length, and only allows the adversary to forward or delete messages input at A.

To use public-key encryption, the receiver initially generates a key pair and transmits the public key to the sender. The sender needs to obtain the correct public key, which corresponds to assuming that the channel from B to A is

[2] The "•" in the notation signifies that the capabilities at the marked interface, i.e., sending or receiving, are exclusive to the respective party. If the "•" is missing, the adversary also has these capabilities. The ⋄-symbol is explained in Section 2.4.

authenticated ($\longleftarrow\!\!\bullet^3$). To transmit a message confidentially, the sender then encrypts the message under the received public key and sends the ciphertext to the receiver over a channel that could be authenticated or completely insecure.

The exact type of channel that is constructed depends on the type of assumed channel used to transmit the ciphertext to the receiver: We show that if the assumed channel is authenticated ($\bullet\!\!-\!\!\diamond\!\!\twoheadrightarrow$) and the PKE scheme is ind-cpa-secure, the constructed channel is a secure channel ($\bullet\!\!-\!\!\diamond\!\!\twoheadrightarrow\!\!\bullet$). If the assumed channel is insecure ($-\,\twoheadrightarrow$) and the PKE scheme is ind-cca-secure, the constructed channel is only confidential ($-\!\!\diamond\!\!\twoheadrightarrow\!\!\bullet$). Using the above notation, for protocols π and π' based on ind-cpa and ind-cca encryption schemes, respectively, these constructions can be written as

$$[\longleftarrow\!\!\bullet, \bullet\!\!-\!\!\diamond\!\!\twoheadrightarrow] \quad \stackrel{\pi}{\models\!\!\Longrightarrow} \quad \bullet\!\!-\!\!\diamond\!\!\twoheadrightarrow\!\!\bullet \quad \text{and} \quad [\longleftarrow\!\!\bullet, -\,\twoheadrightarrow] \quad \stackrel{\pi'}{\models\!\!\Longrightarrow} \quad -\!\!\diamond\!\!\twoheadrightarrow\!\!\bullet,$$

where the bracket notation means that both resources in the brackets are available.

The notion of constructing the confidential (or secure) channel from the two assumed non-confidential ones is made precise in a simulation-based sense [25,24], where the simulator can be interpreted as translating all attacks on the protocol into attacks on the constructed (ideal) channel. As the constructed channel is secure by definition, there are no attacks on the protocol.

The composability of the construction notion then means that the constructed channel can again be used as an assumed resource (possibly along with additional assumed or constructed resources) in other protocols. For instance, if a higher-level protocol uses the confidential channel to transmit a message together with a shared secret value in order to achieve an additionally authenticated (and hence fully secure) transmission of the message, then the proof of this protocol is based on the "idealized" confidential channel and does not (need to) include a reduction to the security of the encryption scheme. In the same spirit, the authenticated channel from B to A could be a physically authenticated channel, but it could also be constructed by using, for instance, a digital signature scheme to authenticate the transmission of the public key (which is done by certificates in practice).

1.2 Constructive Semantics of Game-Based Security Notions

Security properties for PKE are often formalized via a game between a hypothetical challenger and an attacker. We assign constructive semantics to several existing game-based definitions by first characterizing the appropriate assumed and constructed resources and then showing that the "standard use" of a PKE scheme over those channels (as illustrated in Section 1.1) achieves the construction if (and sometimes only if) it has the considered property.[4]

In particular, we show that ind-cpa-security is not only sufficient but also necessary for constructing a secure channel from two authenticated channels. For

[3] The simple arrow indicates that $\longleftarrow\!\!\bullet$ is a single-use channel, i.e., only one message can be transmitted.

[4] We point out that our negative results do *not* rule out the existence of other protocols that are derived from the scheme in some possibly more complicated way; those could still achieve the respective construction.

the construction of a confidential channel from an authenticated and an insecure channel, it turns out that ind-cca-security, while sufficient, is unnecessarily strong. The transformation only requires the weaker notion of ind-rcca-security, which was introduced by Canetti et al. [8] to avoid the artificial strictness of ind-cca. We continue the analysis of ind-cca-security and follow up on work by Bellare et al. [4], where several non-equivalent definitional variants are considered. We show that only the stricter notions they consider are sufficient for the channel construction, leaving the exact semantics of the weaker notions unclear.

We also consider non-adaptive CCA-security (ind-cca1) and non-malleability (nm-cpa). We show that both notions correspond to transformations between somewhat artificial channels, but might still be useful for specific applications.

1.3 Capturing Settings with Potentially Corrupted Senders

Although our security definitions for public-key encryption are phrased in a setting where there is only one legitimate sender (at the A-interface), our treatment can be "lifted" to a setting with multiple senders generically, cf. [29]. In a scenario with multiple senders, it is important to formulate the guarantees that are maintained if one or more of the senders deviate from the protocol because their machines are controlled by some attacker (or virus). This is captured in most security frameworks by considering an external adversary that has the capability of corrupting some of the parties. In the context of PKE and secure communication, the goal is to still provide confidentiality guarantees to non-corrupted senders. (If the receiver is corrupted, then no security can be guaranteed.)

The ability of an attacker to act on behalf of corrupted senders means that it can directly send (bogus) ciphertexts to the receiver, even if the communication to the receiver is authenticated. This capability corresponds exactly to the case of assuming only an unauthenticated channel, where the messages are injected via the E-interface. Hence, our treatment extends to the case of (static) sender corruption by considering the lifting that relates the interfaces of the parties in the multi-party scenario to the A-interface in the three-party setting, and provides all capabilities of the statically corrupted parties also at the E-interface.

In summary, the security of public-key encryption in the presence of potentially (statically) corrupted senders corresponds exactly to the construction of a confidential channel −◇−↠• from one insecure channel − −↠ and one authenticated channel ←−• in the opposite direction, as discussed in Section 1.1. This implies that in the presence of (static) corruption, ind-rcca security is required and sufficient both in the case where the channel from the sender to the receiver is authenticated, and also where it is not authenticated.

1.4 Idealizing Properties vs. Constructing Resources

The security guarantees that one requires from a cryptographic scheme can be modeled in fundamentally different ways, even within a single formal security framework. One approach, which underlies the PKE functionality $\mathcal{F}_{\text{PKE}}$ in [8], is to idealize the properties of the algorithms that comprise the scheme. Such a functionality corresponds to a cryptographic scheme, and its interfaces closely resemble the interfaces of the algorithms (although, e.g., the private key is never output by $\mathcal{F}_{\text{PKE}}$). In such a treatment, elements that are essential for using the scheme, such as the ciphertext or the public key, will still appear in the functionality, but they are idealized in that, e.g., the ciphertext is independent of the corresponding plaintext; the idealized scheme is unbreakable by definition.

Another—fundamentally different—approach is to explicitly model *resources* that are available to one or more parties. The communication channels we describe in Section 1.1 can be considered *network resources*; there are also functionalities in the UC framework, such as $\mathcal{F}_{\text{AUTH}}$ or $\mathcal{F}_{\text{SC}}$ in [7], that can be interpreted in this way. More generally, one can also think of randomness, memory, or even computation as resources of this type. Following the constructive paradigm, the guarantees of a cryptographic scheme are *not* a resource, but modeled as the guarantee that the scheme transforms one (assumed) resource into another (constructed) resource.[5] Compared to ideal functionalities of the above type, the description of resources tends to be simpler and easier to understand. For example, in the case of public-key encryption, the confidential channel does not need to specify implementation artifacts such as ciphertexts or public keys.

While both approaches allow to divide the security proof of a composite protocol into several steps that can be proven independently, only the second approach enables a fully modular protocol design. Each sub-protocol achieves a well-defined construction step transforming a resource R into a resource S, which abstracts from how S is achieved. A higher-level protocol can thus use such a resource S independently of how it is obtained, and the construction of S can be replaced with a different one without affecting the design or proof of the higher-level protocol. Concretely, a protocol using the resource $-\!\diamond\!\!\twoheadrightarrow\!\bullet$ does not depend on whether or not the channel is constructed by a PKE scheme, whereas a protocol using the functionality $\mathcal{F}_{\text{PKE}}$ will always be specific to this step.

1.5 Related Work

We provide here an abridged comparison with related work. A more comprehensive comparison can be found in the full version of this work.

Game-based security. The study of PKE security was initiated by Goldwasser and Micali [17], who introduced the notions of indistinguishability and semantic security. Yao's [36] definition, based on computational entropy, was shown equivalent to variants of [17] by Micali et al. [30]. Goldreich [14,15] made important modifications and also dealt with uniform adversaries. Today's widely-used

[5] By contrast, a typical UC security statement is that a cryptographic scheme implements some functionality. While statements about *hybrid* protocols in UC appear similar to constructive statements, they are less expressive since, e.g., the UC framework technically does not allow to make statements about assuming only *bounded* resources, as protocols that use hybrid functionalities can always instantiate arbitrarily many functionalities of a given type.

variant, *indistinguishability* under chosen-plaintext attack or ind-cpa, has been strengthened by considering more powerful attackers that can additionally obtain decryptions of arbitrary ciphertexts. This lead to the notions of ind-cca1 and ind-cca2 (e.g., [31,37]). Different variants of ind-cca2-security were compared by Bellare et al. [4]. Canetti et al. [8] introduced the weaker notion ind-rcca that suffices for many applications. A second important security property is *non-malleability*, introduced by Dolev et al. [12]. Informally, it requires that an adversary cannot change a ciphertext into one that decrypts to a related message. Variations of this notion have been considered in subsequent work [3,5].

Real-world/ideal-world security. The idea of defining protocol security with respect to an ideal execution was first proposed by Goldreich et al. [16]; the concept of a simulator can be traced back to the seminal work by Goldwasser et al. [18] on zero-knowledge proofs. General security frameworks that allow the formalization of arbitrary functionalities to be realized by cryptographic protocols have been introduced by Canetti [6] as universal composability (UC) as well as by Backes et al. [33,1] as reactive simulatability (RSIM). Treatments of PKE exist in both frameworks. The treatment in UC is with respect to an "ideal PKE" functionality; realizing this functionality is equivalent to ind-cca2-security [8]. Canetti and Krawczyk [7] formulate UC functionalities that model different types of communication channels and can be interpreted as network resources; they do not treat public-key encryption from this perspective. The formalization of the functionalities in [33] is closer to our approach, but less modular and hence formally more complex. In particular, the treatment is restricted to the case where the authenticated transmission of the ciphertexts is achieved by digital signatures instead of using a generic composition statement. More generally, both frameworks [6] and [33] are designed from a bottom-up perspective (starting from a selected machine model), whereas we follow the top-down approach of [25], which leads to simpler, more abstract definitions and statements.

Maurer et al. [26] described symmetric encryption following the constructive cryptography paradigm as the construction of confidential channels from non-confidential channels and shared keys, and compared the security definitions they obtained with game-based definitions. The goal of this work is to provide a comparable treatment for the case of PKE. In the same spirit, specific anonymity-related properties of PKE have been discussed by Kohlweiss et al. [22].

2 Preliminaries

2.1 Systems: Resources, Converters, Distinguishers, and Reductions

At the highest level of abstraction (following the hierarchy in [25]), systems are objects with interfaces by which they connect to (interfaces of) other systems; each interface is labeled with an element of a label set and connects to only a single other interface. This concept of *abstract systems* captures the topological structures that result when multiple systems are connected in this manner.

The abstract systems concept, however, does not model the behavior of systems, i.e., *how* the systems interact via their interfaces. Consequently, statements about cryptographic protocols are statements at the next (lower) abstraction level. In this work, we describe all systems in terms of (probabilistic) discrete systems, which we explain in Section 2.2.

Resources and converters. *Resources* in this work are systems with three interfaces labeled by A, B, and E. A protocol is modeled as a pair of two so-called *converters* (one for each honest party), which are directed in that they have an *inside* and an *outside* interface, denoted by in and out, respectively. As a notational convention, we generally use upper-case, bold-face letters (e.g., $\mathbf{R}$, $\mathbf{S}$) or channel symbols (e.g., •—◇—») to denote resources and lower-case Greek letters (e.g., α, β) or sans-serif fonts (e.g., enc, dec) for converters. We denote by Φ the set of all resources and by Σ the set of all converters.

The topology of a composite system is described using a term algebra, where each expression starts from one (or more) resources on the right-hand side and is subsequently extended with further terms on the left-hand side. An expression is interpreted in the way that all interfaces of the system it describes can be connected to interfaces of systems which are appended on the left. For instance, for a single resource $\mathbf{R} \in \Phi$, all its interfaces A, B, and E are accessible.

For $I \in \{A, B, E\}$, a resource $\mathbf{R} \in \Phi$, and a converter $\alpha \in \Sigma$, the expression $\alpha^I \mathbf{R}$ denotes the composite system obtained by connecting the inside interface of α to interface I of $\mathbf{R}$; the outside interface of α becomes the I-interface of the composite system. The system $\alpha^I \mathbf{R}$ is again a resource (cf. Figure 1 on page 147).

For two resources $\mathbf{R}$ and $\mathbf{S}$, $[\mathbf{R}, \mathbf{S}]$ denotes the parallel composition of $\mathbf{R}$ and $\mathbf{S}$. For each $I \in \{A, B, E\}$, the I-interfaces of $\mathbf{R}$ and $\mathbf{S}$ are merged and become the *sub-interfaces* of the I-interface of $[\mathbf{R}, \mathbf{S}]$, which we denote by $I.1$ and $I.2$. A converter α that connects to the I-interface of $[\mathbf{R}, \mathbf{S}]$ has two inside sub-interfaces, denoted by in.1 and in.2, where the first one connects to $I.1$ of $\mathbf{R}$ and the second one connects to $I.2$ of $\mathbf{S}$.

Any two converters α and β can be composed sequentially by connecting the inside interface of β to the outside interface of α, written $\beta \circ \alpha$, with the effect that $(\beta \circ \alpha)^I \mathbf{R} = \beta^I \alpha^I \mathbf{R}$. Moreover, converters can also be taken in parallel, denoted by $[\alpha, \beta]$, with the effect that $[\alpha, \beta]^I [\mathbf{R}, \mathbf{S}] = [\alpha^I \mathbf{R}, \beta^I \mathbf{S}]$.

We assume the existence of an identity converter $\mathsf{id} \in \Sigma$ with $\mathsf{id}^I \mathbf{R} = \mathbf{R}$ for all resources $\mathbf{R} \in \Phi$ and interfaces $I \in \{A, B, E\}$ and of a special converter $\bot \in \Sigma$ with an inactive outside interface.

Distinguishers. A *distinguisher* is a special type of system $\mathbf{D}$ that connects to all interfaces of a resource $\mathbf{U}$ and outputs a single bit at the end of its interaction with $\mathbf{U}$. In the term algebra, this appears as the expression $\mathbf{DU}$, which defines a binary random variable. The *distinguishing advantage of a distinguisher* $\mathbf{D}$ *on two systems* $\mathbf{U}$ *and* $\mathbf{V}$ is defined as

$$\Delta^{\mathbf{D}}(\mathbf{U}, \mathbf{V}) := |\mathsf{P}[\mathbf{DU} = 1] - \mathsf{P}[\mathbf{DV} = 1]|$$

and as $\Delta^{\mathcal{D}}(\mathbf{U}, \mathbf{V}) := \sup_{\mathbf{D} \in \mathcal{D}} \Delta^{\mathbf{D}}(\mathbf{U}, \mathbf{V})$ for a distinguisher class $\mathcal{D}$.

The distinguishing advantage measures how much the output distribution of **D** differs when it is connected to either **U** or **V**. There is an equivalence notion on systems (which is defined on the discrete systems level), denoted by $\mathbf{U} \equiv \mathbf{V}$, which implies that $\Delta^{\mathbf{D}}(\mathbf{U}, \mathbf{V}) = 0$ for all distinguishers **D**. The distinguishing advantage satisfies the triangle inequality, i.e., $\Delta^{\mathbf{D}}(\mathbf{U}, \mathbf{W}) \leq \Delta^{\mathbf{D}}(\mathbf{U}, \mathbf{V}) + \Delta^{\mathbf{D}}(\mathbf{V}, \mathbf{W})$ for all resources **U**, **V**, and **W** and distinguishers **D**.

Games. We capture games defining security properties as distinguishing problems in which an adversary **A** tries to distinguish between two *game systems* $\mathbf{G}_0$ and $\mathbf{G}_1$. Game systems (or simply *games*) are single-interface systems, which appear, similarly to resources, on the right-hand side of the expressions in the term algebra. The adversary is a distinguisher that connects to a game (instead of a resource). We denote by $\mathcal{A}$ the class of *all* adversaries for games.

Reductions. When relating two distinguishing problems, it is convenient to use a special type of system **C** that translates one setting into the other. Formally, **C** is a converter that has an *inside* and an *outside* interface. When it is connected to a system **S**, which is denoted by **CS**, the inside interface of **C** connects to the (merged) interface(s) of **S** and the outside interface of **C** is the interface of the composed system. **C** is called a *reduction system* (or simply *reduction*).

To reduce distinguishing two systems **S**, **T** to distinguishing two systems **U**, **V**, one exhibits a reduction **C** such that $\mathbf{CS} \equiv \mathbf{U}$ and $\mathbf{CT} \equiv \mathbf{V}$.[6] Then, for all distinguishers **D**, we have $\Delta^{\mathbf{D}}(\mathbf{U}, \mathbf{V}) = \Delta^{\mathbf{D}}(\mathbf{CS}, \mathbf{CT}) = \Delta^{\mathbf{DC}}(\mathbf{S}, \mathbf{T})$. The last equality follows from the fact that **C** can also be thought of as being part of the distinguisher.

2.2 Discrete Systems

Protocols that communicate by passing messages and the respective resources are described as (probabilistic) discrete systems. Their behavior can be formalized by random systems as in [23], i.e., as families of conditional probability distributions of the outputs (as random variables) given all previous inputs and outputs of the system. For systems with multiple interfaces, the interface to which an input or output is associated is specified as part of the input or output. For the restricted (but here sufficient) class of systems that for each input provide (at most) one output, an execution of a collection of systems is defined as the consecutive evaluation of the respective random systems (similarly to the models in [6,20]).

2.3 The Notion of Construction

Recall that we consider resources with interfaces A, B, and E, where A and B are interfaces of honest parties and E is the interface of the adversary. We

[6] For instance, we consider reductions from distinguishing game systems to distinguishing resources. Then, **C** connects to a game on the inside and provides interfaces A, B, and E on the outside.

formalize the security of protocols via the following notion of *construction*, which was introduced in [24] (and is a special case of the abstraction notion from [25]):

Definition 1. *Let Φ and Σ be as in Section 2.1. A protocol $\pi = (\pi_1, \pi_2) \in \Sigma^2$* constructs resource $\mathbf{S} \in \Phi$ from resource $\mathbf{R} \in \Phi$ within ε and with respect to distinguisher class $\mathcal{D}$, *denoted*

$$\mathbf{R} \quad \stackrel{(\pi,\varepsilon)}{\Longmapsto} \quad \mathbf{S},$$

if

$$\begin{cases} & \Delta^{\mathcal{D}}(\pi_1^A \pi_2^B \bot^E \mathbf{R}, \bot^E \mathbf{S}) \leq \varepsilon \quad (\textit{availability}) \\ \exists \sigma \in \Sigma : & \Delta^{\mathcal{D}}(\pi_1^A \pi_2^B \mathbf{R}, \sigma^E \mathbf{S}) \leq \varepsilon \quad (\textit{security}). \end{cases}$$

The availability condition captures that a protocol must correctly implement the functionality of the constructed resource in the absence of the adversary. The security condition models the requirement that everything the adversary can achieve in the *real-world system* (i.e., the assumed resource with the protocol) he can also accomplish in the *ideal-world system* (i.e., the constructed resource with the simulator).

An important property of Definition 1 is its composability. Intuitively, if a resource $\mathbf{S}$ is used in the construction of a larger system, then the composability implies that $\mathbf{S}$ can be replaced by a construction $\pi_1^A \pi_2^B \mathbf{R}$ without affecting the security of the composed system. Security and availability are preserved under composition. More formally, if for some resources $\mathbf{R}$, $\mathbf{S}$, and $\mathbf{T}$ and protocols π and ϕ, $\mathbf{R} \stackrel{(\pi,\varepsilon)}{\Longmapsto} \mathbf{S}$ and $\mathbf{S} \stackrel{(\phi,\varepsilon')}{\Longmapsto} \mathbf{T}$, then

$$\mathbf{R} \quad \stackrel{(\phi\circ\pi,\varepsilon+\varepsilon')}{\Longmapsto} \quad \mathbf{T},$$

as well as

$$[\mathbf{R}, \mathbf{U}] \quad \stackrel{([\pi,\mathsf{id}],\varepsilon)}{\Longmapsto} \quad [\mathbf{S}, \mathbf{U}] \quad \text{and} \quad [\mathbf{U}, \mathbf{R}] \quad \stackrel{([\mathsf{id},\pi],\varepsilon)}{\Longmapsto} \quad [\mathbf{U}, \mathbf{S}]$$

for any resource $\mathbf{U}$. More details can be found in [24].

2.4 Channels

We consider the types of channels shown on the right. Each channel initially expects a special cheating bit $b \in \{0, 1\}$ at interface E, indicating whether the adversary is present and intends to interfere with the transmission of the messages. The special converter $\bot$ (cf. Section 2.1) always sets $b = 0$. For simplicity, we will assume that whenever $\bot$ is not present, all cheating bits are set to 1.

Channel Name	Symbol	$\ell(m)$	inj
Insecure Channel	−−↠	m	✓
Confidential Channel	−◇↠•	$\lvert m\rvert$	✓
Authenticated Channel	•−◇↠	m	×
Secure Channel	•−◇↠•	$\lvert m\rvert$	×

A channel from A to B with leakage ℓ and message space $\mathcal{M} \subseteq \{0,1\}^*$ is a resource with interfaces A, B, and E and behaves as follows:[7] When the ith message $m \in \mathcal{M}$ is input at interface A, it is recorded as (i, m) and $(i, \ell(m))$ is output at interface E. When (dlv, i') is input at interface E, if (i', m') has been recorded, m' is delivered at interface B. If injections are permissible, when (inj, m') is input at interface E, m' is output at interface B.[8]

The security statements in this work are parameterized by the number of messages that are transmitted over the channels. More precisely, for each of the above channels and each $n \in \mathbb{N}$, we define the *n-bounded channel* as the one that processes (only) the first n queries at the A-interface and the first n queries at the E-interface (as described above) and ignores all further queries at these interfaces. We then require from a protocol that it constructs, for all $n \in \mathbb{N}$, the n-bounded "ideal" channel from the n-bounded assumed channel. Wherever the number n is significant, such as in the theorem statements, we denote the n-bounded versions of channels by writing the n on top of the channel symbol (e.g., $\overset{n}{-\!\!\diamond\!\!\twoheadrightarrow\!\!\bullet}$); we omit it in places that are of less formal nature.

Finally, a simple-arrow symbol (e.g., $\bullet\!\!\longrightarrow$) denotes a *single-use* channel. That is, only one message may be transmitted.

2.5 Public-Key Encryption Schemes

A public-key encryption (PKE) scheme with message space $\mathcal{M} \subseteq \{0,1\}^*$ and ciphertext space $\mathcal{C}$ is defined as three algorithms $\Pi = (K, E, D)$, where the key-generation algorithm K outputs a key pair $(\mathsf{pk}, \mathsf{sk})$, the (probabilistic) encryption algorithm E takes a message $m \in \mathcal{M}$ and a public key pk and outputs a ciphertext $c \leftarrow E_{\mathsf{pk}}(m)$, and the decryption algorithm takes a ciphertext $c \in \mathcal{C}$ and a secret key sk and outputs a plaintext $m \leftarrow D_{\mathsf{sk}}(c)$. The output of the decryption algorithm can be the special symbol $\diamond$, indicating an invalid ciphertext.

A PKE scheme is correct if $m = D_{\mathsf{sk}}(E_{\mathsf{pk}}(m))$ (with probability 1 over the randomness in the encryption algorithm) for all messages m and all key pairs $(\mathsf{pk}, \mathsf{sk})$ generated by K.

It will be more convenient to phrase bit-guessing games used in definitions of PKE security properties as a distinguishing problem between two game systems (cf. Section 2.1). We consider the following games, which correspond to the (standard) notions of ind-cpa (cpa for short), ind-cca2 (cca), ind-cca1 (cca1), ind-rcca (rcca), and nm-cpa (nm).[9] Informally, a scheme is secure in the sense of a notion if efficient adversaries have negligible advantage in distinguishing the two corresponding game systems.

---

7 If the cheating bit is set to $b = 0$, all messages input at the sender interface A are immediately delivered to B.

8 Note that none of the channels prevents the adversary from reordering or replaying messages sent over the channel. The $\diamond$-symbol suggests the "internal buffer" in which the channel stores messages input at A.

9 We consider the so-called real-or-random versions of these games, which are equivalent to the more popular left-or-right formulations (as shown in [2] for symmetric encryption). For non-malleability, we use an indistinguishability-based version by [5].

CPA game. Consider the systems $\mathbf{G}_0^{\mathsf{cpa}}$ and $\mathbf{G}_1^{\mathsf{cpa}}$ defined as follows: For a PKE scheme Π, both initially run the key-generation algorithm to obtain $(\mathsf{pk}, \mathsf{sk})$ and output pk. Upon (the first) query (chall, m), $\mathbf{G}_0^{\mathsf{cpa}}$ outputs an encryption $c \leftarrow E_{\mathsf{pk}}(m)$ of m and $\mathbf{G}_1^{\mathsf{cpa}}$ an encryption $c \leftarrow E_{\mathsf{pk}}(\bar{m})$, called the *challenge*, of a randomly chosen message $\bar{m}$ of length $|m|$.

CCA games. For $b \in \{0,1\}$, system $\mathbf{G}_b^{\mathsf{cca1}}$ proceeds as $\mathbf{G}_b^{\mathsf{cpa}}$ but additionally answers decryption queries (dec, c') before the challenge is output by returning $m' \leftarrow D_{\mathsf{sk}}(c')$. $\mathbf{G}_b^{\mathsf{cca}}$ answers decryption queries at any time unless c' equals the challenge c (if defined), in which case the answer is test.

RCCA game. Consider the systems $\mathbf{G}_0^{\mathsf{rcca}}$ and $\mathbf{G}_1^{\mathsf{rcca}}$ defined as follows: Initially, both run the key-generation algorithm to obtain $(\mathsf{pk}, \mathsf{sk})$ and output pk. Upon (the first) query (chall, m), *both* choose a random message $\bar{m}$ of length $|m|$. $\mathbf{G}_0^{\mathsf{rcca}}$ outputs $c \leftarrow E_{\mathsf{pk}}(m)$ and $\mathbf{G}_1^{\mathsf{rcca}}$ outputs $c \leftarrow E_{\mathsf{pk}}(\bar{m})$. Both systems answer decryption queries (dec, c'), but if $D_{\mathsf{sk}}(c') \in \{m, \bar{m}\}$ (if m and $\bar{m}$ are defined), the answer is test.

For more details about RCCA-security, see Section 4.2 or consult [8], where the notion was introduced.

NM game. Consider the systems $\mathbf{G}_0^{\mathsf{nm}}$ and $\mathbf{G}_1^{\mathsf{nm}}$ defined as follows: Both initially run the key-generation algorithm to obtain $(\mathsf{pk}, \mathsf{sk})$ and output pk. Upon (the first) query (chall, m), $\mathbf{G}_0^{\mathsf{nm}}$ outputs an encryption $c \leftarrow E_{\mathsf{pk}}(m)$ of m and $\mathbf{G}_1^{\mathsf{nm}}$ an encryption $c \leftarrow E_{\mathsf{pk}}(\bar{m})$ of a randomly chosen message $\bar{m}$ of length $|m|$. When a query $(\mathsf{dec}, c_1, \ldots, c_\ell)$ is input, both systems decrypt $c_1, \ldots, c_\ell$, return the resulting plaintexts (if any of the ciphertexts equal c, the corresponding plaintexts are replaced by test), and terminate the interaction.

2.6 Asymptotics

To allow for asymptotic security definitions, cryptographic protocols are often equipped with a so-called *security parameter*. We formulate all statements in this paper in a non-asymptotic fashion, but asymptotic statements can be obtained by treating systems $\mathbf{S}$ as asymptotic families $\{\mathbf{S}_\kappa\}_{\kappa \in \mathbb{N}}$ and letting the distinguishing advantage be a real-valued function of κ. Then, for a given notion of efficiency, one can consider security w.r.t. classes of efficient distinguishers and a suitable negligibility notion. All reductions in this work are efficient with respect to the standard polynomial-time notions.

3 Constructing Confidential Channels with PKE

The main purpose of public-key encryption (PKE) is to achieve confidential communication. As a constructive statement, this means that we view a PKE scheme Π as a protocol, a pair of converters $(\mathsf{enc}, \mathsf{dec})$, whose goal is to construct a confidential channel from non-confidential channels. Differentiating between the two cases where the communication from the sender to the receiver is authenticated and unauthenticated, this is written as

$$[\longleftarrow\!\bullet, \bullet\!\!-\!\!\diamond\!\!\twoheadrightarrow] \overset{(\mathsf{enc},\mathsf{dec})}{\Longrightarrow} \bullet\!\!-\!\!\diamond\!\!\twoheadrightarrow\!\bullet \ (1) \quad \text{and} \quad [\longleftarrow\!\bullet, -\!-\!\!\twoheadrightarrow] \overset{(\mathsf{enc},\mathsf{dec})}{\Longrightarrow} -\!\!\diamond\!\!\twoheadrightarrow\!\bullet, \ (2)$$

respectively.

In both cases, the *single-use* channel $\longleftarrow\!\bullet$ captures the ability of the sender to obtain the receiver's public key in an authenticated fashion. In construction (1), the communication from the sender A to the receiver B is authenticated, which is modeled by the channel $\bullet\!\!-\!\!\diamond\!\!\twoheadrightarrow$. The goal is to achieve a secure channel $\bullet\!\!-\!\!\diamond\!\!\twoheadrightarrow\!\bullet$, which only leaks the length of the messages sent at interface A. In construction (2), the communication from A to B is completely insecure, which is captured by the insecure channel $-\!-\!\!\twoheadrightarrow$. Here, the goal is to achieve a confidential channel $-\!\!\diamond\!\!\twoheadrightarrow\!\bullet$, which still hides messages input at the A-interface but also allows to inject arbitrary messages at E.

In the following, we first show how a PKE scheme Π can be seen as a converter pair $(\mathsf{enc}, \mathsf{dec})$. We then prove that $(\mathsf{enc}, \mathsf{dec})$ achieves construction (1) if the underlying PKE scheme is cpa-secure, and construction (2) if the underlying PKE scheme is cca-secure. We also briefly discuss the usefulness of the constructed channels.

3.1 PKE Schemes as Protocols

Let $\Pi = (K, E, D)$ be a PKE scheme. Based on Π, we define a pair of protocol converters $(\mathsf{enc}, \mathsf{dec})$ for constructions (1) and (2). Both converters have two subinterfaces $\mathtt{in}.1$ and $\mathtt{in}.2$ on the inside, as we connect them to a resource that is a parallel composition of two other resources (cf. Section 2.1).

Converter enc works as follows: It initially expects a public key pk at $\mathtt{in}.1$. When a message m is input at the outside interface $\mathtt{out}$, enc outputs $c \leftarrow E_{\mathsf{pk}}(m)$ at $\mathtt{in}.2$. Converter dec initially generates a key pair $(\mathsf{pk}, \mathsf{sk})$ using key-generation algorithm K and outputs pk at $\mathtt{in}.1$. When dec receives c' at $\mathtt{in}.2$, it computes $m' \leftarrow D_{\mathsf{sk}}(c')$ and, if $m' \neq \diamond$, outputs m' at the outside interface $\mathtt{out}$.

3.2 Constructing a Secure from Two Authenticated Channels

Towards proving that the protocol $(\mathsf{enc}, \mathsf{dec})$ indeed achieves construction (1), note first that the correctness of Π implies that the *availability* condition of Definition 1 is satisfied. To prove *security*, we need to exhibit a simulator σ such that the assumed resource $[\longleftarrow\!\bullet, \bullet\!\!-\!\!\diamond\!\!\twoheadrightarrow]$ with the protocol converters is indistinguishable from the constructed resource $\bullet\!\!-\!\!\diamond\!\!\twoheadrightarrow\!\bullet$ with the simulator (cf. Figure 1).

Theorem 1 implies that $(\mathsf{enc}, \mathsf{dec})$ realizes (1) if the underlying PKE scheme is cpa-secure.

Theorem 1. *There exists a simulator* σ *and for any* $n \in \mathbb{N}$ *there exists a (efficient) reduction* $\mathbf{C}$ *such that for every* $\mathbf{D}$,

$$\Delta^{\mathbf{D}}(\mathsf{enc}^A \mathsf{dec}^B [\longleftarrow\!\bullet, \overset{n}{\bullet\!\!-\!\!\diamond\!\!\twoheadrightarrow}], \sigma^E \, \overset{n}{\bullet\!\!-\!\!\diamond\!\!\twoheadrightarrow\!\bullet}) \leq n \cdot \Delta^{\mathbf{DC}}(\mathbf{G}_0^{\mathsf{cpa}}, \mathbf{G}_1^{\mathsf{cpa}}).$$

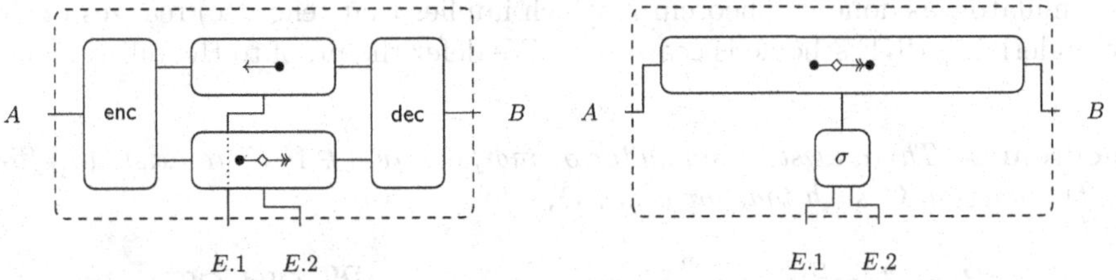

Fig. 1. Left: The assumed resource (two authenticated channels) with protocol converters enc and dec attached to interfaces A and B, denoted $\mathsf{enc}^A\mathsf{dec}^B[\longleftarrow\!\bullet, \bullet\!\!-\!\!\diamond\!\!\twoheadrightarrow]$. Right: The constructed resource (a secure channel) with simulator σ attached to the E-interface, denoted $\sigma^E \bullet\!\!-\!\!\diamond\!\!\twoheadrightarrow\!\bullet$. In particular, σ must simulate the E-interfaces of the two authenticated channels. The protocol is secure if the two systems are indistinguishable.

Proof. First, consider the following simulator σ for interface E of $\bullet\!\!-\!\!\diamond\!\!\twoheadrightarrow\!\bullet$, which has two sub-interfaces, denoted by $\mathtt{out}.1$ and $\mathtt{out}.2$, on the outside (since the real-world system has two sub-interfaces at E): Initially, σ generates a key pair $(\mathsf{pk}, \mathsf{sk})$ and outputs $(1, \mathsf{pk})$ at $\mathtt{out}.1$. When it receives (i, l) at the inside interface $\mathtt{in}$, σ generates an encryption $c \leftarrow E_{\mathsf{pk}}(\bar{m})$ of a randomly chosen message $\bar{m}$ of length l and outputs (i, c) at $\mathtt{out}.2$. When (dlv, i') is input at $\mathtt{out}.2$, σ simply outputs (dlv, i') at $\mathtt{in}$.

Consider the two systems $\mathbf{U} := \mathsf{enc}^A\mathsf{dec}^B[\longleftarrow\!\bullet, \bullet\!\!-\!\!\overset{1}{\diamond}\!\!\twoheadrightarrow]$ and $\mathbf{V} := \sigma^E \bullet\!\!-\!\!\overset{1}{\diamond}\!\!\twoheadrightarrow\!\bullet$. Distinguishing $\mathbf{G}_0^{\mathsf{cpa}}$ from $\mathbf{G}_1^{\mathsf{cpa}}$ can be reduced to distinguishing these two systems via the following reduction system $\mathbf{C}'$, which connects to a game on the inside and provides interfaces A, B, and E on the outside (cf. Section 2.1 for details on reduction systems): Initially, $\mathbf{C}'$ takes a value pk from the game (on the inside) and outputs $(1, \mathsf{pk})$ at the (outside) $E.1$-interface. When a message m is input at the A-interface of $\mathbf{C}'$, it is passed as (chall, m) to the game. The resulting challenge c is output as $(1, c)$ at the $E.2$-interface. When $(\mathsf{dlv}, 1)$ is input at the $E.2$-interface, $\mathbf{C}'$ outputs m at interface B.

We have $\mathbf{C}'\mathbf{G}_0^{\mathsf{cpa}} \equiv \mathbf{U}$ and $\mathbf{C}'\mathbf{G}_1^{\mathsf{cpa}} \equiv \mathbf{V}$, and thus

$$\begin{aligned}\Delta^{\mathbf{D}}(\mathsf{enc}^A\mathsf{dec}^B[\longleftarrow\!\bullet, \bullet\!\!-\!\!\overset{n}{\diamond}\!\!\twoheadrightarrow], \sigma^E \bullet\!\!-\!\!\overset{n}{\diamond}\!\!\twoheadrightarrow\!\bullet) &\leq n \cdot \Delta^{\mathbf{DC}''}(\mathbf{U}, \mathbf{V}) \\ &= n \cdot \Delta^{\mathbf{DC}''}(\mathbf{C}'\mathbf{G}_0^{\mathsf{cpa}}, \mathbf{C}'\mathbf{G}_1^{\mathsf{cpa}}) \\ &= n \cdot \Delta^{\mathbf{DC}}(\mathbf{G}_0^{\mathsf{cpa}}, \mathbf{G}_1^{\mathsf{cpa}}),\end{aligned}$$

where $\mathbf{C} := \mathbf{C}''\mathbf{C}'$ and the first inequality follows from a standard hybrid argument for a reduction system $\mathbf{C}''$ (deferred to the full version). □

3.3 Confidential Channels from Authenticated and Insecure Ones

To prove that the protocol $(\mathsf{enc}, \mathsf{dec})$ achieves construction (2), we need to again exhibit a simulator σ such that the assumed resource $[\longleftarrow\!\bullet, -\!\!\twoheadrightarrow]$ with the protocol converters is indistinguishable from the constructed resource $-\!\!\diamond\!\!\twoheadrightarrow\!\bullet$ with the simulator, as done in Theorem 2, which implies that (enc, dec) realizes (2) if the underlying PKE scheme is cca-secure. We defer the proof to the full version.

Theorem 2. *There exists a simulator σ and for any $n \in \mathbb{N}$ there exists a (efficient) reduction* $\mathbf{C}$ *such that for every* $\mathbf{D}$,

$$\Delta^{\mathbf{D}}(\mathsf{enc}^A\mathsf{dec}^B[\longleftarrow\!\bullet, \overset{n}{-\!\!\twoheadrightarrow}], \sigma^E \overset{n}{-\!\!\diamond\!\!\twoheadrightarrow\!\bullet}) \leq n \cdot \Delta^{\mathbf{DC}}(\mathbf{G}_0^{\mathsf{cca}}, \mathbf{G}_1^{\mathsf{cca}}).$$

The confidential channel $-\!\!\diamond\!\!\twoheadrightarrow\!\bullet$ is the best channel one can construct from the two assumed channels. As the E-interface has the same capabilities as the A-interface at both the authenticated (from B to A) and the insecure channels, it will necessarily also be possible to inject messages to the receiver via the E-interface by simply applying the sender's protocol converter.

3.4 Applicability of the Constructed Channels

The plain use of PKE yields constructions (1) and (2), i.e., one obtains the resources $\bullet\!\!-\!\!\diamond\!\!\twoheadrightarrow\!\bullet$ and $-\!\!\diamond\!\!\twoheadrightarrow\!\bullet$. Both channels allow the adversary to reorder or replace the messages sent by A. In practice, where PKE is often used to encapsulate symmetric keys, it is important, however, that keys used in various protocols by different users are independent. Thus, it is more useful to obtain independent single-use channels $[\bullet\!\!\longrightarrow\!\!\!\!\!\!\rightarrowtail\!\bullet, \ldots, \bullet\!\!\longrightarrow\!\!\!\!\!\!\rightarrowtail\!\bullet]$ and $[\longrightarrow\!\!\!\!\!\!\rightarrowtail\!\bullet, \ldots, \longrightarrow\!\!\!\!\!\!\rightarrowtail\!\bullet]$ instead of $\bullet\!\!-\!\!\diamond\!\!\twoheadrightarrow\!\bullet$ and $-\!\!\diamond\!\!\twoheadrightarrow\!\bullet$, respectively.

In the authenticated setting, given independent authenticated channels, protocol (enc, dec) (with only formal modifications) achieves the construction

$$[\longleftarrow\!\bullet, \bullet\!\!\longrightarrow\!\!\!\!\!\!\rightarrowtail, \ldots, \bullet\!\!\longrightarrow\!\!\!\!\!\!\rightarrowtail] \quad \overset{(\mathsf{enc},\mathsf{dec})}{\Longmapsto} \quad [\bullet\!\!\longrightarrow\!\!\!\!\!\!\rightarrowtail\!\bullet, \ldots, \bullet\!\!\longrightarrow\!\!\!\!\!\!\rightarrowtail\!\bullet].$$

In the unauthenticated setting, however, the analogous construction

$$[\longleftarrow\!\bullet, \longrightarrow\!\!\!\!\!\!\rightarrowtail, \ldots, \longrightarrow\!\!\!\!\!\!\rightarrowtail] \quad \overset{(\mathsf{enc},\mathsf{dec})}{\Longmapsto} \quad [\longrightarrow\!\!\!\!\!\!\rightarrowtail\!\bullet, \ldots, \longrightarrow\!\!\!\!\!\!\rightarrowtail\!\bullet]$$

is not achieved by (enc, dec) since, due to the absence of authenticity, the adversary can freely take a ciphertext it observes on one of the insecure channels $\longrightarrow\!\!\!\!\!\!\rightarrowtail$ and insert it into another one. Thus, the ideal resource cannot consist of independent channels. This issue can be taken care of by (explicitly) introducing session identifiers (SIDs). A systematic treatment of handling multiple sessions and senders can be found in [29].

4 Constructive Semantics of Game-Based Notions

We analyze several game-based security notions from a constructive viewpoint. We complete the analysis of cpa-security from Section 3.2 by showing that it is also necessary to achieve construction (1). Moreover, we explain why the notion of cca is unnecessarily strict for construction (2) and that the construction in fact only requires the weaker notion of rcca introduced in [8].

Then, we follow up on work by Bellare et al. [4], who compared several variants of defining cca-security, and showed that only the stricter notions they consider are sufficient for construction (2). We also provide constructive semantics for non-adaptive chosen-ciphertext security and non-malleability.

4.1 CPA Security Is Necessary for Construction (1)

We prove in Section 3.2 that indistinguishability under chosen-plaintext attacks, cpa-security, suffices to construct a secure channel from two authenticated channels. Here, we show that it is also necessary. That is, if protocol (enc, dec), based on a PKE scheme Π as shown in Section 3.1, achieves the construction, then Π must be cpa-secure.

In the following, let

$$\mathbf{U} := \mathsf{enc}^A \mathsf{dec}^B [\longleftarrow\!\!\bullet, \bullet\!\!-\!\!\diamond\!\!\twoheadrightarrow] \qquad \text{and} \qquad \mathbf{V} := \sigma^E \bullet\!\!-\!\!\diamond\!\!\twoheadrightarrow\!\!\bullet,$$

where σ is an *arbitrary* simulator.

Theorem 3. *There exist (efficient) reductions* $\mathbf{C}_0$ *and* $\mathbf{C}_1$ *such that for all adversaries* $\mathbf{A}$,

$$\Delta^{\mathbf{A}}(\mathbf{G}_0^{\mathsf{cpa}}, \mathbf{G}_1^{\mathsf{cpa}}) \leq \Delta^{\mathbf{A}\mathbf{C}_0}(\mathbf{U}, \mathbf{V}) + \Delta^{\mathbf{A}\mathbf{C}_1}(\mathbf{U}, \mathbf{V}).$$

Proof. Consider the following reduction systems $\mathbf{C}_0$ and $\mathbf{C}_1$, both connecting to an $\{A, B, E\}$-resource on the inside and providing a single interface on the outside (for the adversary): Initially, both obtain $(1, \mathsf{pk})$ at the inside $E.1$-interface and output pk at the outside interface. When (chall, m) is received on the outside, $\mathbf{C}_0$ outputs m at the inside A-interface and $\mathbf{C}_1$ a randomly chosen message $\bar{m}$ of length $|m|$. Subsequently, $(1, c)$ is received at the inside $E.2$-interface, and c is output (as the challenge) on the outside by both systems. We have

$$\mathbf{C}_0\mathbf{U} \equiv \mathbf{G}_0^{\mathsf{cpa}} \qquad \text{and} \qquad \mathbf{C}_1\mathbf{U} \equiv \mathbf{G}_1^{\mathsf{cpa}} \qquad \text{and} \qquad \mathbf{C}_0\mathbf{V} \equiv \mathbf{C}_1\mathbf{V},$$

where the last equivalence follows from the fact that, in $\mathbf{V}$, the input from $\bullet\!\!-\!\!\diamond\!\!\twoheadrightarrow\!\!\bullet$ to σ is the same in both systems (the length of the message input at the A-interface of $\bullet\!\!-\!\!\diamond\!\!\twoheadrightarrow\!\!\bullet$), and therefore they behave identically. Hence,

$$\begin{aligned}
\Delta^{\mathbf{A}}(\mathbf{G}_0^{\mathsf{cpa}}, \mathbf{G}_1^{\mathsf{cpa}}) &= \Delta^{\mathbf{A}}(\mathbf{C}_0\mathbf{U}, \mathbf{C}_1\mathbf{U}) \\
&\leq \Delta^{\mathbf{A}}(\mathbf{C}_0\mathbf{U}, \mathbf{C}_0\mathbf{V}) + \Delta^{\mathbf{A}}(\mathbf{C}_0\mathbf{V}, \mathbf{C}_1\mathbf{V}) + \Delta^{\mathbf{A}}(\mathbf{C}_1\mathbf{V}, \mathbf{C}_1\mathbf{U}) \\
&= \Delta^{\mathbf{A}\mathbf{C}_0}(\mathbf{U}, \mathbf{V}) + \Delta^{\mathbf{A}\mathbf{C}_1}(\mathbf{U}, \mathbf{V}).
\end{aligned}$$

□

4.2 RCCA Security Is Necessary for Construction (2)

Indistinguishability under chosen-ciphertext attacks, cca-security, suffices to construct a confidential channel from an authenticated and an insecure one (cf. Section 3.3). It is, however, unnecessarily strict, as can be seen from the following example, adapted from [8]: Let Π be a PKE scheme and assume it is cca-secure. Consider a modified scheme Π' that works exactly as Π, except that a 0-bit is appended to every encryption, which is ignored during decryption. It is easily seen that Π' is not cca-secure, since the adversary can obtain a decryption of the challenge ciphertext by flipping its last bit and submitting the result to the decryption oracle. PKE scheme Π' can, however, still be used to achieve construction (2) using a simulator that issues the dlv-instruction to —◇—↠• whenever a recorded ciphertext is received at the outside interface or one where flipping the last bit results in a recorded ciphertext (cf. full version for more details).

Canetti et al. [8] introduced the notion of *replayable chosen ciphertext* security, rcca, which is more permissive in that it allows the adversary to transform a ciphertext into one that decrypts to the same message. In the full version of this paper, we show that if protocol (enc, dec), based on a PKE scheme Π (cf. Section 3.1), achieves (2), then Π must be rcca-secure, and that rcca is also sufficient for the construction if the message space of Π is sufficiently large.

4.3 Variants of Chosen-Ciphertext Security

Bellare et al. [4] analyze several ways of enforcing the condition that the adversary must not query the challenge ciphertext c to the decryption oracle. They consider modifications along two axes: First, the condition can be enforced during the entire game (b for *both* phases) or only in the second phase (s for *second* phase), i.e., after the c has been given to the adversary. Second, one can either exclude adversaries with a non-zero probability of violating the condition from consideration (e for *exclusion*) or penalize an adversary (by declaring the game lost) whenever he asks the challenge c (p for *penalty*). The combination of these choices yields four *non-equivalent* notions ind-cca-sp, ind-cca-se, ind-cca-bp, ind-cca-be. The s-notions are equivalent to each other and to our formulation of cca-security (cf. Section 2.5). The e-notions are strictly weaker and do in fact not even imply cca1-security [4]. Since cca1-security is weaker than rcca-security and rcca is needed for construction (2), they are not sufficient for (2).

4.4 Non-malleability

Informally, a non-malleable PKE scheme is such that the adversary cannot transform a ciphertext into one that decrypts to a related message. We consider the notion of non-malleability under chosen-plaintext attacks, nm-cpa, and show that from a PKE scheme with this property we can build a protocol (enc″, dec″) that achieves the construction

$$[\longleftarrow\!\bullet,\ -\!-\!\twoheadrightarrow\!|] \quad \overset{(\mathsf{enc}'',\mathsf{dec}'')}{\Longmapsto} \quad -\!\diamond\!\twoheadrightarrow\!\!\!\!\!|\bullet, \qquad (3)$$

where $-\!\!-\!\!\twoheadrightarrow\!\!\!|$ works like $-\!\!-\!\!\twoheadrightarrow$ but halts when halt is input at B and where the channel $-\!\!\diamond\!\!-\!\!\twoheadrightarrow\!\!\!|\!\!\bullet$ is defined as follows: It internally keeps an initially empty list $\mathcal{L}$ of messages. When the ith message m is input at interface A, it is recorded as (i, m) and $(i, |m|)$ is output at interface E. When (dlv, i') is input at interface E and if (i', m') has been recorded, m' is appended to $\mathcal{L}$. When (inj, m') is input at interface E, m' is appended to $\mathcal{L}$. When $\mathsf{dlv\text{-}all}$ is input at B, all messages in $\mathcal{L}$ are output at B, and the channel halts.

The protocol converters $(\mathsf{enc}'', \mathsf{dec}'')$ are built as $(\mathsf{enc}, \mathsf{dec})$ in Section 3.1, except that dec'' only outputs the messages it received once $\mathsf{dlv\text{-}all}$ is input at the outside interface, at which time it also outputs halt at its inside interface and halts. In the full version of this paper, we prove that $(\mathsf{enc}'', \mathsf{dec}'')$ achieves construction (3) if Π is nm-cpa-secure.

The assumed channel $-\!\!-\!\!\twoheadrightarrow\!\!\!|$ could itself be constructed in a setting where A and B have synchronized clocks and B buffers all messages until an agreed point in time, when A also stops sending. By the composition theorem, the channel that is constructed in this manner can then serve as the assumed channel in construction (3) to construct the channel $-\!\!\diamond\!\!-\!\!\twoheadrightarrow\!\!\!|\!\!\bullet$ using PKE. This channel may then for instance be useful for running a protocol implementing a blind auction.

4.5 Non-adaptive Chosen-Ciphertext Security

ind-cca1-security, is defined via a game $\mathbf{G}^{\mathsf{cca1}}$, which works as $\mathbf{G}^{\mathsf{cca}}$ except that no decryption queries are answered once the adversary has been given the challenge ciphertext. The most natural way to translate this into a constructive statement is to consider the construction of a (type of) confidential channel $\circ\!\!-\!\!\diamond\!\!-\!\!\twoheadrightarrow\!\!\bullet$ where the adversary can inject messages at interface E only as long as no message has been input at A from an insecure channel $\circ\!\!-\!-\!\!\twoheadrightarrow$ with the same property.

In the full version of this paper, we show that protocol $(\mathsf{enc}, \mathsf{dec})$ built from a cca1-secure PKE scheme Π as in Section 3.1 achieves

$$[\longleftarrow\!\!\bullet, \circ\!\!-\!-\!\!\twoheadrightarrow] \quad \overset{(\mathsf{enc}'',\mathsf{dec}'')}{\Longrightarrow} \quad \circ\!\!-\!\!\diamond\!\!-\!\!\twoheadrightarrow\!\!\bullet. \tag{4}$$

Although this construction seems artificial, as with construction (3), it can be used in any setting where the assumed channel is an appropriate modeling of an available physical channel (or can itself be constructed from such a channel).

5 Conclusions

The purpose of this paper is to present the basic ways of applying PKE (within a larger protocol) as constructive steps, to be used for the modular design of complex protocols, thus taming the complexity of security-protocol design. To be ultimately applicable to full-fledged real-world protocols, other relevant cryptographic primitives also need to be modeled in the same way. While for symmetric encryption and MACs this was explained in [28,26], and for commitments in [25], treating digital signatures and other cryptographic schemes and security mechanisms (sequence numbers, session identifiers, etc.) in constructive cryptography is left for future work (cf. [29]).

Acknowledgments. The work was supported by the Swiss National Science Foundation (SNF), project no. 200020-132794.

References

1. Backes, M., Pfitzmann, B., Waidner, M.: The Reactive Simulatability (RSIM) Framework for Asynchronous Systems. Information and Computation 205(12), 1685–1720 (2007)
2. Bellare, M., Desai, A., Jokipii, E., Rogaway, P.: A Concrete Security Treatment of Symmetric Encryption. In: 38th FOCS, pp. 394–403 (1997)
3. Bellare, M., Desai, A., Pointcheval, D., Rogaway, P.: Relations Among Notions of Security for Public-Key Encryption Schemes. In: Krawczyk, H. (ed.) CRYPTO 1998. LNCS, vol. 1462, pp. 26–45. Springer, Heidelberg (1998)
4. Bellare, M., Hofheinz, D., Kiltz, E.: Subtleties in the Definition of IND-CCA: When and How Should Challenge-Decryption be Disallowed? Cryptology ePrint Archive 2009/418 (2009)
5. Bellare, M., Sahai, A.: Non-malleable Encryption: Equivalence between Two Notions, and an Indistinguishability-Based Characterization. In: Wiener, M. (ed.) CRYPTO 1999. LNCS, vol. 1666, pp. 519–536. Springer, Heidelberg (1999)
6. Canetti, R.: Universally Composable Security: A New Paradigm for Cryptographic Protocols. Cryptology ePrint Archive, Report 2000/067 (2000)
7. Canetti, R., Krawczyk, H.: Universally Composable Notions of Key Exchange and Secure Channels. In: Knudsen, L.R. (ed.) EUROCRYPT 2002. LNCS, vol. 2332, pp. 337–351. Springer, Heidelberg (2002)
8. Canetti, R., Krawczyk, H., Nielsen, J.B.: Relaxing Chosen-Ciphertext Security. In: Boneh, D. (ed.) CRYPTO 2003. LNCS, vol. 2729, pp. 565–582. Springer, Heidelberg (2003)
9. Cramer, R., Shoup, V.: A Practical Public Key Cryptosystem Provably Secure Against Adaptive Chosen Ciphertext Attack. In: Krawczyk, H. (ed.) CRYPTO 1998. LNCS, vol. 1462, pp. 13–25. Springer, Heidelberg (1998)
10. Cramer, R., Shoup, V.: Design and Analysis of Practical Public-Key Encryption Schemes Secure against Adaptive Chosen Ciphertext Attack. SIAM Journal on Computing 33, 167–226 (2001)
11. Diffie, W., Hellman, M.E.: New Directions in Cryptography. IEEE Transactions on Information Theory 22(6), 644–654 (1976)
12. Dolev, D., Dwork, C., Naor, M.: Non-Malleable Cryptography (Extended Abstract). In: 23rd ACM STOC, pp. 542–552 (1991)
13. Fujisaki, E., Okamoto, T., Pointcheval, D., Stern, J.: RSA-OAEP Is Secure under the RSA Assumption. In: Kilian, J. (ed.) CRYPTO 2001. LNCS, vol. 2139, pp. 260–274. Springer, Heidelberg (2001)
14. Goldreich, O.: Foundations of Cryptography. Class Notes. Technion University (Spring 1989)
15. Goldreich, O.: A Uniform-Complexity Treatment of Encryption and Zero-Knowledge. Journal of Cryptology 6(1), 21–53 (1993)
16. Goldreich, O., Micali, S., Wigderson, A.: How to Play any Mental Game or A Completeness Theorem for Protocols with Honest Majority. In: 19th ACM STOC, pp. 218–229 (1987)
17. Goldwasser, S., Micali, S.: Probabilistic Encryption. Journal of Computer and System Sciences 28(2), 270–299 (1984)

18. Goldwasser, S., Micali, S., Rackoff, C.: The Knowledge Complexity of Interactive Proof-Systems (Extended Abstract). In: 17th ACM STOC, pp. 291–304 (1985)
19. Hofheinz, D., Kiltz, E.: Practical Chosen Ciphertext Secure Encryption from Factoring. In: Joux, A. (ed.) EUROCRYPT 2009. LNCS, vol. 5479, pp. 313–332. Springer, Heidelberg (2009)
20. Hofheinz, D., Shoup, V.: GNUC: A New Universal Composability Framework. Cryptology ePrint Archive, Report 2011/303 (2011)
21. Kiltz, E., Pietrzak, K., Stam, M., Yung, M.: A New Randomness Extraction Paradigm for Hybrid Encryption. In: Joux, A. (ed.) EUROCRYPT 2009. LNCS, vol. 5479, pp. 590–609. Springer, Heidelberg (2009)
22. Kohlweiss, M., Maurer, U., Onete, C., Tackmann, B., Venturi, D.: Anonymity-Preserving Public-Key Encryption: A Constructive Approach. In: De Cristofaro, E., Wright, M. (eds.) PETS 2013. LNCS, vol. 7981, pp. 19–39. Springer, Heidelberg (2013)
23. Maurer, U.M.: Indistinguishability of Random Systems. In: Knudsen, L.R. (ed.) EUROCRYPT 2002. LNCS, vol. 2332, pp. 110–132. Springer, Heidelberg (2002)
24. Maurer, U.: Constructive Cryptography – A New Paradigm for Security Definitions and Proofs. In: Mödersheim, S., Palamidessi, C. (eds.) TOSCA 2011. LNCS, vol. 6993, pp. 33–56. Springer, Heidelberg (2012)
25. Maurer, U., Renner, R.: Abstract Cryptography. In: Chazelle, B. (ed.) The Second Symposium in Innovations in Computer Science, ICS 2011. pp. 1–21. Tsinghua University Press (January 2011)
26. Maurer, U., Rüedlinger, A., Tackmann, B.: Confidentiality and Integrity: A Constructive Perspective. In: Cramer, R. (ed.) TCC 2012. LNCS, vol. 7194, pp. 209–229. Springer, Heidelberg (2012)
27. Maurer, U., Schmid, P.E.: A Calculus for Security Bootstrapping in Distributed Systems. Journal of Computer Security 4(1), 55–80 (1996)
28. Maurer, U., Tackmann, B.: On the Soundness of Authenticate-then-Encrypt: Formalizing the Malleability of Symmetric Encryption. In: ACM CCS, pp. 505–515 (2010)
29. Maurer, U., Tackmann, B., Coretti, S.: Key Exchange with Unilateral Authentication: Composable Security Definition and Modular Protocol Design. Cryptology ePrint Archive, Report 2013/555 (2013)
30. Micali, S., Rackoff, C., Sloan, B.: The Notion of Security for Probabilistic Cryptosystems. SIAM Journal on Computing 17(2), 412–426 (1988)
31. Naor, M., Yung, M.: Public-key Cryptosystems Provably Secure Against Chosen Ciphertext Attacks. In: 22nd ACM STOC, pp. 427–437 (1990)
32. Peikert, C., Waters, B.: Lossy Trapdoor Functions and Their Applications. In: 40th ACM STOC, pp. 187–196 (2008)
33. Pfitzmann, B., Waidner, M.: A Model for Asynchronous Reactive Systems and its Application to Secure Message Transmission. In: IEEE Symposium on Security and Privacy, pp. 184–200 (2001)
34. Rivest, R.L., Shamir, A., Adleman, L.M.: A Method for Obtaining Digital Signatures and Public-Key Cryptosystems. Communications of the ACM 21(2), 120–126 (1978)
35. Rosen, A., Segev, G.: Chosen-Ciphertext Security via Correlated Products. In: Reingold, O. (ed.) TCC 2009. LNCS, vol. 5444, pp. 419–436. Springer, Heidelberg (2009)
36. Yao, A.C.C.: Theory and Applications of Trapdoor Functions (Extended Abstract). In: 23rd FOCS, pp. 80–91 (1982)
37. Zheng, Y., Seberry, J.: Practical Approaches to Attaining Security against Adaptively Chosen Ciphertext Attacks. In: Brickell, E.F. (ed.) CRYPTO 1992. LNCS, vol. 740, pp. 292–304. Springer, Heidelberg (1993)

Reset Indifferentiability and Its Consequences

Paul Baecher[1], Christina Brzuska[2], and Arno Mittelbach[1]

[1] Darmstadt University of Technology, Germany
[2] Tel-Aviv University, Israel

Abstract. The equivalence of the random-oracle model and the ideal-cipher model has been studied in a long series of results. Holenstein, Künzler, and Tessaro (STOC, 2011) have recently completed the picture positively, assuming that, roughly speaking, equivalence is indifferentiability from each other. However, under the stronger notion of reset indifferentiability this picture changes significantly, as Demay et al. (EUROCRYPT, 2013) and Luykx et al. (ePrint, 2012) demonstrate.

We complement these latter works in several ways. First, we show that any simulator satisfying the reset indifferentiability notion must be stateless and pseudo deterministic. Using this characterization we show that, with respect to reset indifferentiability, two ideal models are either equivalent or incomparable, that is, a model cannot be strictly stronger than the other model. In the case of the random-oracle model and the ideal-cipher model, this implies that the two are incomparable. Finally, we examine weaker notions of reset indifferentiability that, while not being able to allow composition in general, allow composition for a large class of multi-stage games. Here we show that the seemingly much weaker notion of 1-reset indifferentiability proposed by Luykx et al. is equivalent to reset indifferentiability. Hence, the impossibility of coming up with a reset-indifferentiable construction transfers to the setting where only one reset is permitted, thereby re-opening the quest for an achievable and meaningful notion in between the two variants.

1 Introduction

Idealized Models. The standard approach to cryptographic security is to reduce the security of a scheme to a (hopefully) well-studied algebraic or combinatorial complexity assumption. Unfortunately, a large number of cryptographic schemes does not admit a security reduction in the standard model. In these cases, the community often resorts to an idealized model, where we can sometimes obtain a proof of security. It is, of course, highly controversial whether or not proofs in idealized models are acceptable, but there is a tendency to prefer an analysis in an idealized model over the utter absence of any proof at all—in particular, when one is concerned with schemes that are widely deployed in practice [5,6,9].

Arguably the most popular model of this kind is the random-oracle model (ROM) where all parties have oracle access to a public, randomly chosen function [4]. Somewhat related is the ideal-cipher model (ICM) which gives all parties oracle access to a public, randomly chosen (keyed) blockcipher [21]. Knowing

K. Sako and P. Sarkar (Eds.) ASIACRYPT 2013 Part I, LNCS 8269, pp. 154–173, 2013.

that there is a close relation between pseudorandom functions and pseudorandom permutations—namely existential equivalence—one could suspect that the random-oracle model and the ideal-cipher model are equivalent, too. However, formalizing the notion of equivalence is delicate and so are the proofs.

Equivalence of the ROM and ICM under Indifferentiability. Maurer, Renner and Holenstein [19] introduced the concept of indifferentiability, which since then has been regarded as the prevalent and actually only notion of equivalence between ideal primitives. A construction G^{π} with access to some primitive π is called indifferentiable from another ideal primitive Π, if there is a simulator $\mathcal{S}$ such that the construction G^{π} implements an oracle that is indistinguishable from Π, even if the distinguisher $\mathcal{D}$ additionally gets access to π. Now, demanding the distinguisher $\mathcal{D}$ to distinguish (G^{π}, π) from Π is of little sense. Additionally to the oracle Π, the distinguisher gets access to the simulator $\mathcal{S}$ which tries to emulate π's behavior consistently with Π. Thus, the distinguisher tries to distinguish the pair of oracles (G^{π}, π) from the pair of oracles $(\Pi, \mathcal{S}^{\Pi})$.

In the case of the ideal-cipher model and the random-oracle model, considerable effort has led to a proof of equivalence [11,12,17] under indifferentiability. The reason why indifferentiability was considered a suitable notion of equivalence is the appealing composition theorem established by Maurer et al. [19]. Namely, they transform any reductionist argument in the presence of the ideal primitive Π into a proof that relies on the existence of π only. Their theorem, thus, transforms a reduction $\mathcal{R}$ into a reduction $\mathcal{R}'$, where the latter locally implements a single copy of the simulator $\mathcal{S}$. Jumping ahead, it will turn out that in this step, they rely on an implicit assumption.

Multi-Stage Adversaries. Ristenpart et al. [20] were the first to point out scenarios where indifferentiability of G^{π} from Π was not sufficient to replace Π by G^{π}. Their counterexamples involve adversaries that run in multiple stages, i.e., an adversary $\mathcal{A}$ consists of two or more sub-adversaries, say $\mathcal{A} = (\mathcal{A}_1, \mathcal{A}_2)$, that do not share state (or at least not arbitrary state). Now, a reduction $\mathcal{R}$ that reduces to such a *multi-stage game* also needs to be split into two parts $(\mathcal{R}_1, \mathcal{R}_2)$ where the same restriction upon the sharing of state applies. Hence, for the composition theorem by Maurer et al., each part of the reduction $\mathcal{R}_1$ and $\mathcal{R}_2$ needs to implement its own, independent copy of the simulator $\mathcal{S}$. However, in this case, the two copies of the simulator will not necessarily behave in the same way as opposed to the "real" primitive π which is, roughly, what makes the composition theorem collapse in the setting of multi-stage games.

Curiously, their composition holds in the presence of *strong*, colluding adversaries, while it does not in the setting of weaker, non-colluding ones. Usually in cryptography, a conservative approach corresponds to considering the strongest possible adversary, as a primitive that is secure against a strong adversary is also secure against a weaker adversary. However, the indifferentiability composition theorem is not, by itself, a security model or a proof of security. Instead, it is a tool to transform any proof in a security model in the presence of one ideal primitive into a security proof in the same security model in the presence of another ideal primitive. Hence, one tries to cover any type of security model, which, in particular, includes security models where stage-sharing adversaries can mount trivial attacks. And thus, a conservative approach in the setting of indifferentiability demands including also weaker, namely non-colluding state-sharing adversaries. Technically, the composition theorem is *harder* to prove for *weaker* adversaries, because it transforms an adversary of one type into another adversary of the same type. Considering a stronger adversary corresponds to a stronger assumption in the theorem, but also to a harder statement to prove, and vice versa for weaker adversaries.

One might hope that the distinction is of technical interest only. Unfortunately, as we argue, in basically all real-life scenarios, we need to consider multi-stage adversaries. Ristenpart et al. give several examples of multi-stage games for notions such as deterministic encryption [1,2], key-dependent message security [8], related-key attacks [3], and non-malleable hash functions [10]. On the other hand, many classical notions of security seem inherently single stage: IND-CPA or IND-CCA security for encryption, or signature schemes which are existentially unforgeable under (adaptive) chosen message attacks. However, any classical definition of security becomes multi staged if it is augmented with a leakage oracle. The reason is that, in the random oracle model, every party should have access to the random oracle. In particular, this includes the leakage oracle and the adversarially specified leakage function, resulting in an implicit second stage [14]. Hence, whenever side-channel attacks are reflected in a model, adversaries act at least in two stages—and for real-life applications, we cannot discard side-channel attacks.

In order to cope with the new challenge of multi-stage adversaries, Ristenpart et al. put forward a strengthened notion called *reset indifferentiability*. Roughly speaking, in this game, the distinguisher may reset the simulator's internal state between any two queries. Returning to ROM/ICM equivalence, an inspection of the simulators defined in [11] and [17] (as well as [12], for that matter) reveals that their behavior varies substantially with their state and, thus, they are not reset indifferentiable.

Equivalence of the ROM and ICM under Reset Indifferentiability. As plain indifferentiability is not sufficient to argue that two primitives are equivalent, the question regarding the ideal cipher model and the random oracle model is, thus, again open. Building on first negative results from [20], the authors of [13,18] have recently shown that reset-indifferentiable constructions cannot be built via domain extension, thereby ruling out constructions from ideal ciphers that are reset indifferentiable from a random oracle; note that random oracles are usually perceived as having an infinite domain while ideal ciphers have a finite domain. With this result at hand, we thus know that ideal ciphers cannot be used to obtain random oracles via a reset-indifferentiable construction, but it might still be possible to construct an ideal cipher from a random oracle, i.e., either the two models are entirely incomparable, or the random-oracle model is strictly stronger.

We rule out such a possibility. Our so-called duality lemma establishes that if there is no construction G_1^{π} that is reset indifferentiable from primitive Π, then also vice versa, there is no construction G_2^{Π} that is reset indifferentiable from primitive π. Hence, our theorem complements the results by Demay et al. and Luykx et al. [13,18] showing that there can also not be a domain-shrinking construction.

Proving that according to plain indifferentiability, the ICM and ROM are equivalent had been a serious challenge and finally involved a Feistel network with many rounds. A Feistel network is a domain-doubling construction, and is thus ruled out by the previous impossibility results. The few leverages that remain to bypass the current impossibility results possibly require quite new techniques. Firstly, it might still be possible to build a construction that is neither domain shrinking, nor domain extending. However, as we will see later, that means settling either direction (RO from IC and vice versa) simultaneously, and this might be quite challenging. The second leverage is a distinction that has been irrelevant in most works in the area of indifferentiability so far and that we would like to point out. Namely, strong indifferentiability requires the simulator $\mathcal{S}$ to work for any distinguisher $\mathcal{D}$, while weak indifferentiability only demands that for every $\mathcal{D}$, there exists a good simulator $\mathcal{S}$. Known constructions are usually strongly indifferentiable, while most existing impossibility results rule out even weakly indifferentiable constructions. In contrast, we do not rule out weakly indifferentiable constructions. It would be interesting to see techniques that make non-black-box use of the distinguisher $\mathcal{D}$ and establish a reset-indifferentiable construction that is domain shrinking.

Notions between indifferentiability and reset indifferentiability. From the current state-of-the-art, there are two ways to proceed: firstly, we can develop new techniques to exploit the few remaining leverages left to bypass the existing impossibility results. Secondly, we might weaken the notion of reset indifferentiability as introduced by Ristenpart et al., to a notion that is achievable by constructions and which is sufficient for a subclass of multi-stage games.

Demay et al. [13] introduce resource-restricted indifferentiability where adversaries may share a limited amount of state. If a certain amount s of shared state is allowed, then their impossibility result shows that a reset-indifferentiable construction cannot extend the domain by more than $s + \lceil \log(s) \rceil$ bits. Maybe the additional bits allow to bypass the impossibility results more easily, as proving domain extension by a few bits might be easier than requiring equality of the domain sizes—however, in this setting, the composition results accounts for a certain class of games only.

Another approach that has been put forward by Luykx et al. [18] is to reduce the number of resets. Indeed, allowing for a polynomial number of resets/stages seems to be an overkill, as some games such as the security model for deterministic encryption [1,2] and also certain forms of leakage require a constant number of adversarial stages only. To this end, Luykx et al. propose the notion of *single-reset indifferentiability* where a distinguisher can make a single reset call only; naturally, a construction that is single reset indifferentiable would be sufficient in any security game consisting of exactly two distinct adversarial stages such as deterministic encryption. Analogously, one can define n-reset indifferentiability for $n+1$ adversarial stages.

However, as we prove, single-reset indifferentiability is already equivalent to full-reset indifferentiability and so are all notions of n-reset indifferentiability. Hence, reducing the number of allowed reset queries does not help us to establish composition results for a restricted class of games. Thus, if a general indifferentiability result is indeed impossible, then it is a curious open question how to cope with the uncomfortable situation. It might be possible to establish indifferentiability results and composition theorems for a class of games that is restricted in another way than by the number of queries. Indeed, it would be interesting to see how such a class could look like and whether there are games for which, in general, finding a suitable, indifferentiable construction is impossible.

Summary of our Contributions. We first introduce the notion of *pseudo-deterministic* algorithms, which captures, that a probabilistic algorithm almost always returns the same answer on the same queries and thus shares many properties with deterministic algorithms. Essentially, a probabilistic (and possibly stateful) algorithm $\mathcal{A}$ is called pseudo deterministic, if no efficient distinguisher with black-box access to $\mathcal{A}$ can make $\mathcal{A}$ return two different answers on the same input. This notion of pseudo determinism can be seen as an average-case version of the pseudo-deterministic algorithms that were recently introduced by Goldreich, Goldwasser, and Ron [16]. While they require probabilism to be hard to detect on any input, we only require indistinguishability for efficiently generatable inputs, on the average. As stressed by Goldreich et al. [16], pseudo-deterministic algorithms are practically as useful as deterministic algorithms, but they are also easier to construct—which we indeed exploit in our paper.

We will show in Section 3 that simulators for reset indifferentiability need to be stateless and pseudo deterministic. Simplifying pseudo determinism to determinism for the moment, this allows us to establish what we call the duality lemma. Perhaps surprisingly, it states that, with respect to reset indifferentiability, two idealized models are either *equivalent* or *incomparable*. The reason is that a deterministic and stateless simulator can act as a construction and vice versa. Consequently, in order to prove equivalence in terms of reset indifferentiability, this lemma makes it sufficient to prove the "easier" direction, whichever this might be. In turn, for impossibility results, one might use this as a tool to prove impossibility more easily. In fact, we use the duality lemma to establish that not only domain-extending constructions are impossible, but also domain-shrinking constructions (Section 4) thereby complementing the results of [13]. Note that the duality lemma covers strong indifferentiability, leaving non-black-box use of the distinguisher as a potential leverage to bypass this impossibility.

The recently proposed [18] notion of single-reset indifferentiability intends to define a notion of indifferentiability that is easier to achieve and simultaneously covers an interesting class of multi-stage games that has two adversary stages only. Interestingly, as we establish, restricting the number of resets does not yield a weaker notion of equivalence. We prove that single- (and n-) reset indifferentiability is equivalent to reset indifferentiability (Section 5). Maybe surprisingly, our proof does not rely on a hybrid argument; instead, we establish a tight reduction that merely reduces the distinguisher's advantage by a factor of 2.

2 Preliminaries

For a natural number $n \in \mathbb{N}$ we denote by $\{0,1\}^n$ the set of all bit strings of length n. By $\{0,1\}^*$ we denote the set of all bit strings of finite length. As usual $|\mathcal{M}|$ denotes the cardinality of a set $\mathcal{M}$ and logarithms are to base 2. For some probabilistic algorithm $\mathcal{A}$ and input x we denote by $\mathcal{A}(x;R)$ the output of $\mathcal{A}$ on x using randomness R. Throughout this paper we assume that λ is a security parameter (if not explicitly given then implicitly assumed) and that algorithms (resp., Turing machines) run in polynomial time with respect to λ.

In this paper we consider random oracles and ideal ciphers (defined below) which we will collectively refer to as *ideal primitives*. Although we present most of the results directly for ideal ciphers and random oracles, the following more general notion of ideal primitives allows us to generalize some of our results:

Definition 1. *An* ideal primitive Π_λ *is a distribution on functions indexed by the security parameter* λ. *For some algorithm* $\mathcal{A}$, *security parameter* λ *and ideal primitive* Π_λ *we say that* $\mathcal{A}$ has access to Π *if* $\mathcal{A}$ *has oracle access to a function* f *chosen from the distribution* Π_λ.

We simply write Π, i.e., omit the security parameter, if it is clear from the context.

Remark 1. We will usually encounter only single instances of an ideal primitive Π at a time. Unless stated otherwise, if multiple parties have access to Π, then we implicitly assume that the corresponding function f was chosen from the distribution Π using the same randomness for all parties, i.e., all parties have oracle access to the same function f.

Random Oracles and Ideal Ciphers. A random oracle $(\mathcal{R}_{\ell,m})_\lambda$ is the uniform distribution on all functions mapping $\{0,1\}^\ell$ to $\{0,1\}^m$ with $\ell := \ell(\lambda)$ and $m := m(\lambda)$. An ideal cipher $(\mathcal{E}_{k,n})_\lambda$ is the uniform distribution on all keyed permutations of the form $\{0,1\}^k \times \{0,1\}^n \to \{0,1\}^n$ with $k := k(\lambda)$ and $n := n(\lambda)$. That is, for a cipher in the support of $(\mathcal{E}_{k,n})_\lambda$ each key $\kappa \in \{0,1\}^k$ describes a random (independent) permutation $\mathcal{E}_{k,n}(\kappa,\cdot) : \{0,1\}^n \to \{0,1\}^n$. By abuse of notation, the term random oracle (resp., ideal cipher) also refers to a specific instance chosen from the respective distribution.

Keyed vs. unkeyed ciphers. The ideal-cipher model has either been considered as a public *unkeyed* permutation or as a public *keyed* permutation. We present our results in the keyed setting since we feel that the ideal cipher-model is usually perceived in this way. However, we want to point out that the results are equally valid for the unkeyed setting because our proofs do not rely on the presence of a key.

Independently of this, one might be tempted to argue that the settings are interchangeable since we know, for example, constructions of a keyed permutation from an ideal public permutation (Even and Mansour, [15]). Note though, that in order to make this argument work, one needs to show that these constructions are reset indifferentiable. However, the construction by Even and Mansour is a domain extender where the key size is twice the message size and we rule out reset indifferentiability for such extending constructions in Section 4. We note that it is an interesting open problem whether or not such (reset-) indifferentiable non-extending transformations exist.

2.1 Indifferentiability

Let us now recall the indifferentiability notion of Maurer et al. [19] in the version by Coron et al. [11] who replace *random systems* by oracle Turing machines (resp., ideal primitives). Since we are concerned with different types of indifferentiability, we will sometimes use the term *plain* indifferentiability when referring to this original notion of indifferentiability.

Definition 2. *A Turing machine G with black-box access to an ideal primitive π is* strongly indifferentiable *from an ideal primitive Π if there exists a simulator $\mathcal{S}^{\Pi}$, such that for any distinguisher $\mathcal{D}$ there exists negligible function* negl, *such that:*

$$\left|\Pr\left[\mathcal{D}^{G^{\pi},\pi}(1^{\lambda})=1\right]-\Pr\left[\mathcal{D}^{\Pi,\mathcal{S}^{\Pi}}(1^{\lambda})=1\right]\right|\leq\mathsf{negl}(\lambda) \qquad (1)$$

We say that the construction is weakly indifferentiable *if for any $\mathcal{D}$ there exists a simulator $\mathcal{S}$ such that* (1) *holds.*

We will use the term *real world* to denote that the distinguisher $\mathcal{D}$ talks to the construction G^{π} and the primitive π, whereas in the *ideal world*, the distinguisher $\mathcal{D}$ talks to the "target" primitive Π and simulator $\mathcal{S}^{\Pi}$. The goal of the distinguisher is to determine which of the two *pairs* of oracles he is talking to. Towards this goal, the distinguisher $\mathcal{D}$ queries its two oracles, of which one is called the honest interface h which is either G^{π} (in the real world) or Π (in the ideal world). The other oracle is called the adversarial interface a and corresponds to either π (real world) or $\mathcal{S}^{\Pi}$ (ideal world). Thus, $(\mathsf{h},\mathsf{a}) := (G^{\pi},\pi)$ if distinguisher $\mathcal{D}$ is in the real world and $(\mathsf{h},\mathsf{a}) := (\Pi,\mathcal{S}^{\Pi})$ if it is in the ideal world. The names h (honest) and a (adversarial) are in the style of [20] and suggestive: an honest party uses a construction as the designer intended; an adversary could, however, use the underlying building blocks to gain an advantage.

Reset Indifferentiability. Ristenpart et al. show [20] that, in general, we cannot securely replace a primitive Π by a construction G^{π} from primitive π, if the construction is indifferentiable only. Instead, G^{π} needs to be (weakly) *reset* indifferentiable from Π which extends the original indifferentiability definition by giving the distinguisher the power to reset the simulator at arbitrary times:

Definition 3. *Let the setup be as in Definition 2. An oracle Turing machine* G^π *is called* strongly *(resp.* weakly*)* reset indifferentiable *from ideal primitive* Π *if the distinguisher* $\mathcal{D}$ *can reset the simulator* $\mathcal{S}$ *to its initial state arbitrarily many times during the respective experiment.*

For reset indifferentiability the adversarial interface a in the real world simply ignores reset queries. Reset indifferentiability now allows composition in arbitrary games and not only in single-stage games, as does the original indifferentiability notion [20,19].

3 Pseudo-deterministic Stateless Simulators for Indifferentiability

Recall that the composition theorem by Maurer et al. [19] for plain indifferentiability holds for single-stage adversaries only. Their theorem says that if (i) the construction G^π is indifferentiable from the ideal primitive Π and if (ii) there is a reduction $\mathcal{R}$ that transforms a successful adversary $\mathcal{A}$ against some notion of security into an adversary $\mathcal{R}^\mathcal{A}$ against a single-stage game in the presence of the ideal primitive Π, then also in the presence of the construction G^π there is a reduction $\mathcal{R}'$ that transforms a successful adversary $\mathcal{A}$ into an adversary $\mathcal{R}'^\mathcal{A}$ against the single-stage game.

In order to prove a general composition theorem, Ristenpart et al. [20] strengthen the notion of indifferentiability to account for the different stages of the adversary. They introduce the notion of (weak) reset indifferentiability and prove that the aforementioned theorem works for arbitrary games, if the construction G^π is reset indifferentiable from the ideal primitive Π. In contrast to plain indifferentiability, here, the distinguisher gets extra powers, namely to reset the simulator at arbitrary times. Ristenpart et al. [20] and Demay et al. [13] remark that reset indifferentiability is equivalent to plain indifferentiability with stateless simulators. Intuitively, this follows from the observation that the distinguisher in the reset indifferentiability game can simply reset the simulator after each query it asks. We believe that, albeit equivalent, stateless simulators are often easier to handle than reset-resistant simulators and thus explicitly introduce indifferentiability with stateless simulators as *multi-stage indifferentiability* and then prove that it is equivalent to reset indifferentiability.

In Subsection 3.2, we prove that strong multi-stage indifferentiability implies that the simulators are also *pseudo deterministic*, a notion that we put forward in this section. Relative to a random oracle or an ideal cipher, we show how to derandomize pseudo-deterministic simulators, if the simulators are allowed to depend on the number of queries made by the distinguisher.

3.1 Multi-stage Indifferentiability

A stateless interactive algorithm is an algorithm whose behavior is statistically independent from the call/answer history of the algorithm. We now prove that indifferentiability with stateless simulators is equivalent to reset indifferentiability.

Definition 4. *A construction G with black-box access to primitive π is* strongly multi stage indifferentiable *from primitive Π if there exists a stateless probabilistic polynomial-time simulator $\mathcal{S}$ (with access to Π), such that for any probabilistic polynomial-time distinguisher $\mathcal{D}$ there exists negligible function* negl *such that:*

$$\left|\Pr\left[\mathcal{D}^{G^{\pi},\pi}(1^{\lambda})=1\right]-\Pr\left[\mathcal{D}^{\Pi,\mathcal{S}^{\Pi}}(1^{\lambda})=1\right]\right|\leq \mathsf{negl}(\lambda) \tag{2}$$

We say that a construction G^{π} is weakly multi stage indifferentiable *from Π if for any probabilistic polynomial-time distinguisher $\mathcal{D}$ there exists a stateless probabilistic polynomial-time simulator $\mathcal{S}$ such that (2) holds.*

Lemma 1. *A construction G with black-box access to primitive π is weakly (resp., strongly) multi stage indifferentiable from primitive Π if and only if G is weakly (resp., strongly) reset indifferentiable from primitive Π.*

Proof. First note that any stateless simulator is, naturally, indifferent to resets and thus multi-stage indifferentiability implies reset indifferentiability. Moreover, strong reset indifferentiability implies strong multi-stage indifferentiability since the simulator for reset indifferentiability must work for any distinguisher, in particular for those which reset after each query. Hence this stateful simulator can be simply initialized and run by a stateless simulator (the stateless simulator does this for each query it receives).

We now prove the remaining relation, i.e., that weak reset indifferentiability implies weak multi-stage indifferentiability. Assume that reset indifferentiability holds and consider an arbitrary distinguisher $\mathcal{D}$ in the multi-stage indifferentiability game. From this we construct a distinguisher $\mathcal{D}'$ for the reset indifferentiability game which runs $\mathcal{D}$ and sends a reset query to its adversarial a-interface after every a-query issued by $\mathcal{D}$. Let $\mathcal{S}'$ be the simulator for $\mathcal{D}'$ guaranteed to exist by reset indifferentiability. We construct a stateless simulator $\mathcal{S}$ for multi-stage indifferentiability which simply runs (the stateful) $\mathcal{S}'$ and resets its own state after each query. Now the following equations hold for $b \in \{0,1\}$:

$$\Pr\left[\mathcal{D}'^{\Pi,\mathcal{S}'}(1^{\lambda})=b\right]=\Pr\left[\mathcal{D}'^{\Pi,\mathcal{S}}(1^{\lambda})=b\right]=\Pr\left[\mathcal{D}^{\Pi,\mathcal{S}}(1^{\lambda})=b\right].$$

Thus, if equation (2) holds for $(\mathcal{D}',\mathcal{S}')$, then it holds equally for $(\mathcal{D},\mathcal{S})$.

3.2 Pseudo-deterministic Algorithms

Our notion of *pseudo-deterministic* algorithms intuitively captures that no distinguisher can query the algorithm on an input such that it returns something different from the most likely output. That is, the adversary wins if in its set of input/output pairs to the algorithm there is a query for which the algorithm did not return the most likely response. We also introduce a weak notion of this property, where we call $\mathcal{A}$ pseudo deterministic for a specific distinguisher if the probability of the distinguisher winning in the above experiment is negligible.

Our notion of pseudo determinism can be seen as an average-case version of the pseudo-deterministic algorithms as recently introduced by Goldreich et al. [16]. While they require probabilism to be hard to detect on any input, we only require indistinguishability for efficiently generatable inputs, on average.

Definition 5. *Let λ be a security parameter and $\mathcal{A}^{\mathcal{O}}$ a stateless probabilistic polynomial-time oracle Turing machine with access to some oracle $\mathcal{O}$. Let $L[\mathcal{D}, \mathcal{A}, \mathcal{O}]$ denote the induced set of input/output pairs (x, y) of $\mathcal{A}^{\mathcal{O}}$ when queried arbitrarily many times by the distinguisher $\mathcal{D}$, where $\mathcal{A}$ uses fresh coins in each run. We say that $\mathcal{A}^{\mathcal{O}}$ is* pseudo deterministic *if for all probabilistic polynomial-time distinguishers $\mathcal{D}$ there exists a negligible function* negl, *such that*

$$\Pr_{\mathcal{D},\mathcal{A},\mathcal{O}}\left[\forall (x,y) \in L[\mathcal{D},\mathcal{A},\mathcal{O}] \quad y = y_{x,\mathcal{A}^{\mathcal{O}}}\right] \geq 1 - \mathsf{negl}(\lambda). \quad (3)$$

The notation $y_{x,\mathcal{A}^{\mathcal{O}}}$ denotes the most likely output of $\mathcal{A}$ on input x over the randomness of $\mathcal{A}$, *i.e., conditioned on a fixed oracle $\mathcal{O}$. If there are two equally likely answers on input x, we choose $y_{x,\mathcal{A}^{\mathcal{O}}}$ to be the lexicographically smaller one.*

We say algorithm $\mathcal{A}^{\mathcal{O}}$ is pseudo deterministic for distinguisher $\mathcal{D}^{\mathcal{A}^{\mathcal{O}}(1^\lambda,\cdot)}(1^\lambda)$, if there exists negligible function negl, *such that equation* (3) *holds for $\mathcal{D}$.*

Note that the definition of $\mathcal{A}$ being pseudo deterministic for distinguisher $\mathcal{D}$ does not imply that it is hard to distinguish whether $\mathcal{A}$ is probabilistic or deterministic—it is only hard for a particular algorithm $\mathcal{D}$. Although this might sound like a weak and somewhat useless property, it will be sufficient to show that if a simulator is pseudo deterministic for a distinguisher, then the simulator can be entirely derandomized via random oracles/ideal ciphers.

We now show that strong multi-stage indifferentiability implies that the simulators are not only stateless but also pseudo deterministic. This is captured by the following lemma.

Lemma 2. *Let G^π be a construction with black-box access to primitive π which is strongly multi stage indifferentiable from primitive Π. Then there is a stateless* pseudo-deterministic *probabilistic polynomial-time simulator $\mathcal{S}$ such that for all probabilistic polynomial-time distinguishers $\mathcal{D}$ equation* (2) *holds in the strong case.*

Proof. Let us assume there exists stateless simulator $\mathcal{S}$ such that for all distinguishers $\mathcal{D}$ equation (2) holds and such that $\mathcal{S}$ is not pseudo deterministic. The latter implies that there exists distinguisher $\mathcal{D}_{pd}$ against the pseudo determinism of simulator $\mathcal{S}$, i.e., there is a non-negligible probability that $\mathcal{D}_{pd}$ asks a query to $\mathcal{S}$, where $\mathcal{S}$ has a non-negligible probability of returning a different value than the most likely one. We now construct distinguisher $\mathcal{D}'$ against strong multi-stage indifferentiability. Distinguisher $\mathcal{D}'$ runs $\mathcal{D}_{pd}$ on the adversarial a-interface. Let $q_1, \ldots, q_t$ be the queries asked by $\mathcal{D}_{pd}$. Distinguisher $\mathcal{D}'$ then sends the same queries once more to its a-interface and returns 1 if at least one response does not match and 0 otherwise. If $\mathcal{D}'$ is in the real world, talking to G^π and π algorithm $\mathcal{D}'$ will always output 0 as π is a function. If on the other hand, $\mathcal{D}'$ is in the ideal world, then $\mathcal{D}_{pd}$ will succeed with noticeable probability and hence $\mathcal{D}'$ will distinguish both worlds with noticeable probability, a contradiction.

Deterministic Simulators. Bennett and Gill prove in [7] that relative to a random oracle the complexity classes $\mathcal{BPP}$ and $\mathcal{P}$ are equivalent. Let us quickly sketch their idea. Given a probabilistic polynomial time oracle Turing machine $\mathcal{M}^{\mathcal{R}}$ which has access to random oracle $\mathcal{R}$ and which decides a language $\mathcal{L}$ in $\mathcal{BPP}$ we can prove the existence of a *deterministic* polynomial time Turing machine $\mathcal{D}^{\mathcal{R}}$ which also decides $\mathcal{L}$. Let us by $p(|x|)$ denote the runtime of machine $\mathcal{M}^{\mathcal{R}}$ for inputs of length $|x|$. As $\mathcal{M}^{\mathcal{R}}$ runs in polynomial time there exists a polynomial upper bound $p(|x|)$ on the length of queries $\mathcal{M}^{\mathcal{R}}$ can pose to the random oracle. To derandomize $\mathcal{M}^{\mathcal{R}}$ we construct a deterministic machine $\mathcal{D}^{\mathcal{R}}$ which works analogously to $\mathcal{M}^{\mathcal{R}}$ with the single exception that when $\mathcal{M}^{\mathcal{R}}$ requests a random coin then $\mathcal{D}^{\mathcal{R}}$ generates this coin deterministically by querying the random oracle on the next smallest input that cannot have been queried by $\mathcal{M}^{\mathcal{R}}$ due to its runtime restriction. As the random oracle produces perfect randomness, the machines decide the same language with probability 1 over the choice of random oracle.

Using the techniques developed by Bennet and Gill [7] we now show that in the multi-stage indifferentiability setting, if a simulator is pseudo deterministic for a distinguisher $\mathcal{D}$, then it can be derandomized, in case the constructed primitive Π is a random oracle or an ideal cipher. When applied to a simulator $\mathcal{S}$ that is universal for all distinguishers (strong indifferentiability), these derandomization techniques yield a family of simulators that depends only on the number of queries made by the distinguisher (weak indifferentiability).

Lemma 3. *Let $\mathcal{A}^{\Pi}$ be a stateless probabilistic polynomial-time algorithm with oracle access to a random oracle $\mathcal{R}_{\ell,m}$ or an ideal cipher $\mathcal{E}_{k,n}$ for $\ell \in \omega(\log \lambda)$ (resp., $(k+n) \in \omega(\log \lambda)$). Let s be polynomial in λ. From $\mathcal{A}^{\Pi}$, we construct a deterministic algorithm $\mathcal{B}^{\Pi}$ such that the following holds: for all efficient distinguisher $\mathcal{D}$ that make less than s queries to their oracle, it holds that if $\mathcal{A}^{\Pi}$ is pseudo deterministic for $\mathcal{D}$, then*

$$\left|\Pr_{R,\Pi}\left[\mathcal{D}^{\Pi,\mathcal{A}^{\Pi}(R,\cdot)}(1^{\lambda})=1\right]-\Pr_{\Pi}\left[\mathcal{D}^{\Pi,\mathcal{B}^{\Pi}(\cdot)}(1^{\lambda})=1\right]\right|$$

is negligible, where the probability is over the choice of oracle Π and algorithm $\mathcal{A}$'s and distinguisher $\mathcal{D}$'s internal coin tosses for the first case and over the choice of oracle Π and distinguisher $\mathcal{D}$'s internal coin tosses in the second.

Proof. Let $\mathcal{A}^{\Pi}$ be a stateless algorithm with access to ideal primitive Π where Π is either a random oracle $\mathcal{R}_{\ell,m}$ or an ideal cipher $\mathcal{E}_{k,n}$.

Let $\mathcal{D}$ be an efficient distinguisher for which $\mathcal{A}^{\Pi}$ is pseudo deterministic. As distinguisher $\mathcal{D}$ is efficient, there exists an upper bound $p(|\lambda|)$ on the number of queries to the Π-interface by $\mathcal{D}$. We construct a deterministic algorithm $\mathcal{B}$ which works as $\mathcal{A}$ with the only exception that $\mathcal{B}$ deterministically generates "random" bits by querying its random oracle, whenever $\mathcal{A}$ makes use of a random bit. For the jth requested random bit, algorithm $\mathcal{B}$ calls the Π-oracle (either random oracle $\mathcal{R}$ or ideal cipher $\mathcal{E}$ where it uses the encryption interface of $\mathcal{E}$) on $p(|\lambda|)+j$ distinct values xor-ing the result and choosing a bit from this result. Note that as $\ell \in \omega(\log \lambda)$ (resp., $n + k \in \omega(\log \lambda)$) there exist sufficiently many distinct values.

Remember that we denote by $y_{q,\mathcal{A}^{\mathcal{O}}}$ the most likely output of algorithm $\mathcal{A}$ on input q conditioned on fixed oracle $\mathcal{O}$. We want to prove that

$$\left|\Pr_{\Pi,\mathcal{D},\mathcal{A}}\left[\mathcal{D}^{\Pi,\mathcal{A}^{\Pi}}(1^{\lambda}) = 1\right] - \Pr_{\Pi,\mathcal{D}}\left[\mathcal{D}^{\Pi,\mathcal{B}^{\Pi}}(1^{\lambda}) = 1\right]\right|$$

is negligible in λ. We prove a stronger statement, namely, that the outputs of $\mathcal{A}$ and $\mathcal{B}$ are likely to be identical. We define event C capturing that "the outputs of $\mathcal{A}$ and $\mathcal{B}$ agree on all inputs." Towards this goal we define event A as "algorithm $\mathcal{A}$ returns $y_{q_i,\mathcal{A}^{\Pi}}$ for all queries q_i" where $y_{q_i,\mathcal{A}^{\Pi}}$ is the most likely answer of $\mathcal{A}^{\Pi}$ on input q_i, i.e., we set $y_{q_i,\mathcal{A}^{\Pi}} := \arg\max_y \left\{\Pr_R\left[\mathcal{A}^{\Pi}(q_i; R) = y\right]\right\}$ (cf. Definition 5). Likewise, we define event B as "algorithm $\mathcal{B}$ returns $y_{q_i,\mathcal{A}^{\Pi}}$ for all queries q_i." We will show that

$$\Pr_{\Pi,\mathcal{D},\mathcal{A}}[\mathsf{A}] \geq 1 - \mathsf{negl} \tag{4}$$

and

$$\Pr_{\Pi,\mathcal{D}}[\mathsf{B}] \geq 1 - \mathsf{negl}. \tag{5}$$

Clearly, the probability that $\mathcal{A}$ and $\mathcal{B}$ produce the same answers for all q_i is lower bounded by the probability that $\mathcal{A}$ and $\mathcal{B}$ both output $y_{q_i,\mathcal{A}^{\Pi}}$ for all q_i. Thus,

$$\begin{aligned}\Pr_{\Pi,\mathcal{D},\mathcal{A}}[\mathsf{C}] &\geq \Pr_{\Pi,\mathcal{D},\mathcal{A}}[\mathsf{A} \wedge \mathsf{B}] \\ &= 1 - \Pr_{\Pi,\mathcal{D},\mathcal{A}}[\neg\mathsf{A} \vee \neg\mathsf{B}] \\ &\geq 1 - (\Pr_{\Pi,\mathcal{D},\mathcal{A}}[\neg\mathsf{A}] + \Pr_{\Pi,\mathcal{D}}[\neg\mathsf{B}]) \\ &\geq 1 - \mathsf{negl} - \mathsf{negl}.\end{aligned}$$

Let us now make these statements formal as well as prove inequalities (4) and (5). We denote with q_i the queries to $\mathcal{A}$ by $\mathcal{D}$ and by R_i the randomness used by $\mathcal{A}$ on query q_i. We say that event A occurs (over $\Pi, \mathcal{D}, R_1, ..., R_n$), if

$$\forall i\ \mathcal{A}^{\Pi}(q_i; R_i) = y_{q_i,\mathcal{A}^{\Pi}}.$$

Note that the pseudo-determinism of $\mathcal{A}$ for $\mathcal{D}$ directly implies that

$$\Pr_{\Pi,\mathcal{D},R_1,\ldots,R_n}\left[\forall i\ \mathcal{A}^{\Pi}(q_i; R_i) = y_{q_i,\mathcal{A}^{\Pi}}\right] \geq 1 - \mathsf{negl}, \tag{6}$$

which establishes inequality (4). We say that event B occurs (over $\Pi, \mathcal{D}$), if

$$\forall i\ \mathcal{B}^{\Pi}(q_i) = y_{q_i,\mathcal{A}^{\Pi}},$$

where q_i now denotes the queries by $\mathcal{D}$ to algorithm $\mathcal{B}$. Inequality (5) we derive from inequality (4) via an averaging argument. Note that in inequality (6) we consider fresh randomness R_i for every query q_i. If for all queries q_i a random choice of randomness is good with overwhelming probability, then a random choice of randomness is good for all q_i with overwhelming probability:

$$\Pr_{\Pi,\mathcal{D},R}\left[\forall i \; \mathcal{A}^{\Pi}(q_i;R) = y_{q_i,\mathcal{A}^{\Pi}}\right] \geq 1 - \mathsf{negl}. \tag{7}$$

Moreover, when considering the random oracle via lazy sampling, one can observe that the randomness generated by $\mathcal{B}$ from Π is independent from the part of Π that is used in the experiment, which yields that

$$\Pr_{\Pi,\mathcal{D}}\left[\forall i; \mathcal{B}^{\Pi}(q_i) = y_{q_i,\mathcal{A}^{\Pi}}\right] = \Pr_{\Pi,\mathcal{D},R}\left[\forall i; \mathcal{A}^{\Pi}(q_i;R) = y_{q_i,\mathcal{A}^{\Pi}}\right] \geq 1 - \mathsf{negl}$$

as desired.

4 The Random Oracle and Ideal Cipher Model Are Incomparable

In this section we prove that the random oracle-model and the ideal cipher-model are incomparable with respect to strong multi-stage indifferentiability. We start by giving an alternative, simpler proof of the fact that multi-stage indifferentiable constructions cannot be built via domain extension [13,18] (Lemma 4). [13] rule out domain extension even for a single bit of extension. In turn, we obtain an easier proof in the setting where the extension factor is super logarithmic. In Section 4.1 we then present our duality lemma for multi-stage indifferentiability which allows us to conclude that the ROM and the ICM are incomparable with respect to strong multi-stage indifferentiability.

Lemma 4. *Let $\mathcal{R}$ be a random oracle with domain $\{0,1\}^\ell$ (resp., $\mathcal{E}$ be an ideal cipher with domain $\{0,1\}^k \times \{0,1\}^n$) and π be any ideal primitive with domain size 2^v. For $\ell - v \in \omega(\log(\lambda))$ (resp., $k + n - v \in \omega(\log(\lambda))$) there exists no construction G^π that is weakly multi-stage indifferentiable from $\mathcal{R}$ (resp., $\mathcal{E}$).*

We prove Lemma 4 for the random oracle case; the proof for ideal ciphers works analogously. Note that we prove the statement for weak multi-stage indifferentiability, thereby essentially ruling out any (possibly non-black-box) construction.

In the following proof we consider a particular distinguisher that tests for the ideal world by forcing the simulator to query its oracle on a particular value M. We show that no simulator is able to do this with more than negligible probability since M is drawn from a very large set while the simulator, being stateless, is only able to make queries from a negligible fraction of this large set; it thus fails to pass the test.

Proof (Proof of Lemma 4). Assume towards contradiction that there exists construction G^π that is weakly multi stage indifferentiable from random oracle $\mathcal{R}$ and, hence, for every distinguisher $\mathcal{D}$ there exists a stateless simulator $\mathcal{S}$ such that $\mathcal{D}$ cannot distinguish between the real and ideal world.

We consider a distinguisher $\mathcal{D}^{\mathsf{h},\mathsf{a}}$ with access to honest and adversarial interfaces (h,a) which implement the random oracle $\mathcal{R}$ and simulator $\mathcal{S}$ in the ideal world and construction G^{π} and ideal primitive π in the real world. The distinguisher $\mathcal{D}$ chooses a message $M \in \{0,1\}^{\ell}$ uniformly at random and executes construction G via an internal simulation using its adversarial interface a, i.e., it computes $G^{\mathsf{a}}(M)$. Then, the distinguisher asks its honest interface on message M to compute $\mathsf{h}(M)$ and returns 1 if the two results agree and 0 otherwise. Note that in the real world distinguisher $\mathcal{D}$ will always output 1. Thus, the simulator $\mathcal{S}$ has to ensure that $G^{\mathcal{S}^{\mathcal{R}}}(M)$ is equal to $\mathcal{R}(M)$ with overwhelming probability over the choice of the random oracle $\mathcal{R}$. We now prove that, in the ideal world, the two values match only with negligible probability over the choice of the message M and the two settings can thus be distinguished by $\mathcal{D}$.

Let us assume the ideal world and denote the query/response pairs to the a-interface with $(q_i, r_i)_{1\leq i\leq t}$. We analyze the simulator's behavior when it is asked these queries $q_1,\ldots,q_t$. If for none of the q_i the simulator $\mathcal{S}$ asks the random oracle on M, then the answer of $G^{\mathcal{S}^{\mathcal{R}}}(M)$ is independent of $\mathcal{R}(M)$ and thus different with overwhelming probability. By a simple counting argument, we now prove that, with high probability over the choice of M, on *no* query (not even one outside of the set $(q_i, r_i)_{1\leq i\leq t}$), the simulator $\mathcal{S}$ asks $\mathcal{R}$ on M. For this, note that the queries which simulator $\mathcal{S}$ receives are of length v. Hence there are at most 2^v distinct possible queries to $\mathcal{S}$. Denote by c the upper bound on the number of queries that $\mathcal{S}$ asks to its random oracle over all possible queries that $\mathcal{S}$ itself receives. As the simulator $\mathcal{S}$ runs in polynomial time c exists and is polynomial. Noting that $\mathcal{S}$ is stateless, we conclude that $\mathcal{S}$ asks at most $c2^v \ll 2^{\ell}$ queries. Hence the probability that the distinguisher's M is in the set

$$\{M : \exists q\, \mathcal{S}^{\mathcal{R}} \text{ asks } M \text{ on input } q\}$$

is negligible. The probability that the distinguisher $\mathcal{D}$ returns 1 in the ideal world where it is given access to simulator $\mathcal{S}$ and a random oracle $\mathcal{R}$ is therefore also negligible. Thus, the distinguisher $\mathcal{D}$ has a distinguishing advantage of almost 1 which concludes the proof.

4.1 The Duality Lemma for Multi-stage Indifferentiability

We now prove the inverse direction, that is an ideal cipher cannot be build from a random oracle with larger domain. In contrast to the previous section we here give an impossibility result for strong multi-stage indifferentiability. Our result is, however, more general and of independent interest. Strong multi-stage indifferentiability guarantees the existence of a simulator that is stateless and deterministic. Constructions of ideal primitives often need to be stateless and deterministic as well. If for example, the construction, implements a publicly accessible function such as a hash function, it has to be stateless. Note that this is the case both for random oracles and ideal ciphers.

Now, if we assume that constructions are deterministic and stateless, then we show that, in the case of multi-stage indifferentiability, we can exchange the role of the construction and the role of the simulator, if the simulator is also deterministic and stateless. Our *Duality Lemma* establishes that in this case, an impossibility result (resp. feasibility result) in one direction translates into an impossibility result (resp. feasibility result) in the other direction. However, if the simulator is not deterministic, but only pseudo deterministic, then we need to slightly adapt our notion of *constructions* to also allow pseudo-deterministic constructions. For this note that pseudo deterministic constructions are as useful as deterministic ones since inconsistencies due to the pseudo determinism can only be detected with negligible probability. Formally, however, they are not known to be equivalent, in particular, because $\mathcal{P} \neq \mathcal{BPP}$ implies that pseudo-deterministic polynomial-time algorithms are more powerful than deterministic polynomial-time algorithms.

We prove the Duality Lemma in the case of strong multi-stage indifferentiability.

Lemma 5 (Duality Lemma for Multi-stage Indifferentiability). *Let π and π' by two ideal primitives. Assuming constructions are stateless and pseudo deterministic, then one of two following statements holds:*

1. *The two primitives are computationally equivalent, i.e., there exist constructions G_1, G_2 such that G_1^{π} is strongly multi stage indifferentiable from π' and $G_2^{\pi'}$ is strongly multi stage indifferentiable from π, or*
2. *π and π' are incomparable with respect to strong multi-stage indifferentiability.*

In essence this means that a positive or negative result in either direction gives us a result for the other direction. As we have already seen a negative result for domain extenders this gives us the result for the other directions, i.e., going from a large random oracle $\mathcal{R}$ to a small ideal cipher $\mathcal{E}$, or from a large ideal cipher $\mathcal{E}$ to a small random oracle $\mathcal{R}$.

Proof (Proof of Lemma 5). Assume construction G^{π} with black-box access to ideal primitive π is strongly multi stage indifferentiable from π'. Then by definition there exists a (pseudo-)deterministic, stateless simulator $\mathcal{S}$ such that no distinguisher $\mathcal{D}$ can tell apart the ideal world $(\pi', \mathcal{S}^{\pi'})$ from the real world (G^{π}, π). Likewise, by definition, G is stateless and (pseudo-)deterministic. We now exchange the roles of construction G and simulator $\mathcal{S}$, thereby getting a new "construction" $\mathcal{S}^{\pi'}$ implementing primitive π. It remains to show that $\mathcal{S}^{\pi'}$ is strongly multi-stage indifferentiable from π.

Let us assume the contrary. Then there exists distinguisher $\mathcal{D}$ that can distinguish between the settings $(\pi', \mathcal{S}^{\pi'})$ and the setting (G^{π}, π). This, however, contradicts the assumption that G^{π} is strongly multi stage indifferentiable from π'.

An immediate consequence of the duality lemma and Lemma 4 is captured by the following corollary:

Corollary 1. *The ideal cipher model and the random oracle model are incomparable with respect to strong multi-stage indifferentiability.*

Remark 2. One interesting consequence of the duality lemma is best seen by an example: Can a random oracle with smaller domain be constructed from a random oracle with a larger domain? Intuitively, it feels natural to assume that this works. However, Lemma 4 tells us, that the inverse is not possible and, thus, by the duality lemma we can directly conclude that any construction using a large random oracle cannot be strongly multi stage indifferentiable from a small random oracle. So far, we have failed to either prove impossibility for weak multi-stage indifferentiability or to come up with a construction. We leave this for future work.

5 Single versus Multi-reset

Luykx et al. [18] introduce the presumably weaker notion of n-reset indifferentiability, where the distinguisher is allowed to reset the simulator only n times. Naturally, for a construction that is n-reset indifferentiable the composition theorem holds for games that have $n+1$ or less stages. In the following we show that, however, already the extreme single-reset notion implies full reset indifferentiability for simulators that do not depend on the distinguisher (i.e., the *strong* case). This yields that also for n-reset indifferentiability all our separations hold in a black-box fashion.

What we prove is that the advantage of an n-reset distinguisher is bound by the advantage of an $(n-1)$-reset distinguisher and that of a single-reset distinguisher where the advantage of a distinguisher $\mathcal{D}$ in the n-reset indifferentiability game is defined as

$$\mathsf{Adv}^{n\text{-reset}}_{\mathcal{S},\mathcal{D}} := \left|\Pr\left[\mathcal{D}^{\mathcal{R},\mathcal{S}^{\mathcal{R}}}(1^\lambda) = 1\right] - \Pr\left[\mathcal{D}^{G^\pi,\pi}(1^\lambda) = 1\right]\right| .$$

Assuming that a construction is strongly single reset indifferentiable (and thus the advantage for any single-reset distinguisher is negligible) yields the above claim. We use

Lemma 6. *Let G^π be a construction with black-box access to primitive π. Then there exists simulator $\mathcal{S}$ such that for all $n > 1$ and all distinguishers $\mathcal{D}_n$ that make at most n reset queries there exists a distinguisher $\mathcal{D}_{n-1}$ that makes at most $n-1$ reset queries and a distinguisher $\mathcal{D}_1$ that makes a single reset query and*

$$\mathsf{Adv}^{n\text{-reset}}_{\mathcal{S},\mathcal{D}_n}(1^\lambda) \le \mathsf{Adv}^{(n-1)\text{-reset}}_{\mathcal{S},\mathcal{D}_{n-1}}(1^\lambda) + \mathsf{Adv}^{1\text{-reset}}_{\mathcal{S},\mathcal{D}_1}(1^\lambda)$$

is negligible in λ.

The proof idea is simple. Given a distinguisher which makes n resets we construct one that ignores the first reset. Now, either this changes the input/output behavior of the simulator noticeably, which yields a distinguisher that only needs a single reset, or it does not in which case the distinguisher with $n-1$ resets is as good as the n-reset distinguisher.

Proof. Let $\mathcal{D}_n$ be a distinguisher that makes at most n reset queries. We construct a distinguisher $\mathcal{D}_{n-1}$ as follows. The distinguisher $\mathcal{D}_{n-1}$ runs exactly as $\mathcal{D}_n$ but does not perform the first reset query of $\mathcal{D}_n$.

In the real world, where the distinguisher is connected to the construction G^π and π, reset queries have no effect and thus we immediately have that

$$\Pr_{r_\mathcal{D}}\left[\mathcal{D}_n^{G^\pi,\pi}(1^\lambda;r_\mathcal{D})=1\right]=\Pr_{r_\mathcal{D}}\left[\mathcal{D}_{n-1}^{G^\pi,\pi}(1^\lambda;r_\mathcal{D})=1\right] \tag{8}$$

where the probability is over the random coins $r_\mathcal{D}$ of the distinguisher.

Let in the ideal world $L_2[\mathcal{D}_n,\mathcal{S},\mathcal{R},r_\mathcal{D},r_\mathcal{S}]$ denote the ordered list of query-answer pairs of queries by distinguisher $\mathcal{D}_n$ to simulator $\mathcal{S}$ up to the second reset query by $\mathcal{D}_n$ when $\mathcal{D}_n$ runs with randomness $r_\mathcal{D}$ and simulator $\mathcal{S}$ runs with randomness $r_\mathcal{S}$ and $\mathcal{R}$ is the random oracle. Note that after each reset query simulator $\mathcal{S}$ takes a fresh set of random coins. Thus, technically we have that $r_\mathcal{S}:=r_\mathcal{S}^1\|r_\mathcal{S}^2\|\ldots$ where $r_\mathcal{S}^1$ denotes the simulator's coins up to the first reset and $r_\mathcal{S}^2$ its coins after the first and up to the second reset. All further random coins are irrelevant for the definition of L_2 since we only consider queries up to the second reset query.

Similarly, we define $L_1[\mathcal{D}_{n-1},\mathcal{S},\mathcal{R},r_\mathcal{D},r_\mathcal{S}]$ to be the list of query-answer pairs by distinguisher $\mathcal{D}_{n-1}$ to simulator $\mathcal{S}$ up to the first reset query. Note that again $r_\mathcal{S}:=r_\mathcal{S}^1\|r_\mathcal{S}^2\|\ldots$ but this time already the second part $(r_\mathcal{S}^2)$ is irrelevant since we only consider queries up to the first reset query.

Define predicate $\mathsf{E}(\mathcal{R},r_\mathcal{D},r_\mathcal{S})$ to hold, iff

$$L_2[\mathcal{D}_n,\mathcal{S},\mathcal{R},r_\mathcal{D},r_\mathcal{S}]=L_1[\mathcal{D}_{n-1},\mathcal{S},\mathcal{R},r_\mathcal{D},r_\mathcal{S}]$$

for a random oracle $\mathcal{R}$ and randomnesses $r_\mathcal{D}$ and $r_\mathcal{S}$. Note that in case of event $\mathsf{E}(\mathcal{R},r_\mathcal{D},r_\mathcal{S})$ it holds that

$$\begin{aligned}&\Pr_{\mathcal{R},r_\mathcal{D},r_\mathcal{S}}\left[\mathcal{D}_n^{\mathcal{R},\mathcal{S}^\mathcal{R}}(1^\lambda)=1\mid\mathsf{E}(\mathcal{R},r_\mathcal{D},r_\mathcal{S})\right]\\ =&\Pr_{\mathcal{R},r_\mathcal{D},r_\mathcal{S}}\left[\mathcal{D}_{n-1}^{\mathcal{R},\mathcal{S}^\mathcal{R}}(1^\lambda)=1\mid\mathsf{E}(\mathcal{R},r_\mathcal{D},r_\mathcal{S})\right].\end{aligned} \tag{9}$$

In the following we simplify notation and do not make the probability space explicit. That is, the probabilities in the ideal world are always over the random oracle $\mathcal{R}$ the random coins of the distinguisher $r_\mathcal{D}$ and the various random coins of the simulator $r_\mathcal{S}$. Also, we simply write E instead of $\mathsf{E}(\mathcal{R},r_\mathcal{D},r_\mathcal{S})$.

Let $\mathcal{D}_1$ denote a distinguisher which makes only a single reset query and which works as follows: $\mathcal{D}_1$ runs $\mathcal{D}_n$ up to the second reset query, passing on queries to its own oracles but not passing on the two reset queries. Let $\overline{q_1}$ denote the queries to the simulator up to the first (ignored) reset query and $\overline{q_2}$ the queries to the simulator after the first (ignored) reset and up to the second (ignored) reset. Now, after the second ignored reset, distinguisher $\mathcal{D}_1$ makes its single reset query and once more sends the sequence $\overline{q_2}$ to the simulator. It outputs 0 in case the simulator's answers are consistent with the previous $\overline{q_2}$ sequence and else it outputs 1. See Figure 1 for a pictorial representation of this operation.

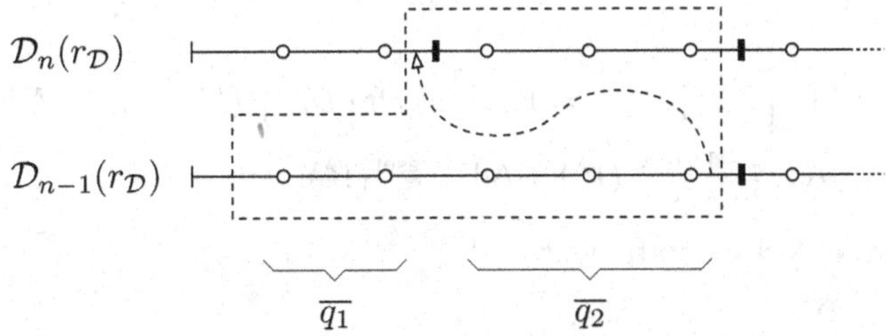

Fig. 1. Illustration of $\mathcal{D}_n$ and $\mathcal{D}_{n-1}$'s operation; circles denote queries and rectangles denote resets. The dashed part resembles the resulting single-reset distinguisher $\mathcal{D}_1$ that asks the queries $\overline{q_2}$ twice (separated by a reset). Whether or not the answer to these two query sequences are identical is captured by the event E.

In the real world, distinguisher $\mathcal{D}_1$ will always output 0 since the answers will always match. Thus, we observe that

$$\begin{aligned}\mathsf{Adv}^{\text{1-reset}}_{\mathcal{S},\mathcal{D}_1}(1^\lambda) &= \Pr\left[\mathcal{D}_1^{\mathcal{R},\mathcal{S}^\mathcal{R}}(1^\lambda)=1\right] - \Pr\left[\mathcal{D}_1^{G^\pi,\pi}(1^\lambda)=1\right]\\ &= \Pr\left[\mathcal{D}_1^{\mathcal{R},\mathcal{S}^\mathcal{R}}(1^\lambda)=1\right]\\ &\geq \Pr\left[\overline{\mathsf{E}}\right]\cdot\Pr\left[\mathcal{D}_1^{\mathcal{R},\mathcal{S}^\mathcal{R}}(1^\lambda)=1\,\middle|\,\overline{\mathsf{E}}\right]\\ &= \Pr\left[\overline{\mathsf{E}}\right]. \end{aligned} \tag{10}$$

For the last equality, note that if $\overline{\mathsf{E}}$ occurs then there is at least one query answer that differs in both runs. This difference must be during $\overline{q_2}$ since, up to $\mathcal{D}_n$'s first reset, both algorithms are identical and operate on the same coins with the same oracles. Hence $\mathcal{D}_1$ always detects this difference and outputs 1. Thus, we have

$$\begin{aligned}\mathsf{Adv}^{n\text{-reset}}_{\mathcal{S},\mathcal{D}_n}(1^\lambda) &= \Pr\left[\mathcal{D}_n^{\mathcal{R},\mathcal{S}^\mathcal{R}}(1^\lambda)=1\right] - \Pr\left[\mathcal{D}_n^{G^\pi,\pi}(1^\lambda)=1\right]\\ &= \Pr[\mathsf{E}]\cdot\Pr\left[\mathcal{D}_n^{\mathcal{R},\mathcal{S}^\mathcal{R}}(1^\lambda)=1\,\middle|\,\mathsf{E}\right]\\ &\quad+\Pr\left[\overline{\mathsf{E}}\right]\cdot\Pr\left[\mathcal{D}_n^{\mathcal{R},\mathcal{S}^\mathcal{R}}(1^\lambda)=1\,\middle|\,\overline{\mathsf{E}}\right] - \Pr\left[\mathcal{D}_n^{G^\pi,\pi}(1^\lambda)=1\right]\\ &\leq \Pr\left[\mathcal{D}_n^{\mathcal{R},\mathcal{S}^\mathcal{R}}(1^\lambda)=1\,\middle|\,\mathsf{E}\right] + \Pr\left[\overline{\mathsf{E}}\right] - \Pr\left[\mathcal{D}_n^{G^\pi,\pi}(1^\lambda)=1\right].\end{aligned}$$

Using equations (8) and (9) we can exchange distinguisher $\mathcal{D}_n$ for distinguisher $\mathcal{D}_{n-1}$ and after reordering we get that

$$= \Pr\left[\mathcal{D}_{n-1}^{\mathcal{R},\mathcal{S}^\mathcal{R}}(1^\lambda)=1\,\middle|\,\mathsf{E}\right] - \Pr\left[\mathcal{D}_{n-1}^{G^\pi,\pi}(1^\lambda)=1\right] + \Pr\left[\overline{\mathsf{E}}\right].$$

Using equation (10)

$$\leq \Pr\left[\mathcal{D}_{n-1}^{\mathcal{R},\mathcal{S}^{\mathcal{R}}}(1^\lambda)=1 \,\middle|\, \mathsf{E}\right] - \Pr\left[\mathcal{D}_{n-1}^{G^\pi,\pi}(1^\lambda)=1\right] + \mathsf{Adv}_{\mathcal{S},\mathcal{D}_1}^{\text{1-reset}}(1^\lambda)$$
$$\leq \mathsf{Adv}_{\mathcal{S},\mathcal{D}_{n-1}}^{(n-1)\text{-reset}}(1^\lambda) + \mathsf{Adv}_{\mathcal{S},\mathcal{D}_1}^{\text{1-reset}}(1^\lambda)$$

which yields the desired statement.

Acknowledgements. We thank the anonymous reviewers, Pooya Farshim, and Giorgia Azzurra Marson for their valuable comments on preliminary versions of this work. Paul Baecher is supported by grant Fi 940/4-1 of the German Research Foundation (DFG). Christina Brzuska is supported by the Israel Science Foundation (grant 1076/11 and 1155/11), the Israel Ministry of Science and Technology grant 3-9094), and the German-Israeli Foundation for Scientific Research and Development (grant 1152/2011). Arno Mittelbach is supported by CASED (`www.cased.de`).

References

1. Bellare, M., Boldyreva, A., O'Neill, A.: Deterministic and efficiently searchable encryption. In: Menezes, A. (ed.) CRYPTO 2007. LNCS, vol. 4622, pp. 535–552. Springer, Heidelberg (2007)
2. Bellare, M., Brakerski, Z., Naor, M., Ristenpart, T., Segev, G., Shacham, H., Yilek, S.: Hedged public-key encryption: How to protect against bad randomness. In: Matsui, M. (ed.) ASIACRYPT 2009. LNCS, vol. 5912, pp. 232–249. Springer, Heidelberg (2009)
3. Bellare, M., Kohno, T.: A theoretical treatment of related-key attacks: RKA-PRPs, RKA-PRFs, and applications. In: Biham, E. (ed.) EUROCRYPT 2003. LNCS, vol. 2656, pp. 491–506. Springer, Heidelberg (2003)
4. Bellare, M., Rogaway, P.: Random oracles are practical: A paradigm for designing efficient protocols. In: Ashby, V. (ed.) ACM CCS 1993: 1st Conference on Computer and Communications Security, November 3-5, pp. 62–73. ACM Press, Fairfax (1993)
5. Bellare, M., Rogaway, P.: Optimal asymmetric encryption. In: De Santis, A. (ed.) EUROCRYPT 1994. LNCS, vol. 950, pp. 92–111. Springer, Heidelberg (1995)
6. Bellare, M., Rogaway, P.: The exact security of digital signatures - how to sign with RSA and Rabin. In: Maurer, U.M. (ed.) EUROCRYPT 1996. LNCS, vol. 1070, pp. 399–416. Springer, Heidelberg (1996)
7. Bennett, C.H., Gill, J.: Relative to a random oracle A, $P^A \neq NP^A \neq coNP^A$ with probability 1. SIAM Journal on Computing 10(1), 96–113 (1981)
8. Black, J., Rogaway, P., Shrimpton, T.: Encryption-scheme security in the presence of key-dependent messages. In: Nyberg, K., Heys, H.M. (eds.) SAC 2002. LNCS, vol. 2595, pp. 62–75. Springer, Heidelberg (2003)
9. Black, J., Rogaway, P., Shrimpton, T., Stam, M.: An analysis of the blockcipher-based hash functions from PGV. Journal of Cryptology 23(4), 519–545 (2010)
10. Boldyreva, A., Cash, D., Fischlin, M., Warinschi, B.: Foundations of non-malleable hash and one-way functions. In: Matsui, M. (ed.) ASIACRYPT 2009. LNCS, vol. 5912, pp. 524–541. Springer, Heidelberg (2009)

11. Coron, J.-S., Dodis, Y., Malinaud, C., Puniya, P.: Merkle-Damgård revisited: How to construct a hash function. In: Shoup, V. (ed.) CRYPTO 2005. LNCS, vol. 3621, pp. 430–448. Springer, Heidelberg (2005)
12. Coron, J.-S., Patarin, J., Seurin, Y.: The random oracle model and the ideal cipher model are equivalent. In: Wagner, D. (ed.) CRYPTO 2008. LNCS, vol. 5157, pp. 1–20. Springer, Heidelberg (2008)
13. Demay, G., Gaži, P., Hirt, M., Maurer, U.: Resource-restricted indifferentiability. In: Johansson, T., Nguyen, P.Q. (eds.) EUROCRYPT 2013. LNCS, vol. 7881, pp. 664–683. Springer, Heidelberg (2013)
14. Dziembowski, S., Kazana, T., Wichs, D.: Key-evolution schemes resilient to space-bounded leakage. In: Rogaway, P. (ed.) CRYPTO 2011. LNCS, vol. 6841, pp. 335–353. Springer, Heidelberg (2011)
15. Even, S., Mansour, Y.: A construction of a cipher from a single pseudorandom permutation. Journal of Cryptology 10(3), 151–162 (1997)
16. Goldreich, O., Goldwasser, S., Ron, D.: On the possibilities and limitations of pseudodeterministic algorithms. Electronic Colloquium on Computational Complexity (ECCC) 19, 101 (2012)
17. Holenstein, T., Künzler, R., Tessaro, S.: The equivalence of the random oracle model and the ideal cipher model, revisited. In: Fortnow, L., Vadhan, S.P. (eds.) 43rd ACM STOC Annual ACM Symposium on Theory of Computing, June 6-8, pp. 89–98. ACM Press, San Jose (2011)
18. Luykx, A., Andreeva, E., Mennink, B., Preneel, B.: Impossibility results for indifferentiability with resets. Cryptology ePrint Archive, Report 2012/644 (2012), `http://eprint.iacr.org/`
19. Maurer, U.M., Renner, R.S., Holenstein, C.: Indifferentiability, impossibility results on reductions, and applications to the random oracle methodology. In: Naor, M. (ed.) TCC 2004. LNCS, vol. 2951, pp. 21–39. Springer, Heidelberg (2004)
20. Ristenpart, T., Shacham, H., Shrimpton, T.: Careful with composition: Limitations of the indifferentiability framework. In: Paterson, K.G. (ed.) EUROCRYPT 2011. LNCS, vol. 6632, pp. 487–506. Springer, Heidelberg (2011)
21. Shannon, C.E.: Communication theory of secrecy systems. Bell Systems Technical Journal 28(4), 656–715 (1949)

Computational Fuzzy Extractors

Benjamin Fuller[1], Xianrui Meng[2], and Leonid Reyzin[2]

[1] Boston University and MIT Lincoln Laboratory
[2] Boston University

Abstract. Fuzzy extractors derive strong keys from noisy sources. Their security is defined information-theoretically, which limits the length of the derived key, sometimes making it too short to be useful. We ask whether it is possible to obtain longer keys by considering computational security, and show the following.

- **Negative Result:** Noise tolerance in fuzzy extractors is usually achieved using an information reconciliation component called a "secure sketch." The security of this component, which directly affects the length of the resulting key, is subject to lower bounds from coding theory. We show that, even when defined computationally, secure sketches are still subject to lower bounds from coding theory. Specifically, we consider two computational relaxations of the information-theoretic security requirement of secure sketches, using conditional HILL entropy and unpredictability entropy. For both cases we show that computational secure sketches cannot outperform the best information-theoretic secure sketches in the case of high-entropy Hamming metric sources.
- **Positive Result:** We show that the negative result can be overcome by analyzing computational fuzzy extractors directly. Namely, we show how to build a computational fuzzy extractor whose output key length equals the entropy of the source (this is impossible in the information-theoretic setting). Our construction is based on the hardness of the Learning with Errors (LWE) problem, and is secure when the noisy source is uniform or symbol-fixing (that is, each dimension is either uniform or fixed). As part of the security proof, we show a result of independent interest, namely that the decision version of LWE is secure even when a small number of dimensions has no error.

Keywords: Fuzzy extractors, secure sketches, key derivation, Learning with Errors, error-correcting codes, computational entropy, randomness extractors.

1 Introduction

Authentication generally requires a secret drawn from some high-entropy source. One of the primary building blocks for authentication is reliable key derivation. Unfortunately, many sources that contain sufficient entropy to derive a key are

K. Sako and P. Sarkar (Eds.) ASIACRYPT 2013 Part I, LNCS 8269, pp. 174–193, 2013.

noisy, and provide similar, but not identical secret values at each reading (examples of such sources include biometrics [14], human memory [37], pictorial passwords [9], measurements of capacitance [35], timing [34], motion [10], quantum information [5] etc.).

Fuzzy extractors [15] achieve reliable key derivation from noisy sources (see [7,16,11] for applications of fuzzy extractors). The setting consists of two algorithms: Generate (used once) and Reproduce (used subsequently). The Generate (Gen) algorithm takes an input w and produces a key r and a public value p. This information allows the Reproduce (Rep) algorithm to reproduce r given p and some value w' that is close to w (according to some predefined metric, such as Hamming distance). Crucially for security, knowledge of p should not reveal r; that is, r should be uniformly distributed conditioned on p. This feature is needed because p is not secret: for example, in a single-user setting (where the user wants to reproduce the key r from a subsequent reading w'), it would be stored in the clear; and in a key agreement application [7] (where two parties have w and w', respectively), it would be transmitted between the parties.

Fuzzy extractors use ideas from information-reconciliation [5] and are defined (traditionally) as information-theoretic objects. The entropy loss of a fuzzy extractor is the difference between the entropy of w and the length of the derived key r. In the information-theoretic setting, some entropy loss is necessary as the value p contains enough information to reproduce r from any close value w'. A goal of fuzzy extractor constructions is to minimize the entropy loss, increasing the security of the resulting application. Indeed, if the entropy loss is too high, the resulting secret key may be too short to be useful.

We ask whether it is possible to obtain longer keys by considering computational, rather than information theoretic, security.

Our Negative Results. We first study (in Section 3) whether it could be fruitful to relax the definition of the main building block of a fuzzy extractor, called a *secure sketch.* A secure sketch is a one-round information reconciliation protocol: it produces a public value s that allows recovery of w from any close value w'. The traditional secrecy requirement of a secure sketch is that w has high min-entropy conditioned on s. This allows the fuzzy extractor of [15] to form the key r by applying a randomness extractor [28] to w, because randomness extractors produce random strings from strings with conditional min-entropy. We call this the *sketch-and-extract* construction.

The most natural relaxation of the min-entropy requirement of the secure sketch is to require HILL entropy [21] (namely, that the distribution of w conditioned on s be *indistinguishable* from a high-min-entropy distribution). Under this definition, we could still use a randomness extractor to obtain r from w, because it would yield a pseudorandom key. Unfortunately, it is unlikely that such a relaxation will yield fruitful results: we prove in Theorem 1 that the entropy loss of such secure sketches is subject to the same coding bounds as the ones that constrain information-theoretic secure sketches.

Another possible relaxation is to require that the value w is unpredictable conditioned on s. This definition would also allow the use of a randomness extractor to get a pseudorandom key, although it would have to be a special extractor—one that has a reconstruction procedure (see [22, Lemma 6]). Unfortunately, this relaxation is also unlikely to be fruitful: we prove in Theorem 2 that the unpredictability is at most log the size of the metric space minus log the volume of the ball of radius t. For high-entropy sources of w over the Hamming metric, this bound matches the best information-theoretic security sketches.

Our Positive Results. Both of the above negative results arise because a secure sketch functions like a decoder of an error-correcting code. To avoid them, we give up on building computational secure sketches and focus directly on the entropy loss in fuzzy extractors. Our goal is to decrease the entropy loss in a fuzzy extractor by allowing the key r to be pseudorandom conditioned on p.

By considering this computational secrecy requirement, we construct the first *lossless* computational fuzzy extractors (Construction 1), where the derived key r is as long as the entropy of the source w. Our construction is for the Hamming metric and uses the code-offset construction [23],[15, Section 5] used in prior work, but with two crucial differences. First, the key r is not extracted from w like in the sketch-and-extract approach; rather w "encrypts" r in a way that is decryptable with the knowledge of some close w' (this idea is similar to the way the code-offset construction is presented in [23] as a "fuzzy commitment"). Our construction uses private randomness, which is allowed in the fuzzy extractor setting but not in noiseless randomness extraction. Second, the code used is a random linear code, which allows us to use the Learning with Errors (LWE) assumption due to Regev [30,31] and derive a longer key r.

Specifically, we use the recent result of Döttling and Müller-Quade [17], which shows the hardness of decoding random linear codes when the error vector comes from the uniform distribution, with each coordinate ranging over a small interval. This allows us to use w as the error vector, assuming it is uniform. We also use a result of Akavia, Goldwasser, and Vaikuntanathan [1], which says that LWE has many hardcore bits, to hide r.

Because we use a random linear code, our decoding is limited to reconciling a logarithmic number of differences. Unfortunately, we cannot utilize the results that improve the decoding radius through the use of trapdoors (such as [30]), because in a fuzzy extractor, there is no secret storage place for the trapdoor. If improved decoding algorithms are obtained for random linear codes, they will improve error-tolerance of our construction. Given the hardness of decoding random linear codes [6], we do not expect significant improvement in the error-tolerance of our construction.

In Section 5, we are able to relax the assumption that w comes from the uniform distribution, and instead allow w to come from a symbol-fixing source [24] (each dimension is either uniform or fixed). This relaxation follows from our results about the hardness of LWE when samples have a fixed (and adversarially known) error vector, which may be of independent interest (Theorem 4).

An Alternative Approach. Computational extractors [26,3,13] have the same goal of obtaining a pseudorandom key r from a source w in the setting without errors.

They can be constructed, for example, by applying a pseudorandom generator to the output of an information-theoretic extractor. One way to build a computational *fuzzy* extractor is by using a computational extractor instead of the information-theoretic extractor in the sketch-and-extract construction of [15]. However, this approach is possible only if conditional min-entropy of w conditioned on the sketch s is high enough. Furthermore, this approach does not allow the use of private randomness; private randomness is a crucial ingredient in our construction. We compare the two approaches in Section 4.4.

2 Preliminaries

For a random variable $X = X_1||...||X_n$ where each X_i is over some alphabet $\mathcal{Z}$, we denote by $X_{1,...,k} = X_1||...||X_k$. The *min-entropy* of X is $\mathrm{H}_\infty(X) = -\log(\max_x \Pr[X = x])$, and the *average (conditional)* min-entropy of X given Y is $\tilde{\mathrm{H}}_\infty(X|Y) = -\log(\mathbb{E}_{y\in Y} \max_x \Pr[X = x|Y = y])$ [15, Section 2.4]. The *statistical distance* between random variables X and Y with the same domain is $\Delta(X, Y) = \frac{1}{2}\sum_x |\Pr[X = x] - \Pr[Y = x]|$. For a distinguisher D (or a class of distinguishers $\mathcal{D}$) we write the *computational distance* between X and Y as $\delta^D(X, Y) = |\mathbb{E}[D(X)] - \mathbb{E}[D(Y)]|$. We denote by $\mathcal{D}_{s_{sec}}$ the class of randomized circuits which output a single bit and have size at most s_{sec}. For a metric space $(\mathcal{M}, \mathsf{dis})$, the *(closed) ball of radius t around x* is the set of all points within radius t, that is, $B_t(x) = \{y|\mathsf{dis}(x, y) \le t\}$. If the size of a ball in a metric space does not depend on x, we denote by $|B_t(\cdot)|$ the size of a ball of radius t. For the Hamming metric over $\mathcal{Z}^n$, $|B_t(\cdot)| = \sum_{i=0}^{t} \binom{n}{t}(|\mathcal{Z}| - 1)^i$. U_n denotes the uniformly distributed random variable on $\{0, 1\}^n$. Usually, we use bold letters for vectors or matrices, capitalized letters for random variables, and lowercase letters for elements in a vector or samples from a random variable.

2.1 Fuzzy Extractors and Secure Sketches

We now recall definitions and lemmas from the work of Dodis et. al. [15, Sections 2.5–4.1], adapted to allow for a small probability of error, as discussed in [15, Sections 8]. Let $\mathcal{M}$ be a metric space with distance function dis.

Definition 1. *An* $(\mathcal{M}, m, \ell, t, \epsilon)$-fuzzy extractor *with error* δ *is a pair of randomized procedures, "generate"* (Gen) *and "reproduce"* (Rep), *with the following properties:*

1. *The generate procedure* Gen *on input* $w \in \mathcal{M}$ *outputs an extracted string* $r \in \{0, 1\}^\ell$ *and a helper string* $p \in \{0, 1\}^*$.
2. *The reproduction procedure* Rep *takes an element* $w' \in \mathcal{M}$ *and a bit string* $p \in \{0, 1\}^*$ *as inputs. The* correctness *property of fuzzy extractors guarantees that for* w *and* w' *such that* $\mathsf{dis}(w, w') \le t$, *if* R, P *were generated by* $(R, P) \leftarrow \mathsf{Gen}(w)$, *then* $\mathsf{Rep}(w', P) = R$ *with probability (over the coins of* Gen, Rep*) at least* $1 - \delta$. *If* $\mathsf{dis}(w, w') > t$, *then no guarantee is provided about the output of* Rep.

3. *The* security *property guarantees that for any distribution W on $\mathcal{M}$ of min-entropy m, the string R is nearly uniform even for those who observe P: if $(R, P) \leftarrow \mathsf{Gen}(W)$, then $\mathbf{SD}((R, P), (U_\ell, P)) \leq \epsilon$.*

A fuzzy extractor is efficient if Gen *and* Rep *run in expected polynomial time.*

Secure sketches are the main technical tool in the construction of fuzzy extractors. Secure sketches produce a string s that does not decrease the entropy of w too much, while allowing recovery of w from a close w':

Definition 2. *An $(\mathcal{M}, m, \tilde{m}, t)$-*secure sketch *with error δ is a pair of randomized procedures, "sketch"* (SS) *and "recover"* (Rec)*, with the following properties:*

1. *The sketching procedure* SS *on input $w \in \mathcal{M}$ returns a bit string $s \in \{0,1\}^*$.*
2. *The recovery procedure* Rec *takes an element $w' \in \mathcal{M}$ and a bit string $s \in \{0,1\}^*$. The* correctness *property of secure sketches guarantees that if $\mathsf{dis}(w, w') \leq t$, then $\Pr[\mathsf{Rec}(w', \mathsf{SS}(w)) = w] \geq 1 - \delta$ where the probability is taken over the coins of* SS *and* Rec*. If $\mathsf{dis}(w, w') > t$, then no guarantee is provided about the output of* Rec.
3. *The* security *property guarantees that for any distribution W over $\mathcal{M}$ with min-entropy m, the value of W can be recovered by the adversary who observes w with probability no greater than $2^{-\tilde{m}}$. That is, $\tilde{\mathrm{H}}_\infty(W|\mathsf{SS}(W)) \geq \tilde{m}$.*

A secure sketch is efficient *if* SS *and* Rec *run in expected polynomial time.*

Note that in the above definition of secure sketches (resp., fuzzy extractors), the errors are chosen before s (resp., P) is known: if the error pattern between w and w' depends on the output of SS (resp., Gen), then there is no guarantee about the probability of correctness.

A fuzzy extractor can be produced from a *secure sketch* and an *average-case randomness extractor*. An average-case extractor is a generalization of a strong randomness extractor [28, Definition 2]) (in particular, Vadhan [36, Problem 6.8] showed that all strong extractors are average-case extractors with a slight loss of parameters):

Definition 3. *Let χ_1, χ_2 be finite sets. A function $\mathtt{ext} : \chi_1 \times \{0,1\}^d \rightarrow \{0,1\}^\ell$ a (m, ϵ)-*average-case extractor *if for all pairs of random variables X, Y over χ_1, χ_2 such that $\tilde{H}_\infty(X|Y) \geq m$, we have $\Delta((\mathtt{ext}(X, U_d), U_d, Y), U_\ell \times U_d \times Y) \leq \epsilon$.*

Lemma 1. *Assume* (SS, Rec) *is an $(\mathcal{M}, m, \tilde{m}, t)$-secure sketch with error δ, and let $\mathtt{ext} : \mathcal{M} \times \{0,1\}^d \rightarrow \{0,1\}^\ell$ be a $(\tilde{m}, \epsilon)$-average-case extractor. Then the following* (Gen, Rep) *is an $(\mathcal{M}, m, \ell, t, \epsilon)$-fuzzy extractor with error δ:*

- $\mathsf{Gen}(w)$: *generate $x \leftarrow \{0,1\}^d$, set $p = (\mathsf{SS}(w), x), r = \mathtt{ext}(w; x)$, and output (r, p).*
- $\mathsf{Rep}(w', (s, x))$: *recover $w = \mathsf{Rec}(w', s)$ and output $r = \mathtt{ext}(w; x)$.*

The main parameter we will be concerned with is the entropy loss of the construction. In this paper, we ask whether a smaller entropy loss can be achieved by considering a fuzzy extractor with a computational security requirement. We therefore relax the security requirement of Definition 1 to require a pseudorandom output instead of a truly random output. Also, for notational convenience, we modify the definition so that we can specify a general class of sources for which the fuzzy extractor is designed to work, rather than limiting ourselves to the class of sources that consists of all sources of a given min-entropy m, as in definitions above (of course, this modification can also be applied to prior definitions of information-theoretic secure sketches and fuzzy extractors).

Definition 4 (Computational Fuzzy Extractor). *Let $\mathcal{W}$ be a family of probability distributions over $\mathcal{M}$. A pair of randomized procedures "generate"* (Gen) *and "reproduce"* (Rep) *is a* $(\mathcal{M}, \mathcal{W}, \ell, t)$-computational fuzzy extractor *that is (ϵ, s_{sec})-hard with error δ if* Gen *and* Rep *satisfy the following properties:*

- *The generate procedure* Gen *on input $w \in \mathcal{M}$ outputs an extracted string $R \in \{0,1\}^\ell$ and a helper string $P \in \{0,1\}^*$.*
- *The reproduction procedure* Rep *takes an element $w' \in \mathcal{M}$ and a bit string $P \in \{0,1\}^*$ as inputs. The* correctness *property guarantees that for all w, w' where* $\mathsf{dis}(w, w') \leq t$, *if $(R, P) \leftarrow$* $\mathsf{Gen}(w)$ *then* $\Pr[\mathsf{Rep}(w', P) = R] \geq 1 - \delta$ *where the probability is over the randomness of* $(\mathsf{Gen}, \mathsf{Rep})$. *If* $\mathsf{dis}(w, w') > t$, *then no guarantee is provided about the output of* Rep.
- *The* security *property guarantees that for any distribution $W \in \mathcal{W}$, the string R is pseudorandom conditioned on P, that is $\delta^{\mathcal{D}_{s_{sec}}}((R, P), (U_\ell, P)) \leq \epsilon$.*

Any efficient fuzzy extractor is also a computational fuzzy extractor with the same parameters.

Remark. Fuzzy extractor definitions make no guarantee about Rep behavior when the distance between w and w' is larger than t. In the information-theoretic setting this seemed inherent as the "correct" R should be information-theoretically unknown conditioned on P. However, in the computationally setting this is not true. Looking ahead, in our construction R is information-theoretically determined conditioned on P (with high probability over the coins of Gen). Our Rep algorithm will never output an incorrect key (with high probability over the coins of Gen) but may not terminate. However, it is not clear this is the desired behavior. For this reason, we leave the behavior of Rep ambiguous when $\mathsf{dis}(w, w') > t$.

3 Impossibility of Computational Secure Sketches

In this section, we consider whether it is possible in build a secure sketch that retains significantly more computational than information-theoretic entropy. We consider two different notions for computational entropy, and for both of them show that corresponding secure sketches are subject to the same upper bounds as those for information-theoretic secure sketches. Thus, it seems that relaxing security of sketches from information-theoretic to computational does not help.

In particular, for the case of the Hamming metric and inputs that have full entropy, our results are as follows. In Section 3.1 we show that a sketch that retains HILL entropy implies a sketch that retains nearly the same amount of min-entropy. In Section 3.2, we show that the computational unpredictability of a sketch is at most $\log|\mathcal{M}| - \log|B_t(\cdot)|$. Dodis et al. [15, Section 8.2] construct sketches with essentially the same information-theoretic security[1] . In Section 3.3, we discuss mechanisms for avoiding these bounds.

3.1 Bounds on Secure Sketches Using HILL Entropy

HILL entropy is a commonly used computational notion of entropy [21]. It was extended to the conditional case by Hsiao, Lu, Reyzin [22]. Here we recall a weaker definition due to Gentry and Wichs [19] (the term relaxed HILL entropy was introduced in [32]); since we show impossibility even for this weaker definition, impossibility for the stronger definition follows immediately.

Definition 5. *Let (W, S) be a pair of random variables. W has* relaxed HILL entropy *at least k conditioned on S, denoted $H^{\text{HILL-rlx}}_{\epsilon,s_{sec}}(W|S) \geq k$ if there exists a joint distribution (X, Y), such that $\tilde{H}_\infty(X|Y) \geq k$ and $\delta^{\mathcal{D}_{s_{sec}}}((W, S), (X, Y)) \leq \epsilon$.*

Intuitively, HILL entropy is as good as average min-entropy for all computationally bounded observers. Thus, redefining secure sketches using HILL entropy is a natural relaxation of the original information-theoretic definition; in particular, the sketch-and-extract construction in Lemma 1 would yield pseudorandom outputs if the secure sketch ensured high HILL entropy. We will consider secure sketches that retain relaxed HILL entropy: that is, we say that $(\mathsf{SS}, \mathsf{Rec})$ is a *HILL-entropy $(\mathcal{M}, m, \tilde{m}, t)$ secure sketch* that is (ϵ, s_{sec})-hard with error δ if it satisfies Definition 2, with the security requirement replaced by $H^{\text{HILL-rlx}}_{\epsilon,s_{sec}}(W|\mathsf{SS}(W)) \geq \tilde{m}$.

Unfortunately, we will show below that such a secure sketch implies an error correcting code with approximately $2^{\tilde{m}}$ points that can correct t random errors (see [15, Lemma C.1] for a similar bound on information-theoretic secure sketches). For the Hamming metric, our result essentially matches the bound on information-theoretic secure sketches of [15, Proposition 8.2]. In fact, we show that, for the Hamming metric, HILL-entropy secure sketches imply information-theoretic ones with similar parameters, and, therefore, the HILL relaxation gives no advantage.

The intuition for building error-correcting codes from HILL-entropy secure sketches is as follows. In order to have $H^{\text{HILL-rlx}}_{\epsilon,s_{sec}}(W|\mathsf{SS}(W)) \geq \tilde{m}$, there must be a distribution X, Y such that $\tilde{\mathrm{H}}_\infty(X|Y) \geq \tilde{m}$ and (X, Y) is computationally indistinguishable from $(W, \mathsf{SS}(W))$. Sample a sketch $s \leftarrow \mathsf{SS}(W)$. We know that SS followed by Rec likely succeeds on $W|s$ (i.e., $\mathsf{Rec}(w', s) = w$ with high probability for $w \leftarrow W|s$ and $w' \leftarrow B_t(w)$). Consider the following experiment: 1) sample $y \leftarrow Y$, 2) draw $x \leftarrow X|y$ and 3) $x' \leftarrow B_t(x)$. By indistinguishability,

[1] The security in [15, Section 8.2] is expressed in terms of entropy of the error rate; recall that $\log B_t(\cdot) \approx H_q(t/n)$, where n is the number of symbols, q is the alphabet size, and H_q is the q-ary entropy function.

$\mathsf{Rec}(x', y) = x$ with high probability. This means we can construct a large set $\mathcal{C}$ from the support of $X|y$. $\mathcal{C}$ will be an error correcting code and Rec an efficient decoder. We can then use standard arguments to turn this code into an information theoretic sketch.

To make this intuition precise, we need an additional technical condition: sampling a random neighbor of a point is efficient.

Definition 6. *We say a metric space* $(\mathcal{M}, \mathsf{dis})$ *is* (s_{neigh}, t)-neighborhood samplable *if there exists a randomized circuit* Neigh *of size* s_{neigh} *that for all* $t' \leq t$, $\mathsf{Neigh}(w, t')$ *outputs a random point at distance* t' *of* w.

We review the definition of a Shannon code [33]:

Definition 7. *Let* $\mathcal{C}$ *be a set over space* $\mathcal{M}$. *We say that* $\mathcal{C}$ *is an* (t, ϵ)-Shannon code *if there exists an efficient procedure* Rec *such that for all* $t' \leq t$ *and for all* $c \in \mathcal{C}$, $\Pr[\mathsf{Rec}(\mathsf{Neigh}(c, t')) \neq c] \leq \epsilon$. *To distinguish it from the average-error Shannon code defined below, we will sometimes call it a* maximal-error *Shannon code.*

This is a slightly stronger formulation than usual, in that for every size $t' < t$ we require the code to correct t' random errors[2]. Shannon codes work for all codewords. We can also consider a formulation that works for an "average" codeword.

Definition 8. *Let* C *be a distribution over space* $\mathcal{M}$. *We say that* C *is an* (t, ϵ)-average error Shannon code *if there exists an efficient procedure* Rec *such that for all* $t' \leq t$ $\Pr_{c \leftarrow C}[\mathsf{Rec}(\mathsf{Neigh}(c, t')) \neq c] \leq \epsilon$.

An average error Shannon code is one whose average probability of error is bounded by ϵ. See [12, Pages 192-194] for definitions of average and maximal error probability. An average-error Shannon code is convertible to a maximal-error Shannon code with a small loss. We use the following pruning argument from [12, Pages 202-204] (we provide a proof in the full version [18]):

Lemma 2. *Let* C *be a* (t, ϵ)*-average error Shannon code with recovery procedure* Rec *such that* $\mathrm{H}_\infty(C) \geq k$. *There is a set* $\mathcal{C}'$ *with* $|\mathcal{C}'| \geq 2^{k-1}$ *that is a* $(t, 2\epsilon)$*-(maximal error) Shannon code with recovery procedure* Rec.

We can now formalize the intuition above and show that a sketch that retains $\tilde{m}$-bits of relaxed HILL entropy implies a good error correcting code with nearly $2^{\tilde{m}}$ points (proof in the full version of this work [18]).

Theorem 1. *Let* $(\mathcal{M}, \mathsf{dis})$ *be a* (s_{neigh}, t)*-neighborhood samplable metric space. Let* $(\mathsf{SS}, \mathsf{Rec})$ *be a HILL-entropy* $(\mathcal{M}, m, \tilde{m}, t)$*-secure sketch that is* (ϵ, s_{sec})*-secure*

[2] In the standard formulation, the code must correct a random error of size up to t, which may not imply that it can correct a random error of a much smaller size t', because the volume of the ball of size t' may be negligible compared to the volume of the ball of size t. For codes that are monotone (if decoding succeeds on a set of errors, it succeeds on all subsets), these formulations are equivalent. However, we work with an arbitrary recover functionality that is not necessarily monotone.

with error δ. Let s_{rec} denote the size of the circuit that computes Rec. *If $s_{sec} \geq (t(s_{neigh} + s_{rec}))$, then there exists a value s and a set $\mathcal{C}$ with $|\mathcal{C}| \geq 2^{\tilde{m}-2}$ that is a $(t, 4(\epsilon + t\delta))$-Shannon code with recovery procedure* $\mathsf{Rec}(\cdot, s)$.

For the Hamming metric, any Shannon code (as defined in Definition 7) can be converted into an information-theoretic secure sketch (as described in [15, Section 8.2] and references therein). The idea is to use the code offset construction, and convert worst-case errors to random errors by randomizing the order of the symbols of w first, via a randomly chosen permutation π (which becomes part of the sketch and is applied to w' during Rec). The formal statement of this result can be expressed in the following Lemma (which is implicit in [15, Section 8.2]).

Lemma 3. *For an alphabet $\mathcal{Z}$, let $\mathcal{C}$ over $\mathcal{Z}^n$ be a (t, δ) Shannon code. Then there exists a $(\mathcal{Z}^n, m, m - (n \log |\mathcal{Z}| - \log |\mathcal{C}|), t)$ secure sketch with error δ for the Hamming metric over $\mathcal{Z}^n$.*

Putting together Theorem 1 and Lemma 3 gives us the negative result for the Hamming metric: a HILL-entropy secure sketch (for the uniform distribution) implies an information-theoretic one with similar parameters:

Corollary 1. *Let $\mathcal{Z}$ be an alphabet. Let* $(\mathsf{SS}', \mathsf{Rec}')$ *be an (ϵ, s_{sec})-HILL-entropy $(\mathcal{Z}^n, n \log |\mathcal{Z}|, \tilde{m}, t)$-secure sketch with error δ for the Hamming metric over $\mathcal{Z}^n$, with* Rec' *of circuit size s_{rec}. If $s_{sec} \geq t(s_{rec} + n \log |\mathcal{Z}|)$, then there exists a $(\mathcal{Z}^n, n \log |\mathcal{Z}|, \tilde{m} - 2, t)$ (information-theoretic) secure sketch with error* $4(\epsilon + t\delta)$.

Note. In Corollary 1 we make no claim about the efficiency of the resulting (SS, Rec), because the proof of Theorem 1 is not constructive.

Corollary 1 extends to non-uniform distributions: if there exists a distribution whose HILL sketch retains $\tilde{m}$ bits of entropy, then for all distributions W, there is an information theoretic sketch that retains $\mathrm{H}_\infty(W) - (n \log |\mathcal{Z}| - \tilde{m}) - 2$ bits of entropy.

3.2 Bounds on Secure Sketches Using Unpredictability Entropy

In the previous section, we showed that any sketch that retained HILL entropy could be transformed into an information theoretic sketch. However, HILL entropy is a strong notion. In this section, we therefore ask whether it is useful to consider a sketch that satisfies a minimal requirement: the value of the input is computationally hard to guess given the sketch. We begin by recalling the definition of conditional unpredictability entropy [22, Definition 7], which captures the notion of "hard to guess" (we relax the definition slightly, similarly to the relaxation of HILL entropy described in the previous section).

Definition 9. *Let (W, S) be a pair of random variables. W has* relaxed unpredictability entropy *at least k conditioned on S, denoted by $H^{\mathtt{unp\text{-}rlx}}_{\epsilon, s_{sec}}(W|S) \geq k$, if there exists a pair of distributions (X, Y) such that $\delta^{\mathcal{D}_{s_{sec}}}((W, S), (X, Y)) \leq \epsilon$, and for all circuits $\mathcal{I}$ of size s_{sec},*

$$\Pr[\mathcal{I}(Y) = X] \leq 2^{-k}.$$

A pair of procedures (SS, Rec) is a *unpredictability-entropy* $(\mathcal{M}, m, \tilde{m}, t)$ *secure sketch* that is (ϵ, s_{sec})-hard with error δ if it satisfies Definition 2, with the security requirement replaced by $H^{\texttt{unp-rlx}}_{\epsilon,s_{sec}}(W|\mathsf{SS}(W)) \geq \tilde{m}$. Note this notion is quite natural: combining such a secure sketch in a sketch-and-extract construction of Lemma 1 with a particular type of extractor (called a *reconstructive* extractor [4]), would yield a computational fuzzy extractor (per [22, Lemma 6]).

Unfortunately, the conditional unpredictability entropy $\tilde{m}$ must decrease as t increases, as the following theorem states. (The proof of the theorem, generalized to more metric spaces, is in the full version [18].)

Theorem 2. *Let $\mathcal{Z}$ be an alphabet. Let* (SS, Rec) *be an unpredictability-entropy* $(\mathcal{Z}^n, m, \tilde{m}, t)$*-secure sketch that is* (ϵ, s_{sec})*-secure with error* δ*, if* $s_{sec} \geq t(|\mathsf{Rec}| + n\log|\mathcal{Z}|)$*, then* $\tilde{m} \leq n\log|\mathcal{Z}| - \log|B_t(\cdot)| + \log(1-\epsilon-t\delta)$.

In particular, if the input is uniform, the entropy loss is about $\log|B_t(\cdot)|$. As mentioned at the beginning of Section 3, essentially the same entropy loss can be achieved with information-theoretic secure sketches, by using the randomized code-offset construction. However, it is conceivable that unpredictability entropy secure sketches could achieve lower entropy loss with greater efficiency for some parameter settings.

3.3 Avoiding Sketch Entropy Upper Bounds

The lower bounds of Corollary 1 and Theorem 2 are strongest for high entropy sources. This is necessary, if a source contains only codewords (of an error correcting code), no sketch is needed, and thus there is no (computational) entropy loss. This same situation occurs when considering lower bounds for information-theoretic sketches [15, Appendix C] .

Both of lower bounds arise because Rec must function as an error-correcting code for many points of any indistinguishable distribution. It may be possible to avoid these bounds if Rec outputs a fresh random variable[3]. Such an algorithm is called a computational fuzzy conductor. See [25] for the definition of a fuzzy conductor. To the best of our knowledge, a computational fuzzy conductor has not been defined in the literature, the natural definition is to replace the pseudorandomness condition in Definition 4 with a HILL entropy requirement.

Our construction (in Section 4) has pseudorandom output and immediately satisfies definition of a computational fuzzy extractor (Definition 4). It may be possible to achieve significantly better parameters with a construction that is a computational fuzzy conductor (but not a computational fuzzy extractor) and then applying an extractor. We leave this as an open problem.

[3] If some efficient algorithm can take the output of Rec and efficiently transform it back to the source W, the bounds of Corollary 1 and Theorem 2 both apply. This means that we need to consider constructions that are hard to invert (either information-theoretically or computationally).

4 Computational Fuzzy Extractor Based on LWE

In this section we describe our main construction. Security of our construction depends on the source W. We first consider a uniform source W; we consider other distributions in Section 5. Our construction uses the code-offset construction [23], [15, Section 5] instantiated with a random linear code over a finite field $\mathbb{F}_q$. Let Decode_t be an algorithm that decodes a random linear code with at most t errors (we will present such an algorithm later, in Section 4.2).

Construction 1. *Let n be a security parameter and let $m \geq n$. Let q be a prime. Define* Gen, Rep *as follows:*

Gen	Rep
1. *Input: $w \leftarrow W$ (where W is some distribution over $\mathbb{F}_q^m$).*	1. *Input: (w', p) (where the Hamming distance between w' and w is at most t).*
2. *Sample $\mathbf{A} \in \mathbb{F}_q^{m\times n}, \mathbf{x} \in \mathbb{F}_q^n$ uniformly.*	2. *Parse p as $(\mathbf{A}, \mathbf{c})$; let $\mathbf{b} = \mathbf{c} - w'$.*
3. *Compute $p = (\mathbf{A}, \mathbf{Ax} + w)$, $r = \mathbf{x}_{1,\dots,n/2}$.*	3. *Let $x = \mathsf{Decode}_t(\mathbf{A}, \mathbf{b})$*
4. *Output (r, p).*	4. *Output $r = x_{1,\dots,n/2}$.*

Intuitively, security comes from the computational hardness of decoding random linear codes with a high number of errors (introduced by w). In fact, we know that decoding a random linear code is NP-hard [6]; however, this statement is not sufficient for our security goal, which is to show

$$\delta^{\mathcal{D}_{s_{sec}}}((X_{1,\dots,n/2}, P), (U_{n/2\log q}, P)) \leq \epsilon.$$

Furthermore, this construction is only useful if Decode_t can be efficiently implemented.

The rest of this section is devoted to making these intuitive statements precise. We describe the LWE problem and the security of our construction in Section 4.1. We describe one possible polynomial-time Decode_t (which corrects more errors than is possible by exhaustive search) in Section 4.2. In Section 4.3, we describe parameter settings that allow us to extract as many bits as the input entropy, resulting in a lossless construction. In Section 4.4, we compare Construction 1 to using a sketch-and-extract approach (Lemma 1) instantiated with a computational extractor.

4.1 Security of Construction 1

The LWE problem was introduced by Regev [30,31] as a generalization of "learning parity with noise." For a complete description of the LWE problem and related lattices problems (which we do not define here) see [30]. We now recall the decisional version of the problem.

Definition 10 (Decisional LWE). *Let n be a security parameter. Let $m = m(n) = \mathtt{poly}(n)$ be an integer and $q = q(n) = \mathtt{poly}(n)$ be a prime*[4]*. Let $\mathbf{A}$ be the uniform distribution over $\mathbb{F}_q^{m \times n}$, X be the uniform distribution over $\mathbb{F}_q^n$ and χ be an arbitrary distribution on $\mathbb{F}_q^m$. The decisional version of the* LWE *problem, denoted* dist-LWE$_{n,m,q,\chi}$*, is to distinguish the distribution $(\mathbf{A}, \mathbf{A}X + \chi)$ from the uniform distribution over $(\mathbb{F}_q^{m \times n}, \mathbb{F}_q^m)$.*

We say that dist-LWE$_{n,m,q,\chi}$ *is (ϵ, s_{sec})-secure if no (probabilistic) distinguisher of size s_{sec} can distinguish the* LWE *instances from uniform except with probability ϵ. If for any $s_{sec} = \mathtt{poly}(n)$, there exists $\epsilon = \mathtt{ngl}(n)$ such that* dist-LWE$_{n,m,q,\chi}$ *is (ϵ, s_{sec})-secure, then we say it is* secure.

Regev [30] and Peikert [29] show that dist-LWE$_{n,m,q,\chi}$ is secure when the distribution χ of errors is Gaussian, as follows. Let $\bar{\Psi}_\rho$ be the discretized Gaussian distribution with variance $(\rho q)^2/2\pi$, where $\rho \in (0,1)$ with $\rho q > 2\sqrt{n}$. If GAPSVP and SIVP are hard to approximate (on lattices of dimension n) within polynomial factors for quantum algorithms, then dist-LWE$_{n,m,q,\bar{\Psi}_\rho^m}$ is secure. (A recent result of Brakerski et al. [8] shows security of LWE based on hardness of approximating lattices problems for classical algorithms. We have not considered how this result can be integrated into our analysis.)

The above formulation of LWE requires the error term to come from the discretized Gaussian distribution, which makes it difficult to use it for constructing fuzzy extractors (because using w and w' to sample Gaussian distributions will increase the distance between the error terms and/or reduce their entropy). Fortunately, recent work Döttling and Müller-Quade [17] shows the security of LWE, under the same assumptions, when errors come from the uniform distribution over a small interval[5]. This allows us to directly encode w as the error term in an LWE problem by splitting it into m blocks. The size of these blocks is dictated by the following result of Döttling and Müller-Quade:

Lemma 4. *[17, Corollary 1] Let n be a security parameter. Let $q = q(n) = \mathtt{poly}(n)$ be a prime and $m = m(n) = \mathtt{poly}(n)$ be an integer with $m \geq 3n$. Let $\sigma \in (0,1)$ be an arbitrarily small constant and let $\rho = \rho(n) \in (0, 1/10)$ be such that $\rho q \geq 2n^{1/2+\sigma} m$. If the approximate decision-version of the shortest vector problem (GAPSVP) and the shortest independent vectors problem (SIVP) are hard within a factor of $\tilde{O}(n^{1+\sigma} m/\rho)$ for quantum algorithms in the worst case, then, for χ the uniform distribution over $[-\rho q, \rho q]^m$,* dist-LWE$_{n,m,q,\chi}$ *is secure.*

To extract pseudorandom bits, we use a result of Akavia, Goldwasser, and Vaikuntanathan [1] to show that X has simultaneously many hardcore bits. The result says that if dist-LWE$_{(n-k,m,q,\chi)}$ is secure then any k variables of X in a dist-LWE$_{(n,m,q,\chi)}$ instance are hardcore. We state their result for a general error distribution (noting that their proof does not depend on the error distribution):

4 Unlike in common formulations of LWE, where q can be any integer, we need q to be prime for decoding.

5 Micciancio and Peikert provide a similar formulation in [27]. The result Döttling and Müller-Quade provides better parameters for our setting.

Lemma 5. *[1, Lemma 2] If* dist-LWE$_{(n-k,m,q,\chi)}$ *is* (ϵ, s_{sec}) *secure, then*

$$\delta^{\mathcal{D}_{s_{sec'}}}((X_{1,\dots,k}, \mathbf{A}, \mathbf{A}X + \chi), (U, \mathbf{A}, \mathbf{A}X + \chi)) \leq \epsilon,$$

where U *denotes the uniform distribution over* $\mathbb{F}_q^k$, $\mathbf{A}$ *denotes the uniform distribution over* $\mathbb{F}_q^{m\times n}$, X *denotes the uniform distribution over* $\mathbb{F}_q^n$, $X_{1,\dots,k}$ *denote the first* k *coordinates of* x, *and* $s'_{sec} \approx s_{sec} - n^3$.

The security of Construction 1 follows from Lemmas 4 and 5 when parameters are set appropriately (see Theorem 3), because we use the hardcore bits of X as our key.

4.2 Efficiency of Construction 1

Construction 1 is useful only if Decode_t can be efficiently implemented. We need a decoding algorithm for a random linear code with t errors that runs in polynomial time. We present a simple Decode_t that runs in polynomial time and can correct correcting $O(\log n)$ errors (note that this corresponds to a superpolynomial number of possible error patterns). This algorithm is a proof of concept, and neither the algorithm nor its analysis have been optimized for constants. An improved decoding algorithm can replace our algorithm, which will increase our correcting capability and improve Construction 1.

Construction 2. *We consider a setting of* (n, m, q, χ) *where* $m \geq 3n$. *We describe* Decode_t*:*

1. *Input* $\mathbf{A}, \mathbf{b} = \mathbf{A}\mathbf{x} + w - w'$
2. *Randomly select rows without replacement* $i_1, ..., i_{2n} \leftarrow [1, m]$.
3. *Restrict* $\mathbf{A}, \mathbf{b}$ *to rows* $i_1, ..., i_{2n}$*; denote these* $\mathbf{A}_{i_1,\dots,i_{2n}}, \mathbf{b}_{i_1,\dots,i_{2n}}$.
4. *Find* n *rows of* $\mathbf{A}_{i_1,\dots,i_{2n}}$ *that are linearly independent. If no such rows exist, output* $\perp$ *and stop.*
5. *Denote by* $\mathbf{A}', \mathbf{b}'$ *the restriction of* $\mathbf{A}_{i_1,\dots,i_{2n}}, \mathbf{b}_{i_1,\dots,i_{2n}}$ *(respectively) to these rows. Compute* $\mathbf{x}' = (\mathbf{A}')^{-1}\mathbf{b}'$.
6. *If* $\mathbf{b} - \mathbf{A}\mathbf{x}'$ *has more than* t *nonzero coordinates, go to step (2).*
7. *Output* $\mathbf{x}'$.

Each step is computable in time $O(n^3)$. For Decode_t to be efficient, we need t to be small enough so that with probability at least $\frac{1}{\mathtt{poly}(n)}$, none of the $2n$ rows selected in step 2 have errors (i.e., so that w and w' agree on those rows). If this happens, and $\mathbf{A}_{i_1,\dots,i_{2n}}$ has rank n (which is highly likely), then $\mathbf{x}' = \mathbf{x}$, and the algorithm terminates. However, we also need to ensure correctness: we need to make sure that if $\mathbf{x}' \neq \mathbf{x}$, we detect it in step 6. This detection will happen if $\mathbf{b} - \mathbf{A}\mathbf{x}' = \mathbf{A}(\mathbf{x} - \mathbf{x}') + (w - w')$ has more than t nonzero coordinates. It suffices to ensure that $\mathbf{A}(\mathbf{x} - \mathbf{x}')$ has at least $2t + 1$ nonzero coordinates (because at most t of those can be zeroed out by $w - w'$), which happens whenever the code generated by $\mathbf{A}$ has distance $2t + 1$.

Setting $t = O(\frac{m}{n} \log n)$ is sufficient to ensure efficiency. Random linear codes have distance at least $O(\frac{m}{n} \log n)$ with probability $1 - e^{-\Omega(n)}$ (the exact statement is in Corollary 2), so this also ensures correctness. The formal statement is below (proof in the full version of this work [18]):

Lemma 6 (Efficiency of Decode_t when $t \leq d(m/n - 2)\log n$). *Let d be a positive constant and assume that $\mathsf{dis}(W, W') \leq t$ where $t \leq d(\frac{m}{n} - 2)\log n$. Then Decode_t runs in expected time $O(n^{4d+3})$ operations in $\mathbb{F}_q$ (this expectation is over the choice of random coins of Decode_t, regardless of the input, as long as $\mathsf{dis}(w, w') \leq t$). It outputs X with probability $1 - e^{-\Omega(n)}$ (this probability is over the choice of the random matrix $\mathbf{A}$ and random choices made by Decode_t).*

4.3 Lossless Computational Fuzzy Extractor

We now state a setting of parameters that yields a lossless construction. The intuition is as follows. We are splitting our source into m blocks each of size $\log \rho q$ (from Lemma 4) for a total input entropy of $m \log \rho q$. Our key is derived from hardcore bits of X: $X_{1,\ldots,k}$ and is of size $k \log q$ (from Lemma 5). Thus, to achieve a lossless construction we need $k \log q = m \log \rho q$. In other words, in order to decode a meaningful number of errors, the vector w is of higher dimension than the vector X, but each coordinate of w is sampled using fewer bits than each coordinate of X. Thus, by increasing the size of q (while keeping ρq fixed) we can set $k \log q = m \log \rho q$, yielding a key of the same size as our source. The formal statement is below.

Theorem 3. *Let n be a security parameter and let the number of errors $t = c \log n$ for some positive constant c. Let d be a positive constant (giving us a tradeoff between running time of Rep and $|w|$). Consider the Hamming metric over the alphabet $\mathcal{Z} = [-2^{b-1}, 2^{b-1}]$, where $b = \log 2(c/d + 2)n^2 = O(\log n)$. Let W be uniform over $\mathcal{M} = \mathcal{Z}^m$, where $m = (c/d + 2)n = O(n)$. If GAPSVP and SIVP are hard to approximate within polynomial factors using quantum algorithms, then there is a setting of $q = \mathtt{poly}(n)$ such that for any polynomial $s_{sec} = \mathtt{poly}(n)$ there exists $\epsilon = \mathtt{ngl}(n)$ such that the following holds: Construction 1 is a $(\mathcal{M}, W, m \log |\mathcal{Z}|, t)$-computational fuzzy extractor that is (ϵ, s_{sec})-hard with error $\delta = e^{-\Omega(n)}$. The generate procedure Gen takes $O(n^2)$ operations over $\mathbb{F}_q$, and the reproduce procedure Rep takes expected time $O(n^{4d+3})$ operations over $\mathbb{F}_q$.*

Proof. Security follows by combining Lemmas 4 and 5; efficiency follows by Lemma 6. For a detailed explanation of the various parameters and constraints see the full version of this work [18].

Theorem 3 shows that a computational fuzzy extractor can be built without incurring any entropy loss. We can essentially think of $\mathbf{A}X + W$ as an encryption of X that where decryption works from any close W'.

4.4 Comparison with Computational-Extractor-Based Constructions

As mentioned in the introduction, an alternative approach to building a computational fuzzy extractor is to use a computational extractor (e.g., [26,3,13]) in place of the information-theoretic extractor in the sketch-and-extract construction. We will call this approach *sketch-and-comp-extract.* (A simple example of a computational extractor is a pseudorandom generator applied to the output of an information-theoretic extractor; note that LWE-based pseudorandom generators exist [2].)

This approach (specifically, its analysis via Lemma 1) works as long as the amount of entropy $\tilde{m}$ of w conditioned on the sketch s remains high enough to run a computational extractor. However, as discussed in Section 3, $\tilde{m}$ decreases with the error parameter t due to coding bounds, and it is conceivable that, if W has barely enough entropy to begin with, it will have too little entropy left to run a computational extractor once s is known.

In contrast, our approach does not require the entropy of w conditioned on $p = (\mathbf{A}, \mathbf{A}X + w)$ to be high enough for a computational extractor. Instead, we require that w is not computationally recoverable given p. This requirement is weaker—in particular, in our construction, w may have no information-theoretic entropy conditioned on p. The key difference in our approach is that instead of extracting from w, we hide secret randomness using w. Computational extractors are not allowed to have private randomness [26, Definition 3].

The main advantage of our analysis (instead of sketch-and-comp-extract) is that security need not depend on the error-tolerance t. In our construction, the error-tolerance depends only on the best available decoding algorithm for random linear codes, because decoding algorithms will not reach the information-theoretic decoding radius.

Unfortunately, LWE parameter sizes require relatively long w. Therefore, in practice, sketch-then-comp-extract will beat our construction if the computational extractor is instantiated efficiently based on assumptions other than LWE (for example, a cryptographic hash function for an extractor and a block cipher for a PRG). However, we believe that our conceptual framework can lead to better constructions. Of particular interest are other codes that are easy to decode up to t errors but become computationally hard as the number of errors increases.

To summarize, the advantage of Construction 1 is that the security of our construction does not depend on the decoding radius t. The disadvantages of Construction 1 are that it supports a limited number of errors and only a uniformly distributed source. We begin to address this second problem in the next section.

5 Computational Fuzzy Extractor for Nonuniform Sources

While showing the security of Construction 1 for arbitrary high-min-entropy distributions is an open problem, in this section we show it for a particular class of distributions called symbol-fixing. First we recall the notion of a symbol fixing source (from [24, Definition 2.3]):

Definition 11. *Let $W = (W_1, ..., W_{m+\alpha})$ be a distribution where each W_i takes values over an alphabet $\mathcal{Z}$. We say that it is a $(m + \alpha, m, |\mathcal{Z}|)$* symbol fixing source *if for α indices $i_1, \ldots, i_\alpha$, the symbols W_{i_α} are fixed, and the remaining m symbols are chosen uniformly at random. Note that $H_\infty(W) = m \log |\mathcal{Z}|$.*

Symbol-fixing sources are a very structured class of distributions. However, extending Construction 1 to such a class is not obvious. Although symbol-fixing sources are deterministically extractible [24], we cannot first run a deterministic extractor before using Construction 1. This is because we need to preserve distance between w and w' and an extractor must not preserve distance between input points. We present an alternative approach, showing security of LWE directly with symbol-fixing sources.

The following theorem states the main technical result of this section, which is of potential interest outside our specific setting. The result is that dist-LWE with symbol-fixing sources is implied by standard dist-LWE (but for n and m reduced by the amount of fixed symbols).

Theorem 4. *Let n be a security parameter, m, α be polynomial in n, and $q = \mathtt{poly}(n)$ be a prime and $\beta \in \mathbb{Z}^+$ be such that $q^{-\beta} = \mathtt{ngl}(n)$. Let U denote the uniform distribution over $\mathcal{Z}^m$ for an alphabet $\mathcal{Z} \subset \mathbb{F}_q$, and let W denote an $(m + \alpha, m, |\mathcal{Z}|)$ symbol fixing source over $\mathcal{Z}^{m+\alpha}$. If $\text{dist-LWE}_{n,m,q,U}$ is secure, then $\text{dist-LWE}_{n+\alpha+\beta,m+\alpha,q,W}$ is also secure.*

Theorem 4 also holds for an arbitrary error distribution (not just uniform error) in the following sense. Let χ' be an arbitrary error distribution. Define χ as the distribution where m dimensions are sampled according to χ' and the remaining dimensions have some fixed error. Then, security of $\text{dist-LWE}_{n,m,q,\chi'}$ implies security of $\text{dist-LWE}_{n+\alpha+\beta,m+\alpha,q,\chi}$. We prove this stronger version of the theorem in the full version of this work [18].

The intuition for this result is as follows. Providing a single sample with no error "fixes" at most a single variable. Thus, if there are significantly more variables than samples with no error, search LWE should still be hard. We are able to show a stronger result that dist-LWE is still hard. The nontrivial part of the reduction is using the additional $\alpha + \beta$ variables to "explain" a random value for the last α samples, without knowing the other variables. The β parameter is the slack needed to ensure that the "free" variables have influence on the last α samples. A similar theorem for the case of a single fixed dimension was shown in concurrent work by Brakerski et al. [8, Lemma 4.3]. The proof techniques of Brakerski et al. can be extended to our setting with multiple fixed dimensions, improving the parameters of Theorem 4 (specifically, removing the need for β).

Theorem 4 allows us to construct a lossless computational fuzzy extractor from block-fixing sources:

Theorem 5. *Let n be a security parameter and let $t = c \log n$ for some positive constant c. Let $d \leq c$ be a positive constant and consider the Hamming metric* *over the alphabet* $\mathcal{Z} = [-2^{b-1}, 2^{b-1}]$, *where* $b \approx \log 2(c/d+2)n^2 = O(\log n)$. *Let* $\mathcal{M} = \mathcal{Z}^{m+\alpha}$ *where* $m = (c/d+2)n = O(n)$ *and* $\alpha \leq n/3$. *Let* $\mathcal{W}$ *be the class of all* $(m+\alpha, m, |\mathcal{Z}|)$*-symbol fixing sources. If GAPSVP and SIVP are hard to approximate within polynomial factors using quantum algorithms, then there is a setting of* $q = \mathtt{poly}(n)$ *such that for any polynomial* $s_{sec} = \mathtt{poly}(n)$ *there exists* $\epsilon = \mathtt{ngl}(n)$ *such that the following holds: Construction 1 is a* $(\mathcal{M}, \mathcal{W}, m \log |\mathcal{Z}|, t)$*-computational fuzzy extractor that is* (ϵ, s_{sec})*-hard with error* $\delta = e^{-\Omega(n)}$. *The generate procedure* Gen *takes* $O(n^2)$ *operations over* $\mathbb{F}_q$, *and the reproduce procedure* Rep *takes expected time* $O(n^{4d+3} \log n)$ *operations over* $\mathbb{F}_q$.

Proof. Security follows by Lemmas 4 and 5 and Theorem 4 . Efficiency follows by Lemma 6. For a more detailed explanation of parameters see the full version of this work [18].

Acknowledgements. The authors are grateful to Jacob Alperin-Sheriff, Ran Canetti, Yevgeniy Dodis, Nico Döttling, Danielle Micciancio, Jörn Müller-Quade, Christopher Peikert, Oded Regev, Adam Smith, and Daniel Wichs for helpful discussions, creative ideas, and important references. In particular, the authors thank Nico Döttling for describing his result on LWE with uniform errors.

This work supported in part by National Science Foundation grants 0831281, 1012910, and 1012798. The work of Benjamin Fuller is sponsored in part by the United States Air Force under Air Force Contract FA8721-05-C-0002. Opinions, interpretations, conclusions and recommendations are those of the authors and are not necessarily endorsed by the United States Government.

References

1. Akavia, A., Goldwasser, S., Vaikuntanathan, V.: Simultaneous hardcore bits and cryptography against memory attacks. In: Reingold, O. (ed.) TCC 2009. LNCS, vol. 5444, pp. 474–495. Springer, Heidelberg (2009), `http://dx.doi.org/10.1007/978-3-642-00457-5_28`
2. Applebaum, B., Ishai, Y., Kushilevitz, E.: On pseudorandom generators with linear stretch in NC 0. Approximation, Randomization, and Combinatorial Optimization. Algorithms and Techniques, pp. 260–271 (2006)
3. Barak, B., Dodis, Y., Krawczyk, H., Pereira, O., Pietrzak, K., Standaert, F.-X., Yu, Y.: Leftover hash lemma, revisited. In: Rogaway, P. (ed.) CRYPTO 2011. LNCS, vol. 6841, pp. 1–20. Springer, Heidelberg (2011)
4. Barak, B., Shaltiel, R., Wigderson, A.: Computational analogues of entropy. In: 11th International Conference on Random Structures and Algorithms, pp. 200–215 (2003)
5. Bennett, C.H., Brassard, G., Robert, J.M.: Privacy amplification by public discussion. SIAM Journal on Computing 17(2), 210–229 (1988)
6. Berlekamp, E., McEliece, R., van Tilborg, H.: On the inherent intractability of certain coding problems. IEEE Transactions on Information Theory 24(3), 384–386 (1978)

7. Boyen, X., Dodis, Y., Katz, J., Ostrovsky, R., Smith, A.: Secure remote authentication using biometric data. In: Cramer, R. (ed.) EUROCRYPT 2005. LNCS, vol. 3494, pp. 147–163. Springer, Heidelberg (2005)
8. Brakerski, Z., Langlois, A., Peikert, C., Regev, O., Stehlé, D.: Classical hardness of learning with errors. In: Proceedings of the 45th Annual ACM Symposium on Symposium on Theory of Computing, pp. 575–584. ACM (2013)
9. Brostoff, S., Sasse, M.: Are passfaces more usable than passwords?: A field trial investigation. People and Computers, 405–424 (2000)
10. Castelluccia, C., Mutaf, P.: Shake them up!: A movement-based pairing protocol for CPU-constrained devices. In: Proceedings of the 3rd International Conference on Mobile Systems, Applications, and Services, pp. 51–64. ACM (2005)
11. Chandran, N., Kanukurthi, B., Ostrovsky, R., Reyzin, L.: Privacy amplification with asymptotically optimal entropy loss. In: Proceedings of the 42nd ACM Symposium on Theory of Computing, pp. 785–794. ACM, New York (2010), `http://doi.acm.org/10.1145/1806689.1806796`
12. Cover, T.M., Thomas, J.A.: Elements of information theory, 2nd edn. Wiley Interscience (2006)
13. Dachman-Soled, D., Gennaro, R., Krawczyk, H., Malkin, T.: Computational extractors and pseudorandomness. In: Cramer, R. (ed.) TCC 2012. LNCS, vol. 7194, pp. 383–403. Springer, Heidelberg (2012)
14. Daugman, J.: How iris recognition works. IEEE Transactions on Circuits and Systems for Video Technology 14(1), 21–30 (2004)
15. Dodis, Y., Ostrovsky, R., Reyzin, L., Smith, A.: Fuzzy extractors: How to generate strong keys from biometrics and other noisy data. SIAM Journal on Computing 38(1), 97–139 (2008)
16. Dodis, Y., Wichs, D.: Non-malleable extractors and symmetric key cryptography from weak secrets. In: Proceedings of the 41st Annual ACM Symposium on Theory of Computing, pp. 601–610. ACM, New York (2009), `http://doi.acm.org/10.1145/1536414.1536496`
17. Döttling, N., Müller-Quade, J.: Lossy codes and a new variant of the learning-with-errors problem. In: Johansson, T., Nguyen, P.Q. (eds.) EUROCRYPT 2013. LNCS, vol. 7881, pp. 18–34. Springer, Heidelberg (2013)
18. Fuller, B., Meng, X., Reyzin, L.: Computational fuzzy extractors. Cryptology ePrint Archive (2013), `http://eprint.iacr.org/2013/416`
19. Gentry, C., Wichs, D.: Separating succinct non-interactive arguments from all falsifiable assumptions. In: STOC, pp. 99–108. ACM, New York (2011)
20. Guruswami, V.: Introduction to coding theory - lecture 2: Gilbert-Varshamov bound. University Lecture (2010)
21. Håstad, J., Impagliazzo, R., Levin, L.A., Luby, M.: A pseudorandom generator from any one-way function. SIAM Journal on Computing 28(4), 1364–1396 (1999)
22. Hsiao, C.-Y., Lu, C.-J., Reyzin, L.: Conditional computational entropy, or toward separating pseudoentropy from compressibility. In: Naor, M. (ed.) EUROCRYPT 2007. LNCS, vol. 4515, pp. 169–186. Springer, Heidelberg (2007)
23. Juels, A., Wattenberg, M.: A fuzzy commitment scheme. In: Sixth ACM Conference on Computer and Communication Security, pp. 28–36. ACM (November 1999)
24. Kamp, J., Zuckerman, D.: Deterministic extractors for bit-fixing sources and exposure-resilient cryptography. SIAM Journal on Computing 36(5), 1231–1247 (2007)
25. Kanukurthi, B., Reyzin, L.: Key agreement from close secrets over unsecured channels. In: Joux, A. (ed.) EUROCRYPT 2009. LNCS, vol. 5479, pp. 206–223. Springer, Heidelberg (2009)

26. Krawczyk, H.: Cryptographic extraction and key derivation: The HKDF scheme. In: Rabin, T. (ed.) CRYPTO 2010. LNCS, vol. 6223, pp. 631–648. Springer, Heidelberg (2010)
27. Micciancio, D., Peikert, C.: Hardness of SIS and LWE with Small Parameters. In: Canetti, R., Garay, J.A. (eds.) CRYPTO 2013, Part I. LNCS, vol. 8042, pp. 21–39. Springer, Heidelberg (2013)
28. Nisan, N., Zuckerman, D.: Randomness is linear in space. Journal of Computer and System Sciences, 43–52 (1993)
29. Peikert, C.: Public-key cryptosystems from the worst-case shortest vector problem: extended abstract. In: Proceedings of the 41st Annual ACM Symposium on Theory of Computing, pp. 333–342. ACM, New York (2009), `http://doi.acm.org/10.1145/1536414.1536461`
30. Regev, O.: On lattices, learning with errors, random linear codes, and cryptography. In: Proceedings of the Thirty-Seventh Annual ACM Symposium on Theory of Computing, pp. 84–93. ACM, New York (2005), `http://doi.acm.org/10.1145/1060590.1060603`
31. Regev, O.: The learning with errors problem (invited survey). In: Annual IEEE Conference on Computational Complexity, pp. 191–204 (2010)
32. Reyzin, L.: Some notions of entropy for cryptography. In: Fehr, S. (ed.) ICITS 2011. LNCS, vol. 6673, pp. 138–142. Springer, Heidelberg (2011)
33. Shannon, C.E., Weaver, W., Blahut, R.E., Hajek, B.: The mathematical theory of communication, vol. 117. University of Illinois press Urbana (1949)
34. Suh, G.E., Devadas, S.: Physical unclonable functions for device authentication and secret key generation. In: Proceedings of the 44th Annual Design Automation Conference, pp. 9–14. ACM (2007)
35. Tuyls, P., Schrijen, G.-J., Škorić, B., van Geloven, J., Verhaegh, N., Wolters, R.: Read-proof hardware from protective coatings. In: Goubin, L., Matsui, M. (eds.) CHES 2006. LNCS, vol. 4249, pp. 369–383. Springer, Heidelberg (2006), `http://dx.doi.org/10.1007/11894063_29`
36. Vadhan, S.: Pseudorandomness. Foundations and Trends in Theoretical Computer Science. Now Publishers (2012)
37. Zviran, M., Haga, W.J.: A comparison of password techniques for multilevel authentication mechanisms. The Computer Journal 36(3), 227–237 (1993)

A Properties of Random Linear Codes

For efficient decoding of Construction 1, we need the LWE instance to have high distance with overwhelming probability. We will use the q-ary entropy function, denoted $H_q(x)$ and defined as $H_q(x) = x\log_q(q-1) - x\log_q x - (1-x)\log_q(1-x)$. Note that $H_2(x) = -x\log x - (1-x)\log(1-x)$. In the region $[0, \frac{1}{2}]$ for any value $q' \geq q$, $H_{q'}(x) \leq H_q(x)$. The following theorem is standard in coding theory:

Theorem 6. *[20, Theorem 8] For prime $q, \delta \in [0, 1-1/q), 0 < \epsilon < 1 - H_q(\delta)$ and sufficiently large m, the following holds for $n = \lceil (1 - H_q(\delta) - \epsilon)m \rceil$. If $\mathbf{A} \in \mathbb{F}_q^{m \times n}$ is drawn uniformly at random, then the linear code with $\mathbf{A}$ as a generator matrix has rate at least $(1 - H_q(\delta) - \epsilon)$ and relative distance at least δ with probability at least $1 - e^{-\Omega(m)}$.*

Our setting is the case where $m = poly(n) \geq 2n$ and $\delta = O(\log n/n)$. This setting of parameters satisfies Theorem 6:

Corollary 2. *Let n be a parameter and let $m = \mathtt{poly}(n) \geq 2n$. Let q be a prime and $\tau = O(\frac{m}{n}\log n)$. For large enough values of n, when $\mathbf{A} \in \mathbb{F}_q^{m\times n}$ is drawn uniformly, the code generated by $\mathbf{A}$ has distance at least τ with probability at least $1 - e^{-\Omega(m)} \geq 1 - e^{-\Omega(n)}$.*

Proof. Let c be some constant. Let $\delta = \tau/m = \frac{c\log n}{n}$. We show the corollary for the case when $m = 2n$ (increasing the size of m only increases the relative distance). It suffices to show that for sufficiently large n, there exists $\epsilon > 0$ where $1 - H_q(\frac{c\log n}{n}) - \epsilon = 1/2$ or equivalently that $H_q(\frac{c\log n}{m}) < 1/2$ as then setting $\epsilon = 1/2 - H_q(\frac{c\log n}{n})$ satisfies Theorem 6. For sufficiently large n:

- $\frac{c\log n}{n} < 1/2$, so we can work with the binary entropy function H_2.
- $\frac{c\log n}{n} < .1 < 1/2$ and thus $H_q(\frac{c\log n}{n}) < H_q(.1)$.

Putting these statements together, for large enough n, $H_q(\frac{c\log n}{n}) < H_q(.1) < H_2(.1) < 1/2$ as desired. This completes the proof.

Efficient One-Way Secret-Key Agreement and Private Channel Coding via Polarization

Joseph M. Renes, Renato Renner, and David Sutter

Institute for Theoretical Phyiscs,
ETH Zurich, Switzerland
{renes,renner,suttedav}@phys.ethz.ch

Abstract. We introduce explicit schemes based on the polarization phenomenon for the task of secret-key agreement from common information and one-way public communication as well as for the task of private channel coding. Our protocols are distinct from previously known schemes in that they combine two practically relevant properties: they achieve the ultimate rate—defined with respect to a strong secrecy condition—and their complexity is essentially linear in the blocklength. However, we are not able to give an efficient algorithm for code construction.

Keywords: One-way secret-key agreement, private channel coding, one-way secret-key rate, secrecy capacity, wiretap channel scenario, more capable, less noisy, degraded, polarization phenomenon, polar codes, practically efficient, strongly secure.

1 Introduction

Consider two parties, Alice and Bob, connected by an authentic but otherwise fully insecure communication channel. It has been shown that without having access to additional resources, it is impossible for them to communicate privately, with respect to an information-theoretic privacy condition [1,2]. In particular they are unable to generate an unconditionally secure key with which to encrypt messages transmitted over the public channel. However, if Alice and Bob have access to correlated randomness about which an adversary (Eve) has only partial knowledge, the situation changes completely: information-theoretically secure secret-key agreement and private communication become possible. Alternatively, if Alice and Bob are connected by a noisy discrete memoryless channel (DMC) to which Eve has only limited access—the so-called *wiretap channel scenario* of Wyner [3], Csiszár and Körner [4], and Maurer [2]—private communication is again possible.

In this paper, we present explicit schemes for efficient one-way secret-key agreement from common randomness and for private channel coding in the wiretap channel scenario. As discussed in Section 2.5, we improve previous work that requires extra assumptions about the structure of the wiretap channel or/and do not achieve strong secrecy. Our schemes are based on *polar codes*, a family of capacity-achieving linear codes, introduced by Arıkan [5], that can be encoded

K. Sako and P. Sarkar (Eds.) ASIACRYPT 2013 Part I, LNCS 8269, pp. 194–213, 2013.

and decoded efficiently. Previous work in a quantum setup [6] already implies that *practically efficient* one-way secret-key agreement and private channel coding in a classical setup is possible, where a practically efficient scheme is one whose computational complexity is essentially linear in the blocklength. The aim of this paper is to explain the schemes in detail and give a purely classical proof that the schemes are reliable, secure, practically efficient and achieve optimal rates.

This paper is structured as follows. Section 2 introduces the problems of performing *one-way secret-key agreement* and *private channel coding*. We summarize known and new results about the optimal rates for these two problems for different wiretap channel scenarios. In Section 3, we explain how to obtain one-way secret-key agreement that is practically efficient, strongly secure, reliable, and achieves the one-way secret-key rate. However, we are not able to give an efficient algorithm for code construction, as discussed in Section 3.3. Section 4 introduces a similar scheme that can be used for strongly secure private channel coding at the secrecy capacity. Finally we conclude in Section 5 and state an open problem that is of interest in the setup of this paper as well as in the quantum mechanical scenario introduced in [6].

2 Background and Contributions

2.1 Notation and Definitions

Let $[k] = \{1, \ldots, k\}$ for $k \in \mathbb{Z}^+$. For $x \in \mathbb{Z}_2^k$ and $\mathcal{I} \subseteq [k]$ we have $x[\mathcal{I}] = [x_i : i \in \mathcal{I}]$, $x^i = [x_1, \ldots, x_i]$ and $x_j^i = [x_j, \ldots, x_i]$ for $j \leqslant i$. The set $\mathcal{A}^{\mathsf{c}}$ denotes the complement of the set $\mathcal{A}$. The uniform distribution on an arbitrary random variable X is denoted by $\overline{P}_X$. For distributions P and Q over the same alphabet $\mathcal{X}$, the variational distance is defined by $\delta(P, Q) := \frac{1}{2} \sum_{x \in \mathcal{X}} |P(x) - Q(x)|$. Let X and Y be two (possibly correlated) random variables. We use standard information theoretic notation, such as $H(X)$ for the (Shannon) entropy of X, $H(X, Y)$ for the joint entropy of (X, Y), $H(X|Y)$ for the conditional entropy of X given Y, and $I(X; Y)$ for the mutual information between X and Y.[1] The notation $X \multimap Y \multimap Z$ means that the random variables X, Y, Z form a Markov chain in the given order.

In this setup we consider a discrete memoryless wiretap channel (DM-WTC) $\mathsf{W} : \mathcal{X} \to \mathcal{Y} \times \mathcal{Z}$, which is characterized by its transition probability distribution $P_{Y,Z|X}$.[2] We assume that the variable X belongs to Alice, Y to Bob and Z to Eve.

According to Körner and Marton [8], a DM-WTC $\mathsf{W} : \mathcal{X} \to \mathcal{Y} \times \mathcal{Z}$ is termed *more capable* if $I(X; Y) \geqslant I(X; Z)$ for every possible distribution on X. The

[1] These quantities are properly defined in [7].

[2] Recall that a *discrete channel* is defined as a system consisting of an input alphabet (here $\mathcal{X}$), an output alphabet (here $\mathcal{Y} \times \mathcal{Z}$) and a transition probability distribution (here $P_{Y,Z|X}$) between the input and the output. A channel is said to be *memoryless* if the probability distribution of the output depends only on the input at that time and is conditionally independent of previous channel inputs or outputs.

channel W is termed *less noisy* if $I(U;Y) \geqslant I(U;Z)$ for every possible distribution on (U,X) where U has finite support and U−∘−X−∘−(Y,Z) form a Markov chain. If X−∘−Y−∘−Z form a Markov chain, W is called *degraded*.[3] It has been shown [8] that being more capable is a strictly weaker condition than being less noisy, which is a strictly weaker condition than being degraded. Hence, having a DM-WTC W which is degraded implies that W is less noisy, which again implies that W is also more capable.

2.2 Polarization Phenomenon

Let X^N be a vector whose entries are i.i.d. Bernoulli(p) distributed for $p \in [0,1]$ and $N = 2^n$ where $n \in \mathbb{Z}^+$. Then define $U^N = G_N X^N$, where G_N denotes the polarization (or polar) transform which can be represented by the matrix

$$G_N := \begin{pmatrix} 1 & 1 \\ 0 & 1 \end{pmatrix}^{\otimes \log N}, \tag{1}$$

where $A^{\otimes k}$ denotes the kth Kronecker power of an arbitrary matrix A. Note that it turns out that G_N is its own inverse. Furthermore, let $Y^N = \mathsf{W}^N X^N$, where W^N denotes N independent uses of a DMC $\mathsf{W} : \mathcal{X} \to \mathcal{Y}$. For $\epsilon \in (0,1)$ we may define the two sets

$$\mathcal{R}_\epsilon^N(X|Y) := \{i \in [N] : H(U_i | U^{i-1}, Y^N) \geqslant 1 - \epsilon\} \quad \text{and} \tag{2}$$

$$\mathcal{D}_\epsilon^N(X|Y) := \{i \in [N] : H(U_i | U^{i-1}, Y^N) \leqslant \epsilon\}. \tag{3}$$

The former consists of outputs U_j which are essentially uniformly random, even given all previous outputs U^{j-1} as well as Y^N, while the latter set consists of the essentially deterministic outputs. The polarization phenomenon is that essentially all outputs are in one of these two subsets, and their sizes are given by the conditional entropy of the input X given Y.

Theorem 1 (Polarization Phenomenon [5,9]). *For any* $\epsilon \in (0,1)$

$$\left|\mathcal{R}_\epsilon^N(X|Y)\right| = NH(X|Y) - o(N) \quad \text{and} \tag{4}$$

$$\left|\mathcal{D}_\epsilon^N(X|Y)\right| = N(1 - H(X|Y)) - o(N). \tag{5}$$

Based on this theorem it is possible to construct a family of linear error correcting codes, called *polar codes*. The logical bits are encoded into the U_i for $i \in \mathcal{D}_\epsilon^N(X|Y)$, whereas the inputs to U_i for $i \in \mathcal{D}_\epsilon^N(X|Y)^\mathsf{c}$ are fixed.[4] It has been shown that polar codes have several desirable attributes [5,10,11,12]: they provably achieve the capacity of any DMC; they have an encoding and decoding

[3] To call a DM-WTC $\mathsf{W} : \mathcal{X} \to \mathcal{Y} \times \mathcal{Z}$ more capable is an abbreviation meaning that the main channel $\mathsf{W}_1 : \mathcal{X} \to \mathcal{Y}$ is more capable than the eavesdropping channel $\mathsf{W}_2 : \mathcal{X} \to \mathcal{Z}$. The same convention is used for less noisy and degraded DM-WTCs.

[4] These are the so-called *frozen bits*.

complexity that is essentially linear in the blocklength N; the error probability decays exponentially in the square root of the blocklength.

Non-binary random variables can be represented by a sequence of correlated binary random variables, which are then encoded separately. Correlated sequences of binary random variables may be polarized using a multilevel construction, as shown in [10].[5] Given M i.i.d. instances of a sequence $X = (X_{(1)}, X_{(2)}, \ldots, X_{(K)})$ and possibly a correlated random variable Y, the basic idea is to first polarize $X_{(1)}^M$ relative to Y^M, then treat $X_{(1)}^M Y^M$ as side information in polarizing $X_{(2)}^M$, and so on. More precisely, defining $U_{(j)}^M = G_M X_{(j)}^M$ for $j = 1, \ldots, K$, we may define the random and deterministic sets for each j as

$$\mathcal{R}_{\epsilon,(j)}^M(X_{(j)}|X_{(j-1)}, \cdots, X_{(1)}, Y) = \{i \in [M] : H\left(U_{(j),i}\middle|U_{(j)}^{i-1}, X_{(j-1)}^M, \cdots, X_{(1)}^M, Y^M\right) \geqslant 1-\epsilon\}, \quad \text{and} \tag{6}$$

$$\mathcal{D}_{\epsilon,(j)}^M(X_{(j)}|X_{(j-1)}, \cdots, X_{(1)}, Y) = \{i \in [M] : H\left(U_{(j),i}\middle|U_{(j)}^{i-1}, X_{(j-1)}^M, \cdots, X_{(1)}^M, Y^M\right) \leqslant \epsilon\}. \tag{7}$$

In principle we could choose different ϵ parameters for each j, but this will not be necessary here. Now, Theorem 1 applies to the random and deterministic sets for every j. The sets $\mathcal{R}_\epsilon^M(X|Y) = \{\mathcal{R}_{\epsilon,(j)}^M(X_{(j)}|X_{(j-1)}, \ldots, X_{(1)}, Y)\}_{j=1}^K$ and $\mathcal{D}_\epsilon^M(X|Y) = \{\mathcal{D}_{\epsilon,(j)}^M(X_{(j)}|X_{(j-1)}, \ldots, X_{(1)}, Y)\}_{j=1}^K$ have sizes given by

$$|\mathcal{R}_\epsilon^M(X|Y)| = \sum_{j=1}^K \left|\mathcal{R}_{\epsilon,(j)}^M(X_{(j)}|X_{(j-1)}, \ldots, X_{(1)}, Y)\right| \tag{8}$$

$$= \sum_{j=1}^K MH(X_{(j)}|X_{(1)}, \ldots, X_{(j-1)}, Y) - o(M) \tag{9}$$

$$= MH(X|Y) - o(KM), \tag{10}$$

and

$$|\mathcal{D}_\epsilon^M(X|Y)| = \sum_{j=1}^K \left|\mathcal{D}_{\epsilon,(j)}^M(X_{(j)}|X_{(j-1)}, \ldots, X_{(1)}, Y)\right| \tag{11}$$

$$= \sum_{j=1}^K M\left(1 - H(X_{(j)}|X_{(1)}, \ldots, X_{(j-1)}, Y)\right) - o(M) \tag{12}$$

$$= M\left(K - H(X|Y)\right) - o(KM). \tag{13}$$

In the following we will make use of both the polarization phenomenon in its original form, Theorem 1, and the multilevel extension. To simplify the presentation, we denote by $\widetilde{G}_M^K$ the K parallel applications of G_M to the K random variables $X_{(j)}^M$.

[5] An alternative approach is given in [13,14], where the polarization phenomenon has been generalized for arbitrary finite fields. We will however focus on the multilevel construction in this paper.

2.3 One-Way Secret-Key Agreement

At the start of the one-way secret-key agreement protocol, Alice, Bob, and Eve share $N = 2^n$, $n \in \mathbb{Z}^+$ i.i.d. copies (X^N, Y^N, Z^N) of a triple of correlated random variables (X, Y, Z) which take values in discrete but otherwise arbitrary alphabets $\mathcal{X}$, $\mathcal{Y}$, $\mathcal{Z}$.[6]

Alice starts the protocol by performing an operation $\tau_A : \mathcal{X}^N \to (\mathcal{S}^J, \mathcal{C})$ on X^N which outputs both her secret key $S_A^J \in \mathcal{S}^J$ and an additional random variable $C \in \mathcal{C}$ which she transmits to Bob over an public but noiseless public channel. Bob then performs an operation $\tau_B : (\mathcal{Y}^N, \mathcal{C}) \to \mathcal{S}^J$ on Y^N and the information C he received from Alice to obtain a vector $S_B^J \in \mathcal{S}^J$; his secret key. The secret-key thus produced should be reliable, i.e., satisfy the

$$\text{reliability condition:} \quad \lim_{N\to\infty} \Pr\left[S_A^J \neq S_B^J\right] = 0, \tag{14}$$

and secure, i.e., satisfy the

$$\text{(strong) secrecy condition:} \quad \lim_{N\to\infty} \left\| P_{S_A^J, Z^N, C} - \overline{P}_{S_A^J} \times P_{Z^N, C} \right\|_1 = 0, \tag{15}$$

where $\overline{P}_{S_A^J}$ denotes the uniform distribution on random variable S_A^J.

Historically, secrecy was first characterized by a (weak) secrecy condition of the form

$$\lim_{N\to\infty} \frac{1}{N} I(S_A^J; Z^N, C) = 0. \tag{16}$$

Maurer and Wolf showed that (16) is not a sufficient secrecy criterion [15,16] and introduced the strong secrecy condition

$$\lim_{N\to\infty} I(S_A^J; Z^N, C) = 0, \tag{17}$$

where in addition it is required that the key is uniformly distributed, i.e.,

$$\lim_{N\to\infty} \delta\left(P_{S_A^J}, \overline{P}_{S_A^J}\right) = 0. \tag{18}$$

In recent years, the strong secrecy condition (17), (18) has often been replaced by (15), since (half) the L_1 distance directly bounds the probability of distinguishing the actual key produced by the protocol with an ideal key. This operational interpretation is particularly helpful in the finite blocklength regime. In the limit $N \to \infty$, the two secrecy conditions (15) and (17) are equivalent, which can be shown using Pinsker's and Fano's inequalities.

Since having weak secrecy is not sufficient, we will only consider strong secrecy in this paper. It has been proven that each secret-key agreement protocol which achieves weak secrecy can be transformed into a strongly secure protocol [16]. However, it is not clear whether the resulting protocol is guaranteed to be practically efficient.

---

[6] The correlation of the random variables (X, Y, Z) is described by their joint probability distribution $P_{X,Y,Z}$.

For *one-way* communication, Csiszár and Körner [4] and later Ahlswede and Csiszár [17] showed that the optimal rate $R := \lim_{N\to\infty} \frac{J}{N}$ of generating a secret key satisfying (14) and (17), called the *secret-key rate* $S_{\rightarrow}(X;Y|Z)$, is characterized by a closed single-letter formula.

Theorem 2 (One-Way Secret-Key Rate [4,17]). *For triples* (X,Y,Z) *described by* $P_{X,Y,Z}$ *as explained above,*

$$S_{\rightarrow}(X;Y|Z) = \begin{cases} \max\limits_{P_{U,V}} H(U|Z,V) - H(U|Y,V) \\ \text{s.t. } V \multimap U \multimap X \multimap (Y,Z), \\ \quad |\mathcal{V}| \leqslant |\mathcal{X}|,\ |\mathcal{U}| \leqslant |\mathcal{X}|^2. \end{cases} \tag{19}$$

The expression for the one-way secret-key rate given in Theorem 2 can be simplified if one makes additional assumptions about $P_{X,Y,Z}$.

Corollary 3. *For* $P_{X,Y,Z}$ *such that the induced DM-WTC* W *described by* $P_{Y,Z|X}$ *is more capable,*

$$S_{\rightarrow}(X;Y|Z) = \begin{cases} \max\limits_{P_V} H(X|Z,V) - H(X|Y,V) \\ \text{s.t. } V \multimap X \multimap (Y,Z), \\ \quad |\mathcal{V}| \leqslant |\mathcal{X}|. \end{cases} \tag{20}$$

Proof. In terms of the mutual information, we have

$$\begin{aligned} &H(U|Z,V) - H(U|Y,V) \\ &\quad = I(U;Y|V) - I(U;Z|V) && (21)\\ &\quad = I(X,U;Y|V) - I(X,U;Z|V) - (I(X;Y|U,V) - I(X;Z|U,V)) && (22)\\ &\quad \leqslant I(X,U;Y|V) - I(X,U;Z|V) && (23)\\ &\quad = I(X;Y|V) - I(X;Z|V), && (24) \end{aligned}$$

using the chain rule, the more capable condition, and the Markov chain properties, respectively. Thus, the maximum in $S_{\rightarrow}(X;Y|Z)$ can be achieved when omitting U. □

Corollary 4. *For* $P_{X,Y,Z}$ *such that the induced DM-WTC* W *described by* $P_{Y,Z|X}$ *is less noisy,*

$$S_{\rightarrow}(X;Y|Z) = H(X|Z) - H(X|Y). \tag{25}$$

Proof. Since W being less noisy implies W being more capable, we know that the one-way secret key rate is given by (20). Using the chain rule we obtain

$$\begin{aligned} &H(X|Z,V) - H(X|Y,V) \\ &\quad = I(X;Y|V) - I(X;Z|V) && (26)\\ &\quad = I(X,V;Y) - I(X,V;Z) - I(V;Y) + I(V;Z) && (27)\\ &\quad = I(X;Y) - I(X;Z) - (I(V;Y) - I(V;Z)) && (28)\\ &\quad \leqslant I(X;Y) - I(X;Z). && (29) \end{aligned}$$

Equation (28) follows from the chain rule and the Markov chain condition. The inequality uses the assumption of being less noisy. □

Note that (25) is also equal to the one-way secret-key rate for the case where W is degraded, as this implies W being less noisy. The proof of Theorem 2 does not imply that there exists an *efficient* one-way secret-key agreement protocol. A computationally efficient scheme was constructed in [18], but is not known to be practically efficient.[7]

For key agreement with two-way communication, no formula comparable to (19) for the optimal rate is known. However, it has been shown that the two-way secret-key rate is strictly larger than the one-way secret-key rate. It is also known that the *intrinsic information* $I(X;Y\!\downarrow Z) := \min_{P_{Z'|Z}} I(X;Y|Z')$ is an upper bound on $S(X;Y|Z)$, but is not tight [17,19,20].

2.4 Private Channel Coding

Private channel coding over a wiretap channel is closely related to the task of one-way secret-key agreement from common randomness (cf. Section 2.5). Here Alice would like to transmit a message $M^J \in \mathcal{M}^J$ privately to Bob. The messages can be distributed according to some arbitrary distribution P_{M^J}. To do so, she first encodes the message by computing $X^N = \text{enc}(M^J)$ for some encoding function $\text{enc} : \mathcal{M}^J \to \mathcal{X}^N$ and then sends X^N over the wiretap channel to Bob (and to Eve), which is represented by $(Y^N, Z^N) = \mathsf{W}^N X^N$. Bob next decodes the received message to obtain a guess for Alice's message $\hat{M}^J = \text{dec}(Y^N)$ for some decoding function $\text{dec} : \mathcal{Y}^N \to \mathcal{M}^J$. As in secret-key agreement, the private channel coding scheme should be reliable, i.e., satisfy the

$$\text{reliability condition:} \quad \lim_{J\to\infty} \Pr\left[M^J \neq \hat{M}^J\right] = 0, \quad \text{for all} \quad M^J \in \mathcal{M}^J \tag{30}$$

and (strongly) secure, i.e., satisfy the

$$\text{(strong) secrecy condition:} \quad \lim_{J\to\infty} \left\|P_{M^J,Z^N,C} - P_{M^J} \times P_{Z^N,C}\right\|_1 = 0. \tag{31}$$

The variable C denotes any additional information made public by the protocol.

As mentioned in Section 2.3, in the limit $J \to \infty$ this strong secrecy condition is equivalent to the historically older (strong) secrecy condition

$$\lim_{J\to\infty} I(M^J; Z^N, C) = 0. \tag{32}$$

The highest achievable rate $R := \lim_{N\to\infty} \frac{J}{N}$ fulfilling (30) and (31) is called the *secrecy capacity*.

Csiszár and Körner showed [4, Corollary 2] that there exists a single-letter formula for the secrecy capacity.[8]

---

[7] As defined in Section 1, we call a scheme practically efficient if its computational complexity is essentially linear in the blocklength.

[8] Maurer and Wolf showed that the single-letter formula remains valid considering strong secrecy [16].

Theorem 5 (Secrecy Capacity [4]). *For an arbitrary DM WTC* W *as introduced above,*

$$C_s = \begin{cases} \max\limits_{P_{V,X}} H(V|Z) - H(V|Y) \\ \text{s.t. } V \multimap X \multimap (Y,Z), \\ \quad |\mathcal{V}| \leqslant |\mathcal{X}|. \end{cases} \tag{33}$$

This expression can be simplified using additional assumptions about W.

Corollary 6 ([8]). *If* W *is more capable,*

$$C_s = \max_{P_X} H(X|Z) - H(X|Y) \,. \tag{34}$$

Proof. A proof can be found in [8] or [21, Section 22.1]. □

2.5 Previous Work and Our Contributions

In Section 3, we present a one-way secret-key agreement scheme based on polar codes that achieves the secret-key rate, is strongly secure, reliable and whose implementation is practically efficient, with complexity $O(N \log N)$ for blocklength N. Our protocol improves previous efficient secret-key constructions [22], where only weak secrecy could be proven and where the eavesdropper has no prior knowledge and/or degradability assumptions are required. Our protocol also improves a very recent efficient secret-key construction [23], which requires to have a small amount of shared key between Alice and Bob and only works for binary *degraded* (symmetric) discrete memoryless sources. However, we note that a possible drawback of our scheme compared to [23] is that its code construction may be more difficult.

In Section 4, we introduce a coding scheme based on polar codes that provably achieves the secrecy capacity for arbitrary discrete memoryless wiretap channels. We show that the complexity of the encoding and decoding operations is $O(N \log N)$ for blocklength N. Our scheme improves previous work on practically efficient private channel coding at the optimal rate [24], where only weak secrecy could be proven under the additional assumption that the channel W is degraded.[9] Recently, Bellare *et al.* introduced a polynomial-time coding scheme that is strongly secure and achieves the secrecy capacity for binary symmetric wiretap channels [25].[10] Several other constructions of private channel coding schemes have been reported [26,27,28], but all achieve only weak secrecy. Very recently, Şaşoğlu and Vardy introduced a new polar coding scheme that

[9] Note that Mahdavifar and Vardy showed that their scheme achieves strong secrecy if the channel to Eve (induced from W) is noiseless. Otherwise their scheme is not provably reliable [24].

[10] They claim that their scheme works for a large class of wiretap channels. However, this class has not been characterized precisely so far. It is therefore not clear whether their scheme requires for example degradability assumptions. Note that to obtain strong secrecy for an arbitrarily distributed message, it is required that the wiretap channel is symmetric [25, Lemma 14].

can be used for private channel coding being strongly secure [29]. However, it still requires the assumption of having a degraded wiretap channel which we do not need for our scheme. In [30], an explicit construction that achieves the secrecy capacity for wiretap channel coding is introduced, but efficiency is not considered.

The tasks of one-way secret-key agreement and private channel coding explained in the previous two subsections are closely related. Maurer showed how a one-way secret-key agreement can be derived from a private channel coding scenario [2]. More precisely, he showed how to obtain the common randomness needed for one-way secret-key agreement by constructing a "virtual" degraded wiretap channel from Alice to Bob. This approach can be used to obtain the one-way secret-key rate from the secrecy capacity result in the wiretap channel scenario [21, Section 22.4.3]. One of the main advantages of the two schemes introduced in this paper is that they are both practically efficient. However, even given a practically efficient private coding scheme, it is not known that Maurer's construction will yield a practically efficient scheme for secret key agreement. For this reason, as well as simplicity of presentation, we treat the one-way secret-key agreement and the private channel coding problem separately in the two sections to follow.

3 One-Way Secret-Key Agreement Scheme

Our key agreement protocol is a concatenation of two subprotocols, an inner and an outer layer, as depicted in Figure 1. The protocol operates on blocks of N i.i.d. triples (X, Y, Z), which are divided into M sub-blocks of size L for input to the inner layer. At the outer layer, we use the multi-level construction introduced in Section 2.2. In the following we assume $\mathcal{X} = \{0, 1\}$, which however is only for convenience; the techniques of [10] and [31] can be used to generalize the schemes to arbitrary alphabets $\mathcal{X}$.

The task of the inner layer is to perform *information reconciliation* and that of the outer layer is to perform *privacy amplification*. Information reconciliation refers to the process of carrying out error correction to ensure that Alice and Bob obtain a shared bit string, and here we only allow communication from Alice to Bob for this purpose. On the other hand, privacy amplification refers to the process of distilling from Alice's and Bob's shared bit string a smaller set of bits whose correlation with the information available to Eve is below a desired threshold.

Each subprotocol in our scheme is based on the polarization phenomenon. For information reconciliation of Alice's random variable X^L relative to Bob's information Y^L, Alice applies a polar transformation to X^L and forwards the bits of the complement of the deterministic set $\mathcal{D}_{\epsilon_1}^L(X|Y)$ to Bob over a public channel, which enables him to recover X^L using the standard polar decoder [5]. Her remaining information is then fed into a multilevel polar transformation and the bits of the random set are kept as the secret key.

Let us now define the protocol more precisely. For $L = 2^\ell$, $\ell \in \mathbb{Z}^+$, let $V^L = G_L X^L$ where G_L is as defined in (1). For $\epsilon_1 > 0$, we define

$$\mathcal{E}_K := \mathcal{D}^L_{\epsilon_1}(X|Y), \tag{35}$$

with $K := |\mathcal{D}^L_{\epsilon_1}(X|Y)|$. Then, let $T_{(j)} = V^L[\mathcal{E}_K]_j$ for $j = 1,\ldots,K$ and $C_{(j)} = V^L[\mathcal{E}^{\mathsf{c}}_K]_j$ for $j = 1,\ldots,L-K$ so that $T = (T_{(1)},\ldots,T_{(K)})$ and $C = (C_{(1)},\ldots,C_{(L-K)})$. For $\epsilon_2 > 0$ and $U^M_{(j)} = G_M T^M_{(j)}$ for $j = 1,\ldots K$ (or, more briefly, $U^M = \widetilde{G}^K_M T^M$), we define

$$\mathcal{F}_J := \mathcal{R}^M_{\epsilon_2}(T|CZ^L), \tag{36}$$

with $J := |R^M_{\epsilon_2}(T|CZ^L)|$.

Protocol 1: One-way secret-key agreement

Given: Index sets $\mathcal{E}_K$ and $\mathcal{F}_J$ (code construction)
Notation: Alice's input: $x^N \in \mathbb{Z}_2^N$ (a realization of X^N)
Bob's / Eve's input: (y^N, z^N) (realizations of Y^N and Z^N)
Alice's output: s^J_A
Bob's output: s^J_B

Step 1: Alice computes $v^{i+L}_{i+1} = G_L x^{i+L}_{i+1}$ for all $i \in \{0, L, 2L, \ldots, (M-1)L\}$.
Step 2: Alice computes $t_i = v^{i+L}_{i+1}[\mathcal{E}_K]$ for all $i \in \{0, L, 2L, \ldots, (M-1)L\}$.
Step 3: Alice sends $c_i = v^{i+L}_{i+1}[\mathcal{E}^{\mathsf{c}}_K]$ for all $i \in \{0, L, 2L, \ldots, (M-1)L\}$ over a public channel to Bob.
Step 4: Alice computes $u^M = \widetilde{G}^K_M t^M$ and obtains $s^J_A = u^M[\mathcal{F}_J]$.[11]
Step 5: Bob applies the standard polar decoder [5,12] to (c_i, y^{i+L}_{i+1}) to obtain $\hat{v}^{i+L}_{i+1}$ and $\hat{t}_i = \hat{v}^{i+L}_{i+1}[\mathcal{E}_K]$, for $i \in \{0, L, 2L, \ldots, (M-1)L\}$.
Step 6: Bob computes $\hat{u}^M = \widetilde{G}^K_M t^M$ and obtains $s^J_B = \hat{u}^M[\mathcal{F}_J]$.

3.1 Rate, Reliability, Secrecy, and Efficiency

Theorem 7. *Protocol 1 allows Alice and Bob to generate a secret key S^J_A respecitvely S^J_B using public one-way communication C^M such that for any $\beta < \frac{1}{2}$:*

$$\text{Reliability:} \quad \Pr\left[S^J_A \neq S^J_B\right] = O\left(M 2^{-L^\beta}\right) \tag{37}$$

$$\text{Secrecy:} \quad \left\| P_{S^J_A, Z^N, C} - \overline{P}_{S^J_A} \times P_{Z^N, C} \right\|_1 = O\left(\sqrt{N} 2^{-\frac{N^\beta}{2}}\right) \tag{38}$$

$$\text{Rate:} \quad R := \frac{J}{N} = H(X|Z) - \frac{1}{L} H\left(V^L[\mathcal{E}^{\mathsf{c}}_K] | Z^L\right) - \frac{o(N)}{N}. \tag{39}$$

All operations by both parties can be performed in $O(N \log N)$ steps.

---

[11] The expression $u^M[\mathcal{F}_J]$ is an abuse of notation, as $\mathcal{F}_J$ is not a subset of [M]. The expression should be understood to be the union of the random bits of $u^M_{(j)}$, for all $j = 1,\ldots,K$, as in the definition of $\mathcal{R}^M_{\epsilon_2}(T|CZ^L)$.

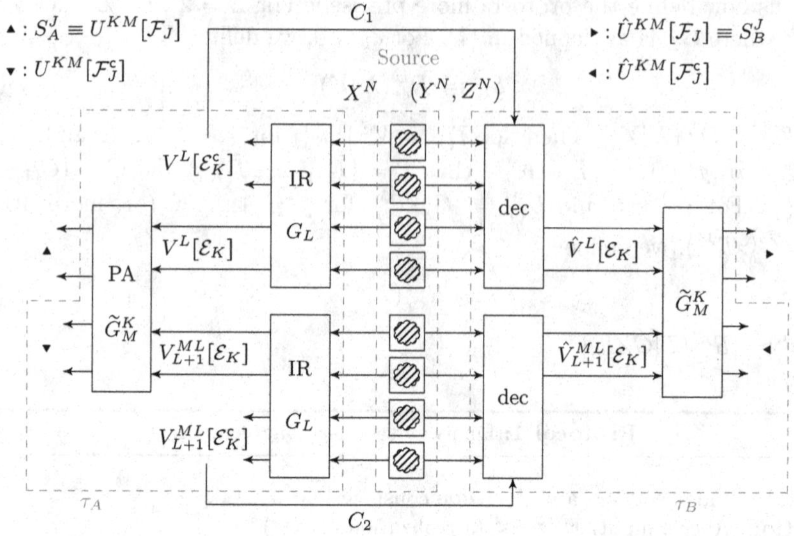

Fig. 1. The secret-key agreement scheme for the setup $N = 8$, $L = 4$, $M = 2$, $K = 2$, and $J = 2$. We consider a source that produces N i.i.d. copies (X^N, Y^N, Z^N) of a triple of correlated random variables (X, Y, Z). Alice performs the operation τ_A, sends $(V^L[\mathcal{E}_K^c])^M$ over a public channel to Bob and obtains S_A^J, her secret key. Bob then performs the operation τ_B which results in his secret key S_B^J.

Proof. The reliability of Alice's and Bob's key follows from the standard polar decoder error probability and the union bound. Each instance of the decoding algorithm employed by Bob has an error probability which scales as $O(2^{-L^\beta})$ for any $\beta < \frac{1}{2}$ [9]; application of the union bound gives the prefactor M. Since G_L as defined in (1) is its own inverse, $\widetilde{G}_M^K$ is its own inverse as well.

The rate of the scheme is

$$
\begin{aligned}
R &= \frac{|\mathcal{F}_J|}{N} && (40)\\
&= \frac{1}{L} H(V^L[\mathcal{E}_K] | V^L[\mathcal{E}_K^c], Z^L) - \frac{o(N)}{N} && (41)\\
&= \frac{1}{L}\left(H(V^L|Z^L) - H(V^L[\mathcal{E}_K^c]|Z^L)\right) - \frac{o(N)}{N} && (42)\\
&= H(X|Z) - \frac{1}{L} H(V^L[\mathcal{E}_K^c]|Z^L) - \frac{o(N)}{N}, && (43)
\end{aligned}
$$

where (41) uses the polarization phenomenon stated in Theorem 1.

To prove the secrecy statement requires more effort. Using Pinsker's inequality we obtain

$$\delta\Big(P_{S_A^J,Z^N,C^M},\overline{P}_{S_A^J}\times P_{Z^N,C^M}\Big)\leqslant\sqrt{\tfrac{\ln 2}{2}D\Big(P_{S_A^J,Z^N,C^M}\Big\|\overline{P}_{S_A^J}\times P_{Z^N,C^M}\Big)}\quad(44)$$

$$=\sqrt{\tfrac{\ln 2}{2}\left(J-H(S_A^J|Z^N,C^M)\right)},\quad(45)$$

where the last step uses the chain rule for relative entropies and that $\overline{P}_{S_A^J}$ denotes the uniform distribution. We can simplify the conditional entropy expression using the chain rule

$$H(S_A^J|Z^N,C^M)$$

$$=H(U^M[\mathcal{F}_J]|Z^N,(V^L[\mathcal{E}_K^{\mathrm{c}}])^M)\quad(46)$$

$$=\sum_{j=1}^{K}H\Big(U_{(j)}^M[\mathcal{F}_{(j)}]\Big|U_{(1)}^M[\mathcal{F}_{(1)}],\ldots,U_{(j-1)}^M[\mathcal{F}_{(j-1)}],Z^N,(V^L[\mathcal{E}_K^{\mathrm{c}}])^M\Big)\quad(47)$$

$$=\sum_{j=1}^{K}\sum_{i=1}^{|\mathcal{F}_{(i)}|}H\Big(U_{(j)}^M[\mathcal{F}_{(j)}]_i\Big|U_{(j)}^M[\mathcal{F}_{(j)}]^{i-1},\big\{U_{(l)}^M[\mathcal{F}_{(l)}]\big\}_{l=1}^{j-1},Z^N,(V^L[\mathcal{E}_K^{\mathrm{c}}])^M\Big)\quad(48)$$

$$\geqslant\sum_{j=1}^{K}\sum_{i\in\mathcal{F}_j}H\Big(U_{(j)i}\Big|U_{(j)}^{i-1},U_{(1)}^M[\mathcal{F}_{(1)}],\ldots,U_{(j-1)}^M[\mathcal{F}_{(j-1)}],Z^N,(V^L[\mathcal{E}_K^{\mathrm{c}}])^M\Big)\quad(49)$$

$$\geqslant J\left(1-\epsilon_2\right),\quad(50)$$

where the first inequality uses the fact that that conditioning cannot increase the entropy and the second inequality follows by the definition of $\mathcal{F}_J$. Recall that we are using the notation introduced in Section 2.2. For $\mathcal{F}_J$ as defined in (36), we have $\mathcal{F}_J=\left\{\mathcal{F}_{(j)}\right\}_{j=1}^{K}$ where $\mathcal{F}_{(j)}=\mathcal{R}_{\epsilon_2}^M\left(T_{(j)}\left|T_{(j-1)},\ldots,T_{(1)},C,Z^L\right.\right)$. The polarization phenomenon, Theorem 1, implies $J=O(N)$, which together with (45) proves the secrecy statement of Theorem 7, since $\epsilon_2=O(2^{-N^\beta})$ for any $\beta<\frac{1}{2}$.

It remains to show that the computational complexity of the scheme is $O(N\log N)$. Alice performs the operation G_L in the first layer M times, each requiring $O(L\log L)$ steps [5]. In the second layer she performs $\tilde{G}_M^K$, or K parallel instances of G_M, requiring $O(KM\log M)$ total steps. From the polarization phenomenon, we have $K=O(L)$, and thus the complexity of Alice's operations is not worse than $O(N\log N)$. Bob runs M standard polar decoders which can be done in $O(ML\log L)$ complexity [5,12]. Bob next performs the polar transform $\tilde{G}_M^K$, whose complexity is not worse than $O(N\log N)$ as justified above. Thus, the complexity of Bob's operations is also not worse than $O(N\log N)$. □

In principle, the two parameters L and M can be chosen freely. However, to maintain the reliability of the scheme (cf.(37)), M may not grow exponentially fast in L. A reasonable choice would be to have both parameters scale comparably fast, i.e., $\frac{M}{L}=O(1)$.

Corollary 8. *The rate of Protocol 1 given in Theorem 7 can be bounded as*

$$R \geqslant \max\left\{0, H(X|Z) - H(X|Y) - \frac{o(N)}{N}\right\}. \tag{51}$$

Proof. According to (43) the rate of Protocol 1 is

$$R = H(X|Z) - \frac{1}{L}H(V^L[\mathcal{E}_K^{\mathsf{c}}]|Z^L) - \frac{o(N)}{N} \tag{52}$$

$$\geqslant \max\left\{0, H(X|Z) - \frac{|\mathcal{E}_K^{\mathsf{c}}|}{L} - \frac{o(N)}{N}\right\} \tag{53}$$

$$= \max\left\{0, H(X|Z) - H(X|Y) - \frac{o(N)}{N}\right\}, \tag{54}$$

where (54) uses the polarization phenomenon stated in Theorem 1. □

3.2 Achieving the Secret-Key Rate of a Given Distribution

Theorem 7 together with Corollaries 4 and 8 immediately imply that Protocol 1 achieves the secret-key rate $S_{\rightarrow}(X;Y|Z)$ if $P_{X,Y,Z}$ is such that the induced DM WTP W described by $P_{Y,Z|X}$ is less noisy. If we can solve the optimization problem (19), i.e., find the optimal auxiliary random variables V and U, our one-way secret-key agreement scheme can achieve $S_{\rightarrow}(X;Y|Z)$ for a general setup. We then make V public, replace X by U and run Protocol 1. Note that finding the optimal random variables V and U might be difficult. It has been shown that for certain distributions the optimal random variables V and U can be found analytically [18].

An open problem discussed in Section 5 addresses the question if Protocol 1 can achieve a rate that is strictly larger than $\max\{0, H(X|Z) - H(X|Y)\}$ if nothing about the optimal auxiliary random variables V and U is known, i.e., if we run the protocol directly for X without making V public.

3.3 Code Construction

To construct the code the index sets $\mathcal{E}_K$ and $\mathcal{F}_J$ need to be determined. The set $\mathcal{E}_K$ can be computed approximately with a linear-time algorithm introduced in [32], given the distributions P_X and $P_{Y|X}$. Alternatively, Tal and Vardy's older algorithm [33] and its adaption to the asymmetric setup [12] can be used.

To approximately compute the outer index set $\mathcal{F}_J$ requires more effort. In principle, we can again use the above algorithms, which require a description of the "super-source" seen by the outer layer, i.e., the source which outputs the triple of random variables $(V^L[\mathcal{E}_K], (Y^L, V^L[\mathcal{E}_K^{\mathsf{c}}]), (Z^L, V^L[\mathcal{E}_K^{\mathsf{c}}]))$. However, its alphabet size is exponential in L, and thus such a direct approach will not be efficient in the overall blocklength N. Nonetheless, due to the structure of the inner layer, it is perhaps possible that the method of approximation by limiting the alphabet size [33,32] can be extended to this case. In particular, a recursive construction motivated by the decoding operation introduced in [6] could potentially lead to an efficient computation of the index set $\mathcal{F}_J$.

4 Private Channel Coding Scheme

Our private channel coding scheme is a simple modification of the secret key agreement protocol of the previous section. Again it consists of two layers, an inner layer which ensures transmitted messages can be reliably decoded by the intended receiver, and an outer layer which guarantees privacy from the unintended receiver. The basic idea is to simply run the key agreement scheme in reverse, *inputting* messages to the protocol where secret key bits would be *output* in key agreement. The immediate problem in doing so is that key agreement also produces outputs besides the secret key, so the procedure is not immediately reversible. To overcome this problem, the encoding operations here simulate the random variables output in the key agreement protocol, and then perform the polar transformations $\widetilde{G}_M^K$ and G_L in reverse.[12]

The scheme is visualized in Figure 2 and described in detail in Protocol 2. Not explicitly shown is the simulation of the bits $U^M[\mathcal{F}_J^{\mathsf{c}}]$ at the outer layer and the bits $V^L[\mathcal{E}_K^{\mathsf{c}}]$ at the inner layer. The outer layer, whose simulated bits are nearly deterministic, makes use of the method described in [34, Definition 1], while the inner layer, whose bits are nearly uniformly-distributed, follows [12, Section 4]. Both proceed by successively sampling from the individual bit distributions given all previous values in the particular block, i.e., constructing V_j by sampling from $P_{V_j|V^{j-1}}$. These distributions can be efficiently constructed, as described in Section 4.3.

Note that a public channel is used to communicate the information reconciliation information to Bob, enabling reliable decoding. However, it is possible to dispense with the public channel and still achieve the same rate and efficiency properties, as will be discussed in Section 4.3.

In the following we assume that the message M^J to be transmitted is uniformly distributed over the message set $\mathcal{M} = \{0,1\}^J$. As mentioned in Section 2.4, it may be desirable to have a private coding scheme that works for an arbitrarily distributed message. This can be achieved by assuming that the wiretap channel W is symmetric—more precisely, by assuming that the two channels $\mathsf{W}_1 : \mathcal{X} \to \mathcal{Y}$ and $\mathsf{W}_2 : \mathcal{X} \to \mathcal{Z}$ induced by W are symmetric. We can define a super-channel $\mathsf{W}' : \mathcal{T} \to \mathcal{Y}^L \times \mathcal{Z}^L \times \mathcal{C}$ which consists of an inner encoding block and L basic channels W. The super-channel W' again induces two channels $\mathsf{W}_1' : \mathcal{T} \to \mathcal{Y}^L \times \mathcal{C}$ and $\mathsf{W}_2' : \mathcal{T} \to \mathcal{Z}^L \times \mathcal{C}$. Arıkan showed that W_1 respectively W_2 being symmetric implies that W_1' respectively W_2' is symmetric [5, Proposition 13]. It has been shown in [24, Proposition 3] that for symmetric channels polar codes remain reliable for an arbitrary distribution of the message bits. We thus conclude that if W_1 is assumed to be symmetric, our coding scheme remains reliable for arbitrarily distributed messages. Assuming having a symmetric channel W_2 implies that W_2' is symmetric which proves that our scheme is strongly secure for arbitrarily distributed messages.[13]

---

[12] As it happens, G_L is its own inverse.

[13] This can be seen easily by the strong secrecy condition given in (31) using that W_2' is symmetric.

Protocol 2: Private channel coding

Given: Index sets $\mathcal{E}_K$ and $\mathcal{F}_J$ (code construction)[14]

Notation: Message to be transmitted: m^J

Outer enc.: Let $u^M[\mathcal{F}_J] = m^J$[15] and $u^M[\mathcal{F}_J^c] = r^{KM-J}$ where r^{KM-J} is (randomly) generated as explained in [34, Definition 1]. Let $t^M = \widetilde{G}_M^K u^M$.

Inner enc.: For all $i \in \{0, L, \dots, L(M-1)\}$, Alice does the following: let $\overline{v}_{i+1}^{i+L}[\mathcal{E}_K] = t_{(i/L)+1}$ and $\overline{v}_{i+1}^{i+L}[\mathcal{E}_K^c] = s_{i+1}^{i+L-K}$ where s_{i+1}^{i+L-K} is (randomly) generated as explained in [12, Section 4]. Send $C_{(i/K)+1} := s_{i+1}^{i+L-K}$ over a public channel to Bob. Finally, compute $x_{i+1}^{i+L} = G_L \overline{v}_{i+1}^{i+L}$.

Transmis.: $(y^N, z^N) = \mathsf{W}^N x^N$

Inner dec.: Bob uses the standard decoder [5,12] with inputs $C_{(i/L)+1}$ and y_{i+1}^{i+L} to obtain $\hat{v}_{i+1}^{i+L}$, and hence $\hat{t}_{(i/L)+1} = \hat{v}_{i+1}^{i+L}[\mathcal{E}_K]$, for each $i \in \{0, L, \dots, L(M-1)\}$.

Outer dec.: Bob computes $\hat{u}^M = \widetilde{G}_M^K \hat{t}^M$ and outputs a guess for the sent message $\hat{m}^J = \hat{u}^M[\mathcal{F}_J]$.

4.1 Rate, Reliability, Secrecy, and Efficiency

Corollary 9. *For any $\beta < \frac{1}{2}$, Protocol 2 satisfies*

$$\text{Reliability:} \quad \Pr\left[M^J \neq \hat{M}^J\right] = O\left(M 2^{-L^\beta}\right) \tag{55}$$

$$\text{Secrecy:} \quad \left\| P_{M^J, Z^N, C} - \overline{P}_{M^J} \times P_{Z^N, C} \right\|_1 = O\left(\sqrt{N} 2^{-\frac{N^\beta}{2}}\right) \tag{56}$$

$$\text{Rate:} \quad R = H(X|Z) - \frac{1}{L} H\left(V^L[\mathcal{E}_K^c] | Z^L\right) - \frac{o(N)}{N} \tag{57}$$

and its computational complexity is $O(N \log N)$.

Proof. Recall that the idea of the private channel coding scheme is to run Protocol 1 backwards. Since Protocol 2 simulates the nearly deterministic bits $U^M[\mathcal{F}_J]$ at the outer encoder as described in [34, Definition 1] and the almost random bits $V^L[\mathcal{E}_K^c]$ at the inner encoder as explained in [12, Section 4], it follows that for large values of L and M the private channel coding scheme approximates the one-way secret-key scheme setup,[16] i.e., $\lim_{N \to \infty} \delta\left(P_{T^M}, P_{(V^L[\mathcal{E}_K])^M}\right) = 0$ and $\lim_{L \to \infty} \delta\left(P_{X^L}, P_{\hat{X}^L}\right) = 0$, where P_{X^L} denotes the distribution of the vector X^L which is sent over the wiretap channel W and $P_{\hat{X}^L}$ denotes the distribution of Alice's random variable $\hat{X}^L$ in the one-way secret-key agreement setup. We

[14] By the code construction the channel input distribution P_X is defined. P_X should be chosen such that it maximizes the scheme's rate.

[15] Again an abuse of notation. See the Footnote 11 of Protocol 1.

[16] This approximation can be made arbitrarily precise for sufficiently large values of L and M.

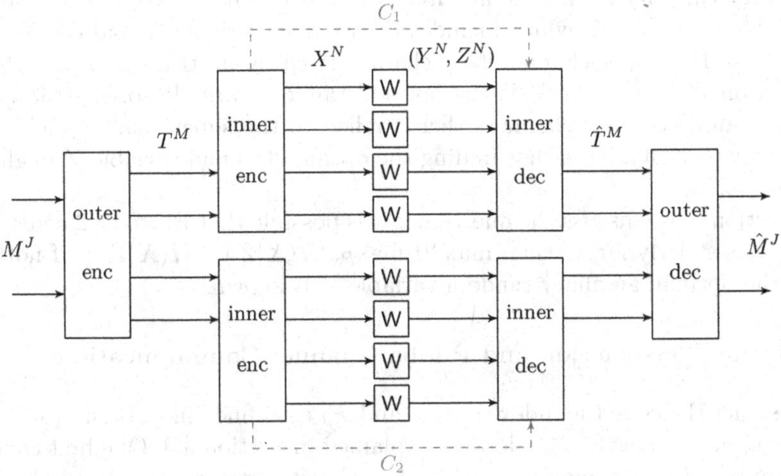

Fig. 2. The private channel coding scheme for the setup $N = 8$, $L = 4$, $M = 2$, $K = 2$, and $J = 2$. The message M^J is first sent through an outer encoder which adds some bits (simulated as explained in [12, Section 4]) and applies the polarization transform $\widetilde{G}_M^K$. The output $T^M = (T_{(1)}, \ldots, T_{(K)})^M$ is then encoded a second time by M independent identical blocks. Note that each block again adds redundancy (as explained in [34, Definition 1]) before applying the polarization transform G_L. Each inner encoding block sends the frozen bits over a public channel to Bob. Note that this extra public communication can be avoided as justified in Section 4.3. The output X^N is then sent over N copies of the wiretap channel W to Bob. Bob then applies a decoding operation as in the key agreement scheme, Section 3.

thus can use the decoder introduced in [9] to decode the inner layer. Since we are using M identical independent inner decoding blocks, by the union bound we obtain the desired reliability condition. The secrecy and rate statement are immediate consequences from Theorem 7. □

As mentioned after Theorem 7, to ensure reliability of the protocol, M may not grow exponentially fast in L.

Corollary 10. *The rate of Protocol 2 given in Corollary 9 can be bounded as*

$$R \geqslant \max\left\{0, H(X|Z) - H(X|Y) - \frac{o(N)}{N}\right\}. \tag{58}$$

Proof. The proof is identical to the proof of Corollary 8. □

4.2 Achieving the Secrecy Capacity of a Wiretap Channel

Corollaries 6 and 10 immediately imply that our private channel coding scheme achieves the secrecy capacity for the setup where W is more capable. If we can find the optimal auxiliary random variable V in (33), Protocol 2 can achieve the secrecy capacity for a general wiretap channel scenario. We define a super-channel $\overline{\mathsf{W}} : \mathcal{V} \to \mathcal{Y} \times \mathcal{Z}$ which includes the random variable X and the wiretap channel W. The super-channel $\overline{\mathsf{W}}$ is characterized by its transition probability distribution $P_{Y,Z|V}$ where V is the optimal random variable solving (33). The private channel coding scheme is then applied to the super-channel, achieving the secrecy capacity. Note that finding the optimal random variable V might be difficult.

In Section 5, we discuss the question if it is possible that Protocol 2 achieves a rate that is strictly larger than $\max\{0, \max_{P_X} H(X|Z) - H(X|Y)\}$, if nothing about the optimal auxiliary random variable V is known.

4.3 Code Construction and Public Channel Communication

To construct the code the index sets $\mathcal{E}_K$ and $\mathcal{F}_J$ as defined in (35) and (36) need to be computed. This can be done as explained in Section 3.3. One first chooses a distribution P_X that maximizes the scheme's rate given in (57), before looking for a code that defines this distribution P_X.

We next explain how the communication $C^M \in \mathcal{C}^M$ from Alice to Bob can be reduced such that it does not affect the rate, i.e., we show that we can choose $|\mathcal{C}| = o(L)$. Recall that we defined the index set $\mathcal{E}_K := \mathcal{D}^L_{\epsilon_1}(X|Y)$ in (35). Let $\mathcal{G} := \mathcal{R}^L_{\epsilon_1}(X|Y)$ using the noation introduced in (2) and $\mathcal{I} := [L]\backslash(\mathcal{E}_K \cup \mathcal{G}) = \mathcal{E}_K^{\mathsf{c}}\backslash\mathcal{G}$. As explained in Section 2.2, $\mathcal{G}$ consists of the outputs V_j which are essentially uniformly random, even given all previous outputs V^{j-1} as well as Y^L, where $V^L = G_L X^L$. The index set $\mathcal{I}$ consists of the outputs V_j which are neither essentially uniformly random nor essentially deterministic given V^{j-1} and Y^L. The polarization phenomenon stated in Theorem 1 ensures that this set is small, i.e., that $|\mathcal{I}| = o(L)$. Since the bits of $\mathcal{G}$ are almost uniformly distributed, we can fix these bits independently of the message—as part of the code construction—without affecting the reliability of the scheme for large blocklengths.[17] We thus only need to communicate the bits belonging to the index set $\mathcal{I}$.

We can send the bits belonging to $\mathcal{I}$ over a seperate public noiseless channel. Alternatively, we could send them over the wiretap channel W that we are using for private channel coding. However since W is assumed to be noisy and it is essential that the bits in $\mathcal{I}$ are recieved by Bob without any errors, we need to protect them using an error correcting code. To not destroy the essentially linear computational complexity of our scheme, the code needs to have an encoder and decoder that are practically efficient. Since $|\mathcal{I}| = o(L)$, we can use any error correcting code that has a non-vanishing rate. For symmetric binary DMCs, polar coding can be used to transmit reliably an arbitrarily distributed message [24, Proposition 3]. We can therefore symmetrize our wiretap channel W and use polar codes to transmit the bits in $\mathcal{I}$.[18]

[17] Recall that we choose $\epsilon_1 = O\left(2^{-L^{\beta}}\right)$ for any $\beta < \frac{1}{2}$, such that for $L \to \infty$ the index set $\mathcal{G}$ contains only uniformly distributed bits.

[18] Note that the symmetrization of the channel will reduce its rate which however does not matter as we need a non-vanishing rate only.

As the reliability of the scheme is the average over the possible assignments of the random bits belonging to $\mathcal{I}$ (or even $\mathcal{E}_K^{\mathsf{c}}$), at least one choice must be as good as the average, meaning a reliable, efficient, and deterministic scheme must exist. However, it might be computationally hard to find this choice. This means that there exists a scheme for private channel coding (having the properties given in Corollary 9) that does not require any extra communication from Alice to Bob, i.e., $\mathcal{C} = \varnothing$, however its code construction might be computationally inefficient.

5 Conclusion and Open Problems

We have constructed practically efficient protocols (with complexity essentially linear in the blocklength) for one-way secret-key agreement from correlated randomness and for private channel coding over discrete memoryless wiretap channels. Each protocol achieves the corresponding optimal rate. Compared to previous methods, we do not require any degradability assumptions and achieve strong (rather than weak) secrecy. Our scheme is formulated for arbitrary discrete memoryless wiretap channels. Using ideas of Şaşoğlu *et al.* [10] the two protocols presented in this paper can also be used for wiretap channels with continuous input alphabets.

Finally, we want to describe an open problem which addresses the question of whether rates beyond $\max\{0, H(X|Z) - H(X|Y)\}$ can be achieved by our key agreement scheme, even if the optimal auxiliary random variables V and U are not given, i.e., if we run Protocol 1 directly for X (instead of U) without making V public. The question could also be formulated in the private coding scenario, whether rates beyond $\max\{0, \max_{P_X} H(X|Z) - H(X|Y)\}$ are possible, but as a positive answer in the former context implies a positive answer in the latter, we shall restrict attention to the key agreement scenario for simplicity.

Question 1 *Does for some distributions $P_{X,Y,Z}$ the rate of Protocol 1 satisfy*

$$R > \max\{0, H(X|Z) - H(X|Y)\}, \quad \text{for} \quad N \to \infty? \tag{59}$$

An equivalent formulation of this question is whether inequality (53) is always tight for large enough N, i.e.,

Question 1' *Is it possible that*

$$\lim_{L\to\infty} \frac{1}{L} H(V^L[\mathcal{E}_K^{\mathsf{c}}]|Z^L) < \lim_{L\to\infty} \frac{1}{L} |\mathcal{E}_K^{\mathsf{c}}|, \quad \text{for} \quad R > 0? \tag{60}$$

From the polarization phenomenon stated in Theorem 1 we obtain $\lim_{L\to\infty} \frac{1}{L} |\mathcal{E}_K^{\mathsf{c}}| = H(X|Y)$, which together with (60) would imply that $R > \max\{0, H(X|Z) - H(X|Y)\}$ for $N \to \infty$ is possible. Relation (60) can only be satisfied if the high-entropy set with respect to Bob's side information, i.e., the set $\mathcal{E}_K^{\mathsf{c}}$, is not always a high-entropy set with respect to Eve's side information. Thus, the question of rates in the key agreement protocol is closely related to fundamental structural properties of the polarization phenomenon.

A positive answer to Question 1 implies that we can send quantum information reliable over a quantum channel at a rate that is beyond the *coherent information* using the scheme introduced in [6].

Acknowledgments. The authors would like to thank Alexander Vardy for useful discussions. This work was supported by the Swiss National Science Foundation (through the National Centre of Competence in Research 'Quantum Science and Technology' and grant No. 200020-135048) and by the European Research Council (grant No. 258932).

References

1. Shannon, C.E.: Communication theory of secrecy systems. Bell System Technical Journal 28, 656–715 (1949)
2. Maurer, U.: Secret key agreement by public discussion from common information. IEEE Trans. on Information Theory 39, 733–742 (1993)
3. Wyner, A.D.: The wire-tap channel. Bell System Technical Journal 54, 1355–1387 (1975)
4. Csiszár, I., Körner, J.: Broadcast channels with confidential messages. IEEE Trans. on Information Theory 24, 339–348 (1978)
5. Arıkan, E.: Channel polarization: A method for constructing capacity-achieving codes for symmetric binary-input memoryless channels. IEEE Trans. on Information Theory 55, 3051–3073 (2009)
6. Sutter, D., Renes, J.M., Dupuis, F., Renner, R.: Efficient quantum channel coding scheme requiring no preshared entanglement. In: Proc. IEEE Int. Symposium on Information Theory (to appear, 2013)
7. Cover, T.M., Thomas, J.A.: Elements of Information Theory. Wiley Interscience (2006)
8. Körner, J., Marton, K.: Comparison of two noisy channels. In: Bolyai, J. (ed.) Topics in Information Theory. Colloquia Mathematica Societatis, pp. 411–424. North-Holland, The Netherlands (1977)
9. Arıkan, E.: Source polarization. In: Proc. IEEE Int. Symposium on Information Theory, pp. 899–903 (2010)
10. Şaşoğlu, E., Telatar, E., Arıkan, E.: Polarization for arbitrary discrete memoryless channels. In: Proc. Information Theory Workshop, pp. 144–148 (2009)
11. Arıkan, E., Telatar, E.: On the rate of channel polarization. In: Proc. IEEE Int. Symposium on Information Theory (2009)
12. Honda, J., Yamamoto, H.: Polar coding without alphabet extension for asymmetric channels. In: Proc. IEEE Int. Symposium on Information Theory, pp. 2147–2151 (2012)
13. Abbe, E.: Randomness and dependencies extraction via polarization. In: Information Theory and Applications Workshop (ITA), pp. 1–7 (2011)
14. Sahebi, A.G., Pradhan, S.S.: Multilevel polarization of polar codes over arbitrary discrete memoryless channels. In: 49th Annual Allerton Conference on Communication, Control, and Computing (Allerton), pp. 1718–1725 (2011)
15. Maurer, U.: The strong secret key rate of discrete random triples. In: Blahut, R.E. (ed.) Communication and Cryptography, pp. 271–285. Kluwer Academic, Boston (1994)
16. Maurer, U., Wolf, S.: Information-theoretic key agreement: From weak to strong secrecy for free. In: Preneel, B. (ed.) EUROCRYPT 2000. LNCS, vol. 1807, pp. 351–368. Springer, Heidelberg (2000)
17. Ahlswede, R., Csiszár, I.: Common randomness in information theory and cryptography. i. secret sharing. IEEE Trans. on Information Theory 39, 1121–1132 (1993)

18. Holenstein, T., Renner, R.: One-way secret-key agreement and applications to circuit polarization and immunization of public-key encryption. In: Shoup, V. (ed.) CRYPTO 2005. LNCS, vol. 3621, pp. 478–493. Springer, Heidelberg (2005)
19. Maurer, U., Wolf, S.: Unconditionally secure key agreement and the intrinsic conditional information. IEEE Trans. on Information Theory 45, 499–514 (1999)
20. Renner, R., Wolf, S.: New bounds in secret-key agreement: The gap between formation and secrecy extraction. In: Biham, E. (ed.) EUROCRYPT 2003. LNCS, vol. 2656, pp. 562–577. Springer, Heidelberg (2003)
21. El Gamal, A., Kim, Y.H.: Network Information Theory. Cambridge University Press (2012)
22. Abbe, E.: Low complexity constructions of secret keys using polar coding. In: Proc. Information Theory Workshop (2012)
23. Chou, R.A., Bloch, M.R., Abbe, E.: Polar coding for secret-key generation (2013), `http://arxiv.org/abs/1305.4746`
24. Mahdavifar, H., Vardy, A.: Achieving the secrecy capacity of wiretap channels using polar codes. IEEE Trans. on Information Theory 57, 6428–6443 (2011)
25. Bellare, M., Tessaro, S., Vardy, A.: Semantic security for the wiretap channel. In: Safavi-Naini, R., Canetti, R. (eds.) CRYPTO 2012. LNCS, vol. 7417, pp. 294–311. Springer, Heidelberg (2012)
26. Andersson, M., Rathi, V., Thobaben, R., Kliewer, J., Skoglund, M.: Nested polar codes for wiretap and relay channels. IEEE Communications Letters 14, 752–754 (2010)
27. Hof, E., Shamai, S.: Secrecy-achieving polar-coding. In: Proc. Information Theory Workshop, pp. 1–5 (2010)
28. Koyluoglu, O.O., El Gamal, H.: Polar coding for secure transmission and key agreement. In: IEEE 21st International Symposium on Personal Indoor and Mobile Radio Communications (PIMRC), pp. 2698–2703 (2010)
29. Şaşoğlu, E., Vardy, A.: A new polar coding scheme for strong security on wiretap channels. In: Proc. IEEE Int. Symposium on Information Theory (to appear, 2013)
30. Hayashi, M., Matsumoto, R.: Construction of wiretap codes from ordinary channel codes. In: Proc. IEEE Int. Symposium on Information Theory, pp. 2538–2542 (2010)
31. Karzand, M., Telatar, E.: Polar codes for q-ary source coding. In: Proc. IEEE Int. Symposium on Information Theory, pp. 909–912 (2010)
32. Tal, I., Sharov, A., Vardy, A.: Constructing polar codes for non-binary alphabets and macs. In: Proc. IEEE Int. Symposium on Information Theory, pp. 2132–2136 (2012)
33. Tal, I., Vardy, A.: How to construct polar codes. Submitted to IEEE Transactions on Information Theory (2011), arXiv:1105.6164
34. Sutter, D., Renes, J.M., Dupuis, F., Renner, R.: Achieving the capacity of any DMC using only polar codes. In: Proc. Information Theory Workshop, pp. 114–118 (2012); extended version available at arXiv:1205.3756

SPHF-Friendly Non-interactive Commitments

Michel Abdalla[1], Fabrice Benhamouda[1], Olivier Blazy[2], Céline Chevalier[3], and David Pointcheval[1]

[1] École Normale Supérieure, CNRS-INRIA, Paris, France
[2] Ruhr-Universität Bochum, Germany
[3] Université Panthéon-Assas, Paris, France

Abstract. In 2009, Abdalla *et al.* proposed a reasonably practical password-authenticated key exchange (PAKE) secure against adaptive adversaries in the universal composability (UC) framework. It exploited the Canetti-Fischlin methodology for commitments and the Cramer-Shoup smooth projective hash functions (SPHFs), following the Gennaro-Lindell approach for PAKE. In this paper, we revisit the notion of non-interactive commitments, with a new formalism that implies UC security. In addition, we provide a quite efficient instantiation. We then extend our formalism to SPHF-friendly commitments. We thereafter show that it allows a blackbox application to one-round PAKE and oblivious transfer (OT), still secure in the UC framework against adaptive adversaries, assuming reliable erasures and a single global common reference string, even for multiple sessions. Our instantiations are more efficient than the Abdalla *et al.* PAKE in Crypto 2009 and the recent OT protocol proposed by Choi *et al.* in PKC 2013. Furthermore, the new PAKE instantiation is the first one-round scheme achieving UC security against adaptive adversaries.

1 Introduction

Commitment schemes are one of the most fundamental primitives in cryptography, serving as a building block for many cryptographic applications such as zero-knowledge proofs [22] and secure multi-party computation [21]. In a typical commitment scheme, there are two main phases. In a *commit* phase, the committer computes a commitment C for some message x and sends it to the receiver. Then, in an *opening* phase, the committer releases some information δ to the receiver which allows the latter to verify that C was indeed a commitment of x. To be useful in practice, a commitment scheme should satisfy two basic security properties. The first one is *hiding*, which informally guarantees that no information about x is leaked through the commitment C. The second one is *binding*, which guarantees that the committer cannot generate a commitment C that can be successfully opened to two different messages.

Smooth Projective Hash Functions (SPHFs) were introduced by Cramer and Shoup [17] as a means to design chosen-ciphertext-secure public-key encryption schemes. In addition to providing a more intuitive abstraction for their

K. Sako and P. Sarkar (Eds.) ASIACRYPT 2013 Part I, LNCS 8269, pp. 214–234, 2013.

original public-key encryption scheme in [16], the notion of SPHF also enabled new efficient instantiations of their scheme under different complexity assumptions, such as quadratic residuosity. Due to its usefulness, the notion of SPHF was later extended to several other contexts, such as password-authenticated key exchange (PAKE) [20], oblivious transfer (OT) [27,15], and blind signatures [7,6].

Password-Authenticated Key Exchange (PAKE) protocols were proposed in 1992 by Bellovin and Merritt [5] where authentication is done using a simple password, possibly drawn from a small space subject to exhaustive search. Since then, many schemes have been proposed and studied. SPHFs have been extensively used, starting with the work of Gennaro and Lindell [20] which generalized an earlier construction by Katz, Ostrovsky, and Yung (KOY) [29], and followed by several other works [11,2]. More recently, a variant of SPHFs proposed by Katz and Vaikuntanathan even allowed the construction of one-round PAKE schemes [30,6].

The first ideal functionality for PAKE protocols in the UC framework [8,12] was proposed by Canetti *et al.* [11], who showed how a simple variant of the Gennaro-Lindell methodology [20] could lead to a secure protocol. Though quite efficient, their protocol was not known to be secure against adaptive adversaries, that are capable of corrupting players at any time, and learn their internal states. The first ones to propose an adaptively secure PAKE in the UC framework were Barak *et al.* [3] using general techniques from multi-party computation (MPC). Though conceptually simple, their solution results in quite inefficient schemes.

The first reasonably practical adaptively secure PAKE was proposed by Abdalla *et al.* [2], following the Gennaro-Lindell methodology with the Canetti-Fischlin commitment [10]. They had to build a complex SPHF to handle the verification of such a commitment. Thus, the communication complexity was high and the protocol required four rounds. No better adaptively secure scheme has been proposed so far.

Oblivious Transfer (OT) was introduced in 1981 by Rabin [34] as a way to allow a receiver to get exactly one out of k messages sent by another party, the sender. In these schemes, the receiver should be oblivious to the other values, and the sender should be oblivious to which value was received. Since then, several instantiations and optimizations of such protocols have appeared in the literature, including proposals in the UC framework [31,13].

More recently, new instantiations have been proposed, trying to reach round-optimality [26], or low communication costs [33]. The 1-out-of-2 OT scheme by Choi *et al.* [15] based on the DDH assumption seems to be the most efficient one among those that are secure against adaptive corruptions in the CRS model with erasures. But it does not scale to 1-out-of-k OT, for $k > 2$.

1.1 Properties of Commitment Schemes

Basic Properties. In addition to the binding and hiding properties, certain applications may require additional properties from a commitment scheme. One such property is *equivocability* [4], which guarantees that a commitment C can be opened in more than a single way when in possession of a certain trapdoor information. Another one is *extractability*, which allows the computation of the message x committed in C when in possession of a certain trapdoor information. Yet another property that may also be useful for cryptographic applications is *non-malleability* [18], which ensures that the receiver of a unopened commitment C for a message x cannot generate a commitment for a message that is related to x.

Though commitment schemes satisfying stronger properties such as *non-malleability*, *equivocability*, and *extractability* may be useful for solving specific problems, they usually stop short of guaranteeing security when composed with arbitrary protocols. To address this problem, Canetti and Fischlin [10] proposed an ideal functionality for commitment schemes in the universal composability (UC) framework [8] which guarantees all these properties simultaneously and remain secure even under concurrent compositions with arbitrary protocols. Unfortunately, they also showed that such commitment schemes can only be realized if one makes additional setup assumptions, such as the existence of a common reference string (CRS) [10], random oracles [25], or secure hardware tokens [28].

Equivocable and Extractable Commitments. As the work of Canetti and Fischlin [10], this work also aims to build *non-interactive* commitment schemes which can simultaneously guarantee *non-malleability*, *equivocability*, and *extractability* properties. To this end, we first define a new notion of commitment scheme, called E^2-commitments, for which there exists an alternative setup algorithm, whose output is computationally indistinguishable from that of a normal setup algorithm and which outputs a common trapdoor that allows for both equivocability and extractability: this trapdoor not only allows for the extraction of a committed message, but it can also be used to create simulated commitments which can be opened to any message.

To define the security of E^2-schemes, we first extend the security notions of standard equivocable commitments and extractable commitments to the E^2-commitment setting: Since the use of a common trapdoor for equivocability and extractability could potentially be exploited by an adversary to break the extractability or equivocability properties of an E^2-commitment scheme, we define stronger versions of these notions, which account for the fact that the same trapdoor is used for both extractability or equivocability. In particular, in these stronger notions, the adversary is given oracle access to the simulated commitment and extractor algorithms.

Finally, after defining the security of E^2-schemes, we further show that these schemes remain secure even under arbitrary composition with other cryptographic protocols. More precisely, we show that any E^2–commitment scheme which meets the strong versions of the equivocability or extraction notions is a non-interactive UC-secure (multiple) commitment scheme in the presence of adaptive adversaries, assuming reliable erasures and a single global CRS.

SPHF-Friendly Commitments. In this work, we are interested in building non-interactive E^2-commitments, to which smooth projective hash functions can be efficiently associated. Unfortunately, achieving this goal is not so easy due to the equivocability property of E^2-commitments. To understand why, let X be the domain of an SPHF function and let L be some underlying NP language such that it is computationally hard to distinguish a random element in L from a random element in $X \setminus L$. A key property of these SPHF functions that makes them so useful for applications such as PAKE and OT is that, for words C in L, their values can be computed using either a *secret* hashing key hk or a *public* projected key hp together a witness w to the fact that C is indeed in L. A typical example of a language in which we are interested is the language L_x corresponding to the set of elements $\{C\}$ such that C is a valid commitment of x. Unfortunately, when commitments are equivocable, the language L_x containing the set of valid commitments of x may not be well defined since a commitment C could potentially be opened to any x. To get around this problem and be able to use SPHFs with E^2-commitments, we show that it suffices for an E^2-commitment scheme to satisfy two properties. The first one is the stronger version of the equivocability notion, which guarantees that equivocable commitments are computationally indistinguishable from normal commitments, even when given oracle access to the simulated commitment and extractor algorithms. The second one, which is called *robustness*, is new and guarantees that commitments generated by polynomially-bounded adversaries are perfectly binding. Finally, we say that a commitment scheme is *SPHF-friendly* if it satisfies both properties and if it admits an SPHF on the languages L_x.

1.2 Contributions

A New SPHF-friendly E^2-commitment Construction. First, we define the notion of SPHF-friendly E^2-commitment together with an instantiation. The new construction, which is called $\mathcal{E}^2\mathcal{C}$ and described in Section 4, is inspired by the commitment schemes in [10,13,2]. Like the construction in [2], it combines a variant of the Cramer-Shoup encryption scheme (as an extractable commitment scheme) and an equivocable commitment scheme to be able to simultaneously achieve both equivocability and extractability. However, unlike the construction in [2], we rely on Haralambiev's perfectly hiding commitment [24, Section 4.1.4], instead of the Pedersen commitment [32].

Since the opening value of Haralambiev's scheme is a group element that can be encrypted in one ElGamal-like ciphertext to allow extractability, this globally leads to a better communication and computational complexity for the commitment. The former is linear in $m \cdot \mathfrak{K}$, where m is the bit-length of the committed value and $\mathfrak{K}$, the security parameter. This is significantly better than the extractable commitment construction in [2] which was linear in $m \cdot \mathfrak{K}^2$, but asymptotically worse than the two proposals in [19] that are linear in $\mathfrak{K}$, and thus independent of m. However, we point out the latter proposals in [19] are not SPHF-friendly since they are not robust.

We then show in Theorem 4 that a labeled E^2-commitment satisfying stronger notions of equivocability and extractability is a non-interactive UC-secure commitment scheme in the presence of adaptive adversaries, assuming reliable erasures and a single global CRS, and we apply this result to our new construction.

One-Round Adaptively Secure PAKE. Second, we provide a generic construction of a one-round UC-secure PAKE from any SPHF-friendly commitment. The UC-security holds against adaptive adversaries, assuming reliable erasures and a single global CRS, as shown in Section 6. In addition to being the first one-round adaptively secure PAKE, our new scheme also enjoys a much better communication complexity than previous adaptively secure PAKE schemes. For instance, in comparison to the PAKE in [2], which is currently the most efficient adaptively secure PAKE, the new scheme gains a factor of $\mathfrak{K}$ in the overall communication complexity, where $\mathfrak{K}$ is the security parameter. However, unlike their scheme, our new construction requires pairing-friendly groups.

Three-round Adaptively Secure 1-out-of-k OT. Third, we provide a generic construction of a three-round UC-secure 1-out-of-k OT from any SPHF-friendly commitment. The UC-security holds against adaptive adversaries, assuming reliable erasures and a single global CRS, as shown in Section 7. Besides decreasing the total number of rounds with respect to existing OT schemes with similar security levels, our resulting protocol also has a better communication complexity than the best known solution so far [15]. Moreover, our construction is more general and provides a solution for 1-out-of-k OT schemes while the solution in [15] only works for $k = 2$.

Due to space restrictions, complete proofs and some details were postponed to the full version [1].

2 Basic Notions for Commitments

We first review the basic definitions of non-interactive commitments, with some examples. Then, we consider the classical additional notions of equivocability and extractability. In this paper, the qualities of adversaries will be measured by their successes and advantages in certain experiments $\mathsf{Exp}^{\mathsf{sec}}$ or $\mathsf{Exp}^{\mathsf{sec}\text{-}b}$ (between the cases $b = 0$ and $b = 1$), denoted $\mathsf{Succ}^{\mathsf{sec}}(\mathcal{A}, \mathfrak{K})$ and $\mathsf{Adv}^{\mathsf{sec}}(\mathcal{A}, \mathfrak{K})$ respectively, while the security of a primitive will be measured by the maximal successes or advantages of any adversary running within a time bounded by some t in the appropriate experiments, denoted $\mathsf{Succ}^{\mathsf{sec}}(t)$ and $\mathsf{Adv}^{\mathsf{sec}}(t)$ respectively. Adversaries can keep state during the different phases. We denote $\stackrel{\$}{\leftarrow}$ the outcome of a probabilistic algorithm or the sampling from a uniform distribution.

2.1 Non-interactive Labeled Commitments

A non-interactive labeled commitment scheme $\mathcal{C}$ is defined by three algorithms:

- $\mathsf{SetupCom}(1^{\mathfrak{K}})$ takes as input the security parameter $\mathfrak{K}$ and outputs the global parameters, passed through the CRS ρ to all other algorithms;

$\mathsf{Exp}^{\mathtt{hid}\text{-}b}_{\mathcal{A}}(\mathfrak{K})$	$\mathsf{Exp}^{\mathtt{bind}}_{\mathcal{A}}(\mathfrak{K})$
$\rho \overset{\$}{\leftarrow} \mathsf{SetupCom}(1^{\mathfrak{K}})$ $(\ell, x_0, x_1, \mathsf{state}) \overset{\$}{\leftarrow} \mathcal{A}(\rho)$ $(C, \delta) \overset{\$}{\leftarrow} \mathsf{Com}^{\ell}(x_b)$ **return** $\mathcal{A}(\mathsf{state}, C)$	$\rho \overset{\$}{\leftarrow} \mathsf{SetupCom}(1^{\mathfrak{K}})$ $(C, \ell, x_0, \delta_0, x_1, \delta_1) \overset{\$}{\leftarrow} \mathcal{A}(\rho)$ **if** $\neg\mathsf{VerCom}^{\ell}(C, x_0, \delta_0)$ **then return** 0 **if** $\neg\mathsf{VerCom}^{\ell}(C, x_1, \delta_1)$ **then return** 0 **return** $x_0 \neq x_1$

Fig. 1. Hiding and Binding Properties

- $\mathsf{Com}^{\ell}(x)$ takes as input a label ℓ and a message x, and outputs a pair (C, δ), where C is the commitment of x for the label ℓ, and δ is the corresponding opening data (a.k.a. decommitment information). This is a probabilistic algorithm;
- $\mathsf{VerCom}^{\ell}(C, x, \delta)$ takes as input a commitment C, a label ℓ, a message x, and the opening data δ and outputs 1 (true) if δ is a valid opening data for C, x and ℓ. It always outputs 0 (false) on $x = \perp$.

Using the experiments $\mathsf{Exp}^{\mathtt{hid}}_{\mathcal{A}}(\mathfrak{K})$ and $\mathsf{Exp}^{\mathtt{bind}}_{\mathcal{A}}(\mathfrak{K})$ defined in Figure 1, one can state the basic properties:

- *Correctness*: for all correctly generated CRS ρ, all commitments and opening data honestly generated pass the verification VerCom test: for all ℓ, x, if $(C, \delta) \overset{\$}{\leftarrow} \mathsf{Com}^{\ell}(x)$, then $\mathsf{VerCom}^{\ell}(C, x, \delta) = 1$;
- *Hiding Property*: the commitment does not leak any information about the committed value. $\mathcal{C}$ is said (t, ε)-hiding if $\mathsf{Adv}^{\mathtt{hid}}_{\mathcal{C}}(t) \leq \varepsilon$.
- *Binding Property*: no adversary can open a commitment in two different ways. $\mathcal{C}$ is said (t, ε)-binding if $\mathsf{Succ}^{\mathtt{bind}}_{\mathcal{C}}(t) \leq \varepsilon$.

Correctness is always perfectly required, and one can also require either the binding or the hiding property to be perfect.

The reader can remark that labels are actually useless in the hiding and the binding properties. But they will become useful in $\mathtt{E}^2$-commitment schemes introduced in the next section. This is somehow similar to encryption scheme: labels are useless with encryption schemes which are just IND-CPA, but are very useful with IND-CCA encryption schemes.

2.2 Perfectly Binding Commitments: Public-Key Encryption

To get perfectly binding commitments, classical instantiations are public-key encryption schemes, which additionally provide extractability (see below). The encryption algorithm is indeed the commitment algorithm, and the random coins become the opening data that allow to check the correct procedure of the commit phase. The hiding property relies on the indistinguishability (IND-CPA), which is computationally achieved, whereas the binding property relies on the correctness of the encryption scheme and is perfect.

Let us define the ElGamal-based commitment scheme:

- $\mathsf{SetupCom}(1^{\mathfrak{K}})$ chooses a cyclic group $\mathbb{G}$ of prime order p, g a generator for this group and a random scalar $z \xleftarrow{\$} \mathbb{Z}_p$. It sets the CRS $\rho = (\mathbb{G}, g, h = g^z)$;
- $\mathsf{Com}(M)$, for $M \in \mathbb{G}$, chooses a random element $r \xleftarrow{\$} \mathbb{Z}_p$ and outputs the pair $(C = (u = g^r, e = h^r \cdot M), \delta = r)$;
- $\mathsf{VerCom}(C = (u, e), M, \delta = r)$ checks whether $C = (u = g^r, e = h^r \cdot M)$.

This commitment scheme is hiding under the DDH assumption and perfectly binding. It is even extractable using the decryption key z: $M = e/u^z$. However, it is not labeled. The Cramer-Shoup encryption scheme [16] admits labels and is extractable and non-malleable, thanks to the IND-CCA security level.

2.3 Perfectly Hiding Commitments

The Pedersen scheme [32] is the most famous perfectly hiding commitment: $\mathsf{Com}(m) = g^m h^r$ for a random scalar $r \xleftarrow{\$} \mathbb{Z}_p$ and a fixed basis $h \in \mathbb{G}$. The binding property relies on the DL assumption. Unfortunately, the opening value is the scalar r, which makes it hard to encrypt/decrypt efficiently, as required in our construction below. Haralambiev [24, Section 4.1.4] recently proposed a new commitment scheme, called TC4 (without label), with a group element as opening value:

- $\mathsf{SetupCom}(1^{\mathfrak{K}})$ chooses an asymmetric pairing-friendly setting $(\mathbb{G}_1, g_1, \mathbb{G}_2, g_2, \mathbb{G}_T, p, e)$, with an additional independent generator $T \in \mathbb{G}_2$. It sets the CRS $\rho = (\mathbb{G}_1, g_1, \mathbb{G}_2, g_2, T, \mathbb{G}_T, p, e)$;
- $\mathsf{Com}(x)$, for $x \in \mathbb{Z}_p$, chooses a random element $r \xleftarrow{\$} \mathbb{Z}_p$ and outputs the pair $(C = g_2^r T^x, \delta = g_1^r)$;
- $\mathsf{VerCom}(C, x, \delta)$ checks whether $e(g_1, C/T^x) = e(\delta, g_2)$.

This commitment scheme is clearly perfectly hiding, since the groups are cyclic, and for any $C \in \mathbb{G}_2$, $x \in \mathbb{Z}_p$, there exists $\delta \in \mathbb{G}_1$ that satisfies $e(g_1, C/T^x) = e(\delta, g_2)$. More precisely, if $C = g_2^u$ and $T = g_2^t$, then $\delta = g_1^{u-tx}$ opens C to any x. The binding property holds under the DDH assumption in $\mathbb{G}_2$, as proven in [24, Section 4.1.4].

2.4 Equivocable Commitments

An equivocable commitment scheme $\mathcal{C}$ extends on the previous definition, with $\mathsf{SetupCom}$, Com, VerCom, and a second setup $\mathsf{SetupComT}(1^{\mathfrak{K}})$ that additionally outputs a trapdoor τ, and

- $\mathsf{SimCom}^{\ell}(\tau)$ that takes as input the trapdoor τ and a label ℓ and outputs a pair (C, eqk), where C is a commitment and eqk an equivocation key;
- $\mathsf{OpenCom}^{\ell}(\mathsf{eqk}, C, x)$ that takes as input a commitment C, a label ℓ, a message x, and an equivocation key eqk for this commitment, and outputs an opening data δ for C and ℓ on x.

$\mathsf{Exp}_{\mathcal{A}}^{\mathtt{sim\text{-}ind\text{-}}b}(\mathfrak{K})$
$\quad (\rho, \tau) \stackrel{\$}{\leftarrow} \mathsf{SetupComT}(1^{\mathfrak{K}})$
$\quad (\ell, x, \mathsf{state}) \stackrel{\$}{\leftarrow} \mathcal{A}^{\mathsf{SCom}^{\cdot}(\tau,\cdot)}(\rho)$
$\quad$ **if** $b = 0$ **then** $(C, \delta) \stackrel{\$}{\leftarrow} \mathsf{Com}^{\ell}(x)$
$\quad$ **else** $(C, \delta) \stackrel{\$}{\leftarrow} \mathsf{SCom}^{\ell}(\tau, x)$
$\quad$ **return** $\mathcal{A}^{\mathsf{SCom}^{\cdot}(\tau,\cdot)}(\mathsf{state}, C, \delta)$

$\mathsf{Exp}_{\mathcal{A}}^{\mathtt{bind\text{-}ext}}(\mathfrak{K})$
$\quad (\rho, \tau) \stackrel{\$}{\leftarrow} \mathsf{SetupComT}(1^{\mathfrak{K}})$
$\quad (C, \ell, x, \delta) \stackrel{\$}{\leftarrow} \mathcal{A}^{\mathsf{ExtCom}^{\cdot}(\tau,\cdot)}(\rho)$
$\quad x' \leftarrow \mathsf{ExtCom}^{\ell}(\tau, C)$
$\quad$ **if** $x' = x$ **then return** 0
$\quad$ **else return** $\mathsf{VerCom}^{\ell}(C, x, \delta)$

Fig. 2. Simulation Indistinguishability and Binding Extractability

Let us denote SCom the algorithm that takes as input the trapdoor τ, a label ℓ and a message x and which outputs $(C, \delta) \stackrel{\$}{\leftarrow} \mathsf{SCom}^{\ell}(\tau, x)$, computed as $(C, \mathsf{eqk}) \stackrel{\$}{\leftarrow} \mathsf{SimCom}^{\ell}(\tau)$ and $\delta \leftarrow \mathsf{OpenCom}^{\ell}(\mathsf{eqk}, C, x)$. Three additional properties are then associated: a *correctness* property, and two *indistinguishability* properties, which all together imply the *hiding* property.

- *Trapdoor Correctness*: all simulated commitments can be opened on any message: for all ℓ, x, if $(C, \mathsf{eqk}) \stackrel{\$}{\leftarrow} \mathsf{SimCom}^{\ell}(\tau)$ and $\delta \leftarrow \mathsf{OpenCom}^{\ell}(\mathsf{eqk}, C, x)$, then $\mathsf{VerCom}^{\ell}(C, x, \delta) = 1$;
- *Setup Indistinguishability*: one cannot distinguish the CRS ρ generated by $\mathsf{SetupCom}$ from the one generated by $\mathsf{SetupComT}$. $\mathcal{C}$ is said (t, ε)-setup-indistinguishable if the two distributions for ρ are (t, ε)-computationally indistinguishable. We denote $\mathsf{Adv}_{\mathcal{C}}^{\mathtt{setup\text{-}ind}}(t)$ the distance between the two distributions.
- *Simulation Indistinguishability*: one cannot distinguish a real commitment (generated by Com) from a fake commitment (generated by SCom), even with oracle access to fake commitments. $\mathcal{C}$ is said (t, ε)-simulation-indistinguishable if $\mathsf{Adv}_{\mathcal{C}}^{\mathtt{sim\text{-}ind}}(t) \leq \varepsilon$ (see the experiments $\mathsf{Exp}_{\mathcal{A}}^{\mathtt{sim\text{-}ind\text{-}}b}(\mathfrak{K})$ in Figure 2).

More precisely, when the trapdoor correctness is satisfied, since commitments generated by SimCom are perfectly hiding (they can be opened in any way using $\mathsf{OpenCom}$), $\mathsf{Adv}_{\mathcal{C}}^{\mathtt{hid}}(t) \leq \mathsf{Adv}_{\mathcal{C}}^{\mathtt{setup\text{-}ind}}(t) + \mathsf{Adv}_{\mathcal{C}}^{\mathtt{sim\text{-}ind}}(t)$.

Definition 1 (Equivocable Commitment). *A commitment scheme $\mathcal{C}$ is said (t, ε)-equivocable if, first, the basic commitment scheme satisfies the correctness property and is both (t, ε)-binding and (t, ε)-hiding, and, secondly, the additional algorithms guarantee the trapdoor correctness and make it both (t, ε)-setup-indistinguishable and (t, ε)-simulation-indistinguishable.*

2.5 Extractable Commitments

An extractable commitment scheme $\mathcal{C}$ also extends on the initial definition, with $\mathsf{SetupCom}$, Com, VerCom, as well as the second setup $\mathsf{SetupComT}(1^{\mathfrak{K}})$ that additionally outputs a trapdoor τ, and

- $\mathsf{ExtCom}^{\ell}(\tau, C)$ which takes as input the trapdoor τ, a commitment C, and a label ℓ, and outputs the committed message x, or $\perp$ if the commitment is invalid.

As above, three additional properties are then associated: a *correctness* property, and the *setup indistinguishability*, but also an *extractability* property, which implies, together with the setup indistinguishability, the *binding* property:

- *Trapdoor Correctness*: all commitments honestly generated can be correctly extracted: for all ℓ, x, if $(C, \delta) \stackrel{\$}{\leftarrow} \mathsf{Com}^{\ell}(x)$ then $\mathsf{ExtCom}^{\ell}(C, \tau) = x$;
- *Setup Indistinguishability*: as above;
- *Binding Extractability*: one cannot fool the extractor, *i.e.*, produce a commitment and a valid opening data to an input x while the commitment does not extract to x. $\mathcal{C}$ is said (t, ε)-binding-extractable if $\mathsf{Succ}_{\mathcal{C}}^{\mathtt{bind\text{-}ext}}(t) \leq \varepsilon$ (see the experiment $\mathsf{Exp}_{\mathcal{A}}^{\mathtt{bind\text{-}ext}}(\mathfrak{K})$ in Figure 2).

More precisely, when one breaks the binding property with $(C, \ell, x_0, \delta_0, x_1, \delta_1)$, if the extraction oracle outputs $x' = x_0$, then one can output (C, ℓ, x_1, δ_1), otherwise one can output (C, ℓ, x_0, δ_0). In both cases, this breaks the binding-extractability: $\mathsf{Adv}_{\mathcal{C}}^{\mathtt{bind}}(t) \leq \mathsf{Adv}_{\mathcal{C}}^{\mathtt{setup\text{-}ind}}(t) + \mathsf{Succ}_{\mathcal{C}}^{\mathtt{bind\text{-}ext}}(t)$.

Definition 2 (Extractable Commitment). *A commitment scheme $\mathcal{C}$ is said (t, ε)-extractable if, first, the basic commitment scheme satisfies the correctness property and is both (t, ε)-binding and (t, ε)-hiding, and, secondly, the additional algorithms guarantee the trapdoor correctness and make it both (t, ε)-setup-indistinguishable and (t, ε)-binding-extractable.*

3 Equivocable and Extractable Commitments

3.1 E^2-Commitments: Equivocable and Extractable

Public-key encryption schemes are perfectly binding commitments that are additionally extractable. The Pedersen and Haralambiev commitments are perfectly hiding commitments that are additionally equivocable. But none of them have the two properties at the same time. This is now our goal.

Definition 3 (E^2-Commitment). *A commitment scheme $\mathcal{C}$ is said (t, ε)-E^2 (equivocable and extractable) if the indistinguishable setup algorithm outputs a common trapdoor that allows both equivocability and extractability. If one denotes $\mathsf{Adv}_{\mathcal{C}}^{\mathtt{e}^2}(t)$ the maximum of $\mathsf{Adv}_{\mathcal{C}}^{\mathtt{setup\text{-}ind}}(t)$, $\mathsf{Adv}_{\mathcal{C}}^{\mathtt{sim\text{-}ind}}(t)$, and $\mathsf{Succ}_{\mathcal{C}}^{\mathtt{bind\text{-}ext}}(t)$, then it should be upper-bounded by ε.*

But with such a common trapdoor, the adversary could exploit the equivocation queries to break extractability and extraction queries to break equivocability. Stronger notions can thus be defined, using the experiments $\mathsf{Exp}_{\mathcal{A}}^{\mathtt{s\text{-}sim\text{-}ind\text{-}}b}(\mathfrak{K})$ and $\mathsf{Exp}_{\mathcal{A}}^{\mathtt{s\text{-}bind\text{-}ext}}(\mathfrak{K})$ in Figure 3, in which SCom is supposed to store each query/answer (ℓ, x, C) in a list Λ and ExtCom-queries on such an SCom-output (ℓ, C) are answered by x (as it would be when using Com instead of SCom).

- *Strong Simulation Indistinguishability*: one cannot distinguish a real commitment (generated by Com) from a fake commitment (generated by SCom), even with oracle access to the extraction oracle (ExtCom) and to fake commitments (using SCom). $\mathcal{C}$ is said (t, ε)-strongly-simulation-indistinguishable if $\mathsf{Adv}_{\mathcal{C}}^{\mathtt{s\text{-}sim\text{-}ind}}(t) \leq \varepsilon$;

$\mathsf{Exp}^{\texttt{s-sim-ind-}b}_{\mathcal{A}}(\mathfrak{K})$	$\mathsf{Exp}^{\texttt{s-bind-ext}}_{\mathcal{A}}(\mathfrak{K})$
$(\rho,\tau) \stackrel{\$}{\leftarrow} \mathsf{SetupComT}(1^{\mathfrak{K}})$; $(\ell, x, \mathsf{state}) \stackrel{\$}{\leftarrow} \mathcal{A}^{\mathsf{SCom}^\cdot(\tau,\cdot),\mathsf{ExtCom}^\cdot(\tau,\cdot)}(\rho)$ **if** $b=0$ **then** $(C,\delta) \stackrel{\$}{\leftarrow} \mathsf{Com}^\ell(x)$ **else** $(C,\delta) \stackrel{\$}{\leftarrow} \mathsf{SCom}^\ell(\tau,x)$ **return** $\mathcal{A}^{\mathsf{SCom}^\cdot(\tau,\cdot),\mathsf{ExtCom}^\cdot(\tau,\cdot)}(\mathsf{state}, C, \delta)$	$(\rho,\tau) \stackrel{\$}{\leftarrow} \mathsf{SetupComT}(1^{\mathfrak{K}})$ $(C, \ell, x, \delta) \stackrel{\$}{\leftarrow} \mathcal{A}^{\mathsf{SCom}^\cdot(\tau,\cdot),\mathsf{ExtCom}^\cdot(\tau,\cdot)}(\rho)$ $x' \leftarrow \mathsf{ExtCom}^\ell(\tau, C)$ **if** $(\ell, x', C) \in \Lambda$ **then return** 0 **if** $x' = x$ **then return** 0 **else return** $\mathsf{VerCom}^\ell(C, x, \delta)$

Fig. 3. Strong Simulation Indistinguishability and Strong Binding Extractability

– *Strong Binding Extractability* (informally introduced in [13] as "simulation extractability"): one cannot fool the extractor, *i.e.*, produce a commitment and a valid opening data (not given by SCom) to an input x while the commitment does not extract to x, even with oracle access to the extraction oracle (ExtCom) and to fake commitments (using SCom). $\mathcal{C}$ is said (t,ε)-strongly-binding-extractable if $\mathsf{Succ}^{\texttt{s-bind-ext}}_{\mathcal{C}}(t) \le \varepsilon$.

They both imply the respective weaker notions since they just differ by giving access to the ExtCom-oracle in the former game, and to the SCom oracle in the latter. We insist that ExtCom-queries on SCom-outputs are answered by the related SCom-inputs. Otherwise, the former game would be void. In addition, VerCom always rejects inputs with $x = \bot$, which is useful in the latter game.

3.2 UC-Secure Commitments

The security definition for commitment schemes in the UC framework was presented by Canetti and Fischlin [10], refined by Canetti [9]. The ideal functionality is presented in Figure 4, where a *public delayed output* is an output first sent to the adversary $\mathcal{S}$ that eventually decides if and when the message is actually delivered to the recipient. In case of corruption of the committer, if this is before the `Receipt`-message for the receiver, the adversary chooses the committed value, otherwise it is provided by the ideal functionality, according to the `Commit`-message. Note this is actually the multiple-commitment functionality that allows multiple executions of the commitment protocol (multiple ssid's) for the same functionality instance (one sid). This avoids the use of joint-state UC [14].

Theorem 4. *A labeled* E^2*-commitment scheme* $\mathcal{C}$*, that is in addition strongly-simulation-indistinguishable or strongly-binding-extractable, is a non-interactive UC-secure commitment scheme in the presence of adaptive adversaries, assuming reliable erasures and authenticated channels.*

4 A Construction of Labeled E^2-Commitment Scheme

4.1 Labeled Cramer-Shoup Encryption on Vectors

For our construction we use a variant of the Cramer-Shoup encryption scheme for vectors of messages. Let $\mathbb{G}$ be a cyclic group of order p, with two independent

The functionality $\mathcal{F}_{\text{com}}$ is parametrized by a security parameter k. It interacts with an adversary $\mathcal{S}$ and a set of parties $P_1,\ldots,P_n$ via the following queries: **Commit phase: Upon receiving a query $(\texttt{Commit}, \text{sid}, \text{ssid}, P_i, P_j, x)$ from party P_i**: record the tuple $(\text{sid}, \text{ssid}, P_i, P_j, x)$ and generate a *public delayed output* $(\texttt{Receipt}, \text{sid}, \text{ssid}, P_i, P_j)$ to P_j. Ignore further Commit-message with the same ssid from P_i. **Decommit phase. Upon receiving a query $(\texttt{Reveal}, \text{sid}, \text{ssid}, P_i, P_j)$ from party P_i**: ignore the message if $(\text{sid}, \text{ssid}, P_i, P_j, x)$ is not recorded; otherwise mark the record $(\text{sid}, \text{ssid}, P_i, P_j)$ as revealed and generate a *public delayed output* $(\texttt{Revealed}, \text{sid}, \text{ssid}, P_i, P_j, x)$ to P_j. Ignore further Reveal-message with the same ssid from P_i.

Fig. 4. Ideal Functionality for Commitment Scheme $\mathcal{F}_{\text{com}}$

generators g and h. The secret decryption key is a random vector $\mathsf{sk} = (x_1, x_2, y_1, y_2, z) \stackrel{\$}{\leftarrow} \mathbb{Z}_p^5$ and the public encryption key is $\mathsf{pk} = (g, h, c = g^{x_1}h^{x_2}, d = g^{y_1}h^{y_2}, f = g^z, H)$, where H is randomly chosen in a collision-resistant hash function family $\mathcal{H}$ (actually, second-preimage resistance is enough). For a message-vector $\boldsymbol{M} = (M_i)_{i=1,\ldots,m} \in \mathbb{G}^m$, the multi-Cramer-Shoup encryption is defined as $m\text{-}\mathsf{MCS}^\ell_{\mathsf{pk}}(\boldsymbol{M}; (r_i)_i) = (\mathsf{CS}^\ell_{\mathsf{pk}}(M_i, \theta; r_i) = (u_i = g^{r_i}, v_i = h^{r_i}, e_i = f^{r_i} \cdot M_i, w_i = (cd^\theta)^{r_i}))_i$, where $\theta = H(\ell, (u_i, v_i, e_i)_i)$ is the same for all the w_i's to ensure non-malleability contrary to what we would have if we had just concatenated Cramer-Shoup ciphertexts of the M_i's. Such a ciphertext $C = (u_i, v_i, e_i, w_i)_i$ is decrypted by $M_i = e_i/u_i^z$, after having checked the validity of the ciphertext, $w_i \stackrel{?}{=} u_i^{x_1+\theta y_1} v_i^{x_2+\theta y_2}$, for $i = 1, \ldots, m$. This multi-Cramer-Shoup encryption scheme, denoted MCS, is IND-CCA under the DDH assumption. It even verifies a stronger property VIND-PO-CCA (for Vector-Indistinguishability with Partial Opening under Chosen-Ciphertext Attacks), useful for the security proof of our commitment $\mathcal{E}^2\mathcal{C}$.

4.2 Construction

In this section, we provide a concrete construction $\mathcal{E}^2\mathcal{C}$, inspired from [10,13,2], with the above multi-Cramer-Shoup encryption (as an extractable commitment scheme) and the TC4 Haralambiev's equivocable commitment scheme [24, Section 4.1.4]. The latter will allow equivocability while the former will provide extractability:

- $\mathsf{SetupComT}(1^{\mathfrak{K}})$ generates a pairing-friendly setting $(\mathbb{G}_1, g_1, \mathbb{G}_2, g_2, \mathbb{G}_T, p, e)$, with another independent generator h_1 of $\mathbb{G}_1$. It then generates the parameters of a Cramer-Shoup-based commitment in $\mathbb{G}_1$: $x_1, x_2, y_1, y_2, z \stackrel{\$}{\leftarrow} \mathbb{Z}_p$ and $H \stackrel{\$}{\leftarrow} \mathcal{H}$, and sets $\mathsf{pk} = (g_1, h_1, c = g_1^{x_1}h_1^{x_2}, d = g_1^{y_1}h_1^{y_2}, f_1 = g_1^z, H)$. It then chooses a random scalar $t \stackrel{\$}{\leftarrow} \mathbb{Z}_p$, and sets $T = g_2^t$. The CRS ρ is set as (pk, T) and the trapdoor τ is the decryption key (x_1, x_2, y_1, y_2, z) (a.k.a. extraction trapdoor) together with t (a.k.a. equivocation trapdoor).

For $\mathsf{SetupCom}(1^{\mathfrak{K}})$, the CRS is generated the same way, but forgetting the scalars, and thus without any trapdoor;

- $\mathsf{Com}^{\ell}(\boldsymbol{M})$, for $\boldsymbol{M} = (M_i)_i \in \{0,1\}^m$ and a label ℓ, works as follows:
  - For $i = 1, \ldots, m$, it chooses a random scalar $r_{i,M_i} \stackrel{\$}{\leftarrow} \mathbb{Z}_p$, sets $r_{i,1-M_i} = 0$, and commits to M_i, using the TC4 commitment scheme with r_{i,M_i} as randomness: $a_i = g_2^{r_{i,M_i}} T^{M_i}$, and sets $d_{i,j} = g_1^{r_{i,j}}$ for $j = 0, 1$, which makes d_{i,M_i} the opening value for a_i to M_i; Let us also write $\boldsymbol{a} = (a_1, \ldots, a_m)$, the tuple of commitments.
  - For $i = 1, \ldots, m$ and $j = 0, 1$, it gets $\boldsymbol{b} = (b_{i,j})_{i,j} = 2m\text{-}\mathsf{MCS}^{\ell'}_{\mathsf{pk}}(\boldsymbol{d}; \boldsymbol{s})$, that is $(u_{i,j}, v_{i,j}, e_{i,j}, w_{i,j})_{i,j}$, where $\boldsymbol{d} = (d_{i,j})_{i,j}$ computed above, $\boldsymbol{s} = (s_{i,j})_{i,j} \stackrel{\$}{\leftarrow} \mathbb{Z}_p^{2m}$, and $\ell' = (\ell, \boldsymbol{a})$.

  The commitment is $C = (\boldsymbol{a}, \boldsymbol{b})$, and the opening information is the m-tuple $\delta = (s_{1,M_1}, \ldots, s_{m,M_m})$.
- $\mathsf{VerCom}^{\ell}(C, \boldsymbol{M}, \delta)$ checks the validity of the ciphertexts b_{i,M_i} with s_{i,M_i} and θ computed on the full ciphertext C, extracts d_{i,M_i} from b_{i,M_i} and s_{i,M_i}, and checks whether $e(g_1, a_i/T^{M_i}) = e(d_{i,M_i}, g_2)$, for $i = 1, \ldots, m$.
- $\mathsf{SimCom}^{\ell}(\tau)$ takes as input the equivocation trapdoor, namely t, and outputs $C = (\boldsymbol{a}, \boldsymbol{b})$ and $\mathsf{eqk} = \boldsymbol{s}$, where
  - For $i = 1, \ldots, m$, it chooses a random scalar $r_{i,0} \stackrel{\$}{\leftarrow} \mathbb{Z}_p$, sets $r_{i,1} = r_{i,0} - t$, and commits to both 0 and 1, using the TC4 commitment scheme with $r_{i,0}$ and $r_{i,1}$ as respective randomness: $a_i = g_2^{r_{i,0}} = g_2^{r_{i,1}} T$, and $d_{i,j} = g_1^{r_{i,j}}$ for $j = 0, 1$, which makes $d_{i,j}$ the opening value for a_i to the value $j \in \{0, 1\}$. This leads to $\boldsymbol{a}$;
  - $\boldsymbol{b}$ is built as above: $\boldsymbol{b} = (b_{i,j})_{i,j} = 2m\text{-}\mathsf{MCS}^{\ell'}_{\mathsf{pk}}(\boldsymbol{d}; \boldsymbol{s})$, with random scalars $(s_{i,j})_{i,j}$.
- $\mathsf{OpenCom}^{\ell}(\mathsf{eqk}, C, \boldsymbol{M})$ simply extracts the useful values from $\mathsf{eqk} = \boldsymbol{s}$ to make the opening value $\delta = (s_{1,M_1}, \ldots, s_{m,M_m})$ in order to open to $\boldsymbol{M} = (M_i)_i$.
- $\mathsf{ExtCom}^{\ell}(\tau, C)$ takes as input the extraction trapdoor, namely the Cramer-Shoup decryption key. Given $\boldsymbol{b}$, it can decrypt all the $b_{i,j}$ into $d_{i,j}$ and check whether $e(g_1, a_i/T^j) = e(d_{i,j}, g_2)$ or not. If, for each i, exactly one $j = M_i$ satisfies the equality, then the extraction algorithm outputs $(M_i)_i$, otherwise (no correct decryption or ambiguity with several possibilities) it outputs $\perp$.

4.3 Security Properties

The above commitment scheme $\mathcal{E}^2\mathcal{C}$ is a labeled $\mathtt{E}^2$-commitment, with both strong-simulation-indistinguishability and strong-binding-extractability, under the DDH assumptions in both $\mathbb{G}_1$ and $\mathbb{G}_2$. It is thus a UC-secure commitment scheme. The stronger `VIND-PO-CCA` security notion for the encryption scheme is required because the SCom/Com oracle does not only output the commitment (and thus the ciphertexts) but also the opening values which include the random coins of the encryption, but just for the plaintext components that are the same in the two vectors, since the two vectors only differ for unnecessary data (namely the $d_{i,1-M_i}$'s) in the security proof. More details can be found in the full version [1].

5 SPHF-Friendly Commitments

5.1 Smooth Projective Hash Functions

Projective hash function families were first introduced by Cramer and Shoup [17], but we here use the definitions of Gennaro and Lindell [20], provided to build secure password-based authenticated key exchange protocols, together with non-malleable commitments.

Let X be the domain of these functions and let L be a certain subset of this domain (a language). A key property of these functions is that, for words C in L, their values can be computed by using either a *secret* hashing key hk or a *public* projection key hp but with a witness w of the fact that C is indeed in L:

- $\mathsf{HashKG}(L)$ generates a hashing key hk for the language L;
- $\mathsf{ProjKG}(\mathsf{hk}, L, C)$ derives the projection key hp, possibly depending on the word C;
- $\mathsf{Hash}(\mathsf{hk}, L, C)$ outputs the hash value from the hashing key, on any word $C \in X$;
- $\mathsf{ProjHash}(\mathsf{hp}, L, C, w)$ outputs the hash value from the projection key hp, and the witness w, for $C \in L$.

The *correctness* of the SPHF assures that if $C \in L$ with w a witness of this fact, then $\mathsf{Hash}(\mathsf{hk}, L, C) = \mathsf{ProjHash}(\mathsf{hp}, L, C, w)$. On the other hand, the security is defined through the *smoothness*, which guarantees that, if $C \notin L$, $\mathsf{Hash}(\mathsf{hk}, L, C)$ is *statistically* indistinguishable from a random element, even knowing hp.

Note that HashKG and ProjKG can just depend partially on L (a superset L') and not at all on C: we then note $\mathsf{HashKG}(L')$ and $\mathsf{ProjKG}(\mathsf{hk}, L', \perp)$ (see [6] for more details on GL-SPHF and KV-SPHF and language definitions).

5.2 Robust Commitments

For a long time, SPHFs have been used to implicitly check some statements, on language membership, such as "C indeed encrypts x". This easily extends to perfectly binding commitments with labels: $L_x = \{(\ell, C) \mid \exists\delta, \mathsf{VerCom}^{\ell}(C, x, \delta) = 1\}$. But when commitments are equivocable, this intuitively means that a commitment C with the label ℓ contains any x and is thus in all the languages L_x. In order to be able to use SPHFs with E^2-commitments, we want the commitments generated by polynomially-bounded adversaries to be perfectly binding, and thus to belong to at most one language L_x. We thus need a *robust verification* property for such *E^2-commitments.*

Definition 5 (Robustness). *One cannot produce a commitment and a label that extracts to x' (possibly $x' = \perp$) such that there exists a valid opening data to a different input x, even with oracle access to the extraction oracle (*ExtCom*) and to fake commitments (using* SCom*).* $\mathcal{C}$ *is said (t, ε)-robust if $\mathsf{Succ}_{\mathcal{C}}^{\mathsf{robust}}(t) \leq \varepsilon$, according to the experiment $\mathsf{Exp}_{\mathcal{A}}^{\mathsf{robust}}(\mathfrak{K})$ in Figure 5.*

It is important to note that the latter experiment $\mathsf{Exp}_{\mathcal{A}}^{\mathsf{robust}}(\mathfrak{K})$ may not be run in polynomial time. Robustness implies strong-binding-extractability.

$\mathsf{Exp}_{\mathcal{A}}^{\mathtt{robust}}(\mathfrak{K})$
$(\rho, \tau) \stackrel{\$}{\leftarrow} \mathsf{SetupComT}(1^{\mathfrak{K}})$
$(C, \ell) \stackrel{\$}{\leftarrow} \mathcal{A}^{\mathsf{SCom}^{\cdot}(\tau,\cdot),\mathsf{ExtCom}^{\cdot}(\tau,\cdot)}(\rho)$
$x' \leftarrow \mathsf{ExtCom}^{\ell}(\tau, C)$
if $(\ell, x', C) \in \Lambda$ **then return** 0
if $\exists x \neq x'$, $\exists \delta$, $\mathsf{VerCom}^{\ell}(C, x, \delta)$ **then return** 1
else return 0

Fig. 5. Robustness

5.3 Properties of SPHF-Friendly Commitments

We are now ready to define SPHF-friendly commitments, which admit an SPHF on the languages $L_x = \{(\ell, C) | \exists \delta, \mathsf{VerCom}^{\ell}(C, x, \delta) = 1\}$, and to discuss about them:

Definition 6 (SPHF-Friendly Commitments). *An SPHF-friendly commitment is an $\mathbb{E}^2$-commitment that admits an SPHF on the languages L_x, and that is both strongly-simulation-indistinguishable and robust.*

Let us consider such a family $\mathcal{F}$ of SPHFs on languages L_x for $x \in X$, with X a non trivial set (with at least two elements), with hash values in the set G. From the smoothness of the SPHF on L_x, one can derive the two following properties on SPHF-friendly commitments, modeled by the experiments in Figure 6. The first notion of *smoothness* deals with adversary-generated commitments, that are likely perfectly binding from the robustness, while the second notion of *pseudo-randomness* deals with simulated commitments, that are perfectly hiding. They are inspired by the security games from [20].

In both security games, note that when hk and hp do not depend on x nor on C, and when the smoothness holds even if the adversary can choose C after having seen hp (*i.e.*, the SPHF is actually a KV-SPHF [6]), they can be generated from the beginning of the games, with hp given to the adversary much earlier.

Smoothness of SPHF-Friendly Commitments. If the adversary $\mathcal{A}$, with access to the oracles SCom and ExtCom, outputs a fresh commitment (ℓ, C) that extracts to $x' \leftarrow \mathsf{ExtCom}^{\ell}(\tau, C)$, then the robustness guarantees that for any $x \neq x'$, $(\ell, C) \notin L_x$ (excepted with small probability), and thus the distribution of the hash value is statistically indistinguishable from the random distribution, even when knowing hp. In the experiment $\mathsf{Exp}_{\mathcal{A}}^{\mathtt{c\text{-}smooth}}(\mathfrak{K})$, we let the adversary choose x, and we have: $\mathsf{Adv}_{\mathcal{C},\mathcal{F}}^{\mathtt{c\text{-}smooth}}(t) \leq \mathsf{Succ}_{\mathcal{C}}^{\mathtt{robust}}(t) + \mathsf{Adv}_{\mathcal{F}}^{\mathtt{smooth}}$.

Pseudo-Randomness of SPHF on Robust Commitments. If the adversary $\mathcal{A}$ is given a commitment C by SCom on x' with label ℓ, both adversary-chosen, even with access to the oracles SCom and ExtCom, then for any x, it cannot distinguish the hash value of (ℓ, C) on language L_x from a random value, even being given hp, since C could have been generated as $\mathsf{Com}^{\ell}(x'')$ for some $x'' \neq x$,

$\mathsf{Exp}_{\mathcal{A}}^{\texttt{c-smooth-}b}(\mathfrak{K})$
$(\rho, \tau) \stackrel{\$}{\leftarrow} \mathsf{SetupComT}(1^{\mathfrak{K}})$
$(C, \ell, x, \mathsf{state}) \stackrel{\$}{\leftarrow} \mathcal{A}^{\mathsf{SCom}^{\cdot}(\tau,\cdot),\mathsf{ExtCom}^{\cdot}(\tau,\cdot)}(\rho)$; $x' \leftarrow \mathsf{ExtCom}^{\ell}(\tau, C)$
if $(\ell, x', C) \in \Lambda$ **then return** 0
$\mathsf{hk} \stackrel{\$}{\leftarrow} \mathsf{HashKG}(L_x)$; $\mathsf{hp} \leftarrow \mathsf{ProjKG}(\mathsf{hk}, L_x, (\ell, C))$
if $b = 0 \vee x' = x$ **then** $H \leftarrow \mathsf{Hash}(\mathsf{hk}, L_x, (\ell, C))$ **else** $H \stackrel{\$}{\leftarrow} G$
return $\mathcal{A}^{\mathsf{SCom}^{\cdot}(\tau,\cdot),\mathsf{ExtCom}^{\cdot}(\tau,\cdot)}(\mathsf{state}, \mathsf{hp}, H)$

$\mathsf{Exp}_{\mathcal{A}}^{\texttt{c-ps-rand-}b}(\mathfrak{K})$
$(\rho, \tau) \stackrel{\$}{\leftarrow} \mathsf{SetupComT}(1^{\mathfrak{K}})$
$(\ell, x, x', \mathsf{state}) \stackrel{\$}{\leftarrow} \mathcal{A}^{\mathsf{SCom}^{\cdot}(\tau,\cdot),\mathsf{ExtCom}^{\cdot}(\tau,\cdot)}(\rho)$; $(C, \delta) \stackrel{\$}{\leftarrow} \mathsf{SCom}^{\ell}(\tau, x')$
$\mathsf{hk} \stackrel{\$}{\leftarrow} \mathsf{HashKG}(L_x)$; $\mathsf{hp} \leftarrow \mathsf{ProjKG}(\mathsf{hk}, L_x, (\ell, C))$
if $b = 0$ **then** $H \leftarrow \mathsf{Hash}(\mathsf{hk}, L_x, (\ell, C))$ **else** $H \stackrel{\$}{\leftarrow} G$
return $\mathcal{A}^{\mathsf{SCom}^{\cdot}(\tau,\cdot),\mathsf{ExtCom}^{\cdot}(\tau,\cdot)}(\mathsf{state}, C, \mathsf{hp}, H)$

Fig. 6. Smoothness and Pseudo-Randomness

which excludes it to belong to L_x, under the robustness. In the experiment $\mathsf{Exp}_{\mathcal{A}}^{\texttt{c-ps-rand}}(\mathfrak{K})$, we let the adversary choose (ℓ, x), and we have: $\mathsf{Adv}_{\mathcal{C},\mathcal{F}}^{\texttt{c-ps-rand}}(t) \leq \mathsf{Adv}_{\mathcal{C}}^{\texttt{s-sim-ind}}(t) + \mathsf{Succ}_{\mathcal{C}}^{\texttt{robust}}(t) + \mathsf{Adv}_{\mathcal{F}}^{\texttt{smooth}}$.

5.4 Our Commitment Scheme $\mathcal{E}^2\mathcal{C}$ is SPHF-Friendly

In order to be *SPHF-friendly*, the commitment first needs to be *strongly-simulation-indistinguishable* and *robust*. We have already shown the former property, and the latter is also proven in the full version [1]. One additionally needs an SPHF able to check the verification equation: using the notations from Section 4.2, $C = (\boldsymbol{a}, \boldsymbol{b})$ is a commitment of $\boldsymbol{M} = (M_i)_i$, if there exist $\delta = (s_{1,M_1}, \ldots, s_{m,M_m})$ and $(d_{1,M_1}, \ldots, d_{m,M_m})$ such that $b_{i,M_i} = (u_{i,M_i}, v_{i,M_i}, e_{i,M_i}, w_{i,M_i}) = \mathsf{CS}_{\mathsf{pk}}^{\ell'}(d_{i,M_i}, \theta; s_{i,M_i})$ (with a particular θ) and $e(g_1, a_i/T^{M_i}) = e(d_{i,M_i}, g_2)$, for $i = 1, \ldots, m$. Since e is non-degenerated, we can eliminate the need of d_{i,M_i}, by lifting everything in $\mathbb{G}_T$, and checking that, first, the ciphertexts are all valid:

$$e(u_{i,M_i}, g_2) = e(g_1^{s_{i,M_i}}, g_2) \quad e(v_{i,M_i}, g_2) = e(h_1^{s_{i,M_i}}, g_2)$$
$$e(w_{i,M_i}, g_2) = e((cd^{\theta})^{s_{i,M_i}}, g_2)$$

and, second, the plaintexts satisfy the appropriate relations:

$$e(e_{i,M_i}, g_2) = e(f_1^{s_{i,M_i}}, g_2) \cdot e(g_1, a_i/T^{M_i}).$$

From these expressions we derive several constructions of such SPHFs in the full version [1], and focus here on the most interesting ones for the following applications:

– First, when C is sent in advance (known when generating hp), as in the OT protocol described in Section 7, for $\mathsf{hk} = (\eta, \alpha, \beta, \mu, \varepsilon) \stackrel{\$}{\leftarrow} \mathbb{Z}_p^5$, and $\mathsf{hp} = (\varepsilon, \mathsf{hp}_1 = g_1^\eta h_1^\alpha f_1^\beta (cd^\theta)^\mu) \in \mathbb{Z}_p \times \mathbb{G}_1$:

$$H = \mathsf{Hash}(\mathsf{hk}, \boldsymbol{M}, C)$$
$$\stackrel{\text{def}}{=} \prod_i \left(e(u_{i,M_i}^\eta \cdot v_{i,M_i}^\alpha, g_2) \cdot (e(e_{i,M_i}, g_2)/e(g_1, a_i/T^{M_i}))^\beta \cdot e(w_{i,M_i}^\mu, g_2)\right)^{\varepsilon^{i-1}}$$
$$= e(\textstyle\prod_i \mathsf{hp}_1^{s_{i,M_i}\varepsilon^{i-1}}, g_2) \stackrel{\text{def}}{=} \mathsf{ProjHash}(\mathsf{hp}, \boldsymbol{M}, C, \delta) = H'.$$

– Then, when C is not necessarily known for computing hp, as in the one-round PAKE, described in Section 6, for $\mathsf{hk} = (\eta_{i,1}, \eta_{i,2}, \alpha_i, \beta_i, \mu_i)_i \stackrel{\$}{\leftarrow} \mathbb{Z}_p^{5m}$, and $\mathsf{hp} = (\mathsf{hp}_{i,1} = g_1^{\eta_{i,1}} h_1^{\alpha_i} f_1^{\beta_i} c^{\mu_i}, \mathsf{hp}_{i,2} = g_1^{\eta_{i,2}} d^{\mu_i})_i \in \mathbb{G}_1^{2m}$:

$$H = \mathsf{Hash}(\mathsf{hk}, \boldsymbol{M}, C)$$
$$\stackrel{\text{def}}{=} \prod_i \left(e(u_{i,M_i}^{(\eta_{i,1}+\theta\eta_{i,2})} \cdot v_{i,M_i}^{\alpha_i}, g_2) \cdot (e(e_{i,M_i}, g_2)/e(g_1, a_i/T^{M_i}))^{\beta_i} \cdot e(w_{i,M_i}^{\mu_i}, g_2)\right)$$
$$= e(\textstyle\prod_i (\mathsf{hp}_{i,1} \mathsf{hp}_{i,2}^\theta)^{s_{i,M_i}}, g_2) \stackrel{\text{def}}{=} \mathsf{ProjHash}(\mathsf{hp}, \boldsymbol{M}, C, \delta) = H'.$$

5.5 Complexity and Comparisons

As summarized in Table 1, the communication complexity is linear in $m \cdot \mathfrak{K}$ (where m is the bit-length of the committed value and $\mathfrak{K}$ is the security parameter), which is much better than [2] that was linear in $m \cdot \mathfrak{K}^2$, but asymptotically worse than the two proposals in [19] that are linear in $\mathfrak{K}$, and thus independent of m (as long as $m = \mathcal{O}(\mathfrak{K})$).

Basically, the first scheme in [19] consists of a Cramer-Shoup-like encryption C of the message x, and a perfectly-sound Groth-Sahai [23] NIZK π that C contains x. The actual commitment is C and the opening value on x is $\delta = \pi$. The trapdoor-setup provides the Cramer-Shoup decryption key and changes the Groth-Sahai setup to the perfectly-hiding setting. The indistinguishable setups of the Groth-Sahai mixed commitments ensure the setup-indistinguishability. The extraction algorithm uses the Cramer-Shoup decryption algorithm, while the equivocation uses the simulator of the NIZK. The IND-CCA security notion for C and the computational soundness of π make it strongly-binding-extractable, the IND-CCA security notion and the zero-knowledge property of the NIZK provide the strong-simulation-indistinguishability. It is thus UC-secure. However, the verification is not robust: because of the perfectly-hiding setting of Groth-Sahai proofs, for any ciphertext C and for any message x, there exists a proof π that makes the verification of C on x. As a consequence, it is not SPHF-friendly. The second construction is in the same vein: they cannot be used in the following applications.

6 Password-Authenticated Key Exchange

6.1 A Generic Construction

The ideal functionality of a Password-Authenticated Key Exchange (PAKE) has been proposed in [11]. In Figure 7, we describe a one-round PAKE that

Table 1. Comparison with existing non-interactive UC-secure commitments with a single global CRS (m = bit-length of the committed value, $\mathfrak{K}$ = security parameter)

	SPHF-Friendly	Commitment C	Decommitment δ	Assumption
[2][a]	yes	$(m + 16m\mathfrak{K}) \times \mathbb{G}$	$2m\mathfrak{K} \times \mathbb{Z}_p$	DDH
[19], 1	no	$5 \times \mathbb{G}$	$16 \times \mathbb{G}$	DLIN
[19], 2	no	$37 \times \mathbb{G}$	$3 \times \mathbb{G}$	DLIN
this paper	yes	$8m \times \mathbb{G}_1 + m \times \mathbb{G}_2$	$m \times \mathbb{Z}_p$	SXDH

[a] slight variant without one-time signature but using labels for the IND-CCA security of the multi-Cramer-Shoup ciphertexts, as in our new scheme, and supposing that an element in the cyclic group $\mathbb{G}$ has size $2\mathfrak{K}$, to withstand generic attacks.

is UC-secure against adaptive adversaries, assuming erasures. It can be built from any SPHF-friendly commitment scheme (that is E^2, strongly-simulation-indistinguishable, and robust as described in Section 5), if the SPHF is actually a KV-SPHF [6] and the algorithms HashKG and ProjKG do not need to know the committed value π (nor the word (ℓ, C) itself). We thus denote L_π the language of the pairs (ℓ, C), where C is a commitment that opens to π under the label ℓ, and L the union of all the L_π (L does not depend on π).

Theorem 7. *The Password-Authenticated Key-Exchange described on Figure 7 is UC-secure in the presence of adaptive adversaries, assuming erasures, as soon as the commitment scheme is SPHF-friendly with a KV-SPHF.*

6.2 Concrete Instantiation

Using our commitment $\mathcal{E}^2\mathcal{C}$ introduced Section 4 together with the second SPHF described Section 5 (which satisfies the above requirements for HashKG and ProjKG), one gets a quite efficient protocol, described in the full version [1]. More precisely, for m-bit passwords, each player has to send $\mathsf{hp} \in \mathbb{G}_1^{2m}$ and

CRS: $\rho \stackrel{\$}{\leftarrow} \mathsf{SetupCom}(1^{\mathfrak{K}})$.

Protocol execution by P_i with π_i:

1. P_i generates $\mathsf{hk}_i \stackrel{\$}{\leftarrow} \mathsf{HashKG}(L)$, $\mathsf{hp}_i \leftarrow \mathsf{ProjKG}(\mathsf{hk}_i, L, \perp)$ and erases any random coins used for the generation
2. P_i computes $(C_i, \delta_i) \stackrel{\$}{\leftarrow} \mathsf{Com}^{\ell_i}(\pi_i)$ with $\ell_i = (\mathsf{sid}, P_i, P_j, \mathsf{hp}_i)$
3. P_i stores δ_i, completely erases random coins used by Com and sends hp_i, C_i to P_j

Key computation: Upon receiving hp_j, C_j from P_j

1. P_i computes $H'_i \leftarrow \mathsf{ProjHash}(\mathsf{hp}_j, L_{\pi_i}, (\ell_i, C_i), \delta_i)$ and $H_j \leftarrow \mathsf{Hash}(\mathsf{hk}_i, L_{\pi_i}, (\ell_j, C_j))$ with $\ell_j = (\mathsf{sid}, P_j, P_i, \mathsf{hp}_j)$
2. P_i computes $\mathsf{sk}_i = H'_i \cdot H_j$.

Fig. 7. UC-Secure PAKE from an SPHF-Friendly Commitment

Table 2. Comparison with existing UC-secure PAKE schemes

	Adaptive	One-round	Communication complexity	Assumption
[2][a]	yes	no	$2 \times (2m + 22m\mathfrak{K}) \times \mathbb{G} + \mathrm{OTS}^{b}$	DDH
[30]	no	yes	$\approx 2 \times 70 \times \mathbb{G}$	DLIN
[6]	no	yes	$2 \times 6 \times \mathbb{G}_1 + 2 \times 5 \times \mathbb{G}_2$	SXDH
this paper	yes	yes	$2 \times 10m \times \mathbb{G}_1 + 2 \times m \times \mathbb{G}_2$	SXDH

[a] with the commitment variant of note "a" of Table 1.
[b] OTS: one-time signature (public key size and signature size) to link the flows in the PAKE protocol.

$C \in \mathbb{G}_1^{8m} \times \mathbb{G}_2^{m}$, which means $10m$ elements from $\mathbb{G}_1$ and m elements from $\mathbb{G}_2$. In Table 2, we compare our new scheme with some previous UC-secure PAKE.

7 Oblivious Transfer

7.1 A Generic Construction

The ideal functionality of an Oblivious Transfer (OT) protocol is depicted in the full version [1]. It is inspired from [15]. In Figure 8, we describe a 3-round OT that is UC-secure against adaptive adversaries, and a 2-round variant which is UC-secure against static adversaries. They can be built from any SPHF-friendly commitment scheme, where L_t is the language of the commitments that open to t under the associated label ℓ, and from any IND-CPA encryption scheme $\mathcal{E} = (\mathsf{Setup}, \mathsf{KeyGen}, \mathsf{Encrypt}, \mathsf{Decrypt})$ with plaintext size at least $\mathfrak{K}$, and from any Pseudo-Random Generator (PRG) F with input size equal to plaintext size, and output size equal to the size of the messages in the database. Details on encryption schemes and PRGs can be found in the full version [1]. Notice the adaptive version can be seen as a variant of the static version where the last flow is sent over a somewhat secure channel, as in [15]; and the preflow and pk and c are used to create this somewhat secure channel.

Theorem 8. *The two Oblivious Transfer schemes described in Figure 8 are UC-secure in the presence of adaptive adversaries and static adversaries respectively, assuming reliable erasures and authenticated channels, as soon as the commitment scheme is SPHF-friendly.*

7.2 Concrete Instantiation and Comparison

Using our commitment $\mathcal{E}^2\mathcal{C}$ introduced Section 4 together with the first SPHF described Section 5, one gets the protocol described in the full version [1], where the number of bits of the commited value is $m = \lceil \log k \rceil$. For the statically secure version, the communication cost is, in addition to the database $\boldsymbol{m}$ that is sent in $\boldsymbol{M}$ in a masked way, 1 element of $\mathbb{Z}_p$ and k elements of $\mathbb{G}_1$ (for $\mathbf{hp}$, by using the same scalar ε for all hp_t's) for the sender, while the receiver sends $\lceil \log k \rceil$

CRS: $\rho \stackrel{\$}{\leftarrow} \mathsf{SetupCom}(1^{\mathfrak{K}}), \mathtt{param} \stackrel{\$}{\leftarrow} \mathsf{Setup}(1^{\mathfrak{K}})$.

Pre-flow (for adaptive security only):

1. P_i generates a key pair $(\mathsf{pk}, \mathsf{sk}) \stackrel{\$}{\leftarrow} \mathsf{KeyGen}(\mathtt{param})$ for $\mathcal{E}$
2. P_i stores sk, completely erase random coins used by KeyGen, and sends pk to P_i

Index query on s**:**

1. P_j chooses a random value S, computes $R \leftarrow F(S)$ and encrypts S under pk: $c \stackrel{\$}{\leftarrow} \mathsf{Encrypt}(\mathsf{pk}, S)$ (for adaptive security only; for static security: $c = \perp, R = 0$)
2. P_j computes $(C, \delta) \stackrel{\$}{\leftarrow} \mathsf{Com}^{\ell}(s)$ with $\ell = (\mathrm{sid}, \mathrm{ssid}, P_i, P_j)$
3. P_j stores δ and completely erase R, S and random coins used by Com and $\mathsf{Encrypt}$ and sends C and c to P_i

Database input $(m_1, \ldots, m_k)$**:**

1. P_i decrypts $S \leftarrow \mathsf{Decrypt}(\mathsf{sk}, c)$ and gets $R \leftarrow F(S)$ (for static security: $R = 0$)
2. P_i computes $\mathsf{hk}_t \stackrel{\$}{\leftarrow} \mathsf{HashKG}(L_t)$, $\mathsf{hp}_t \leftarrow \mathsf{ProjKG}(\mathsf{hk}_t, L_t, (\ell, C))$, $K_t \leftarrow \mathsf{Hash}(\mathsf{hk}_t, L_t, (\ell, C))$, and $M_t \leftarrow R \oplus K_t \oplus m_t$, for $t = 1, \ldots, k$
3. P_i erases everything except $(\mathsf{hp}_t, M_t)_{t=1,\ldots,k}$ and sends them over a secure channel

Data recovery:

Upon receiving $(\mathsf{hp}_t, M_t)_{t=1,\ldots,k}$, P_j computes $K_s \leftarrow \mathsf{ProjHash}(\mathsf{hp}_s, L_s, (\ell, C), \delta)$ and gets $m_s \leftarrow R \oplus K_s \oplus M_s$.

Fig. 8. UC-Secure 1-out-of-k OT from an SPHF-Friendly Commitment (for Adaptive and Static Security)

elements of $\mathbb{G}_2$ (for $\boldsymbol{a}$) and $\lceil 8 \log k \rceil$ elements of $\mathbb{G}_1$ (for $\boldsymbol{b}$), in only two rounds. In the particular case of $k = 2$, the scalar can be avoided since the message consists of 1 bit, so our construction just requires: 2 elements from $\mathbb{G}_1$ for the sender, and 1 from $\mathbb{G}_2$ and 8 from $\mathbb{G}_1$ for the receiver, in two rounds. For the same security level (static corruptions in the CRS, with erasures), the best known solution from [15] required to send at least 23 group elements and 7 scalars, in 4 rounds. If adaptive security is required, our construction requires 3 additional elements in $\mathbb{G}_1$ and 1 additional round, which gives a total of 13 elements in $\mathbb{G}_1$, in 3 rounds. This is also more efficient then the best known solution from [15], which requires 26 group elements and 7 scalars, in 4 rounds.

Acknowledgments. We thank Ralf Küsters for his comments on a preliminary version. This work was supported in part by the French ANR-12-INSE-0014 SIMPATIC Project and in part by the European Commission through the FP7-ICT-2011-EU-Brazil Program under Contract 288349 SecFuNet. The third author was funded by a Sofja Kovalevskaja Award of the Alexander von Humboldt Foundation and the German Federal Ministry for Education and Research.

References

1. Abdalla, M., Benhamouda, F., Blazy, O., Chevalier, C., Pointcheval, D.: SPHF-friendly non-interactive commitments. In: Sako, K., Sarkar, P. (eds.) ASIACRYPT 2013 Part I. LNCS, vol. 8269, pp. 214–234. Springer, Heidelberg (2013), Full version available on the Cryptology ePrint Archive as Report 2013/588

2. Abdalla, M., Chevalier, C., Pointcheval, D.: Smooth projective hashing for conditionally extractable commitments. In: Halevi, S. (ed.) CRYPTO 2009. LNCS, vol. 5677, pp. 671–689. Springer, Heidelberg (2009)
3. Barak, B., Canetti, R., Lindell, Y., Pass, R., Rabin, T.: Secure computation without authentication. In: Shoup, V. (ed.) CRYPTO 2005. LNCS, vol. 3621, pp. 361–377. Springer, Heidelberg (2005)
4. Beaver, D.: Adaptive zero knowledge and computational equivocation (extended abstract). In: 28th ACM STOC, pp. 629–638. ACM Press (May 1996)
5. Bellovin, S.M., Merritt, M.: Encrypted key exchange: Password-based protocols secure against dictionary attacks. In: 1992 IEEE Symposium on Security and Privacy, pp. 72–84. IEEE Computer Society Press (May 1992)
6. Benhamouda, F., Blazy, O., Chevalier, C., Pointcheval, D., Vergnaud, D.: New techniques for SPHFs and efficient one-round PAKE protocols. In: Canetti, R., Garay, J.A. (eds.) CRYPTO 2013, Part I. LNCS, vol. 8042, pp. 449–475. Springer, Heidelberg (2013); full version available on the Cryptology ePrint Archive as reports 2013/034 and 2013/341
7. Blazy, O., Pointcheval, D., Vergnaud, D.: Round-optimal privacy-preserving protocols with smooth projective hash functions. In: Cramer, R. (ed.) TCC 2012. LNCS, vol. 7194, pp. 94–111. Springer, Heidelberg (2012)
8. Canetti, R.: Universally composable security: A new paradigm for cryptographic protocols. In: 42nd FOCS, pp. 136–145. IEEE Computer Society Press (October 2001)
9. Canetti, R.: Universally composable security: A new paradigm for cryptographic protocols. Cryptology ePrint Archive, Report 2000/067 (2005), `http://eprint.iacr.org/`
10. Canetti, R., Fischlin, M.: Universally composable commitments. In: Kilian, J. (ed.) CRYPTO 2001. LNCS, vol. 2139, pp. 19–40. Springer, Heidelberg (2001)
11. Canetti, R., Halevi, S., Katz, J., Lindell, Y., MacKenzie, P.: Universally composable password-based key exchange. In: Cramer, R. (ed.) EUROCRYPT 2005. LNCS, vol. 3494, pp. 404–421. Springer, Heidelberg (2005)
12. Canetti, R., Krawczyk, H.: Universally composable notions of key exchange and secure channels. In: Knudsen, L.R. (ed.) EUROCRYPT 2002. LNCS, vol. 2332, pp. 337–351. Springer, Heidelberg (2002)
13. Canetti, R., Lindell, Y., Ostrovsky, R., Sahai, A.: Universally composable two-party and multi-party secure computation. In: 34th ACM STOC, pp. 494–503. ACM Press (May 2002)
14. Canetti, R., Rabin, T.: Universal composition with joint state. In: Boneh, D. (ed.) CRYPTO 2003. LNCS, vol. 2729, pp. 265–281. Springer, Heidelberg (2003)
15. Choi, S.G., Katz, J., Wee, H., Zhou, H.-S.: Efficient, adaptively secure, and composable oblivious transfer with a single, global CRS. In: Kurosawa, K., Hanaoka, G. (eds.) PKC 2013. LNCS, vol. 7778, pp. 73–88. Springer, Heidelberg (2013)
16. Cramer, R., Shoup, V.: A practical public key cryptosystem provably secure against adaptive chosen ciphertext attack. In: Krawczyk, H. (ed.) CRYPTO 1998. LNCS, vol. 1462, pp. 13–25. Springer, Heidelberg (1998)
17. Canetti, R., Krawczyk, H.: Universally composable notions of key exchange and secure channels. In: Knudsen, L.R. (ed.) EUROCRYPT 2002. LNCS, vol. 2332, pp. 45–64. Springer, Heidelberg (2002)
18. Dolev, D., Dwork, C., Naor, M.: Nonmalleable cryptography. SIAM Journal on Computing 30(2), 391–437 (2000)

19. Fischlin, M., Libert, B., Manulis, M.: Non-interactive and re-usable universally composable string commitments with adaptive security. In: Lee, D.H., Wang, X. (eds.) ASIACRYPT 2011. LNCS, vol. 7073, pp. 468–485. Springer, Heidelberg (2011)
20. Gennaro, R., Lindell, Y.: A framework for password-based authenticated key exchange. In: Biham, E. (ed.) EUROCRYPT 2003. LNCS, vol. 2656, pp. 524–543. Springer, Heidelberg (2003), http://eprint.iacr.org/2003/032.ps.gz
21. Goldreich, O., Micali, S., Wigderson, A.: How to play any mental game, or a completeness theorem for protocols with honest majority. In: Aho, A. (ed.) 19th ACM STOC, pp. 218–229. ACM Press (May 1987)
22. Goldreich, O., Micali, S., Wigderson, A.: Proofs that yield nothing but their validity or all languages in NP have zero-knowledge proof systems. Journal of the ACM 38(3), 691–729 (1991)
23. Groth, J., Sahai, A.: Efficient non-interactive proof systems for bilinear groups. In: Smart, N.P. (ed.) EUROCRYPT 2008. LNCS, vol. 4965, pp. 415–432. Springer, Heidelberg (2008)
24. Haralambiev, K.: Efficient Cryptographic Primitives for Non-Interactive Zero-Knowledge Proofs and Applications. Ph.D. thesis, New York University (2011)
25. Hofheinz, D., Müller-Quade, J.: Universally composable commitments using random oracles. In: Naor, M. (ed.) TCC 2004. LNCS, vol. 2951, pp. 58–76. Springer, Heidelberg (2004)
26. Horvitz, O., Katz, J.: Universally-composable two-party computation in two rounds. In: Menezes, A. (ed.) CRYPTO 2007. LNCS, vol. 4622, pp. 111–129. Springer, Heidelberg (2007)
27. Kalai, Y.T.: Smooth projective hashing and two-message oblivious transfer. In: Cramer, R. (ed.) EUROCRYPT 2005. LNCS, vol. 3494, pp. 78–95. Springer, Heidelberg (2005)
28. Katz, J.: Universally composable multi-party computation using tamper-proof hardware. In: Naor, M. (ed.) EUROCRYPT 2007. LNCS, vol. 4515, pp. 115–128. Springer, Heidelberg (2007)
29. Katz, J., Ostrovsky, R., Yung, M.: Efficient password-authenticated key exchange using human-memorable passwords. In: Pfitzmann, B. (ed.) EUROCRYPT 2001. LNCS, vol. 2045, pp. 475–494. Springer, Heidelberg (2001)
30. Katz, J., Vaikuntanathan, V.: Round-optimal password-based authenticated key exchange. In: Ishai, Y. (ed.) TCC 2011. LNCS, vol. 6597, pp. 293–310. Springer, Heidelberg (2011)
31. Naor, M., Pinkas, B.: Efficient oblivious transfer protocols. In: 12th SODA, pp. 448–457. ACM-SIAM (January 2001)
32. Pedersen, T.P.: Non-interactive and information-theoretic secure verifiable secret sharing. In: Feigenbaum, J. (ed.) CRYPTO 1991. LNCS, vol. 576, pp. 129–140. Springer, Heidelberg (1992)
33. Peikert, C., Vaikuntanathan, V., Waters, B.: A framework for efficient and composable oblivious transfer. In: Wagner, D. (ed.) CRYPTO 2008. LNCS, vol. 5157, pp. 554–571. Springer, Heidelberg (2008)
34. Rabin, M.O.: How to exchange secrets with oblivious transfer. Technical Report TR81, Harvard University (1981)

Self-Updatable Encryption: Time Constrained Access Control with Hidden Attributes and Better Efficiency

Kwangsu Lee[1], Seung Geol Choi[2], Dong Hoon Lee[1], Jong Hwan Park[1,3], and Moti Yung[4]

[1] CIST, Korea University, Korea
[2] US Naval Academy, USA
[3] Sangmyung University, Korea
[4] Google Inc. and Columbia University, USA

Abstract. Revocation and key evolving paradigms are central issues in cryptography, and in PKI in particular. A novel concern related to these areas was raised in the recent work of Sahai, Seyalioglu, and Waters (Crypto 2012) who noticed that revoking past keys should at times (e.g., the scenario of cloud storage) be accompanied by revocation of past ciphertexts (to prevent unread ciphertexts from being read by revoked users). They introduced revocable-storage attribute-based encryption (RS-ABE) as a good access control mechanism for cloud storage. RS-ABE protects against the revoked users not only the future data by supporting key-revocation but also the past data by supporting ciphertext-update, through which a ciphertext at time T can be updated to a new ciphertext at time $T+1$ *using only the public key*. Motivated by this pioneering work, we ask whether it is possible to have a modular approach, which includes a primitive for time managed ciphertext update as a primitive. We call encryption which supports this primitive a "self-updatable encryption" (SUE). We then suggest a modular cryptosystems design methodology based on three sub-components: a primary encryption scheme, a key-revocation mechanism, and a time-evolution mechanism which controls the ciphertext self-updating via an SUE method, coordinated with the revocation (when needed). Our goal in this is to allow the self-updating ciphertext component to take part in the design of new and improved cryptosystems and protocols in a flexible fashion. Specifically, we achieve the following results:

- We first introduce a new cryptographic primitive called *self-updatable encryption (SUE)*, realizing a time-evolution mechanism. We also construct an SUE scheme and prove its full security under static assumptions.
- Following our modular approach, we present a new RS-ABE scheme with shorter ciphertexts than that of Sahai et al. and prove its security. The length efficiency is mainly due to our SUE scheme and the underlying modularity.
- We apply our approach to predicate encryption (PE) supporting attribute-hiding property, and obtain a revocable-storage PE (RS-PE) scheme that is selectively-secure.
- We further demonstrate that SUE is of independent interest, by showing it can be used for timed-release encryption (and its applications), and for augmenting key-insulated encryption with forward-secure storage.

Keywords: Public-key encryption, Attribute-based encryption, Predicate encryption, Self-updatable encryption, Revocation, Key evolving systems, Cloud storage.

K. Sako and P. Sarkar (Eds.) ASIACRYPT 2013 Part I, LNCS 8269, pp. 235–254, 2013.

1 Introduction

Cloud data storage has many advantages: A virtually unlimited amount of space can be flexibly allocated with very low costs, and storage management, including back-up and recovery, has never been easier. More importantly, it provides great accessibility: users in any geographic location can access their data through the Internet. However, when an organization is to store *privacy-sensitive data*, existing cloud services do not seem to provide a good security guarantee yet (since the area is in its infancy). In particular, access control is one of the greatest concerns, that is, the sensitive data items have to be protected from any illegal access, whether it comes from outsiders or even from insiders without proper access rights.

One possible approach for this problem is to use attribute-based encryption (ABE) that provides cryptographically enhanced access control functionality in encrypted data [14, 18, 30]. In ABE, each user in the system is issued a private key from an authority that reflects their attributes (or credentials), and each ciphertext specifies access to itself as a boolean formula over a set of attributes. A user will be able to decrypt a ciphertext if the attributes associated with their private key satisfy the boolean formula associated with the ciphertext. To deal with the change of user's credentials that takes place over time, revocable ABE (R-ABE) [3] has been suggested, in which a user's private key can be revoked. In R-ABE, a key generation authority uses broadcast encryption to allow legitimate users to update their keys. Therefore, a revoked user cannot learn any partial information about the messages encrypted when the ciphertext is created after the time of revocation (or after the user's credential has expired).

As pointed out by Sahai, Seyalioglu, and Waters [29], R-ABE alone does not suffice in managing dynamic credentials for cloud storage. In fact, R-ABE cannot prevent *a revoked user from accessing ciphertexts that were created before the revocation*, since the old private key of the revoked user is enough to decrypt these ciphertexts. To overcome this, they introduced a novel revocable-storage ABE (RS-ABE) which solves this issue by supporting not only the revocation functionality but also the ciphertext update functionality such that a ciphertext at any arbitrary time T can be updated to a new ciphertext at time $T+1$ by any party *just using the public key* (in particular, by the cloud servers).

Key-revocation and key evolution are general sub-area in cryptosystems design, and ciphertext-update is a new concern which may be useful elsewhere. So, in this paper, we ask natural questions:

> Can we achieve key-revocation and ciphertext-update in other encryption schemes?
> Can we use ciphertext-update as an underlying primitive by itself?

We note that, in contrast to our questions, the methodology that Sahai et al. [29] used to achieve ciphertext-update is customized to the context of ABE. In particular, they first added ciphertext-delegation to ABE, and then, they *represented time as a set of attributes*, and by doing so they reduced ciphertext-update to ciphertext-delegation.

1.1 Our Results

We address the questions by taking a modular approach, that is, by actually constructing a cryptographic component realizing each of the two functionalities: key revocation and ciphertext update. In particular, our design approach is as follows:

- The overall system has three components: a primary encryption scheme (i.e., ABE or some other encryption scheme), a key-revocation mechanism, and a time-evolution mechanism.
- We combine the components by putting the key-revocation mechanism in the center and connecting it with the other two. This is because the revoked users need to be taken into account both in the decryption of the primary scheme and in the time-evolution of ciphertexts.

There are a few potential benefits to this approach. First, we may be able to achieve key-revocation and time-evolution mechanisms, *independently of the primary encryption scheme*. Secondly, each mechanism may be of independent interest and be used in other interesting scenarios. Thirdly, looking at each mechanism alone may open the door to various optimizations and flexibilities of implementations.

Time-Evolution Mechanism: Self-Updatable Encryption. We first formulate a new cryptographic primitive called *self-updatable encryption (SUE)*, realizing a time-evolution mechanism. In SUE, a ciphertext and a private key are associated with time T_c and T_k respectively. A user who has a private key with time T_k can decrypt the ciphertext with time T_c if $T_c \leq T_k$. Additionally, *anyone can update the ciphertext* with time T_c to a new ciphertext with new time T_c' such that $T_c < T_c'$. We construct an SUE scheme in composite order bilinear groups. In our SUE scheme, a ciphertext consists of $O(\log T_{max})$ group elements, and a private key consists of $O(\log T_{max})$ group elements, where T_{max} is the maximum time period in the system. Our SUE scheme is fully secure under static assumptions by using the dual system encryption technique of Waters [19,31].

RS-ABE with Shorter Ciphertexts. Following the general approach above, we construct a new RS-ABE scheme and prove that it is fully secure under static assumptions. In particular, we take the ciphertext-policy ABE (CP-ABE) scheme of Lewko et al. [18] as the primary encryption scheme, and combine it with our SUE scheme and a revocation mechanism. The revocation mechanism follows the design principle of Boldyreva, Goyal, and Kumar [3] that uses the complete subtree method to securely update the keys of the non-revoked users. Compared with the scheme of Sahai et al. [29], our scheme has a shorter ciphertext length consisting of $O(l + \log T_{max})$ groups elements where l is the size of row in the ABE access structure; a ciphertext in their scheme consists of $O(l \log T_{max} + \log^2 T_{max})$ group elements (reflecting the fact that time is dealt with in a less modular fashion there, while we employ the more separated SUE component which is length efficient).

Revocable-Storage Predicate Encryption. We apply our approach to predicate encryption (PE) and give the first RS-PE scheme. In particular, taking the PE scheme of Park [26] as the primary encryption scheme, we combine it with the same revocation functionality and (a variant of) our SUE scheme. The scheme is in prime-order groups and is shown to be selectively secure (a previously used weaker notion than (full) security, where the adversary selects the target of attack at the start). Obviously, compared with the RS-ABE scheme, the RS-PE scheme is a PE system and, thus, additionally supports the attribute-hiding property: even a decryptor cannot obtain information about the attributes x of a ciphertext except $f(x)$, where f is the predicate of its private key.

Other Systems. These are discussed below in this section.

1.2 Our Technique

To devise our SUE scheme, we use a full binary tree structure to represent time. The idea of using the full binary tree for time was already used by Canetti et al. [8] to construct a forward-secure public-key encryption (FSE) scheme. However, our scheme greatly differs on a technical level from their approach; in our scheme, *a ciphertext is updated* from time T_i to time $T_j > T_i$, whereas in their scheme *a private key is updated* from time T_i to time $T_j > T_i$. We start from the HIBE scheme of Boneh and Boyen [4], and then construct a *ciphertext delegatable encryption (CDE)* scheme, by switching the structure of private keys with that of ciphertexts; our goal is to support ciphertext delegation instead of private key delegation. In CDE, each ciphertext is associated with a tree node, so is each private key. A ciphertext at a tree node v_c can be decrypted by any keys with a tree node v_k where v_k is a descendant (or self) of v_c. We note that the CDE scheme may be of independent interest. The ciphertext delegation property of CDE allows us to construct an SUE scheme. An SUE ciphertext at time T_i consists of multiple CDE ciphertexts in order to support ciphertext-update for every T_j such that $T_j > T_i$. We were able to reduce the number of group elements in the SUE ciphertext *by carefully reusing the randomness of CDE ciphertexts.*

Our key-revocation mechanism, as mentioned above, uses a symmetric-key broadcast encryption scheme to periodically broadcast update keys to non-revoked users. A set of non-revoked users is represented as a node (more exactly the leaves of the subtree rooted at the node) in a tree, following the complete subset (CS) scheme of Naor et al. [22]. So, we use two different trees in this paper, i.e., one for representing time in the ciphertext domain, and the other for managing non-revoked users in the key domain.

In the RS-ABE/RS-PE setting, a user u who has a private key with attributes x and an update key with a revoked set R at time T' can decrypt a ciphertext with a policy f and time T if the attribute satisfies the policy $(f(x) = 1)$ and the user is not revoked $(u \notin R)$, and $T \leq T'$. The main challenge in combining all the components is protecting the overall scheme against a collusion attack, e.g., a non-revoked user with a few attributes should not decrypt more ciphertexts than he is allowed to, given the help of a revoked user with many attributes. To achieve this, we use a secret sharing scheme as suggested in [3]. Roughly speaking, the overall scheme is associated with a secret key α. For each node v_i in the revocation tree, this secret key α is split into γ_i for ABE/PE, and $\alpha - \gamma_i$ for SUE, where γ_i is random. Initially, each user will have some tree nodes v_is according to the revocation mechanism, and get ABE/PE private keys subject to his attributes at each of v_is (associated with the ABE/PE master secret γ_i). In key-update at time T, only non-revoked users receive SUE private keys with time T at a tree node v_j representing a set of non-revoked users (associated with the SUE master secret $\alpha - \gamma_j$).

1.3 Other Applications

Timed-Release Encryption. One application of SUE is timed-release encryption (TRE) and its variants [27, 28]. TRE is a specific type of PKE such that a ciphertext specified with time T can only be decrypted after time T. In TRE, a semi-trusted time server periodically broadcasts a time instant key (TIK) with time T' to all users. A sender creates a ciphertext by specifying time T, and a receiver can decrypt the ciphertext if he has a TIK with time T' such that $T' \geq T$. TRE can be used for electronic auctions, key escrow, on-line gaming, and press releases. TRE and its variants can be realizable by using IBE, certificateless encryption, or forward-secure PKE (FSE) [10, 27]. An SUE scheme can be used for a TRE scheme with augmented properties, since a ciphertext with time T can be decrypted by a private key with time $T' \geq T$ from using the ciphertext update functionality, and, in addition, we have flexibility of having a public ciphertext server which can tune the ciphertext time forward before final public release. In this scheme, a ciphertext consists of $O(\log T_{max})$ and a TIK consists of $O(\log T_{max})$. TRE, in turn, can help in designing synchronized protocols, like fair exchanges in some mediated but protocol-oblivious server model.

Key-Insulated Encryption with Ciphertext Forward Security. SUE can be used to enhance the security of key-insulated encryption (KIE) [12]. KIE is a type of PKE that additionally provides tolerance against key exposures. For a component of KIE, a master secret key MK is stored on a physically secure device, and a temporal key SK_T for time T is stored on an insecure device. At a time period T, a sender encrypts a message with the time T and the public key PK, and then a receiver who obtains SK_T by interacting with the physically secure device can decrypt the ciphertext. KIE provides the security of all time periods except those in which the compromise of temporal keys occurred. KIE can be obtained from IBE. Though KIE provides strong level of security, it does not provide security of ciphertexts available in compromised time periods, even if these ciphertexts are to be read in a future time period. To enhance the security and prevent this premature disclosure, we can build a KIE scheme with forward-secure storage by combining KIE and SUE schemes. Having cryptosystems with key-insulated key and forward-secure storage is different from intrusion-resilient cryptosystems [11].

1.4 Related Work

Attribute-Based Encryption. As mentioned, ABE extends IBE, such that a ciphertext is associated with an attribute x and a private key is associated with an access structure f. When a user has a private key with f, only then he can decrypt a ciphertext with x that satisfies $f(x) = 1$. Sahai and Waters [30] introduced fuzzy IBE (F-IBE) that is a special type of ABE. Goyal et al. [14] proposed a key-policy ABE (KP-ABE) scheme that supports flexible access structures in private keys. Bethencourt et al. [2] proposed a ciphertext-policy ABE (CP-ABE) scheme such that a ciphertext is associated with an access structure f and a private key is associated with an attribute x. After that, numerous ABE schemes with various properties were proposed [9, 18, 20, 25, 32].

Predicate Encryption. PE is also an extension of IBE that additionally provides an attribute-hiding property in ciphertexts: A ciphertext is associated with an attribute x

and a private key is associated with a predicate f. Boneh and Waters [7] introduced the concept of PE and proposed a hidden vector encryption (HVE) scheme that supports conjunctive queries on encrypted data. Katz et al. [15] proposed a PE scheme that supports inner-product queries on encrypted data. After that, many PE schemes with different properties were proposed [17, 23, 24, 26]. Boneh, Sahai, and Waters [6] formalized the concept of functional encryption (FE) by generalizing ABE and PE.

Revocation. Boneh and Franklin [5] proposed a revocation method for IBE that periodically re-issues the private key of users. That is, the identity ID of a user contains time information, and a user cannot obtain a valid private key for new time from a key generation center if he is revoked. However, this method requires for all users to establish secure channels to the server and prove their identities every time. To solve this problem, Boldyreva et al. [3] proposed an R-IBE scheme by combining an F-IBE scheme and a full binary tree structure. Libert and Vergnaud [21] proposed a fully secure R-IBE scheme.

2 Preliminaries

2.1 Notation

We let λ be a security parameter. Let $[n]$ denote the set $\{1,\ldots,n\}$ for $n \in \mathbb{N}$. For a string $L \in \{0,1\}^n$, let $L[i]$ be the ith bit of L, and $L|_i$ be the prefix of L with i-bit length. For example, if $L = 010$, then $L[1] = 0, L[2] = 1, L[3] = 0$, and $L|_1 = 0, L|_2 = 01, L|_3 = 010$. Concatenation of two strings L and L' is denoted by $L||L'$.

2.2 Full Binary Tree

A full binary tree $\mathcal{BT}$ is a tree data structure where each node except the leaf nodes has two child nodes. Let N be the number of leaf nodes in $\mathcal{BT}$. The number of all nodes in $\mathcal{BT}$ is $2N-1$. For any index $0 \leq i < 2N-1$, we denote by v_i a node in $\mathcal{BT}$. We assign the index 0 to the root node and assign other indices to other nodes by using breadth-first search. The depth of a node v_i is the length of the path from the root node to the node. The root node is at depth zero. Siblings are nodes that share the same parent node.

For any node $v_i \in \mathcal{BT}$, L is defined as a label that is a fixed and unique string. The label of each node in the tree is assigned as follows: Each edge in the tree is assigned with 0 or 1 depending on whether the edge is connected to its left or right child node. The label L of a node v_i is defined as the bitstring obtained by reading all the labels of edges in the path from the root node to the node v_i. Note that we assign a special empty string to the root node as a label.

2.3 Subset Cover Framework

The subset cover (SC) framework introduced by Naor, Naor, and Lotspiech [22] is a general methodology for the construction of efficient revocation systems. The SC framework consists of the subset-assigning part and key-assigning part for the subset.

We define the SC scheme by including only the subset-assigning part. The formal definition of SC is given in the full version of this paper [16].

We use the complete subset (CS) scheme proposed by Naor et al. [22] as a building block for our schemes. The CS scheme uses a full binary tree $\mathcal{BT}$ to define the subsets S_i. For any node $v_i \in \mathcal{BT}$, $\mathcal{T}_i$ is defined as a subtree that is rooted at v_i and S_i is defined as the set of leaf nodes in $\mathcal{T}_i$. For the tree $\mathcal{BT}$ and a subset R of leaf nodes, $ST(\mathcal{BT},R)$ is defined as the Steiner Tree induced by the set R and the root node, that is, the minimal subtree of $\mathcal{BT}$ that connects all the leaf nodes in R and the root node. we simply denote $ST(\mathcal{BT},R)$ by $ST(R)$. The CS scheme is described as follows:

CS.Setup(N_{max}): This algorithm takes as input the maximum number of users N_{max}. Let $N_{max} = 2^d$ for simplicity. It first sets a full binary tree $\mathcal{BT}$ of depth d. Each user is assigned to a different leaf node in $\mathcal{BT}$. The collection $\mathcal{S}$ of CS is $\{S_i : v_i \in \mathcal{BT}\}$. Recall that S_i is the set of all the leaves in the subtree $\mathcal{T}_i$. It outputs the full binary tree $\mathcal{BT}$.

CS.Assign($\mathcal{BT},u$): This algorithm takes as input the tree $\mathcal{BT}$ and a user $u \in \mathcal{N}$. Let v_u be the leaf node of $\mathcal{BT}$ that is assigned to the user u. Let $(v_{j_0}, v_{j_1}, \ldots, v_{j_d})$ be the path from the root node $v_{j_0} = v_0$ to the leaf node $v_{j_n} = v_u$. It sets $PV_u = \{S_{j_0}, \ldots, S_{j_d}\}$, and outputs the private set PV_u.

CS.Cover($\mathcal{BT},R$): This algorithm takes as input the tree $\mathcal{BT}$ and a revoked set R of users. It first computes the Steiner tree $ST(R)$. Let $\mathcal{T}_{i_1}, \ldots \mathcal{T}_{i_m}$ be all the subtrees of $\mathcal{BT}$ that hang off $ST(R)$, that is all subtrees whose roots $v_{i_1}, \ldots v_{i_m}$ are not in $ST(R)$ but adjacent to nodes of outdegree 1 in $ST(R)$. It outputs a covering set $CV_R = \{S_{i_1}, \ldots, S_{i_m}\}$.

CS.Match(CV_R, PV_u): This algorithm takes input as a covering set $CV_R = \{S_{i_1}, \ldots, S_{i_m}\}$ and a private set $PV_u = \{S_{j_0}, \ldots, S_{j_d}\}$. It finds a subset S_k such that $S_k \in CV_R$ and $S_k \in PV_u$. If there is such a subset, it outputs (S_k, S_k). Otherwise, it outputs $\perp$.

Lemma 1 ([22]). *Let N_{max} be the number of leaf nodes in a full binary tree and r be the size of a revoked set. In the CS scheme, the size of a private set is $O(\log N_{max})$ and the size of a covering set is at most $r\log(N_{max}/r)$.*

3 Self-Updatable Encryption

3.1 Definitions

Ciphertext Delegatable Encryption (CDE). Before introducing self-updatable encryption, we first introduce ciphertext delegatable encryption. Ciphertext delegatable encryption (CDE) is a special type of public-key encryption (PKE) with the ciphertext delegation property such that a ciphertext can be easily converted to a new ciphertext under a more restricted label string by using public values. The following is the syntax of CDE.

Definition 1 (Ciphertext Delegatable Encryption). *A ciphertext delegatable encryption (CDE) scheme for the set $\mathcal{L}$ of labels consists of seven PPT algorithms* **Init**, **Setup**, **GenKey**, **Encrypt**, **DelegateCT**, **RandCT**, *and* **Decrypt**, *which are defined as follows:*

Init(1^{λ}). *The initialization algorithm takes as input a security parameter* 1^{λ}, *and it outputs a group description string GDS.*

Setup(GDS, d_{max}). *The setup algorithm takes as input a group description string GDS and the maximum length* d_{max} *of the label strings, and it outputs public parameters PP and a master secret key MK.*

GenKey(L, MK, PP). *The key generation algorithm takes as input a label string* $L \in \{0,1\}^k$ *with* $k \leq d_{max}$, *the master secret key MK, and the public parameters PP, and it outputs a private key* SK_L.

Encrypt$(L, s, \mathbf{s}, PP)$. *The encryption algorithm takes as input a label string* $L \in \{0,1\}^d$ *with* $d \leq d_{max}$, *a random exponent s, an exponent vector* $\mathbf{s}$, *and the public parameters PP, and it outputs a ciphertext header* CH_L *and a session key EK.*

DelegateCT(CH_L, c, PP). *The ciphertext delegation algorithm takes as input a ciphertext header* CH_L *for a label string* $L \in \{0,1\}^d$ *with* $d < d_{max}$, *a bit value* $c \in \{0,1\}$, *and the public parameters PP, and it outputs a delegated ciphertext header* $CH_{L'}$ *for the label string* $L' = L||c$.

RandCT$(CH_L, s', \mathbf{s}, PP)$. *The ciphertext randomization algorithm takes as input a ciphertext header* CH_L *for a label string* $L \in \{0,1\}^d$ *with* $d < d_{max}$, *a new random exponent* s', *a new vector* $\mathbf{s}$, *and the public parameters PP, and it outputs a re-randomized ciphertext header* CH'_L *and a partial session key* EK'.

Decrypt$(CH_L, SK_{L'}, PP)$. *The decryption algorithm takes as input a ciphertext header* CH_L, *a private key* $SK_{L'}$, *and the public parameters PP, and it outputs a session key EK or the distinguished symbol* $\perp$.

The correctness property of CDE is defined as follows: For all PP, MK generated by ***Setup***, *all* L, L', *any* $SK_{L'}$ *generated by* ***GenKey***, *any* CH_L *and EK generated by* ***Encrypt*** *or* ***DelegateCT***, *it is required that:*

- *If L is a prefix of* L', *then* ***Decrypt***$(CH_L, SK_{L'}, PP) = EK$.
- *If L is not a prefix of* L', *then* ***Decrypt***$(CH_L, SK_{L'}, PP) = \perp$ *with all but negligible probability.*

Additionally, it requires that the ciphertext distribution of ***RandCT*** *is statistically equal to that of* ***Encrypt***.

Remark 1. The syntax of CDE is different with the usual syntax of encryption since the encryption algorithm additionally takes input random values instead of selecting its own randomness. Because of this difference, we cannot show the security of SUE under the security of CDE, but this syntax difference is essential for the ciphertext efficiency of SUE.

Self-Updatable Encryption (SUE). Self-updatable encryption (SUE) is a new type of PKE with the ciphertext updating property such that a time is associated with private keys and ciphertexts and a ciphertext with a time can be easily updatable to a new ciphertext with a future time. In SUE, the private key of a user is associated with a time T' and a ciphertext is also associated with a time T. If $T \leq T'$, then a user who has a private key with a time T' can decrypt a ciphertext with a time T. That is, a user who has a private key for a time T' can decrypt any ciphertexts attached a past time T such that

$T \leq T'$, but he cannot decrypt a ciphertext attached a future time T such that $T' < T$. Additionally, the SUE scheme has the ciphertext update algorithm that updates the time T of a ciphertext to a new time $T+1$ by using public parameters. The following is the syntax of SUE.

Definition 2 (Self-Updatable Encryption). *A self-updatable encryption (SUE) scheme consists of seven PPT algorithms* ***Init***, ***Setup***, ***GenKey***, ***Encrypt***, ***UpdateCT***, ***RandCT***, *and* ***Decrypt***, *which are defined as follows:*

Init(1^λ). *The initialization algorithm takes as input a security parameter 1^λ, and it outputs a group description string GDS.*

Setup(GDS, T_{max}). *The setup algorithm takes as input a group description string GDS and the maximum time T_{max}, and it outputs public parameters PP and a master secret key MK.*

GenKey(T, MK, PP). *The key generation algorithm takes as input a time T, the master secret key MK, and the public parameters PP, and it outputs a private key SK_T.*

Encrypt(T, s, PP). *The encryption algorithm takes as input a time T, a random value s, and the public parameters PP, and it outputs a ciphertext header CH_T and a session key EK.*

UpdateCT$(CH_T, T+1, PP)$. *The ciphertext update algorithm takes as input a ciphertext header CH_T for a time T, a next time $T+1$, and the public parameters PP, and it outputs an updated ciphertext header CH_{T+1}.*

RandCT(CH_T, s', PP). *The ciphertext randomization algorithm takes as input a ciphertext header CH_T for a time T, a new random exponent s', and the public parameters PP, and it outputs an re-randomized ciphertext header CH'_T and a partial session key EK'.*

Decrypt$(CH_T, SK_{T'}, PP)$. *The decryption algorithm takes as input a ciphertext header CH_T, a private key $SK_{T'}$, and the public parameters PP, and it outputs a session key EK or the distinguished symbol $\perp$.*

The correctness property of SUE is defined as follows: For all PP, MK generated by ***Setup***, *all T, T', any $SK_{T'}$ generated by* ***GenKey***, *and any CH_T and EK generated by* ***Encrypt*** *or* ***UpdateCT***, *it is required that:*

- *If $T \leq T'$, then* $\textbf{Decrypt}(CH_T, SK_{T'}, PP) = EK$.
- *If $T > T'$, then* $\textbf{Decrypt}(CH_T, SK_{T'}, PP) = \perp$ *with all but negligible probability.*

Additionally, it requires that the ciphertext distribution of ***RandCT*** *is statistically equal to that of* ***Encrypt***.

Remark 2. For the definition of SUE, we follow the syntax of key encapsulation mechanisms instead of following that of standard encryption schemes since the session key of SUE serves as the partial share of a real session key in other schemes.

Definition 3 (Security). *The security property for SUE schemes is defined in terms of the indistinguishability under a chosen plaintext attack (IND-CPA). The security game for this property is defined as the following game between a challenger $\mathcal{C}$ and a PPT adversary $\mathcal{A}$:*

1. ***Setup***: *C runs* ***Init*** *and* ***Setup*** *to generate the public parameters PP and the master secret key MK, and it gives PP to* $\mathcal{A}$.
2. ***Query 1***: $\mathcal{A}$ *may adaptively request a polynomial number of private keys for times* $T_1,\ldots,T_{q'}$, *and C gives the corresponding private keys* $SK_{T_1},\ldots,SK_{T_{q'}}$ *to* $\mathcal{A}$ *by running* ***GenKey***(T_i, MK, PP).
3. ***Challenge***: $\mathcal{A}$ *outputs a challenge time* T^* *subject to the following restriction: For all times* $\{T_i\}$ *of private key queries, it is required that* $T_i < T^*$. *C chooses a random bit* $b \in \{0,1\}$ *and computes a ciphertext header* CH^* *and a session key* EK^* *by running* ***Encrypt***(T^*, s, PP). *If* $b = 0$, *then it gives* CH^* *and* EK^* *to* $\mathcal{A}$. *Otherwise, it gives* CH^* *and a random session key to* $\mathcal{A}$.
4. ***Query 2***: $\mathcal{A}$ *may continue to request private keys for additional times* $T_{q'+1},\ldots,T_q$ *subject to the same restriction as before, and C gives the corresponding private keys to* $\mathcal{A}$.
5. ***Guess***: *Finally* $\mathcal{A}$ *outputs a bit* b'.

The advantage of $\mathcal{A}$ *is defined as* $\mathbf{Adv}_{\mathcal{A}}^{SUE}(\lambda) = \left|\Pr[b = b'] - \frac{1}{2}\right|$ *where the probability is taken over all the randomness of the game. A SUE scheme is fully secure under a chosen plaintext attack if for all PPT adversaries* $\mathcal{A}$, *the advantage of* $\mathcal{A}$ *in the above game is negligible in the security parameter* λ.

Remark 3. In the above security game, it is not needed to explicitly describe **UpdateCT** since the adversary can run **UpdateCT** to the challenge ciphertext header by just using *PP*. Note that the use of **UpdateCT** does not violate the security game since the adversary only can request a private key query for T_i such that $T_i < T^*$.

3.2 Bilinear Groups of Composite Order

Let $N = p_1 p_2 p_3$ where p_1, p_2, and p_3 are distinct prime numbers. Let $\mathbb{G}$ and $\mathbb{G}_T$ be two multiplicative cyclic groups of same composite order n and g be a generator of $\mathbb{G}$. The bilinear map $e : \mathbb{G} \times \mathbb{G} \rightarrow \mathbb{G}_T$ has the following properties:

1. Bilinearity: $\forall u, v \in \mathbb{G}$ and $\forall a, b \in \mathbb{Z}_n$, $e(u^a, v^b) = e(u,v)^{ab}$.
2. Non-degeneracy: $\exists g$ such that $e(g,g)$ has order N, that is, $e(g,g)$ is a generator of $\mathbb{G}_T$.

We say that $\mathbb{G}$ is a bilinear group if the group operations in $\mathbb{G}$ and $\mathbb{G}_T$ as well as the bilinear map e are all efficiently computable. Furthermore, we assume that the description of $\mathbb{G}$ and $\mathbb{G}_T$ includes generators of $\mathbb{G}$ and $\mathbb{G}_T$ respectively. We use the notation $\mathbb{G}_{p_i}$ to denote the subgroups of order p_i of $\mathbb{G}$ respectively. Similarly, we use the notation $\mathbb{G}_{T,p_i}$ to denote the subgroups of order p_i of $\mathbb{G}_T$ respectively.

3.3 Complexity Assumptions

We give three static assumptions in bilinear groups of composite order that were introduced by Lewko and Waters [19]. The Assumption 1 (Subgroup Decision), the Assumption 2 (General Subgroup Decision), and the Assumption 3 (Composite Diffie-Hellman) are described in the the full version of this paper [16].

3.4 Design Principle

We use a full binary tree to represent time in our SUE scheme by assigning time periods to all tree nodes instead of assigning time periods to leaf nodes only. The use of binary trees to construct key-evolving schemes dates back to the work of Bellare and Miner [1], and the idea of using all tree nodes to represent time periods was introduced by Canetti, Halevi, and Katz [8]. They used a full binary tree for private key update in forward-secure PKE schemes, but we use the full binary tree for ciphertext update.

In the full binary tree $\mathcal{BT}$, each node v (internal node or leaf node) is assigned a unique time value by using the pre-order tree traversal that recursively visits the root node, the left subtree, and the right subtree. Note that we use breadth-first search for index assignment, but we use pre-order traversal for time assignment. Let $\mathbf{Path}(v)$ be the set of path nodes from the root node to a node v, $\mathbf{RightSibling}(\mathbf{Path}(v))$[1] be the set of right sibling nodes of $\mathbf{Path}(v)$, and $\mathbf{TimeNodes}(v)$ be the set of nodes that consists of v and $\mathbf{RightSibling}(\mathbf{Path}(v))$ excluding the parent's path nodes. That is, $\mathbf{TimeNodes}(v) = \{v\} \cup \mathbf{RightSibling}(\mathbf{Path}(v)) \setminus \mathbf{Path}(\mathbf{Parent}(v))$. Pre-order traversal has the property such that if a node v is associated with time T and a node v' is associated with time T', then we have

$$\mathbf{TimeNodes}(v) \cap \mathbf{Path}(v') \neq \varnothing \text{ if and only if } T \leq T'.$$

Thus if a ciphertext has the delegation property such that it's association can be changed from a node to its child node, then a ciphertext for the time T can be easily delegated to a ciphertext for the time T' such that $T \leq T'$ by providing the ciphertexts of its own and right sibling nodes of path nodes excluding path nodes.

For the construction of an SUE scheme that uses a full binary tree, we need a CDE scheme that has the ciphertext delegation property in the tree such that a ciphertext associated with a node can be converted to another ciphertext associated with its child node. Hierarchical identity-based encryption (HIBE) has the similar delegation property in the tree, but the private keys of HIBE can be delegated [4, 13]. To construct a CDE scheme that supports the ciphertext delegation property, we start from the HIBE scheme of Boneh and Boyen [4] and interchange the private key structure with the ciphertext structure of their HIBE scheme. To use the structure of HIBE, we associate each node with a unique label string $L \in \{0,1\}^*$. The ciphertext delegation property in CDE is easily obtained from the private-key delegation property of HIBE.

To build an SUE scheme from the CDE scheme, we define a mapping function ψ that maps time T to a label L in the tree nodes since these two scheme uses the same full binary tree. The SUE ciphertext for time T consists of all CDE ciphertexts for all nodes in $\mathbf{TimeNodes}(v)$ where time T is associated with a node v. Although the ciphertext of SUE just consists of $O(\log T_{max})$ number of CDE ciphertexts, the ciphertext of SUE can be $O(\log^2 T_{max})$ group elements since the ciphertext of a naive CDE scheme from the HIBE scheme has $O(\log T_{max})$ number of group elements. To improve the efficiency of the ciphertext size, we use the randomness reuse technique for CDE ciphertexts. In this case, we obtain an SUE scheme with $O(\log T_{max})$ group elements in ciphertexts.

[1] Note that we have $\mathbf{RightSibling}(\mathbf{Path}(v)) = \mathbf{RightChild}(\mathbf{Path}(\mathbf{Parent}(v)))$ where $\mathbf{RightChild}(\mathbf{Path}(v))$ be the set of right child nodes of $\mathbf{Path}(v)$ and $\mathbf{Parent}(v)$ be the parent node of v.

3.5 Construction

CDE.Init(1^λ): This algorithm takes as input a security parameter 1^λ. It generates a bilinear group $\mathbb{G}$ of composite order $N = p_1 p_2 p_3$ where p_1, p_2, and p_3 are random primes. It chooses a random generator $g_1 \in \mathbb{G}_{p_1}$ and outputs a group description string as $GDS = ((N, \mathbb{G}, \mathbb{G}_T, e), g_1, p_1, p_2, p_3)$.

CDE.Setup(GDS, d_{max}): This algorithm takes as input the string GDS and the maximum length d_{max} of the label strings. Let $l = d_{max}$. It chooses random elements w, $\{u_{i,0}, u_{i,1}\}_{i=1}^{l}, \{h_{i,0}, h_{i,1}\}_{i=1}^{l} \in \mathbb{G}_{p_1}$, a random exponent $\beta \in \mathbb{Z}_N$, and a random element $Y \in \mathbb{G}_{p_3}$. We define $F_{i,b}(L) = u_{i,b}^{L} h_{i,b}$ where $i \in [l]$ and $b \in \{0,1\}$. It outputs the master secret key $MK = (\beta, Y)$ and the public parameters as

$$PP = \Big((N, \mathbb{G}, \mathbb{G}_T, e), g = g_1,\ w,\ \{u_{i,0}, u_{i,1}\}_{i=1}^{l},\ \{h_{i,0}, h_{i,1}\}_{i=1}^{l},\ \Lambda = e(g,g)^{\beta}\Big).$$

CDE.GenKey(L, MK, PP): This algorithm takes as input a label string $L \in \{0,1\}^n$, the master secret key MK, and the public parameters PP. It first selects a random exponent $r \in \mathbb{Z}_N$ and random elements $Y_0, Y_1, Y_{2,1}, \ldots, Y_{2,n} \in \mathbb{G}_{p_3}$. It outputs a private key as

$$SK_L = \Big(K_0 = g^{\beta} w^{-r} Y_0,\ K_1 = g^r Y_1,\ K_{2,1} = F_{1,L[1]}(L|_1)^r Y_{2,1},\ \ldots,\ K_{2,n} = F_{n,L[n]}(L|_n)^r Y_{2,n}\Big).$$

CDE.Encrypt(L, s, s, PP): This algorithm takes as input a label string $L \in \{0,1\}^d$, a random exponent $s \in \mathbb{Z}_N$, a vector $s = (s_1, \ldots, s_d) \in \mathbb{Z}_N^d$ of random exponents, and PP. It outputs a ciphertext header as

$$CH_L = \Big(C_0 = g^s,\ C_1 = w^s \prod_{i=1}^{d} F_{i,L[i]}(L|_i)^{s_i},\ C_{2,1} = g^{-s_1},\ \ldots,\ C_{2,d} = g^{-s_d}\Big)$$

and a session key as $EK = \Lambda^s$.

CDE.DelegateCT(CH_L, c, PP): This algorithm takes as input a ciphertext header $CH_L = (C_0, \ldots, C_{2,d})$ for a label string $L \in \{0,1\}^d$, a bit value $c \in \{0,1\}$, and PP. It selects a random exponent $s_{d+1} \in \mathbb{Z}_N$ and outputs a delegated ciphertext header for the new label string $L' = L||c$ as

$$CH_{L'} = \Big(C_0,\ C_1' = C_1 \cdot F_{d+1,c}(L')^{s_{d+1}},\ C_{2,1},\ \ldots,\ C_{2,d},\ C_{2,d+1}' = g^{-s_{d+1}}\Big).$$

CDE.RandCT(CH_L, s', s', PP): This algorithm takes as input a ciphertext header $CH_L = (C_0, \ldots, C_{2,d})$ for a label string $L \in \{0,1\}^d$, a new random exponent $s' \in \mathbb{Z}_N$, a new vector $s' = (s_1', \ldots, s_d') \in \mathbb{Z}_N^d$, and PP. It outputs a re-randomized ciphertext header as

$$CH_L' = \Big(C_0' = C_0 \cdot g^{s'},\ C_1' = C_1 \cdot w^{s'} \prod_{i=1}^{d} F_{i,L[i]}(L|_i)^{s_i'},\ C_{2,1}' = C_{2,1} \cdot g^{-s_1'},\ \ldots,$$
$$C_{2,d}' = C_{2,d} \cdot g^{-s_d'}\Big).$$

and a partial session key $EK' = \Lambda^{s'}$ that will be multiplied with the session key EK of CH_L to produce a re-randomized session key.

CDE.Decrypt($CH_L, SK_{L'}, PP$)**:** This algorithm takes as input a ciphertext header CH_L for a label string $L \in \{0,1\}^d$, a private key $SK_{L'}$ for a label string $L' \in \{0,1\}^n$, and PP. If L is a prefix of L', then it computes $CH'_{L'} = (C'_0, \ldots, C'_{2,n})$ by running **DelegateCT** and outputs a session key as $EK = e(C'_0, K_0) \cdot e(C'_1, K_1) \cdot \prod_{i=1}^{n} e(C'_{2,i}, K_{2,i})$. Otherwise, it outputs $\perp$.

Let ψ be a mapping from time T to a label L[2]. Our SUE scheme that uses our CDE scheme as a building block is described as follows:

SUE.Init(1^λ)**:** This algorithm outputs GDS by running **CDE.Init**(1^λ).
SUE.Setup(GDS, T_{max})**:** This algorithm outputs MK and PP by running **CDE.Setup** (GDS, d_{max}) where $T_{max} = 2^{d_{max}+1} - 1$.
SUE.GenKey(T, MK, PP)**:** This algorithm outputs SK_T by running **CDE.GenKey** $(\psi(T), MK, PP)$.
SUE.Encrypt(T, s, PP)**:** This algorithm takes as input a time T, a random exponent $s \in \mathbb{Z}_N$, and PP. It proceeds as follows:

1. It first sets a label string $L \in \{0,1\}^d$ by computing $\psi(T)$. It sets an exponent vector $\mathbf{s} = (s_1, \ldots, s_d)$ by selecting random exponents $s_1, \ldots, s_d \in \mathbb{Z}_N$, and obtains $CH^{(0)}$ by running **CDE.Encrypt**($L, s, \mathbf{s}, PP$).
2. For $1 \le j \le d$, it sets $L^{(j)} = L|_{d-j}||1$ and proceeds the following steps:
   (a) If $L^{(j)} = L|_{d-j+1}$, then it sets $CH^{(j)}$ as an empty one.
   (b) Otherwise, it sets a new exponent vector $\mathbf{s}' = (s'_1, \ldots, s'_{d-j+1})$ where $s'_1, \ldots s'_{d-j}$ are copied from $\mathbf{s}$ and s'_{d-j+1} is randomly selected in $\mathbb{Z}_N$ since $L^{(j)}$ and L have the same prefix string. It obtains $CH^{(j)} = (C'_0, \ldots, C'_{2,d-j+1})$ by running **CDE.Encrypt**($L^{(j)}, s, \mathbf{s}', PP$). It also prunes the redundant elements $C'_0, C'_{2,1}, \ldots, C'_{2,d-j}$ from $CH^{(j)}$, which are already contained in $CH^{(0)}$.
3. It removes all empty $CH^{(j)}$ and sets $CH_T = \left(CH^{(0)}, CH^{(1)}, \ldots, CH^{(d')}\right)$ for some $d' \le d$ that consists of non-empty $CH^{(j)}$.
4. It outputs a ciphertext header as CH_T and a session key as $EK = \Lambda^s$.

SUE.UpdateCT($CH_T, T+1, PP$)**:** This algorithm takes as input a ciphertext header $CH_T = (CH^{(0)}, \ldots, CH^{(d)})$ for a time T, a next time $T+1$, and PP. Let $L^{(j)}$ be the label of $CH^{(j)}$. It proceeds as follows:

1. If the length d of $L^{(0)}$ is less than d_{max}, then it first obtains $CH_{L^{(0)}||0}$ and $CH_{L^{(0)}||1}$ by running **CDE.DelegateCT**($CH^{(0)}, c, PP$) for all $c \in \{0,1\}$ since $CH_{L^{(0)}||0}$ is the ciphertext header for the next time $T+1$ by pre-order traversal. It also prunes the redundant elements in $CH_{L^{(0)}||1}$. It outputs an updated ciphertext header as $CH_{T+1} = \left(CH'^{(0)} = CH_{L^{(0)}||0}, CH'^{(1)} = CH_{L^{(0)}||1}, CH'^{(2)} = CH^{(1)}, \ldots, CH'^{(d+1)} = CH^{(d)}\right)$.

[2] In a full binary tree, each node is associated with a unique time T by the pre-order traversal and a unique label L by the label assignment. Thus there exist a unique mapping function ψ from a time T to a label L.

2. Otherwise, it copies the common elements in $CH^{(0)}$ to $CH^{(1)}$ and simply remove $CH^{(0)}$ since $CH^{(1)}$ is the ciphertext header for the next time $T+1$ by pre-order traversal. It outputs an updated ciphertext header as $CH_{T+1} = \big(CH'^{(0)} = CH^{(1)}, \ldots, CH'^{(d-1)} = CH^{(d)}\big)$.

SUE.RandCT(CH_T, s', PP): This algorithm takes as input a ciphertext header $CH_T = (CH^{(0)}, \ldots, CH^{(d)})$ for a time T, a new random exponent $s' \in \mathbb{Z}_N$, and PP. Let $L^{(j)}$ be the label of $CH^{(j)}$ and $d^{(j)}$ be the length of the label $L^{(j)}$. It proceeds as follows:
1. It first sets a vector $s' = (s'_1, \ldots, s'_{d^{(0)}})$ by selecting random exponents $s'_1, \ldots, s'_{d^{(0)}} \in \mathbb{Z}_N$, and obtains $CH'^{(0)}$ by running **CDE.RandCT**$(CH^{(0)}, s', s', PP)$.
2. For $1 \leq j \leq d$, it sets a new vector $s'' = (s'_1, \ldots, s'_{d^{(j)}})$ where $s'_1, \ldots s'_{d^{(j)}-1}$ are copied from s' and $s'_{d^{(j)}}$ is randomly chosen in $\mathbb{Z}_N$, and obtains $CH'^{(j)}$ by running **CDE.RandCT**$(CH^{(j)}, s', s'', PP)$.
3. It outputs a re-randomized ciphertext header as $CH'_T = \big(CH'^{(0)}, \ldots, CH'^{(d)}\big)$ and a partial session key as $EK' = \Lambda^{s'}$ that will be multiplied with the session key EK of CH_T to produce a re-randomized session key.

SUE.Decrypt($CH_T, SK_{T'}, PP$): This algorithm takes as input a ciphertext header CH_T, a private key $SK_{T'}$, and PP. If $T \leq T'$, then it finds $CH^{(j)}$ from CH_T such that $L^{(j)}$ is a prefix of $L' = \psi(T')$ and outputs EK by running **CDE.Decrypt**$(CH^{(j)}, SK_{T'}, PP)$. Otherwise, it outputs $\perp$.

Remark 4. The ciphertext delegation (or update) algorithm of CDE (or SUE) just outputs a valid ciphertext header. However, we can easily modify it to output a ciphertext header that is identically distributed with that of the encrypt algorithm of CDE (or SUE) by applying the ciphertext randomization algorithm.

3.6 Correctness

In CDE, if the label string L of a ciphertext is a prefix of the label string L' of a private key, then the ciphertext can be changed to a new ciphertext for the label string L' by using the ciphertext delegation algorithm. Thus the correctness of CDE is easily obtained from the following equation.

$$
\begin{aligned}
& e(C_0, K_0) \cdot e(C_1, K_1) \cdot \prod_{i=1}^{n} e(C_{2,i}, K_{2,i}) \\
&= e(g^s, g^\beta w^{-r} Y_0) \cdot e(w^s \prod_{i=1}^{n} F_{i,L[i]}(L|_i)^{s_i}, g^r Y_1) \cdot \prod_{i=1}^{n} e(g^{-s_i}, F_{i,L[i]}(L|_i)^r Y_{2,i}) \\
&= e(g^s, g^\beta) \cdot e(g^s, w^{-r}) \cdot e(w^s, g^r) = e(g,g)^{\beta s}
\end{aligned}
$$

The SUE ciphertext header of a time T consists of the CDE ciphertext headers $CH^{(0)}, CH^{(1)}, \ldots, CH^{(d)}$ that are associated with the nodes in **TimeNodes**(v). If the SUE private key of a time T' associated with a node v' satisfies $T \leq T'$, then we can find a unique node v'' such that **TimeNodes**$(v) \cap$ **Path**$(v') = v''$ since the property of the pre-order tree traversal. Let CH'' be the CDE ciphertext header that is associated with the node v''. The correctness of SUE is easily obtained from the correctness of CDE since the label string L'' of CH'' is a prefix of the label string L' of the private key.

In CDE, the output of **CDE.DelegateCT** is a valid ciphertext header since the function $F_{d+1,c}(L')$ is used with a new random exponent s_{d+1} for the new label string L' with depth $d+1$. The output of **CDE.RandCT** is statistically indistinguishable from that of **CDE.Encrypt** since it has a random exponent $s'' = s + s'$ and a random vector $s'' = (s_1 + s'_1, \ldots, s_d + s'_d)$ where $s, s_1, \ldots, s_d$ are original values in the ciphertext header and $s', s'_1, \ldots, s'_d$ are newly selected random values.

In SUE, the output of **SUE.UpdateCT** is a valid ciphertext header since the output of **CDE.DelegateCT** is a valid ciphertext header and the CDE ciphertext headers $CH^{(0)}, \ldots CH^{(d)}$ are still associated with the nodes in **TimeNodes**(v) where v is a node for the time $T+1$. The output of **SUE.RandCT** is statistically indistinguishable from that of the encryption algorithm since new random exponents $s', s'_1, \ldots, s'_{d^{(0)}}$ are chosen and these random exponents are reused among the CDE ciphertext headers.

3.7 Security Analysis

Theorem 1. *The above SUE scheme is fully secure under a chosen plaintext attack if Assumptions 1, 2, and 3 hold. That is, for any PPT adversary $\mathcal{A}$, we have that $\boldsymbol{Adv}_{\mathcal{A}}^{SUE}(\lambda) \leq \boldsymbol{Adv}_{\mathcal{B}}^{A1}(\lambda) + 2q\boldsymbol{Adv}_{\mathcal{B}}^{A2}(\lambda) + \boldsymbol{Adv}_{\mathcal{B}}^{A3}(\lambda)$ where q is the maximum number of private key queries of $\mathcal{A}$.*

The proof of this theorem is given in the full version of this paper [16].

4 Revocable-Storage Attribute-Based Encryption

4.1 Definitions

Revocable-storage attribute-based encryption (RS-ABE) is attribute-based encryption (ABE) that additionally supports the revocation functionality and the ciphertext update functionality. Boldyreva, Goyal, and Kumar introduced the concept of revocable ABE (R-ABE) that provides the revocation functionality [3], and Sahai, Seyalioglu, and Waters introduced the concept of RS-ABE that provides the ciphertext update functionality in R-ABE [29].

Definition 4 (Revocable-Storage Attribute-Based Encryption). *A revocable-storage (ciphertext-policy) attribute-based encryption (RS-ABE) scheme consists of seven PPT algorithms **Setup**, **GenKey**, **UpdateKey**, **Encrypt**, **UpdateCT**, **RandCT**, and **Decrypt**, which are defined as follows:*

Setup$(1^\lambda, \mathcal{U}, T_{max}, N_{max})$. *The setup algorithm takes as input a security parameter 1^λ, the universe of attributes $\mathcal{U}$, the maximum time T_{max}, and the maximum number of users N_{max}, and it outputs public parameters PP and a master secret key MK.*

GenKey(S, u, MK, PP). *The key generation algorithm takes as input a set of attributes $S \subseteq \mathcal{U}$, a user index $u \in \mathcal{N}$, the master secret key MK, and the public parameters PP, and it outputs a private key $SK_{S,u}$.*

UpdateKey(T, R, MK, PP). *The key update algorithm takes as input a time $T \leq T_{max}$, a set of revoked users $R \subseteq \mathcal{N}$, the master secret key MK, and the public parameters PP, and it outputs an update key $UK_{T,R}$.*

Encrypt$(\mathbb{A}, T, M, PP)$. *The encryption algorithm takes as input an access structure* $\mathbb{A}$, *a time* $T \leq T_{max}$, *a message* M, *and the public parameters* PP, *and it outputs a ciphertext* $CT_{\mathbb{A},T}$.

UpdateCT$(CT_{\mathbb{A},T}, T+1, PP)$. *The ciphertext update algorithm takes as input a ciphertext* $CT_{\mathbb{A},T}$ *for an access structure* $\mathbb{A}$ *and a time* T, *a new time* $T+1$ *such that* $T+1 \leq T_{max}$, *and the public parameters* PP, *and it outputs an updated ciphertext* $CT_{\mathbb{A},T+1}$.

RandCT$(CT_{\mathbb{A},T}, PP)$. *The ciphertext randomization algorithm takes as input a ciphertext* $CT_{\mathbb{A},T}$ *for an access structure* $\mathbb{A}$ *and a time* T, *and the public parameters* PP, *and it outputs a re-randomized ciphertext* $CT'_{\mathbb{A},T}$.

Decrypt$(CT_{\mathbb{A},T}, SK_{S,u}, UK_{T',R}, PP)$. *The decryption algorithm takes as input a ciphertext* $CT_{\mathbb{A},T}$, *a private key* $SK_{S,u}$, *an update key* $UK_{T',R}$, *and the public parameters* PP, *and it outputs a message* M *or the distinguished symbol* $\perp$.

The correctness property of RS-ABE is defined as follows: For all PP,MK generated by ***Setup***, *all* S *and* u, *any* $SK_{S,u}$ *generated by* ***GenKey***, *all* $\mathbb{A}$, T, *and* M, *any* $CT_{\mathbb{A},T}$ *generated by* ***Encrypt*** *or* ***UpdateCT***, *all* T' *and* R, *any* $UK_{T',R}$ *generated by* ***UpdateKey***, *it is required that:*

- *If* $(S \in \mathbb{A}) \wedge (u \notin R) \wedge (T \leq T')$, *then* ***Decrypt***$(CT_{\mathbb{A},T}, SK_{S,u}, UK_{T',R}, PP) = M$.
- *If* $(S \notin \mathbb{A}) \vee (u \in R) \vee (T' < T)$, *then* ***Decrypt***$(CT_{\mathbb{A},T}, SK_{S,u}, UK_{T',R}, PP) = \perp$ *with all but negligible probability.*

Additionally, it requires that the ciphertext distribution of ***RandCT*** *is statistically equal to that of* ***Encrypt***.

Definition 5 (Security). *The security property for RS-ABE is defined in terms of the indistinguishability under a chosen plaintext attack (IND-CPA). The security game for this property is defined as the following game between a challenger* $\mathcal{C}$ *and a PPT adversary* $\mathcal{A}$:

1. ***Setup***: $\mathcal{C}$ *runs* ***Setup*** *to generate the public parameters* PP *and the master secret key* MK, *and it gives* PP *to* $\mathcal{A}$.
2. ***Query 1***: $\mathcal{A}$ *may adaptively request a polynomial number of private keys and update keys.* $\mathcal{C}$ *proceeds as follows:*
   - *If this is a private key query for a set of attributes* S *and a user index* u, *then it gives the corresponding private key* $SK_{S,u}$ *to* $\mathcal{A}$ *by running* ***GenKey***(S, u, MK, PP). *Note that the adversary is allowed to query only one private key for each user* u.
   - *If this is an update key query for an update time* T *and a set of revoked users* R, *then it gives the corresponding update key* $UK_{T,R}$ *to* $\mathcal{A}$ *by running* ***UpdateKey***(T, R, MK, PP). *Note that the adversary is allowed to query only one update key for each time* T.
3. ***Challenge***: $\mathcal{A}$ *outputs a challenge access structure* $\mathbb{A}^*$, *a challenge time* T^*, *and challenge messages* $M_0^*, M_1^* \in \mathcal{M}$ *of equal length subject to the following restriction:*
   - *It is required that* $(S_i \notin \mathbb{A}^*) \vee (u_i \in R_j) \vee (T_j < T^*)$ *for all* $\{(S_i, u_i)\}$ *of private key queries and all* $\{(T_j, R_j)\}$ *of update key queries.*

$\mathcal{C}$ chooses a random bit b and gives the ciphertext CT^* to $\mathcal{A}$ by running ***Encrypt***$(\mathbb{A}^*, T^*, M_b^*, PP)$.

4. ***Query 2**: $\mathcal{A}$ may continue to request private keys and update keys subject to the same restrictions as before, and $\mathcal{C}$ gives the corresponding private keys and update keys to $\mathcal{A}$.*
5. ***Guess**: Finally $\mathcal{A}$ outputs a bit b'.*

The advantage of $\mathcal{A}$ is defined as $\mathbf{Adv}_{\mathcal{A}}^{RS\text{-}ABE}(\lambda) = \left|\Pr[b=b'] - \frac{1}{2}\right|$ *where the probability is taken over all the randomness of the game. A RS-ABE scheme is fully secure under a chosen plaintext attack if for all PPT adversaries $\mathcal{A}$, the advantage of $\mathcal{A}$ in the above game is negligible in the security parameter λ.*

Remark 5. In the above security game, it is not needed to explicitly describe **UpdateCT** since the adversary can run **UpdateCT** to the challenge ciphertext by just using PP. Note that the use of **UpdateCT** does not violate the security game because of the restrictions in the game.

4.2 Construction

For our RS-ABE scheme, we use the (ciphertext-policy) ABE scheme of Lewko et al. [18] as a primary encryption scheme with slight modifications. That is, we use the key encapsulation mechanism version of CP-ABE and the encryption algorithm additionally takes input a random exponent for a session key. The detailed description of CP-ABE is given in the full version of this paper [16]. Our RS-ABE scheme is described as follows:

RS-ABE.Setup$(1^\lambda, \mathcal{U}, T_{max}, N_{max})$**:** This algorithm takes as input a security parameter 1^λ, the universe of attributes $\mathcal{U}$, the maximum time T_{max}, and the maximum number of users N_{max}.

1. It first generates bilinear groups $\mathbb{G}, \mathbb{G}_T$ of composite order $N = p_1p_2p_3$ where p_1, p_2, and p_3 are random primes. Let g_1 be the generator of $\mathbb{G}_{p_1}$. It sets $GDS = ((N, \mathbb{G}, \mathbb{G}_T, e), g_1, p_1, p_2, p_3)$.
2. It obtains MK_{ABE}, PP_{ABE} and MK_{SUE}, PP_{SUE} by running **ABE.Setup**$(GDS, \mathcal{U})$ and **SUE.Setup**(GDS, T_{max}) respectively. It also obtains $\mathcal{BT}$ by running **CS.Setup**(N_{max}) and assigns a random exponent $\gamma_i \in \mathbb{Z}_N$ to each node v_i in $\mathcal{BT}$.
3. It selects a random exponent $\alpha \in \mathbb{Z}_N$, and then it outputs $MK = (MK_{ABE}, MK_{SUE}, \alpha, \mathcal{BT})$ and $PP = \left(PP_{ABE}, PP_{SUE}, g = g_1, \Omega = e(g,g)^\alpha\right)$.

RS-ABE.GenKey(S, u, MK, PP)**:** This algorithm takes as input a set of attributes S, a user index u, $MK = (MK_{ABE}, MK_{SUE}, \alpha, \mathcal{BT})$, and PP.

1. It first obtains $PV_u = \{S_{j_0}, \ldots, S_{j_d}\}$ by running **CS.Assign**$(\mathcal{BT}, u)$ and retrieves $\{\gamma_{j_0}, \ldots, \gamma_{j_d}\}$ from $\mathcal{BT}$ where γ_i is assigned to the node v_i.
2. For $0 \leq k \leq d$, it sets $MK'_{ABE} = (\gamma_{j_k}, Y)$ and obtains $SK_{ABE,k}$ by running **ABE.GenKey**(S, MK'_{ABE}, PP_{ABE}).
3. It outputs $SK_{S,u} = \left(PV_u, SK_{ABE,0}, \ldots, SK_{ABE,d}\right)$.

RS-ABE.UpdateKey(T, R, MK, PP)**:** This algorithm takes as input an update time T, a set of revoked users R, MK, and PP.

1. It first obtains $CV_R = \{S_{i_1}, \dots, S_{i_m}\}$ by running **CS.Cover**$(\mathcal{BT}, R)$ and retrieves $\{\gamma_{i_1}, \dots, \gamma_{i_m}\}$ from $\mathcal{BT}$.
2. For $1 \leq k \leq m$, it sets $MK'_{SUE} = (\alpha - \gamma_{i_k}, Y)$ and obtains $SK_{SUE,k}$ by running **SUE.GenKey**(T, MK'_{SUE}, PP_{SUE}).
3. It outputs $UK_{T,R} = \big(CV_R, SK_{SUE,1}, \dots, SK_{SUE,m}\big)$.

RS-ABE.Encrypt($\mathbb{A}, T, M, PP$): This algorithm takes as input an LSSS access structure $\mathbb{A}$, a time T, a message M, and PP. It selects a random exponent $s \in \mathbb{Z}_N$ and obtains CH_{ABE} and CH_{SUE} by running **ABE.Encrypt**$(\mathbb{A}, s, PP_{ABE})$ and **SUE.Encrypt** (T, s, PP_{SUE}) respectively. It outputs as $CT_{\mathbb{A},T} = \big(CH_{ABE}, CH_{SUE}, C = \Omega^s \cdot M\big)$.

RS-ABE.UpdateCT($CT_{\mathbb{A},T}, T+1, PP$): This algorithm takes as input a ciphertext $CT_{\mathbb{A},T} = (CH_{ABE}, CH_{SUE}, C)$ for an LSSS access structure $\mathbb{A}$ and a time T, a new time $T+1$, and PP. It obtains CH'_{SUE} by running **SUE.UpdateCT**$(CH_{SUE}, T+1, PP_{SUE})$. It outputs $CT_{\mathbb{A},T+1} = \big(CH_{ABE}, CH'_{SUE}, C\big)$.

RS-ABE.RandCT($CT_{\mathbb{A},T}, PP$): This algorithm takes as input a ciphertext $CT_{\mathbb{A},T} = (CH_{ABE}, CH_{SUE}, C)$ and PP. It first selects a random exponent $s' \in \mathbb{Z}_N$. It obtains CH'_{ABE} and CH'_{SUE} by running **ABE.RandCT**(CH_{ABE}, s', PP_{ABE}) and **SUE.RandCT** (CH_{SUE}, s', PP_{SUE}), respectively. It outputs $CT'_{\mathbb{A},T} = \big(CH'_{ABE}, CH'_{SUE}, C' = C \cdot \Omega^{s'}\big)$.

RS-ABE.Decrypt($CT_{\mathbb{A},T}, SK_{S,u}, UK_{T',R}, PP$): This algorithm takes as input a ciphertext $CT_{\mathbb{A},T} = (CH_{ABE}, CH_{SUE}, C)$, a private key $SK_{S,u} = (PV_u, SK_{ABE,0}, \dots, SK_{ABE,d})$, an update key $UK_{T',R} = (CV_R, SK_{SUE,1}, \dots, SK_{SUE,m})$, and PP.
1. If $u \notin R$, then it obtains (S_i, S_j) by running **CS.Match**(CV_R, PV_u). Otherwise, it outputs $\perp$.
2. If $S \in \mathbb{A}$ and $T \leq T'$, then it can obtain EK_{ABE} and EK_{SUE} by running **ABE.Decrypt**$(CH_{ABE}, SK_{ABE,j}, PP_{ABE})$ and **SUE.Decrypt**$(CH_{SUE}, SK_{SUE,i}, PP_{SUE})$ respectively and outputs M by computing $C \cdot \big(EK_{ABE} \cdot EK_{SUE}\big)^{-1}$. Otherwise, it outputs $\perp$.

Remark 6. The ciphertext update algorithm of our scheme just outputs a valid updated ciphertext since a past ciphertext will be erased in most applications. However, the definition of Sahai et al. [29] requires that the output of **UpdateCT** should be equally distributed with that of **Encrypt**. Our scheme also can meet this strong requirement by applying **RandCT** to the output of **UpdateCT**.

Theorem 2. *The above RS-ABE scheme is fully secure under a chosen plaintext attack if Assumptions 1, 2, and 3 hold. That is, for any PPT adversary $\mathcal{A}$, we have that $\boldsymbol{Adv}_{\mathcal{A}}^{RS\text{-}ABE}(\lambda) \leq \boldsymbol{Adv}_{\mathcal{B}}^{A1}(\lambda) + O(q) \cdot \boldsymbol{Adv}_{\mathcal{B}}^{A2}(\lambda) + \boldsymbol{Adv}_{\mathcal{B}}^{A3}(\lambda)$ where q is the maximum number of private key and update key queries of $\mathcal{A}$.*

The proof of this theorem is given in the full version of this paper [16].

4.3 Discussions and RS-PE Results

Efficiency. In our RS-ABE scheme, the number of group elements in a ciphertext is $2l + 3\log T_{max}$ where l is the row size of an access structure. In the RS-ABE scheme of Sahai et al. [29], the number of group elements in a ciphertext is $2\log T_{max} \cdot (l + 2\log T_{max})$ since a piecewise CP-ABE scheme was used.

Revocable-Storage Predicate Encryption. If we use the PE scheme of Park [26] as a primary encryption scheme, then we can build an RS-PE scheme in prime order bilinear groups that additionally supports attribute-hiding property. The definition, construction, and proof of RS-PE are given in the full version of this paper [16].

Theorem 3. *The RS-PE scheme is selectively secure under a chosen plaintext attack if the DBDH and the DLIN assumptions hold. That is, for any PPT adversary $\mathcal{A}$, we have that* $\boldsymbol{Adv}_{\mathcal{A}}^{RS\text{-}PE}(\lambda) \leq 2\boldsymbol{Adv}_{\mathcal{B}}^{DLIN}(\lambda) + \boldsymbol{Adv}_{\mathcal{B}}^{DBDH}(\lambda)$.

Acknowledgements. We thank the anonymous reviewers of ASIACRYPT 2013 for their helpful comments. This work was partially done while Kwangsu Lee and Seung Geol Choi were at Columbia University. Kwangsu Lee was supported by Basic Science Research Program through NRF funded by the Ministry of Education (2013R1A1A2008394). Dong Hoon Lee was supported by Mid-career Researcher Program through NRF grant funded by the MEST (2010-0029121). Jong Hwan Park was supported by Basic Science Research Program through NRF funded by the Ministry of Education (2013R1A1A2009524).

References

1. Bellare, M., Miner, S.K.: A forward-secure digital signature scheme. In: Wiener, M. (ed.) CRYPTO 1999. LNCS, vol. 1666, pp. 431–448. Springer, Heidelberg (1999)
2. Bethencourt, J., Sahai, A., Waters, B.: Ciphertext-policy attribute-based encryption. In: IEEE Symposium on Security and Privacy, pp. 321–334. IEEE Computer Society (2007)
3. Boldyreva, A., Goyal, V., Kumar, V.: Identity-based encryption with efficient revocation. In: Ning, P., Syverson, P.F., Jha, S. (eds.) ACM Conference on Computer and Communications Security, pp. 417–426. ACM (2008)
4. Boneh, D., Boyen, X.: Efficient selective-ID secure identity-based encryption without random oracles. In: Cachin, C., Camenisch, J.L. (eds.) EUROCRYPT 2004. LNCS, vol. 3027, pp. 223–238. Springer, Heidelberg (2004)
5. Boneh, D., Franklin, M.: Identity-based encryption from the weil pairing. In: Kilian, J. (ed.) CRYPTO 2001. LNCS, vol. 2139, pp. 213–229. Springer, Heidelberg (2001)
6. Boneh, D., Sahai, A., Waters, B.: Functional encryption: Definitions and challenges. In: Ishai, Y. (ed.) TCC 2011. LNCS, vol. 6597, pp. 253–273. Springer, Heidelberg (2011)
7. Boneh, D., Waters, B.: Conjunctive, subset, and range queries on encrypted data. In: Vadhan, S.P. (ed.) TCC 2007. LNCS, vol. 4392, pp. 535–554. Springer, Heidelberg (2007)
8. Canetti, R., Halevi, S., Katz, J.: A forward-secure public-key encryption scheme. In: Biham, E. (ed.) EUROCRYPT 2003. LNCS, vol. 2656, pp. 255–271. Springer, Heidelberg (2003)
9. Chase, M.: Multi-authority attribute based encryption. In: Vadhan, S.P. (ed.) TCC 2007. LNCS, vol. 4392, pp. 515–534. Springer, Heidelberg (2007)
10. Chow, S.S.M., Roth, V., Rieffel, E.G.: General certificateless encryption and timed-release encryption. In: Ostrovsky, R., De Prisco, R., Visconti, I. (eds.) SCN 2008. LNCS, vol. 5229, pp. 126–143. Springer, Heidelberg (2008)
11. Dodis, Y., Franklin, M., Katz, J., Miyaji, A.: Intrusion-resilient public-key encryption. In: Joye, M. (ed.) CT-RSA 2003. LNCS, vol. 2612, pp. 19–32. Springer, Heidelberg (2003)
12. Dodis, Y., Katz, J., Xu, S., Yung, M.: Key-insulated public key cryptosystems. In: Knudsen, L.R. (ed.) EUROCRYPT 2002. LNCS, vol. 2332, pp. 65–82. Springer, Heidelberg (2002)

13. Gentry, C., Silverberg, A.: Hierarchical ID-based cryptography. In: Zheng, Y. (ed.) ASIACRYPT 2002. LNCS, vol. 2501, pp. 548–566. Springer, Heidelberg (2002)
14. Goyal, V., Pandey, O., Sahai, A., Waters, B.: Attribute-based encryption for fine-grained access control of encrypted data. In: Juels, A., Wright, R.N., di Vimercati, S.D.C. (eds.) ACM Conference on Computer and Communications Security, pp. 89–98. ACM (2006)
15. Katz, J., Sahai, A., Waters, B.: Predicate encryption supporting disjunctions, polynomial equations, and inner products. In: Smart, N.P. (ed.) EUROCRYPT 2008. LNCS, vol. 4965, pp. 146–162. Springer, Heidelberg (2008)
16. Lee, K., Choi, S.G., Lee, D.H., Park, J.H., Yung, M.: Self-updatable encryption: Time constrained access control with hidden attributes and better efficiency. Cryptology ePrint Archive (2013), `http://eprint.iacr.org/2013/`
17. Lee, K., Lee, D.H.: Improved hidden vector encryption with short ciphertexts and tokens. Des. Codes Cryptography 58(3), 297–319 (2011)
18. Lewko, A., Okamoto, T., Sahai, A., Takashima, K., Waters, B.: Fully secure functional encryption: Attribute-based encryption and (Hierarchical) inner product encryption. In: Gilbert, H. (ed.) EUROCRYPT 2010. LNCS, vol. 6110, pp. 62–91. Springer, Heidelberg (2010)
19. Lewko, A., Waters, B.: New techniques for dual system encryption and fully secure HIBE with short ciphertexts. In: Micciancio, D. (ed.) TCC 2010. LNCS, vol. 5978, pp. 455–479. Springer, Heidelberg (2010)
20. Lewko, A., Waters, B.: Decentralizing attribute-based encryption. In: Paterson, K.G. (ed.) EUROCRYPT 2011. LNCS, vol. 6632, pp. 568–588. Springer, Heidelberg (2011)
21. Libert, B., Vergnaud, D.: Adaptive-ID secure revocable identity-based encryption. In: Fischlin, M. (ed.) CT-RSA 2009. LNCS, vol. 5473, pp. 1–15. Springer, Heidelberg (2009)
22. Naor, D., Naor, M., Lotspiech, J.: Revocation and tracing schemes for stateless receivers. In: Kilian, J. (ed.) CRYPTO 2001. LNCS, vol. 2139, pp. 41–62. Springer, Heidelberg (2001)
23. Okamoto, T., Takashima, K.: Hierarchical predicate encryption for inner-products. In: Matsui, M. (ed.) ASIACRYPT 2009. LNCS, vol. 5912, pp. 214–231. Springer, Heidelberg (2009)
24. Okamoto, T., Takashima, K.: Adaptively attribute-hiding (Hierarchical) inner product encryption. In: Pointcheval, D., Johansson, T. (eds.) EUROCRYPT 2012. LNCS, vol. 7237, pp. 591–608. Springer, Heidelberg (2012)
25. Ostrovsky, R., Sahai, A., Waters, B.: Attribute-based encryption with non-monotonic access structures. In: Ning, P., di Vimercati, S.D.C., Syverson, P.F. (eds.) ACM Conference on Computer and Communications Security, pp. 195–203. ACM (2007)
26. Park, J.H.: Inner-product encryption under standard assumptions. Des. Codes Cryptography 58(3), 235–257 (2011)
27. Paterson, K.G., Quaglia, E.A.: Time-specific encryption. In: Garay, J.A., De Prisco, R. (eds.) SCN 2010. LNCS, vol. 6280, pp. 1–16. Springer, Heidelberg (2010)
28. Rivest, R.L., Shamir, A., Wagner, D.A.: Time-lock puzzles and timed-release crypto. Technical Report MIT/LCS/TR-684 (1996)
29. Sahai, A., Seyalioglu, H., Waters, B.: Dynamic credentials and ciphertext delegation for attribute-based encryption. In: Safavi-Naini, R., Canetti, R. (eds.) CRYPTO 2012. LNCS, vol. 7417, pp. 199–217. Springer, Heidelberg (2012)
30. Sahai, A., Waters, B.: Fuzzy identity-based encryption. In: Cramer, R. (ed.) EUROCRYPT 2005. LNCS, vol. 3494, pp. 457–473. Springer, Heidelberg (2005)
31. Waters, B.: Dual system encryption: Realizing fully secure IBE and HIBE under simple assumptions. In: Halevi, S. (ed.) CRYPTO 2009. LNCS, vol. 5677, pp. 619–636. Springer, Heidelberg (2009)
32. Waters, B.: Ciphertext-policy attribute-based encryption: An expressive, efficient, and provably secure realization. In: Catalano, D., Fazio, N., Gennaro, R., Nicolosi, A. (eds.) PKC 2011. LNCS, vol. 6571, pp. 53–70. Springer, Heidelberg (2011)

Function-Private Subspace-Membership Encryption and Its Applications*

Dan Boneh[1], Ananth Raghunathan[1], and Gil Segev[2,**]

[1] Stanford University
{dabo,ananthr}@cs.stanford.edu
[2] Hebrew University
segev@cs.huji.ac.il

Abstract. Boneh, Raghunathan, and Segev (CRYPTO '13) have recently put forward the notion of *function privacy* and applied it to identity-based encryption, motivated by the need for providing predicate privacy in public-key searchable encryption. Intuitively, their notion asks that decryption keys reveal essentially no information on their corresponding identities, beyond the absolute minimum necessary. While Boneh et al. showed how to construct function-private identity-based encryption (which implies predicate-private encrypted keyword search), searchable encryption typically requires a richer set of predicates.

In this paper we significantly extend the function privacy framework. First, we consider the notion of *subspace-membership* encryption, a generalization of inner-product encryption, and formalize a meaningful and realistic notion for capturing its function privacy. Then, we present a generic construction of a *function-private* subspace-membership encryption scheme based on *any* inner-product encryption scheme. This is the first generic construction that yields a function-private encryption scheme based on a non-function-private one.

Finally, we present various applications of function-private subspace-membership encryption. Among our applications, we significantly improve the function privacy of the identity-based encryption schemes of Boneh et al.: whereas their schemes are function private only for identities that are highly unpredictable (with min-entropy of at least $\lambda + \omega(\log \lambda)$ bits, where λ is the security parameter), we obtain function-private schemes assuming only the *minimal* required unpredictability (i.e., min-entropy of only $\omega(\log \lambda)$ bits). This improvement offers a much more realistic function privacy guarantee.

Keywords: Function privacy, functional encryption.

1 Introduction

Predicate encryption systems [13,23] are public-key schemes where a single public encryption key has *many* corresponding secret keys: every secret key corresponds

* The full version is available as Cryptology ePrint Archive, Report 2013/403 [11].
** This work was done while the author was visiting Stanford University.

K. Sako and P. Sarkar (Eds.) ASIACRYPT 2013 Part I, LNCS 8269, pp. 255–275, 2013.

to a predicate $p : \Sigma \to \{0,1\}$ where Σ is some pre-defined set of indices (or attributes). Plaintext messages are pairs (x, m) where $x \in \Sigma$ and m is in some message space. A secret key sk_p for a predicate p has the following semantics: if c is an encryption of the pair (x, m) then sk_p can be used to decrypt c only if the "index" x satisfies the predicate p. More precisely, attempting to decrypt c using sk_p will output m if $p(x) = 1$ and output $\perp$ otherwise. A predicate encryption system is secure if it provides semantic security for the pair (x, m) even if the adversary has a few benign secret keys.

The simplest example of predicate encryption is a system supporting the set of equality predicates, that is, predicates $p_{\mathsf{id}} : \Sigma \to \{0,1\}$ defined as $p_{\mathsf{id}}(x) = 1$ iff $x = \mathsf{id}$. In such a system there is a secret key $\mathsf{sk}_{\mathsf{id}}$ for every $\mathsf{id} \in \Sigma$ and given the encryption c of a pair (x, m) the key $\mathsf{sk}_{\mathsf{id}}$ can decrypt c and recover m only when $x = \mathsf{id}$. It is easy to see that predicate encryption for the set of equality predicates is the same thing as (anonymous) identity-based encryption [8,1].

Currently the most expressive collusion-resistant predicate encryption systems [23,3] support the family of inner product predicates: for a vector space $\Sigma = \mathbb{F}_q^\ell$ this is the set of predicates $p_v : \Sigma \to \{0,1\}$ where $v \in \Sigma$ and $p_v(x) = 1$ iff $x \perp v$. This family of predicates includes the set of equality predicates and others.

Searching on Encrypted Data. Predicate encryption systems provide a general framework for searching on encrypted data. Consider a mail gateway whose function is to route incoming user email based on characteristics of the email. For example, emails from "boss" that are marked "urgent" are routed to the user's cell phone as are all emails from "spouse." All other emails are routed to the user's desktop. When the emails are transmitted in the clear the gateway's job is straight forward. However, when the emails are encrypted with the user's public key the gateway cannot see data needed for the routing decision. The simplest solution is to give the gateway the user's secret key, but this enables the gateway to decrypt all emails and exposes more information than the gateway needs.

A better solution is to encrypt emails using predicate encryption. The email header functions as the index x and the the routing instructions are used as m. The gateway is given a secret key sk_p corresponding to the "route to cell phone" predicate. This secret key enables the gateway to learn the routing instructions for messages satisfying the predicate p, but learn nothing else about emails.

Function Privacy. A limitation of many existing predicate encryption systems is that the secret key sk_p reveals information about the predicate p. As a result, the gateway, and anyone else who has access to sk_p, learns the predicate p. Since in many practical settings it is important to keep the predicate p secret, our goal is to provide *function privacy*: sk_p should reveal as little information about p as possible.

At first glance it seems that hiding p is impossible: given sk_p the gateway can itself encrypt messages (x, m) and then apply sk_p to the resulting ciphertext. In doing so the gateway learns if $p(x) = 1$ which reveals some information about p. Nevertheless, despite this inherent limitation, function privacy can still be achieved.

Towards a Solution. In recent work Boneh, Raghunathan, and Segev [10] put forward a new notion of function privacy and applied it to identity-based encryption systems (i.e. to predicate encryption supporting equality predicates). They observe that if the identity id is chosen from a distribution with super-logarithmic min-entropy then the inherent limitation above is not a problem since the attacker cannot learn id from $\mathsf{sk}_{\mathsf{id}}$ by a brute force search since there are too many potential identities to test. They define function privacy for IBE systems by requiring that when id has sufficient min-entropy then $\mathsf{sk}_{\mathsf{id}}$ is indistinguishable from a secret key derived for an independently and uniformly distributed identity. This enables function private keyword searching on encrypted data. They then construct several IBE systems supporting function-private keyword searching.

While Boneh et al. [10] showed how to achieve function privacy for equality predicates, encrypted search typically requires a richer set of searching predicates, including conjunctions, disjunctions, and many others. The authors left open the important question of achieving function privacy for a larger family of predicates.

Our Contributions. In this paper we extend the framework and techniques of Boneh et al. [10] for constructing function-private encryption schemes. We put forward a generalization of inner-product predicate encryption [23,18,3], which we denote subspace-membership encryption, and present a definitional framework for capturing its function privacy. Our framework identifies the minimal restrictions under which a strong and meaningful notion of function privacy can be obtained for subspace-membership encryption schemes.

Then, we present a generic construction of a *function-private* subspace membership encryption scheme based on any underlying inner-product encryption scheme (even when the underlying scheme is *not* function private). Our construction is efficient, and in addition to providing function privacy, it preserves the security properties of the underlying scheme. This is the first generic construction that yields a function-private encryption scheme based on a non-function-private one. Recall that even for the simpler case of identity-based encryption, Boneh et al. [10] were not able to provide a generic construction, and had to individually modify various existing schemes.

Finally, we present various applications of function-private subspace membership encryption (we refer the reader to Section 1.1 for an overview of these applications). Among our applications, we significantly improve the function privacy of the identity-based encryption schemes of Boneh et al. [10]. Specifically, whereas their schemes guarantee function privacy only for identity distributions that are highly unpredictable (with min-entropy of at least $\lambda + \omega(\log \lambda)$ bits, where λ is the security parameter), we construct schemes that guarantee function privacy assuming only *minimal* unpredictability (i.e., min-entropy of $\omega(\log \lambda)$ bits). This improvement presents a much more realistic function privacy guarantee.

1.1 Overview of Our Contributions

A subspace-membership encryption scheme is a predicate encryption scheme supporting subspace-membership predicates. That is, an encryption of a message is associated with an attribute $\mathbf{x} \in \mathbb{S}^\ell$, and secret keys are derived for subspaces defined by all vectors in $\mathbb{S}^\ell$ orthogonal to a matrix $\mathbf{W} \in \mathbb{S}^{m \times \ell}$ (for integers $m, \ell \in \mathbb{N}$ and an additive group $\mathbb{S}$).[1] Decryption recovers the message iff $\mathbf{W} \cdot \mathbf{x} = \mathbf{0}$. We refer the reader to [11, Section 2.3] for the standard definitions of the functionality and data security of predicate encryption (following [23,3]).

Function Privacy for Subspace-Membership Encryption. Our goal is to design subspace-membership encryption schemes in which a secret key, $\mathsf{sk}_{\mathbf{W}}$, does not reveal any information, beyond the absolute minimum necessary, on the matrix $\mathbf{W}$. Formalizing a realistic notion of function privacy, however, is not straightforward due to the actual functionality of subspace-membership encryption encryption. Specifically, assuming that an adversary who is given a secret key $\mathsf{sk}_{\mathbf{W}}$ has some a-priori information that the matrix $\mathbf{W}$ belongs to a small set of matrices (e.g., $\{\mathbf{W}_0, \mathbf{W}_1\}$), then the adversary may be able to fully recover $\mathbf{W}$: The adversary simply needs to encrypt a (possibly random) message m for some attribute $\mathbf{x}$ that is orthogonal to $\mathbf{W}_0$ but not to $\mathbf{W}_1$, and then run the decryption algorithm on the given secret key $\mathsf{sk}_{\mathbf{W}}$ and the resulting ciphertext to identify the one that decrypts correctly. In fact, as in [10], as long as the adversary has some a-priori information according to which the matrix $\mathbf{W}$ is sampled from a distribution whose min-entropy is at most logarithmic in the security parameter, there is a non-negligible probability for a full recovery.

In the setting of subspace-membership encryption (unlike that of identity-based encryption [10]), however, the requirement that $\mathbf{W}$ is sampled from a source of high min-entropy does not suffice for obtaining a meaningful notion of function privacy. In Section 2 we show that even if $\mathbf{W}$ has nearly full min-entropy, but two of its columns may be correlated, then a meaningful notion of function privacy is not within reach.

In this light, our notion of function privacy for subspace-encryption schemes focuses on secret key $\mathsf{sk}_{\mathbf{W}}$ for which the columns of $\mathbf{W}$ form a block source. That is, each column of $\mathbf{W}$ should have a reasonable amount of min-entropy even given all previous columns. Our notion of function privacy requires that such a secret key $\mathsf{sk}_{\mathbf{W}}$ (where $\mathbf{W}$ is sampled from an *adversarially*-chosen distribution) be indistinguishable from a secret key for a subspace chosen uniformly at random.

A Function-Private Construction from Inner-product Encryption. Given any underlying inner-product encryption scheme we construction a function-private subspace-membership encryption scheme quite naturally. We modify the key-generation algorithm as follows: for generating a secret key for a subspace described by $\mathbf{W}$, we first sample a uniform $\mathbf{s} \leftarrow \mathbb{S}^m$ and use the key-generation algorithm of the underlying scheme for generating a secret key for the vector $\mathbf{v} = \mathbf{W}^\mathsf{T}\mathbf{s}$. Observe that as long as the columns of $\mathbf{W}$ form a block source, then the leftover

[1] Note that by setting $m = 1$ one obtains the notion of an inner-product encryption scheme [23,18,3].

hash lemma for block sources guarantees that $\mathbf{v}$ is statistically close to uniform. In particular, essentially no information on $\mathbf{W}$ is revealed.

We also observe that extracting from the columns of $\mathbf{W}$ using the same seed for the extractor $\langle \mathbf{s}, \cdot \rangle$ interacts nicely with the subspace-membership functionality. Indeed, if $\mathbf{W} \cdot \mathbf{x} = \mathbf{0}$, it holds that $\mathbf{v}^\mathsf{T}\mathbf{x} = 0$ and vice-versa with high probability. We note that the case where the attribute set is small requires some additional refinement that we omit from this overview, and we refer the reader to Section 3 for more details.

Application 1: Function Privacy When Encrypting to Roots of Polynomials. We consider predicate encryption schemes supporting polynomial evaluation where secret keys correspond to polynomials $p \in \mathbb{S}[\mathsf{X}]$ and messages are encrypted to an attribute $x \in \mathbb{S}$. Given a secret key sk_p and a ciphertext with an attribute x, decryption recovers the message iff $p(x)$ evaluates to 0. Our work constructs such schemes from any underlying subspace-membership scheme.

We also explore the notion of function privacy for such polynomial encryption schemes. We require that secret keys for degree-d polynomials $p(x)$ with coefficients $(p_0, p_1, \ldots, p_d) \in \mathbb{S}^{d+1}$ coming from a sufficiently unpredictable adversarially chosen (joint) distribution be indistinguishable from secret keys for degree-d polynomials where each coefficient is sampled uniformly from the underlying set. Unlike the case of subspace membership, we do not restrict our security to those distributions whose unpredictability holds even when conditioned on all previous (i.e., here we obtain security for any min-entropy source and not only for block sources).

Our function-private construction maps attributes x to Vandermonde vectors $\mathbf{x} = (1, x, x^2, \ldots)$ and a polynomial $p(x)$ to a subspace $\mathbf{W}$ as follows. We sample $d+1$ polynomials $r_1(x), \ldots, r_{d+1}(x)$ in a particular manner (as a product of d uniformly random linear polynomials) and construct the subspace $\mathbf{W}$ whose i^{th} row comprises the coefficients of $p(x) \cdot r_i(x)$. In section 4.1, we elaborate on the details and prove that our choice of randomizing polynomials allows us to show that for polynomials whose coefficients come from an unpredictable distribution, $\mathbf{W}$'s columns have conditional unpredictability. And similarly, for polynomials with uniformly distributed coefficients, $\mathbf{W}$'s columns are uniformly distributed. This allows us to infer the function privacy of the polynomial encryption scheme from the function privacy of the underlying subspace-membership encryption scheme.

Application 2: Function-Private IBE with Minimal Unpredictability. As another interesting application of predicate encryption supporting polynomial evaluation, we consider the question of constructing function-private IBE schemes whose function privacy requires only the minimal necessary unpredictability assumption. It is easy to see (and as was shown in [10]) that if the adversary has some a-priori information according to which identities are sampled from a distribution with only logarithmic bits of entropy, then a simple adversary recovers id from $\mathsf{sk}_{\mathsf{id}}$ with non-negligible probability by simply encrypting a messages to a guessed id and checking if decryption recovers the messages successfully.

Their constructions use a technique of preprocessing the id with a randomness extractor to recover $\mathsf{id}_{\mathsf{Ext}}$ that is statistically close to uniform and thus hides any information about the underlying distribution of identities. As the extracted identity is roughly λ bits long, the distribution of identities must have min-entropy at least $\lambda+\omega(\log\lambda)$ bits to guarantee that extraction works. The identity space is much larger and this is still a meaningful notion of function privacy but the question of designing schemes that require the minimal min-entropy of $\omega(\log\lambda)$ bits was left open.

Starting from encryption schemes supporting polynomial evaluation (for our construction, linear polynomials suffice), this work shows how to construct function-private IBE schemes with the only restriction on identities being that they are unpredictable. We consider identities in a set $\mathbb{S}$ and consider a polynomial $p_{\mathsf{id}}(x) = (x - \mathsf{id})$. By first randomizing the polynomial with uniformly chosen r in $\mathbb{S}$, we observe that if id has the minimal super-logarithmic unpredictability, then the coefficients of the polynomial $r \cdot (x - \mathsf{id})$ have sufficient unpredictability. Thus, considering polynomial encryption schemes where secret keys correspond to such polynomials and attributes correspond to $x = \mathsf{id}$, we construct IBE schemes that are function private against distributions that only have the minimum necessary unpredictability.

1.2 Related Work

As discussed above, the notion of function privacy was recently put forward by Boneh, Raghunathan, and Segev [10]. One of the main motivations of Boneh et al. was that of designing public-key searchable encryption schemes [8,20,1,13,28,23,5,14,2,3] that are keyword private. That is, public-key searchable encryption schemes in which search tokens hide, as much as possible, their corresponding predicates. They presented a framework for modeling function privacy, and constructed various function-private anonymous identity-based encryption schemes (which, in particular, imply public-key keyword-private searchable encryption schemes).

More generally, the work of Boneh et al. initiated the study of function privacy in functional encryption [12,26,6,21,4,19], where a functional secret key sk_f corresponding to a function f enables to compute $f(m)$ given an encryption $c = \mathsf{Enc}_{\mathsf{pk}}(m)$. Intuitively, in this setting function privacy guarantees that a functional secret key sk_f does not reveal information about f beyond what is already known and what can be obtained by running the decryption algorithm on test ciphertexts. In [10], the authors also discuss connections of function privacy to program obfuscation.

Our notion of subspace-membership encryption generalizes that of inner-product encryption introduced by Katz, Sahai, and Waters [23]. They defined and constructed predicate encryption schemes for predicates corresponding to inner products over $\mathbb{Z}_N$ (for some large N). Informally, this class of predicates corresponds to functions $f_{\mathbf{v}}$ where $f_{\mathbf{v}}(\mathbf{x}) = 1$ if and only if $\langle \mathbf{v}, \mathbf{x} \rangle = 0$. Subsequently, Freeman [18] modified their construction to inner products over groups of prime order p, and Agrawal, Freeman, and Vaikuntanathan [3] constructed an inner-product encryption scheme over $\mathbb{Z}_p$ for a small prime p. Other results on inner product encryption study adaptive security [25], delegation in the context of hierarchies [24], and generalized IBE [9].

Finally, we note that function privacy in the symmetric-key setting, where the encryptor and decryptor have a shared secret key, was studied by Shen, Shi, and Waters [27]. They designed a function-private inner-product encryption scheme. As noted by Boneh et al. [10], achieving function privacy in the public-key setting is a more subtle task due to the inherent conflict between privacy and functionality.

1.3 Notation

For an integer $n \in \mathbb{N}$ we denote by $[n]$ the set $\{1, \ldots, n\}$, and by U_n the uniform distribution over the set $\{0,1\}^n$. For a random variable X we denote by $x \leftarrow X$ the process of sampling a value x according to the distribution of X. Similarly, for a finite set S we denote by $x \leftarrow S$ the process of sampling a value x according to the uniform distribution over S. We denote by $\mathbf{x}$ (and sometimes $\boldsymbol{x}$) a vector $(x_1, \ldots, x_{|\mathbf{x}|})$. We denote by $\mathbf{X} = (X_1, \ldots, X_T)$ a joint distribution of T random variables. A non-negative function $f : \mathbb{N} \to \mathbb{R}$ is *negligible* if it vanishes faster than any inverse polynomial. A non-negative function $f : \mathbb{N} \to \mathbb{R}$ is *super-polynomial* if it grows faster than any polynomial.

The *min-entropy* of a random variable X is $\mathbf{H}_\infty(X) = -\log(\max_x \Pr[X = x])$. A *$k$-source* is a random variable X with $\mathbf{H}_\infty(X) \geq k$. A *$(T,k)$-block source* is a random variable $\mathbf{X} = (X_1, \ldots, X_T)$ where for every $i \in [T]$ and $x_1, \ldots, x_{i-1}$ it holds that $\mathbf{H}_\infty(X_i | X_1 = x_1, \ldots, X_{i-1} = x_{i-1}) \geq k$. The *statistical distance* between two random variables X and Y over a finite domain Ω is $\mathbf{SD}(X,Y) = \frac{1}{2}\sum_{\omega \in \Omega} |\Pr[X = \omega] - \Pr[Y = \omega]|$. Two random variables X and Y are *δ-close* if $\mathbf{SD}(X,Y) \leq \delta$. Two distribution ensembles $\{X_\lambda\}_{\lambda \in \mathbb{N}}$ and $\{Y_\lambda\}_{\lambda \in \mathbb{N}}$ are *statistically indistinguishable* if it holds that $\mathbf{SD}(X_\lambda, Y_\lambda)$ is negligible in λ. They are *computationally indistinguishable* if for every probabilistic polynomial-time algorithm $\mathcal{A}$ it holds that $\left|\Pr\left[\mathcal{A}(1^\lambda, x) = 1\right] - \Pr\left[\mathcal{A}(1^\lambda, y) = 1\right]\right|$ is negligible in λ, where $x \leftarrow X_\lambda$ and $y \leftarrow Y_\lambda$.

1.4 Paper Organization

The remainder of this paper is organized as follows. Due to space constraints, we refer the reader to the full version [11, Section 2] for standard definitions and tools. In Section 2 we introduce the notions of subspace-membership encryption and function privacy for subspace-membership encryption. In Section 3 we present a generic construction of a function-private subspace-membership encryption scheme based on any inner-product encryption scheme. In Section 4 we present various applications of function-private subspace-membership encryption. In Section 5 we discuss several open problems that arise from this work.

2 Subspace-Membership Encryption and Its Function Privacy

In this section we formalize the notion of subspace-membership encryption and its function privacy within the framework of Boneh, Raghunathan and Segev [10]. A subspace-membership encryption scheme is a predicate encryption scheme [13,23] supporting the class of predicates $\mathcal{F}$, over an attribute space $\Sigma = \mathbb{S}^\ell$, defined as

$$\mathcal{F} = \left\{ f_{\mathbf{W}} : \mathbf{W} \in \mathbb{S}^{m \times \ell} \right\} \qquad \text{with} \qquad f_{\mathbf{W}}(\mathbf{x}) = \begin{cases} 1 \ \mathbf{W} \cdot \mathbf{x} = \mathbf{0} \in \mathbb{S}^m \\ 0 \text{ otherwise} \end{cases}$$

for integers $m, \ell \in \mathbb{N}$, and an additive group $\mathbb{S}$. Informally, in a subspace-membership encryption, an encryption of a message is associated with an attribute $\mathbf{x} \in \mathbb{S}^\ell$, and secret keys are derived for subspaces defined by all vectors in $\mathbb{S}^\ell$ orthogonal to a matrix $\mathbf{W} \in \mathbb{S}^{m \times \ell}$. Decryption recovers the message if and only if $\mathbf{W} \cdot \mathbf{x} = \mathbf{0}$. (See [11, Section 2.3] for the standard definitions of the functionality and data security of predicate encryption.) Subspace-membership encryption with delegation was also studied in [24,25]. Here we do not need the delegation property.

Based on the framework introduced by Boneh, Raghunathan, and Segev [10], our notion of function privacy for subspace-membership encryption considers adversaries that are given the public parameters of the scheme and can interact with a "real-or-random" function-privacy oracle $\mathsf{RoR}^{\mathsf{FP}}$ defined as follows, and with a key-generation oracle.

Definition 2.1 (Real-or-random function-privacy oracle). *The real-or-random function-privacy oracle* RoR^{FP} *takes as input triplets of the form* $(\mathsf{mode}, \mathsf{msk}, V)$, *where* $\mathsf{mode} \in \{\mathsf{real}, \mathsf{rand}\}$, msk *is a master secret key, and* $V = (V_1, \ldots, V_\ell) \in \mathbb{S}^{m \times \ell}$ *is a circuit representing a joint distribution over* $\mathbb{S}^{m \times \ell}$ *(i.e., each* V_i *is a distribution over* $\mathbb{S}^m$*). If* $\mathsf{mode} = \mathsf{real}$ *then the oracle samples* $\mathbf{W} \leftarrow V$ *and if* $\mathsf{mode} = \mathsf{rand}$ *then the oracle samples* $\mathbf{W} \leftarrow \mathbb{S}^{m \times \ell}$ *uniformly. It then invokes the algorithm* $\mathsf{KeyGen}(\mathsf{msk}, \cdot)$ *on* $\mathbf{W}$ *for outputting a secret key* $\mathsf{sk}_{\mathbf{W}}$.

Definition 2.2 (Function-privacy adversary). *An* (ℓ, k)*-block-source function-privacy adversary* $\mathcal{A}$ *is an algorithm that is given as input a pair* $(1^\lambda, \mathsf{pp})$ *and oracle access to* $\mathsf{RoR}^{FP}(\mathsf{mode}, \mathsf{msk}, \cdot)$ *for some* $\mathsf{mode} \in \{\mathsf{real}, \mathsf{rand}\}$, *and to* $\mathsf{KeyGen}(\mathsf{msk}, \cdot)$. *It is required that each of* $\mathcal{A}$*'s queries to* RoR^{FP} *be an* (ℓ, k)*-block-source.*

Definition 2.3 (Function-private subspace-membership encryption). *A subspace-membership encryption scheme* $\Pi = (\mathsf{Setup}, \mathsf{KeyGen}, \mathsf{Enc}, \mathsf{Dec})$ *is* (ℓ, k)-block-source function private *if for any probabilistic polynomial-time* (ℓ, k)*-block-source function-privacy adversary* $\mathcal{A}$, *there exists a negligible function* $\nu(\lambda)$ *such that*

$$\mathbf{Adv}^{FP}_{\Pi,\mathcal{A}}(\lambda) \stackrel{\mathsf{def}}{=} \left| \Pr\left[\mathsf{Expt}^{\mathsf{real}}_{FP,\Pi,\mathcal{A}}(\lambda) = 1 \right] - \Pr\left[\mathsf{Expt}^{\mathsf{rand}}_{FP,\Pi,\mathcal{A}}(\lambda) = 1 \right] \right| \leq \nu(\lambda),$$

where for each $\mathsf{mode} \in \{\mathsf{real}, \mathsf{rand}\}$ *and* $\lambda \in \mathbb{N}$ *the experiment* $\mathsf{Expt}^{\mathsf{mode}}_{FP,\Pi,\mathcal{A}}(\lambda)$ *is defined as follows:*

1. $(\mathsf{pp}, \mathsf{msk}) \leftarrow \mathsf{Setup}(1^\lambda)$.
2. $b \leftarrow \mathcal{A}^{\mathsf{RoR}^{FP}(\mathsf{mode},\mathsf{msk},\cdot),\mathsf{KeyGen}(\mathsf{msk},\cdot)}(1^\lambda, \mathsf{pp})$.
3. *Output* b.

In addition, such a scheme is statistically (ℓ, k)*-block-source function private if the above holds for all* computationally-unbounded (ℓ, k)*-block-source function-privacy adversary making a polynomial number of queries to the* RoR^{FP} *oracle.*

We note here that a security model that allows the adversary to receive the master secret key msk in place of the oracle $\mathsf{KeyGen}(\mathsf{msk}, \cdot)$ leads to a seemingly stronger notion of function privacy. However, such a notion is subsumed by *statistical* function privacy and the schemes constructed in this paper actually satisfy this stronger notion.

Multi-shot vs. Single-shot Adversaries. Note that Definition 2.3 considers adversaries that query the function-privacy oracle for any polynomial number of times. In fact, as adversaries are also given access to the key-generation oracle, this "multi-shot" definition is polynomially equivalent to its "single-shot" variant in which adversaries query the real-or-random function-privacy oracle $\mathsf{RoR}^{\mathsf{FP}}$ at most once. This is proved via a straightforward hybrid argument, where the hybrids are constructed such that only one query is forwarded to the function-privacy oracle, and all other queries are answered using the key-generation oracle.

The Block-source Requirement on the Columns of W. Our definition of function privacy for subspace-membership encryption requires that a secret key $\mathsf{sk}_{\mathbf{W}}$ reveals no unnecessary information about $\mathbf{W}$ as long as the columns of $\mathbf{W}$ form a block source (i.e., each column is unpredictable even given the previous columns). One might consider a stronger definition, in which the columns of $\mathbf{W}$ may be arbitrarily correlated, as long as each column of $\mathbf{W}$ is sufficiently unpredictable. Such a definition, however, is impossible to satisfy.

Specifically, consider the special case of inner-product encryption (i.e., $m = 1$), and an adversary that queries the real-or-random oracle with a distribution over vectors $\mathbf{w} \in \mathbb{S}^\ell$ defined as follows: sample $\ell - 1$ independent and uniform values $u_1, \ldots, u_{\ell-1} \leftarrow \mathbb{S}$ and output $\mathbf{w} = (u_1, 2u_1, u_2, \ldots, u_{\ell-1})$. Such a distribution clearly has high min-entropy (specifically, $(\ell-1)\log|\mathbb{S}|$ bits), and each coordinate of $\mathbf{w}$ has min-entropy $\log|\mathbb{S}|$ bits. However, secret keys for vectors drawn from this distribution can be easily distinguished from secret keys for vectors drawn from the uniform distribution over $\mathbb{S}^\ell$: encrypt a message M to the attribute $\mathbf{x} = (-2, 1, 0, \ldots, 0) \in \mathbb{S}^\ell$ and check to see if decryption succeeds in recovering M. For a random vector $\mathbf{w} \in \mathbb{S}^\ell$ the decryption succeeds only with probability $1/|\mathbb{S}|$ giving the adversary an overwhelming advantage.

Therefore, restricting function privacy adversaries to query the $\mathsf{RoR}^{\mathsf{FP}}$ oracle only with sources whose columns form block sources is essential for achieving a meaningful notion of function privacy.

On Correlated $\mathsf{RoR}^{\mathsf{FP}}$ Queries. In Definition 2.2 we consider adversaries that receives only a single secret key $\mathsf{sk}_{\mathbf{W}}$ for each query to the $\mathsf{RoR}^{\mathsf{FP}}$ oracle. Our definition easily generalizes to include adversaries that are allowed to query the $\mathsf{RoR}^{\mathsf{FP}}$ oracle with *correlated* queries. More specifically, an adversary can receive secret keys $\mathsf{sk}_{\mathbf{W}_1}, \ldots, \mathsf{sk}_{\mathbf{W}_T}$ for any parameter T that is polynomial in the security parameter. The $\mathsf{RoR}^{\mathsf{FP}}$ oracle samples subspaces $\mathbf{W}_1, \ldots, \mathbf{W}_T$ from an adversarially chosen joint distribution over $\left(\mathbb{S}^{m \times \ell}\right)^T$ with the restriction that for every $1 \leq i \leq T$, the columns of $\mathbf{W}_i$ come from a (ℓ, k)-block-source even conditioned on any fixed values for $\mathbf{W}_1, \ldots, \mathbf{W}_{i-1}$.[2]

Function Privacy of Existing Inner-product Encryption Schemes. The inner-product predicate encryption scheme from lattices [3] is trivially not function private as the secret key includes the corresponding function $f_{\mathbf{v}}$ as part of it (this is necessary for the decryption algorithm to work correctly). The scheme constructed from bilinear groups with composite order [23] however presents no such obvious attack, but we were not able to prove its function privacy based on any standard cryptographic assumption.

3 A Generic Construction Based on Inner-Product Encryption

In this section we present a generic construction of a function-private subspace-membership encryption scheme starting from any inner-product encryption scheme. Due to space constraints, we deal with a large attribute space $\mathbb{S}$ of size super-polynomial in the security parameter λ here, and explain our idea of extending our construction to the case when $|\mathbb{S}|$ is small (see [11, Section 4.2] for the details).

Our Construction. Let $\mathcal{IP} = (\mathsf{IP.Setup}, \mathsf{IP.KeyGen}, \mathsf{IP.Enc}, \mathsf{IP.Dec})$ be an inner-product encryption scheme with attribute set $\Sigma = \mathbb{S}^\ell$. We construct a subspace-membership encryption scheme $\mathcal{SM} = (\mathsf{SM.Setup}, \mathsf{SM.KeyGen}, \mathsf{SM.Enc}, \mathsf{SM.Dec})$ as follows.

- **Setup:** $\mathsf{SM.Setup}$ is identical to $\mathsf{IP.Setup}$. On input the security parameter it outputs public parameters pp and the master secret key msk by running $\mathsf{IP.Setup}(1^\lambda)$.
- **Key generation:** $\mathsf{SM.KeyGen}$ takes as input the master secret key msk and a function $f_{\mathbf{W}}$ where $\mathbf{W} \in \mathbb{S}^{m \times \ell}$ and proceeds as follows. It samples uniform $\mathbf{s} \leftarrow \mathbb{S}^m$ and computes $\mathbf{v} = \mathbf{W}^\intercal \mathbf{s} \in \mathbb{S}^\ell$. Next, it samples a secret key $\mathsf{sk}_{\mathbf{v}} \leftarrow \mathsf{IP.KeyGen}(\mathsf{msk}, \mathbf{v})$ and outputs $\mathsf{sk}_{\mathbf{W}} \stackrel{\text{def}}{=} \mathsf{sk}_{\mathbf{v}}$.
- **Encryption:** $\mathsf{SM.Enc}$ is identical to $\mathsf{IP.Enc}$. On input the public parameters, an attribute $\mathbf{x} \in \mathbb{S}^\ell$, and a message M, the algorithm outputs a ciphertext $c \leftarrow \mathsf{IP.Enc}(\mathsf{pp}, \mathbf{x}, \mathsf{M})$.

[2] Or equivalently, the columns of $[\, \mathbf{W}_1 \mid \mathbf{W}_2 \mid \cdots \mid \mathbf{W}_T \,]$ are distributed according to a $(T\ell, k)$-block-source.

– **Decryption:** SM.Dec is identical to IP.Dec. On input the public parameters pp, a secret key $sk_{\mathbf{W}}$, and a ciphertext c, it outputs $\mathsf{M} \leftarrow \mathsf{IP.Dec}(\mathsf{pp}, \mathsf{sk}_{\mathbf{W}}, c)$.

Correctness. Correctness of the construction follows from the correctness of the underlying inner-product encryption scheme. For every $\mathbf{W} \in \mathbb{S}^{m \times \ell}$ and every $\mathbf{x} \in \mathbb{S}^{\ell}$, it suffices to show the following:

– If $f(I) = 1$, then it holds that $\mathbf{W} \cdot \mathbf{x} = \mathbf{0}$. This implies $\mathbf{x}^{\mathsf{T}}\mathbf{v} = \mathbf{x}^{\mathsf{T}}(\mathbf{W}^{\mathsf{T}}\mathbf{s}) = 0$ and therefore SM.Dec correctly outputs M as required.
– If $f(I) = 0$, then it holds that $\mathbf{e} \stackrel{\mathsf{def}}{=} \mathbf{W} \cdot \mathbf{x} \neq \mathbf{0} \in \mathbb{S}^m$. As $\mathbf{x}^{\mathsf{T}}\mathbf{v} = \mathbf{x}^{\mathsf{T}}(\mathbf{W}^{\mathsf{T}}\mathbf{s}) = \mathbf{e}^{\mathsf{T}}\mathbf{s}$, for any $\mathbf{e} \neq \mathbf{0}$ the quantity $\mathbf{x}^{\mathsf{T}}\mathbf{v}$ is zero with probability $1/|\mathbb{S}|$ over choices of $\mathbf{s}$. As $1/|\mathbb{S}|$ is negligible in λ whenever $|\mathbb{S}|$ is super-polynomial in λ, the proof of correctness follows.

Security. We state the following theorem about the security of our construction.

Theorem 3.1. *If $\mathcal{IP}$ is an attribute hiding (resp. weakly attribute hiding) inner-product encryption scheme for an attribute set $\mathbb{S}$ of size super-polynomial in the security parameter, then it holds that:*

1. *The scheme $\mathcal{SM}$ is an attribute hiding (resp. weakly attribute hiding) subspace-membership encryption scheme under the same assumption as the security of the underlying inner-product encryption scheme.*
2. *The scheme $\mathcal{SM}$ when $m \geq 2$ is statistically function private for (ℓ, k)-block-sources for any $\ell = \mathrm{poly}(\lambda)$ and $k \geq \log|\mathbb{S}| + \omega(\log\lambda)$.*

Proof. We first prove the attribute-hiding property of the scheme, and then prove its function privacy.

Attribute Hiding. Attribute-hiding property of $\mathcal{SM}$ follows from the attribute-hiding property of $\mathcal{IP}$ in a rather straightforward manner. Given a challenger for the attribute-hiding property of $\mathcal{IP}$, an $\mathcal{SM}$ adversary $\mathcal{A}$ can be simulated by algorithm $\mathcal{B}$ as follows: $\mathcal{A}$'s challenge attributes are forwarded to the $\mathcal{IP}$-challenger and the resulting public parameterers are published. Secret key queries can be simulated by first sampling uniform $\mathbf{s} \leftarrow \mathbb{S}^m$, then computing $\mathbf{v} = \mathbf{W}^{\mathsf{T}}\mathbf{s}$ and forwarding $\mathbf{v}$ to the $\mathcal{IP}$ key generation oracle. Similarly, the challenge messages from the adversary are answered by forwarding them to the challenger. In the full version [11, Section 4.1], we elaborate on the details and show that if Q denotes the number of secret key queries by $\mathcal{A}$, it holds that

$$\mathbf{Adv}_{\mathcal{IP},\mathcal{B}}(\lambda) \geq \mathbf{Adv}_{\mathcal{SM},\mathcal{A}}(\lambda) - 2Q/|\mathbb{S}|, \qquad (1)$$

thus completing the proof of the attribute hiding property of $\mathcal{SM}$.

Function Privacy. Let $\mathcal{A}$ be a computationally unbounded (ℓ, k)-block-source function-privacy adversary that makes a polynomial number $Q = Q(\lambda)$ of queries to the $\mathsf{RoR}^{\mathsf{FP}}$ oracle. We prove that the distribution of $\mathcal{A}$'s view in the experiment $\mathsf{Expt}^{\mathsf{real}}_{\mathsf{FP},\mathcal{SM},\mathcal{A}}$ is statistically close to the distribution of $\mathcal{A}$'s view in the experiment $\mathsf{Expt}^{\mathsf{rand}}_{\mathsf{FP},\mathcal{SM},\mathcal{A}}$ (we refer the reader to Definition 2.3 for the descriptions of these experiments). We denote these two distributions by $\mathsf{View}_{\mathsf{real}}$ and $\mathsf{View}_{\mathsf{rand}}$, respectively.

As the adversary $\mathcal{A}$ is computationally unbounded, we assume without loss of generality that $\mathcal{A}$ does not query the $\mathsf{KeyGen}(\mathsf{msk}, \cdot)$ oracle—such queries can be internally simulated by $\mathcal{A}$. Moreover, as discussed in Section 2, it suffices to focus on adversaries $\mathcal{A}$ that query the $\mathsf{RoR}^{\mathsf{FP}}$ oracle exactly once. From this point on we fix the public parameters pp chosen by the setup algorithm, and show that the two distributions $\mathsf{View}_{\mathsf{real}}$ and $\mathsf{View}_{\mathsf{rand}}$ are statistically close for any such pp.

Denote by $V = (V_1, \dots, V_\ell)$ the random variable corresponding to the (ℓ, k)-source with which $\mathcal{A}$ queries the $\mathsf{RoR}^{\mathsf{FP}}$ oracle. For each $i \in [\ell]$, let $(w_{i,1}, \dots, w_{i,m})$ denote a sample from V_i. Also, let $\mathbf{s} = (s_1, \dots, s_m) \in \mathbb{S}^m$. As $\mathcal{A}$ is computationally unbounded, and having fixed the public parameters, we can in fact assume that

$$\mathsf{View}_{\mathsf{mode}} = \left(\left(\sum_{i=1}^{m} s_i \cdot w_{i,1} \right), \dots, \left(\sum_{i=1}^{m} s_i \cdot w_{i,\ell} \right) \right) \tag{2}$$

for $\mathsf{mode} \in \{\mathsf{real}, \mathsf{rand}\}$, where $\mathbf{W} = \{w_{i,j}\}_{i \in [m], j \in [\ell]}$ is drawn from V for $\mathsf{mode} = \mathsf{real}$, $\mathbf{W}$ is uniformly distributed over $\mathbb{S}^{m \times \ell}$ for $\mathsf{mode} = \mathsf{rand}$, and $s_i \leftarrow \mathbb{S}$ for every $i \in [\ell]$. For $\mathsf{mode} \in \{\mathsf{real}, \mathsf{rand}\}$ we prove that the distribution $\mathsf{View}_{\mathsf{mode}}$ is statistically close to a uniform distribution over $\mathbb{S}^m$.

Note that the collection of functions $\{g_{s_1,\dots,s_m} : \mathbb{S}^m \to \mathbb{S}\}_{s_1,\dots,s_m \in \mathbb{S}}$ defined by $g_{s_1,\dots,s_m}(w_1, \dots, w_m) = \sum_{j=1}^{m} s_j \cdot w_j$ is universal. This enables us to directly apply the Leftover Hash Lemma for block-sources [16,22,29,17] implying that for our choice of parameters m, ℓ and k the statistical distance between $\mathsf{View}_{\mathsf{real}}$ and the uniform distribution is negligible in λ.[3] The same clearly holds also for $\mathsf{View}_{\mathsf{rand}}$, as the uniform distribution over $\mathbb{S}^{m \times \ell}$ is, in particular, a (ℓ, k)-block-source. This completes the proof of function privacy. ∎

Theorem 3.1 for correlated $\mathsf{RoR}^{\mathsf{FP}}$ queries. Recollect that the definition of function privacy for subspace membership (Definition 2.3) extends to adversaries that query the $\mathsf{RoR}^{\mathsf{FP}}$ oracle with secret keys for T correlated subspaces $\mathbf{W}_1, \dots, \mathbf{W}_T$ for any $T = \mathrm{poly}(\lambda)$. If the columns of the jointly sampled subspaces $[\mathbf{W}_1 | \mathbf{W}_2 | \cdots | \mathbf{W}_T]$ form a block source, we can extend the proof of function privacy to consider such correlated queries. The adversaries view comprises T terms as in Equation (2) with randomly sampled vectrs $\mathbf{s}_1, \dots, \mathbf{s}_T$ in place of $\mathbf{s}$. The collection of functions g remains universal and a simple variant of the Leftover Hash Lemma implies that for our choice of parameters, the statistical distance between $\mathsf{View}_{\mathsf{real}}$ and the uniform distribution is negligible in λ (and similarly for $\mathsf{View}_{\mathsf{rand}}$).

Dealing with Small Attribute Spaces. We also consider constructing subspace-membership encryption schemes where we do not place any restrictions on the size of the underlying attribute space $\mathbb{S}$. In our generic construction,

[3] We note here that a weaker version of the Leftover Hash Lemma will suffice as the adversary's view does not include $(s_1, \dots, s_m)$.

observe that correctness requires that $1/|\mathbb{S}|$ be negligible in λ. If $|\mathbb{S}|$ is not super-polynomial in the security parameter, then correctness fails with a non-negligible probability. Additionally, this breaks the proof of attribute-hiding security in Theorem 3.1: In Equation (1), if the quantity $2Q/|\mathbb{S}|$ is non-negligible, then a non-negligible advantage of an adversary $\mathcal{A}$ *does not* translate to a non-negligible advantage for the reduction algorithm $\mathcal{B}$ against the inner-product encryption scheme.

To overcome this difficulty, we refine the construction as follows using a parameter $\tau = \tau(\lambda) \in \mathbb{N}$. We split the message into τ secret shares and apply parallel repetition of τ copies of the underlying inner-product encryption scheme, where each copy uses independent public parameters and master secret keys. For the proof of security, it suffices to have τ such that the quantity $\tau/|\mathbb{S}|^\tau$ is negligible in λ. Due to space constraints, a formal description of the scheme and a statement of its security is deferred to [11, Section 4.2].

4 Applications of Function-Private Subspace-Membership Encryption

4.1 Roots of a Polynomial Equation

We can construct a predicate encryption scheme for predicates corresponding to polynomial evaluation. Let $\Phi^{\mathsf{poly}}_{<d} \stackrel{\mathsf{def}}{=} \{f_p : p \in \mathbb{S}[\mathsf{X}], \deg(p) < d\}$, where

$$f_p(x) = \begin{cases} 1 \text{ if } p(x) = 0 \in \mathbb{S} \\ 0 \quad \text{otherwise} \end{cases} \qquad \text{for } x \in \mathbb{S}.$$

Correctness and attribute hiding properties of the predicate encryption scheme for the class of predicates $\Phi^{\mathsf{poly}}_{<d}$ are defined as in the case of a generic predicate encryption scheme in a natural manner (see [11, Definition 2.3]).

Function-Private Polynomial Encryption. For the class $\Phi^{\mathsf{poly}}_{<d}$, consider a real-or-random function privacy oracle $\mathsf{RoR}^{\mathsf{FP}\text{-}\Phi}$ (along the lines of Definition 2.1) that takes as input triplets of the form $(\mathsf{mode}, \mathsf{msk}, \mathbf{P})$, where $\mathsf{mode} \in \{\mathsf{real}, \mathsf{rand}\}$, msk is a master secret key, and $\mathbf{P} = (P_0, \ldots, P_{d-1}) \in \mathbb{S}^d$ is a circuit representing a joint distribution over coefficients of polynomials p with $\deg(p) < d$. If $\mathsf{mode} = \mathsf{real}$ then the oracle samples $p \leftarrow \mathbf{P}$ and if $\mathsf{mode} = \mathsf{rand}$ then the oracle samples $p \leftarrow \mathbb{S}^d$ uniformly. It then invokes the algorithm $\mathsf{KeyGen}(\mathsf{msk}, \cdot)$ on p and outputs secret key sk_p.

Along the lines of Definition 2.2, we consider a k-source $\Phi^{\mathsf{poly}}_{<d}$ function-privacy adversary $\mathcal{A}$. Such an adversary is given inputs $(1^\lambda, \mathsf{pp})$ and oracle access to $\mathsf{RoR}^{\mathsf{FP}\text{-}\Phi}$ and each query to the oracle is a k-source (over the coefficients of the polynomial).

Definition 4.1 ($\Phi^{\mathrm{poly}}_{<d}$ Function privacy). *A predicate encryption scheme for the class of predicates $\Phi^{\mathrm{poly}}_{<d}$ denoted $\Pi = (\mathsf{Setup}, \mathsf{KeyGen}, \mathsf{Enc}, \mathsf{Dec})$ is k-source* function-private *if for any probabilistic polynomial-time k-source $\Phi^{\mathrm{poly}}_{<d}$ function-privacy adversary $\mathcal{A}$, there exists a negligible function $\nu(\lambda)$ such that*

$$\mathbf{Adv}^{\mathsf{FP}\text{-}\Phi}_{\Pi,\mathcal{A}}(\lambda) \stackrel{\mathsf{def}}{=} \left| \Pr\left[\mathsf{Expt}^{\mathsf{real}}_{\mathsf{FP}\text{-}\Phi,\Pi,\mathcal{A}}(\lambda) = 1\right] - \Pr\left[\mathsf{Expt}^{\mathsf{rand}}_{\mathsf{FP}\text{-}\Phi,\Pi,\mathcal{A}}(\lambda) = 1\right] \right| \leq \nu(\lambda),$$

where for each $\mathsf{mode} \in \{\mathsf{real}, \mathsf{rand}\}$ *and* $\lambda \in \mathbb{N}$ *the experiment* $\mathsf{Expt}^{\mathsf{mode}}_{\mathsf{FP}\text{-}\Phi,\Pi,\mathcal{A}}(\lambda)$ *is defined as follows:*

1. $(\mathsf{pp}, \mathsf{msk}) \leftarrow \mathsf{Setup}(1^\lambda)$.
2. $b \leftarrow \mathcal{A}^{\mathsf{RoR}^{\mathsf{FP}\text{-}\Phi}(\mathsf{mode},\mathsf{msk},\cdot),\mathsf{KeyGen}(\mathsf{msk},\cdot)}(1^\lambda, \mathsf{pp})$.
3. *Output b.*

In addition, such a scheme is statistically *k-source function private if the above holds for any* computationally-unbounded *k-source $\Phi^{\mathrm{poly}}_{<d}$ function privacy adversary making a polynomial number of queries to the $\mathsf{RoR}^{\mathsf{FP}\text{-}\Phi}$ oracle.*

Correlated $\mathsf{RoR}^{\mathsf{FP}\text{-}\Phi}$ Queries. Definition 4.1 extends to adversaries that query the $\mathsf{RoR}^{\mathsf{FP}\text{-}\Phi}$ oracle on T correlated queries. A scheme Π is said to be (T, k)-source (resp. (T, k)-block-source) function private if each query $(\mathbf{P}_1, \ldots, \mathbf{P}_T)$ of a joint distribution over T polynomials is a (T, k)-source (resp. (T, k)-block-source).

Constructing Function-Private Predicate Encryption Schemes Supporting Polynomial Evaluation. Given a subspace membership encryption scheme $(\mathsf{Setup}, \mathsf{KeyGen}, \mathsf{Enc}, \mathsf{Dec})$ with parameters $m = d$ and $\ell = 2d - 1$, we can construct a predicate encryption scheme for $\Phi^{\mathrm{poly}}_{<d}$ as follows (for simplicity, we consider the instructive case $d = 3$ and subsequently explain how our technique generalizes):

- **Setup:** The Setup algorithm remains unchanged.
- **Encryption:** To encrypt a message M for the attribute $x \in \mathbb{S}$, the encryption algorithm sets $\mathbf{x} = \left(x^4, x^3, x^2, x, 1\right)^\mathsf{T}$ and outputs the ciphertext $\mathsf{Enc}(\mathsf{pp}, \mathbf{x}, \mathsf{M})$.
- **Key generation:** To generate a secret key corresponding to the polynomial $p = p_2 \cdot x^2 + p_1 \cdot x + p_0$, the key-generation algorithm constructs a vector $\mathbf{p} = (p_2, p_1, p_0)^\mathsf{T} \in \mathbb{S}^3$. Next, it "blinds" the polynomial $p(x)$ with two linear polynomials $r(x) = r_1 \cdot x + r_0$ and $s(x) = s_1 \cdot x + s_0$ and computes the coefficients of the polynomial $p(x) \cdot r(x) \cdot s(x)$. The coefficients r_1, r_0, s_1, s_0 are sampled independently and uniformly at random from $\mathbb{S}$. The key generation algorithm repeats this step with two more sets of polynomials (we refer to them as "randomizing" polynomials) $r'(x), s'(x)$ and $r''(x), s''(x)$ whose coefficients are also sampled uniformly at random. It constructs

$$\mathbf{W} = \begin{bmatrix} \text{— coefficients of } p(x) \cdot r(x) \cdot s(x) \text{ —} \\ \text{— coefficients of } p(x) \cdot r'(x) \cdot s'(x) \text{ —} \\ \text{— coefficients of } p(x) \cdot r''(x) \cdot s''(x) \text{ —} \end{bmatrix} \in \mathbb{S}^{3\times 5}. \quad (3)$$

$$= \begin{bmatrix} p_2r_1s_1 & \begin{matrix} p_2r_1s_0 + p_2r_0s_1 \\ +p_1r_1s_1 \end{matrix} & \begin{matrix} p_2r_0s_0 + p_1r_1s_0 \\ +p_1r_0s_1 + p_0r_1s_1 \end{matrix} & \begin{matrix} p_1r_0s_0 + p_0r_0s_1 \\ +p_0r_1s_0 \end{matrix} & p_0r_0s_0 \\ p_2r'_1s'_1 & \begin{matrix} p_2r'_1s'_0 + p_2r'_0s'_1 \\ +p_1r'_1s'_1 \end{matrix} & \begin{matrix} p_2r'_0s'_0 + p_1r'_1s'_0 \\ +p_1r'_0s'_1 + p_0r'_1s'_1 \end{matrix} & \begin{matrix} p_1r'_0s'_0 + p_0r'_0s'_1 \\ +p_0r'_1s'_0 \end{matrix} & p_0r'_0s'_0 \\ p_2r''_1s_1 & \begin{matrix} p_2r''_1s_0 + p_2r''_0s_1 \\ +p_1r''_1s_1 \end{matrix} & \begin{matrix} p_2r''_0s_0 + p_1r''_1s_0 \\ +p_1r''_0s_1 + p_0r''_1s_1 \end{matrix} & \begin{matrix} p_1r''_0s_0 + p_0r''_0s_1 \\ +p_0r''_1s_0 \end{matrix} & p_0r''_0s_0 \end{bmatrix}.$$

The algorithm then runs $\mathsf{KeyGen}(\mathsf{msk}, \mathbf{W})$ and outputs $\mathsf{sk}_{\mathbf{W}}$.

- **Decryption:** The decryption algorithm remains unchanged.

Correctness and Attribute Hiding. Given a ciphertext c for attribute x and a secret key for polynomial p, if $p(x) = 0$ then it follows that $\mathbf{W} \cdot \mathbf{x} = \mathbf{0}$. If $\mathbf{W} \cdot \mathbf{x} = \mathbf{0}$, then x is a root of polynomials $p \cdot r \cdot s$, $p \cdot r' \cdot s'$, and $p \cdot r'' \cdot s''$ which implies that x is a root of $p(x)$ with overwhelming probability over the choices of polynomials $r, r', r'', s, s', s'' \in \mathbb{S}[\mathsf{X}]$.[4] The attribute hiding property of the scheme follows in a fairly straightforward manner from the attribute hiding property of the subspace membership encryption scheme.

Function Privacy. We show that with overwhelming probability over the choices of the randomizing polynomials: (a) if the coefficients of p, namely (p_2, p_1, p_0) are sampled from a k-source, then $\mathbf{W}$ is distributed according to a $(5, k)$-block source, and (b) if the coefficients of p are sampled uniformly at random from $\mathbb{S}^3$, then $\mathbf{W}$ is distributed uniformly over $\mathbb{S}^{3\times 5}$. Given the above two claims, a straightforward reduction allows us to simulate a $\mathsf{RoR}^{\mathsf{FP}\text{-}\Phi}$ oracle given access to a RoR oracle for the subspace membership predicate with parameters $m = 3$ and $\ell = 5$. Thus, we can state the following theorem.

Theorem 4.2. *If $\mathcal{SM}$ is a subspace membership encryption scheme with parameters $m = 3$ and $\ell = 5$ that satisfies function privacy against $(5, k)$-block-source adversaries, then the predicate encryption scheme for the class of predicates $\Phi^{\mathsf{poly}}_{<3}$ constructed above is statistically function private against k-source adversaries.*

Applying Theorem 3.1 for adversaries that query the $\mathsf{RoR}^{\mathsf{FP}}$ oracle with T correlated queries immediately gives us the following corollary.

Corollary 4.3. *Given any large attribute space inner-product encryption scheme with $\ell = 3$, there exists a predicate encryption scheme for the class of predicates $\Phi^{\mathsf{poly}}_{<3}$ that is statistically function-private against (T, k)-block-sources for any $T = \mathsf{poly}(\lambda)$ and $k \geq \log|\mathbb{S}| + \omega(\log \lambda)$.*

---

[4] From a simple union bound over the events where three linear polynomials share a root, this probability works out to be $\geq 1 - 8/|\mathbb{S}|^2$ which is indeed overwhelming.

Proof of Claims (a) and (b). Consider column $\mathbf{w}_1 = (p_2 r_1 s_1, p_2 r'_1 s'_1, p_2 r''_1 s''_1)^\mathsf{T}$. We observe that over choices of s_1, s'_1, and s''_1, the column $\mathbf{w}_1$ is distributed uniformly over $\mathbb{S}^3$. The second column $\mathbf{w}_2$ is also distributed uniformly at random by noting that the elements $p_2 r_1 s_0$, $p_2 r'_1 s'_0$, and $p_2 r''_1 s''_0$ are distributed uniformly in $\mathbb{S}^3$ over choices of r_1, r'_1, and r''_1 (which are themselves information theoretically hidden in $\mathbf{w}_1$). An identical argument shows that over choices of r_0, r'_0, and r''_0, and s_0, s'_0, and s''_0, the fourth and fifth columns, $\mathbf{w}_4$ and $\mathbf{w}_5$, are distributed uniformly in $\mathbb{S}^3$. This is true even conditioned on all the other columns. It suffices to show that conditioned on $\mathbf{w}_1$, $\mathbf{w}_2$, $\mathbf{w}_4$, and $\mathbf{w}_5$, column $\mathbf{w}_3$ has entropy at least $\log|\mathbb{S}| + \omega(\log \lambda)$.

We re-write $\mathbf{w}_3$ as $\mathbf{R} \cdot \mathbf{p}$ where

$$\mathbf{R} = \begin{bmatrix} r_0 s_0 & r_1 s_0 + r_0 s_1 & r_1 s_1 \\ r'_0 s'_0 & r'_1 s'_0 + r'_0 s'_1 & r'_1 s'_1 \\ r''_0 s''_0 & r''_1 s''_0 + r''_0 s''_1 & r''_1 s''_1 \end{bmatrix} \in \mathbb{S}^{3 \times 3}. \tag{4}$$

With overwhelming probability over random choices of all the coefficients in the polynomials r, s, r', s', r'', and s'', the matrix $\mathbf{R}$ is full-rank over $\mathbb{S}$. Therefore, the distribution of $\mathbf{w}_3$ has a one-one correspondence with the distribution of $\mathbf{p}$. Therefore, $\mathbf{w}_3$ has entropy at least k even given $\mathbf{R}$ if p is sampled from a k-source and $\mathbf{w}_3$ is uniform over $\mathbb{S}^3$ even given $\mathbf{R}$ if p is sampled uniformly from $\mathbb{S}^3$. This concludes the proof of claims (a) and (b). ∎

A General Technique for $\Phi^{\text{poly}}_{<d}$. As stated earlier, we can construct predicate encryption for the class of predicates $\Phi^{\text{poly}}_{<d}$ starting with a subspace membership encryption scheme with parameters $m = d$ and $\ell = 2d - 1$. The main idea in extending beyond $d = 3$ is to construct d randomized "blindings" of $p(x)$. For $i \in [d]$, the ith row of $\mathbf{W}$ now comprises coefficients of a polynomial $p(x) \cdot r_{i,1}(x) \cdots r_{i,d-1}(x)$ where each of the $r_{i,j}(x)$'s are random linear polynomials sampled as $r(x)$ and $s(x)$ are sampled in the $d = 3$ construction. The details of our construction are as follows. Due to space constraints the details about the construction are deferred to the full version [11, Section 5.1].

Comparing Entropy Requirements. In Definition 4.1 and Corollary 4.3 it suffices to consider function-privacy adversaries that query the "real-or-random" oracle with polynomials whose coefficients come from a k-source. We *do not* require the sources have conditional min-entropy in contrast to subspace membership function privacy (see Definition 2.3 and the discussion in Section 2). The reason this weaker restriction on $\Phi^{\text{poly}}_{<d}$ function-privacy adversaries suffices when it does not suffice against subspace membership function-privacy adversaries is that the class of predicates $\Phi^{\text{poly}}_{<d}$ offers a weaker functionality than is offered by subspace membership. In particular, if the adversary evaluates ciphertexts with attributes corresponding to "ill-formed" non-Vandermonde vectors, i.e., vectors not of the form $(1, x, x^2, \ldots)$, correctness of decryption is not guaranteed and the particular attack outlined in Section 2 fails. It is easy to see this in our construction as well—the randomizing polynomials ensure correctness only holds when the subspace membership predicate is evaluated on Vandermonde vectors.

4.2 Function-Private IBE with Minimal Unpredictability

As discussed in Section 1.1, the IBE schemes of Boneh et al. [10] are function private only for identity distributions with min-entropy at least $\lambda + \omega(\log \lambda)$. However, the only inherent restriction required for a meaningful notion of security is that identity distributions have min-entropy $\omega(\log \lambda)$. In this section, starting with predicate encryption schemes for polynomial evaluation constructed in Section 4.1, we construct an IBE scheme satisfying function privacy with only a super-logarithmic min-entropy restriction on identity distributions.

Scheme. Consider a predicate encryption scheme for the class of *linear* predicates $\Phi^{\text{poly}}_{<2}$ comprising algorithms $(\mathsf{Setup}, \mathsf{KeyGen}, \mathsf{Enc}, \mathsf{Dec})$. From Section 4.1, such a predicate encryption scheme can be built from any underlying subspace membership scheme for parameters $m = 2$ and $\ell = 3$. Given such a scheme, we construct an IBE scheme $\mathcal{IBE}^{\mathsf{OPT}}$ for the space of identities $\mathbb{S}$ as follows.

- **Setup:** On input 1^λ, the IBE setup algorithm runs $\mathsf{Setup}(1^\lambda)$ to receive $(\mathsf{pp}, \mathsf{msk})$ and publishes pp.
- **Key generation:** On input msk and an identity $\mathsf{id} \in \mathbb{S}$, the key generation algorithm constructs a (randomized) polynomial $p_{\mathsf{id}}(x)$ such that $p_{\mathsf{id}}(x) = 0$ if and only if $x = \mathsf{id}$ as follows. The algorithm samples uniform $r \leftarrow \mathbb{S}$ and computes $p_{\mathsf{id}}(x) = r(x - \mathsf{id})$. It then runs the underlying KeyGen algorithm to output $\mathsf{sk}_{\mathsf{id}} \leftarrow \mathsf{KeyGen}(\mathsf{msk}, p_{\mathsf{id}})$.
- **Encryption:** On input pp, an identity id, and a message M, the encryption algorithm computes $\mathsf{Enc}(\mathsf{pp}, \mathsf{id}, \mathsf{M})$.
- **Decryption:** On input pp, a ciphertext c, and a secret key sk, the decryption algorithm simply computes the underlying decryption algorithm to output $\mathsf{M} \leftarrow \mathsf{Dec}(\mathsf{pp}, \mathsf{sk}, c)$.

Correctness of the IBE scheme follows from the correctness of the underlying $\Phi^{\text{poly}}_{<2}$-predicate encryption scheme. Data privacy and anonymity of the IBE scheme (see [11, Definition 2.5]) follows directly from the attribute hiding property of the underlying $\Phi^{\text{poly}}_{<2}$-predicate encryption scheme. In the theorem that follows, we prove that $\mathcal{IBE}^{\mathsf{OPT}}$ is function-private against minimally unpredictable sources.

Theorem 4.4. *Given any large attribute space inner-product encryption scheme for dimension $\ell = 3$, there exists an IBE scheme function private against (T, k)-block-sources for any $T = \text{poly}(\lambda)$ and $k \geq \omega(\log \lambda)$.*

Proof Outline. For simplicity, consider adversaries that query the real-or-random oracle with k-sources (i.e., $T = 1$). As outlined in Section 4.1 we first construct a predicate encryption scheme for $\Phi^{\text{poly}}_{<2}$ that is function private against k'-sources for $k' \geq \log |\mathbb{S}| + \omega(\log \lambda)$. We instantiate $\mathcal{IBE}^{\mathsf{OPT}}$ described above with this predicate encryption scheme.

The proof proceeds by showing that $\mathsf{RoR}^{\mathsf{FP\text{-}IBE}}$ queries (see [11, Definition 2.6]) ID can be compiled to distributions over coefficients of linear polynomials

$\mathbf{P} = (P_1, P_0)$ such that if $\mathbf{H}_\infty(ID) = k$, then $\mathbf{H}_\infty(\mathbf{P}) = k + \log|\mathbb{S}|$. This allows us to simulate a $\mathsf{RoR}^{\mathsf{FP\text{-}IBE}}$ oracle given an oracle $\mathsf{RoR}^{\mathsf{FP}\text{-}\Phi}$ for linear polynomials thus showing that $\mathcal{IBE}^{\mathsf{OPT}}$ is function-private against k-sources if the encryption scheme for $\Phi^{\mathsf{poly}}_{<2}$ is function-private against k'-sources. Due to space constraints, the reader is referred to the full version for details [11, Section 5.1].

Fully-Secure Function-Private IBE. Current constructions of inner-product encryption schemes [23,3] satisfy a selective notion of security where the challenge attributes are chosen by the adversary before seeing the public parameters. Our transformation of inner-product encryption schemes to function-private IBE schemes with minimal unpredictability is not limited to selective security. Starting from an inner-product encryption scheme satisfying an adaptive version of attribute hiding, we can construct fully-secure IBE schemes. We also note that the standard complexity leveraging approach (see [7, Section 7.1]) gives a generic transformation from selectively-secure IBE to fully-secure IBE. This approach does not modify the key generation algorithm and therefore preserves function privacy.

5 Conclusions and Open Problems

Our work proposes subspace-membership encryption and constructs the first such function-private schemes from any inner-product encryption scheme. We also show its application to constructing function-private polynomial encryption schemes and function-private IBE schemes with minimal unpredictability. In this section, we discuss a few extensions and open problems that arise from this work.

Function Privacy from Computational Assumptions. In this work we construct subspace-membership schemes that are *statistically* function private. Although the construction of inner-product encryption schemes from lattices [3] presents an immediate function-privacy attack, we were unable to find such attacks for the construction from composite-order groups [23] (or its prime order variant [18]). We conjecture that suitable "min-entropy" variants of the decisional Diffie-Hellman assumption [15] have a potential for yielding a proof of computational function privacy for these schemes.

Other Predicates. A pre-cursor to the work on predicate encryption supporting inner-products was work on predicate encryption supporting comparison and range queries by Boneh and Waters [13]. They achieve this by constructing predicate encryption supporting an interesting primitive, denoted Hidden-Vector Encryption (HVE). Briefly, in HVE, attributes correspond to vectors over an alphabet Σ and secret keys correspond to vectors over the *augmented* alphabet $\Sigma \cup \{\star\}$. Decryption works if the attributes and secret key match for every coordinate that is not a $\star$.

HVE can be implemented using inner-product encryption schemes [23] but it breaks function privacy in a rather trivial manner. Formalizing function privacy for HVE does not immediately follow from the notion of function privacy for inner-products because of the role played by $\star$. The questions of formalizing function privacy (which in turn will imply realistic notions also for encryption supporting range and comparison queries) and designing function-private HVE schemes are left as open problems. It is also open to formalize security and design function-private encryption schemes that support multivariate polynomial evaluation.

Enhanced Function Privacy. A stronger notion of function privacy, denoted enhanced function privacy [10], asks that an adversary learn nothing more than the minimum necessary from a secret key even given corresponding ciphertexts with attributes that allow successful decryption. Constructing enhanced function-private schemes for subspace membership and inner products is an interesting line of research that may require new ideas and techniques.

Acknowledgments. This work was supported by NSF, the DARPA PROCEED program, an AFO SR MURI award, a grant from ONR, an IARPA project provided via DoI/NBC, and by Samsung. Opinions, findings and conclusions or recommendations expressed in this material are those of the author(s) and do not necessarily reflect the views of DARPA or IARPA.

References

1. Abdalla, M., Bellare, M., Catalano, D., Kiltz, E., Kohno, T., Lange, T., Malone-Lee, J., Neven, G., Paillier, P., Shi, H.: Searchable encryption revisited: Consistency properties, relation to anonymous IBE, and extensions. Journal of Cryptology 21(3), 350–391 (2008)
2. Abdalla, M., Bellare, M., Neven, G.: Robust encryption. In: Proceedings of the 7th Theory of Cryptography Conference, pp. 480–497 (2010)
3. Agrawal, S., Freeman, D.M., Vaikuntanathan, V.: Functional encryption for inner product predicates from learning with errors. In: Lee, D.H., Wang, X. (eds.) ASIACRYPT 2011. LNCS, vol. 7073, pp. 21–40. Springer, Heidelberg (2011)
4. Agrawal, S., Gorbunov, S., Vaikuntanathan, V., Wee, H.: Functional encryption: New perspectives and lower bounds. In: Canetti, R., Garay, J.A. (eds.) CRYPTO 2013, Part II. LNCS, vol. 8043, pp. 500–518. Springer, Heidelberg (2013)
5. Baek, J., Safavi-Naini, R., Susilo, W.: Public key encryption with keyword search revisited. In: Gervasi, O., Murgante, B., Laganà, A., Taniar, D., Mun, Y., Gavrilova, M.L. (eds.) ICCSA 2008, Part I. LNCS, vol. 5072, pp. 1249–1259. Springer, Heidelberg (2008)
6. Bellare, M., O'Neill, A.: Semantically-secure functional encryption: Possibility results, impossibility results and the quest for a general definition. Cryptology ePrint Archive, Report 2012/515 (2012)
7. Boneh, D., Boyen, X.: Efficient selective identity-based encryption without random oracles. Journal of Cryptology 24(4), 659–693 (2011)
8. Boneh, D., Di Crescenzo, G., Ostrovsky, R., Persiano, G.: Public key encryption with keyword search. In: Cachin, C., Camenisch, J.L. (eds.) EUROCRYPT 2004. LNCS, vol. 3027, pp. 506–522. Springer, Heidelberg (2004)
9. Boneh, D., Hamburg, M.: Generalized identity based and broadcast encryption schemes. In: Pieprzyk, J. (ed.) ASIACRYPT 2008. LNCS, vol. 5350, pp. 455–470. Springer, Heidelberg (2008)

10. Boneh, D., Raghunathan, A., Segev, G.: Function-private identity-based encryption: Hiding the function in functional encryption. In: Canetti, R., Garay, J.A. (eds.) CRYPTO 2013, Part II. LNCS, vol. 8043, pp. 461–478. Springer, Heidelberg (2013)
11. Boneh, D., Raghunathan, A., Segev, G.: Function-private subspace-membership encryption and its applications. Cryptology ePrint Archive, Report 2013/403 (2013)
12. Boneh, D., Sahai, A., Waters, B.: Functional encryption: Definitions and challenges. In: Ishai, Y. (ed.) TCC 2011. LNCS, vol. 6597, pp. 253–273. Springer, Heidelberg (2011)
13. Boneh, D., Waters, B.: Conjunctive, subset, and range queries on encrypted data. In: Vadhan, S.P. (ed.) TCC 2007. LNCS, vol. 4392, pp. 535–554. Springer, Heidelberg (2007)
14. Camenisch, J., Kohlweiss, M., Rial, A., Sheedy, C.: Blind and anonymous identity-based encryption and authorised private searches on public key encrypted data. In: Jarecki, S., Tsudik, G. (eds.) PKC 2009. LNCS, vol. 5443, pp. 196–214. Springer, Heidelberg (2009)
15. Canetti, R.: Towards realizing random oracles: Hash functions that hide all partial information. In: Kaliski Jr., B.S. (ed.) CRYPTO 1997. LNCS, vol. 1294, pp. 455–469. Springer, Heidelberg (1997)
16. Chor, B., Goldreich, O.: Unbiased bits from sources of weak randomness and probabilistic communication complexity. SIAM Journal on Computing 17(2), 230–261 (1988)
17. Chung, K.-m., Vadhan, S.P.: Tight bounds for hashing block sources. In: Goel, A., Jansen, K., Rolim, J.D.P., Rubinfeld, R. (eds.) APPROX and RANDOM 2008. LNCS, vol. 5171, pp. 357–370. Springer, Heidelberg (2008)
18. Freeman, D.M.: Converting pairing-based cryptosystems from composite-order groups to prime-order groups. In: Gilbert, H. (ed.) EUROCRYPT 2010. LNCS, vol. 6110, pp. 44–61. Springer, Heidelberg (2010)
19. Goldwasser, S., Kalai, Y.T., Popa, R.A., Vaikuntanathan, V., Zeldovich, N.: Reusable garbled circuits and succinct functional encryption. In: Proceedings of the 45th Annual ACM Symposium on the Theory of Computing, pp. 555–564 (2013)
20. Golle, P., Staddon, J., Waters, B.: Secure conjunctive keyword search over encrypted data. In: Jakobsson, M., Yung, M., Zhou, J. (eds.) ACNS 2004. LNCS, vol. 3089, pp. 31–45. Springer, Heidelberg (2004)
21. Gorbunov, S., Vaikuntanathan, V., Wee, H.: Functional encryption with bounded collusions via multi-party computation. In: Safavi-Naini, R., Canetti, R. (eds.) CRYPTO 2012. LNCS, vol. 7417, pp. 162–179. Springer, Heidelberg (2012)
22. Håstad, J., Impagliazzo, R., Levin, L.A., Luby, M.: A pseudorandom generator from any one-way function. SIAM Journal on Computing 28(4), 1364–1396 (1999)
23. Katz, J., Sahai, A., Waters, B.: Predicate encryption supporting disjunctions, polynomial equations, and inner products. In: Smart, N.P. (ed.) EUROCRYPT 2008. LNCS, vol. 4965, pp. 146–162. Springer, Heidelberg (2008)
24. Okamoto, T., Takashima, K.: Hierarchical predicate encryption for inner-products. In: Matsui, M. (ed.) ASIACRYPT 2009. LNCS, vol. 5912, pp. 214–231. Springer, Heidelberg (2009)
25. Okamoto, T., Takashima, K.: Adaptively attribute-hiding (Hierarchical) inner product encryption. In: Pointcheval, D., Johansson, T. (eds.) EUROCRYPT 2012. LNCS, vol. 7237, pp. 591–608. Springer, Heidelberg (2012)

26. O'Neill, A.: Definitional issues in functional encryption. IACR Cryptology ePrint Archive, Report 2010/556 (2010)
27. Shen, E., Shi, E., Waters, B.: Predicate privacy in encryption systems. In: Reingold, O. (ed.) TCC 2009. LNCS, vol. 5444, pp. 457–473. Springer, Heidelberg (2009)
28. Shi, E., Bethencourt, J., Chan, H.T.-H., Song, D., Perrig, A.: Multi-dimensional range query over encrypted data. In: IEEE Symposium on Security and Privacy, pp. 350–364 (2007)
29. Zuckerman, D.: Simulating BPP using a general weak random source. Algorithmica 16(4/5), 367–391 (1996)

Random Projections, Graph Sparsification, and Differential Privacy

Jalaj Upadhyay

David R. Cheriton School of Computer Science
University of Waterloo,
200, University Avenue West
Waterloo, ON, Canada–N2L 3G1
jkupadhy@cs.uwaterloo.ca

Abstract. This paper initiates the study of preserving *differential privacy* (DP) when the data-set is sparse. We study the problem of constructing efficient sanitizer that preserves DP and guarantees high utility for answering cut-queries on graphs. The main motivation for studying sparse graphs arises from the empirical evidences that social networking sites are sparse graphs. We also motivate and advocate the necessity to include the efficiency of sanitizers, in addition to the utility guarantee, if one wishes to have a practical deployment of privacy preserving sanitizers.

We show that the technique of Blocki et al. [3] (BBDS) can be adapted to preserve DP for answering cut-queries on sparse graphs, with an asymptotically efficient sanitizer than BBDS. We use this as the base technique to construct an efficient sanitizer for arbitrary graphs. In particular, we use a preconditioning step that preserves the spectral properties (and therefore, size of any cut is preserved), and then apply our basic sanitizer. We first prove that our sanitizer preserves DP for graphs with high conductance. We then carefully compose our basic technique with the modified sanitizer to prove the result for arbitrary graphs. In certain sense, our approach is complementary to the Randomized sanitization for answering cut queries [17]: we use graph sparsification, while Randomized sanitization uses graph densification.

Our sanitizers almost achieves the best of both the worlds with the same privacy guarantee, i.e., it is almost as efficient as the most efficient sanitizer and it has utility guarantee almost as strong as the utility guarantee of the best sanitization algorithm.

We also make some progress in answering few open problems by BBDS. We make a combinatorial observation that allows us to argue that the sanitized graph can also answer (S,T)-cut queries with same asymptotic efficiency, utility, and DP guarantee as our sanitization algorithm for $S,\bar{S}$-cuts. Moreover, we achieve a better utility guarantee than Gupta, Roth, and Ullman [17]. We give further optimization by showing that fast Johnson-Lindenstrauss transform of Ailon and Chazelle [2] also preserves DP.

Keywords: Differential privacy, Graph sparsification, (S,T)-cut queries, Fast Johnson-Lindenstrauss transform.

K. Sako and P. Sarkar (Eds.) ASIACRYPT 2013 Part I, LNCS 8269, pp. 276–295, 2013.

1 Introduction

The privacy of a data is a fundamental problem in today's age of information. Many agencies collect enormous amount of data and store it in its database. These data may contain sensitive informations about an individual. However, given the benefits of analyzing these data, the problem that curators of such a database face is to provide useful information in such a manner so that no personal or sensitive information about an individual is leaked. A trivial way to guarantee this is to add a lot of noise to the database; however, nothing useful could be harnessed from such noisy database. Most of the research in this area is geared towards providing a tight utility and privacy tradeoff and only consider the query generator in mind. In this paper, we take a conceptual review and ask the practical question: what would a firm, that is going to deploy these sanitizers, demand from the group that develops these algorithms?

The question one expect to get from real firms or agencies is what extra resources they have to invest to provide this facility. This is expected in the real-world because a curator would prefer to deploy its resources to facilitate other interfaces that are primary to its business if differential private sanitizer uses a lot of resource. In general, sanitizers are polynomial time, but the exact bound on this polynomial is never made explicit in earlier works. In fact, Exponential sanitization [17, 19] may be intractable! We initiate the study of the question whether it is possible to guarantee DP that has high utility guarantee with an efficient sanitizer, emphasizing on a *concrete bound* on the efficiency parameter.

Motivation of Our Problem. Our motivation of studying cut queries on sparse graphs arises from a natural problem in social networks. One of the question that is commonly asked in social network is, given a set of individuals, how many friends/acquaintance do a set of people have outside their circle? The natural approach to solve this problem is to construct a *friendship graph*, where each vertex is labeled by an individual and there is an edge between two vertices if they are friends. These graphs on social networks are usually sparse, i.e., the average degree of the graph is very small in comparison to the number of vertices.

For a concrete example, consider the *friendship graph* on Facebook. According to the recent data released by Facebook, it has around one billion active users! It is not outrageous to assume that only a small fraction of users on Facebook have more than a thousand friends. Therefore, this graph is highly sparse. The friendship graph is undirected; however, this might not always be the case. For example, consider the *following graph* based on the networking of Twitter. It is a directed graph with nodes labeled by an individual. A node is the tail of an edge if the individual follows the head of the edge. The number of active users on Twitter is few million; however, it is less likely that an individual follows more than a few hundred fellow users. Thus, the *following graph* is very sparse. In these scenarios, the difference between performing $10^{9\times 2.38}$ and 10^{18} algebraic operations is huge. Any firm, like Facebook and Twitter, which is motivated by economics is less likely to invest in the former sanitization algorithm and may consider investing in the latter one.

It could be argued that if a sanitizer works for dense graphs (and therefore, also for sparse graphs), then there is no need for a specialized sanitizer for sparse graphs. The reason why we feel it is important to study sparse graphs exclusively is that sanitizers for dense graphs do not use the structural properties present in sparse graphs. In general, sparse graphs provides faster algorithms [7, 15, 28]. For example, consider the Johnson-Lindenstrauss (JL) sanitizer [3]. The sanitization algorithm of BBDS first overlays a complete graph on top of the input graph and then applies JL transform to the columns of the Laplacian of the modified graph. The step in which we overlay the complete graph destroys all the structural properties the input graph might have.

Now consider the situation when the input graph is sparse, and a hypothetical sanitizer that does not overlay K_n on top of it. When we perform random projection on this graph, the number of operations would depend on the number of edges of the graph (more concretely, on the non-zero entries in the representative matrix of the graph which could be Laplacian or adjacency matrix). Unfortunately, this sanitizer is not differentially private if the graph is weakly connected (in graph theoretic terms, has low conductance).

To see why this hypothetical sanitization does not provide DP, consider an n-vertices graph with two connected components. If the query is to find the cut of all the set of vertices in one component, the answer is 0 with probability 1. However, for a graph that has an edge joining two vertices present in different connected components, the probability with which the response to the query is non-zero is 1. This gives an easy way to differentiate the two cases. To resolve this particular problem, BBDS overlaid an n-vertex complete graph on top of the input graph.

Unfortunately, if we overlay the complete graph on a sparse graph, then we destroy the sparsity, and lose any (possible) gain in the computational time. On the other hand, even if the graph is connected and we do not perturb the graph, chances of privacy leakage are still present. More specifically, adding a single edge in a sparse graph can potentially have more privacy leak than a corresponding change in dense graphs. For example, consider a line graph or tree. They are acyclic; however adding any edge introduces a cycle. A slight modification of the differentiating algorithm used in the case of two component graph could be used to break the DP.

Our Contributions. This work is motivated by practical scenarios in which a sanitizer might be deployed. One of the objective of this paper is to advocate that, in addition to the utility and privacy guarantee, a design methodology for sanitizers should also give a concrete analysis of the efficiency of sanitizers. We initiate this line of work by studying differentially private sanitizer for cut queries on graphs.

As mentioned above, in practice, sparse graphs are more likely to occur than dense graphs. Every sanitizer that are proposed in the literature for dense graphs also works for sparse graphs, but they are not efficient. Moreover, there are examples of sparse graphs that could leak more information in DP sense than dense graphs, mainly because an addition or deletion of an edge could change the graph properties more dramatically in sparse graphs than in dense graphs. Thus, the problem is non-trivial, especially when we wish to construct efficient sanitizer.

On the fundamental level, we advocate the need of considering the efficiency of the sanitizer in the design methodology and give an explicit bound on the running time. The reason why we believe this is an important parameter is that, in many practical scenarios, the data-sets are held by agencies who might not have the privacy of an individual as their biggest priority. Therefore, unless a sanitizer is efficient, they might not have any incentive to perform the required sanitization. The technical contributions of this paper are as follows.

1. We show that it is possible to adapt the JL sanitizer of BBDS to answer cut queries when the graph is sparse. Additionally, our sanitizer has to perform only $O(n^{2+o(1)})$ algebraic operations. On the other hand, irrespective of whether or not the graph is sparse, the sanitizer of BBDS needs $O(rn^{2.38})$ operations[1], where r is the dimension of the subspace to which the JL transform projects the columns of the Laplacian. This improvement is *asymptotically significant.*
2. A natural question that arises next is whether our basic sanitizer is useful when the graph is fairly dense. We answer this question in affirmative. More precisely, we show that if we precondition a graph by reassigning the weights to the edges such that the transformed graph is guaranteed to be sparse and maintain the spectral properties of the graph, then applying the basic sanitizer on the conditioned graph preserves DP. This can be seen as a complementary approach to the Randomized sanitizer [17].
3. We make a simple combinatorial observation to argue that our sanitizer also preserves (S, T)-cut queries. This answers an open problem raised by BBDS.
4. Our last contribution is directed towards the optimization of the algebraic computations. We show that DP is maintained even if we replace the standard JL transform by the fast JL transform of Ailon and Chazelle [2]. This *partially answers another open problem of* BBDS.

Remark 1. An important characteristics of our preconditioning step, in item 2 above, is that it preserves the spectral properties. Any sanitizer that answers the queries based on spectral property of a graph could be transformed to first apply the preconditioning step before the sanitization step to improve the efficiency.

We note that none of our sanitizer randomly projects the vector corresponding to the column vector of the graph to a smaller dimension r. The main observation is that the mechanism is non-interactive, and in order to preserve the privacy for all set of queries, the dimension of the projected space has to be at least the dimension of the input space.

Overview of our Techniques. We first give a brief overview of the sanitizer of BBDS. In BBDS, the sanitization algorithm first reweighs the graph by overlaying

[1] Assuming that the matrix multiplication is done using Coppersmith-Winograd's algorithm.

a complete graph on top of the input graph, i.e., every edge, $e = (a, b)$, with weight w_e is replaced by an edge with weight $w'_e := \frac{w}{n} w_e + \left(1 - \frac{w}{n}\right)$. In other words, the weight on the edges are redistributed such that the overlaid complete graph has an equal weight, $\frac{w}{n}$, on all its edges. This makes the graph connected, and from Lemma 4, the smallest non-zero eigenvalues of this modified graph is greater than w/n. The JL transform is then applied on the columns of the Laplacian of the modified graph.

The major challenge that we face is that sparse graphs have low conductance. When we overlay the complete graph, it increases the conductance, but simultaneously makes it inefficient to answer the cut-queries by destroying the structural properties. As an extreme, consider a line graph. The cut-queries are fairly straightforward to answer; however, if we overlay a complete graph on top of it, we destroy the structural property and need to do some extra arithmetic to answer the same set of queries. An alternative is to overlay a sparse graph that increases the conductance, but does not destroy the structure of the underlying graph by much. The most natural candidate for this is an expander graph. As we show in our analysis, this suits our purpose very well.

We modify our basic sanitization technique, as outlined above, to construct an efficient sanitization technique for dense graphs. The key idea is to use graph sparsification at an appropriate step. As a warm-up, we assume that the input graph has high conductance. Our technique in this case is simple: apply the graph sparsification algorithms followed by the JL transform. The key observation here is that conductance helps in proving that, when the sparsification technique is applied on two neighboring graphs, the corresponding sparse graphs differ on at most one edge. This allows us to use the proof of BBDS for DP. On the other hand, due to the spectral guarantee provided by the sparsification technique, we know that all the cuts of the graph is maintained within a multiplicative factor. The utility guarantee then follows using simple arithmetic.

In order to apply the above analysis to arbitrary graph, we need a high conductance graph. This directs the order of the steps we follow for arbitrary graph, i.e, we first overlay a complete graph (or an expander) on the input graph before applying the sparsification algorithm. Finally, we apply the JL transform.

Related Work. Differential privacy, introduced by Dwork et al. [9], provides a robust guarantee of privacy. Informally speaking, if a curator sanitizes a data-set, then even if an individual's data is removed from the database, none of the responses to a query is more or less likely than the others. The key idea used in Dwork et al. [9] is to add noise to an output of the query according to a Laplace distribution, where the distribution is parameterized by the *sensitivity* of the query function. The Gaussian variant of this basic sanitizer was proven to preserve DP by Dwork et al. [8] in a follow-up work. Since then, many sanitizers for preserving DP have been proposed in the literature, including the Exponential sanitizer [4, 23, 27], the Multiplicative Update sanitizer [16–19], the Median sanitizer [30], the Boosting sanitizer [10], and the Random Projection sanitizer [25]. All these sanitizers have a common theme: they perturb the output before responding to queries. Blocki et al. [3] took a complementary approach. They perturb the input by performing a random projection of the input and show that existing algorithms preserves DP if the input is perturbed in a reversible manner.

The first work to explicitly study DP when the underlying data-set is a graph or a social network was by Hay et al. [20]. They presented a differentially private sanitizer for answering the degree of a node in a graph. They were followed by works of Nissim et al. [29] and Karwa et al. [24]. Gupta et al. [17] first studied the question of answering (S, T)-cut queries. The literature of studying faster computations on a sparse variant of any mathematical objects is so extensive that we cannot hope to cover it in all details here. An extensive study of faster methods of doing linear algebra on a sparse matrix is covered in standard textbooks [7, 15]. We refer the readers to an excellent book by Nesetril and de Mendez [28] for the properties and algorithms on sparse graph.

Organization of the Paper. In Section 2, we cover the basic preliminaries and definitions to the level required to understand the presentation of this paper. In Section 3, we give our basic sanitizer for sparse graph that serves as the building blocks for the sanitizers in Section 4. We conclude the paper by showing in Section 5 that the fast JL transform of Ailon and Chazelle [2] also preserves DP.

2 Preliminaries, Notations, and Basic Definitions

2.1 Privacy and Utility

In this work, we deal with privacy-preserving sanitizers for answering cut queries on graphs. The notion of differential privacy requires a definition of neighboring data-sets. Two data-sets (graphs, respectively) are *neighboring* if they differ on at most one entry (edge, respectively).

Definition 1. *A randomized algorithm $\mathcal{K}$, also called a* sanitizer, *gives* ε-DP, *if for all neighboring data-sets D_1 and D_2, and all range $S \subset Range(\mathcal{K})$,* $\mathsf{Pr}[\mathcal{K}(D_1) \in S] \leq \exp(e^{\epsilon})\mathsf{Pr}[\mathcal{K}(D_2) \in S]$, *where the probability is over the coin tosses of the sanitizer $\mathcal{K}$.*

In this paper, we study a natural relaxation of DP, called *approximate* DP.

Definition 2. *A randomized algorithm, $\mathcal{K}$ also called a* sanitizer, *gives* (ε, δ)-DP, *if for all neighboring data-sets D_1 and D_2, and all range $S \subset Range(\mathcal{K})$,* $\mathsf{Pr}[\mathcal{K}(D_1) \in S] \leq \exp(e^{\epsilon})\mathsf{Pr}[\mathcal{K}(D_2) \in S] + \delta$, *where the probability is over the coin tosses of the sanitizer $\mathcal{K}$.*

2.2 Linear Algebra

Let A be an $n \times m$ matrix. We let $\mathsf{rk}(A)$ denote the rank and $\mathsf{Tr}(A)$ denote the trace norm of the matrix A. The singular value decomposition of A is $A = V \Lambda U^T$, where U, V are unitary matrices and Λ is a diagonal matrix. The entries of Λ, denoted by $\lambda_1(A), \cdots, \lambda_{\mathsf{rk}(A)}(A)$, are called the *singular values* of A. Since U, V are unitary matrices, one can write $A^i = V \Lambda^i U^T$ for any real value i. If A is not a full rank matrix, then its inverse is called *Moore-Penrose* inverse and is denoted by $A^{\dagger}$ and its determinant is called *pseudo-determinant* and is defined as $\tilde{\Delta}(A) = \Pi_{i=1}^{\mathsf{rk}(A)} \lambda_i(A)$. We let $\chi_S \in \{0,1\}^n$ denote the characteristic vector for a subset $S \subseteq \mathcal{V}$.

2.3 Gaussian Distribution

Given a random variable, X, we denote by $X \sim \mathcal{N}(\mu, \sigma^2)$ the fact that X is distributed according to a Gaussian distribution with the probability density function, $\mathsf{PDF}_X(x) = \frac{1}{\sqrt{2\pi}\sigma} \exp\left(-\frac{(x-\mu)^2}{2\sigma^2}\right)$. The Gaussian distribution is invariant under affine transformation, i.e., if $X \sim \mathcal{N}(\mu_x, \sigma_x)$ and $Y \sim \mathcal{N}(\mu_y, \sigma_y)$, then $Z = aX + bY$ has the distribution $Z \sim \mathcal{N}(a\mu_x + b\mu_y, a\sigma_x^2 + b\sigma_y^2)$.

Multivariate Gaussian Distribution. The multivariate Gaussian distribution is a generalization of univariate Gaussian distribution. Given an m-dimensional multivariate random variable, $X \sim \mathcal{N}(\mu, \Sigma)$ with mean $\mu \in \mathbb{R}^m$ and covariance matrix $\Sigma = \mathbb{E}[(X - \mu)(X - \mu)^T]$, the PDF of a multivariate Gaussian is given by $\mathsf{PDF}_{\mathbf{X}}(\mathbf{x}) := \frac{1}{\sqrt{2\pi\tilde{\Delta}(\Sigma)}} \exp\left(-\frac{1}{2}\mathbf{x}^T \Sigma^{\dagger} \mathbf{x}\right)$. It is easy to see from the description of the PDF that, in order to define the PDF corresponding to a multivariate Gaussian distribution, Σ has to have full rank. If Σ has a non-trivial kernel space, then the PDF is undefined. However, in this paper, we only need to compare the probability distribution of two random variables which are defined over the same subspace. Therefore, in those scenarios, we would restrict our attention to the subspace orthogonal to the kernel space of Σ.

Multivariate Gaussian distribution maintains many key properties of univariate Gaussian distribution. For example, any (non-empty) subset of multivariate normals is multivariate normal. Another key property that is important in our analysis is that linearly independent linear functions of multivariate normal random variables are multivariate normal random variables, i.e., if $Y = AX + \mathbf{b}$, where A is an $n \times n$ non-singular matrix and $\mathbf{b}$ is a (column) n-vector of constants, then $Y \sim \mathcal{N}(A\mu + \mathbf{b}, A\Sigma A^T)$.

2.4 Graph Theory

We reserve the symbol $\mathcal{G}$ and $\mathcal{H}$ to denote a graph. We denote by $\mathcal{G}'$ the graph formed by adding an edge to the graph $\mathcal{G}$. In the case when $\mathcal{H}$ is formed from $\mathcal{G}$ using some transformation, we denote by $\mathcal{H}'$ the graph formed by performing the same transformation on $\mathcal{G}'$. For any $S \subseteq \mathcal{V}(\mathcal{G})$, the *cut* of the set of vertices S, denoted it by $\Phi_{\mathcal{G}}(S)$, is the weight of the edges that are present between S and $V \backslash S$.

We follow the same terminology of BBDS to define the utility guarantee.

Definition 3. *We say a sanitizer* $\mathcal{K}$ *gives a* (η, τ, ν)-approximation for cut queries, *if for every non-empty set* $S \subseteq \mathcal{V}$, *it holds that*

$$\mathsf{Pr}[(1-\eta)\Phi_{\mathcal{G}}(S) - \tau \leq \mathcal{K}(S, \mathcal{G}) \leq (1+\eta)\Phi_{\mathcal{G}}(S) + \tau] \geq 1 - \nu.$$

For the entire paper, we fix $w = \Theta\left(\frac{\log(1/\delta)+\log(1/\nu)}{\varepsilon}\right)$.

Laplacian of a Graph. For a weighted graph $\mathcal{G} := (\mathcal{V}, \mathcal{E}, w)$, its adjacency matrix $A_\mathcal{G}$ is given by $A_\mathcal{G}(i,j) = w_{ij}$ if $(i,j) \in \mathcal{E}$. The degree matrix of a weighted graph $\mathcal{G}$ is given by a diagonal matrix $D_\mathcal{G}$ such that the diagonal entries (i,i) is $\sum_j A_\mathcal{G}(i,j)$. The signed-edge matrix, $B_\mathcal{G}$, is constructed in the similar fashion as in BBDS: let $\mathcal{O}$ be an arbitrary orientation of edges. For an edge $e = (u,v)$, place $\sqrt{w_e}$ at position (e,v) if the edge e has v as its head and $-\sqrt{w_e}$ if it has v as its tail. For the other (e,i) when $i \neq u, v$, place 0.

The matrix for the Laplacian of a weighted graph, denoted by $L_\mathcal{G}$, is defined as $D_\mathcal{G} - A_\mathcal{G}$. One of the most useful form of Laplacian of a graph is the following form: $L_\mathcal{G} = \sum_{(a,b)\in E} w_{ab} L_{ab} = B_\mathcal{G}^T B_\mathcal{G}$, where L_{ab} is the Laplacian of a graph with a single edge (a,b). Many interesting properties of the Laplacian of a graph follows from this representation. For example, Laplacian of a graph is positive semi-definite, i.e., all the eigenvalues are non-negative. For a set S of vertices, its cut-set is $\Phi_\mathcal{G}(S) = \chi_S^T L_\mathcal{G} \chi_S$. Moreover, for $S, T \subseteq \mathcal{V}$, the sum of the weights of the edges with one end in S and other in T is denoted by $\Phi_\mathcal{G}(S,T)$. We explore this in detail later in Section 4.3.

We let $\lambda_i(\mathcal{G})$ denote the eigenvalues of $L_\mathcal{G}$ for $1 \leq i \leq n$. Next we present few lemmata that are useful in our analysis. In our analysis, we analyze multivariate Gaussian distributions that are linear combination of the Laplacian of two graphs. In order to analyze the two distributions, the corresponding covariance matrices must span the same subspace. The first lemma allows us to work on the same subspace, that is, the subspace orthogonal to $\mathsf{Span}\{\mathbf{1}\}$.

Lemma 1. [11, 12] *Let* $0 = \lambda_1(\mathcal{G}) \leq \lambda_2(\mathcal{G}) \cdots \leq \lambda_n(\mathcal{G})$ *be the* n *eigenvalues of* $L_\mathcal{G}$. *Then* $\mathcal{G}$ *is connected iff* $\lambda_2 > 0$ *and the kernel space of a connected graph is* $\mathsf{Span}\{\mathbf{1}\}$. *More generally, if a graph has* k *components, then the multiplicity of eigenvalue* 0 *is* k.

The following two lemmata are useful in giving the upper bound while proving the DP of our sanitizer.

Lemma 2. *Let* $\mathcal{G}$ *and* $\mathcal{G}'$ *be two graphs, where* $\mathcal{G}'$ *is obtained from* $\mathcal{G}$ *by adding one edge joining two distinct vertices of* $\mathcal{G}$. *Then*

$$\lambda_2(\mathcal{G}) \leq \lambda_2(\mathcal{G}') \leq \lambda_2(\mathcal{G}) + 2.$$

Lemma 3. *Let* $\mathcal{G}'$ *be formed by adding an edge* (u,v) *to* $\mathcal{G}$. *For any vector* $\boldsymbol{x} \in \mathbb{R}^n$, *we have* $\mathsf{Tr}(L_{\mathcal{G}'}) \leq \mathsf{Tr}(L_\mathcal{G}) + 2$.

The following lemma is particularly useful in arguing that the lowest non-zero eigenvalues of all the graphs is bounded from below by a constant (which is the second smallest eigenvalue of an expander).

Lemma 4. (Eigenvalue Interlacing). *Let* $\mathcal{G}$ *and* $\mathcal{G}'$ *be two graphs, where* $\mathcal{G}'$ *is obtained from* $\mathcal{G}$ *by adding one edge joining two distinct vertices of* $\mathcal{G}$. *Then*

$$\lambda_i(\mathcal{G}) \leq \lambda_i(\mathcal{G}') \leq \lambda_{i+1}(\mathcal{G}).$$

In particular, if $\mathcal{H}$ be a subgraph of $\mathcal{G}$, then $\lambda_i(\mathcal{H}) \leq \lambda_i(\mathcal{G}) \forall 1 \leq i \leq n$.

We refer the readers to the excellent book by Godsil and Royle [14] for a comprehensive treatment of the algebraic properties of graphs.

Graph Approximation. A graph $\mathcal{H}$ is said to ϵ-approximate a graph $\mathcal{G}$ if $\mathcal{H}$ approximates the spectral properties of $\mathcal{G}$, i.e.,

$$(1-\epsilon)\mathbf{x}^T L_{\mathcal{G}} \mathbf{x} \leq \mathbf{x}^T L_{\mathcal{H}} \mathbf{x} \leq (1+\epsilon)\mathbf{x}^T L_{\mathcal{G}} \mathbf{x} \qquad \forall \mathbf{x} \in \mathbb{R}^n.$$

We denote it by $(1-\epsilon)L_{\mathcal{G}} \preceq L_{\mathcal{H}} \preceq (1+\epsilon)L_{\mathcal{G}}$.

Electrical Flows and Resistance. We need the concept of electrical flow in graphs at various points for the analysis of Theorem 7. Intuitively, electrical flow of a graph measures how easy or difficult it is to move from one vertex to the other. If the "resistance" (as described later) between two vertices is high, then it is more difficult to reach from one vertex to the other, and vice versa. We give a brief exposition of the electrical flow that is required to understand this paper. Let $\mathbf{i}$ be the vector of current injected at the vertices of the graph $\mathcal{G}$. Then the effective resistance between two vertices u and v is defined as the potential difference induced between them when a unit current is injected at one vertex and extracted from the other. For any pair of vertices u and v, the effective resistance,

$$R_{uv} = (\chi_u - \chi_v)^T L_{\mathcal{G}}^{\dagger}(\chi_u - \chi_v) = \|B_{\mathcal{G}} L_{\mathcal{G}}^{\dagger}(\chi_u - \chi_v)^2\|_2^2.$$

Conductance. At the intuitive level, the conductance of a graph is the inverse of the resistance. For a graph, $\mathcal{G} = (V, E)$, let d_v denote the degree of vertex $v \in V$. Let $Vol(S) = \sum_{i \in S} d_i$, then the conductance of a set of vertex S, denoted by $\mathsf{cond}_S(G)$ is defined by

$$\mathsf{cond}_S(G) := \frac{|\Phi_{\mathcal{G}}(S)|}{\min\{Vol(S), Vol(V-S)\}}.$$

The conductance of a graph $\mathcal{G}$ is then given by $\mathsf{cond}(G) := \min_{S \subset V, |S| \geq 1} \mathsf{cond}_S(G)$. The conductance of a graph has a strong relation to the smallest non-zero eigenvalue of its Laplacian and we use it implicitly or explicitly in all of our analyses.

Theorem 1. (Cheeger's Inequality). *For a graph $\mathcal{G}$, $\mathsf{cond}(G)^2/2 \leq \lambda_2(L_{\mathcal{G}}) \leq 2\mathsf{cond}(G)$.*

2.5 JL Transform

The famous JL transform [1, 2, 5, 6, 21, 22] can be seen as a random projection of d points from a n-dimensional space to a lower dimensional space such that the Euclidean distance between any two pairs of points is maintained.

Theorem 2. *Fix any $\eta \in (0, 1/2)$ and M be a $k \times n$ matrix, whose entries are chosen from $\mathcal{N}(0,1)$. Then $\forall \boldsymbol{x} \in \mathbb{R}^n$, we have*

$$\mathsf{Pr}_M \left[(1-\eta)\|\boldsymbol{x}\|^2 \leq \frac{1}{k}\|M\boldsymbol{x}\|^2 \leq (1+\eta)\|\boldsymbol{x}\|^2 \right] \geq 1 - 2\exp(-\eta^2 k/8). \quad (1)$$

Fast JL *Transform.* Ailon and Chazelle [2] gave an elegant transform that is asymptotically faster than the traditional JL transform. It involves preconditioning the input. In this section, m denotes the number of n-dimensional points on which the transform is applied and k denotes the target dimension. More specifically, the fast JL transform is $M = PWD$, where (i) W is a $n \times n$ normalized Walsh-Hadamard matrix, (ii) D is a $n \times n$ diagonal matrix, where $\mathsf{Pr}[D_{ij} = 1] = \mathsf{Pr}[D_{ij} = -1] = 1/2$, and (iii) P is a $k \times n$ matrix whose elements are independently distributed as follows. With probability $1 - q$, set $P_{ij} = 0$; otherwise draw P_{ij} from a normal distribution of expectation 0 and variance $1/q$. The constant q is called the *sparsity* constant and is set to $q = \Theta\left(\frac{\eta^{p-2}\log^p n}{n}\right)$, where p is the norm we wish to preserve. Since W encodes the discrete Fourier transform, using fast Fourier transform, Ailon and Chazelle [2] proved that the transform satisfies equation (1), and takes time $\tilde{O}(n + qm/\eta^2))$.

3 Sanitizer for Cut Queries

In this section, we give our basic sanitizer for sparse graphs. Our key observation is that, in the sanitizer of BBDS, the overlay of the complete graph is required to maintain high conductance, and the result regarding the second smallest eigenvalue of the Laplacian follows immediately from the fact that a complete graph is a subgraph of the resulting graph. Unfortunately, this perturbation, when applied to sparse graphs, destroys the structural benefits of sparsity.

We get the same two objectives by overlaying an expander graph. An expander graph makes the graph connected while the smallest non-zero eigenvalue has the desired lower bound if we chose our expander graph with care. Recently, Friedman [13] proved that a random graph is an expander graph with high probability. In fact, he showed that such graphs are Ramanujan graphs. Marcus, Spielman, and Srivastava [26] recently proved the existential result for bipartite expander that matches the Ramanujan bound for every degree d.

We first derive a connection between the spectral properties of an expander graph and a complete graph. The most useful relation for this derivation is an alternate definition of an expander graph, i.e., *a d-regular graph $\mathcal{G}$ is an expander if $\lambda_2(\mathcal{G}) \geq (1 - \epsilon')d$ for some arbitrary constant ϵ'*. We give our basic sanitizer for sparse graphs in Figure 1.

Theorem 3. *The basic algorithm in Figure 1 preserves (ε, δ)-*DP*, provides an utility of (η, τ, ν)-approximation, where $\tau \leq O((\eta + \epsilon')ws)$, and runs in time $O(n^{2+o(1)})$.*

Input. A n-vertex sparse graph $\mathcal{G}$, and parameters $\varepsilon, \delta, \eta$.
Output A Laplacian of a graph $\tilde{L}$.

1. Pick a d-regular expander graph E.
2. Set $L_{\mathcal{H}} \leftarrow \frac{w}{d} L_E + \left(1 - \frac{w}{d}\right) L_{\mathcal{G}}$
3. Pick an $n \times n$ matrix M with each entries picked from $\mathcal{N}(0,1)$.
4. Output $\tilde{L} = (M^T L_{\mathcal{H}} M)/n$

The sanitizer publishes $\tilde{L}$ and anyone with a set of vertices S as input can compute $\Phi_{\mathcal{G}}(S)$ as below

$$\Phi_{\mathcal{G}}(S) = \frac{1}{1 - \frac{w}{d}} \left(\chi_S^T \tilde{L} \chi_S - \frac{ws(n-s)}{n} \right),$$

where $|S| = s$.

Fig. 1. The Basic Sanitizer

Proof. We first perform the complexity analysis of the above sanitizer. For a sparse graph, $m = O(n)$; therefore, using [34], it takes $\tilde{O}(n)$ time to compute the JL transform (since every column in the Laplacian of a sparse graph has $\tilde{O}(1)$ entries). Since, matrix multiplication takes $\Omega(n^2)$, this is almost tight for any sanitizer design that uses noise multiplication for answering cut queries.

The proof of DP proceeds in the similar manner as in BBDS. This is because our change still fulfills the requirement for which BBDS introduced the complete graph, i.e., it makes the graph connected. Using Lemma 1, the kernel space is $\mathsf{Span}\{\mathbf{1}\}$. Also, by a suitable choice of the expander graph, i.e., one with $1 - \epsilon' \geq w/d$, Lemma 4 guarantees that all the eigenvalues of $\mathcal{H}$ is greater that w, which is required in the privacy analysis of BBDS.

Our proof of utility guarantee develops on a useful relation between an expander graph and a complete graph. Let E be a d-regular expander graph such that the eigenvalues of A_E are $\leq \epsilon' d$. From the expression, $L_E = D_E - A_E$, where D_E is the degree matrix of E, we have that all the non-zero eigenvalues of L_E are between $(1-\epsilon')d$ and $(1+\epsilon')d$. Therefore, from Courant-Fischer formula,

$$(1-\epsilon')d\mathbf{x}^T\mathbf{x} \leq \mathbf{x}^T L_E \mathbf{x} \leq (1+\epsilon')d\mathbf{x}^T\mathbf{x} \qquad \forall \mathbf{x} \in \mathbb{R}^n. \tag{2}$$

We wish to relate this to the complete graph. For the complete graph, K_n, the eigenvalues are 0 with multiplicity 1 and n with multiplicity $n-1$. Therefore,

$$\mathbf{x}^T L_{K_n} \mathbf{x} = n\mathbf{x}^T\mathbf{x} \qquad \Rightarrow \frac{d}{n}\mathbf{x}^T L_{K_n}\mathbf{x} = d\mathbf{x}^T\mathbf{x}. \tag{3}$$

Plugging equation (3) in equation (2), we have

$$(1-\epsilon')\frac{d}{n}\mathbf{x}^T L_{K_n}\mathbf{x} \leq \mathbf{x}^T L_E \mathbf{x} \leq (1+\epsilon')\frac{d}{n}\mathbf{x}^T L_{K_n}\mathbf{x} \qquad \forall \mathbf{x} \in \mathbb{R}^n. \tag{4}$$

One way to look at equation 4 is that an expander graph is a sparsified complete graph. Using equation 4, we have the following approximation:

$$\left((1-\epsilon')\frac{w}{n}L_{K_n} + \left(1-\frac{w}{d}\right)L_{\mathcal{G}}\right) \preceq \left(\frac{w}{d}L_E + \left(1-\frac{w}{d}\right)L_{\mathcal{G}}\right) \preceq \left((1+\epsilon')\frac{w}{n}L_{K_n} + \left(1-\frac{w}{d}\right)L_{\mathcal{G}}\right). \quad (5)$$

We can now calculate the utility guarantee using equation (5) and similar arithmetic as in BBDS. More specifically, the upper bound on the utility guarantee can be calculated as below.

$$\begin{aligned}(1+\eta)\chi_S^T L_{\mathcal{H}}\chi_S &= (1+\eta)\chi_S^T\left(\frac{w}{d}L_E + \left(1-\frac{w}{d}\right)L_{\mathcal{G}}\right)\chi_S \\ &\leq (1+\eta)\chi_S^T\left((1+\epsilon')\frac{w}{n}L_{K_n} + \left(1-\frac{w}{d}\right)L_{\mathcal{G}}\right)\chi_S \\ &\leq (1+\eta)(1+\epsilon')\frac{w}{n}s(n-s) + \left(1-\frac{w}{d}\right)(1+\eta)\Phi_{\mathcal{G}}(S).\end{aligned}$$

Therefore,

$$\begin{aligned}\frac{1}{1-\frac{w}{d}}\left(\chi_S^T\tilde{L}\chi_S - \frac{ws(n-s)}{n}\right) &= \left(\frac{w(\eta+\epsilon'+\eta\epsilon')s}{(1-\frac{w}{d})}\right)\left(1-\frac{s}{n}\right) + (1+\eta)\Phi_{\mathcal{G}}(S) \\ &\leq 2(\eta+\epsilon'+\eta\epsilon')ws + (1+\eta)\Phi_{\mathcal{G}}(S).\end{aligned}$$

This gives an additive approximation of $O((\eta+\epsilon')ws)$. The proof of the lower bound is similar.

4 Differential Privacy by Sparsification

In Section 3, we showed that a simple change to the sanitizer of BBDS gives an efficient sanitizer for sparse graphs. In this section, we consider the case of an arbitrary graph. In particular, we show that various graph sparsification techniques also preserves DP. This serves as the second main contribution of this paper. Intuitively, the result in this section follows from the observation that, for large enough n, the sparsification techniques can be seen as a random projection. Thus, sparsification composed with our basic scheme should preserve DP by the composition theorem [10] and Theorem 3.

In certain sense, our approach is complementary to the approach used in Randomized sanitization. The Randomized sanitization [17] constructs a weighted graph, $\mathcal{H} = (\mathcal{V}, \mathcal{E}', w')$, such that $\forall u, y \in \mathcal{V}$, the weight of edge (u,v) in $\mathcal{H}$ is distributed as per the following distribution: $\mathsf{Pr}[w'_{uv} = 1] = (1+\varepsilon w_{uv})/2$ and $\mathsf{Pr}[w'_{uv} = -1] = (1-\varepsilon w_{uv})/2$.

4.1 Sanitization of Graphs with High Conductance

Every randomized sparsification technique picks an edge to be included in a sparsified graph with some specified probability distribution. At a high level,

Input. A n-vertex graph $\mathcal{G}$ with high conductance, and parameters $\varepsilon, \delta, \eta$.
Output A Laplacian of a graph $\tilde{L}$.

1. Convert $\mathcal{G}$ to a $\mathcal{H}$ using [31] or [32], such that $\mathcal{H}$ is an ϵ-sparsification of $\mathcal{G}$.
2. Pick an $n \times n$ matrix M with each entries picked from $\mathcal{N}(0,1)$.
3. Output $\tilde{L} = (M^T L_{\mathcal{H}} M)/n$.

The sanitizer publishes $\tilde{L}$ and anyone who has as input the set of vertices S can compute $\Phi_{\mathcal{G}}(S) = \left(1 - \frac{w}{n}\right) \chi_S^T \tilde{L} \chi_S$.

Fig. 2. Sanitizer for Graph With High Conductance

the distribution can be defined either dependent on the local structure or on the global structure of the graph. We give our sanitizer for both types of distribution, picking the most efficient one for instantiation. In this section, we analyze our sanitizer, stated in Figure 2, for graphs with high conductance.

The utility guarantee follows from the sparsification guarantee provided by the respective sparsifiers and the JL transform. The efficiency guarantee is straightforward from the observation that there are $\tilde{O}(1)$ entries in every row or column of the Laplacian of sparse graphs, and the run time of the step 3 in the Figure 2 is governed by the number of non-zero entries in the Laplacian. More concretely, assume that the sparsification algorithm takes $\tilde{O}(m)$ time to output a graph with $\tilde{O}(n)$ edges (as we will see, both the techniques, Spielman and Teng [32] based on local properties of the graph (Theorem 4), and Spielman and Srivastava [31] based on global properties of the graph (Theorem 6), satisfies these two conditions). Therefore, even if the graph is dense, i.e., $m = O(n^2)$, the run time for sparsification is $\tilde{O}(n^2)$. Therefore, the time taken by the sanitizer is bounded by $\tilde{O}(n^2)$ (since, in expectation, every column in the Laplacian of the sparse matrix has $\tilde{O}(1)$ entries).

The tricky part is to prove the privacy guarantee. For the privacy guarantee, we prove that for two neighboring graphs $\mathcal{G}$ and $\mathcal{G}'$, the respective sparse graphs differ on at most one edge. We can then apply Lemma 3, and the rest of the proof follows along the same line as BBDS. For both the sparsifier, we prove that if the graph has high enough conductance, then the probability distribution on edges with which the sparsification algorithm picks an edge does not differ by a lot. We then analyze two types of edges: (i) the edge (a, b) that is present in $\mathcal{G}'$ but not in $\mathcal{G}$ and (ii) the edges that are in both $\mathcal{G}$ and $\mathcal{G}'$. In the first case, the probability that the edge (a, b) is present in $\mathcal{H}'$ is non-zero and is identically zero in $\mathcal{H}$. We then prove that if the probability distribution on the edges does not differ by "lot", then with all but negligible probability, the respective sparse graphs will differ on at most one edge, i.e., only due to the (possible) presence of the edge (a, b) in $\mathcal{H}'$. The privacy guarantee follows using the proof of BBDS.

Using Local Sparsification Techniques: Construction of Spielman and Teng [32]. Spielman and Teng [32] proved the following result for any graph.

Theorem 4. [32] *There exists an $\tilde{O}(m)$ time algorithm which on input $0 < \epsilon, p < 1/2$ and a graph $\mathcal{G}$ with n-vertices and m-edges, outputs a sparse graph $\mathcal{H}$ with $\tilde{O}(n/\epsilon^2)$ edges such that $\mathcal{H}$ is an ϵ-approximation of $\mathcal{G}$.*

The main construction of Spielman and Teng [32] for arbitrary graphs is little complicated and uses techniques of graph decomposition and contraction. In this section, we will use the construction that works for a graph with high conductance. In this construction, every edge, $e = (i,j)$, is picked with probability, $p_{ij} := \frac{144k^2}{\epsilon^2 \lambda_2(G)^2 \min\{d_i, d_j\}}$, where $k = \max\{\log_2(3/p), \log_2 n\}$, d_i denotes the degree of the vertex i, and p is an arbitrary constant between $(0, 1/2)$. It is an easy exercise to check that the probability distribution on the edges that are already present does not change by a lot when a new edge is added due to the dependence only on the local structure of the graph (the eigenvalue changes by at most two by Lemma 2, and degree of only the two end vertices changes).

Using the proof outline mentioned above, we have the following theorem for the sanitizer in Figure 2 when we use the sparsifier of Spielman and Teng [32].

Theorem 5. *The algorithm in Figure 2 preserves (ε, δ) DP, provides an answer that is $((1+\eta)(1+\epsilon), 0, \nu)$-approximation, and runs in time $O(n^{2+o(1)})$ when using [32] sparsifier.*

Using Global Sparsification Techniques: Construction of Spielman and Srivastava [31]. We first recall the spectral sparsifier of Spielman and Srivastava [31]. One alternative way to see their sparsification is that $\mathcal{H}$ is a random projection of the edge matrix of $\mathcal{G}$, where edges are picked according to their importance in the original graph. The sparsifier construct a graph $\mathcal{H}$ by picking every edge, $e \in \mathcal{E}(\mathcal{G})$, with probability $p_e = w_e R_e/(n-1)$ to be included in $\mathcal{H}$, where R_e is the effective resistance across the edge e and w_e is the weight of the edge e. The effective resistance on an edge can be computed as

$$R_e = b_e L_{\mathcal{G}}^{\dagger} b_e^T \qquad \Rightarrow \qquad R = B_{\mathcal{G}} L_{\mathcal{G}}^{\dagger} B_{\mathcal{G}}^T. \tag{6}$$

Using the above probability distribution, Spielman and Srivastava [31] proved the following for any arbitrary graph.

Theorem 6. [31] *There exists an $\tilde{O}(m(\log r)/\epsilon^2)$ algorithm which on input $\epsilon > 1/\sqrt{n}$ and an n vertex, m edges graph $\mathcal{G}$, with the ratio of maximum weight to minimum weight r, outputs a sparse graph $\mathcal{H}$ such that $\mathcal{H}$ is an ϵ-approximation of $\mathcal{G}$.*

Consider the matrix, $\Pi = BL_{\mathcal{G}}^{+}B^T$, where B is the signed edge-vertex matrix. It is easy to see that Π is a projection matrix and has a well defined spectrum: eigenvalue 1 with multiplicity $(n-1)$ and 0 otherwise. Also, it has a nice relation to the probability with which an edge is picked to be placed in $\mathcal{H}$: $\Pi_{e,e} = W_{e,e}^{1/2} R_e W_{e,e}^{1/2} = w_e R_e$ for every edge $e = (a,b)$. Moreover, since the trace of Π is $n-1$; therefore, $p_e = \Pi_{e,e}/(n-1) = \|BL^{\dagger}(\chi_a - \chi_b)\|/(n-1)$, where χ_a is the characteristic vector of the vertex a.

We have the following theorem for DP when using the sparsification technique of Spielman and Srivastava [31].

Theorem 7. *If the input graph has high conductance, then the algorithm in Figure 2 runs in time $O(n^{2+o(1)})$ and preserves (ε, δ) differential privacy with an utility of $((1+\eta)(1+\epsilon), 0, \nu)$ approximation when using [31] sparsifier.*

Few remarks are in order regarding Theorems 5 and 7. The theorems state that the sanitizer allows zero additive error. Therefore, if we have a graph with high conductance, we can get an answer with only multiplicative approximation for as low tolerance as possible. This stands in stark contrast with Theorem 3 and Theorem 9 where there is an additive approximation that governs the tolerance achieved by the sanitizer. The reason is that, as the underlying graph has high conductance, the smallest non-zero eigenvalue is large. This allows us to remove the step where we overlay an expander graph!

4.2 Sanitizer for Arbitrary Graph

Before we move to arbitrary graphs, we give an alternative for the local sparsification technique of Spielman and Teng [31]. They (and Trevisan [33], independently) proved an important combinatorial property of an arbitrary graph, which can be used, in composition to their basic technique for high conductance graph, to prove the sparsification result for any arbitrary graph.

Theorem 8. [32, 33] *Let $\mathcal{G} = (V, E)$ be an arbitrary graph. Then there exists a set $\mathcal{E}' \subset \mathcal{E}$ of $\kappa|\mathcal{E}|$ edges, such that removal of these edges decomposes the graph in some components, each of which have an smallest non-zero eigenvalue at least $\kappa^2/72 \cdot (\log|\mathcal{E}|)^2$. Furthermore, these edges can be found in polynomial time.*

The above theorem could be used to get sparsification algorithm for an arbitrary graph. Let $\mathcal{E}'$ be the set of edges found by the algorithm guaranteed in Theorem 8. First apply the sparsification algorithm of [32] on all the components with high conductance, neglecting the edges in the set $\mathcal{E}'$. We recursively apply the sparsification algorithm on $\mathcal{E}'$ until we get a sparse graph, i.e, one with $|\mathcal{E}'| \leq \tilde{O}(n)$. The recursion depth is at most $O(\log n)$ rounds, so the overall run time of the algorithm is still under the bound guaranteed by Theorem 4. Therefore, one could use the complete sparsification technique that decomposes the graph in to graphs of high conductance with few bridge edges between the components before the third step of Figure 1. The DP of the sanitizer would follow the same idea as in the proof of Theorem 5 because the probability of picking edges depends on local graph structure, and the utility guarantee would follow from the partition-then-sample lemma of Spielman and Teng [32] and the guarantee of JL transform.

The idea of constructing sparse graphs using recursion also applies to the sparsifier of Spielman and Srivastava [31], but does not help us in proving the DP because of a subtle reason: the probability distribution in [32] depend locally on the graph structure, i.e., only on the degree of the end-points of the edges (it also depends on the smallest non-zero eigenvalue of the Laplacian, but from Lemma 3 and Lemma 4, it is easy to prove that the eigenvalue change by at most 2). Thus, the probability distribution changes by any significant amount only for the edge that is added.

On the other hand, the probability distribution on an edge can change drastically for Spielman and Srivastava's sparsifier [31]. This is because the probability distribution depends globally on the whole graph and if an edge is added, it is possible that the effective resistance along a different edge changes by a lot. For example, if a unit weight edge (a, b) is added, then any edge (u, v) that is parallel to (a, b) sees an effective drop to less than 1. If the conductance of the graph is not large, then this drop is significant.

The key idea to work around this problem is to maintain the conductance of the graph. We do this by using appropriate order of composition: we first overlay a expander (or complete) graph on top of the input graph and then apply the Spielman-Srivastava's sparsifier [31].

The utility guarantee follows by incorporating the approximation guarantee provided by Spielman and Teng [32] or Spielman and Srivastava [31] in the analysis of Theorem 3.

Theorem 9. *The algorithm in Figure 3 preserves* (ε, δ)*-*DP*, provides an utility of* (η, τ, ν) *approximation, where* $\tau \leq O((\eta+\epsilon)ws)$*, and runs in time* $O(n^{2+o(1)})$.

Input. A n-vertex graph $\mathcal{G}$, and parameters $\varepsilon, \delta, \eta$.
Output A Laplacian of a graph $\tilde{L}$.

1. Pick a d-regular expander (or n-vertices complete) graph E.
2. Set $L_{\mathcal{G}} \leftarrow \frac{w}{d} L_E + \left(1 - \frac{w}{d}\right) L_{\mathcal{G}}$.
3. Convert $\mathcal{G}$ to a $\mathcal{H}$ using [31] or [32], such that $\mathcal{H}$ is an ϵ-sparsification of $\mathcal{G}$.
4. Pick an $n \times n$ matrix M with each entries picked from $\mathcal{N}(0, 1)$.
5. Output $\tilde{L} = (M^T L_{\mathcal{H}} M)/n$

The sanitizer publishes the matrix $\tilde{L}$. For an input $S \subset V$, one can compute the number of vertices that crosses the cut as below

$$\Phi_S(\mathcal{G}) = \frac{1}{1 - \frac{w}{n}} \left(\chi_S^T \tilde{L} \chi_S - \frac{ws(n-s)}{n} \right).$$

Fig. 3. Sanitizer for Arbitrary Graph

Remark. Note that we can perform the complexity analysis of all the sanitizer mentioned in Sections 3 and 4 using the optimization mentioned in Section 5.

4.3 Answering (S, T)-cut Queries

One of the open problems listed by BBDS was to construct a sanitizer that answers (S, T)-cut queries on arbitrary graphs. Their concern for using the JL transform based mechanism is related to the inner product problem in JL transform. We get around this problem by making a simple combinatorial observation.

Let $S, T \subseteq \mathcal{V}$ be the set of vertices and we wish to find $\Phi_{\mathcal{G}}(S, T) = \sum_{s \in S, t \in T} w_{st}$. Note that

$$\Phi_{\mathcal{G}}(S) = \sum_{s \in S, u \in V \setminus S} w_{su} \quad \text{and} \quad \Phi_{\mathcal{G}}(T) = \sum_{t \in T, v \in V \setminus T} w_{st}.$$

Therefore, $\Phi_{\mathcal{G}}(S) + \Phi_{\mathcal{G}}(T)$ counts the weight of the edges that are crossing the boundaries of either S or T. These edges includes two types of edges: one that are crossing the boundaries of either S or T but do not have end vertices in both the sets and the one that have one end in S and the other in T. Note that we are interested in counting the weight of the edges of the latter form. Therefore, $\Phi_{\mathcal{G}}(S) + \Phi_{\mathcal{G}}(T) - \Phi_{\mathcal{G}}(S \cup T)$ is the sum of the weights of the edges between S and T, counted twice. This observation gives us that $\Phi_{\mathcal{G}}(S,T) = (\Phi_{\mathcal{G}}(S) + \Phi_{\mathcal{G}}(T) - \Phi_{\mathcal{G}}(S \cup T))/2$. Therefore, anyone with a set of vertices S and T as input and the sanitized graph from any of the mechanisms in this paper can compute $\Phi_{\mathcal{G}}(S,T)$ as below

$$\Phi_{\mathcal{G}}(S,T) = \frac{1}{2(1-\frac{w}{n})}\left(\left(\chi_S^T \tilde{L} \chi_S + \chi_T^T \tilde{L} \chi_T - \chi_{S\cup T}^T \tilde{L} \chi_{S\cup T}\right)\right) - \frac{1}{2(1-\frac{w}{n})}\left(\frac{ws(n-s)}{n} + \frac{wt(n-t)}{n} - \frac{w(s+t)(n-(s+t))}{n}\right).$$

Since computing the (S,T)-cut is three sequential applications of our basic sanitizer, DP follows from Theorem 3 (Theorem 5 and 7, respectively) for sparse graphs (high conductance graphs and arbitrary graphs, respectively) and the composition theorem of Dwork, Rothblum, and Vadhan [10, Theorem III.1]. The utility guarantee and efficiency guarantee are straightforward, giving the following theorem.

Theorem 10. *We can preserves (ε, δ) DP and provides an utility of (η, τ, ν) approximation, where $\tau \leq O((\eta + \epsilon) w \max\{s,t\})$ in time $O(n^{2+o(1)})$ for answering (S,T)-cut queries.*

4.4 Comparison with Other Algorithms

In Table 1, we compare our sanitizer algorithms with other sanitizers that are proposed in the literature. It is not clear how to compare interactive and non-interactive sanitizers; therefore, for the additive errors, we have a column when total number of cut queries are at most k.

Table 1. Comparison Between our Sanitizers and Other Sanitizers

Method	τ for any k	Curator's Run Time
Randomized Response [17]	$O(\sqrt{sn \log k/\varepsilon})$	$O(n^2)$
Exponential Sanitizer [4, 27]	$O(n \log n/\varepsilon)$	Intractable
Multiplicative Weight [17, 19]	$\tilde{O}(\sqrt{\lvert\mathcal{E}\rvert \log k/\varepsilon})$	$O(n^2)$
JL [3]	$O(s\sqrt{\log k}/\varepsilon)$	$O(rn^{2.38})$
Basic Scheme	$O(s(\eta + \epsilon')\sqrt{\log k}/\varepsilon)$	$O(n^{2+o(1)})$
Using Sparsifier	$O(s(\eta + \epsilon')\sqrt{\log k}/\varepsilon)$	$O(n^{2+o(1)})$

It is easy to see from Table 1 that our sanitizer almost matches the best of both the worlds; it is almost as efficient as Randomized Response and the utility guarantee is as high as JL transform for constant ϵ', ϵ. For JL mechanism, we assume that for publishing the matrix, the sanitizer uses the Coppersmith-Winograd's matrix multiplication algorithm.

5 Optimization Using Fast-JL Transform

Our last major contribution is to explore whether some other variants of JL transform also preserve DP. We show a positive result for fast JL-transform of Ailon and Chazelle [2]; thereby, partially answering an open problem of BBDS.

Due to lack of space, we just give an overview of our proof. Recall that the fast-JL transform is the product PWD. The intuitive reason why fast JL transform preserves privacy is that fast-JL transform preconditions the input by performing a random projection by matrix D. This is a random projection by the result of Achlioptas [1]. It then applies an unitary matrix that is a FFT and then another random projection matrix P. Thus, it can be seen as the application of two random projections. Using Theorem III.1 of Dwork, Rothblum, and Vadhan [10] and the main observation of BBDS, it preserves DP. This is our intuition behind the proof.

The exact proof uses case analysis. Consider the edge (a, b) that is present in $\mathcal{G}'$ and absent in $\mathcal{G}$. Let $d_1, \cdots, d_n$ be the diagonal entries of the matrix D. The proof proceeds by consider two cases: when $d_a = d_b$ and when $d_a \neq d_b$. The first case is almost the same as in BBDS because WD is an unitary matrix. The upper bound when $d_a \neq d_b$ is also immediate. However, for lower bound, we need to analyze the terms in the decomposition of matrix $WDL_{ab}D^TW^T$, and the eigenvalues of the projection of $WDL_{\mathcal{G}}$ on the co-ordinates a, b.

6 Open Problems

Our technique of using spectral sparsification is very general. We believe it could be used as a subroutine in many sanitization algorithms, which are designed to answer queries based on the spectral properties, to improve their run time. It would be interesting to investigate other such spectral properties. For example, one possible candidate for this improvement could be the differentially private low rank approximation algorithm of Kapralov and Talwar [23]. This is because Kapralov and Talwar [23] assume that their private matrices are covariance matrices and publish the low rank approximation by computing the singular vectors. Since covariance matrices are symmetric, one could compute the spectral sparsification.

Another aspect that is still open is to investigate whether other off-the-shelf JL transforms also preserve privacy or not. We have partially answered this question by studying fast JL transform, but there are many other variants that have applicability in different domains of computer science. In particular, we believe that any positive result for sparse JL transforms will be a significant step in improving the efficiency of our sanitizers and help in better understanding the relation between JL transforms and DP.

Acknowledgements. The author would like to thank the anonymous reviewers of ASIACRYPT, 2013 for many useful comments. The author would also like to thank the attendees of the C&O reading group and the members of CrySP and CACR for the useful discussions during the author's informal presentations.

References

1. Achlioptas, D.: Database-friendly random projections: Johnson-lindenstrauss with binary coins. J. Comput. Syst. Sci. 66(4), 671–687 (2003)
2. Ailon, N., Chazelle, B.: The fast johnson–lindenstrauss transform and approximate nearest neighbors. SIAM J. Comput. 39(1), 302–322 (2009)
3. Blocki, J., Blum, A., Datta, A., Sheffet, O.: The johnson-lindenstrauss transform itself preserves differential privacy. In: FOCS, pp. 410–419 (2012)
4. Blum, A., Ligett, K., Roth, A.: A learning theory approach to non-interactive database privacy. In: STOC, pp. 609–618 (2008)
5. Dasgupta, A., Kumar, R., Sarlós, T.: A sparse johnson: Lindenstrauss transform. In: STOC, pp. 341–350 (2010)
6. Dasgupta, S., Gupta, A.: An elementary proof of a theorem of johnson and lindenstrauss. Random Struct. Algorithms 22(1), 60–65 (2003)
7. Davis, T.: Direct Methods for Sparse Linear Systems. Part of the SIAM Book Series on the Fundamentals of Algorithms. SIAM, Philadelphia (2008)
8. Dwork, C., Kenthapadi, K., McSherry, F., Mironov, I., Naor, M.: Our data, ourselves: Privacy via distributed noise generation. In: Vaudenay, S. (ed.) EUROCRYPT 2006. LNCS, vol. 4004, pp. 486–503. Springer, Heidelberg (2006)
9. Dwork, C., McSherry, F., Nissim, K., Smith, A.: Calibrating noise to sensitivity in private data analysis. In: Halevi, S., Rabin, T. (eds.) TCC 2006. LNCS, vol. 3876, pp. 265–284. Springer, Heidelberg (2006)
10. Dwork, C., Rothblum, G., Vadhan, S.: Boosting and DP. In: FOCS, pp. 51–60 (2010)
11. Fiedler, M.: Algebraic connectivity of graphs. Czechoslovak Mathematical Journal 23(98), 298–305 (1973)
12. Fiedler, M.: A property of eigenvectors of nonnegative symmetric matrices and its applications to graph theory. Czechoslovak Mathematical Journal 25(100), 618–633 (1975)
13. Friedman, J.: A proof of alon's second eigenvalue conjecture. In: STOC, pp. 720–724 (2003)
14. Godsil, C., Royle, G.: Algebraic Graph Theory. Springer (2001)
15. Golub, G., Loan, C.: Matrix Computations. John Hopkins University Press, Maryland (1996)
16. Gupta, A., Hardt, M., Roth, A., Ullman, J.: Privately releasing conjunctions and the statistical query barrier. In: STOC, pp. 803–812 (2011)
17. Gupta, A., Roth, A., Ullman, J.: Iterative constructions and private data release. In: Cramer, R. (ed.) TCC 2012. LNCS, vol. 7194, pp. 339–356. Springer, Heidelberg (2012)
18. Hardt, M., Ligett, K., McSherry, F.: A simple and practical algorithm for differentially private data release. In: NIPS, pp. 2348–2356 (2012)
19. Hardt, M., Roth, A.: Beating randomized response on incoherent matrices. In: STOC, pp. 1255–1268 (2012)

20. Hay, M., Li, C., Miklau, G., Jensen, C.: Accurate estimation of the degree distribution of private networks. In: ICDM, pp. 169–178 (2009)
21. Johnson, W., Lindenstrauss, J.: Extensions of Lipschitz mapping into Hilbert space. In: Conf. in Modern Analysis and Probability. Contemporary Mathematics, vol. 26, pp. 189–206. American Mathematical Society (1984)
22. Kane, D., Nelson, J.: Sparser johnson-lindenstrauss transforms. In: SODA, pp. 1195–1206 (2012)
23. Kapralov, M., Talwar, K.: On differentially private low rank approximation. In: SODA, pp. 1395–1414 (2013)
24. Karwa, K., Raskhodnikova, S., Smith, A., Yaroslavtsev, G.: Private analysis of graph structure. PVLDB 4(11), 1146–1157 (2011)
25. Kenthapadi, K., Korolova, A., Mironov, I., Mishra, N.: Privacy via the johnson-lindenstrauss transform. CoRR, abs/1204.2606 (2012)
26. Marcus, A., Spielman, D., Srivastava, N.: Interlacing families i: Bipartite ramanujan graphs of all degrees. To Appear in FOCS (2013)
27. McSherry, F., Talwar, K.: Mechanism design via DP. In: FOCS, pp. 94–103 (2007)
28. Nesetril, J., Mendez, P.: Sparsity - Graphs, Structures, and Algorithms. Algorithms and combinatorics, vol. 28. Springer (2012)
29. Nissim, K., Raskhodnikova, S., Smith, A.: Smooth sensitivity and sampling in private data analysis. In: STOC, pp. 75–84 (2007)
30. Roth, A., Roughgarden, T.: Interactive privacy via the median mechanism. In: STOC, pp. 765–774 (2010)
31. Spielman, D., Srivastava, N.: Graph sparsification by effective resistances. SIAM J. Comput. 40(6), 1913–1926 (2011)
32. Spielman, D., Teng, S.: Spectral sparsification of graphs. SIAM J. Comput. 40(4), 981–1025 (2011)
33. Trevisan, L.: Approximation algorithms for unique games. Theory of Computing 4(1), 111–128 (2008)
34. Yuster, R., Zwick, U.: Fast sparse matrix multiplication. ACM Transactions on Algorithms 1(1), 2–13 (2005)

Notions of Black-Box Reductions, Revisited

Paul Baecher[1], Christina Brzuska[2], and Marc Fischlin[1]

[1] Department of Computer Science, Darmstadt University of Technology, Germany
[2] Tel-Aviv University, Israel

Abstract. Reductions are the common technique to prove security of cryptographic constructions based on a primitive. They take an allegedly successful adversary against the construction and turn it into a successful adversary against the underlying primitive. To a large extent, these reductions are black-box in the sense that they consider the primitive and/or the adversary against the construction only via the input-output behavior, but do not depend on internals like the code of the primitive or of the adversary. Reingold, Trevisan, and Vadhan (TCC, 2004) provided a widely adopted framework, called the RTV framework from hereon, to classify and relate different notions of black-box reductions.

Having precise notions for such reductions is very important when it comes to black-box separations, where one shows that black-box reductions cannot exist. An impossibility result, which clearly specifies the type of reduction it rules out, enables us to identify the potential leverages to bypass the separation. We acknowledge this by extending the RTV framework in several respects using a more fine-grained approach. First, we capture a type of reduction—frequently ruled out by so-called meta-reductions—which escapes the RTV framework so far. Second, we consider notions that are "almost black-box", i.e., where the reduction receives additional information about the adversary, such as its success probability. Third, we distinguish explicitly between efficient and inefficient primitives and adversaries, allowing us to determine how relativizing reductions in the sense of Impagliazzo and Rudich (STOC, 1989) fit into the picture.

1 Introduction

A fundamental question in cryptography refers to the possibility of constructing one primitive from another one. For some important primitives like one-way functions, pseudorandom generators, pseudorandom functions, and signature schemes it has been shown that one can be built from the other one [24, 17, 34]. For other primitives, however, there are results separating primitives like key agreement or collision-resistant hash functions from one-way functions [26, 36].

Separations between cryptographic primitives usually refer to a special kind of reductions called *black-box* reductions. These reductions from a primitive $\mathcal{P}$ to another primitive $\mathcal{Q}$ treat the underlying primitive $\mathcal{Q}$ and/or the adversary as a black box. Reingold et al. [33] suggested a taxonomy for such reductions which can be divided roughly into three categories:

K. Sako and P. Sarkar (Eds.) ASIACRYPT 2013 Part I, LNCS 8269, pp. 296–315, 2013.

Fully Black-Box Reductions: A fully black-box reduction $\mathcal{S}$ is an efficient algorithm that transforms any (even inefficient) adversary $\mathcal{A}$, breaking any instance G^f of primitive $\mathcal{P}$, into an algorithm $\mathcal{S}^{\mathcal{A},f}$ breaking the instance f of $\mathcal{Q}$. Here, the reduction treats both the adversary as well as the primitive as a black box, and G is the (black-box) construction out of f.

Semi Black-Box Reductions: In a semi black-box reduction, for any instance G^f of $\mathcal{P}$, if an efficient adversary $\mathcal{A}^f$ breaks G^f, then there is an algorithm $\mathcal{S}^f$ breaking the instance f of $\mathcal{Q}$. Here, S^f can be tailor-made for $\mathcal{A}$ and f.

Weakly Black-Box Reductions: In a weakly black-box reduction, for any instance G^f of $\mathcal{P}$, if an efficient adversary $\mathcal{A}$ (now without access to f) breaks G^f, then there is an algorithm $\mathcal{S}^f$ breaking the instance f of $\mathcal{Q}$.

Reingold et al. [33] indicate that the notion of weakly black-box reductions is close to free reductions (with no restrictions), such that separation results for this type of reduction are presumably hard to find. They discuss further notions like "$\forall\exists$ versions" of the above definitions, where the construction G does not make black-box use of f but may depend arbitrarily on f, and relativizing reductions where security of the primitives should hold relative to any oracle. We discuss these notions later in more detail.

1.1 Black-Box Separation Techniques

Known black-box separations usually obey the following two-oracle approach: to separate $\mathcal{P}$ from $\mathcal{Q}$ one oracle essentially makes any instance of $\mathcal{P}$ insecure, whereas the other oracle implements an instance of $\mathcal{Q}$. It follows that one cannot build (in a black-box way) $\mathcal{P}$ out of $\mathcal{Q}$. For example, Impagliazzo and Rudich [26] separate key agreement from one-way permutations by using a PSPACE-complete oracle to break any key agreement, and a random permutation oracle to realize the one-way permutation. This type of separation rules out so-called relativizing reductions, and are in this case equivalent to semi black-box reductions via embedding of the PSPACE-complete oracle into the black-box primitive [33].

Later, Hsiao and Reyzin [25] consider simplified separations for fully black-box reductions. Roughly speaking, they move the breaking oracle into the adversary such that the reduction can only access this oracle through the adversary (instead of directly, as in [26]). Because this makes separations often much more elegant this technique has been applied successfully for many other primitives, e.g., [11, 20, 21, 27, 5, 13, 29, 28, 3].

Interestingly, recently there has been another type of separations based on so-called meta-reduction techniques, originally introduced by Boneh and Venkatesan [6], and subsequently used in many other places [9, 30, 22, 14, 31, 15, 10, 35, 12]. Such meta-reductions take an alleged reduction from $\mathcal{P}$ to $\mathcal{Q}$ and show how to use such a reduction to break the primitive $\mathcal{P}$ directly, simulating the adversary for the reduction usually via rewinding techniques. It turns out that meta-reductions are somewhat dual to the above notions for black-box reductions. They usually work against reductions which use the adversary only in a black-box way, whereas the reduction often receives the description of the primitive f. This notion then escapes the treatment in [33].

An interesting side effect when the reduction is given the description of f is that then the separation technique still applies to concrete problems like RSA or discrete logarithms, and to constructions which use zero-knowledge proofs relative to f. Such zero-knowledge proofs often rely on Karp reductions of f to an NP-complete language and therefore on the description of f. In contrast, for black-box use of the primitive f such constructions do not work in general, although some of them can still be rescued by augmenting the setup through a zero-knowledge oracle which allows to prove statements relative to f (see [7]). We also remark that in some cases, such as Barak's ingenious result about non-black-box zero-knowledge and related results [2, 4], the security relies on the code of the adversary instead, though.

1.2 Our Results

The purpose of this paper is to complement the notions of fully, semi, and weakly black-box reductions. We also introduce a more fine-grained view on the involved algorithms, such as the distinction between efficient and non-efficient adversaries, or the question in how far the framework can deal with the reduction having partial knowledge about the adversary. We also formalize meta-reductions in the new framework and thus enable classification of this type of separation results. We give a comprehensive picture of the relationship of all reduction types. Next we discuss these results in more detail.

As explained above, we extend the classification of black-box reductions to other types, like meta-reductions relying on black-box access to the adversary but allowing to depend on the primitive's representation. This, interestingly, also affects the question of efficiency of the involved algorithms. That is, we believe that reductions for inefficient and efficient adversaries and primitives should in general not be resumed under a single paradigm, if efficiently computable primitives like one-way functions are concerned. For this class, classical separations techniques such as the embedding of the adversarially exploited PSPACE-complete oracle into the primitive do not work anymore. Hence, in this case one would need to additionally rely on a complexity assumption, such as for example in the work by Pass et al. [32]. To testify the importance of the distinction between efficient and inefficient adversaries in black-box reductions we show for example that black-box use of efficient adversaries is equivalent to non-black-box use, for constructions and reductions which are non-black-box for the primitive. Another example where the non-black-box use of the primitive turned out to be crucial is in the work by Mahmoody and Pass [29] where non-interactive commitments are built from non-black-box one-way functions, whereas constructions out of black-box one-way functions provably fail.

Another issue we address is the question in how far information about the adversary available to the reduction may be considered as covered by black-box notions. Technically speaking, the running time of an efficient fully black-box reduction must not depend on the adversary's running time, and thus for example on the number of queries the adversary makes to the primitive. Else, one would need to use a non-standard cost model for the reduction's oracle queries to the adversary. We overcome this dilemma by allowing the reduction's running time (or other parameters) to depend on adversarial parameters, such as the number of queries the adversary makes when attacking primitive $\mathcal{P}$. We call this a parameter-dependent reduction.

We can go even one step further and give the reduction the adversarial parameters as input. This is for example necessary to allow the reduction to depend on the adversary's success probability, but otherwise treating the adversary as a black box. A well-known example of such an "almost" fully black box reduction is the security proof of the Goldreich–Levin hardcore predicate [19], attributed to Rackoff in [16]. This reduction depends on the adversary's success probability for a majority decision, but does not rely on any specifics of the adversary nor the function to be inverted itself. We call such reductions parameter-aware.

We note that it is up to the designer of the reduction or separation to precisely specify the parameters. Such parametrized black-box reductions potentially allow authors to counteract the idea behind black-box reductions by placing the adversary's code in the parameters and thus making the reduction depend on the adversary again (via a universal Turing machine). But we assume that such trivial cases can be easily detected *if the dependency is signalized clearly*, just as in the case of a trivial reduction of a cryptographic protocol to its own security. So far, however, literature seems to be often less explicit on which parameters the reduction is based upon, and if the reduction should really count as black box. Stating reductions clearly as parametrized black-box reduction should make this more prominent.

In summary, we thus provide a more comprehensive and fine-grained view on black-box constructions and separations, allowing to identify and relate separations more clearly. In our view, two important results are that we can place relativizing reductions between non-black box constructions for inefficient and for efficient adversaries, and that for efficient adversaries the question of the reduction having black-box access to the adversary, or allowing full dependency on the adversary, is irrelevant. This holds as long as the construction and reduction itself make non-black-box use of the primitive. From a technical point of view, one of the interesting results is clearly that any reduction from the indistinguishability of hardcore bits to one-wayness, such as in the Goldreich–Levin case [19], must depend on the adversary's success probability (and thus needs to be parametrized).

Nevertheless, we view the contributions in this paper to be primarily on the conceptual side. Given the central role that reductions play in modern cryptography, our impression is that a fundamental—but rather coarse—work like [33] leaves some potential for refinement. Let us demonstrate this by the following two examples.

The Hsiao-Reyzin separation [25] is often termed fully black-box (according to [33]) and considered to be a rather "weak" separation. Our more fine-grained picture shows that the separation is actually of the NNN type and thus rather a low-level (i.e., strong) separation which cannot be bypassed through, say, any non-black-box technique in either direction of the CAP dimensions. Hence, non-black-box techniques cannot be used to sidestep this impossibility result; looking at efficient adversaries/primitives may help, though.

Similarly, according to [33], meta-reductions only rule out BBB reductions. So, the framework does not make any distinction between the strength of meta-reductions and some oracle separations. However, most meta-reductions today rely on unbounded adversaries. As our paper exhibits one might circumvent such meta-reductions by switching to the "parallel universe" of efficient adversaries, identifying exactly what kind of black-boxness is still admissible according to our implications (e.g., if the meta-reduction rules out NBN reductions, then one may still manage to find an NBNa reduction).

Thus, our framework reveals that some impossibility results actually rule out a great class of reductions and points exactly to the remaining few leverages to give positive results.

2 Notions of Reducibility

We extend the original framework for notions of reducibility by Reingold, Trevisan and Vadhan [33]. Since we augment the basic notions in various directions, we find it useful to use a different terminology for the reduction types. Instead of referring the original terms fully, semi, weakly, and their $\forall\exists$ variants, we use a more descriptive three-character "CAP" notation with words from the language $\{\mathrm{B}, \mathrm{N}\}^3$, with the meaning that a 'B' in the first position (the C-position) refers to the fact that the Construction is black-box, in the second A-position that the Adversary is treated as a black-box by the reduction, and in the third P-position the Primitive is treated as black-box by the reduction. Accordingly, an entry 'N' stands for a non-black-box use. From each combination of constraints, we then derive the order of quantification to obtain the actual definitions.

Hence, a fully black-box reduction in the RTV framework corresponds to a BBB-reduction in our notation, and a $\forall\exists$ fully black-box reduction is an NBB-reduction in our sense. The CAP notation will later turn out to be handy when showing implications from an XYZ-reduction to an $\widehat{X}\widehat{Y}\widehat{Z}$-reduction, whenever $\widehat{X}\widehat{Y}\widehat{Z}$ is pointwise at most as large as XYZ (with N being smaller than B). It also allows to see immediately that the RTV framework only covers a fraction of all 8 possibilities for the CAP choices (although the NNB type is actually not meaningful, as we discuss later), and that we fill in the missing types BBN, as often ruled out by meta-reductions, and the dual BNB type where the primitive but not the adversary is treated as a black-box.

Extending the RTV framework in another dimension, we differentiate further based on the (in)efficiency of the primitives and adversaries. We append the suffix 'a' to denote an efficiency requirement on the adversary, i.e., a BBBa-reduction only works for all probabilistic polynomial-time (PPT) adversaries $\mathcal{A}$, while a BBB-reduction is a fully black-box reduction that transforms *any* adversary $\mathcal{A}$ into an adversary against another primitive. Likewise, we use 'p' to indicate that we restrict primitives to those which are efficiently computable; the suffix 'ap' naturally combines both restrictions.

2.1 Overview

At the top of the RTV hierarchy there are fully black-box reductions—or, BBB-reductions in our CAP terminology. These BBB-reductions from a primitive $\mathcal{P}$ to a primitive $\mathcal{Q}$ is a pair $(G, \mathcal{S})$ consisting of a construction G and a reduction algorithm $\mathcal{S}$. Both treat the primitive in a black-box way and the reduction treats the adversary in a black-box way. So, for *all* adversaries $\mathcal{A}$ and *all* instantiations f of the primitive $\mathcal{Q}$, we have that, if the adversary $\mathcal{A}^f$ breaks G^f, then the reduction $\mathcal{S}^{\mathcal{A},f}$ with black-box access to the adversary $\mathcal{A}$ and f breaks the implementation f. As a consequence, the existence of primitive $\mathcal{Q}$ implies the existence of the primitive $\mathcal{P}$.

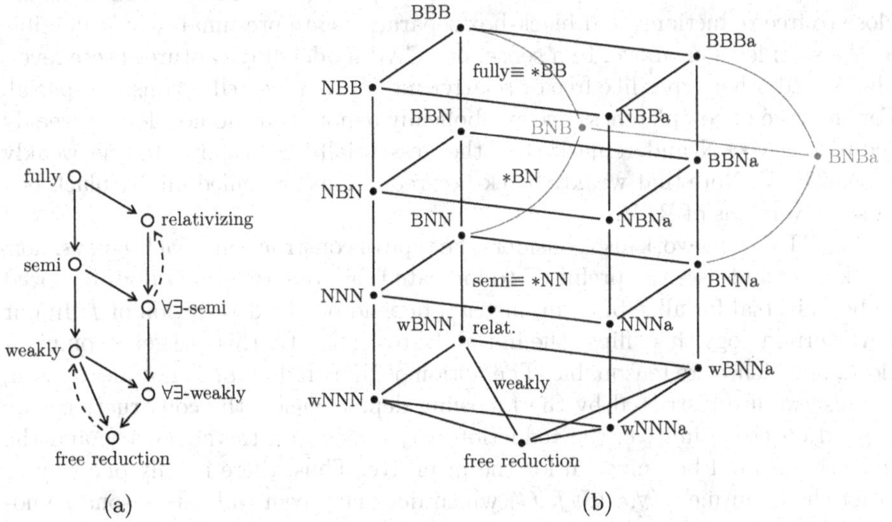

Fig. 1. (a) shows the relation of notions in the RTV framework. The dashed arrows indicate equivalence for a restricted class of reductions. In our framework (b), it is instructive to look at the vertical planes for fully, ∗BN, semi, and weakly. The left side corresponds to inefficient adversaries, the right side to efficient ones. The front is the ∀∃ layer, i.e., non-black-box constructions, and the back corresponds to black-box constructions. As NNB-reductions are not meaningful, we only need the BNB type (in gray). The w∗NN notions are equivalent to the weakly notions of RTV. A notion A implies notion B if there is a path of edges between both notions and notion A is located above notion B.

The RTV framework discusses several variants and relaxations of fully black-box reductions, called semi, weakly, and relativizing reductions. For semi black-box reductions (aka. BNN-reductions) $\mathcal{S}$ can depend on both, the description of the adversary $\mathcal{A}$ and of the instantiation f, and only the construction is black-box. For weakly black-box reductions (which are also of the BNN type) the adversary is additionally restricted to be efficient and does not get access oracle to the primitive (but may depend on it). There is a relativizing reduction between the primitives $\mathcal{P}$ and $\mathcal{Q}$, if for all oracles, the primitive $\mathcal{P}$ exists relative to an oracle whenever $\mathcal{Q}$ exists relative to this oracle. Figure 1a illustrates the relationships between these classes.

We augment the RTV framework by new classes which represent, among others, reductions that are ruled out by certain meta-reductions. That is, we first introduce the notion of BBN-reductions where $\mathcal{S}$ has to work for all (black-box) adversaries, but may depend on the code of f. The other case, where $\mathcal{S}$ is universal for all black-box f but may depend on $\mathcal{A}$, is called BNB-reduction. In both cases the initial 'B' indicates that the construction still makes black-box calls to the primitive. We remark that semi black-box and weakly black-box reductions are of the same BNN type in our notation as they only differ in regard to the adversary's access to f. As pointed out in [33] weakly black-box reductions are close to free reductions, and black-box separations are presumably only possible at the semi level or above. In a sense, our CAP model only captures these levels above, and other types like free or relativizing (or weakly) reductions are special. For the sake of completeness, we symbolically denote (but do not define) weakly reductions w$*$NN and remark that they essentially correspond to the weakly type of RTV. Note that weakly black-box reductions are called mildly black-box in some versions of RTV.

The RTV framework also considers the type of construction (black-box vs. non-black-box) and uses the prefix $\forall\exists$ to indicate that construction G does not need to be universal for all f but can, instead, depend on the description of f. In our CAP terminology this "flips" the initial 'B' to an 'N'. By this, we get 8 combinations, of which 7 are reasonable. The notion of NNB-reduction is not meaningful, because we are restricted by the following dependencies: the construction may depend on the primitive, the reduction may depend on the adversary, and the reduction should be universal for the primitive. Thus, there is only one way to order the quantifiers ($\forall\mathcal{A}\exists\mathcal{S}\forall f\exists G$) which does not seem to be a reasonable notion of security, because the construction can now depend on the adversary (and if it does not, we are in the other cases).

We note that the notion of an NBB-reduction is debatable, because it relies on a universal reduction which works for arbitrary constructions. That is, the order of quantifiers is $\exists\mathcal{S}\forall f\exists G\forall\mathcal{A}$. But since there may indeed be such reductions, say, a trivial reduction from a primitive to itself, we do not exclude this type of reduction here.

2.2 Definitions of Reductions

We next provide definitions of BBB (aka. fully black-box) reductions, BNB and BBN reductions; the remaining definitions are delegated to the full version of this paper [1].

A primitive $\mathcal{Q} = (\mathcal{F}_\mathcal{Q}, \mathcal{R}_\mathcal{Q})$ is represented as a set $\mathcal{F}_\mathcal{Q}$ of random variables, corresponding to the set of implementations, and a relation $\mathcal{R}_\mathcal{Q}$ that describes the security of the primitive as tuples of random variables, i.e., a random variable $\mathcal{A}$ is said to break an instantiation $f \in \mathcal{F}_\mathcal{Q}$, if and only if $(f, \mathcal{A}) \in \mathcal{R}_\mathcal{Q}$. Following [33], we say that a primitive exists if there is a polynomial-time

CAP	[33] name	Remark(s)
BBB	fully	known meta reductions: [8, 22]
BBN		
BNB		known reduction: [19]
BNN	semi (weakly)	
NBB	∀∃-fully	formally not defined in [33], only "trivial" reductions
NBN		known meta reductions: [6, 22, 14, 31]
NNB		not meaningful
NNN	∀∃-semi (∀∃-weakly)	

Fig. 2. CAP indicates whether the construction (C), the adversary in the reduction (A), or the primitive in the reduction (P) is treated in a black-box (B) or non-black-box (N) way

computable instantiation $f \in \mathcal{F}_Q$ such that no polynomial-time random variable breaks the primitive. Indeed, [33] demand that primitive sets $\mathcal{F}_Q$ are non-empty, but do not motivate this further. We drop this requirement here as reductions explicitly depend on primitives, such that one can enforce such non-empty sets by investigating only such primitives if necessary. Still, we remark that all our implications and separations would work in this case as well.

For efficient primitives or adversaries we stipulate that the random variable is efficiently computable in the underlying machine model which, unless mentioned differently, is assumed to be Turing machines; the results remain valid for other computational models like circuit families. Considering security as a general relation allows to cover various (if not all) notions of security: games such as CMA-UNF for unforgeability of signature schemes, simulation-based notions such as implementing a UC commitment functionality, and even less common notions such as distributional one-way functions. In the full version of this paper [1] we define as examples the DDH assumption (cast as a primitive) and the indistinguishability of the ElGamal encryption scheme . We also present the reduction from the ElGamal encryption to the DDH assumption and identify its type according to our terminology. Note that a "black-boxness" consideration in this particular setting is indeed meaningful, because the DDH assumption can hold in a variety of group distributions and the concrete procedures that sample from these group distributions can be abstracted away. In the full version we discuss another example of weak one-way functions (and the construction of strong one-way functions [37]) to highlight that the type of reduction hinges on the exact formulation of the underlying primitive: the construction and the reduction is then either of the NBN type or of the BBB kind.

We stress that the distinction between the *mathematical object* describing the adversary as a random variable, and its *implementation* through, say, a Turing machine is important here; else one can find counter examples to implications among black-box reduction types proven in [33]. The problem is roughly that the relation may simply be secure because it syntactically excludes all oracle Turing machines $\mathcal{A}^f$. We note that Reingold et al. [33] indeed define the relations for adversarial *machines*. Our discussion in [1] shows that only interpreting such adversaries as abstract objects sustains the implications in [33]. However, for sake of convenience, we too often refer to $\mathcal{A}^f$ by the machine implementing it, even when considering the mathematical random process for relations $\mathcal{R}_\mathcal{Q}$. In this case it is understood that we actually mean the abstract random variable instead. The same holds for the constructions of the form G^f and the first component of the security relations. An alternative approach, also presented in the full version is to rely on machines, but to formally introduce semantical relations. These relations roughly require that, for any algorithm $\mathcal{A}$ in $\mathcal{R}_\mathcal{Q}$, any oracle machine $\mathcal{A}^f$ with the same output behavior is also in $\mathcal{R}_\mathcal{Q}$.

We now turn to the actual definitions. Many (but not all) reductions in cryptography fall into the class of so-called fully black-box reductions, a very restrictive notion, where the reduction algorithm is only provided with black-box access to the primitive and the adversary. Throughout the paper, if there is a XYZ-reduction from primitive $\mathcal{P}$ to a primitive $\mathcal{Q}$, we notate this as $(\mathcal{P} \hookrightarrow \mathcal{Q})$-$XYZ$-reduction. Note that the correctness is requirement is the same for all definitions. Therefore, the shorthand notation towards the end of each definition covers the security requirement only.

Definition 1 (($\mathcal{P} \hookrightarrow \mathcal{Q}$)-BBB or Fully Black-Box Reduction). *There exists a* fully black-box (or BBB-)reduction *from a primitive $\mathcal{P} = (\mathcal{F}_\mathcal{P}, \mathcal{R}_\mathcal{P})$ to a primitive $\mathcal{Q} = (\mathcal{F}_\mathcal{Q}, \mathcal{R}_\mathcal{Q})$ if there exist probabilistic polynomial-time oracle algorithms G and $\mathcal{S}$ such that:*

Correctness. *For every $f \in \mathcal{F}_\mathcal{Q}$, it holds that $G^f \in \mathcal{F}_\mathcal{P}$.*
Security. *For every implementation $f \in \mathcal{F}_\mathcal{Q}$ and every machine $\mathcal{A}$, if $(G^f, \mathcal{A}^f) \in \mathcal{R}_\mathcal{P}$, then $(f, \mathcal{S}^{\mathcal{A},f}) \in \mathcal{R}_\mathcal{Q}$, i.e.,*

$$\exists \mathsf{PPT} G\ \exists \mathsf{PPT} \mathcal{S}\ \forall f \in \mathcal{F}_\mathcal{Q}\ \forall \mathcal{A}\ \ ((G^f, \mathcal{A}^f) \in \mathcal{R}_\mathcal{P} \Rightarrow (f, \mathcal{S}^{\mathcal{A},f}) \in \mathcal{R}_\mathcal{Q}).$$

Definition 2 (($\mathcal{P} \hookrightarrow \mathcal{Q}$)-BNB-reduction). *There exists a* BNB-reduction *from a primitive $\mathcal{P} = (\mathcal{F}_\mathcal{P}, \mathcal{R}_\mathcal{P})$ to a primitive $\mathcal{Q} = (\mathcal{F}_\mathcal{Q}, \mathcal{R}_\mathcal{Q})$ if there exists a probabilistic polynomial-time oracle machine G such that:*

Correctness. *For every $f \in \mathcal{F}_\mathcal{Q}$, it holds that $G^f \in \mathcal{F}_\mathcal{P}$.*
Security. *For every machine $\mathcal{A}$, there is a probabilistic polynomial-time oracle algorithm $\mathcal{S}$ such that: for every implementation $f \in \mathcal{F}_\mathcal{Q}$, if $(G^f, \mathcal{A}^f) \in \mathcal{R}_\mathcal{P}$, then $(f, \mathcal{S}^{\mathcal{A},f}) \in \mathcal{R}_\mathcal{Q}$, i.e.,*

$$\exists \mathsf{PPT} G\ \forall \mathcal{A}\ \exists \mathsf{PPT} \mathcal{S}\ \forall f \in \mathcal{F}_\mathcal{Q}\ \ ((G^f, \mathcal{A}^f) \in \mathcal{R}_\mathcal{P} \Rightarrow (f, \mathcal{S}^{\mathcal{A},f}) \in \mathcal{R}_\mathcal{Q}).$$

Definition 3 (($\mathcal{P} \hookrightarrow \mathcal{Q}$)-BBN-reduction). *There exists a* BBN-reduction *from a primitive $\mathcal{P} = (\mathcal{F}_\mathcal{P}, \mathcal{R}_\mathcal{P})$ to a primitive $\mathcal{Q} = (\mathcal{F}_\mathcal{Q}, \mathcal{R}_\mathcal{Q})$ if there exists a probabilistic polynomial-time oracle machine G such that:*

Correctness. *For every $f \in \mathcal{F}_\mathcal{Q}$, it holds that $G^f \in \mathcal{F}_\mathcal{P}$.*
Security. *For every implementation $f \in \mathcal{F}_\mathcal{Q}$, there is a probabilistic polynomial-time oracle algorithm $\mathcal{S}$ such that for every machine $\mathcal{A}$, if $(G^f, \mathcal{A}) \in \mathcal{R}_\mathcal{P}$, then $(f, \mathcal{S}^{\mathcal{A},f}) \in \mathcal{R}_\mathcal{Q}$, i.e.,*

$$\exists \mathsf{PPT} G\ \forall f \in \mathcal{F}_\mathcal{Q}\ \exists \mathsf{PPT} \mathcal{S}\ \forall \mathcal{A}\ \ ((G^f, \mathcal{A}^f) \in \mathcal{R}_\mathcal{P} \Rightarrow (f, \mathcal{S}^{\mathcal{A},f}) \in \mathcal{R}_\mathcal{Q}).$$

Name	Summary of definition				
BBB	$\exists$PPT$\mathcal{G}$	$\exists$PPT$\mathcal{S}$	$\forall f \in \mathcal{F_Q}$	$\forall \mathcal{A}$	$((G^f, A^f) \in \mathcal{R_P} \Rightarrow (f, \mathcal{S}^{A,f}) \in \mathcal{R_Q})$
BNB	$\exists$PPT$\mathcal{G}$	$\forall \mathcal{A}$	$\exists$PPT$\mathcal{S}$	$\forall f \in \mathcal{F_Q}$	$((G^f, A^f) \in \mathcal{R_P} \Rightarrow (f, \mathcal{S}^{A,f}) \in \mathcal{R_Q})$
BBN	$\exists$PPT$\mathcal{G}$	$\forall f \in \mathcal{F_Q}$	$\exists$PPT$\mathcal{S}$	$\forall \mathcal{A}$	$((G^f, A^f) \in \mathcal{R_P} \Rightarrow (f, \mathcal{S}^{A,f}) \in \mathcal{R_Q})$
BNN	$\exists$PPT$\mathcal{G}$	$\forall f \in \mathcal{F_Q}$	$\forall \mathcal{A}$	$\exists$PPT$\mathcal{S}$	$((G^f, A^f) \in \mathcal{R_P} \Rightarrow (f, \mathcal{S}^{A,f}) \in \mathcal{R_Q})$
NBB	$\exists$PPT$\mathcal{S}$	$\forall f \in \mathcal{F_Q}$	$\exists$PPT$\mathcal{G}$	$\forall \mathcal{A}$	$((G^f, A^f) \in \mathcal{R_P} \Rightarrow (f, \mathcal{S}^{A,f}) \in \mathcal{R_Q})$
NBN	$\forall f \in \mathcal{F_Q}$	$\exists$PPT$\mathcal{G}$	$\exists$PPT$\mathcal{S}$	$\forall \mathcal{A}$	$((G^f, A^f) \in \mathcal{R_P} \Rightarrow (f, \mathcal{S}^{A,f}) \in \mathcal{R_Q})$
NNN	$\forall f \in \mathcal{F_Q}$	$\exists$PPT$\mathcal{G}$	$\forall \mathcal{A}$	$\exists$PPT$\mathcal{S}$	$((G^f, A^f) \in \mathcal{R_P} \Rightarrow (f, \mathcal{S}^{A,f}) \in \mathcal{R_Q})$
weakly-BB	$\exists$PPT$\mathcal{G}$	$\forall \mathcal{A}$	$\forall f \in \mathcal{F_Q}$	$\exists$PPT$\mathcal{S}$	$((G^f, \mathcal{A}) \in \mathcal{R_P} \Rightarrow (f, \mathcal{S}^{A,f}) \in \mathcal{R_Q})$
$\forall\exists$-weakly-BB	$\forall f \in \mathcal{F_Q}$	$\exists$PPT$\mathcal{G}$	$\forall \mathcal{A}$	$\exists$PPT$\mathcal{S}$	$((G^f, \mathcal{A}) \in \mathcal{R_P} \Rightarrow (f, \mathcal{S}^{A,f}) \in \mathcal{R_Q})$

Fig. 3. Overview of notions of reducibility

Note that we always grant $\mathcal{S}$ black-box access to f and $\mathcal{A}$, as they may not be efficiently computable so that the probabilistic polynomial-time reduction algorithm $\mathcal{S}$ cannot efficiently simulate them, even if it knows the code of f, respectively, of $\mathcal{A}$. For a compact summary of all definitions, see Figure 3; the full definitions omitted above appear in the full version of this paper [1].

2.3 Efficient versus Inefficient Algorithms

Reductions usually run the original adversary as a subroutine. However, in many cases, the reduction does not use the code of the original adversary, but instead only transforms the adversary's inputs and outputs. Thus, one might consider the reduction algorithm as having black-box access to the adversary only. An efficient reduction can then also be given black-box access to an inefficient adversary, and, maybe surprisingly, most reductions even work for inefficient adversaries. Imagine, for example, the case that one extracts a forgery against a signature scheme from a successful intrusion attack against an authenticated channel. Then, the extraction usually still works for inefficient adversaries. On the other hand, (unconditional) impossibility results often require the reduction algorithm to be able to deal with inefficient adversaries.

When designing a fine-grained framework for notions of reducibility, one thus needs to decide whether one considers efficient or inefficient adversaries. Reingold et al. [33] defined their most restrictive notion of reductions, the fully-BB-reductions (aka. BBB), for inefficient adversaries. In contrast, their notion of semi-BB-reduction treats only efficient adversaries thus making it easier to find such a reduction. Surprisingly, even for such a weak notion, they were able to give impossibility results. The reason is that they used inefficient primitives, which allow to embed arbitrary oracles so that they could make use of two-oracle separation techniques. Hence, the efficiency question does not only apply to adversaries, but also to the primitives (and, consequently, to the combination of both). We postpone the treatment of the case of primitives for now and refer the reader to Section 2.6.

We now define the efficient adversary analogues of the notions of reduction introduced in the previous section. Note that we still give the reduction $\mathcal{S}$ oracle access to the adversary $\mathcal{A}$ in *all* notions, even though the latter can be dropped for all cases where $\mathcal{S}$ depends on $\mathcal{A}$ in a non-black-box way. In these cases, a probabilistic polynomial-time reduction $\mathcal{S}$ can simulate the now likewise efficient adversarial algorithm $\mathcal{A}$. For consistency, though, we keep the $\mathcal{A}$ oracles in the definitions. To distinguish the two cases of efficient and unbounded adversaries, denote by BBBa-reduction a reduction only dealing with efficient adversaries.

Definition 4 (($\mathcal{P} \hookrightarrow \mathcal{Q}$)-BBBa-reduction for Efficient Adversaries). *There exists a BBBa-reduction from a primitive $\mathcal{P} = (\mathcal{F}_\mathcal{P}, \mathcal{R}_\mathcal{P})$ to a primitive $\mathcal{Q} = (\mathcal{F}_\mathcal{Q}, \mathcal{R}_\mathcal{Q})$ if there exist probabilistic polynomial-time oracle machines G and $\mathcal{S}$ such that:*

Correctness. *For every $f \in \mathcal{F}_\mathcal{Q}$, it holds that $G^f \in \mathcal{F}_\mathcal{P}$.*

Security. *For every implementation $f \in \mathcal{F}_\mathcal{Q}$ and every probabilistic polynomial-time machine $\mathcal{A}$, if $(G^f, \mathcal{A}) \in \mathcal{R}_\mathcal{P}$, then $(f, \mathcal{S}^{\mathcal{A},f}) \in \mathcal{R}_\mathcal{Q}$, i.e.,*

$$\exists \mathsf{PPT} G\ \exists \mathsf{PPT} \mathcal{S}\ \forall f \in \mathcal{F}_\mathcal{Q}\ \forall \mathsf{PPT} \mathcal{A}\ \ ((G^f, \mathcal{A}^f) \in \mathcal{R}_\mathcal{P} \Rightarrow (f, \mathcal{S}^{\mathcal{A},f}) \in \mathcal{R}_\mathcal{Q}).$$

Again, the definitions for the remaining types of reductions are presented in the full version of this paper [1].

2.4 Relations amongst the Definitions

We first note that a number of implications among the reductions is immediately clear by simply shifting quantifiers, that is, if we have an for-all quantifier, there is certainly an existential version of the reduction in question. The next proposition states this formally, we omit the proof because it is only syntactical.

Theorem 1. *Let XYZ and $\widehat{X}\widehat{Y}\widehat{Z}$ be two types of CAP reductions such that $\widehat{X}\widehat{Y}\widehat{Z} \leq XYZ$ point-wise (where $N \leq B$) and let $\mathcal{P}$ and $\mathcal{Q}$ be two primitives. If there is a ($\mathcal{P} \hookrightarrow \mathcal{Q}$)-$XYZ$-reduction, then there is a ($\mathcal{P} \hookrightarrow \mathcal{Q}$)-$\widehat{X}\widehat{Y}\widehat{Z}$ reduction. Also, if there is a ($\mathcal{P} \hookrightarrow \mathcal{Q}$)-$XYZ$a-reduction, then there is a ($\mathcal{P} \hookrightarrow \mathcal{Q}$)-$\widehat{X}\widehat{Y}\widehat{Z}$a reduction.*

In the full version of this paper [1], we prove via means of counterexamples that for all notions for inefficient adversaries, almost all the above implications are, indeed, strict. These separations are split into a number of interesting observations. For example, we prove that the Goldreich–Levin hardcore bit reduction [19] has to depend on the success probability of the adversary (Theorem D.3 of [1]). Moreover, we show that the construction of one-way functions out of weak one-way functions ([37, 18]) needs to depend on the weakness parameter of the weak one-way function (Theorem D.2 of [1]). Interestingly, some of the implications of Theorem 1 are not strict when one is concerned with reductions for efficient adversaries. Maybe surprisingly, NNNa-reductions and NBNa-reductions are, indeed, equivalent. Note that this means that knowledge of the code of the adversary does not lend additional power to the reduction:

Theorem 2 (Equivalence of NNNa and NBNa). *For all primitives $\mathcal{P}$ and $\mathcal{Q}$, there is a $(\mathcal{P} \hookrightarrow \mathcal{Q})$-NBNa-reduction if and only if there is a $(\mathcal{P} \hookrightarrow \mathcal{Q})$-NNNa-reduction.*

Proof. Using straightforward logical deductions, it follows that NBNa-reductions imply NNNa-reductions. For the converse direction, assume that we have two primitives $\mathcal{P}$ and $\mathcal{Q}$ such that there is a $(\mathcal{P} \hookrightarrow \mathcal{Q})$-NNNa-reduction. We now have to show that there also is a $(\mathcal{P} \hookrightarrow \mathcal{Q})$-NBNa-reduction, that is, we have to give a reduction algorithm $\mathcal{S}$ that depends on f in a non-black-box-way, and yet $\mathcal{S}$ depends on $\mathcal{A}$ only in a black-box way. We proceed by case distinction over f.

Case I: Suppose $f \in \mathcal{F}_{\mathcal{Q}}$ such that for all constructions G, the primitive G^f is a secure implementation of $\mathcal{P}$, i.e., for all polynomial-time adversaries $\mathcal{A}$ it holds that $(G^f, \mathcal{A}^f) \notin \mathcal{R}_{\mathcal{P}}$. Then proving the existence of a reduction satisfying the implication $(G^f, \mathcal{A}^f) \in \mathcal{R}_{\mathcal{P}} \Rightarrow (f, \mathcal{S}^{\mathcal{A},f}) \in \mathcal{R}_{\mathcal{Q}}$ is trivial, as the premise of the implication is never satisfied.

Case II: For any $f \in \mathcal{F}_{\mathcal{Q}}$ outside the class described in Case I, we know that there exists a PPT construction G such that for all $\mathcal{A}$ there is a reduction algorithm $\mathcal{S}$ that satisfies $(G^f, \mathcal{A}^f) \in \mathcal{R}_{\mathcal{P}} \Rightarrow (f, \mathcal{S}^{\mathcal{A},f}) \in \mathcal{R}_{\mathcal{Q}}$, and such an efficient $\mathcal{A}$ with $(G^f, \mathcal{A}^f) \in \mathcal{R}_{\mathcal{P}}$ exists. For any such f, we now fix a unique adversary $\mathcal{A}_f$, say, by taking the random variable $\mathcal{A}_f$ with the shortest description according to a particular encoding, such that it satisfies $(G^f, \mathcal{A}_f^f) \in \mathcal{R}_{\mathcal{P}}$. For such an $\mathcal{A}_f$ let $\mathcal{S}$ be a probabilistic polynomial-time reduction making black-box use of $\mathcal{A}_f$ such that $(f, \mathcal{S}^{\mathcal{A}_f,f}) \in \mathcal{R}_{\mathcal{Q}}$. Consider the oracle algorithm $\mathcal{S}_f^f$ that has the same behavior as $\mathcal{S}^{\mathcal{A}_f,f}$, but it incorporates $\mathcal{A}_f$ and only has an f-oracle. The algorithm $\mathcal{S}_f^f$ only depends on f, satisfies $(\mathcal{S}_f^f, f) \in \mathcal{R}_{\mathcal{Q}}$, and is implementable in probabilistic polynomial time, as $\mathcal{S}$ and $\mathcal{A}_f$ are both polynomial time algorithms. Thus, regardless of construction G, we showed that for all f there is an efficient reduction $\mathcal{S}$ such that $(\mathcal{S}^f, f) \in \mathcal{R}_{\mathcal{Q}}$, namely by choosing $\mathcal{S}^f = \mathcal{S}_f^f$. Thus, we also know that for all f, there is a reduction $\mathcal{S}$ such that for all $\mathcal{A}$, if $(\mathcal{A}, G^f) \in \mathcal{R}_P$ then $(\mathcal{S}^f, f) \in \mathcal{R}_{\mathcal{Q}}$. If now, we add an adversary oracle $\mathcal{A}$ that is ignored[1] by $\mathcal{S}$, we also obtain that $(\mathcal{S}^f, f) \in \mathcal{R}_{\mathcal{Q}}$. And thus, there is a $(\mathcal{P} \hookrightarrow \mathcal{Q})$-NBNa-reduction. □

We now show that, while a reduction for inefficient adversaries always implies a reduction for efficient adversaries of the same type, the converse is not true in general.

Theorem 3. *For each XYZ ∈ {BBB, BNB, BBN, NBB, BNN, NBN, NNN}, there are primitives $\mathcal{P}$ and $\mathcal{Q}$ such that there is a $(\mathcal{P} \hookrightarrow \mathcal{Q})$-XYZa-reduction, but no $(\mathcal{P} \hookrightarrow \mathcal{Q})$-XYZ-reduction.*

Proof. For the primitive $\mathcal{P}$ we consider a trivial primitive, namely the constant all-zero function, denoted f_0. Let $\mathcal{L}$ be an EXPTIME-complete problem. The pair $(f_0, \mathcal{A})$ is in the relation $\mathcal{R}_{\mathcal{P}}$ if and only if the adversary $\mathcal{A}$ is a deterministic function that decides $\mathcal{L}$. Let $\mathcal{F}_{\mathcal{Q}}$ also consist of the set that only contains the all-zero function f_0. The relation $\mathcal{R}_{\mathcal{Q}}$ is empty. Observe that, for efficient adversaries,

[1] Here, we require the relation to be machine-independent.

the primitive $\mathcal{P}$ is secure because EXPTIME strictly contains the complexity class P [23]. Thus, there is a trivial reduction since the premise of the implication

$$(G^f, \mathcal{A}^f) \in \mathcal{R}_{\mathcal{P}} \Rightarrow (f, \mathcal{S}^{\mathcal{A},f}) \in \mathcal{R}_{\mathcal{Q}}$$

is never satisfied for any efficient adversary $\mathcal{A}$. Hence, for all $XYZ \neq$ NNB, there is a $(\mathcal{P} \hookrightarrow \mathcal{Q})$-$XYZ$a-reduction. In contrast, inefficient adversaries can break the primitive $\mathcal{P}$, while, as $\mathcal{R}_{\mathcal{Q}}$ is empty, no reduction $\mathcal{S}$ can break $\mathcal{R}_{\mathcal{Q}}$, even oracle $\mathcal{A}$. Thus, for all $XYZ \in \{\text{BBB}, \text{BNB}, \text{BBN}, \text{NBB}, \text{BNN}, \text{NBN}, \text{NNN}\}$, there is no $(\mathcal{P} \hookrightarrow \mathcal{Q})$-$XYZ$-reduction. □

2.5 Relativizing Reductions

In complexity theory as in cryptography, most reductions relativize in the presence of oracles, i.e., if a (secure instantiation of the) primitive $\mathcal{P}$ can be built from a (secure instantiation of the) primitive $\mathcal{Q}$, then the construction still works, if additionally, all parties get access to a random oracle (or any other oracle). We say that there is a *relativizing* reduction from $\mathcal{P}$ to $\mathcal{Q}$, if for all oracles Π, the primitive $\mathcal{P}$ exists relative to Π, whenever $\mathcal{Q}$ exists relative to Π. Often, separation results rule out such reductions.

Definition 5 (Relativizing Reduction). *There exists a* relativizing *reduction from a primitive $\mathcal{P}$ to a primitive $\mathcal{Q}$, if for all oracles Π, the primitive $\mathcal{P}$ exists relative to Π whenever $\mathcal{Q}$ exists relative to Π. A primitive $\mathcal{P}$ is said to exist relative to Π if there is an $f \in \mathcal{F}_{\mathcal{P}}$ which has an efficient implementation when having access to the oracle Π such that there is no probabilistic polynomial-time algorithm $\mathcal{A}$ with $(f, \mathcal{A}^{\Pi,f}) \in \mathcal{R}_{\mathcal{P}}$.*

We remark that, since we define security relations over random variables and not their implementations, it is understood that the implementation of f may actually depend on Π, too. According to Reingold et al. [33], relativizing reductions are a relatively restrictive notion of reducibility that they place between BBB-reductions and NNNa-reductions. Jumping ahead, we note this is due their treatment of (in-)efficient adversaries: they require BBB-reductions to also work for inefficient adversaries $\mathcal{A}$, and so do we. In contrast, for NNNa-reductions, Reingold et al. allow the reduction algorithm to fail for inefficient adversaries $\mathcal{A}$. As we can show, *all* notions of reducibility for inefficient adversaries, including NNN-reductions, imply relativizing reductions, i.e., we can place relativizing reductions between NNN- and NNNa-reductions showing that, in fact, the notion is very liberal compared to notions of reductions that treat inefficient adversaries. In contrast, for efficient adversaries, relativizing reductions imply NNNa- and (the equivalent) NBNa-reductions and are incomparable to all stronger notions that treat efficient adversaries.

We now prove that relativizing reductions are implied by NNN-reductions for inefficient adversaries, i.e., according to Definition C.4 of [1]. The proof is inspired by Reingold et al. [33] who show that BBB-reductions imply relativizing reductions.

Theorem 4. *If there is a $(P \hookrightarrow Q)$-NNN-reduction, then there is a relativizing reduction from $\mathcal{P}$ to $\mathcal{Q}$.*

Proof. Assume there is an NNN-reduction between two primitives $\mathcal{P}$ and $\mathcal{Q}$ and assume towards contradiction that there is an oracle Π such that $\mathcal{Q}$ exists relative to this oracle, but $\mathcal{P}$ does not. Let $f \in \mathcal{F}_{\mathcal{Q}}$ be an instantiation of $\mathcal{Q}$ that is efficiently computable by an algorithm that has oracle access to Π and such that f is secure against all efficient oracle machines $\mathcal{S}$, i.e., for all probabilistic polynomial-time machines $\mathcal{S}$, one has $(f, \mathcal{S}^{\Pi}) \notin \mathcal{R}_{\mathcal{Q}}$. By assumption of a $(\mathcal{P} \hookrightarrow \mathcal{Q})$-NNN-reduction, there exists a PPT oracle algorithm G for f, such that for all (possibly unbounded) adversaries $\mathcal{A}$ there is a PPT reduction algorithm $\mathcal{S}$ such that $(G^f, \mathcal{A}^f) \in \mathcal{R}_{\mathcal{P}}$ implies $(f, \mathcal{S}^{f,\mathcal{A}}) \in \mathcal{R}_{\mathcal{Q}}$. Now, G^f is efficiently computable relative to the oracle Π, because G is PPT and f is efficiently computable relative to Π. Since $\mathcal{P}$ does not exist relative to Π, there is an efficient adversary $\mathcal{A}$ such that $(G^f, \mathcal{A}^{\Pi}) \in \mathcal{R}_{\mathcal{P}}$, i.e., by considering that the relations are defined over random variables, setting $\mathcal{A}' := \mathcal{A}^{\Pi}$ one also has $(G^f, \mathcal{A}'^f) \in \mathcal{R}_{\mathcal{P}}$. Thus, the NNN-reduction gives an efficient reduction $\mathcal{S}$ such that $(f, \mathcal{S}^{\mathcal{A}',f}) \in \mathcal{R}_{\mathcal{Q}}$. As $\mathcal{S}$ is PPT and as f and $\mathcal{A}'$ are efficiently computable relative to oracle Π, one has that $\mathcal{S}^{\mathcal{A}',f}$ is efficiently computable relative to Π. Thus, f is not "$\mathcal{Q}$-secure" against all efficient oracle machines with oracle access to Π, yielding a contradiction. □

This proves that for inefficient adversaries, relativizing reductions are implied by NNN-reductions, the most liberal notion of reductions for inefficient adversaries. Conversely, for efficient adversaries, relativizing reductions imply NNNa and NBNa reductions, but they are not implied by any of the stronger notions. We adapt the proof due to Reingold et al. [33] for the following theorem.

Theorem 5. *If there is a relativizing reduction from $\mathcal{P}$ to $\mathcal{Q}$, then there is a $(\mathcal{P} \hookrightarrow \mathcal{Q})$-NNNa-reduction, and a $(\mathcal{P} \hookrightarrow \mathcal{Q})$-NBNa-reduction.*

Proof. It suffices to show that relativizing reductions imply NNNa-reductions for efficient adversaries, as Theorem 2 proves that NBNa-reductions and NNNa-reductions are equivalent. Assume that there is a relativizing reduction between the primitives $\mathcal{P}$ and $\mathcal{Q}$, and assume towards contradiction that there is an $f \in \mathcal{F}_{\mathcal{Q}}$ such that for all constructions G, there is an efficient adversary $\mathcal{A}$ such that for all efficient reductions algorithms $\mathcal{S}$, it holds that $(G^f, \mathcal{A}^f) \in \mathcal{R}_{\mathcal{P}}$ but, simultaneously, $(f, \mathcal{S}^{\mathcal{A},f}) \notin \mathcal{R}_{\mathcal{Q}}$. Then, by definition, relative to oracle f, the primitive $\mathcal{Q}$ exists, as no efficient algorithm with oracle access to f can break f. Note that we can view $\mathcal{S}^f$ as an algorithm $\mathcal{S}'^{\mathcal{A},f}$ which does not query $\mathcal{A}$ but has the same output distribution, if viewed as random variables. By assumption, there exists a relativizing reduction between $\mathcal{P}$ and $\mathcal{Q}$, and thus, relative to the oracle f, not only $\mathcal{Q}$ exists but also the primitive $\mathcal{P}$. In particular, there is a probabilistic polynomial-time oracle machine G such that G^f implements $\mathcal{P}$ and such that for all efficient oracle machines $\mathcal{A}$, one has $(G^f, \mathcal{A}^f) \notin \mathcal{R}_{\mathcal{P}}$, i.e., $\mathcal{P}$ is secure against all efficient adversaries that get f as an oracle, a contradiction. □

Theorem 6. *For $XYZ \in \{BBB, NBB, BBN, BNB, BNN, NBN, NNN\}$, there are primitives $\mathcal{P}$ and $\mathcal{Q}$ such that there is a $(\mathcal{P} \hookrightarrow \mathcal{Q})$-$XYZa$-reduction for efficient adversaries, but no relativizing reduction.*

Proof. We show that BBBa-reductions do not imply relativizing reductions; as BBBa-reductions imply the "lower level" reductions, the other cases follow. We use the same approach as for Theorem 3.

Let $\mathcal{Q}$ be the primitive that contains the constant 0-function f_0. We define the relation $\mathcal{R}_\mathcal{P}$ such that $\mathcal{P}$ is trivially secure against all *efficient* adversaries, namely, let $\mathcal{L}$ be an EXPTIME-complete language, then $(f_0, \mathcal{A})$ is in $\mathcal{R}_\mathcal{P}$ if $\mathcal{A}$ is a deterministic function and decides $\mathcal{L}$. As the complexity class P is strictly contained in EXPTIME, no efficient adversary can break $\mathcal{P}$. Let $\mathcal{Q}$ also be the primitive that contains the constant 0-function f_0, but with a different relation, namely $\mathcal{R}_\mathcal{Q}$ is empty. In particular, no adversary can break $\mathcal{Q}$. Hence, there is a trivial $(\mathcal{P} \hookrightarrow \mathcal{Q})$-BBBa-reduction, because the premise of the implication

$$(G^f, \mathcal{A}^f) \in \mathcal{R}_\mathcal{P} \Rightarrow (f, \mathcal{S}^{\mathcal{A},f}) \in \mathcal{R}_\mathcal{Q}$$

is never satisfied for efficient adversaries and the implication is thus trivially true. In contrast, there is no relativizing reduction between the two primitives. That is, assume, we add an oracle that decides the EXPTIME-complete language $\mathcal{L}$, then relative to this oracle, there are suddenly efficient adversaries that break $\mathcal{P}$. However, as $\mathcal{R}_\mathcal{Q}$ is still empty, there cannot be a reduction $\mathcal{S}$ in this oracle world, giving us a contradiction. □

Reingold et al. [33] note that BNNa-reductions for efficient adversaries and relativizing reductions are often equivalent. In particular, they prove that if a primitive $\mathcal{Q}$ allows any oracle Π to be embedded into it, then a $(\mathcal{P} \hookrightarrow \mathcal{Q})$-BNNa-reduction implies a $(\mathcal{P} \hookrightarrow \mathcal{Q})$-relativizing reduction. However, *efficient* primitives $\mathcal{Q}$ such as one-way functions (as opposed to random oracles, for example), are not known to satisfy this property. We discuss this issue in more detail in the following section about efficient primitives.

2.6 Efficient Primitives versus Inefficient Primitives

A reduction for *efficient* primitives is a reduction that only works if $f \in \mathcal{F}_\mathcal{Q}$ is efficiently implementable, i.e., in probabilistic polynomial-time. If we make this distinction then, according to Figure 1, we unfold another dimension (analogously to the case of efficient adversaries). As we discuss below our results for non-efficient primitives hold in this "parallel universe" of efficient primitives as well, and between the two universes there are straightforward implications and separations (as in the case of efficient and inefficient adversaries).

Technically, one derives the efficient primitive version XYZp of an XYZ-reduction by replacing all universal quantifiers over primitives f in $\mathcal{F}_\mathcal{Q}$ by universal quantifiers that are restricted to efficiently implementable f in $\mathcal{F}_\mathcal{Q}$. More concretely, we replace $\forall f \in \mathcal{F}_\mathcal{Q}$ by the term $\forall \mathsf{PPT} f \in \mathcal{F}_\mathcal{Q}$. For example, the notion of a BBBp-reduction then reads as follows:

Definition 6 (($\mathcal{P} \hookrightarrow \mathcal{Q}$)-BBBp or Fully Black-Box Reduction for Efficient Primitives). *There exists a* fully black-box (or BBBp-)reduction *for efficient primitives from* $\mathcal{P} = (\mathcal{F}_\mathcal{P}, \mathcal{R}_\mathcal{P})$ *to* $\mathcal{Q} = (\mathcal{F}_\mathcal{Q}, \mathcal{R}_\mathcal{Q})$ *if there exist probabilistic polynomial-time oracle algorithms* G *and* $\mathcal{S}$ *such that:*

Correctness. *For every polynomial-time computable function* $f \in \mathcal{F}_\mathcal{Q}$*, it holds that* $G^f \in \mathcal{F}_\mathcal{P}$*.*
Security. *For every polynomial-time computable function* $f \in \mathcal{F}_\mathcal{Q}$ *and every machine* $\mathcal{A}$*, if* $(G^f, \mathcal{A}) \in \mathcal{R}_\mathcal{P}$*, then* $(f, \mathcal{S}^{\mathcal{A},f}) \in \mathcal{R}_\mathcal{Q}$*, i.e.,*

$$\exists \mathsf{PPT} G\ \exists \mathsf{PPT} \mathcal{S}\ \forall \mathsf{PPT} f \in \mathcal{F}_\mathcal{Q}\ \forall \mathcal{A}\ \ ((G^f, \mathcal{A}^f) \in \mathcal{R}_\mathcal{P} \Rightarrow (f, \mathcal{S}^{\mathcal{A},f}) \in \mathcal{R}_\mathcal{Q}).$$

In the same manner, for any XYZ-reduction, we can define the corresponding XYZp-reduction. Similarly, one can transform all reduction types XYZa for efficient adversaries into reduction types XYZap for efficient adversaries *and efficient primitives.* Most relations that this paper establishes for XYZ-reductions and XYZa-reductions also hold for XYZp- and XYZap-reductions, except for the relation to relativizing reductions, where only some of the results carry over, see Theorem 2.15 of [1]. Building on proof ideas of Theorem 3, we also establish in Theorem 2.14 of [1] that the implication from reductions for arbitrary primitives to reductions for efficient primitives is strict. We refer the reader to the full version of this paper [1] for formal theorem statements, proofs and further discussion of the relations of reductions for efficient primitives.

3 Parametrized Black-Box Reductions

Many reductions in cryptography commonly classified as "black box" technically do not fall in this class, as a black box reduction algorithm must not have any information about the adversary beyond the input/output behavior, except for the sole guarantee that it breaks security with non-negligible probability. Strictly speaking, this excludes information such as running time, number of queries, or the actual success probability of a given adversary. This prompts the question of what the "natural" notion of a black-box reduction should be. Not surprisingly, the answer is a matter of taste, just like the question whether fully black-box or semi black-box is the "right" notion of a black-box reduction. As in the case of different notions of black-box reductions, we can nonetheless give a technically profound, and yet easy-to-use notion of *parametrized* black-box reductions (of any type). In the full version [1] we motivate and formalize two different degrees of parameterization by distinguishing between parameter-*aware* and parameter-*dependent* reductions. The difference is essentially whether or not the reduction algorithm receives the parameter values as input.

We note that parametrized black-box reductions and separations rely critically on the specific parameters. In particular, some of our separations consider reductions that are required to depend on, say, the success probability of the adversary, as in the case of the Goldreich–Levin hardcore bit. This separation

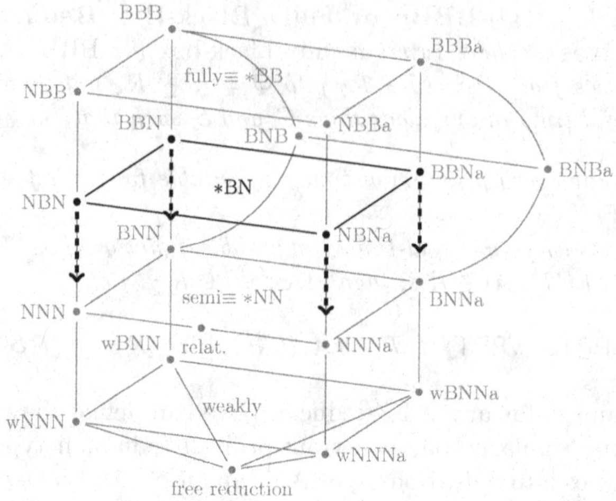

Fig. 4. The effect of parametrization (in the case of ∗BN-reductions). Parametrized counterparts of each type partly descend towards the corresponding ∗NN-reduction with full dependency on the adversary.

does not carry over to the parametrized case. In contrast, separations for efficient/inefficient adversaries as well as the theorems on relativized reductions still apply.

More pictorially, one can imagine parametrized black-box reductions in light of our Figure 1 as descending from the ∗B∗ plane for black-box adversaries towards the ∗N∗ plane, where the reduction can completely depend on the adversary, see Figure 4. The parameters and the distinction between awareness and dependency determines how far one descends. Analogously, parametrization for BBB-reductions means to descend from the top node BBB to BNB (also in the case of efficient adversaries). As such, it is clear that implications along edge paths remain valid, e.g., a parametrized NBN-reduction still implies a NNN-reduction.

The case of NBB-reductions, however, shows that parametrization cannot fully bridge the gap to NNB-reductions. As explained before, the latter type with quantification $\forall \mathcal{A} \exists \mathcal{S} \forall f \exists G$ does not seem to be meaningful, because the construction G would now depend on the adversary $\mathcal{A}$. Parametrization of NBB-reductions (with quantification $\exists \mathcal{S} \forall f \exists G \forall \mathcal{A}$) still makes sense, though, because the dependency of $\mathcal{S}$ on the adversary is only through the running time or the input. Put differently, the parametrization allows for the "admissible non-black-boxness" for the NBB type of reduction. If one parametrizes the black-box access to the primitive, either for the construction or the reduction, then this parametrization corresponds to a (partial) shift from back plane to the front plane resp. from the top ∗BB plane to the ∗BN plane. In the full version of this paper [1], we establish formal relationships between parameter-awareness and parameter-depedency.

4 Conclusion

We provide a comprehensive framework to classify black-box reductions more precisely. We believe that this is important to fully understand and appreciate the implications and limitations of black-box separation results. In particular, we point out how subtleties such as different possibilities to define a primitive, the distinction between efficient and non-efficient adversaries and primitives, or parameterization, affect the results. Such details have previously been often neglected, and our work draws more attention to these issues.

Acknowledgements. We thank the anonymous reviewers and Pooya Farshim for valuable comments on previous versions of this work. Paul Baecher is supported by grant Fi 940/4-1 of the German Research Foundation (DFG). Christina Brzuska was supported in part by the Israel Ministry of Science and Technology (grant 3-9094) and by the Israel Science Foundation (grant 1155/11 and grant 1076/11); parts of the work done while being at TU Darmstadt and CASED (`www.cased.de`). Marc Fischlin is supported by the Heisenberg Program of the DFG under grant Fi 940/3-1.

References

1. Baecher, P., Brzuska, C., Fischlin, M.: Notions of black-box reductions, revisited. Cryptology ePrint Archive, Report 2013/101 (2013), `http://eprint.iacr.org/`
2. Barak, B.: How to go beyond the black-box simulation barrier. In: 42nd FOCS, pp. 106–115. IEEE Computer Society Press (October 2001)
3. Barhum, K., Holenstein, T.: A cookbook for black-box separations and a recipe for UOWHFs. In: Sahai, A. (ed.) TCC 2013. LNCS, vol. 7785, pp. 662–679. Springer, Heidelberg (2013)
4. Bitansky, N., Paneth, O.: From the impossibility of obfuscation to a new non-black-box simulation technique. In: Proceedings of the Annual Symposium on Foundations of Computer Science (FOCS) 2012, pp. 223–232. IEEE Computer Society Press (2012)
5. Boldyreva, A., Cash, D., Fischlin, M., Warinschi, B.: Foundations of non-malleable hash and one-way functions. In: Matsui, M. (ed.) ASIACRYPT 2009. LNCS, vol. 5912, pp. 524–541. Springer, Heidelberg (2009)
6. Boneh, D., Venkatesan, R.: Breaking RSA may not be equivalent to factoring. In: Nyberg, K. (ed.) EUROCRYPT 1998. LNCS, vol. 1403, pp. 59–71. Springer, Heidelberg (1998)
7. Brakerski, Z., Katz, J., Segev, G., Yerukhimovich, A.: Limits on the power of zero-knowledge proofs in cryptographic constructions. In: Ishai, Y. (ed.) TCC 2011. LNCS, vol. 6597, pp. 559–578. Springer, Heidelberg (2011)
8. Bresson, E., Monnerat, J., Vergnaud, D.: Separation results on the "one-more" computational problems. In: Malkin, T. (ed.) CT-RSA 2008. LNCS, vol. 4964, pp. 71–87. Springer, Heidelberg (2008)
9. Coron, J.-S.: Optimal security proofs for PSS and other signature schemes. In: Knudsen, L.R. (ed.) EUROCRYPT 2002. LNCS, vol. 2332, pp. 272–287. Springer, Heidelberg (2002)

10. Dodis, Y., Haitner, I., Tentes, A.: On the instantiability of hash-and-sign RSA signatures. In: Cramer, R. (ed.) TCC 2012. LNCS, vol. 7194, pp. 112–132. Springer, Heidelberg (2012)
11. Dodis, Y., Oliveira, R., Pietrzak, K.: On the generic insecurity of the full domain hash. In: Shoup, V. (ed.) CRYPTO 2005. LNCS, vol. 3621, pp. 449–466. Springer, Heidelberg (2005)
12. Fischlin, M., Fleischhacker, N.: Limitations of the meta-reduction technique: The case of schnorr signatures. In: Johansson, T., Nguyen, P.Q. (eds.) EUROCRYPT 2013. LNCS, vol. 7881, pp. 444–460. Springer, Heidelberg (2013)
13. Fischlin, M., Lehmann, A., Ristenpart, T., Shrimpton, T., Stam, M., Tessaro, S.: Random oracles with(out) programmability. In: Abe, M. (ed.) ASIACRYPT 2010. LNCS, vol. 6477, pp. 303–320. Springer, Heidelberg (2010)
14. Fischlin, M., Schröder, D.: On the impossibility of three-move blind signature schemes. In: Gilbert, H. (ed.) EUROCRYPT 2010. LNCS, vol. 6110, pp. 197–215. Springer, Heidelberg (2010)
15. Gentry, C., Wichs, D.: Separating succinct non-interactive arguments from all falsifiable assumptions. In: Fortnow, L., Vadhan, S.P. (eds.) 43rd ACM STOC, pp. 99–108. ACM Press (June 2011)
16. Goldreich, O.: Foundations of Cryptography: Basic Applications, vol. 2. Cambridge University Press, Cambridge (2004)
17. Goldreich, O., Goldwasser, S., Micali, S.: How to construct random functions. Journal of the ACM 33, 792–807 (1986)
18. Goldreich, O., Impagliazzo, R., Levin, L.A., Venkatesan, R., Zuckerman, D.: Security preserving amplification of hardness. In: FOCS, pp. 318–326. IEEE Computer Society (1990)
19. Goldreich, O., Levin, L.A.: A hard-core predicate for all one-way functions. In: 21st ACM STOC, pp. 25–32. ACM Press (May 1989)
20. Haitner, I., Hoch, J.J., Reingold, O., Segev, G.: Finding collisions in interactive protocols - a tight lower bound on the round complexity of statistically-hiding commitments. In: 48th FOCS, pp. 669–679. IEEE Computer Society Press (October 2007)
21. Haitner, I., Holenstein, T.: On the (Im)Possibility of key dependent encryption. In: Reingold, O. (ed.) TCC 2009. LNCS, vol. 5444, pp. 202–219. Springer, Heidelberg (2009)
22. Haitner, I., Holenstein, T.: On the (Im)Possibility of key dependent encryption. In: Reingold, O. (ed.) TCC 2009. LNCS, vol. 5444, pp. 202–219. Springer, Heidelberg (2009)
23. Hartmanis, J., Stearns, R.E.: On the computational complexity of algorithms. Transactions of the American Mathematical Society 117, 285–306 (1965)
24. Håstad, J., Impagliazzo, R., Levin, L.A., Luby, M.: A pseudorandom generator from any one-way function. SIAM Journal on Computing 28(4), 1364–1396 (1999)
25. Hsiao, C.-Y., Reyzin, L.: Finding collisions on a public road, or do secure hash functions need secret coins? In: Franklin, M. (ed.) CRYPTO 2004. LNCS, vol. 3152, pp. 92–105. Springer, Heidelberg (2004)
26. Impagliazzo, R., Rudich, S.: Limits on the provable consequences of one-way permutations. In: 21st ACM STOC, pp. 44–61. ACM Press (May 1989)
27. Kiltz, E., Pietrzak, K.: On the security of padding-based encryption schemes – or – why we cannot prove OAEP secure in the standard model. In: Joux, A. (ed.) EUROCRYPT 2009. LNCS, vol. 5479, pp. 389–406. Springer, Heidelberg (2009)

28. Lindell, Y., Omri, E., Zarosim, H.: Completeness for symmetric two-party functionalities - revisited. In: Wang, X., Sako, K. (eds.) ASIACRYPT 2012. LNCS, vol. 7658, pp. 116–133. Springer, Heidelberg (2012)
29. Mahmoody, M., Pass, R.: The curious case of non-interactive commitments – on the power of black-box vs. non-black-box use of primitives. In: Safavi-Naini, R., Canetti, R. (eds.) CRYPTO 2012. LNCS, vol. 7417, pp. 701–718. Springer, Heidelberg (2012)
30. Paillier, P., Villar, J.L.: Trading one-wayness against chosen-ciphertext security in factoring-based encryption. In: Lai, X., Chen, K. (eds.) ASIACRYPT 2006. LNCS, vol. 4284, pp. 252–266. Springer, Heidelberg (2006)
31. Pass, R.: Limits of provable security from standard assumptions. In: Fortnow, L., Vadhan, S.P. (eds.) 43rd ACM STOC, pp. 109–118. ACM Press (June 2011)
32. Pass, R., Tseng, W.-L.D., Venkitasubramaniam, M.: Towards non-black-box lower bounds in cryptography. In: Ishai, Y. (ed.) TCC 2011. LNCS, vol. 6597, pp. 579–596. Springer, Heidelberg (2011)
33. Reingold, O., Trevisan, L., Vadhan, S.P.: Notions of reducibility between cryptographic primitives. In: Naor, M. (ed.) TCC 2004. LNCS, vol. 2951, pp. 1–20. Springer, Heidelberg (2004)
34. Rompel, J.: One-way functions are necessary and sufficient for secure signatures. In: 22nd ACM STOC, pp. 387–394. ACM Press (May 1990)
35. Seurin, Y.: On the exact security of schnorr-type signatures in the random oracle model. In: Pointcheval, D., Johansson, T. (eds.) EUROCRYPT 2012. LNCS, vol. 7237, pp. 554–571. Springer, Heidelberg (2012)
36. Simon, D.R.: Findings collisions on a one-way street: Can secure hash functions be based on general assumptions? In: Nyberg, K. (ed.) EUROCRYPT 1998. LNCS, vol. 1403, pp. 334–345. Springer, Heidelberg (1998)
37. Yao, A.C.: Theory and applications of trapdoor functions. In: 23rd FOCS, pp. 80–91. IEEE Computer Society Press (November 1982)

Adaptive and Concurrent Secure Computation from New Adaptive, Non-malleable Commitments

Dana Dachman-Soled[1,⋆], Tal Malkin[2], Mariana Raykova[3,⋆⋆], and Muthuramakrishnan Venkitasubramaniam[4]

[1] University of Maryland, College Park, MD 20742, USA
danadach@ece.umd.edu
[2] Columbia University, New York, NY 10027, USA
tal@cs.columbia.edu
[3] IBM T.J. Watson Research Center, Yorktown Heights, NY 10598, USA, and SRI, Menlo Park, CA 94025, USA
mariana@cs.columbia.edu
[4] University of Rochester, Rochester, NY 14627, USA
muthuv@cs.rochester.edu

Abstract. We present a unified approach for obtaining general secure computation that achieves adaptive-Universally Composable (UC)-security. Using our approach we essentially obtain all previous results on adaptive concurrent secure computation, both in relaxed models (e.g., quasi-polynomial time simulation), as well as trusted setup models (e.g., the CRS model, the imperfect CRS model). This provides conceptual simplicity and insight into what is required for adaptive and concurrent security, as well as yielding improvements to set-up assumptions and/or computational assumptions in known models. Additionally, we provide the first constructions of concurrent secure computation protocols that are adaptively secure in the timing model, and the non-uniform simulation model. As a corollary we also obtain the first adaptively secure multiparty computation protocol in the plain model that is secure under bounded-concurrency.

Conceptually, our approach can be viewed as an adaptive analogue to the recent work of Lin, Pass and Venkitasubramaniam [STOC '09], who considered only non-adaptive adversaries. Their main insight was that the non-malleability requirement could be decoupled from the simulation requirement to achieve UC-security. A main conceptual contribution of this work is, quite surprisingly, that it is still the case even when considering adaptive security.

A key element in our construction is a commitment scheme that satisfies a strong definition of non-malleability. Our new primitive of *concurrent equivocal non-malleable commitments*, intuitively, guarantees that even when a man-in-the-middle adversary observes concurrent equivocal commitments and decommitments, the binding property of the commitments continues to hold for commitments made by the adversary. This definition is stronger than previous ones, and may be of independent interest. Previous constructions that satisfy our definition have been constructed in setup models, but either require existence of stronger encryption schemes such as CCA-secure encryption or require independent "trapdoors" provided by the setup for every pair of parties to ensure non-malleability. A main technical contribution of this work is to provide a construction that eliminates these requirements and requires *only* a single trapdoor.

⋆ Most of this work was done while at Columbia University and Microsoft Research.
⋆⋆ Supported by NSF Grant No.1017660.

K. Sako and P. Sarkar (Eds.) ASIACRYPT 2013 Part I, LNCS 8269, pp. 316–336, 2013.

1 Introduction

The notion of *secure multi-party computation* allows mutually distrustful parties to securely compute a function on their inputs, such that only the (correct) output is obtained, and no other information is leaked, even if the adversary controls an arbitrary subset of parties. This security is formalized via the real/ideal simulation paradigm, requiring that whatever the adversary can do in a real execution of the protocol, can be simulated by an adversary ("simulator") working in the ideal model, where the parties submit their inputs to a trusted party who then computes and hands back the output. Properly formalizing this intuitive definition and providing protocols to realize it requires care, and has been the subject of a long line of research starting in the 1980s.

In what is recognized as one of the major breakthroughs in cryptography, strong feasibility results were provided, essentially showing that *any function that can be efficiently computed, can be efficiently computed securely,* assuming the existence of enhanced trapdoor permutations (eTDP) [46,27]. However, these results were originally investigated in the *stand-alone setting*, where a single instance of the protocol is run in isolation. A stronger notion is that of *concurrent security*, which guarantees security even when many different protocol executions are carried out concurrently. In this work, we focus on the strongest (and most widely used) notion of concurrent security, namely universally-composable (UC) security [6]. This notion guarantees security even when an unbounded number of different protocol executions are run concurrently in an arbitrary interleaving schedule and is critical for maintaining security in an uncontrolled environment that allows concurrent executions (e.g., the Internet). Moreover, this notion also facilitates modular design and analysis of protocols, by allowing the design and security analysis of small protocol components, which may then be composed to obtain a secure protocol for a complex functionality.

Unfortunately, achieving these strong notions of concurrent security is far more challenging than achieving stand-alone security, and we do not have general feasibility results for concurrently secure computation of every function. In fact, there are lower bounds showing that concurrent security (which is implied by UC security) cannot be achieved for general functions, unless trusted setup is assumed [8,9,35]. Previous works overcome this barrier either by using some trusted setup infrastructure [8,11,2,7,30,12], or by relaxing the definition of security [39,45,3,10,25] (we will see examples below).

Another aspect of defining secure computation, is the power given to the adversary. A *static* (or non-adaptive) adversary is one who has to decide which parties to corrupt at the outset, before the execution of the protocol begins. A stronger notion is one that allows for an *adaptive* adversary, who may corrupt parties at any time, based on its current view of the protocol. It turns out that achieving security in the adaptive setting is much more challenging than in the static one. The intuitive reason for this is that the simulator needs to simulate messages from uncorrupted parties, but may later need to explain the messages (i.e. produce the randomness used to generate those messages) when that party is corrupted. Moreover, the simulator must simulate messages from uncorrupted parties *without knowing their inputs*, but when corrupted, must explain the messages according to the actual input that the party holds. On the other hand, in the real protocol execution, messages must information-theoretically determine the actual inputs of the party, both for correctness as well as to ensure that an adversary is committed to its inputs and cannot cheat. We note that although the setting of adaptive corruptions *with erasures* has been considered in the literature, in our work we assume adaptive corruptions *without erasures*. Here we assume that honest parties cannot reliably erase randomness used to generate messages of the protocol and thus when corrupted, the adversary learns the randomness used by that party to generate previous protocol messages. Clearly, this is the more general and challenging setting. Canetti, Lindell, Ostrovsky and Sahai [11] provided the first constructions of UC-secure protocols with static and adaptive security in the *common reference string* model (CRS)[1]. Subsequently, several results were obtained for both the static and adaptive case in other trusted-setup models and relaxed-security models. The techniques for achieving security against adaptive adversaries are generally quite different than the techniques needed to achieve security against static adversaries, and many results for concurrent secure computation do not readily extend to the adaptive setting. In fact, several of the previous results allowing general concurrent secure computation (e.g., using a trusted setup) were only proved for the static case [33,34,42,40,22,30], and extending them to the adaptive setting has remained an open problem.

In this paper we focus on the strongest notions of security, and study their fundamental power and limitations. The main question we ask is:

Under which circumstances is adaptive concurrent security generally feasible?

In particular, we refine this question to ask:

What is the minimum setup required to achieve adaptive concurrent security?

We address these questions on both a conceptual and technical level. In particular, we unify and generalize essentially all previous results in the generic adaptive concurrent setting, as well as providing completely new results (constructions with weaker trusted setup requirements, weaker computational assumptions, or in relaxed models of security), conceptual simplicity, and insight into what is required for adaptive and concurrent secure computation. Our main technical tool is a new primitive of equivocal non-malleable commitment. We describe our results in more detail below.

1.1 Our Results

We extend the general framework of [33], to obtain a composition theorem that allows us to establish adaptive UC-security in models both with, and without, trusted set-up. With this theorem, essentially all general UC-feasibility results for adaptive adversaries follow as simple corollaries, often improving the set-up and/or complexity theoretic assumptions; moreover, we obtain adaptive UC secure computation in new models (such as the timing model). Additionally, our work is the first to achieve bounded-concurrent adaptively-secure multiparty computation without setup assumptions. As such, similar to [33], our theorem takes a step towards characterizing those models in which adaptive UC security is realizable, and also at what cost.

Although technically quite different, as mentioned previously, our theorem can be viewed as an adaptive analogue of the work of Lin, Pass and Venkitasubramaniam

[1] In the CRS model, all parties have access to public reference string sampled from a prespecified distribution.

[33], who study the *static* case. Their work puts forward the very general notion of a "UC-puzzle" to capture the models (or setup assumptions) that admit general static UC-security. More precisely, they prove that if we assume the existence of enhanced trapdoor permutations and stand-alone non-malleable commitments, static UC-security is achievable in any model that admits a UC-puzzle. In this work, we establish an analogous result for the more difficult case of *adaptive* UC-security, as we outline below.

We start by introducing the notion of an *Adaptive UC-Puzzle*. Next, we define the new primitive (which may be of independent interest), *equivocal non-malleable commitment* or EQNMCom, which is a commitment with the property that a man-in-the-middle observing concurrent equivocal commitments and decommitments cannot break the binding property. We then present a construction of equivocal non-malleable commitment for any model that admits an adaptive UC-puzzle (thus, requiring this primitive does not introduce an additional complexity-theoretic assumption). Finally, we rely on a computational assumption that is known to imply adaptively secure OT (analogous to the eTDP used by [33], which implies statically secure OT). Specifically, we use *simulatable public key encryption* [18,13]. Although a weaker assumption, *trapdoor* simulatable public key encryption is known to imply semi-honest adaptively secure OT, it is unknown how to achieve malicious, adaptive, UC secure OT (in any setup model) from only trapdoor simulatable public key encryption. We remark here that, more recently, for the static case, Lin et al. show how to extend their framework and rely on the minimal assumptions of stand-alone semi-honest oblivious-transfer and static UC-puzzle [41]. More concretely, we show the following:

Theorem 1 (Main Theorem (Informal)). *Assume the existence of an adaptive UC-secure puzzle Σ using some setup $\mathcal{T}$, the existence of an* EQNMCom *primitive, and the existence of a simulatable public-key encryption scheme. Then, for every m-ary functionality f, there exists a protocol Π using the same set-up $\mathcal{T}$ that adaptively, UC-realizes f.*

As an immediate corollary of our theorem, it follows that to establish feasibility of adaptive UC-secure computation in any set-up model, it suffices to construct an adaptive UC-puzzle in that model. Complementing the main theorem, we show that in many previously studied models, adaptive UC-puzzles are easy to construct. Indeed, in many models the straightforward puzzle constructions for the static case (cf. [33]) are sufficient to obtain adaptive puzzles; some models require puzzle constructions that are more complex (see the full version [17] for details). We highlight some results below.

Adaptive UC in the "imperfect" String Model. Canetti, Pass and shelat [12] consider adaptive UC security where parties have access to an "imperfect" reference string–called a "sunspot"–that is generated by an arbitrary efficient min-entropy source (obtained e.g., by measurement of some physical phenomenon). The CPS-protocol requires m communicating parties to share m reference strings, each of them generated using fresh entropy. We show that a *single* reference string is sufficient for UC and adaptively-secure MPC (regardless of the number of parties m).

Adaptive UC in the Timing Model. Dwork, Naor and Sahai [22] introduced the *timing model* in the context of concurrent zero-knowledge, where all players are assumed to have access to clocks with a certain drift. Kalai, Lindell and Prabhakaran [30] subsequently presented a concurrent secure computation protocol in the timing model;

whereas the timing model of [22] does not impose a maximal upper-bound on the clock drift, the protocol of [30] requires the clock-drift to be "small"; furthermore, it requires extensive use of delays (roughly $n\Delta$, where Δ is the latency of the network). Finally, [33] showed that UC security against *static* adversaries is possible also in the *unrestricted* timing model (where the clock drift can be "large"); additionally, they reduce the use of delays to only $O(\Delta)$. To the best of our knowledge, our work is the first to consider security against adaptive adversaries in the timing model, giving the first feasibility results for UC and adaptively-secure MPC in the timing model; moreover, our results also hold in the unrestricted timing model.

Adaptive UC with Quasi-polynomial Simulation. Pass [39] proposed a relaxation of the standard simulation-based definition of security, allowing for super polynomial-time or Quasi-polynomial simulation (QPS). In the static and adaptive setting, Prabhakaran and Sahai [45] and Barak and Sahai [3] obtained general MPC protocols that are concurrently QPS-secure without any trusted set-up, but rely on strong complexity assumptions. We achieve adaptive security in the QPS model under relatively weak complexity assumptions. Moreover, we achieve a stronger notion of security, which (in analogy with [39]) requires that indistinguishability of simulated and real executions holds for all of quasi-polynomial time; in contrast, [3] only achieves indistinguishability w.r.t. distinguishers with running-time smaller than that of the simulator.

Adaptive UC with Non-uniform Simulation. Lin et al. [33] introduced the non-uniform UC model, which considers environments that are $\mathcal{PPT}$ machines and ideal-model adversaries that are non-uniform $\mathcal{PPT}$ machines and prove feasibility of MPC in the same model. Relying on the same assumptions as those introduced by [33] to construct a puzzle in non-uniform model (along with the assumption of the existence of simulatable PKE), we show feasibility results for secure MPC in the adaptive, non-uniform UC model.

Adaptive Bounded-Concurrent Secure Multiparty Computation. Several works [34,42,40] consider the notion of bounded-concurrency for general functionalities where a single secure protocol Π implementing a functionality f is run concurrently, and there is an *a priori* bound on the number of concurrent executions. In our work, we show how to construct an adaptive puzzle in the bounded-concurrent setting (with no setup assumptions). Thus, we achieve the first results showing feasibility of bounded-concurrency of general functionalities under adaptive corruptions.

In addition to these models, we obtain feasibility of adaptive UC in existing models such as the common reference string (CRS) model [11], uniform reference string (URS) model [11], key registration model [2], tamper-proof hardware model [31], and partially isolated adversaries model [20] (see the full version [17]). For relaxed security models, we obtain UC in the quasi-polynomial time model [39,45,3].

Beyond the specific instantiations, our framework provides conceptual simplicity, technical insight, and the potential to facilitate "translation" of results in the static setting into corresponding (and much stronger) adaptive security results. For example, in recent work of Garg et al. [24], one of the results—constructing UC protocols using multiple setups when the parties share an arbitrary belief about the setups—can be translated to the adaptive model by replacing (static) puzzles with our notion of adaptive puzzles. Other results may require more work to prove, but again are facilitated by our framework.

1.2 Technical Approach and Comparison with Previous Work

There are two basic properties that must be satisfied in order to achieve adaptive UC secure computation: (1) concurrent simulation and (2) concurrent non-malleability. The former requirement amounts to providing the simulator with a trapdoor while the latter requirement amounts to establishing independence of executions. The simulation part is usually "easy" to achieve. Consider, for instance, the *common random string* (CRS) or *Uniform Reference String* (URS) model where the players have access to a public reference string that is sampled uniformly at random. A trapdoor can be easily provided to the simulator as the inverse of the reference string under a pseudo-random generator. Concurrent non-malleability on the other hand is significantly harder to achieve. For the specific case of the CRS model, Canetti et al. [11] and subsequent works [23,37] show that adaptive security can be achieved using a single trapdoor. However, more general setup models require either strong computational assumptions, or provide the simulator with *different* and *independent* trapdoors for different executions. For example, in the URS model, [11] interpret the random string as a public-key for a CCA-secure encryption scheme, and need to assume dense cryptosystems, while in the imperfect random string (sunspot) model, [12] require multiple trapdoors. Other models follow a similar pattern, where concurrent non-malleability is difficult.

In the static case, [33] provided a framework that allowed to decouple the concurrent simulation requirement from the concurrent non-malleability. More precisely, they show that providing a (single) trapdoor to achieve concurrent simulation is sufficient, and once a trapdoor is established concurrent non-malleability can be obtained for free. This allows them to obtain significant improvement in computational/set-up assumptions since no additional assumptions are required to establish non-malleability.

A fundamental question is whether the requirement of concurrent simulation and concurrent non-malleability can be decoupled in the case of adaptive UC-security. Unfortunately, the techniques used in the static case are not applicable in the adaptive case. Let us explain the intuition. [33] and subsequent works rely on *stand-alone non-malleable* primitives to achieve concurrent non-malleability. An important reason this was possible in the static case is because non-malleable primitives can be constructed in the plain-model (i.e. assuming no trapdoor). Furthermore, these primitives inherently admit black-box simulation, i.e. involve the simulator *rewinding* the adversary. Unfortunately, in the adaptive case both these properties are difficult to achieve. First, primitives cannot be constructed in the plain model since adaptive security requires the simulator to be able to simultaneously *equivocate* the simulated messages for honest parties for different inputs and demostrate their validity at any point in the execution by revealing the random coins for the honest parties consistent with the messages. Second, as demostrated in [26], black-box rewinding techniques cannot be employed since the adversary can, in between messages, corrupt an arbitrary subset of the players (some not even participating in the execution) whose inputs are not available to the simulator.

In this work, we show, somewhat surprisingly that a single trapdoor is still sufficient to achieve concurrent non-malleability. Although we do not decouple the requirements, this establishes that even for the case of adaptive security no additional setup, and therefore, no additional assumptions, are required to achieve concurrent non-malleability, thereby yielding similar improvements to complexity and set-up assumptions to [33].

The basic approach we take resembles closely the unified framework of [33]. By relying on previous works [40,42,35,11,27], Lin et. al in [33] argue that to construct a UC protocol for realizing any multi-party functionality, it suffices to construct a zero-knowledge protocol that is concurrently simulatable and concurrently simulation-sound[2]. To formalize concurrent-simulation, they introduce the notion of a UC-puzzle that captures the property that no adversary can successfully complete the puzzle and also obtain a trapdoor, but a simulator exists that can generate (correctly distributed) puzzles together with trapdoors. To achieve simulation-soundness, they introduce the notion of strong non-malleable witness indistinguishability and show how a protocol satisfying this notion can be based on stand-alone non-malleable commitments.

A first approach for the adaptive case, would be to extend the techniques from [33], by replacing the individual components with analogues that are adaptively secure and rely on a similar composition theorem. While the notion of UC-puzzle can be strengthened to the adaptive setting, the composition theorem does not hold for stand-alone non-malleable commitments. This is because, in the static case, it is enough to consider a commitment scheme that is statistically-binding for which an indistinguishability-based notion of non-malleability is sufficient; such a notion, when defined properly, is concurrently composable. However, when we consider adaptive security, commitments need to be equivocable (i.e., the simulator must be capable of producing a fake commitment transcript and inputs for honest committers that allow the transcript to be decommitted to both 0 and 1) and such commitments cannot be statistically-binding. Therefore, we need to consider a stronger simulation-based notion of non-malleability. Furthermore, as mentioned before, an equivocal commitment, even in the stand-alone case, requires the simulator to have a trapdoor, which in turn requires some sort of a trusted set-up.

Our approach here is to consider a "strong" commitment scheme, one that is both equivocable and concurrently non-malleable at the same time, but relies on a UC-puzzle (i.e. single trapdoor) and then establish a new composition theorem that essentially establishes feasibility of UC-secure protocol in any setup that admits a UC-puzzle. While the core contribution of [33] was in identifying the right notion of UC-puzzle and providing a modular analysis, in this work, the main technical novelty is in identifying the right notion of commitment that will allow feasibility with a single trapdoor. Once this is established the results from [33] can be extended analogously by constructing an adaptively secure UC-puzzle for each model. In fact, in most of the models considered in this work, the puzzle constructions are essentially the same as the static case and thus we obtain similar corollaries to [33]. While the general framework for our work resembles [33], as we explain in the next section, the commitment scheme and the composition theorem are quite different and requires an intricate and subtle analysis.

1.3 Main Tool: Equivocal Non-malleable Commitments

We define and construct a new primitive called *equivocal non-malleable commitments* or EQNMCom. Such commitments have previously been defined in the works of [15,16] but only for the limited case of bounded concurrency and non-interactive commitments. In our setting, we consider the more general case of unbounded concurrency as well as

[2] Simulation-soundness is a stronger property that implies and is closely related to non-malleability.

interactive commitments. Intuitively, the property we require from these commitments is that even when a man-in-the-middle receives concurrent equivocal commitments and concurrent equivocal decommitments, the man-in-the-middle cannot break the binding property of the commitment. Thus, the man-in-the-middle receives equivocal commitments and decommitments, but cannot equivocate himself. Formalizing this notions seems to be tricky and has not been considered in literature before. Previously, non-malleability of commitments has been dealt with in two scenarios:

Non-malleability w.r.t commitment:[21,43,32] This requires that no adversary that receives a commitment to value v be able to commit to a related value (even without being able to later decommit to this value).

Non-malleability w.r.t decommitment (or opening):[15,43,19] This requires that no adversary that receives a commitment and decommitment to a value v be able to commit and decommit to a related value.

While the former is applicable only in the case the of statistically-binding commitments the latter is useful even for statistically-hiding commitments. In this work, we need a definition that ensures independence of commitments schemes that additionally are equivocable and adaptively secure. Equivocability means that there is a way to commit to the protocol without knowing the value being committed to and later open to any value. Such a scheme cannot be statistically-binding. Furthermore, since we consider the setting where the adversary receives concurrent equivocal decommitments, our definition needs to consider non-malleability w.r.t decommitment. Unfortunately, current definitions for non-malleability w.r.t decommitment in literature are defined only in the scenario where the commitment phase and decommitment phases are decoupled, i.e. in a first phase, a man-in-the-middle adversary receives commitments and sends commitments, then, in a second phase, the adversary requests decommitments of the commitments received in the first phase, followed by it decommitting its own commitments. For our construction, we need to define concurrent non-malleability w.r.t decommitments and such a two phase scenario is not applicable as the adversary can arbitrarily and adaptively decide when to obtain decommitments. Furthermore, it is not clear how to extend the traditional definition to the general case, as at any point, only a subset of the commitments received by the adversary could be decommitted and the adversary could selectively decommit based on the values seen so far and hence it is hard to define a "related" value.

We instead propose a new definition, along the lines of *simulation-extractability* that has been defined in the context of constructing non-malleable zero-knowledge proofs [44]. Loosely speaking, an interactive protocol is said to be simulation extractable if for any man-in-the-middle adversary A, there exists a probabilistic polynomial time machine (called the simulator-extractor) that can simulate both the left and the right interaction for A, while outputting a witness for the statement proved by the adversary in the right interaction. Roughly speaking, we say that a tag-based commitment scheme (i.e., commitment scheme that takes an identifier—called the tag—as an additional input) is *concurrent non-malleable w.r.t opening* if for every man-in-the-middle adversary A that participates in several interactions with honest committers as a receiver (called *left* interactions) as well as several interactions with honest receivers as a committer (called *right* interactions), there exists a simulator S that can simulate the left interactions, while extracting the commitments made by the adversary in the right interactions (whose identifiers are different from all the left identifiers) before the adversary decommits.

A related definition in literature is that of *simulation-sound trapdoor commitments* from [23,37] which considers *non-interactive* equivocable commitments and require that no adversary be able to equivocate even when it has access to an oracle that provides equivocal commitments and decommitments. This can be thought of as the CCA analogue for equivocal commitments. We believe that such a scheme would suffice for our construction, however, it is not clear how to construct such commitments from any trapdoor (i.e. any set-up) even if we relax the definition to consider interactive commitments.

It is not hard to construct equivocal commitments using trusted set-up. The idea here is to provide the simulator with a trapdoor with which it can equivocate as wells as extract the commitments on the right. (by e.g., relying on encryption). However, to ensure non-malleability, most constructions in literature additionally impose CCA-security or provide independent trapdoors for every interaction. Our main technical contribution consists of showing how to construct a concurrent non-malleable commitment scheme in any trusted set-up by providing the simulator with just one trapdoor, i.e. we show how to construct a concurrent non-malleable commitment scheme w.r.t opening using any UC-puzzle. We remark here that, in the static case, a stand-alone non-malleable commitment was sufficient, since the indistinguishability based notion of non-malleability allowed for some form of concurrent composition. However, in the adaptive case, it is not clear if our definition yields a similar composition and hence we construct a scheme and prove non-malleability directly in the concurrent setting.

Although our main application of equivocal non-malleable commitments is achieving UC-security, these commitments may also be useful for other applications such as concurrent non-malleable zero knowledge secure under adaptive corruptions. We believe that an interesting open question is to explore other applications of equivocal non-malleable commitments and non-malleable commitments with respect to decommitment.

2 Equivocal Non-malleable Commitments

In this section, we define Equivocal Non-malleable Commitments. Intuitively, these are equivocal commitments such that even when a man-in-the-middle adversary receives equivocal commitments and openings from a simulator, the adversary himself remains unable to equivocate. Since we are interested in constructing equivocal commitments from any trapdoor (i.e. setup), we will capture trapdoors, more generally, as witnesses to NP-statements. First, we provide definitions of language-based commitments.

Language-Based Commitment Schemes: We adopt a variant of language-based commitment schemes introduced by Lindell et. al [36] which in turn is a variant of [4,29]. Roughly speaking, in such commitments the sender and receiver share a common input, a statement x from an NP language L. The properties of the commitment scheme depend on the whether x is in L or not and the binding property of the scheme asserts that any adversary violating the binding can be used to extract an NP-witness for the statement. We present the formal definition below.

Definition 1 (Language-Based Commitment Schemes). *Let L be an* NP*-Language and $\mathcal{R}$, the associated* NP*-relation. A language-based commitment scheme (LBCS) for L is commitment scheme $\langle S, R\rangle$ such that:*

Computational hiding: *For every (expected) PPT machine R^*, it holds that, the following ensembles are computationally indistinguishable over $n \in N$.*
- $\{\mathsf{sta}^{R^*}_{\langle S,R\rangle}(x, v_1, z)\}_{n\in N, x\in\{0,1\}^n, v_1,v_2\in\{0,1\}^n, z\in\{0,1\}^*}$
- $\{\mathsf{sta}^{R^*}_{\langle S,R\rangle}(x, v_2, z)\}_{n\in N, x\in\{0,1\}^n, v_1,v_2\in\{0,1\}^n, z\in\{0,1\}^*}$

where $\mathsf{sta}^{R^*}_{\langle S,R\rangle}(x, v, z)$ *denotes the random variable describing the output of* $R^*(x, z)$ *after receiving a commitment to* v *using* $\langle S, R\rangle$.

Computational binding: *The binding property asserts that, there exists an polynomial-time witness-extractor algorithm Ext, such that for any cheating committer S^*, that can decommit a commitment to two different values v_1, v_2 on common input $x \in \{0,1\}^n$, outputs w such that $w \in \mathcal{R}(x)$.*

We now extend the definition to include equivocability.

Definition 2 (Language-Based Equivocal Commitments). *Let L be an* NP*-Language and $\mathcal{R}$, the associated* NP*-relation. A language-based commitment scheme $\langle S, R\rangle$ for L is said to be equivocal, if there exists a tuple of algorithms $(\tilde{S}, \mathsf{Adap})$ such that the following holds:*

Special-Hiding: *For every (expected) PPT machine R^*, it holds that, the following ensembles are computationally indistinguishable over $n \in N$.*
- $\{\mathsf{sta}^{R^*}_{\langle S,R\rangle}(x, v_1, z)\}_{n\in N, x\in L\cap\{0,1\}^n, w\in\mathcal{R}(x), v_1\in\{0,1\}^n, z\in\{0,1\}^*}$
- $\{\mathsf{sta}^{R^*}_{\langle \tilde{S},R\rangle}(x, w, z)\}_{n\in N, x\in L\cap\{0,1\}^n, w\in\mathcal{R}(x), v_1\in\{0,1\}^n, z\in\{0,1\}^*}$

where $\mathsf{sta}^{R^*}_{\langle \tilde{S},R\rangle}(x, w, z)$ *denotes the random variable describing the output of* $R^*(x, z)$ *after receiving a commitment using* $\langle \tilde{S}, R\rangle$.

Equivocability: *Let τ be the transcript of the interaction between R and $\tilde{S}$ on common input $x \in L \cap \{0,1\}^n$ and private input $w \in \mathcal{R}(x)$ and random tape $r \in \{0,1\}^*$ for $\tilde{S}$. Then for any $v \in \{0,1\}^n$, $\mathsf{Adap}(x, w, r, \tau, v)$ produces a random tape r' such that (r', v) serves as a valid decommitment for C on transcript τ.*

2.1 Definition of Equivocal Non-malleable Commitments

Let $\langle C, R\rangle$ be a commitment scheme, and let $n \in N$ be a security parameter. Consider man-in-the-middle adversaries that are participating in left and right interactions in which $m = \text{poly}(n)$ commitments take place[3]. We compare between a *man-in-the-middle* and a *simulated* execution. In the man-in-the-middle execution, the adversary A is simultaneously participating in m left and right interactions. In the left interactions the man-in-the-middle adversary A interacts with C receiving commitments to values $v_1, \ldots, v_m$, using identities $\mathsf{id}_1, \ldots, \mathsf{id}_m$ of its choice. It must be noted here that values $v_1, \ldots, v_m$ are provided to committer on the left prior to the interaction. In the right interaction A interacts with R attempting to commit to a sequence of related values again

[3] We may also consider relaxed notions of concurrent non-malleability: one-many, many-one and one-one secure non-malleable commitments. In a one-one (i.e., a stand-alone secure) non-malleable commitment, we consider only adversaries A that participate in one left and one right interaction; in one-many, A participates in one left and many right, and in many-one, A participates in many left and one right.

using identities of its choice $\tilde{\mathsf{id}}_1, \ldots, \tilde{\mathsf{id}}_m$; $\tilde{v}_i$ is set to the value decommitted by A in the jth right interaction. If any of the right commitments are invalid its committed value is set to $\perp$. For any i such that $\tilde{\mathsf{id}}_i = \mathsf{id}_j$ for some j, set $\tilde{v}_i = \perp$—i.e., any commitment where the adversary uses the same identity as one of the honest committers is considered invalid. Let $\mathsf{MIM}^A_{\langle C,R\rangle}(v_1, \ldots, v_m, z)$ denote a random variable that describes the values $\tilde{v}_1, \ldots, \tilde{v}_m$ and the view of A, in the above experiment.

In the simulated execution, a simulator S interacts only with receivers on the right as follows:

1. Whenever the commitment phase of jth interaction with a receiver on the right is completed, S outputs a value $\tilde{v}_j$ as the alleged committed value in a special-output tape.
2. During the interaction, S may output a partial view for a man-in-the-middle adversary whose right-interactions are identical to S's interaction so far. If the view contains a left interaction where the ith commitment phase is completed and the decommitment is requested, then a value v_i is provided as the decommitment.
3. Finally, S outputs a view and values $\tilde{v}_1, \ldots, \tilde{v}_m$. Let $\mathsf{sim}^S_{\langle C,R\rangle}(1^n, v_1, \ldots, v_m, z)$ denote the random variable describing the view output by the simulation and values $\tilde{v}_1, \ldots, \tilde{v}_m$.

Definition 3. *A commitment scheme $\langle C, R\rangle$ is said to be* concurrent non-malleable w.r.t. opening *if for every polynomial $p(\cdot)$, and every probabilistic polynomial-time man-in-the-middle adversary A that participates in at most $m = p(n)$ concurrent executions, there exists a probabilistic polynomial time simulator S such that the following ensembles are computationally indistinguishable over $n \in N$:*

$$\left\{\mathsf{MIM}^A_{\langle C,R\rangle}(v_1, \ldots, v_m, z)\right\}_{n\in N, v_1,\ldots,v_m\in\{0,1\}^n, z\in\{0,1\}^*}$$
$$\left\{\mathsf{sim}^S_{\langle C,R\rangle}(1^n, v_1, \ldots, v_m, z)\right\}_{n\in N, v_1,\ldots,v_m\in\{0,1\}^n, z\in\{0,1\}^*}$$

A slightly relaxed definition considers all the values committed to the adversary in the left interaction to be sampled independently from an arbitrary distribution D. We show how to construct a commitment satisfying only this weaker definition. However, this will be sufficient to establish our results.

Definition 4. *A commitment scheme $\langle C, R\rangle$ is said to be* concurrent non-malleable w.r.t. opening with independent and identically distributed (i.i.d.) commitments *if for every polynomial $p(\cdot)$ and polynomial time samplable distribution D, and every probabilistic polynomial-time man-in-the-middle adversary A that participates in at most $m = p(n)$ concurrent executions, there exists a probabilistic polynomial time simulator S such that the following ensembles are computationally indistinguishable over $n \in N$:*

$$\left\{(v_1 \ldots, v_m) \leftarrow D^n : \mathsf{MIM}^A_{\langle C,R\rangle}(v_1, \ldots, v_m, z)\right\}_{n\in N, z\in\{0,1\}^*}$$
$$\left\{(v_1 \ldots, v_m) \leftarrow D^n : \mathsf{sim}^S_{\langle C,R\rangle}(1^n, v_1, \ldots, v_m, z)\right\}_{n\in N, z\in\{0,1\}^*}$$

Remark 1. Any scheme that satisfies our definition with a straight-line simulator in essence realizes the ideal commitment functionality with UC-security as it acheives equivocation and straight-line extraction. If the simulator is not straight-line, then the requirement that the left commitments are sampled from i.i.d distributions is seemingly inherent. This is because our definition implies security against *selective opening attacks* (SOA) and as proved in [38], achieving fully concurrent SOA-security with (black-box) rewinding techniques is impossible when the distributions of the commitments are not sampleable (or unknown).

Finally, we consider commitment schemes that are both non-malleable w.r.t opening and language-based equivocal. In a *setup model*, the simulator will obtain a trapdoor via the setup procedure and the witness relation will satisfy that language requirement.

Definition 5. *A commitment scheme* $\langle C, R \rangle$ *is said to be an* equivocal non-malleable commitment scheme *if it is both a language-based equivocal commitment scheme (see Definition 2) and is concurrent non-malleable w.r.t. opening (see Definition 4).*

3 Adaptive UC-Puzzles

Informally, an adaptive UC-puzzle is a protocol $\langle \mathsf{S}, \mathsf{R} \rangle$ between two players–a *sender* S and a *receiver* R – and a PPT computable relation $\mathcal{R}$, such that the following two properties hold:

Soundness: No efficient receiver R^* can successfully complete an interaction with S and also obtain a "trapdoor" y, such that $\mathcal{R}(\mathsf{TRANS}, y) = 1$, where TRANS is the transcript of the interaction.

Statistical UC-simulation with adaptive corruptions: For every efficient adversary $\mathcal{A}$ participating in a polynomial number of concurrent executions with receivers R (i.e., $\mathcal{A}$ is acting as a puzzle sender in all these executions) and at the same time communicating with an environment $\mathcal{Z}$, there exists a simulator $\mathcal{S}$ that is able to statistically simulate the view of $\mathcal{A}$ for $\mathcal{Z}$, while at the same time outputting trapdoors to all successfully completed puzzles. Moreover, $\mathcal{S}$ successfully simulates the view even when $\mathcal{A}$ may adaptively corrupt the receivers.

We provide a formal definition in the full version [17]. In essence, it is the same definition as in [33] with the additional requirement of adaptive security in the simulation. We remark that our analysis will require the puzzle to be **straight-line simulatable**. In fact, for almost all models considered in this work, this is indeed the case, with the exception of the timing and partially-isolated adversaries model (for which we argue the result independently). Using the result of [26], it is possible to argue that straight-line simulation is necessary to achieve adaptive security (except when we consider restricted adversaries, such as the timing or partially-isolated adversaries model).

4 Achieving Adaptive UC-Security

In this section, we give a high-level overview of the construction of an $\mathsf{EQNMCom}$ scheme and the proof of Theorem 1, which relies on the existence of an $\mathsf{EQNMCom}$

scheme. For the formal construction and analysis of our EQNMCom scheme, see the full version [17]. A formal proof of Theorem 1 can be found in the full version [17].

By relying on previous results [11,18,28,14,13], the construction of an adaptive UC-secure protocol for realizing any multiparty functionality can be reduced to the task of constructing a UC-commitment assuming the existence of simulatable PKE. First, we show how to construct an equivocal non-malleable commitment scheme based on any adaptive UC-puzzle. Then combining the equivocal non-malleable commitment scheme with a simulatable PKE scheme we show how to realize the UC-commitment.

4.1 Constructing EQNMCom Based on Adaptive UC-Puzzles

Our protocol on a very high-level is a variant of the non-malleable commitment protocol from [32] which in turn is a variant of the protocol from [21]. While non-malleability relies on the message-scheduling technique of [21,32] protocol, the equivocability is obtained by relying on a variant of Feige-Shamir's trapdoor commitment scheme[4] and adaptively secure witness-indistinguishable proof of knowledge (WIPOK) protocol (see the full version [17]) for a formal definition and construction) of Lindell-Zarosim[36]. More precisely, our protocol proceeds in two phases: in the preamble phase, the Committer and Receiver exchange a UC-puzzle where the Receiver is the sender of the puzzle and the Committer is the receiver of the puzzle (this phase establishes a trapdoor through which an equivocal commitment can be generated). This is followed by the commitment phase: here the Committer first commits to its value using a language-based (non-interactive) equivocal commitment scheme, where the NP-language is the one corresponding to the UC-puzzle and the particular statement is the puzzle exchanged in the preamble (this relies on the Feige-Shamir trapdoor commitment scheme). This is followed by several invocations of an (adaptively-secure) WIPOK where the Committer proves the statement that either it knows the value committed to in phase 2 or possesses a solution to the puzzle from phase 1. Here we rely on the *adaptively-secure* (without erasures) WIPOK of [36] where the messages are scheduled based on the Committers id using the scheduling of [21]. This phase allows for any Committer that possess a solution to the puzzle from the preamble phase to generate a commitment that can be equivocated (i.e. later be opened to any value). Conversely, any adversary that can equivocate the non-interactive commitment of the second phase can be used to obtain a solution to the puzzle. The decommitment information is simply the value and the random tape of an honest committer consistent with the commitment phase. More specifically, our protocol proceeds as follows:

1. In the Preamble Phase, the Committer and Receiver exchange a UC-puzzle where the Receiver is the sender of the puzzle and the Committer is the receiver of the puzzle. Let x be the transcript of the interaction.
2. In the Committing Phase, the Committer sends $c = \mathsf{EQCom}^x(v)$, where EQCom^x is a language-based equivocal commitment scheme as in Definition 2 with common input x. This is followed by the Committer proving that c is a valid commitment

[4] Let x be an NP-statement. The sender commits to bit b by running the honest-verifier simulator for Blum's Hamiltonian Circuit protocol [5] on input the statement x and the verifier message b, generating the transcript (a, b, z), and finally outputting a as its commitment. In the decommitment phase, the sender reveals the bit b by providing both b, z.

for v. This is proved by 4ℓ invocations of an adaptively-secure (without erasures) WIPOK where the messages are scheduled based on the id (as in [21,32]). More precisely, there are ℓ rounds, where in round i, the schedule $\mathsf{design}_{\mathsf{id}_i}$ is followed by $\mathsf{design}_{1-\mathsf{id}_i}$ (See Figure 1).

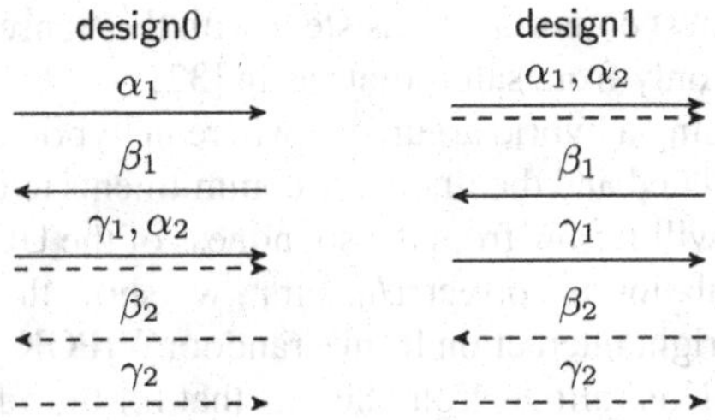

Fig. 1. Message schedule in a round in adaptively-secure WIPOK

While the protocol is an adaptation of the [32] commitment scheme, where the individual components are replaced by adaptively-secure alternatives, proving security requires a substantially different analysis. It is easy to see that concurrent equivocability of our scheme follows from the UC-Puzzle simulation. However proving concurrent non-malleability w.r.t opening with i.i.d commitments is the hard part and the core of our contribution. Recall that, achieving this, essentially entails constructing a simulator for any man-in-the-middle adversary, that while equivocating all commitments to the adversary (in the left interactions), can extract all the values the value committed to by the adversary (in the right interactions) before the decommitment phase.

Towards extracting from the right interactions, we first recall the basic idea in [32,21]. Their scheduling ensures that for every right interaction with a tag that is different from a left interaction, there exists a point—called a safe-point—from which we can *rewind* the right interaction (and extract the committed value), without violating the hiding property of the left interaction. It now follows from the hiding property of the left interaction that the values committed to on the right do not depend on the value committed to on the left. However, this technique only allows for extraction from a right interaction without violating the hiding property of *one* left interaction. However, here we need to extract without violating the hiding property of all the left interactions.

Our simulator-extractor as follows: In a main execution with the man-in-the-middle adversary, the simulator simulates all puzzles to obtain trapdoors and equivocates the left interactions using the solution of the puzzle and simulates the right interactions honestly. Whenever a decommitment on the left is requested, the simulator obtains a value externally (a value sampled independently from distribution D) which it decommits to the adversary (this is achieved since the protocol is adaptively secure). After the adversary completes the commitment phase of a right interaction in the main execution, the simulator switches to a rewinding phase, where it tries to extract the value committed to by the adversary in that right interaction. Towards this, it chooses a random WIPOK (instead of a safe point) from the commitment phase and rewinds the adversary to obtain the witness used in the WIPOK (using the proof-of-knowledge extractor). In the rewinding phase, the left interactions are now simulated using the honest committer strategy (as opposed to equivocating using the solution to the puzzle). More precisely, in the rewinding phase, for every left interaction that has already been opened (i.e. decommitment phase has occurred in the main execution), the simulator has a value and random tape for an honest committer and for those that have not yet been opened, using the adaptive-security of the protocol, the simulator simply samples a random value from distribution D (since we consider only i.i.d. values for left interactions) and generates a random tape for an honest committer consistent with the transcript so far. This stands in contrast of extracting only from safe-points as in [32].

The proof proceeds using a hybrid argument, where in hybrid experiment H_i all puzzle interactions are simulated and the first i left commitments to complete the preamble phase is equivocated. It will follow from the soundness of the UC-puzzle and statistical simulation that the simulation is correct H_0. First, we show that in H_0, the value extracted in any particular right interaction from a random WIPOK is the value decommitted to by the adversary. This follows from the fact that for the adversary to equivocate, it must know the solution to the UC-puzzle and this violates the statistical simulation and soundness condition of the puzzle. We show the following properties for every i, and the proof of correctness follows using a standard hybrid argument.

- *If the value extracted in any particular right interaction from a random WIPOK is the value decommitted to by the adversary in H_{i-1}, then the value extracted from a random WIPOK and the safe point of that right interaction w.r.t to i^{th} left interaction are the same and equal to the decommitment.* We show this by carefully considering another sequence of hybrids that yields an adversary that violates the soundness of the UC-puzzle in an execution where the puzzles are not simulated. This will rely on fact that the simulator simulates the left interactions in the rewindings using the honest committer strategy and the pseudo-randomness of the non-interactive commitment scheme used in the Commitment phase.
- *If the value extracted from the safe point is the decommitment in H_{i-1} then the same holds in H_i.* We rely on the proof technique of [32] through safe-points to establish this. In slightly more detail, we show that for any particular right interaction, the value extracted from the safe-point w.r.t i^{th} left interaction does not change when the ith left commitment is changed from an honest commitment to an equivocal commitment. Recall that a safe-point can be used to extract the value committed to in the right without rewinding the particular left interaction. Since, the non-interactive commitment scheme used has pseudo-random commitments, an adversary cannot distinguish if it is receiving an honest or equivocal commitment in the ith interaction.
- *If the value extracted in the right interaction from the safe point is the value decommitted to by the adversary in H_i, then the value extracted from a random WIPOK and the safe point are the same and equal to the decommitment in H_i.* This is established exactly as the first property.

See the full version [17] for the formal construction and proof.

4.2 Adaptive UC-Secure Commitment Scheme

We now provide the construction of a UC-commitment scheme. First, we recall the construction of the adaptive UC-secure commitment in the common reference string model (CRS) from [11] to motivate our construction. In the [11] construction, the CRS contains two strings. The first string consists of a random image $y = f(x)$ of a one-way function f and the second string consists of a public key for a cca-secure encryption scheme. The former allows a simulator to equivocate the commitment when it knows x and the public key allows the simulator to extract committed values from the adversary using its knowledge of the corresponding secret-key. The additional CCA requirement ensures non-malleability.

Our construction follows a similar approach, with the exception that instead of having a common reference string generated by a trusted party, we use the equivocal non-malleable commitment to generate two common-reference strings between every pair of parties: one for equivocation and the other for extraction. This is achieved by running the following "non-malleable" coin-tossing protocol between an initiator and a responder. Let $\langle \mathsf{S}_{\mathsf{com}}, \mathsf{R}_{\mathsf{com}} \rangle$ be a concurrent equivocal non-malleable commitment scheme and $\langle \mathsf{S}_{\mathsf{puz}}, \mathsf{R}_{\mathsf{puz}} \rangle$ be a UC-puzzle.

1. The initiator commits to a random string r^0 using $\langle \mathsf{S}_{\mathsf{com}}, \mathsf{R}_{\mathsf{com}} \rangle$ to the responder.
2. The responder chooses a random string r^1 and sends to the Initiator.
3. The initiator opens its commitment and reveals r^0.
4. The output of the coin toss is: $r = r^0 \oplus r^1$.

The coin-tossing protocol is run between an initiator and responder and satisfies the following two properties: (1) For all interactions where the initiator is honest, there is a way to simulate the coin-toss. This follows directly from the equivocability of the commitment scheme $\langle \mathsf{S}_{\mathsf{com}}, \mathsf{R}_{\mathsf{com}} \rangle$. (2) For all interactions where the initiator is controlled by the adversary, the coin-toss generated is uniformly-random. This follows from the simulation-extractability of the commitment scheme.

Using the coin-tossing protocol we construct an adaptive UC-commitment scheme. First, the sender and receiver interact in two coin-tossing protocols, one where the sender is the initiator, with outcome $coin_1$ and the other, where the receiver is the initiator, with outcome $coin_2$. Let x be the statement that $coin_1$ is in the image of a pseudorandom generator G. Also let, $\text{PK} = \mathsf{oGen}(coin_2)$ be a public key for the simulatable encryption scheme $(\mathsf{Gen}, \mathsf{Enc}, \mathsf{Dec}, \mathsf{oGen}, \mathsf{oRndEnc}, \mathsf{rGen}, \mathsf{rRndEnc})$. To commit to a string β, the sender sends a commitment to β using the non-interactive language-based commitment scheme with statement x along with strings S_0 and S_1 where one of the two strings (chosen at random) is an encryption of decommitment information to β and the other string is outputted by $\mathsf{oRndEnc}$. In fact, this is identical to the construction in [11], with the exception that a simulatable encryption scheme is used instead of a CCA-secure scheme.

Binding follows from the soundness of the adaptive UC-puzzle and hiding follows from the hiding property of the non-interactive commitment scheme and the semantic security of the encryption scheme. It only remains to show that the scheme is concurrently equivocable and extractable. To equivocate a commitment from a honest committer, the simulator manipulates $coin_1$ (as the honest party is the initiator) so that $coin_1 = G(s)$ for a random string s and then equivocates by equivocating the non-interactive commitment and encrypting the decommitment information for both bits 0 and 1 in S_b and S_{1-b} (where b is chosen at random). To extract a commitment made by the adversary, the simulator manipulates $coin_2$ so that $coin_2 = \mathsf{rGen}(r)$ and

$(\mathsf{PK}, \mathsf{SK}) = \mathsf{Gen}(r)$ for a random string r. Then it extracts the decommitment information in the encryptions sent by the adversary using SK.

The procedure described above works only if the adversary does not encrypt the decommitment information for both 0 and 1 even when the simulator is equivocating. On a high-level, this follows since, if the coin-toss $coin_1$ cannot be manipulated by the adversary when it is the initiator, then the $coin_1$ is not in the range of G with very high probability and hence the adversary cannot equivocate (equivocating implies a witness can be extracted that proves that $coin_1$ is in the range of G). Proving this turns out to be subtle and an intricate analysis relying on the simulation-extractability of the $\langle \mathsf{S_{com}}, \mathsf{R_{com}} \rangle$-scheme is required.

We use a "non-malleable" coin-toss protocol to generate two keys, one for equivocation and another for extraction. Such an idea has been pursued before, for instance, in [19], they use a coin-toss to generate keys for extraction and equivocation. However, they use a single coin-toss and depending on which party is corrupt, the simulation yields an extraction or equivocation key. In recent and independent work, Garg and Sahai [26], show how to achieve stand-alone adaptively-secure multiparty computation in the plain model (assuming no-setup) using non black-box simulation. They rely on a coin-tossing protocol using equivocal commitments to generate a common random string and then rely on previous techniques used in the uniform reference string model [11] to securely realize any functionality. An important difference between their approach and ours is that while our construction relies on a single trapdoor they require the trapdoors to be non-malleable.[5] See Figure 2 for a formal description of our protocol (For further details and the proof, we refer the reader to the full version [17]).

5 Puzzle Instantiations

By Theorem 1, it suffices to present an adaptive UC puzzle in a given model to demonstrate feasibility of adaptive and UC secure computation. We first give some brief intuition on the construction of adaptive UC-puzzles in various models. Formal constructions and proofs are found in the full version [17].

In the Common reference string (CRS) model, the Uniform reference string (URS) model and the Key registration model the puzzles are identical to the ones presented in [33] for the static case, where the puzzle interactions essentially consists of a call to the corresponding ideal setup functionalities. Thus, in these models, the simulator is essentially handed the trapdoor for the puzzle via its simulation of the ideal functionality and the puzzles are non-interactive. In the Timing model and the Partially Isolated Adversaries model, we rely on essentially the same puzzles as [33], however, we need to modify the simulator to accommodate adaptive corruption by the adversary.

Constructing adaptive UC-puzzles in the Sunspots model is less straightforward and so we give more detail here. Simulated reference strings r in the Sunspots model have Kolmogorov complexity smaller than k. Thus, as in [33], the puzzle sender and receiver exchange 4 strings (v_1, c_2, v_2, c_2). We then let Φ' denote the statement that c_1, c_2 are commitments to messages p_1, p_2 such that (v_1, p_1, v_2, p_2) is an accepting transcript of

[5] In [19], they use separate keys for each party and in [26], the trapdoors admit a "simulation-soundness" property.

Protocol $\langle \mathsf{S}, \mathsf{R} \rangle$**: Input:** The sender S has a bit β to be committed to.
Preamble:

- An adaptive UC-Puzzle interaction $\langle \mathsf{S}_{\mathsf{puz}}, \mathsf{R}_{\mathsf{puz}} \rangle$ on input 1^n where R is the receiver and S is the sender. Let TRANS_1 be the transcript of the messages exchanged.
- An adaptive UC-Puzzle interaction $\langle \mathsf{S}_{\mathsf{puz}}, \mathsf{R}_{\mathsf{puz}} \rangle$ on input 1^n where S is the receiver and R is the sender. Let TRANS_2 be the transcript of the messages exchanged.

Commit phase:

Stage 1: S and R run a coin-tossing protocol to agree on strings PK and CRS:

Coin-toss to generate PK:

1. The parties run protocol $\langle \mathsf{S}_{\mathsf{com}}, \mathsf{R}_{\mathsf{com}} \rangle$ with common input TRANS_1. R plays the part of sender with input a random string r_{R}^0.
2. S chooses a random string r_{S}^0 and sends to R.
3. R opens its commitment and reveals r_{R}^0.
4. The output of the coin toss is: $r = r_{\mathsf{S}}^0 \oplus r_{\mathsf{R}}^0$. S and R run $\mathsf{oGen}(r)$ to obtain public key PK.

Coin-toss to generate CRS**:**

1. The parties run protocol $\langle \mathsf{S}_{\mathsf{com}}, \mathsf{R}_{\mathsf{com}} \rangle$ with common input TRANS_2. S plays the part of sender with input a random string r_{S}^1.
2. R chooses a random string r_{R}^1 and sends to S.
3. S opens its commitment and reveals r_{S}^1.
4. The output of the coin-toss is: $x = r_{\mathsf{S}}^1 \oplus r_{\mathsf{R}}^1$.

Stage 2:

1. The parties run $\langle \mathsf{S}_{\mathsf{eq}}, \mathsf{R}_{\mathsf{eq}} \rangle$ with common input x to generate a commitment $C = \mathsf{EQCom}^x(\beta; r)$ where S plays the part of S_{eq} with input bit β.
2. S chooses $b \in \{0, 1\}$ at random and sends to R the strings (S_0, S_1) to where:
   - S_b is an encryption of the decommitment information of C (to bit β) under PK.
   - S_{1-b} is generated by running $\mathsf{oRndEnc}(\mathrm{PK}, r_{\mathsf{Enc}})$ where r_{Enc} is chosen uniformly at random.

Reveal phase:

1. S sends β, b, and the randomness used to generate S_0, S_1 to R.
2. R checks that S_0, S_1 can be reconstructed using β, b and the randomness produced by S.

Fig. 2. The Adaptive Commitment Protocol $\langle \mathsf{S}, \mathsf{R} \rangle$

a Universal argument of the statement $\Phi = \mathsf{KOL}(r) \leq k$. Note that since we require *statistical* and *adaptive* simulation of puzzles, the commitment scheme used must be both statistically-hiding and "obliviously samplable" (i.e. there is a way to generate strings that are statistically indistinguishable from commitments, without "knowing" the committed value).

To construct an adaptive puzzle for the bounded-concurrent model we follow an approach similar to the sunspots model combined with the bounded-concurrent non black-box zero-knowledge protocol of Barak[1]. In fact this is inspired by the stand alone adaptive secure multiparty computation construction of Garg, et al, [26].

References

1. Barak, B.: How to go beyond the black-box simulation barrier. In: FOCS, pp. 106–115 (2001)
2. Barak, B., Canetti, R., Nielsen, J.B., Pass, R.: Universally composable protocols with relaxed set-up assumptions. In: FOCS, pp. 186–195 (2004)
3. Barak, B., Sahai, A.: How to play almost any mental game over the net - concurrent composition via super-polynomial simulation. In: FOCS, pp. 543–552 (2005)
4. Bellare, M., Micali, S., Ostrovsky, R.: The (true) complexity of statistical zero knowledge. In: STOC, pp. 494–502 (1990)
5. Blum, M.: How to prove a theorem so no one else can claim it. In: Proceedings of the International Congress of Mathematicians, pp. 1444–1451 (1986)
6. Canetti, R.: Universally composable security: A new paradigm for cryptographic protocols. In: FOCS, pp. 136–145 (2001)
7. Canetti, R., Dodis, Y., Pass, R., Walfish, S.: Universally composable security with global setup. In: Vadhan, S.P. (ed.) TCC 2007. LNCS, vol. 4392, pp. 61–85. Springer, Heidelberg (2007)
8. Canetti, R., Fischlin, M.: Universally composable commitments. In: Kilian, J. (ed.) CRYPTO 2001. LNCS, vol. 2139, pp. 19–40. Springer, Heidelberg (2001)
9. Canetti, R., Kushilevitz, E., Lindell, Y.: On the limitations of universally composable two-party computation without set-up assumptions. In: Biham, E. (ed.) EUROCRYPT 2003. LNCS, vol. 2656, pp. 68–86. Springer, Heidelberg (2003)
10. Canetti, R., Lin, H., Pass, R.: Adaptive hardness and composable security in the plain model from standard assumptions. In: FOCS, pp. 541–550 (2010)
11. Canetti, R., Lindell, Y., Ostrovsky, R., Sahai, A.: Universally composable two-party and multi-party secure computation. In: STOC, pp. 494–503 (2002)
12. Canetti, R., Pass, R., Shelat, A.: Cryptography from sunspots: How to use an imperfect reference string. In: FOCS, pp. 249–259 (2007)
13. Choi, S.G., Dachman-Soled, D., Malkin, T., Wee, H.: Improved non-committing encryption with applications to adaptively secure protocols. In: Matsui, M. (ed.) ASIACRYPT 2009. LNCS, vol. 5912, pp. 287–302. Springer, Heidelberg (2009)
14. Choi, S.G., Dachman-Soled, D., Malkin, T., Wee, H.: Simple, black-box constructions of adaptively secure protocols. In: Reingold, O. (ed.) TCC 2009. LNCS, vol. 5444, pp. 387–402. Springer, Heidelberg (2009)
15. Di Crescenzo, G., Ishai, Y., Ostrovsky, R.: Non-interactive and non-malleable commitment. In: STOC, pp. 141–150 (1998)
16. Di Crescenzo, G., Katz, J., Ostrovsky, R., Smith, A.: Efficient and non-interactive non-malleable commitment. In: Pfitzmann, B. (ed.) EUROCRYPT 2001. LNCS, vol. 2045, p. 40. Springer, Heidelberg (2001)
17. Dachman-Soled, D., Malkin, T., Raykova, M., Venkitasubramaniam, M.: Adaptive and concurrent secure computation from new notions of non-malleability. IACR Cryptology ePrint Archive, 2011:611 (2011)
18. Damgård, I.B., Nielsen, J.B.: Improved non-committing encryption schemes based on a general complexity assumption. In: Bellare, M. (ed.) CRYPTO 2000. LNCS, vol. 1880, pp. 432–450. Springer, Heidelberg (2000)
19. Damgård, I.B., Nielsen, J.B.: Perfect hiding and perfect binding universally composable commitment schemes with constant expansion factor. In: Yung, M. (ed.) CRYPTO 2002. LNCS, vol. 2442, pp. 581–596. Springer, Heidelberg (2002)
20. Damgård, I., Nielsen, J.B., Wichs, D.: Universally composable multiparty computation with partially isolated parties. In: Reingold, O. (ed.) TCC 2009. LNCS, vol. 5444, pp. 315–331. Springer, Heidelberg (2009)

21. Dolev, D., Dwork, C., Naor, M.: Nonmalleable cryptography. SIAM J. Comput. 30(2), 391–437 (2000)
22. Dwork, C., Naor, M., Sahai, A.: Concurrent zero-knowledge. In: IN 30TH STOC, pp. 409–418 (1999)
23. Garay, J.A., MacKenzie, P.D., Yang, K.: Strengthening zero-knowledge protocols using signatures. In: Biham, E. (ed.) EUROCRYPT 2003. LNCS, vol. 2656, pp. 177–194. Springer, Heidelberg (2003)
24. Garg, S., Goyal, V., Jain, A., Sahai, A.: Bringing people of different beliefs together to do UC. In: Ishai, Y. (ed.) TCC 2011. LNCS, vol. 6597, pp. 311–328. Springer, Heidelberg (2011)
25. Garg, S., Goyal, V., Jain, A., Sahai, A.: Concurrently secure computation in constant rounds. In: Pointcheval, D., Johansson, T. (eds.) EUROCRYPT 2012. LNCS, vol. 7237, pp. 99–116. Springer, Heidelberg (2012)
26. Garg, S., Sahai, A.: Adaptively secure multi-party computation with dishonest majority. In: Safavi-Naini, R., Canetti, R. (eds.) CRYPTO 2012. LNCS, vol. 7417, pp. 105–123. Springer, Heidelberg (2012)
27. Goldreich, O., Micali, S., Wigderson, A.: How to play any mental game or a completeness theorem for protocols with honest majority. In: STOC, pp. 218–229 (1987)
28. Ishai, Y., Prabhakaran, M., Sahai, A.: Founding cryptography on oblivious transfer – efficiently. In: Wagner, D. (ed.) CRYPTO 2008. LNCS, vol. 5157, pp. 572–591. Springer, Heidelberg (2008)
29. Itoh, T., Ohta, Y., Shizuya, H.: A language-dependent cryptographic primitive. J. Cryptology 10, 37–50 (1997)
30. Kalai, Y.T., Lindell, Y., Prabhakaran, M.: Concurrent composition of secure protocols in the timing model. J. Cryptology 20(4), 431–492 (2007)
31. Katz, J.: Universally composable multi-party computation using tamper-proof hardware. In: Naor, M. (ed.) EUROCRYPT 2007. LNCS, vol. 4515, pp. 115–128. Springer, Heidelberg (2007)
32. Lin, H., Pass, R., Venkitasubramaniam, M.: Concurrent non-malleable commitments from any one-way function. In: Canetti, R. (ed.) TCC 2008. LNCS, vol. 4948, pp. 571–588. Springer, Heidelberg (2008)
33. Lin, H., Pass, R., Venkitasubramaniam, M.: A unified framework for concurrent security: universal composability from stand-alone non-malleability. In: STOC, pp. 179–188 (2009)
34. Lindell, Y.: Protocols for bounded-concurrent secure two-party computation. Chicago J. Theor. Comput. Sci. (2006)
35. Lindell, Y.: Bounded-concurrent secure two-party computation without setup assumptions. In: STOC, pp. 683–692 (2003)
36. Lindell, Y., Zarosim, H.: Adaptive zero-knowledge proofs and adaptively secure oblivious transfer. In: Reingold, O. (ed.) TCC 2009. LNCS, vol. 5444, pp. 183–201. Springer, Heidelberg (2009)
37. MacKenzie, P.D., Yang, K.: On simulation-sound trapdoor commitments. In: Cachin, C., Camenisch, J.L. (eds.) EUROCRYPT 2004. LNCS, vol. 3027, pp. 382–400. Springer, Heidelberg (2004)
38. Ostrovsky, R., Rao, V., Scafuro, A., Visconti, I.: Revisiting lower and upper bounds for selective decommitments. In: Sahai, A. (ed.) TCC 2013. LNCS, vol. 7785, pp. 559–578. Springer, Heidelberg (2013)
39. Pass, R.: Simulation in quasi-polynomial time, and its application to protocol composition. In: Biham, E. (ed.) EUROCRYPT 2003. LNCS, vol. 2656, pp. 160–176. Springer, Heidelberg (2003)
40. Pass, R.: Bounded-concurrent secure multi-party computation with a dishonest majority. In: STOC, pp. 232–241 (2004)

41. Pass, R., Lin, H., Venkitasubramaniam, M.: A unified framework for UC from only OT. In: Wang, X., Sako, K. (eds.) ASIACRYPT 2012. LNCS, vol. 7658, pp. 699–717. Springer, Heidelberg (2012)
42. Pass, R., Rosen, A.: Bounded-concurrent secure two-party computation in a constant number of rounds. In: FOCS, pp. 404–413 (2003)
43. Pass, R., Rosen, A.: Concurrent non-malleable commitments. In: Proceedings of the 46th Annual IEEE Symposium on Foundations of Computer Science, FOCS 2005, pp. 563–572 (2005)
44. Pass, R., Rosen, A.: New and improved constructions of non-malleable cryptographic protocols. In: STOC, pp. 533–542 (2005)
45. Prabhakaran, M., Sahai, A.: New notions of security: achieving universal composability without trusted setup. In: STOC, pp. 242–251 (2004)
46. Yao, A.C.-C.: How to generate and exchange secrets (extended abstract). In: FOCS, pp. 162–167 (1986)

Key Recovery Attacks on 3-round Even-Mansour, 8-step LED-128, and Full AES^2

Itai Dinur[1], Orr Dunkelman[1,2,⋆], Nathan Keller[1,3,⋆⋆], and Adi Shamir[1]

[1] Computer Science department, The Weizmann Institute, Rehovot, Israel
[2] Computer Science Department, University of Haifa, Israel
[3] Department of Mathematics, Bar-Ilan University, Israel

Abstract. The Even-Mansour (EM) encryption scheme received a lot of attention in the last couple of years due to its exceptional simplicity and tight security proofs. The original 1-round construction was naturally generalized into r-round structures with one key, two alternating keys, and completely independent keys. In this paper we describe the first key recovery attack on the one-key 3-round version of EM which is asymptotically faster than exhaustive search (in the sense that its running time is $o(2^n)$ rather than $O(2^n)$ for an n-bit key). We then use the new cryptanalytic techniques in order to improve the best known attacks on several concrete EM-like schemes. In the case of LED-128, the best previously known attack could only be applied to 6 of its 12 steps. In this paper we develop a new attack which increases the number of attacked steps to 8, is slightly faster than the previous attack on 6 steps, and uses about a thousand times less data. Finally, we describe the first attack on the full AES^2 (which uses two complete AES-128 encryptions and three independent 128-bit keys, and looks exceptionally strong) which is about 7 times faster than a standard meet-in-the-middle attack, thus violating its security claim.

Keywords: Cryptanalysis, key recovery attacks, iterated Even-Mansour, LED encryption scheme, AES^2 encryption scheme.

1 Introduction

The Even-Mansour (EM) block cipher was first proposed at Asiacrypt'1991 [9]. It uses a single publicly known random permutation P on n-bit values and two secret n-bit keys K_1 and K_2, and defines the encryption of the n-bit plaintext m as $E(m) = P(m \oplus K_1) \oplus K_2$. The decryption of the n-bit ciphertext c is similarly defined as $D(c) = P^{-1}(c \oplus K_2) \oplus K_1$. It can be naturally generalized into an r-round iterated EM encryption function (a.k.a. a key-alternating scheme in [1, 5]), which is defined using r permutations $P_1, P_2, \ldots, P_r$ and $r + 1$ keys

⋆ The second author was supported in part by the German-Israeli Foundation for Scientific Research and Development through grant No. 2282-2222.6/2011.
⋆⋆ The third author was supported by the Alon Fellowship.

K. Sako and P. Sarkar (Eds.) ASIACRYPT 2013 Part I, LNCS 8269, pp. 337–356, 2013.

$K_1, K_2, \ldots K_{r+1}$ as $E(m) = P_r(\ldots P_2(P_1(m \oplus K_1) \oplus K_2) \oplus K_3 \ldots \oplus K_r) \oplus K_{r+1}$, where decryption is defined in an analogous way.

For about 20 years this scheme received very little attention in the cryptographic literature, but in the last couple of years it became a very active research area: multiple papers about this scheme appeared at Crypto, Eurocrypt, Asiacrypt, CHES and FSE [1, 5, 8, 12, 14–16], analyzing its theoretical properties, generalizing it in various ways, and proposing concrete constructions of block ciphers which are based on the EM structure.

In this paper we describe several new key recovery attacks on iterated EM schemes, analyze their complexity, and apply them to some concrete proposals of block ciphers which have this structure. The origin of the observations used in our attacks goes back to the first paper which attacked EM, by Daemen [6] in 1991. Daemen observed that in single-round one-key EM, an attacker can use the fact that the XOR of the unknown input and output of the permutation P is equal to the known XOR of the plaintext and the ciphertext. This observation can be used to break 1-round EM significantly faster than exhaustive search.

At FSE'13, Nicolic et al. [16] extended the basic observation considerably. They considered the graph of the function $P'(x) = x \oplus P(x)$,[1] and showed that vertices with a large in-degree in this graph can be exploited to bypass an additional round of EM, but at the expense of enlarging the time complexity to slightly less than exhaustive key search.

In this paper, we develop the techniques one step further, and show that graphs of the functions P_1' and P_3' (corresponding to the permutations P_1 and P_3) can be deployed simultaneously, resulting in an attack on 3-round EM. However, this enhancement is not sufficient by itself, since the time complexity becomes very close to that of exhaustive key search. Nevertheless, a surprising feature of our 3 round attack is that it has about the same time complexity as the 2-round attack. This feature is due to a novel filtering technique based on tailor-made linear subspaces that we develop in Section 2.2, and allows us to quickly dispose of data which is useless for our attack. Another novel technique that we develop in this paper allows us to adapt the differential-based attack of [15] (which was originally applied to 2-round iterated EM with one key) to 2-round iterated EM with completely independent keys, and thus to attack the full AES^2 scheme. While the attack of [15] makes use of plaintext pairs with a fixed difference, we notice that in its original form it cannot improve the standard meet-in-the-middle attack on this scheme. In our attack, we work on non-standard structures of plaintext triplets which allow us to filter out wrong guesses for the key more efficiently.

Throughout the paper we follow the standard conventions in the analysis of time and memory complexities. Our basic unit of memory is an n-bit block. Our basic unit of time is a single evaluation of the encryption or the decryption function, i.e., the full r-round iterated EM scheme. The scheme requires the evaluation of the r permutations P_i (which are assumed to be heavy operations)

[1] In [16], the permutation P is actually the full encryption function, and thus x is a message and $P(x)$ is its corresponding ciphertext.

and a small number of simple operations (such as XORs) which are assumed to require negligible time.[2] Thus, an invocation of a single permutation P_i (or its inverse) costs $1/r$ time units. For the sake of convenience, we often partition the attack into an *offline preprocessing phase* which analyzes the properties of the public P_i's, and an *online attack phase* which analyzes the given plaintexts and ciphertexts. However, we always define the time complexity of the attack as the sum of the complexities of its offline and online phases. This is different from the model used by Hellman in his time/memory tradeoff attack, which allowed unlimited free preprocessing and considered only the online complexity (note that in our model, Hellman's attack is no better than exhaustive search). To prevent other types of "cheating", we always add the time required to generate the data to the final time complexity, and add the space required to hold the data to the final space complexity.

All our attacks are only slightly better than exhaustive search, which raises the natural question whether they should be considered as legitimate attacks. This is a general problem in cryptanalysis, since it is difficult to decide whether an attack such as the Biclique attack on AES-128 [4] which requires 2^{126} time really "breaks" a scheme whose exhaustive search requires 2^{128} time. Some researchers suggested that this issue should be decided by the nature of the attack: If an attack on an n-bit scheme has an outer loop which tries 2^n different possibilities, but performs for each one of them an operation which is cheaper than a single encryption, then the attack should be called an "improvement of exhaustive search" rather than a "real attack", and the scheme is not said to be "broken" by it. However, this is a fragile definition since the same attack can be described in multiple ways, and it is not always clear whether it tries 2^n or fewer possibilities.

Fortunately, in cryptographic schemes such as EM which can be naturally defined for arbitrarily large key sizes n, we can avoid this fragility by analyzing the asymptotic complexity of the attack. As we show in this paper, our attacks are about $n/\log(n)$ times faster than exhaustive search. Since this ratio is unbounded when n increases, our attacks are asymptotically better than any standard or improved version of exhaustive search, and this is a robust statement since it ignores all the multiplicative constants which are associated with a particular model of computation.

Some of the concrete schemes we consider in this paper (such as LED and AES^2) pose the following problem: they use the general EM framework, but instantiate P with a fixed-key AES-like permutation which is defined only for a few values of n, and thus it is difficult to define their asymptotic security. We solve this problem in two ways. First, we observe that all our attacks are completely generic, and do not exploit any particular properties of P besides its randomness. We can thus analyze the performance of our attacks assuming

[2] This complexity gap is typically large for normal choices of n, and likely to grow even larger as n increases: the number of 2-bit to 1-bit gates in the Boolean circuit of P_i which are needed to thoroughly and independently mix the n input bits into n output bits is expected to grow super-linearly with n, whereas the number of gates in the Boolean circuit of XOR grows only linearly with n.

that AES is replaced in these schemes by a random permutation over n-bit values, and show that their asymptotic time complexity is $o(2^n)$. In addition, we carefully analyze the exact complexity of our attacks for the particular values of n recommended for these schemes, and show that they are between 7 and 20 times faster than exhaustive search, depending on which scheme we attack.

We would like to point out that some of the previously published attacks on these schemes (such as [5]) are distinguishing attacks, and thus they are incomparable to our key recovery attacks. In addition, our attacks may fail to find the key or require longer than expected time for a small fraction of "bad" permutations, since we only analyze their expected behavior when the P_i's are randomly chosen permutations.

The paper is organized as follows. In Section 2, we introduce our new cryptanalytic techniques, and use them to attack the one-key, three-round version of EM (the best previous attack could only handle the two-round version of EM). In particular, our new attack influenced the decision of the designers of the Zorro block cipher [11] to increase its number of steps from 3 to 6. In Section 3, we consider the LED block cipher, which was proposed at CHES 2011 [12]. It has two flavors: a one-key version called LED-64, and a two-key version (in which the two keys are alternately used) called LED-128. In the case of LED-64, the best previously published attack [13] appeared at ACISP 2012, and could only handle 2 steps. We increase the number of steps we can attack from 2 to 3. In the case of LED-128, the best previously published attack [16] appeared at FSE 2013, and could handle 6 steps out of the 12 steps of full LED-128. We increase this number to 8, using smaller time and data complexities. In Section 4, we consider the generalized version of EM in which all the keys are completely independent, and show how to attack the 2-round version of this scheme. We then use the new techniques in order to describe the first published attack on the full version of the block cipher AES^2, which was presented at Eurocrpyt 2012 by [5]. The scheme looks exceptionally strong, using two complete AES encryptions and three independent 128-bit keys. In fact, the designers of AES^2 conjectured that the best attack on their scheme is a meet-in-the-middle attack, but our new attack disproves this claim since it is about 7 times faster.

2 Attacks on Iterated Even-Mansour with One Key

We first consider iterated EM schemes with one key K and r permutations $P_1, P_2, \ldots, P_r$, as shown in Figure 1 (note that if all the permutations are also the same, the scheme is extremely vulnerable to slide attacks [3]). Our goal is to use properties of one of the public permutations $P \in \{P_1, P_2, \ldots, P_r\}$ in order to deduce properties of the associated keyed permutation[3] $Q(K, x) = K \oplus P(x \oplus K)$ (used inside the EM construction), which hold for any value of K. As Daeman pointed out in 1991 [6], for any value of K and in any invocation of $Q(K, x)$, the XOR of its input and output is equal to the XOR of the input and output of the internal P function in the same invocation, i.e., $x \oplus Q(K, x) = (x \oplus K) \oplus P(x \oplus K)$.

[3] In general, given some public permutation P_i, we denote $Q_i(K, x) = K \oplus P_i(x \oplus K)$.

Another interesting observation is that when K is unknown we cannot determine $x \oplus K$, but the addition of K just renames the input vertices in the bipartite graph of $P'(x) = x \oplus P(x)$, and thus it preserves the distribution of in-degrees of its output vertices. In particular, if some output values of P' are more likely than expected (i.e., appear more than the average), then we can predict the value $Q(K, x)$ with a higher probability than expected even when K is unknown. More specifically, any t-way collision on the value v in P', namely $x_1, x_2, \ldots, x_t$ such that $x_1 \oplus P(x_1) = x_2 \oplus P(x_2) = \ldots = x_t \oplus P(x_t) = v$ for some value of v, yields a t-way collision on the value v in the function $Q'(K, x) = x \oplus Q(K, x) = x \oplus K \oplus P(x \oplus K)$. Assume that indeed we manage to find during a preprocessing phase a large t-way collision in the public $P'(x)$ on the output value v. Since it also yields a t-way collision on the value v in the keyed function $Q'(K, x)$, there are at least t values of x for which $Q'(K, x) = v$, and thus $Q(K, x) = x \oplus v$. Consequently, we can guess $Q(K, x)$ with a probability which is t times higher than the expected $1/2^n$ even when we know nothing about K.

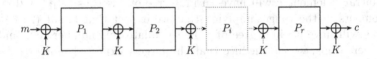

Fig. 1. An iterated EM with one key

This graph theoretic property is strongly related to the one used in [16], but we use it in a different way. Whereas we use properties of the public permutations (which can be observed during a preprocessing phase), [16] exploits properties of the given plaintext-ciphertext pairs: assume that $m_j \oplus c_j = v$ for multiple plaintext-ciphertext pairs (m_j, c_j). Then, for all of these pairs, they know that $(m_j \oplus K) \oplus (c_j \oplus K) = v$. Thus, the attack of [16] is based on the property that the XOR of the inputs to the first and last public permutations P_1 and P_r^{-1} attain the value v more than the expected number of times. In particular, in their attack it is not clear how to compute such a v during a preprocessing phase, and they have to wait for the actual data in order to search for the best v in it. Our attacks, on the other hand, are based on the property that the XOR of the input and output of a single public permutation attains some value v more than the expected number of times, and thus we can find the best v once and for all, before any data is given for a particular key.

In order to estimate the highest expected in-degree in the bipartite graph of $P'(x) = x \oplus P(x)$, we assume that for a random choice of the permutation P, the function P' behaves as a random function. This is not completely true, since there are some extremely expensive ways to distinguish between such cases (for example, the XOR of all the 2^n values of P' is zero, whereas the XOR of all the outputs of a truly random function is unlikely to be zero). However, it is easy to verify with appropriate simulations that the in-degree distributions of the two models behave almost identically, which is all we need in our attack.[4]

The main problem in applying this attack is that going over all the 2^n possible values of x in order to find the most popular v will make our attack slower than exhaustive search (since we do not allow free preprocessing in our model). Fortunately, we can find vertices v' which are almost as popular by trying only a small subset $X \subseteq \{0,1\}^n$ of possible inputs. We denote this restricted function by $f_{|X}$, and note that it induces a subgraph in the bipartite graph associated with f, in which the left side of the graph contains only the vertices in X. Our goal now is to analyze the expected distribution of the in-degrees in random subgraphs of random functions.

Random functions have been extensively analyzed in the literature (e.g., see [10]). It is well-known that the in-degree of an element in the range of $f_{|X}$ is distributed according to the Poisson distribution with an expectation λ, which is equal to the average in-degree (i.e., $\lambda = |X|/2^n$, which is the ratio between the sizes of the domain and range of $f_{|X}$). Given a parameter t, the probability that an arbitrary element v will have an in-degree of t is thus $(\lambda^t e^{-\lambda})/t!$. We have 2^n elements in the range, implying that we expect that about $(2^n \cdot \lambda^t e^{-\lambda})/t!$ vertices will have an in-degree of t. If we equate this number to 1 and ignore low order terms, we can deduce that the largest expected in-degree t satisfies $t \cdot \log(t) = n$, and thus t is approximately equal to $n/\log(n)$. The crucial point is that this highest in-degree grows in an unbounded way as n increases, and thus any complexity of the form $O(2^n/t)$ behaves asymptotically as $o(2^n)$. If we reduce this maximal t to $t-i$ for a small i, we expect to find about $(t/\lambda)^i$ vertices which have this reduced in-degree. Since $t > 1$ and $\lambda < 1$, this number grows exponentially with i, and we can thus find a huge number of vertices which have almost maximal in-degrees.

To get a sense of the concrete values implied by this distribution, consider the recommended value of $n = 64$ in the LED block cipher. If we consider all the 2^{64} possible inputs, we expect to see 2 or 3 vertices of degree 20, 55 vertices of degree 19, and 1060 vertices of degree 18. If we reduce the number of possible inputs to 2^{63}, we expect to see 1 vertex of degree 17, 8 vertices of degree 16, and 260 vertices of degree 15. If we further reduce the number of possible inputs to 2^{60}, we expect to see 4 vertices of degree 10, 695 vertices of degree 9, and $100130 \approx 2^{16.6}$ vertices of degree 8.

The attacks in this paper are described in terms of several parameters, and it is usually possible to obtain various tradeoffs between their time, data and memory complexities by tweaking the parameter values. However, since there is no simple formula which describes the exact tradeoff curves, one needs to determine favorable tradeoff points on the curves by plugging in a few values for the parameters

[4] In fact, collisions in P'(x) are slightly less likely to occur when P is a random function, since if $P(x) = P(y)$ (for $x \neq y$) then $P'(x) \neq P'(y)$, whereas if P is random permutation then $x \neq y$ implies $P(x) \neq P(y)$, and the probability for $P'(x) = P'(y)$ is a bit higher. As a result, our analysis slightly underestimates the highest expected in-degree, and thus the attacks that we describe are actually (negligibly) faster.

and calculating the resultant complexities of the algorithms. This is demonstrated in our attacks, where we suggest concrete points on the curves which minimize the time complexity, but stress that there are other options as well.

2.1 Attacks on 2-Round Iterated Even-Mansour with One Key

We start by describing a very basic attack, $2Round1KeyBasic$. Let S and D be parameters.

Preprocessing:

PR1. Evaluate P_1' on an arbitrary subset of inputs X, such that $|X| = S$, and store the output values (without their associated input values) in a sorted list.

PR2. Traverse the sorted list and find the output v_1 which occurs the maximal number of times (in t_1 consecutive locations).

Online:

O1. Ask for the encryption of D arbitrary plaintexts.

O2. For each plaintext-ciphertext pair (m_i, c_i):
  (a) Assume that $Q_1(K, m_i) = m_i \oplus v_1 \triangleq z_i$ and calculate $P_2(z_i)$.
  (b) Test the suggestion for the key $K' = P_2(z_i) \oplus c_i$ by checking whether indeed $Q_1(K', m_i) = m_i \oplus v_1$. If the test fails, increment i and return to Step O2. Otherwise, return the suggested key.

The time complexity of the preprocessing phase is S evaluations of P_1, and its memory complexity is also S. Note that the output of the preprocessing phase is only the value v_1 and the corresponding number t_1, and we can discard the rest of the sorted list (In our model, we can ignore the sorting time of the list, since sorting uses only cheap comparison operations.[5]). In addition, since we can execute the online phase in streaming mode by working on each given plaintext-ciphertext pair independently and discarding it afterwards, its memory complexity is negligible. The expected time and data complexities of the online phase depend on the value of t_1: we know that there are at least t_1 values of x such that $Q_1(K, x) = x \oplus v_1$. According to the birthday paradox, after trying about $2^n/t_1$ arbitrary messages we expect that at least one m_i will satisfy $Q_1(K, m_i) = m_i \oplus v_1$ and suggest the correct value of K. Thus, the expected data complexity of the online algorithm is $2^n/t_1$, and in order to compute its time complexity, we need to sum $2^n/t_1$ evaluations of P_2 in Step O2.(b), and $2^n/t_1$ encryptions in order to generate the data.

[5] One may notice that since sorting requires $O(n \log(n))$ basic operations, then our algorithm actually requires about 2^n basic operations. However, as mentioned before we expect the circuit size of any reasonable choice of P_1 to grow at least as $n^{1+\epsilon}$ (for some $\epsilon > 0$) when n increases, and thus the real time complexity of exhaustive search is in fact $O(n^{1+\epsilon} \cdot 2^n)$ basic operations, which is asymptotically larger than the number of basic operations performed by our algorithm when we take the sorting time of $\tilde{O}(2^n/n)$ values into account.

Optimizing the Basic Algorithm. We now describe several useful optimizations of the *2Round1KeyBasic* algorithm. The first optimization is to use the freedom to choose the subset X during the preprocessing phase in order to immediately filter out most of the wrong key suggestions that are now filtered only in Step O2.(b) of the online algorithm, and thus avoid the Q_1 evaluations in these cases. The idea uses a technique that resembles (but is not the same as) splice-and-cut [2]: assume that we choose the set X of size S as the subspace of values x in which the $n - \log(S)$ LSBs are zero (or any other constant). Then the value of these $n - \log(S)$ LSBs in all the t_1 inputs x that satisfy $P'_{1|X}(x) = v_1$ is zero. Consequently, we know that for any plaintext m_i, if $m_i \oplus K$ is one of these t_1 inputs, then the $n - \log(S)$ LSBs of K are equal to those of m_i. Thus, before testing the suggested key in Step O2.(b), we can check whether its $n - \log(S)$ LSBs are equal to those of m_i, and otherwise discard it without evaluating Q_1. We note that in this attack, the saving in time complexity due to this optimization is small, however, in Section 2.2 we show that a similar idea yields a more significant saving in our attacks on 3-round iterated EM. We alert the reader that even though the values in X are now chosen in a specific way, the attack remains a known plaintext attack since there is no restriction on the choice of the m_i's.

The second optimization is to consider $\ell > 1$ outputs of P'_1 with a high in-degree instead of just one. This allows us to reduce the data complexity of the attack at the expense of using more memory and slightly more time during the online phase of the attack. Since the original online algorithm required only negligible memory, this tradeoff seems favorable. Our optimized algorithm *2Round1KeyOpt* is described below, using S, D and ℓ as parameters.

Preprocessing:

PR1. Evaluate P'_1 on a subset of S inputs, X, such that the $n - \log(S)$ LSBs of each $x \in X$ are zero. Store the output values in a sorted list.

PR2. Traverse the sorted list and store the outputs $v_1, v_2, \ldots, v_\ell$ which have the highest in-degrees. Denote the in-degrees of the outputs $v_1, v_2, \ldots, v_\ell$ by $t_1, t_2, \ldots, t_\ell$, respectively.

Online:

O1. Ask for the encryption of D arbitrary plaintexts.

O2. For each plaintext-ciphertext pair (m_i, c_i):
   (a) For $j \in \{1, 2, \ldots, \ell\}$:
      i. Assume that $Q_1(K, m_i) = m_i \oplus v_j \triangleq z_{ij}$ and calculate $P_2(z_{ij})$.
      ii. Let $K' = P_2(z_{ij}) \oplus c_i$. If the $n - \log(S)$ LSBs of K' are different from those of c_i, discard it and return to Step O2.(a) (if $j = \ell$ return to Step O2). Otherwise, test K' by checking whether $Q_1(K', m_i) = m_i \oplus v_j$. If the test succeeds, return K', otherwise, if $j < \ell$ return to Step O2.(a) and if $j = \ell$ return to Step O2.

As in the *2Round1KeyBasic*, the time complexity of the preprocessing phase is S evaluations of P_1, and its memory complexity is also S. However, in

$2Round1KeyOpt$, a bigger list of size ℓ is carried over to the online algorithm, and thus its memory complexity is increased to ℓ. In order to calculate the time and data complexities, we denote by $\bar{t}$ the average value of $t_1, t_2, \ldots, t_\ell$, and thus there are $\bar{t}\ell$ values of x for which $Q_1(K,x) = x \oplus v_j$ for $j \in \{1, 2, \ldots, \ell\}$. According to the birthday paradox, after trying about $2^n/(\bar{t}\ell)$ arbitrary messages, we expect that at least one m_i will satisfy $Q_1(K, m_i) = m_i \oplus v_j$ and suggest the correct value of K. Thus, the expected data complexity of the attack is $2^n/(\bar{t}\ell)$. Since we perform ℓ evaluations of P_2 per given message, the expected time complexity of the online algorithm is about $\ell \cdot D = 2^n/\bar{t}$ evaluations of P_2, $S/2^n \cdot 2^n/\bar{t} = S/\bar{t}$ evaluations of P_1 in Step O2.(a).ii, and $2^n/(\bar{t}\ell)$ time to generate the data.

Concrete Parameters. For $n = 64$, let $S = 2^{60}$, which implies $\lambda = 2^{60}/2^{64} = 2^{-4}$. As shown before, by using the formula $(2^n \cdot \lambda^t e^{-\lambda})/t! = 2^{64} \cdot (2^{-4t} e^{-1/16})/t!$ with $t = 10$, it is easy to check that in such an evaluated subgraph of a random function we expect to see at least $\ell = 4$ vertices with an in-degree of 10. With these parameters, the time complexity of the preprocessing phase is 2^{60} evaluations of P_1 (which is equivalent to 2^{59} evaluations of the 2-round scheme), and its memory complexity is 2^{60}. The memory complexity of the online algorithm is negligible, its data complexity is $2^{64}/(10 \cdot 4) = 2^{58.7}$ known plaintexts and its time complexity is $2^{64}/10$ evaluations of P_2 and $2^{60}/10$ evaluations of P_1, which is equivalent to about $2^{59.8}$ time units. Adding the $2^{58.7}$ time required to generate the data, we obtain a total time complexity of about $2^{60.4}$, which is about 12 times faster than exhaustive search.

We can significantly reduce the data complexity by considering all the vertices with an in-degree of at least 8, whose number ℓ is expected to exceed 2^{16}. This does not affect the time and memory complexities of the preprocessing phase. The memory complexity of the online algorithm is now 2^{16} (which is still quite small), its data complexity is $2^{64}/(8 \cdot 2^{16}) = 2^{45}$ known plaintexts and its time complexity is now $2^{64}/8$ evaluations of P_2 and $2^{60}/8$ evaluations of P_1, which is equivalent in total to about $2^{60.1}$ time units, or about 15 times faster than exhaustive search (note that we actually gain in time complexity since we use significantly less data).

2.2 Attacks on 3-Round Iterated Even-Mansour with One Key

In the attacks on 2-round iterated EM with one key, we use properties of P_1 in order to guess a value of $Q_1(K,x)$ with a higher probability than expected. We then apply to this guess the public permutation P_2, which immediately gives us a suggestion for the key by XORing the obtained value with the ciphertext. In order to attack 3-round iterated EM with one key, we start with the same idea. However, after the evaluation of P_2, we cannot immediately get a suggestion for the key, as we still have to apply the complex operation of XOR'ing the unknown key, applying P_3, and XOR'ing the unknown key again, before we can compare the result to the ciphertext. Nevertheless, we notice that given the value at the output of P_2, we reduce the key recovery problem to attacking a single-round EM scheme with one key, to which we can apply the simple attack of [8]. Thus, we run an additional preprocessing step which evaluates and stores in a sorted list of values of $P'_3(x) = x \oplus P_3(x)$ for various inputs x. The sorted list is used in the online algorithm in order to obtain suggestions for the key, as described in the basic algorithm $3Round1KeyBasic$ below, which uses S_1, S_3 and D as parameters.

Preprocessing:

PR1. Evaluate P'_1 on an arbitrary subset of inputs X_1 such that $|X_1| = S_1$, and store the output values in a sorted list.
PR2. Traverse the sorted list and find an output v_1 with a maximal in-degree, denoted by t_1.
PR3. Evaluate P'_3 on an arbitrary subset of inputs X_3, such that $|X_3| = S_3$, and store the output values $P'_3(x)$ in a sorted list L_3 next to the corresponding value of $P_3(x)$.

Online:

O1. Ask for the encryption of D arbitrary plaintexts.
O2. For each plaintext-ciphertext pair (m_i, c_i):
  (a) Assume that $Q_1(K, m_i) = m_i \oplus v_1 \triangleq z_i$ and calculate $P_2(z_i)$.
  (b) Look for the value of $P_2(z_i) \oplus c_i$ in L_3. If there is no match, return to Step O2 and increment i.
  (c) For each match of $P_2(z_i) \oplus c_i$, obtain the value of $P_3(x)$ (for which $P_2(z_i) \oplus c_i = P'_3(x) = x \oplus P_3(x)$), and test the key suggestion $K' = P_3(x) \oplus c_i$ by checking whether $Q_1(K', m_i) = m_i \oplus v_1$. If the test fails, continue with the next match (if none remain, return to Step O2). Otherwise, return the key.

The time complexity of the preprocessing phase is S_1 evaluations of P_1 and S_3 evaluations of P_3, and its memory complexity is $max(S_1, S_3)$. Note that we do not need to store any of the values generated in the first step of the preprocessing after Step PR2 terminates. The memory complexity of the online algorithm is S_3. In order to calculate the expected time and data complexities of the online algorithm, we notice that after we process D pairs (m_i, c_i), we expect that at least $(t_1 \cdot D)/2^n$ of them satisfy $Q_1(K, m_i) = m_i \oplus v_1$, and consequently at least $(t_1 \cdot D \cdot S_3)/2^{2n}$ pairs will be matched and suggest the correct value for the key in Step O2.(c). Thus, in order to obtain a correct suggestion for the key, we require $(t_1 \cdot D \cdot S_3)/2^{2n} = 1$, implying that the data complexity of the attack is $D = 2^{2n}/(t_1 \cdot S_3)$. We expect a match in Step O2.(c) for a fraction of $S_3/2^n$ of the (m_i, c_i) pairs. Thus, we estimate the time complexity of the online algorithm as $D = 2^{2n}/(t_1 \cdot S_3)$ evaluations of P_2, $S_3/2^n \cdot 2^{2n}/(t_1 \cdot S_3) = 2^n/t_1$ evaluations of P_1, and $2^{2n}/(t_1 \cdot S_3)$ time required to generate the data.

Optimizing the Basic Algorithm. Similarly to our $2Round1KeyOpt$ attack, we would like to use the freedom to choose the subset X_1 during preprocessing in order to reduce the time complexity of the attack. However, in this attack we will use this freedom in a different way: we "synchronize" the sets X_1 and X_3 such that we can instantly rule out most pairs (m_i, c_i) (just by comparing bits of m_i and c_i) that do not simultaneously satisfy both $Q_1(K, m_i) = m_i \oplus v_1$ and $P_3^{-1}(c_i \oplus K) \in X_3$. Thus, we can discard most pairs (m_i, c_i) which will suggest a wrong key (or suggest no key at all) with negligible computation.

We now assume that $|X_1| = |X_3| = S$. Similarly to the *2Round1KeyOpt* algorithm, we choose X_1 as a subspace of values x in which the $n - \log(S)$ LSBs are zero (or any other constant). This implies that for any plaintext m_i, if $m_i \oplus K$ is one of the t_1 inputs that satisfy $P'_{1|X_1}(x) = v_1$, then the $n - \log(S)$ LSBs of K are equal to those of m_i. As for $x \in X_3$, we store the values of $P'_3(x) = x \oplus P_3(x)$, and set the additional condition that the $n - \log(S)$ LSBs of $P_3(x)$ are zero (or any other constant). In fact, during preprocessing, we do not evaluate $P_3(x)$ on $x \in X_3$, but rather evaluate $P_3^{-1}(y)$ for each $y \in Y_3$, where Y_3 contains all n-bit vectors whose $n - \log(S)$ LSBs are zero. Thus, we know that if $c_i \oplus K \in Y_3$, then the $n - \log(S)$ LSBs of K are equal to those of c_i. Combining the conditions on m_i and c_i, we know that a pair (m_i, c_i) will suggest a correct key in our algorithm only if the $n - \log(S)$ LSBs of m_i and c_i are equal.

Similarly to the *2Round1KeyOpt* attack, the second optimization is to consider $\ell > 1$ outputs of P'_1 with a high in-degree (instead of just one), which allows us to reduce the data complexity of the attack. Our optimized algorithm *3Round1KeyOpt* is described below, and Figure 2 illustrates its online part. Let S, D and ℓ be parameters.

Preprocessing:

PR1. Evaluate P'_1 on a subset of S inputs, X, such that the $n - \log(S)$ LSBs of each $x \in X$ are zero. Store the output values in a sorted list.

PR2. Traverse the sorted list and store the outputs $v_1, v_2, \ldots, v_\ell$ with the highest in-degrees. Denote the in-degrees of outputs $v_1, v_2, \ldots, v_\ell$ by $t_1, t_2, \ldots, t_\ell$, respectively.

PR3. Let Y_3 be the subspace of the $|S|$ n-bit vectors in which the $n - \log(S)$ LSBs are zero. For each $y \in Y_3$, store $P_3^{-1}(y) \oplus y = P'_3(P_3^{-1}(y))$ in a sorted list L_3 next to y.

Online:

O1. Ask for the encryption of D arbitrary plaintexts.

O2. For each plaintext-ciphertext pair (m_i, c_i), if the $n - \log(S)$ LSBs of m_i and c_i are not equal, discard it. Otherwise:

   (a) For $j \in \{1, 2, \ldots, \ell\}$:

      i. Assume that $Q_1(K, m_i) = m_i \oplus v_j \triangleq z_{ij}$ and calculate $P_2(z_{ij})$.

      ii. Look for the value of $P_2(z_{ij}) \oplus c_i$ in L_3. If there is no match: if $j < \ell$ return to Step O2.(a), otherwise ($j = \ell$) return to Step O2.

      iii. For each match of $P_2(z_{ij}) \oplus c_i$, obtain the value of y (such that $P_2(z_i) \oplus c_i = P_3^{-1}(y) \oplus y = P'_3(P_3^{-1}(y))$), and test the key suggestion $K' = y \oplus c_i$ by checking whether $Q_1(K, m_i) = m_i \oplus v_j$. If the test succeeds, return K', otherwise, if $j < \ell$ return to Step O2.(a), and if $j = \ell$ return to Step O2.

The time complexity of the preprocessing phase is S evaluations of P_1 and P_3^{-1}, and its memory complexity is $S + \ell$. The memory complexity of the online algorithm is also $S + \ell$. We denote by $\bar{t}$ the average value of $t_1, t_2, \ldots, t_\ell$, and thus there are $\bar{t}\ell$ values of x for which $Q_1(K, x) = x \oplus v_j$ for $j \in \{1, 2, \ldots, \ell\}$. Consequently, in order to obtain a correct suggestion for the key, we require that $(\bar{t}\ell \cdot D \cdot S)/2^{2n} = 1$, implying that the data complexity of the attack is $D = 2^{2n}/(\bar{t}\ell \cdot S)$. We process a pair (m_i, c_i) (i.e., we do not discard it in step 2) with probability $S/2^n$, and for each such pair we perform ℓ evaluations of P_2 and for a $S/2^n$ fraction of those we also evaluate Q_1 (or P_1). The expected time complexity of the online algorithm is thus $\ell \cdot S/2^n \cdot D = 2^n/\bar{t}$ evaluations of P_2, $S/\bar{t}$ evaluations of P_1, and $2^{2n}/(\bar{t}\ell \cdot S)$ time required to generate the data.

Thus, the attack has about the same time complexity as the 2*Round*1*KeyOpt* attack, and for $\ell = 1$ it is more efficient than the 3*Round*1*KeyBasic* attack by a factor of about $2^n/S$.

Concrete Parameters. For $n = 64$, let $S = 2^{60}$, i.e., $\lambda = 2^{60}/2^{64} = 2^{-4}$. Again, we use the formula $(2^n \cdot \lambda^t e^{-\lambda})/t! = (2^{64} \cdot 2^{-4t} e^{-1/16})/t!$ with $t = 8$, such that we expect at least $\ell = 2^{16}$ vertices with an in-degree of 8. With these parameters, the time complexity of the preprocessing phase is 2^{60} evaluations of P_1 (equivalent to about $2^{58.5}$ evaluations of the 3-round scheme), and its memory complexity is 2^{60}. The memory complexity of the online algorithm is 2^{60}, its expected data complexity is $2^{128}/(8 \cdot 2^{16} \cdot 2^{60}) = 2^{49}$ known plaintexts and its expected time complexity is $2^{64}/8$ evaluations of P_2 and $2^{60}/8$ evaluations of P_1, whose sum is equivalent to about $2^{59.6}$ time units (the time required to generate the data is negligible). Note that it is possible to reduce the data complexity further at the expense of increasing the time complexity by considering vertices of a lower in-degree.[6]

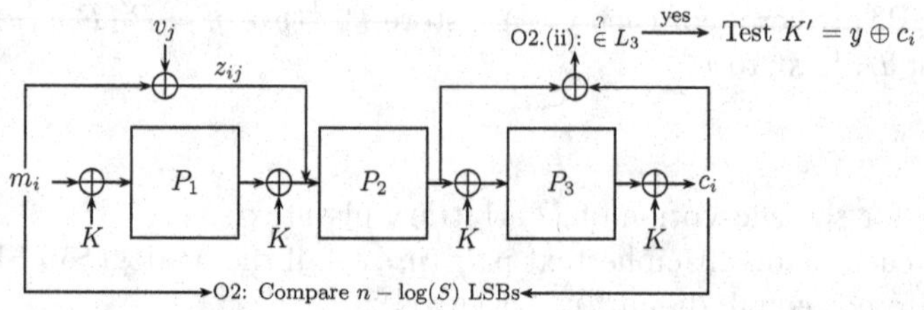

Fig. 2. The online algorithm of 3*Round*1*KeyOpt*

[6] For example, we expect more than 2^{23} vertices with an in-degree of at least 7, and thus if we use only $2^{128}/(8 \cdot 2^{23} \cdot 2^{60}) = 2^{42}$ known plaintexts for the attack, the time complexity of the online algorithm slightly increases from $2^{59.6}$ to about $2^{59.8}$.

3 Applications to Step-Reduced LED

LED is a 64-bit block cipher designed for resource-constrained environments, proposed by Guo et al. at CHES 2011 [12]. The two main variants of LED are LED-64 (which supports 64-bit keys) and LED-128 (which supports 128-bit keys). The design of LED can be viewed as a special case of iterated EM schemes: LED-64 is in fact an 8-step iterated EM scheme[7] with one key and LED-128 is a 12-step iterated EM scheme with alternating keys K_1 and K_2. The inner permutations of LED are based on the AES design framework, however, since our attacks do not exploit any properties of these permutations, we do not specify them here and refer the reader to [12] for further details.

In the single-key model, the best attack published so far on reduced LED-64 breaks 2 steps of this cipher [13]. For LED-128, the largest number of attacked steps was 6 (see [16]). In this paper, we use our generic attacks in order to improve the data complexity of the attack on 6-step LED-128 from 2^{59} to 2^{45}, while keeping the time and memory complexities similar to the original attack. More significantly, we present the first single-key attacks which are faster than exhaustive search on 3-step LED-64 and on 8-step LED-128. The previous attacks on LED (which are in the single-key rather than in the related-key model) and our new attacks are summarized in Table 1. Note in particular that our new attack on 8-step LED-128 actually has a slightly better time complexity and requires about a thousand times less data than the best previous attack which could only be applied to 6 steps of LED-128, out of the full 12.

Table 1. Single-Key Attacks of Step-Reduced LED

Reference	Cipher	Steps	Time	Data	Memory
[13]	LED-64	2	2^{56}	2^{8} CP	2^{11}
This paper	LED-64	3	$2^{60.2}$	2^{49} KP	2^{60}
[13]	LED-128	4	2^{112}	2^{16} CP	2^{19}
[15]	LED-128	4	2^{96}	2^{64} KP	2^{64}
[16]	LED-128	4	2^{96}	2^{32} KP	2^{32}
[16]	LED-128	6	$2^{124.4}$	2^{59} KP	2^{59}
This paper	LED-128	6	$2^{124.5}$	2^{45} KP	2^{60}
This paper	LED-128	8	$2^{123.8}$	2^{49} KP	2^{60}

The data complexity is given in chosen plaintexts (CP), or in known plaintexts (KP).

[7] In the design of LED, the term "step" is used in order to describe what we refer to as a "round" of an iterated EM scheme. On the other hand, a "round" of LED is used in order to describe a smaller component of its internal permutation. Thus, in order to avoid confusion, we will use the term "step" in this section.

3.1 An Attack on 6 Steps of LED-128

As was pointed out in [15, 16], it is easy to reduce $2r+2$-steps of LED-128 (with its alternating use of two keys) into an iterated EM scheme variant with one key by guessing K_1 and combining consecutive pairs of permutations (along with the XOR'ed key between them) into a single known permutation. In particular, [16] reduced 6-step LED-128 into a 2-step iterated EM, which was relatively easy to attack. Similarly, we guess K_1, and for each guess, we partially encrypt and decrypt the given plaintext-ciphertext pairs and remain with a 2-step iterated EM scheme with a single key (K_2). Thus, we can apply our 2-step iterated EM attack (presented in Section 2.1) for each guess of K_1. However, we note that the preprocessing phase of our $2Round1KeyOpt$ attack should be executed for each guess of K_1, and it is thus now a part of the online algorithm of the attack on LED-128. Moreover, the algorithm can no longer be performed in streaming mode, as we need to reuse each plaintext-ciphertext pair for each guess of K_1. The general framework of the algorithm is given below.

1. Ask for the encryption of D arbitrary plaintexts and store them.
2. For each value of K_1:
   (a) Apply the $2Round1KeyOpt$ attack (including the preprocessing steps) on the resultant scheme, with plaintext-ciphertext pairs $(P_1(m_i \oplus K_1), P_6^{-1}(c_i \oplus K_1))$. Test each returned key using another pair (m_j, c_j).

Using the parameters of our $2Round1KeyOpt$ attack (presented in Section 2.1), the expected data complexity of the attack is 2^{45} known plaintexts and its memory complexity is 2^{60} (required for preprocessing, which is now part of the online algorithm). We calculate the expected time complexity of the algorithm as follows: adding the preprocessing and online time complexities, the main procedure of the attack performed for each guess of K_1 requires about $2^{60.1}+2^{60} \approx 2^{61.1}$ evaluations of 4 out of the 6 permutations, which is equivalent to about $2^{60.5}$ evaluations of the full scheme. Compared to this complexity, the partial encryption and decryption of each (m_j, c_j) pair, and the trial encryptions using (m_j, c_j) (performed on average once per guess of K_1) are negligible. Thus, the expected time complexity of the attack is about $2^{64+60.5} = 2^{124.5}$, which is about 11 times better than exhaustive search.

3.2 An Attack on 3 Steps of LED-64

We can attack 3-step LED-64 by directly applying $3Round1KeyOpt$ attack with $n = 64$, presented in Section 2.2. Thus, the preprocessing phase has a time complexity of about $2^{58.5}$ and memory complexity of 2^{60}. The online algorithm has a memory complexity of 2^{60}, data complexity of 2^{49} known plaintexts and time complexity of $2^{59.6}$. Since in this paper we consider the preprocessing time as part of the attack (i.e., we assume that we are trying to attack the scheme for the first time), the total time complexity of the algorithm is about $2^{60.2}$, which is about 14 times better than exhaustive search.

3.3 An Attack on 8 Steps of LED-128

We use the same framework of our 6 step attack on LED-128 in order to attack 8 steps of LED-128 (shown in Figure 3). Namely, we guess K_1, and for each guess, we partially encrypt and decrypt the given plaintext-ciphertext pairs and remain with a 3-step iterated EM scheme with a single key (K_2). We then apply our *3Round1KeyOpt* attack (presented in Section 2.2) for each guess of K_1. Thus, the memory complexity of the attack is 2^{60} and its data complexity is 2^{49} known plaintexts. We calculate the expected time complexity of the algorithm as follows: adding the preprocessing and online time complexities, the main procedure of the algorithm performed for each guess of K_1 requires about $2^{58.5} + 2^{59.6} \approx 2^{60.2}$ evaluations of 6 out of the 8 permutations, equivalent to about $2^{59.8}$ evaluations of the full scheme. Thus, the expected time complexity of the attack is about $2^{64+59.8} = 2^{123.8}$, which is about 18 times better than exhaustive search.

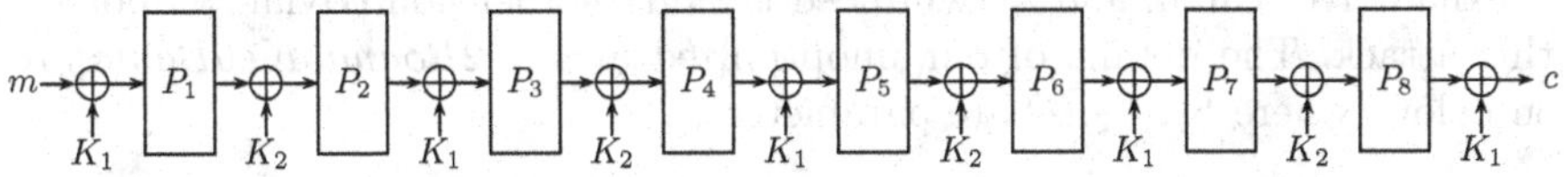

Fig. 3. 8-step LED-128

4 Attacks on 2-Round Iterated Even-Mansour with Independent Keys

The best known generic attack on 2-Round iterated EM with independent keys (see Figure 4) is a MITM attack. This attack is described in the full version of this paper [7] and it requires 2^n memory and has a time complexity of about $2^{n+1.6}$ full cipher evaluations.

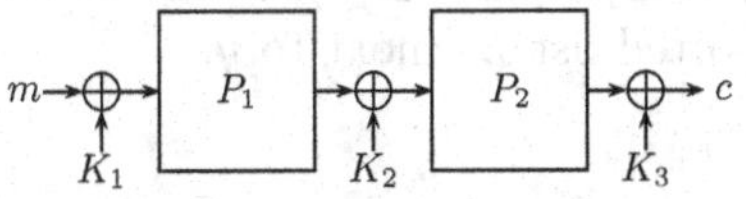

Fig. 4. A 2-round iterated EM with independent keys

In this attack, we use a property of the permutation P_i, which is shared by the keyed permutation $Q_i(K_i, K_{i+1}, x) = P_i(x \oplus K_i) \oplus K_{i+1}$ for any value of K_i and K_{i+1}: these permutations have the same difference distribution table. In order to demonstrate this, consider an entry with the value of t in the difference distribution table of P_i, and denote its input and output differences by Δ_1 and Δ_2, respectively. Let us denote the t corresponding input-output pairs[8]

[8] In this paper, we consider unordered pairs, i.e., $((x, y), (u, v))$ and $((u, v), (x, y))$ are considered the same pair.

by $((x_1, y_1), (x_1 \oplus \Delta_1, y_1 \oplus \Delta_2)), \ldots, ((x_t, y_t), (x_t \oplus \Delta_1, y_t \oplus \Delta_2))$. Then, the t input-output pairs $((x_1 \oplus K_1, y_1 \oplus K_2), (x_1 \oplus K_1 \oplus \Delta_1, y_1 \oplus K_2 \oplus \Delta_2)), \ldots, ((x_t \oplus K_1, y_t \oplus K_2), (x_t \oplus K_1 \oplus \Delta_1, y_t \oplus K_2 \oplus \Delta_2))$ correspond to the same entry in the difference distribution table of Q_i (i.e., the entry with input and output differences Δ_1 and Δ_2, respectively).

Using the property above, if we find an entry $[\Delta_1, \Delta_2]$ in the difference distribution table of P_i with a large value, then we can use a similar attack to the one given in [15] on 2-round iterated EM,[9] in order to break the scheme. However, our main observation is that we can find such an entry by preprocessing the public function P_i, which does not need to admit any special property in order to attack the scheme. Thus, our attack adds a preprocessing algorithm to the online algorithm of the attack of [15] (which assumes that we have an entry in the difference distribution table of P_i with a large value). In addition (as we will see later), in the case of independent keys, the basic attack of [15] is not better than exhaustive search, and we will need to add another non-trivial component to this attack. The details of our unoptimized attack *2Round3KeyBasic* are given below, where S_1, S_2, D are parameters:

Preprocessing:

PR1. Choose an arbitrary input difference $\Delta_1 \neq 0$ and evaluate P_1 on S_1 arbitrary input pairs with input difference Δ_1. For each pair $(x, P_1(x)), (x \oplus \Delta_1, P_1(x \oplus \Delta_1))$, store the output difference $P_1(x) \oplus P_1(x \oplus \Delta_1)$ in a sorted list, next to x.

PR2. Traverse the sorted list and find the most common output difference Δ_2 (if there are several options for Δ_2, choose one arbitrarily). Keep only the entries of the list which correspond to pairs with the output difference of Δ_2 (assume that we have t such pairs). For each such entry, recalculate and store the full pair $(x, P_1(x)), (x \oplus \Delta_1, P_1(x \oplus \Delta_1))$.

PR3. Evaluate P_2 on S_2 arbitrary input pairs with input difference Δ_2. For each pair $(y, P_2(y)), (y \oplus \Delta_2, P_2(y \oplus \Delta_2))$, store the output difference $P_2(y) \oplus P_2(y \oplus \Delta_2)$ in a sorted list L_2, next to y.

Online:

O1. Ask for the encryption of D arbitrary input pairs with difference Δ_1.

O2. For each pair of plaintext-ciphertext pairs $((m_i^1, c_i^1), (m_i^2 = m_i^1 \oplus \Delta_1, c_i^2))$:

(a) Search for the output difference $c_i^1 \oplus c_i^2$ in L_2, (if there is no match, discard the pair and return to Step O2).

(b) For each match $(y, P_2(y)), (y \oplus \Delta_2, P_2(y \oplus \Delta_2))$, we have 2 candidates for K_3: $P_2(y) \oplus c_i^1$ and $P_2(y) \oplus c_i^2$. We also have $2t$ candidates for K_1: the candidates $x \oplus m_i^1$ and $x \oplus m_i^2$ for each of the t values of x. As each pair of values for K_1 and K_3 suggests a value for K_2, we have $4t$ suggestions of the full key to test using another plaintext-ciphertext pair.

[9] Although the attack of [15] was previously applied to 2-round iterated EM with one key, it can be adapted to work for the case of independent keys.

Similarly to our analysis of random functions, assuming that P_1 is a random permutation, then each entry in its difference distribution table is distributed according to the Poisson distribution [17].[10] This will allow us to easily determine the expected value of t and use it in order to analyze the expected complexity of our algorithm.

The memory complexity of the preprocessing phase is $max(S_1, S_2)$, and its time complexity is $2 \cdot S_1$ evaluations of P_1 and $2 \cdot S_2$ evaluations of P_2, or $S_1 + S_2$ evaluations of the full scheme. The memory complexity of the online algorithm is S_2. Using the birthday paradox, out of the D plaintext-ciphertext pairs evaluated in the online phase, at least $(D \cdot t)/2^{n-1}$ are expected to have a difference of Δ_2 after P_1 (note that we have 2^{n-1} unordered pairs with a given difference). Using the same argument, we expect that $(D \cdot t \cdot S_2)/2^{2(n-1)}$ of them will match the pairs evaluated for P_2 during proprocessing. Thus, we require that $(D \cdot t \cdot S_2)/2^{2(n-1)} = 1$, or $D = 2^{2(n-1)}/(t \cdot S_2)$ in order to find the key with high probability. Without going into the details of the time complexity analysis, note that we are using only two plaintext-ciphertext pairs to filter the key suggestions, tested in Step O2.(b). As we have $3n$ bits of key and $2n$ bits of filtering, we need to test at least 2^n keys in Step O2.(b), and thus the attack is not faster than the simple MITM attack on this scheme.

4.1 A Time-Optimized Attack on 2-Round Iterated Even-Mansour

In order to improve the attack, we need to add more filtering conditions, and thus we actually work on triplets, as described in the improved algorithm *2Round3KeyOpt*:

Preprocessing:

PR1. Choose an arbitrary input difference $\Delta_1 \neq 0$ and evaluate P_1 on S_1 arbitrary input pairs with input difference Δ_1. For each pair $(x, P_1(x)), (x \oplus \Delta_1, P_1(x \oplus \Delta_1))$, store the output difference $P_1(x) \oplus P_1(x \oplus \Delta_1)$ in a sorted list, next to x.

PR2. Traverse the sorted list and find the most common output difference Δ_2 (if there are several options for Δ_2, choose one arbitrarily). Keep only the entries of the list which correspond to pairs with the output difference of Δ_2 (assume that we have t such pairs). For each such entry recalculate and store the full pair $(x, P_1(x)), (x \oplus \Delta_1, P_1(x \oplus \Delta_1))$ in a list L_1.

PR3. Choose another non-zero input difference Δ'_1. For each value x stored in L_1, evaluate P_1 an additional time to obtain the pair $(x \oplus \Delta'_1, P_1(x \oplus \Delta'_1))$. Store the (total of) additional t output differences $P_1(x) \oplus P_1(x \oplus \Delta'_1)$ in a separate sorted list of differences, L'_1.

PR4. Evaluate P_2 on S_2 arbitrary input pairs with input difference Δ_2. For each pair $(y, P_2(y)), (y \oplus \Delta_2, P_2(y \oplus \Delta_2))$, store the output difference $P_2(y) \oplus P_2(y \oplus \Delta_2)$ in a sorted list L_2, next to y.

[10] However, we note that since we consider unordered pairs, then we have only 2^{n-1} possible pairs of a given difference, and each pair can attain (almost) all 2^n output differences

Online:

O1. Ask for the encryption of D arbitrary input triplets of the form m, $m \oplus \Delta_1$ and $m \oplus \Delta'_1$ (for D arbitrary values of m).
O2. For each pair of plaintext-ciphertext pairs $((m_i^1, c_i^1), (m_i^2 = m_i^1 \oplus \Delta_1, c_i^2))$:
  (a) Search for the output difference $c_i^1 \oplus c_i^2$ in the list L_2 (if there is no match, discard the pair and return to Step O2).
  (b) For each match $(y, P_2(y)), (y \oplus \Delta_2, P_2(y \oplus \Delta_2))$, compute the 2 candidates for K_3: $K'_3 = P_2(y) \oplus c_i^1$ and $K''_3 = P_2(y) \oplus c_i^2$.
  (c) Denote the third plaintext-ciphertext pair in the triplet by $(m_i^3 = m_i^1 \oplus \Delta'_1, c_i^3)$. Compute $y' = P_2^{-1}(c_i^3 \oplus K'_3)$ and $y'' = P_2^{-1}(c_i^3 \oplus K''_3)$.
  (d) Search L'_1 for the four possibilities of the third difference obtained at this stage: $y' \oplus y$, $y' \oplus \Delta_2 \oplus y$, $y'' \oplus y$, $y'' \oplus \Delta_2 \oplus y$ (if there is no match, discard the pair and return to Step O2).
  (e) Test the $4t$ suggestions of the full key using (m_i^3, c_i^3). If the test succeeds, return the key.

The time and memory complexities of the preprocessing phase are similar to those of the $2Round3KeyBasic$ attack (the additional t evaluations of P_1 and t units of storage are negligible). Using the calculation done for $2Round3KeyBasic$, the online algorithm requires $D = 2^{2(n-1)}/(t \cdot S_2)$ plaintext-ciphertext triplets. For each processed triplet, we expect to find a match in L_2 with probability $S_2/2^n$. For each such matched triplet, we need to compute $P_2(y)$ (in order to compute K'_3 and K''_3) and evaluate P_3^{-1} twice in order to compute y' and y''. Once we do so, the probability of a match in L'_1 in Step O2.(d) is proportional to $t/2^n$. This is a negligible probability, and thus we can neglect the complexity of the trial encryptions in Step O2.(e). Thus, the online time complexity (without counting the data) is about $3 \cdot D \cdot S_2/2^n = 0.75 \cdot 2^n/t$ evaluations of P_2, or $0.375 \cdot 2^n/t$ evaluations of the full scheme.

The data complexity of the attack is D triplets, or $3D$ chosen plaintexts. However, we can easily reduce it to $2D$ by requesting encryptions of structures containing the messages m, $m \oplus \Delta_1$, $m \oplus \Delta'_1$ and $m \oplus \Delta_1 \oplus \Delta'_1$. Each such structure of 4 plaintexts contains two triplets which we can exploit, implying that the data complexity of the attack is indeed $2D$. If we add the time to generate the data to the time complexity, we get that the total time complexity of the online attack is about $2D + 0.375 \cdot 2^n/t$ evaluations of the full scheme.

4.2 Applications to Full AES^2

AES^2 is a 128-bit block cipher presented at Eurocrpyt 2012 [5]. The cipher is a 2-round iterated EM construction, where each of the public permutations P_1 and P_2 is based on an invocation of full AES-128 with a pre-fixed and publicly known key. The designers of the scheme claim that its security is 2^{128}. However, the best attack known to the designers (as claimed in [5]) is the MITM attack presented in [7], and based on our analysis, it has a slightly higher time complexity of $3 \cdot 2^{128} \approx 2^{129.6}$ and a memory complexity of 2^{128}.

In order to attack AES^2, we use our *2Round3KeyOpt* attack with $S_1 = 2^{124}$ and $S_2 = 2^{125.4}$. This implies that the memory complexities of both the preprocessing and online phases is $2^{125.4}$. The time complexity of the preprocessing phase is $S_1 + S_2 = 2^{124} + 2^{125.4} \approx 2^{125.9}$ evaluations of the full scheme. Using the formula $(2^n \cdot \lambda^t \cdot e^{-\lambda})/t!$ with $\lambda = 2^{124}/2^{128} = 1/16$ and $t = 18$, it is easy to check that we expect to find at least 10 entries in the difference distribution table with a value of 18 (we need only one). Plugging in these values into the formula $D = 2^{2(n-1)}/(t \cdot S_2)$, we obtain $D \approx 2^{124.4}$, implying that the data complexity of the attack is $2^{125.4}$ chosen plaintexts. The time complexity of the online attack is $2D + 0.375 \cdot 2^n/t \approx 2^{125.6}$, and adding the preprocessing time, the total time complexity of the algorithm is about $2^{125.9} + 2^{125.6} \approx 2^{126.8}$. This is better than the $2^{129.6}$ time complexity of the MITM attack by a factor of about 7, and clearly violates the 128-bit security claimed for AES^2 in [5]. We also note that the memory complexity is improved from 2^{128} to about $2^{125.4}$, however the data complexity is greatly increased to $2^{125.4}$.

5 Conclusions

In this paper we considered several schemes which are based on the iterated Even-Mansour scheme, and improved their best known attacks. For the recommended values of n our attacks are between 7 and 20 times faster than exhaustive search, but they differ from other improvements of exhaustive search since their improvement factor is about $n/\log(n)$, which increases to infinity as n grows. In particular, we described the first attack on the full AES^2, and improved the number of steps which can be attacked in the well known LED-128 block cipher from 6 to 8. Even though our attacks are not likely to be practically significant, they indicate that block ciphers based on the EM scheme with one key should have at least 4 rounds, regardless of how strong we make the internal permutations.

References

1. Andreeva, E., Bogdanov, A., Dodis, Y., Mennink, B., Steinberger, J.P.: On the Indifferentiability of Key-Alternating Ciphers. In: Canetti, R., Garay, J.A. (eds.) CRYPTO 2013, Part I. LNCS, vol. 8042, pp. 531–550. Springer, Heidelberg (2013)
2. Aoki, K., Sasaki, Y.: Preimage Attacks on One-Block MD4, 63-Step MD5 and More. In: Avanzi, R.M., Keliher, L., Sica, F. (eds.) SAC 2008. LNCS, vol. 5381, pp. 103–119. Springer, Heidelberg (2009)
3. Biryukov, A., Wagner, D.: Slide Attacks. In: Knudsen, L.R. (ed.) FSE 1999. LNCS, vol. 1636, pp. 245–259. Springer, Heidelberg (1999)
4. Bogdanov, A., Khovratovich, D., Rechberger, C.: Biclique Cryptanalysis of the Full AES. In: Lee, D.H., Wang, X. (eds.) ASIACRYPT 2011. LNCS, vol. 7073, pp. 344–371. Springer, Heidelberg (2011)
5. Bogdanov, A., Knudsen, L.R., Leander, G., Standaert, F.-X., Steinberger, J.P., Tischhauser, E.: Key-Alternating Ciphers in a Provable Setting: Encryption Using a Small Number of Public Permutations - (Extended Abstract). In Pointcheval, Johansson (eds.) [18], pp. 45–62

6. Daemen, J.: Limitations of the Even-Mansour Construction. In: Matsumoto, T., Imai, H., Rivest, R.L. (eds.) ASIACRYPT 1991. LNCS, vol. 739, pp. 495–498. Springer, Heidelberg (1993)
7. Dinur, I., Dunkelman, O., Keller, N., Shamir, A.: Key Recovery Attacks on 3-round Even-Mansour, 8-step LED-128, and Full AES^2. IACR Cryptology ePrint Archive, 2013:391 (2013)
8. Dunkelman, O., Keller, N., Shamir, A.: Minimalism in Cryptography: The Even-Mansour Scheme Revisited. In: Pointcheval, Johansson (eds.) [18], pp. 336–354
9. Even, S., Mansour, Y.: A Construction of a Cipher from a Single Pseudorandom Permutation. J. Cryptology 10(3), 151–162 (1997)
10. Flajolet, P., Odlyzko, A.M.: Random Mapping Statistics. In: Quisquater, J.-J., Vandewalle, J. (eds.) EUROCRYPT 1989. LNCS, vol. 434, pp. 329–354. Springer, Heidelberg (1990)
11. Gérard, B., Grosso, V., Naya-Plasencia, M., Standaert, F.-X.: Block Ciphers That Are Easier to Mask: How Far Can We Go? In: Bertoni, G., Coron, J.-S. (eds.) CHES 2013. LNCS, vol. 8086, pp. 383–399. Springer, Heidelberg (2013)
12. Guo, J., Peyrin, T., Poschmann, A., Robshaw, M.: The LED Block Cipher. In: Preneel, B., Takagi, T. (eds.) CHES 2011. LNCS, vol. 6917, pp. 326–341. Springer, Heidelberg (2011)
13. Isobe, T., Shibutani, K.: Security Analysis of the Lightweight Block Ciphers XTEA, LED and Piccolo. In: Susilo, W., Mu, Y., Seberry, J. (eds.) ACISP 2012. LNCS, vol. 7372, pp. 71–86. Springer, Heidelberg (2012)
14. Lampe, R., Patarin, J., Seurin, Y.: An Asymptotically Tight Security Analysis of the Iterated Even-Mansour Cipher. In: Wang, Sako (eds.) [19], pp. 278–295
15. Mendel, F., Rijmen, V., Toz, D., Varici, K.: Differential Analysis of the LED Block Cipher. In: Wang, Sako (eds.) [19], pp. 190–207
16. Nikolić, I., Wang, L., Wu, S.: Cryptanalysis of Round-Reduced LED. In: FSE. To appear in Lecture Notes in Computer Science (2013)
17. O'Connor, L.: On the Distribution of Characteristics in Bijective Mappings. In: Helleseth, T. (ed.) EUROCRYPT 1993. LNCS, vol. 765, pp. 360–370. Springer, Heidelberg (1994)
18. Pointcheval, D., Johansson, T. (eds.): EUROCRYPT 2012. LNCS, vol. 7237. Springer, Heidelberg (2012)
19. Wang, X., Sako, K. (eds.): ASIACRYPT 2012. LNCS, vol. 7658. Springer, Heidelberg (2012)

Key Difference Invariant Bias in Block Ciphers

Andrey Bogdanov[1,*] and Christina Boura[1,*], Vincent Rijmen[2,*], Meiqin Wang[3,*], Long Wen[3,*], Jingyuan Zhao[3,*]

[1] Technical University of Denmark, Denmark
[2] KU Leuven ESAT/SCD/COSIC and iMinds, Belgium
[3] Shandong University, Key Laboratory of Cryptologic Technology and Information Security, Ministry of Education, Shandong University, Jinan 250100, China

Abstract. In this paper, we reveal a fundamental property of block ciphers: There can exist linear approximations such that their biases ε are deterministically *invariant under key difference.* This behaviour is highly unlikely to occur in idealized ciphers but persists, for instance, in 5-round AES. Interestingly, the property of key difference invariant bias is independent of the bias value ε itself and only depends on the form of linear characteristics comprising the linear approximation in question as well as on the key schedule of the cipher.

We propose a statistical distinguisher for this property and turn it into an key recovery. As an illustration, we apply our novel cryptanalytic technique to mount related-key attacks on two recent block ciphers — LBlock and TWINE. In these cases, we break 2 and 3 more rounds, respectively, than the best previous attacks.

Keywords: block ciphers, key difference invariant bias, linear cryptanalysis, linear hull, key-alternating ciphers, LBlock, TWINE.

1 Introduction

1.1 Linear Cryptanalysis, Linear Approximations, and Bias

Linear cryptanalysis is a central and indispensable attack on block ciphers. Having been proposed as early as in 1992 [23–25], it forms an established research field within symmetric-key cryptology. Since then, many interesting results have been obtained in the area, among others including correlation matrices by Daemen et al. [8], multiple linear cryptanalysis by Kaliski and Robshaw [15], linear hull effect by Nyberg [29], multidimensional cryptanalysis by Hermelin et al. [13], comprehensive bounds on linear properties by Keliher and Sui [18], as well as success probability estimations by Selçuk [35].

The basis of linear cryptanalysis is a *linear approximation* of a function f. If the linear approximation holds with probability $1/2+\varepsilon$, ε is called its *bias*. A linear approximation can comprise numerous linear characteristics θ, each contributing their linear characteristic bias ε_θ to the linear approximation bias ε.

* Corresponding authors.

K. Sako and P. Sarkar (Eds.) ASIACRYPT 2013 Part I, LNCS 8269, pp. 357–376, 2013.

There are essentially two standard approaches to deal with the key-dependency of these biases: they are either averaged over all keys or evaluated for a fixed key. Both cases have been studied in great detail and these approaches have turned out to be very fruitful: While the average behaviour of the bias is vital to the foundations of linear cryptanalysis and the demonstration of the linear hull effect, Murphy has demonstrated [27] that there can be keys for which the linear distinguisher might not apply. The latter observation is more inline with the fixed-key correlation-matrix approach, which also, among others, has lead to zero-correlation attacks by Bogdanov et al. [3–5] and improved linear attacks on PRESENT by Cho [6].

Apart from the average case and the fixed-key case, recently, Abdelraheem et al. [1] have managed to compute the distribution of linear characteristic bias for several interesting examples. Moreover, there has been quite some interest towards deducing key information from the value of the bias [7,28,30]. Kim [19] studies the combined related-key linear-differential attacks on block ciphers. Interestingly, a linear-hull version of Matsui's Algorithm 1 by Röck and Nyberg [32] uses the fact that, in some ciphers, the linear characteristic biases ε_θ are the same for different keys.

At the same time, much less is known about the even more fundamental question of how *the bias ε of the entire linear approximation behaves under a change of key.* This is not least due to the fact that the entire linear hull is notoriously difficult to analyze for the immense number of linear characteristics θ comprising it. In this paper, we tackle this problem and reveal a property for many block ciphers, namely, that *the bias ε of a linear approximation can be actually invariant under the modification of the key.*

1.2 Our Contributions

The contributions of this paper are as follows.

Bias Invariant under Key Difference in Iterative Block Ciphers. We investigate the bias of a linear approximation in *key-alternating ciphers* (iterative SPN ciphers with XOR addition of subkeys) under a change of the key. By looking at the composition of the fixed-key linear hull from individual characteristics, we derive a sufficient condition on the keys and linear approximations such that *the bias remains unaffected by a change of key.* The class of key-alternating ciphers is already broad enough to include AES, most of the other SPN ciphers, and some Feistel ciphers. After recalling some background on linear cryptanalysis in Section 2, we describe these findings in Section 3.

An Instructive Example with AES. With our technique, the key difference invariant bias property is easy to construct over (a part of) susceptible ciphers since it mainly depends on the differential diffusion in the key schedule and on the linear diffusion in the data transform of a cipher. We use AES to show how the property can be derived. Namely, we demonstrate a key difference invariant bias property holding deterministically over 5 rounds of the original AES-256. This serves as a pedagogical illustration. See Section 3.3.

Statistical Distinguisher and Generic Key Recovery. The probability to have the key difference invariant bias property in an idealized block cipher with block size n, is about $\frac{1}{\sqrt{2\pi}} 2^{\frac{3-n}{2}}$. This forms the basis for a statistical distinguisher that can be used for key recovery. Here, we use the fact that the *key difference invariant bias property is actually truncated*, i.e., there are many linear approximations with key difference invariant bias in most susceptible ciphers. In our distinguisher, for two keys, we compute the sample biases of a set of approximations with this property (using the part of the plaintext-ciphertext pairs available to the adversary) and test their collective proximity. We demonstrate that it is possible to efficiently distinguish this from an idealized cipher, under some basic independency assumptions. The distinguisher can be used for hash functions and block ciphers. In the related-key setting, we propose a key recovery procedure for block ciphers which is similar to Matsui's Algorithm 2. These techniques are given in Section 4.

Applications to Block Ciphers LBlock and TWINE. As an illustration, we apply our new cryptanalytic technique of key difference invariant bias to the recently proposed block ciphers LBlock [39] and TWINE [37] . LBlock was designed by Wu et al. and presented in ACNS 2011. Its state and key sizes are 64 and 80 bits respectively. LBlock has received the attention of many cryptographers and various attacks have been published so far on some reduced versions [16, 20–22, 26, 33, 34]. The best attack breaks 22 rounds of the cipher. TWINE is a block cipher proposed in SAC 2012 by Suzaki et al. that is operating on a 64-bit state that is parameterized by keys of length 80 or 128 bits. The total number of rounds is 36. The best known attack on TWINE-128, is an impossible differential attack given in [37], that breaks 24 rounds of the cipher.

We identify key difference invariant bias properties over 16 rounds of LBlock and 17 rounds of TWINE-128. This allows us to attack 24-round LBlock and 27-round TWINE-128 in the classical related-key model with differences in the user-supplied master keys. Thus, our attacks improve upon the state-of-the-art cryptanalysis for both LBlock and TWINE by breaking 2 and 3 more rounds, respectively, than the best previous attacks. Our cryptanalysis is provided in Sections 5 and Section 6.

2 Preliminaries

2.1 Key-Alternating Ciphers

A *block cipher* operating on n-bit blocks with a k-bit key can be seen as a subset of cardinality 2^k of the set of all $2^n!$ permutations over the space of n-bit strings. In an *idealized block cipher*, this subset is randomly chosen. In all practical settings, however, one is concerned with efficiently implementable block ciphers. So all block ciphers used in practice contain at their core the iterative application of r similar invertible transformations (called *rounds*). *Key-alternating block ciphers* form a special but important subset of the modern block ciphers (see Figure 1):

Definition 1 (Key-alternating block cipher [9]). *Let each round i, $1 \leq i \leq r$, of a block cipher have its own n-bit subkey k_i. This block cipher is* key-alternating, *if the key material in round i is introduced by XORing the subkey k_i to the state at the end of the round. Additionally, the subkey k_0 is XORed with the plaintext before the first round.*

The $r+1$ round subkeys $k_0, k_1, \ldots, k_{r-1}, k_r$ build the *expanded key* K (of length $n(r+1)$ bits) which is derived from the user-supplied key κ using a key-schedule algorithm φ. Numerous popular and widely used block ciphers belong to the class of key-alternating block ciphers. Among others, almost all SPNs (including AES) and some Feistel ciphers are key-alternating [11].

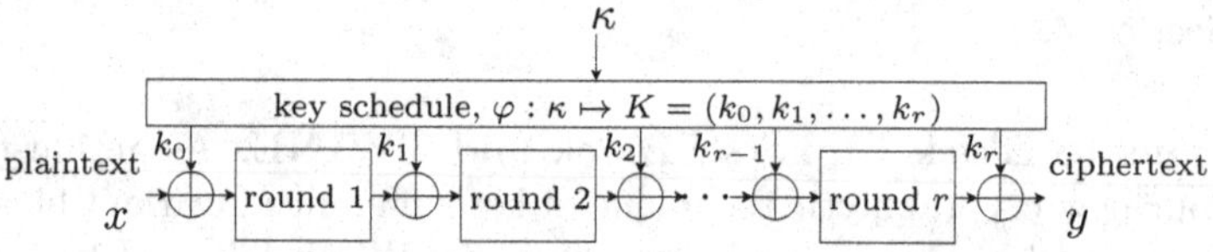

Fig. 1. Key-alternating cipher

2.2 Linear Approximations and Bias

We briefly recall the concepts of linear approximations and bias. We denote the scalar product of binary vectors by $a^t x = \bigoplus_{i=1}^{n} a_i x_i$. Linear cryptanalysis is based on *linear approximations* determined by input mask a and output mask b. A linear approximation (a, b) of a vectorial function f has a *bias* defined by

$$\varepsilon^f_{a,b} = \Pr_x\{b^t f(x) \oplus a^t x\} - 1/2$$

to which we also refer simply as ε if its assignment to function and linear approximation is clear from the context. We call a linear approximation *trivial* if both a and b are zero. Otherwise, with both $a \neq 0$ and $b \neq 0$, it is *non-trivial.*

2.3 Linear Characteristics and Linear Hulls

A linear approximation (a, b) of an iterative block cipher (e.g. a key-alternating block cipher of Definition 1) is called a *linear hull* in [29]. The linear hull contains all possible sequences of the linear approximations over individual rounds, with input mask a and output mask b. These sequences are called *linear characteristics* which we denote by θ. Now we recall the relations between the bias of a linear characteristic and the bias of the entire linear hull it belongs to, for key-alternating block ciphers.

Given a linear hull (a, b), a linear characteristic θ is the concatenation of an input mask $a = \theta_0$ before the first round, an output mask $b = \theta_r$ after the last round, and $r-1$ intermediate masks θ_i between rounds $i-1$ and i:

$$\theta = (\theta_0, \theta_1, \ldots, \theta_{r-1}, \theta_r). \tag{1}$$

Thus, each linear characteristic consists of $n(r+1)$ bits (cf. the length of the expanded key K). The *bias* ε_θ *of the linear characteristic* θ is defined as the scaled product of the individual biases $\varepsilon_{\theta_{i-1},\theta_i}$ over each round:

$$\varepsilon_\theta = 2^{r-1} \prod_{i=1}^{r} \varepsilon_{\theta_{i-1},\theta_i}.$$

In a key-alternating cipher, only the sign of ε_θ depends on the key value, while the absolute bias value $|\varepsilon_\theta|$ remains exactly the same for all keys. As a reference point, we denote by $d_\theta \in \{0, 1\}$ the sign of the linear characteristic bias with expanded key $K = 0$:

$$\varepsilon_\theta[0] = (-1)^{d_\theta} |\varepsilon_\theta|.$$

Now we formulate the following central proposition that deterministically connects the linear approximation bias with the individual linear characteristic biases through a fixed key value:

Proposition 1 ([9, Subsection 7.9.2]). *For a key-alternating block cipher, the bias* ε *of a non-trivial linear hull* (a, b) *is*

$$\varepsilon = \sum_{\theta:\theta_0=a,\theta_r=b} (-1)^{d_\theta + \theta^t K} |\varepsilon_\theta|.$$

We will be relying on Proposition 1 in the next section to determine when ε is invariant under a change of key.

3 Towards Bias Invariant under Key Difference

For a non-trivial linear hull (a, b) of a block cipher, let ε and ε' be two biases under two distinct keys κ and κ', respectively. Now we consider when $\varepsilon = \varepsilon'$ with $\kappa \neq \kappa'$, that is, when the bias is invariant under a change of key.

3.1 Key Difference Invariant Bias in Key-Alternating Ciphers

In a key-alternating block cipher, let K and K' be the expanded keys corresponding to two user-supplied keys κ and κ', $K = \varphi(\kappa)$ and $K' = \varphi(\kappa')$ for key schedule φ as in Section 2, such that $K' = K \oplus \Delta$ where the difference Δ describes a connection between K and K'. We will now derive a condition on Δ and θ such that the value of linear approximation bias $\varepsilon = \varepsilon'$ is unaffected by the key change $\kappa \neq \kappa'$.

In a key-alternating cipher, the bias for an expanded key can be computed due to Proposition 1. That is:

$$\varepsilon = \sum_{\theta:\theta_0=a,\theta_r=b} (-1)^{d_\theta+\theta^t K}|\varepsilon_\theta| \text{ and } \varepsilon' = \sum_{\theta:\theta_0=a,\theta_r=b} (-1)^{d_\theta+\theta^t K'}|\varepsilon_\theta|. \quad (2)$$

We want to attain the equality $\varepsilon = \varepsilon'$, so we study when both sides of (2) are equal: One can observe that the only part that is different are *the signs of the individual linear characteristic biases.* Therefore, the equation will hold if all the signs are equal, that is, if the following is satisfied for each θ:

$$d_\theta + \theta^t K = d_\theta + \theta^t K'. \quad (3)$$

Since d_θ is the same, (3) holds if and only if $\theta^t(K \oplus K') = 0$. Recalling that we denote $K \oplus K'$ by Δ, we have the following statement:

Theorem 1 (Key difference invariant bias for key-alternating ciphers). *Let (a, b) be a non-trivial linear hull of a key-alternating block cipher. Its biases ε for expanded key K and ε' for expanded key K' with $K = K' \oplus \Delta$ have exactly equal values $\varepsilon = \varepsilon'$, if $\theta^t\Delta = 0$ for each linear characteristic θ of the linear hull (a, b) with $\varepsilon_\theta \neq 0$.*

Theorem 1 yields a sufficient condition on the relation between the masks of linear characteristics and the expanded key difference for the key difference invariant bias property to hold. We will deal with this in the next subsection.

3.2 Sufficient Condition for Key Difference Invariant Bias

For a fixed pair of keys K and K', the difference Δ connecting them is also constant. At the same time, the linear masks θ will be different for each linear characteristic in the given linear hull (a, b). Thus, Δ can be seen as a linear mask itself on θ that *chooses* certain positions in characteristics θ, cf. (1).

In a linear characteristic θ, we address each of the $n(r + 1)$ bits by $\theta(j)$, $j = 1, \ldots, n(r + 1)$. We focus on bit positions $\theta(j)$ in linear characteristics θ such that $\theta(j) = 0$ for all θ with $\varepsilon_\theta \neq 0$. We call such positions **zero positions.** Otherwise, a position is called a **nonzero position.**

Now we are ready to formulate a more explicit sufficient condition for deterministically keeping $\theta^t\Delta = 0$:

Condition 1 (Sufficient condition for key difference invariant bias). *For a fixed non-trivial linear approximation (a, b) of a key-alternating block cipher, the relation between a pair of the user-supplied keys κ and κ' is such that the expanded key difference $\Delta = K \oplus K'$ chooses an arbitrary number of zero positions and no nonzero positions in the linear characteristics θ of the linear hull, with $\varepsilon_\theta \neq 0$.*

Once Condition 1 is fulfilled, Theorem 1 becomes applicable with $\theta^t\Delta = 0$ and yields $\varepsilon = \varepsilon'$.

In the next subsection, for instructive and pedagogical purposes, we show one example of key difference invariant bias property using Condition 1 with AES.

3.3 The Instructive Example of AES

Here we provide an illustration of the key difference invariant bias property for AES. The goal of this section is mainly pedagogical and we simply aim to show how such a property can be derived in practice. We demonstrate a key difference invariant bias property for reduced-round AES-256. We provide an example where Condition 1 is satisfied, which in turn makes Theorem 1 applicable.

For AES-256, let the two user-supplied 32-byte keys be connected by

$$\kappa \oplus \kappa' = \begin{bmatrix} 0&0&0&0&0&0&\delta&0 \\ 0&0&0&0&0&0&0&0 \\ 0&0&0&0&0&0&0&0 \\ 0&0&0&0&0&0&0&0 \end{bmatrix} \tag{4}$$

with the first byte $\delta \neq 0$ of the 7-th column being the only non-zero byte. Furthermore, let the (truncated) linear approximation be defined by the 16-byte input/output masks:

$$a = \begin{bmatrix} a&0&0&0 \\ 0&0&0&0 \\ 0&0&0&0 \\ 0&0&0&0 \end{bmatrix} \text{ and } b = \begin{bmatrix} b&0&0&0 \\ 0&0&0&0 \\ 0&0&0&0 \\ 0&0&0&0 \end{bmatrix}. \tag{5}$$

The masks define a linear hull for any non-zero byte values a and b. We show that the key difference (4) and the linear hulls (5) result in the key difference invariant bias property for 5 rounds of AES-256.

The AES data transform diffuses a single-byte input mask to the full state only after two rounds. Analogously, a single-byte output mask applies to the full state only after three rounds of backward computations. This fact makes Condition 1 applicable to AES. The byte positions involved into the propagation of linear patterns over 5 rounds of AES with a and b above as input/output masks are shown as ■ in Figure 2. Correspondingly, byte positions not involved are depicted as □. Since AddRoundKey is addition with constant and MixColumns is an affine operation, one can exchange their order under the suitable modification of the subkey value. In this case, ShiftRows is followed directly by the modified AddRoundKey (AK') which is the case in the last round of Figure 2.

We track the propagation of the difference in the user-supplied key to the expanded key difference which is shown as × in Figure 2. $\kappa \oplus \kappa'$ specified above satisfies Condition 1. In Figure 2, all non-zero bytes × of Δ are only concentrated in impossible positions □ of θ and do not interfere with ■.
Thus, $\varepsilon = \varepsilon'$ is fulfilled with probability 1 and the key difference invariant bias property holds deterministically.

3.4 Key Difference Invariant Bias and Idealized Cipher

In random block ciphers, the bias ε under a fixed key is the bias for a fixed randomly drawn permutation. Using [10, Theorem 4.7], one can demostrate that

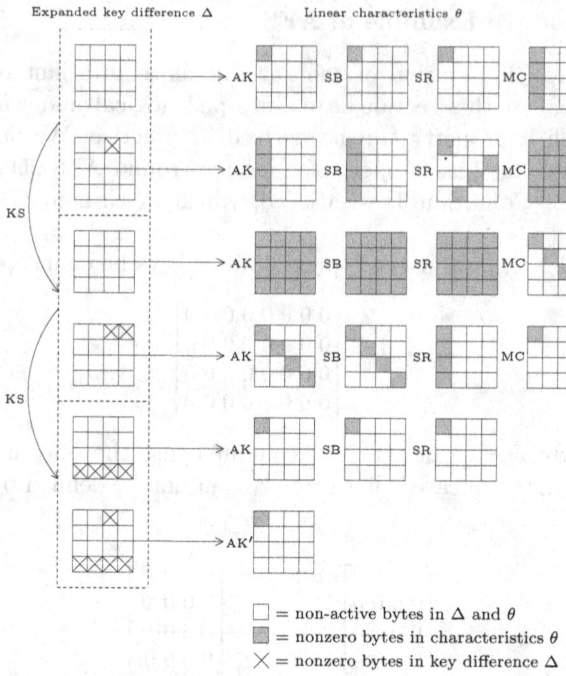

Fig. 2. Key difference invariant bias for 5 rounds of AES-256

the probability for the biases with two different keys to be exactly equal is $\Pr\{\varepsilon = \varepsilon' | \kappa \neq \kappa'\} \approx \frac{1}{\sqrt{2\pi}} 2^{\frac{3-n}{2}}$ for block sizes $n \geq 5$. Thus, the key difference invariant property for idealized block ciphers is a rare event, which yields a distinguisher for susceptible ciphers outlined in the next section.

4 Statistical Distinguisher and Key Recovery with Key Difference Invariant Bias

In this section, we present the statistical distinguisher based on the key difference invariant bias for an n-bit block cipher, followed by a generic key recovery procedure.

4.1 Distinguisher

In the distinguisher, our aim is to tell if we deal with the target cipher featuring the property or an idealized cipher. The setup for the statistical test is as follows. Suppose that we are given N plaintext-ciphertext pairs and λ linear approximations under a pair of expanded keys (K, K') connected by Δ in the way described in Condition 1. Then, for each one of these linear approximations we compute and store in counters S_i and S'_i, $1 \leq i \leq \lambda$, which account for the number of times these approximations are satisfied for K and K' with the N texts.

The counters S_i and S'_i suggest empirical biases $\hat{\varepsilon}_i = \frac{S_i}{N} - \frac{1}{2}$ and $\hat{\varepsilon}'_i = \frac{S'_i}{N} - \frac{1}{2}$ respectively. We evaluate consequently the following statistic s:

$$s = \sum_{i=1}^{\lambda} \left[\left(\frac{S_i}{N} - \frac{1}{2} \right) - \left(\frac{S'_i}{N} - \frac{1}{2} \right) \right]^2 .$$

We expect the statistic s to be lower for the target cipher, featuring the key difference invariant bias property, than for a random cipher. As we aim to perform key-recovery with this test, we will derive the distribution of this statistic for the right key guess (assuming the target structure) and for the wrong key guess (assuming a random cipher).

Right Key Guess. The empirical bias value $\hat{\varepsilon}_i$ for the i-th linear approximation approximately follows the normal distribution with the exact value of bias ε_i as mean and variance $1/4N$ with good precision (cf., e.g., [14, 35]) for sufficiently large N:

$$\hat{\varepsilon}_i \sim \mathcal{N}(\varepsilon_i, 1/4N).$$

In this case, the following proposition holds:

Proposition 2 (Distribution of statistic s for the right key). *Consider λ nontrivial linear approximations for a block cipher under a pair of expanded keys (K, K') connected by Δ conforming to Condition 1. If N is the number of known plaintext-ciphertext pairs, S_i and S'_i are the numbers of times such a linear approximation is fulfilled for K and K', respectively, $i \in \{1, \ldots, \lambda\}$, and λ is high enough, then, assuming the counters S_i and S'_i are all independent, the following approximate distribution holds for sufficiently large N and n:*

$$s \sim \mathcal{N}\left(\frac{\lambda}{2N}, \frac{\lambda}{2N^2} \right).$$

Proof. See the full version of this paper [2].

Wrong Key Guess. In this case, we base upon the hypothesis that for a wrong key, we deal with a random cipher consisting of permutations drawn at random. Then, each of the values $\hat{\varepsilon}_i$ can be approximated by a normal distribution with mean ε_i and variance $1/4N$ for sufficiently large N:

$$\hat{\varepsilon}_i \sim \mathcal{N}(\varepsilon_i, 1/4N) \text{ with } \varepsilon_i \sim \mathcal{N}(0, 1/2^{n+2}),$$

where ε_i is the exact value of the bias which is itself distributed over n-bit random permutations for $n \geq 5$ [10, 31].

Then we have then the following proposition for the distribution of the statistic s:

Proposition 3 (Distribution of statistic s for the wrong key). *Consider λ nontrivial linear approximations for two randomly drawn permutations. If N is* *the number of known plaintext-ciphertext pairs, S_i and S_i' are the numbers of times a linear approximation is fulfilled for these two permutations, $i \in \{1, \ldots, \lambda\}$, and λ is high enough, then, assuming the independency of all S_i and S_i', the following approximate distribution holds for sufficiently large N and n:*

$$s \sim \mathcal{N}\left(\frac{\lambda}{2N} + \frac{\lambda}{2^{n+1}}, \frac{\lambda}{2N^2} + \frac{\lambda}{2^{2n+1}} + \frac{\lambda}{N2^n}\right).$$

Proof. See the full version of this paper [2].

Data Complexity of Distinguisher. In the two above cases, we have seen that the statistic s will follow, depending on if we deal with the right or the wrong key, two different normal distributions. In the first case, it follows the normal distribution with mean $\mu_0 = \frac{\lambda}{2N}$ and variance $\sigma_0^2 = \frac{\lambda}{2N^2}$, while in the second case it follows the normal distribution with mean $\mu_1 = \frac{\lambda}{2N} + \frac{\lambda}{2^{n+1}}$ and variance $\sigma_1^2 = \frac{\lambda}{2N^2} + \frac{\lambda}{2^{2n+1}} + \frac{\lambda}{N2^n}$. It has to be decided if the obtained statistic s is from $\mathcal{N}(\mu_0, \sigma_0^2)$ or from $\mathcal{N}(\mu_1, \sigma_1^2)$. To do that, we perform a test that compares the statistic s to a threshold value τ. This test says that s belongs to $\mathcal{N}(\mu_0, \sigma_0^2)$ if $s \leq \tau$ and that s belongs to $\mathcal{N}(\mu_1, \sigma_1^2)$, otherwise.

As in any statistical test, one has to deal with two types of error probabilities here. The first one – denoted by α_0 – is the probability to reject the right key, whereas the second one – denoted by α_1 – is the probability to accept a wrong key. The decision threshold used is $\tau = \mu_0 + \sigma_0 q_{1-\alpha_0} = \mu_1 - \sigma_1 q_{1-\alpha_1}$, where $q_{1-\alpha_1}$ and $q_{1-\alpha_0}$ are the quantiles of the standard normal distribution $\mathcal{N}(0,1)$. This simple test is visualized in Figure 3.

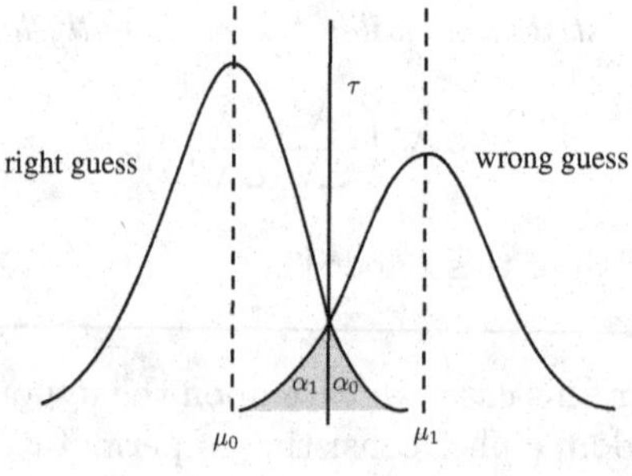

Fig. 3. Statistical test for key difference invariant bias in key recovery

It is well known [12] that in order for such a test to have error probabilities of at most α_0 and α_1, the parameters μ_0, σ_0^2, μ_1 and σ_1^2 should be such that $q_{1-\alpha_1}\sigma_1 + q_{1-\alpha_0}\sigma_0 = |\mu_1 - \mu_0|$.

Now, using Proposition 2 and Proposition 3, we obtain the following equation that determines the amount of data needed by the distinguisher:

$$N = \frac{2^{n+0.5}}{\sqrt{\lambda} - q_{1-\alpha_1}\sqrt{2}}(q_{1-\alpha_0} + q_{1-\alpha_1}). \tag{6}$$

4.2 How to Recover the Key with Key Difference Invariant Bias

Here, we describe a generic key recovery attack approach that can be applied to block ciphers for which a key difference invariant bias property for r rounds has been identified. This procedure is described in Algorithm 1. We will feed this algorithm with the related key differential paths that are going to be used for the attack. Other entries to the algorithm will be the number of rounds of the distinguisher r, the number of rounds r_{top} that we are going to append at the top of the distinguisher and the number of rounds r_{bot} that we are going to add at the bottom of the distinguisher. In Algorithm 1, $V[x]$ and $V'[x']$ are the counters containing the number of times the partial state values x and x' (values corresponding to non-zero mask of linear approximations) occur for N plaintext-ciphertext pairs under the key pair.

Algorithm 1. Generic Attack Procedure

Require: A set of linear approximations (a, b) and master key difference $\delta = \kappa \oplus \kappa'$ with the key difference invariant bias property holding.

for all related-key differential paths with a difference δ on the master-key **do**
  Collect N plaintext-ciphertext pairs (P, C) under a key κ.
  Collect N plaintext-ciphertext pairs (P', C') under $\kappa' = \kappa \oplus \delta$.
  Partially encrypt r_{top} rounds and partially decrypt r_{bot} rounds, obtain partial state values x and x' covered by the input/output masks of (a, b) and compute $V[x]$ and $V'[x']$ (number of times these partial state values occur).
  Allocate a counter s.
  for all linear approximations (a, b) **do**
    Allocate counters S and S' and set them to zero.
    for all values of x and x' **do**
      if the linear approximation holds **then**
        Add $V[x]$ and $V[x']$ to S and S', respectively.
      end if
    end for
    Compute $s = s + \left[\left(\frac{S}{N} - \frac{1}{2}\right) - \left(\frac{S'}{N} - \frac{1}{2}\right)\right]^2$.
  end for
  if $s \leq \tau$ **then**
    The guessed subkey is a possible subkey value.
    Check exhaustively the remaining keys against several plaintext-ciphertext pairs.
  end if
end for
return encryption key.

5 Attack on 24-round LBlock

LBlock is a lightweight block cipher presented at ACNS 2011 by Wu and Zhang [39]. It uses 64-bit block and 80-bit key and is based on a modified 32-round Feistel structure. Its description is provided in the full version of this paper [2].

5.1 Previous Cryptanalysis

Despite its recent proposal, LBlock has already been extensively analyzed. For example, impossible differential attacks have been mounted in the single-key model [16, 21, 39] as well as attacks in the related-key model [26]. A related-key truncated differential attack on 22-round LBlock was given in [22]. Some other results concern integral cryptanalysis [20,33,34,39]. A zero-correlation linear attack was equally mounted against 22 rounds of LBlock [36]. Finally, biclique attacks [17,40] provide only a small gain against exhaustive search. So the currently best non-exhaustive attacks against LBlock can break at most 22 rounds.

In this paper, we propose an attack on 24 rounds of LBlock. Our results are summarized and compared to previous cryptanalysis in Table 1.

Table 1. Summary of attacks on LBlock

Model	Attack	#Rounds	#keys	Data per key	Time	Memory	Ref.
SK	Imp. Diff	20	1	2^{63} CP	$2^{72.7}$	2^{68}	[39]
	Imp. Diff	21	1	$2^{62.5}$CP	$2^{73.7}$	$2^{55.5}$	[21]
	Imp. Diff	21	1	2^{63}CP	$2^{69.5}$	2^{75}	[16]
	Imp. Diff	22	1	2^{58}CP	$2^{79.28}$	2^{76}	[16]
	Integral	20	1	$2^{63.7}$CP	$2^{63.7}$	N/A	[39]
	Integral	20	1	$2^{63.6}$CP	$2^{39.6}$	2^{35}	[33]
	Integral	22	1	$2^{61.6}$CP	$2^{71.2}$	N/A	[20]
	Integral	21	1	$2^{61.6}$CP	$2^{54.1}$	$2^{51.58}$	[34]
	Integral	22	1	2^{61}CP	2^{70}	2^{63}	[34]
	Zero-Correlation	22	1	2^{64}KP	$2^{70.54}$	2^{64}	[36]
	Zero-Correlation	22	1	$2^{62.1}$KP	$2^{71.27}$	2^{64}	[36]
	Zero-Correlation	22	1	2^{60}KP	2^{79}	2^{64}	[36]
RK	Imp. Diff	22	8	2^{47}RKCP	2^{70}	N/A	[26]
	Differential	22	2	$2^{63.1}$RKCP	2^{67}	N/A	[22]
	Key Diff Inv Bias	**24**	32	$2^{62.29}$ **RKKP**	$2^{74.59}$	2^{61}	**Here**
	Key Diff Inv Bias	**24**	32	$2^{62.95}$ **RKKP**	$2^{70.67}$	2^{61}	**Here**

5.2 Linear Approximations with Key Difference Invariant Bias for LBlock

We start by presenting the linear approximations with key difference invariant bias under two keys related by a difference on a single nibble of the master key. These linear approximations depicted in Figure 4, hold for 16 rounds (from round 5 to round 20) under the related-key differential paths depicted in the full version of this paper [2]. The input mask of the 5-th round is $(0000\alpha00000000000)$ and the output mask of the 20-th round is $(000000000\beta000000)$, $\alpha \neq 0, \beta \neq 0$. There are in total $(2^4 - 1) \cdot (2^4 - 1) \approx 2^{7.81}$ such linear approximations.

We can see from Figure 4 that the relations $\Gamma_r \cdot \Delta K_r = 0$, for $5 \leq r \leq 20$ hold for all the related-key differential paths listed in the full version of this paper [2].

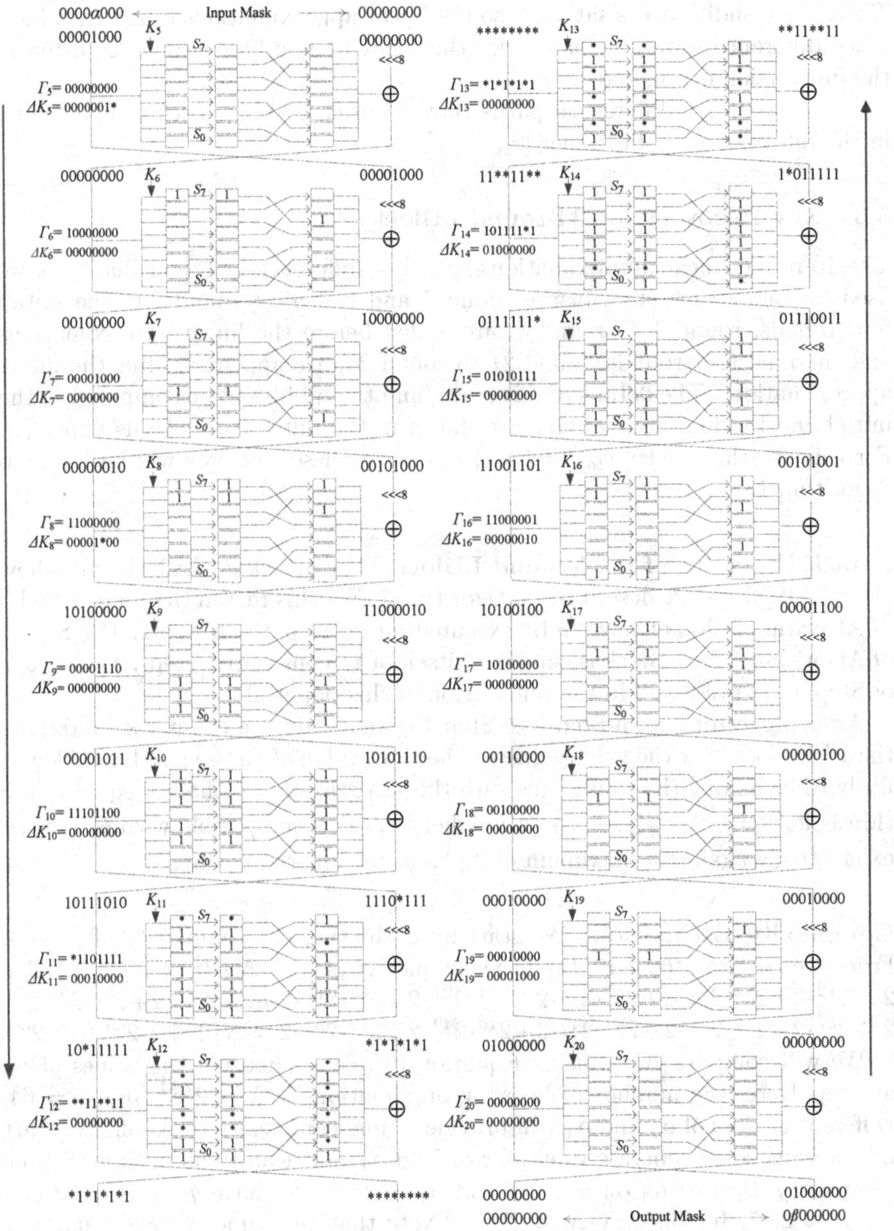

$\Gamma_r, 5 \leq r \leq 20$: input mask value for the S-boxes in round r.
$\Delta K_r, 5 \leq r \leq 20$: the subkey difference in round r.
In masks, '0', '1' and '*': zero, nonzero and arbitrary mask for a nibble, resp.
In differences, '0' , '1' and '*': zero, nonzero and arbitrary difference for a nibble, resp.

Fig. 4. 16-round linear approximations with key difference invariant bias for LBlock

Therefore Condition 1 is satisfied, so the linear approximations in Figure 4 have a key difference invariant bias under the related-key differential paths listed in the full version of this paper [2].

The related-key differential paths that we used for our attack are presented in the full version of this paper [2].

5.3 Key Recovery for 24-round LBlock

The 16-round linear approximations with key difference invariant bias that we used for our attack start before round 5 and end after round 20. The initial four rounds, round 1 to round 4, are added before the linear approximations and the final four rounds, round 21 to round 24, are appended after the linear approximations. The details of this stage, and the nibbles to be computed in the initial and the final four rounds are shown in the full version of this paper [2]. For this attack, $r = 16$, $r_{top} = 4$ and $r_{bot} = 4$. These elements will be input to Algorithm 1.

Attack Procedure for 24-round LBlock. The attack for LBlock will follow the attack procedure described in Algorithm 1. For this reason the Steps 2 and 3 of Algorithm 1 do not have to be executed for every path of Step 1. The Step 4 of Algorithm 1 for LBlock is composed itself of 14 consecutive steps. The details of Step 4 are presented in the full version of this paper [2].

After proceeding from Step 5 to Step 15, we obtain the counter s containing the χ^2 statistics for the subkey guess. The right value of guessed 53-bit subkey is likely to be among the candidates with the statistic s lower than or equal to the threshold $\tau = \frac{\sqrt{\lambda}}{N\sqrt{2}}q_{1-\alpha_0} + \frac{\lambda}{2N}$. All cipher keys it is compatible with are tested exhaustively against a maximum of 2 plaintext-ciphertext pairs.

Complexity Estimation. We start by evaluating the complexity of Step 4. From Step 4.1 to Step 4.14, the time complexity is $T_1 = N \cdot 2^4 \cdot 2 + 2^{60} \cdot 2^8 \cdot 2 + 2^{56} \cdot 2^{12} \cdot 2 + 2^{52} \cdot 2^{13} \cdot 2 + 2^{48} \cdot 2^{17} \cdot 2 + 2^{44} \cdot 2^{21} \cdot 2 + 2^{40} \cdot 2^{25} \cdot 2 + 2^{36} \cdot 2^{29} \cdot 2 + 2^{32} \cdot 2^{33} \cdot 2 + 2^{28} \cdot 2^{37} \cdot 2 + 2^{24} \cdot 2^{41} \cdot 2 + 2^{20} \cdot 2^{45} \cdot 2 + 2^{16} \cdot 2^{49} \cdot 2 + 2^{12} \cdot 2^{53} \cdot 2 = N \cdot 2^5 + 2 \cdot 2^{69} + 11 \cdot 2^{66}$.

We will compute N by using Equation (6), after choosing the values of α_0 and α_1. Here, the number of linear approximations is $\lambda = 2^{7.81}$ and $n = 64$. Different choices of α_0 and α_1 will provide a time-complexity trade-off. We start by choosing some concrete values for α_0 and α_1 that lead to an optimized time complexity. By setting $\alpha_0 = 2^{-2.7}$ and $\alpha_1 = 2^{-8.5}$, we have $q_{1-\alpha_0} \approx 1.02$ and $q_{1-\alpha_1} \approx 2.77$. In this way $N \approx 2^{62.95}$ (Note that the same N (P, C) pairs or N (P', C') pairs can be reused for different related-key differential paths under the condition that $\Delta\kappa_{14\sim17}$ remains the same.) and the threshold value gets $\tau \approx 2^{-55.02}$. Then, $T_1 \approx 2^{70.95}$ times of $\frac{1}{8}$ round encryption which is equivalent to $2^{63.37}$ times of 24-round encryptions. Note that the time complexity of the procedure described in Steps 6$\sim$14 is negligible. Under each related-key differential path, the value of $\kappa_{14\sim17}$ is already known, so the time complexity of Steps 16-19 is about $2^{76} \cdot 2^{-8.5} = 2^{67.5}$ times of 24-round encryption. Therefore, the total complexity from Step 2 to Step 18 is about $2^{67.58}$ encryptions. After proceeding from Step 2 to Step 18, if we can not succeed, this means that the value of the right key does not belong to the values corresponding to the related-key differential path tested. We can then use another related-key differential path to proceed the above attack. All possible values of the master key bits $\kappa_{4\sim21}$ are covered by the related-key differential paths, so we could always find the right key where in the worst case, all the related-key differential paths have to be tested. So the expected time complexity of our attack on 24-round LBlock is about $2^{67.58}\cdot[1+(1-\frac{1}{16})+\cdots+(1-\frac{15}{16})] \approx 2^{70.67}$ 24-round encryptions. The data complexity is $2^{62.95}$ known plaintexts under each master key, while $2^{60}\cdot 2 = 2^{61}$ bytes of memory are required to store the counters.

Another possible choice of α_0 and α_1 can lead to a different time-data complexity trade-off. For example, if we set $\alpha_0 = 2^{-2.7}$ and $\alpha_1 = 2^{-4.5}$, then $q_{1-\alpha_0} \approx 1.01$ and $q_{1-\alpha_1} \approx 1.70$, we get $N \approx 2^{62.29}$. For these parameters the expected time complexity is about $2^{74.59}$ encryptions and the expected data complexity is $2^{62.29}$ known plaintexts for each master key. The memory requirements are the same as in the previous attack.

Other possible time-data trade-offs with $\beta_0 = 2^{-2.7}$ for the attack on LBlock can be visualized in Figure 6.

6 Attack on 27-round TWINE-128

TWINE is a lightweight block cipher proposed by Suzaki, Minematsu, Morioka and Kobayashi in [37]. Its structure is based on a modified Type-2 generalized Feistel scheme. The cipher's description is given in the full version of this paper.

6.1 Previous Cryptanalysis

In the original proposal of TWINE [37], the authors analyze the resistance of TWINE against various types of attacks, such as impossible differential and saturation attacks. The best analysis in this proposal is an impossible differential attack against 23 rounds of TWINE-80 and against 24 rounds of TWINE-128. Moreover, biclique attacks have been mounted in [17] for both full-round versions of TWINE, but the time complexity of these attacks is only marginally lower than exhaustive search.

6.2 Linear Approximations with Key Difference Invariant Bias for TWINE-128

We present 17-round (from round 6 to round 22) linear approximations with key difference invariant bias under related-key differential paths for TWINE-128 in Figure 5. In our attack, the input mask of the 6-th round is $00000000000\alpha000$ and the output mask of the 22-th round is $0000000\beta00000000$, $\alpha, \beta \neq 0$. Thus, there are $15 * 15 \approx 2^{7.81}$ such linear approximations, exactly as in the case of LBlock. We start by describing the related-key truncated differential path that we use in our attack. This differential path was found by considering only differences in only one nibble of the master key and by searching exhaustively over all such configurations.

This path is described in the full version of this paper [2]. More precisely, we consider a difference equal to 1 in the 22nd nibble of the master key. This differential path covers all the possible key values and is sufficient to recover the right key value. From Figure 5, we can see that $\Gamma_r \cdot \Delta K_r = 0$, $6 \leq r \leq 22$ (where K_r and ΔK_r denote the subkey value and the subkey difference for the round r respectively) and thus Condition 1 is satisfied.

6.3 Key Recovery for 27-round TWINE-128

We utilize the 17-round distinguisher in Figure 5 to attack 27 rounds of TWINE-128. The initial five rounds from round 1 to round 5 are added before the distinguisher and the final five rounds from round 23 to round 27 are appended after the distinguisher, as shown in the full version of this paper. In such a way, the first 27 rounds of TWINE-128 are covered. The attack is proceeded by following Algorithm 1. The parameters are $r = 17$, $r_{top} = 5$, $r_{bot} = 5$, see the full version of this paper.

After proceeding from Step 5 to Step 15, we obtain the counter s containing the χ^2 statistics for the subkey guess. The right value of guessed 96-bit subkey is likely to be among the candidates with the statistic s lower than or equal to the threshold $\tau = \frac{\sqrt{\lambda}}{N\sqrt{2}} q_{1-\alpha_0} + \frac{\lambda}{2N}$. All cipher keys it is compatible with are tested exhaustively against a maximum of 2 plaintext-ciphertext pairs.

Complexity Estimation. We start by evaluating the complexity T_1 of Steps 4.1-4.17. $T_1 = N \cdot 2^{20} \cdot 2 + N \cdot 2^{32} \cdot 15 \cdot 2 + N \cdot 2^{40} \cdot 15 \cdot 2 + 2^{60} \cdot 2^{44} \cdot 2 \cdot 15 + 2^{56} \cdot 2^{48} \cdot 2 \cdot 15 + 2^{52} \cdot 2^{52} \cdot 2 \cdot 15 + 2^{48} \cdot 2^{56} \cdot 2 \cdot 15 + 2^{44} \cdot 2^{60} \cdot 2 \cdot 15 + 2^{40} \cdot 2^{64} \cdot 2 \cdot 15 + 2^{36} \cdot 2^{68} \cdot 2 \cdot 15 + 2^{36} \cdot 2^{72} \cdot 2 \cdot 15 + 2^{32} \cdot 2^{76} \cdot 2 \cdot 15 + 2^{28} \cdot 2^{80} \cdot 2 \cdot 15 + 2^{24} \cdot 2^{84} \cdot 2 \cdot 15 + 2^{20} \cdot 2^{88} \cdot 2 \cdot 15 + 2^{16} \cdot 2^{92} \cdot 2 \cdot 15 + 2^{12} \cdot 2^{96} \cdot 2 \cdot 15 = N \cdot 2^{20} \cdot 2 + N \cdot 2^{32} \cdot 15 \cdot 2 + N \cdot 2^{40} \cdot 15 \cdot 2 + 7 \cdot 2^{104} \cdot 2 \cdot 15 + 7 \cdot 2^{108} \cdot 2 \cdot 15$.

To compute N, we will use Equation (6). Here, the number of linear approximations is $\lambda = 2^{7.81}$ and $n = 64$. Therefore N will be computed after choosing the values of α_0 and α_1. Different choices of these values will provide a data-time trade-off. We start by choosing some concrete values for α_0 and α_1 that lead to an optimized time complexity.

Consider for example $\alpha_0 = 2^{-2.7}$ and $\alpha_1 = 2^{-8.5}$. Then $q_{1-\alpha_0} \approx 1.02$ and $q_{1-\alpha_1} \approx 2.77$. By replacing these values to Equation (6), we obtain $N \approx 2^{62.95}$. The threshold value gets $\tau = \frac{\sqrt{\lambda}}{N\sqrt{2}} q_{1-\alpha_0} + \frac{\lambda}{2N} \approx 2^{-55.02}$. Thus $T_1 \approx 2^{115.81}$ times of 1/8 encryption, which is equivalent to $2^{108.05}$ times of 27-round encryption. The complexity of computing the counters S and S' is negligible. The complexity of the last step is $2^{128} \cdot 2^{-8.5} = 2^{119.5}$ times of 27-round encryption. Thus the total time complexity of the attack is about $2^{119.5}$ 27-round TWINE-128 encryptions. The data complexity is $N \approx 2^{62.95}$ known plaintexts per key and the memory requirements are 2^{61} bytes to store the counters.

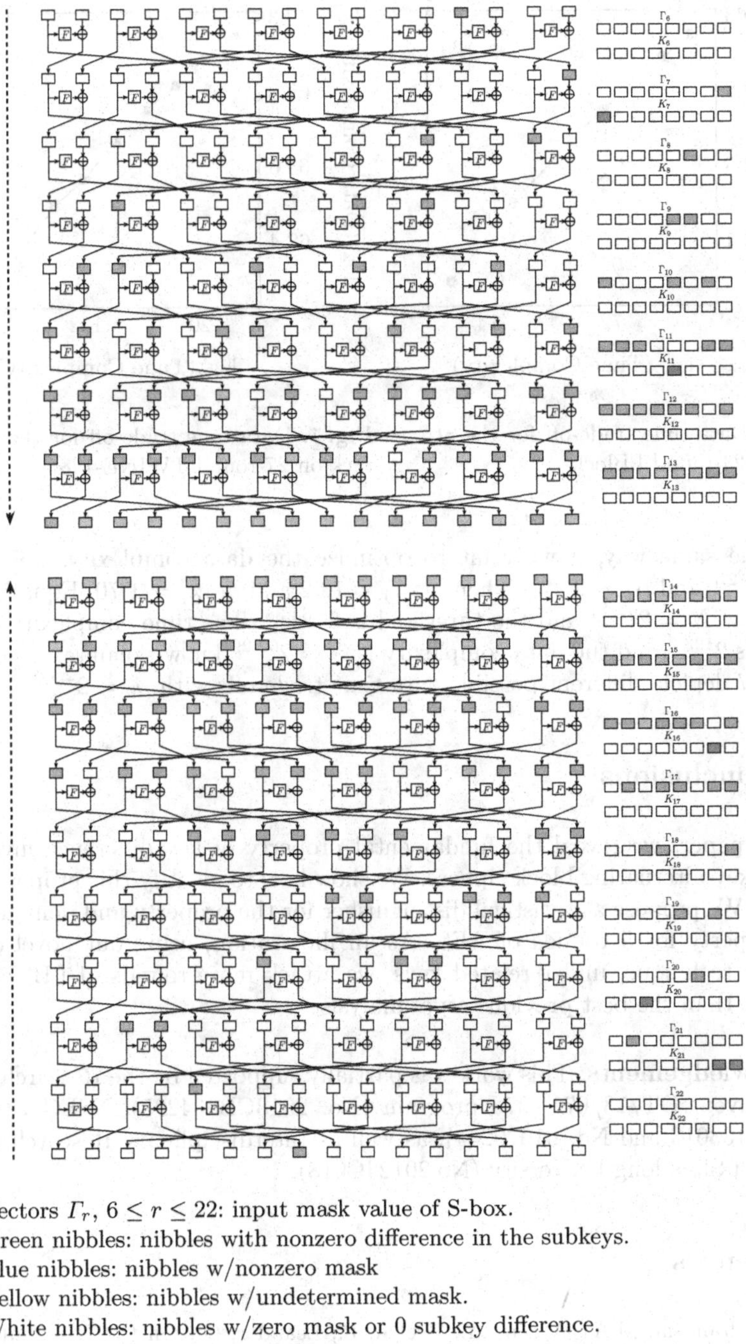
Vectors Γ_r, $6 \leq r \leq 22$: input mask value of S-box.
Green nibbles: nibbles with nonzero difference in the subkeys.
Blue nibbles: nibbles w/nonzero mask
Yellow nibbles: nibbles w/undetermined mask.
White nibbles: nibbles w/zero mask or 0 subkey difference.

Fig. 5. 17-round linear approximations for key difference invariant bias for TWINE-128

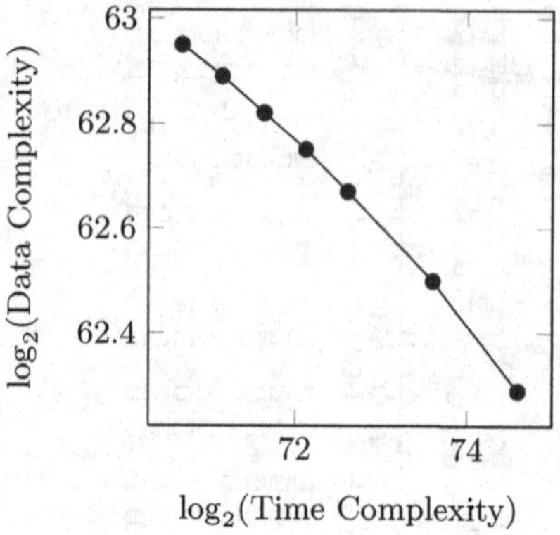

Fig. 6. Data-time trade-off for the attack on 24-round LBlock

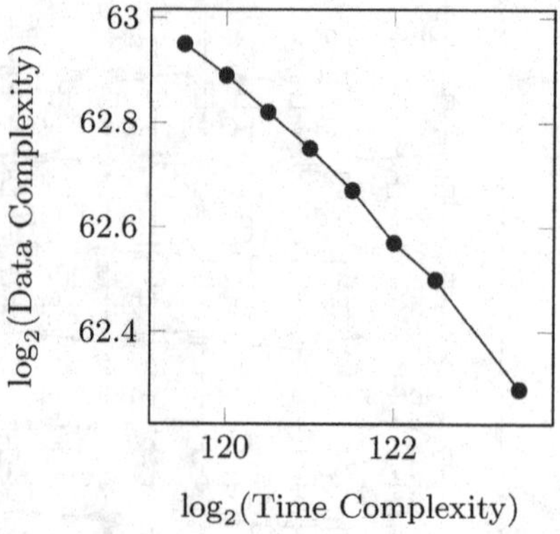

Fig. 7. Data-time trade-off for the attack on 27-round TWINE-128

In the same way, if we want to optimize the data complexity, we choose $\alpha_0 = 2^{-2.7}$ and $\alpha_1 = 2^{-4.5}$. Then $q_{1-\alpha_0} \approx 1.02$ and $q_{1-\alpha_1} = 1.70$. Equation (6) gives now $N = 2^{62.29}$ and the threshold is $2^{-54.38}$. The time complexity of the attack is $2^{123.5}$ and the data complexity is $N = 2^{62.29}$ known plaintexts per key. Figure 7 depicts different possible data-time trade-offs with $\beta_0 = 2^{-2.7}$.

7 Conclusions

In this paper, we reveal the fundamental property of key difference invariant bias in key-alternating block ciphers. We show how to identify this property efficiently. We propose a statistical distinguisher for the property and demonstrate the property for 5 rounds of AES. As an illustration, using our novel cryptanalytic technique, under related keys, we attack more rounds of LBlock and TWINE than the best previous cryptanalysis.

Acknowledgements. This work was partially supported by the Research Fund KU Leuven, OT/13/071, 973 program (No. 2013CB834205), NSFC Projects (No. 61133013 and No. 61070244), as well as Interdisciplinary Research Foundation of Shandong University (No.2012JC018).

References

1. Abdelraheem, M.A., Ågren, M., Beelen, P., Leander, G.: On the Distribution of Linear Biases: Three Instructive Examples. In: Safavi-Naini, R., Canetti, R. (eds.) CRYPTO 2012. LNCS, vol. 7417, pp. 50–67. Springer, Heidelberg (2012)
2. Bogdanov, A., Boura, C., Rijmen, V., Wang, M., Wen, L., Zhao, J.: Key Difference Invariant Bias in Block Ciphers. IACR Eprint report (2013)

3. Bogdanov, A., Rijmen, V.: Linear Hulls with Correlation Zero and Linear Cryptanalysis of Block Ciphers. Accepted to Designs, Codes and Cryptography. Springer (2012) (in press)
4. Bogdanov, A., Wang, M.: Zero Correlation Linear Cryptanalysis with Reduced Data Complexity. In: Canteaut, A. (ed.) FSE 2012. LNCS, vol. 7549, pp. 29–48. Springer, Heidelberg (2012)
5. Bogdanov, A., Leander, G., Nyberg, K., Wang, M.: Integral and Multidimensional Linear Distinguishers with Correlation Zero. In: Wang, X., Sako, K. (eds.) ASIACRYPT 2012. LNCS, vol. 7658, pp. 244–261. Springer, Heidelberg (2012)
6. Cho, J.Y.: Linear Cryptanalysis of Reduced-Round PRESENT. In: Pieprzyk, J. (ed.) CT-RSA 2010. LNCS, vol. 5985, pp. 302–317. Springer, Heidelberg (2010)
7. Collard, B., Standaert, F.-X.: Experimenting Linear Cryptanalysis. In: Junod, P., Canteaut, A. (eds.) Advanced Linear Cryptanalysis of Block and Stream Ciphers. ISO Press (2011)
8. Daemen, J., Govaerts, R., Vandewalle, J.: Correlation Matrices. In: Preneel, B. (ed.) FSE 1994. LNCS, vol. 1008, pp. 275–285. Springer, Heidelberg (1995)
9. Daemen, J., Rijmen, V.: The Design of Rijndael: AES - The Advanced Encryption Standard. Springer (2002) ISBN 3-540-42580-2
10. Daemen, J., Rijmen, V.: Probability Distributions of Correlations and Differentials in Block Ciphers. Journal of Mathematical Cryptology 1(3), 221–242 (2007)
11. Daemen, J., Rijmen, V.: Probability Distributions of Correlation and Differentials in Block Ciphers. Tech. Rep. 212, IACR ePrint Report 2005/212 (2005), http://eprint.iacr.org/2005/212
12. Feller, W.: An Introduction to Probability Theory and Its Applications (1971)
13. Hermelin, M., Cho, J.Y., Nyberg, K.: Multidimensional Extension of Matsui's Algorithm 2. In: Dunkelman, O. (ed.) FSE 2009. LNCS, vol. 5665, pp. 209–227. Springer, Heidelberg (2009)
14. Junod, P.: On the Complexity of Matsui's Attack. In: Vaudenay, S., Youssef, A.M. (eds.) SAC 2001. LNCS, vol. 2259, pp. 199–211. Springer, Heidelberg (2001)
15. Kaliski Jr., B.S., Robshaw, M.: Linear Cryptanalysis Using Multiple Approximations. In: Desmedt, Y.G. (ed.) CRYPTO 1994. LNCS, vol. 839, pp. 26–39. Springer, Heidelberg (1994)
16. Karakoç, F., Demirci, H., Harmancı, A.E.: Impossible Differential Cryptanalysis of Reduced-Round LBlock. In: Askoxylakis, I., Pöhls, H.C., Posegga, J. (eds.) WISTP 2012. LNCS, vol. 7322, pp. 179–188. Springer, Heidelberg (2012)
17. Karakoç, F., Demirci, H., Harmanci, A.: Biclique Cryptanalysis of LBlock and TWINE. Inf. Process. Lett. 113(12), 423–429 (2013)
18. Keliher, L., Sui, J.: Exact Maximum Expected Differential and Linear Probability for Two-Round Advanced Encryption Standard. IET Information Security 1(2), 53–57 (2007)
19. Kim, J.: Combined Differential, Linear and Related-Key Attacks on Block Ciphers and MAC Algorithms. Ph.D. thesis, K.U.Leuven (2006)
20. Li, Y.: Integral Cryptanalysis on Block Ciphers. Institute of Software, Chinese Academy of Sciences, Beijing (2012) (in Chinese)
21. Liu, Y., Gu, D., Liu, Z., Li, W.: Impossible Differential Attacks on Reduced-Round LBlock. In: Ryan, M.D., Smyth, B., Wang, G. (eds.) ISPEC 2012. LNCS, vol. 7232, pp. 97–108. Springer, Heidelberg (2012)
22. Liu, S., Gong, Z., Wang, L.: Improved Related-Key Differential Attacks on Reduced-Round LBlock. In: Chim, T.W., Yuen, T.H. (eds.) ICICS 2012. LNCS, vol. 7618, pp. 58–69. Springer, Heidelberg (2012)

23. Matsui, M.: Linear Cryptanalysis Method for DES Cipher. In: Helleseth, T. (ed.) EUROCRYPT 1993. LNCS, vol. 765, pp. 386–397. Springer, Heidelberg (1994)
24. Matsui, M.: The First Experimental Cryptanalysis of the Data Encryption Standard. In: Desmedt, Y.G. (ed.) CRYPTO 1994. LNCS, vol. 839, pp. 1–11. Springer, Heidelberg (1994)
25. Matsui, M., Yamagishi, A.: A New Method for Known Plaintext Attack of FEAL Cipher. In: Rueppel, R.A. (ed.) EUROCRYPT 1992. LNCS, vol. 658, pp. 81–91. Springer, Heidelberg (1993)
26. Minier, M., Naya-Plasencia, M.: A Related Key Impossible Differential Attack against 22 Rounds of the Lightweight Block Cipher LBlock. Inf. Process. Lett. 112(16), 624–629 (2012)
27. Murphy, S.: The Effectiveness of the Linear Hull Effect. J. Mathematical Cryptology 6(2), 137–147 (2012)
28. Nyberg, K., Hakala, R.: A Key-Recovery Attack on SOBER-128, Symmetric Cryptography Dagstuhl Seminar No. 07021 (2007)
29. Nyberg, K.: Linear Approximation of Block Ciphers. In: De Santis, A. (ed.) EUROCRYPT 1994. LNCS, vol. 950, pp. 439–444. Springer, Heidelberg (1995)
30. Nyberg, K.: Linear Cryptanalysis Using Multiple Linear Approximations. In: Early Symmetric Crypto (ESC 2010) Seminar, Remich, Luxembourg (2011), https://cryptolux.org/mediawiki.esc/images/5/52/Escnyberg.pdf
31. O'Connor, L.: Properties of Linear Approximation Tables. In: Preneel, B. (ed.) FSE 1994. LNCS, vol. 1008, pp. 131–136. Springer, Heidelberg (1995)
32. Röck, A., Nyberg, K.: Generalization of Matsui's Algorithm 1 to Linear Hull for Key-Alternating Block Ciphers. Designs, Codes and Cryptography 66(1-3), 175–193 (2013)
33. Sasaki, Y., Wang, L.: Meet-in-the-Middle Technique for Integral Attacks against Feistel Ciphers. In: Knudsen, L.R., Wu, H. (eds.) SAC 2012. LNCS, vol. 7707, pp. 234–251. Springer, Heidelberg (2013)
34. Sasaki, Y., Wang, L.: Comprehensive Study of Integral Analysis on 22-Round LBlock. In: Kwon, T., Lee, M.-K., Kwon, D. (eds.) ICISC 2012. LNCS, vol. 7839, pp. 156–169. Springer, Heidelberg (2013)
35. Selçuk, A.A.: On Probability of Success in Linear and Differential Cryptanalysis. Journal of Cryptology 21(1), 131–147 (2008)
36. Soleimany, H., Nyberg, K.: Zero-Correlation Linear Cryptanalysis of Reduced-Round LBlock. Accepted to WCC 2013 (2012) (to appear), http://eprint.iacr.org/2012/570.pdf
37. Suzaki, T., Minematsu, K., Morioka, S., Kobayashi, E.: TWINE: A Lightweight Block Cipher for Multiple Platforms. In: Knudsen, L.R., Wu, H. (eds.) SAC 2012. LNCS, vol. 7707, pp. 339–354. Springer, Heidelberg (2013)
38. Wagner, D.: The Boomerang Attack. In: Knudsen, L.R. (ed.) FSE 1999. LNCS, vol. 1636, pp. 156–170. Springer, Heidelberg (1999)
39. Wu, W., Zhang, L.: LBlock: A Lightweight Block Cipher. In: Lopez, J., Tsudik, G. (eds.) ACNS 2011. LNCS, vol. 6715, pp. 327–344. Springer, Heidelberg (2011)
40. Wang, Y., Wu, W., Yu, X., Zhang, L.: Security on LBlock against Biclique Cryptanalysis. In: Lee, D.H., Yung, M. (eds.) WISA 2012. LNCS, vol. 7690, pp. 1–14. Springer, Heidelberg (2012)

Leaked-State-Forgery Attack against the Authenticated Encryption Algorithm ALE

Shengbao Wu[1,3], Hongjun Wu[2], Tao Huang[2], Mingsheng Wang[4], and Wenling Wu[1]

[1] Trusted Computing and Information Assurance Laboratory, Institute of Software, Chinese Academy of Sciences, Beijing 100190, P.O. Box 8718, China
[2] Division of Mathematical Sciences, School of Physical and Mathematical Sciences, Nanyang Technological University, Singapore
[3] Graduate School of Chinese Academy of Sciences, Beijing 100190, China
[4] State Key Laboratory of Information Security, Institute of Information Engineering, Chinese Academy of Sciences, Beijing, China
{wushengbao,wwl}@tca.iscas.ac.cn
{wuhj,huangtao}@ntu.edu.sg
wangmingsheng@iie.ac.cn

Abstract. ALE is a new authenticated encryption algorithm published at FSE 2013. The authentication component of ALE is based on the strong Pelican MAC, and the authentication security of ALE is claimed to be 128-bit. In this paper, we propose the leaked-state-forgery attack (LSFA) against ALE by exploiting the state information leaked from the encryption of ALE. The LSFA is a new type of differential cryptanalysis in which part of the state information is known and exploited to improve the differential probability. Our attack shows that the authentication security of ALE is only 97-bit. And the results may be further improved to around 93-bit if the whitening key layer is removed. We implemented our attacks against a small version of ALE (using 64-bit block size instead of 128-bit block size). The experimental results match well with the theoretical results.

Keywords: authenticated encryption, forgery attack, ALE.

1 Introduction

Confidentiality and message authentication are two fundamental goals in cryptography. In symmetric key cryptography, a block cipher/stream cipher is used to protect the confidentiality of messages; and a message authentication code (MAC) is used to authenticate messages. In the widely used Transport Layer Security (TLS), the MAC-then-Encrypt approach is used: HMAC [27] is applied to authenticate the TCP packets, and AES [9] in CBC mode [26] can be used to encrypt the payload of TCP packets.

In many applications, both confidentiality and message authentication are required. The authenticated encryption algorithm can achieve encryption and authentication simultaneously, and its performance is much better than the

K. Sako and P. Sarkar (Eds.) ASIACRYPT 2013 Part I, LNCS 8269, pp. 377–404, 2013.

combination of separate encryption and authentication. Authenticated encryption has received considerable research interests in recent years. A number of block cipher based authenticated encryption modes have been proposed, *e.g.*, IAPM [21], OCB [28], CCM [29], CWC [23], GCM [24], EAX [4], HBS [19], BTM [18] and McOE [15]. The ISO/IEC 19772:2009 [17] standardized several modes, including EAX, CCM, GCM and OCB 2.0. Besides the authenticated encryption modes, several authenticated encryption algorithms have been proposed, such as Helix [14], Phelix [30], Hummingbird-2 [13], ASC-1 [20], the 3GPP algorithm 128-EIA3 [2] and Grain-128a [3]. The coming competition CAESAR (Competition for Authenticated Encryption: Security, Applicability and Robustness) [7] is expected to attract many new authenticated encryption algorithms.

ALE. ALE (*A*uthenticated *L*ightweight *E*ncryption) is an AES-based authenticated encryption algorithm proposed by Bogdanov *et al.* at FSE 2013 [6]. It is designed for the low-cost embedded systems (such as RFID tags and smart cards) and provides single-pass authenticated encryption with associated data. The keystream generation of ALE uses the idea of the LEX stream cipher [5], and the tag generation uses the idea of Pelican MAC [10]. It has 256-bit internal state and aims to have a probability of success at most 2^{-128} for a forgery attack.

Pelican MAC is an extremely simple MAC based on AES. In Pelican MAC, any difference being introduced in the forgery attack passes through at least four AES rounds. It ensures that the success rate of a forgery attack is at most 2^{-128}. The state size of Pelican MAC is only 128 bits. The small state size means that the number of messages being authenticated under the same key should be less than 2^{64}. Yuan *et al.* delivered a state recovery attack against the Pelican MAC by exploiting the state collision when more than 2^{64} authentication tags are generated from the same key [33]. The attack given in [33] cannot be applied to ALE. In ALE, the state size is increased to 256 bits, and a new nonce is needed for generating each authentication tag when the same key is used.

The stream cipher LEX is based on AES, and four keystream bytes are extracted from the AES state after each round. LEX suffers from two attacks. The slide attack against LEX recovers the key with negligible complexity when around 2^{60} nonces are used with the same secret key [31]. Another attack recovers the key with around 2^{100} simple operations and 2^{40} keystream bytes [11,12]. ALE is not vulnerable to these two attacks due to its large state and the changing AES round keys (the round keys in LEX are fixed for the same key).

The design of ALE is similar to the authenticated encryption algorithm ASC-1. In ASC-1, a leaked byte is protected by an additional key byte before it is extracted as keystream byte. However, the additional key byte is not used in ALE for better hardware efficiency. Unfortunately, the lacking of additional key bytes in ALE allows part of the AES state being leaked as keystream, and such leaked state information can be exploited to improve the forgery attack, as demonstrated in this paper.

In this paper, we propose a new attack – leaked-state-forgery attack (LSFA) against ALE. The general idea of this attack is to exploit the leaked state information so as to increase the differential probability. For ALE, there exists four-round AES differential characteristics with probability much larger than 2^{-128} after taking into account the leaked state information. The forgery attack against ALE can reach the success rate of 2^{-97}, which is 2^{31} higher than the claimed probability. We show that the results may be further improved if the whitening key layer is removed. We implemented our attack on a small version of ALE, in which 64-bit block and 4-bit-to-4-bit S-box are used. The experimental results match well with the theoretical results.

Very recently, Khovratovich and Rechberger independently proposed an attack against ALE in SAC 2013 [22] which also exploits the weakness of the ALE scheme. However, we notice that their attack is applied to a variant of ALE which the four bytes are leaked after `SubByte`. And in this work, we optimized the differential characteristics used in our attacks so that lower complexities can be obtained in this paper.

This paper is organized as follows. The specification of ALE is given in Sect. 2. Section 3 describes a basic forgery attack against ALE. Section 4 optimizes the forgery attack. Section 5 discusses the effect of removing the whitening key layer of four-round AES. Section 6 gives the experimental results on ALE with reduced block size. Section 7 concludes this paper.

2 The Specification of ALE

In this section, we give a brief description of the ALE. The full specifications of ALE can be found in the original paper [6].

AES round function. AES-128 is used as an underlying primitive of ALE. A full specification of AES can be found in [9]. There are four operations in an AES round: `SubBytes(SB)`, `ShiftRows(SR)`, `MixColumns(MC)` and `AddRoundKey(ARK)`.

```
AESRound(State, ExpandedKey[i])
{
  SubBytes(State);
  ShiftRows(State);
  MixColumns(State);
  AddRoundKey(State,ExpandedKey[i]);
}
```

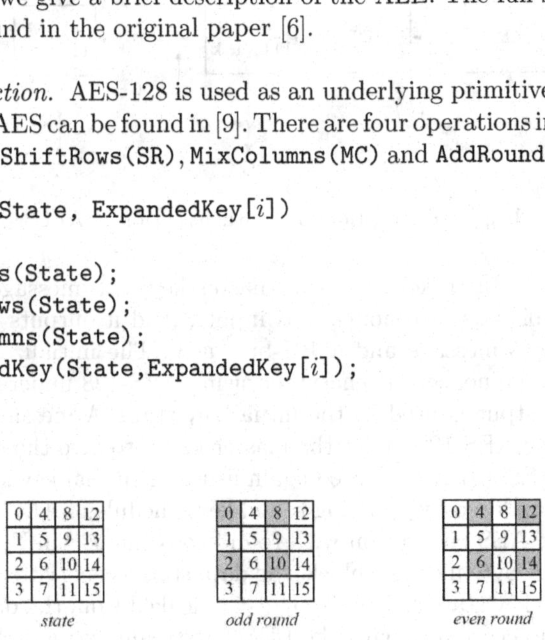

Fig. 1. The positions of the leaked bytes in the even and odd rounds of LEX

LEX keystream extraction. In the stream cipher LEX, AES round functions are repeatedly applied to a state (the subkeys are fixed). At the end of each AES round, 4 bytes from the state are extracted as the keystream [5]. The positions of leaked bytes are shown in Fig. 1.

Pelican MAC. In the Pelican MAC, each 128-bit message block is xored to a secret 128-bit state, then the state passes through 4 AES rounds. In Pelican MAC, each difference passes through at least 25 active S-boxes (following directly from the analysis of AES), thus Pelican MAC provides strong security against forgery attack.

Specification of ALE. The encryption/authentication of ALE is shown in Fig. 2. The process of associated data and last partial block are omitted here. The encryption component of ALE is based on LEX, and its authentication component is based on Pelican MAC. A different nonce is used in ALE for the protection of every message. When the verification fails, the plaintext from the decryption should be kept secret so as to prevent state recovery attack. To encrypt/authen-

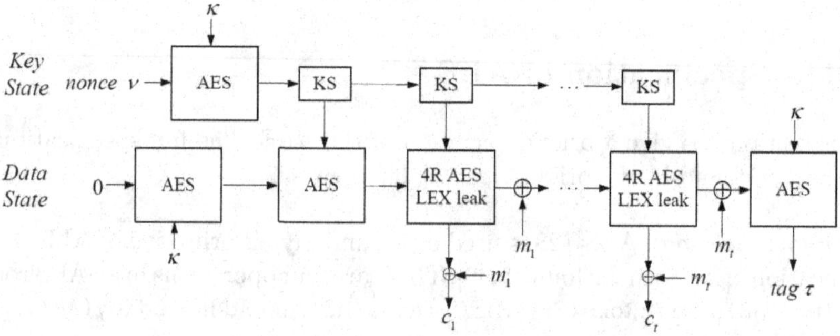

Fig. 2. Encryption and authentication of ALE

ticate a message, ALE takes a 128-bit master key κ, a message μ, associated data α and 128-bit non-zero nonce ν as inputs. And it outputs ciphertext γ of the same length as message and a 128-bit tag τ. The initialization of ALE is given as follows: the nonce ν is encrypted using AES-128 under the master key κ. The 128-bit output is used as the initial key state. A message with value 0 is encrypted using AES-128 under the master key κ to give the data state. The 128-bit output $\text{AES}_\kappa(0)$ is encrypted again using the initial key state as the key. The key state is updated by applying round key schedule of AES-128 to the final round key of last AES encryption with round constant x^{10} in $\mathbb{F}_{2^8}$.

To process a 16-byte message block, the data state is encrypted with 4 rounds of AES using the key state as key. 16 bytes are leaked from the data state in the 4 AES rounds in accordance with the LEX keystream extraction. According to the code provided by the authors of ALE, five round keys are used during the 4 AES rounds, namely, an initial whitening key is used. And at each AES round, four bytes are leaked after the `AddRoundKey()` function. The leak is xored to the current 16-byte block M for encryption. The final round subkey is updated one more time using the AES round key schedule with byte round constant x^4 in $\mathbb{F}_{2^8}$ to get the key state. The current message block M is xored to the data state so that it would pass through the next 4 AES rounds for authentication purpose (similar to that in Pelican MAC).

The decryption/verification is similar to the encryption/authentication, except that the ciphertext block is xored to the keystream to get the message, as shown in Fig. 3. We provide this figure here since the decryption/verification is important in our attack.

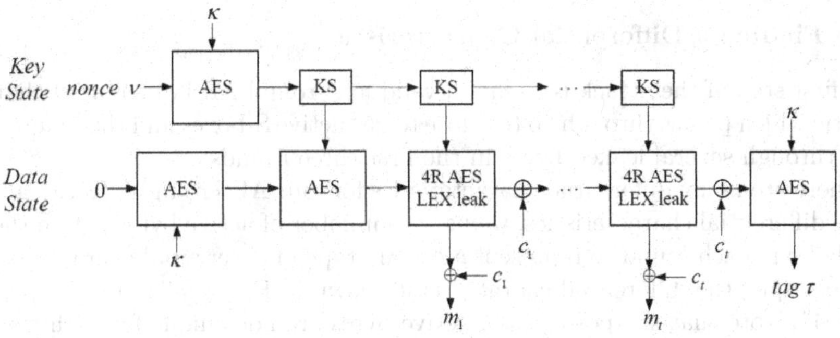

Fig. 3. Decryption and verification of ALE

The designers of ALE claim that any forgery attack not involving key recovery/internal state recovery has a success probability at most 2^{-128}. It is stated that each secret key is used to protect at most 2^{48} message bits. Such restriction on message bits does not affect the success rate of our forgery attack.

3 A Basic Leaked-State Forgery Attack on ALE

In this section, we present a basic forgery attack against ALE. The chance of successful forgery attack is 2^{-106}, which is 2^{22} larger than the claimed success rate 2^{-128}. This attack requires 2^{41} known plaintext blocks.

3.1 The Main Idea of the Attack

The following property of active S-box will be used in our attack:

Property 1. *For an active S-box, if the values of an input and the input/output difference are known, the output/input difference is known with probability 1.*

Here the active S-box is the S-box with non-zero input difference. In the rest of the paper, we will use a new term *active leaked byte* to denote a leaked byte with difference on it.

In the security analysis of Pelican MAC [10] and ALE [6], the probability of four-round differential characteristic of ALE follows the analysis of AES. It has been shown that for any four-round AES differential characteristic, the number of active S-boxes is at least 25 [8]. For each S-box, the differential probability is at most 2^{-6}. Hence, there is a trivial upper bound for the four-round AES differential probability which is 2^{-150}. However, different from the Pelican MAC, 4 state bytes are leaked at the end of every round in ALE. Using Property 1, it is possible to bypass some active S-boxes with probability 1 when the input bytes to those active S-boxes are leaked. It means that the overall differential probability could be significantly increased.

3.2 Finding a Differential Characteristic

The first step of the attack is to find a valid four-round AES differential characteristic which passes through 25 (or close to 25) active S-boxes and the differences pass through several leaked bytes in the first three rounds.

There are many differential characteristics for four AES rounds. To categorize those differential characteristics, we use the number of active bytes before the S-box layer in each round to represent a certain type of differential characteristics. For example, the differential characteristic shown in Fig. 4 falls in the type "1–4–16–4". Note that the positions of active bytes are not unique for each type.

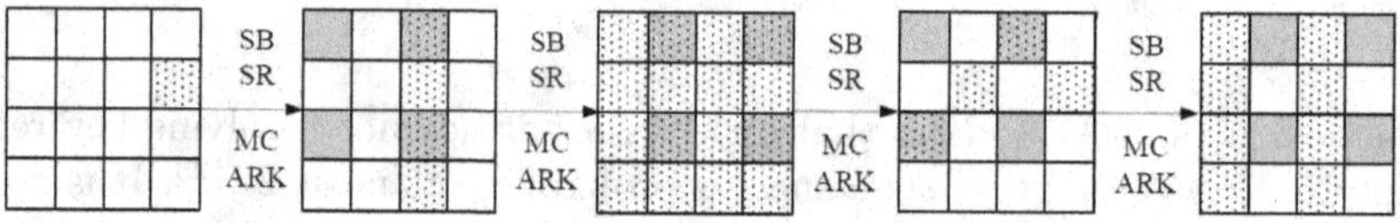

Fig. 4. An example of 1-4-16-4 differential characteristic. Gray squares denote leaked bytes. Squares marked with broken line denote active bytes.

In our basic attack, we use the type of differential characteristic shown in Fig. 4. There are 25 active S-boxes in the differential characteristic, and 8 active leaked bytes are located in the first three rounds.

Next we need to find a differential characteristic with high probability. Note that it is not always guaranteed that the differential probability of each active S-box can reach the maximum value 2^{-6}. The AES S-box has a property that for any input difference δ_1 and output difference δ_2, the probability that equation $S(x) \oplus S(x \oplus \delta_1) = \delta_2$ has a solution is 127/256. Among the 127 solutions, there are 126 solutions have probability 2^{-7} and only one solution has probability 2^{-6}. Hence, for an active S-box, there is a unique output difference reaches the probability 2^{-6} for difference propagation. It shows the conditions to set active S-boxes with difference propagation probability 2^{-6} will limit the number of choices for the possible differential characteristics.

It is thus not surprising that we found no differential characteristic such that every active S-box (except those involving the leaked ones) has the maximum

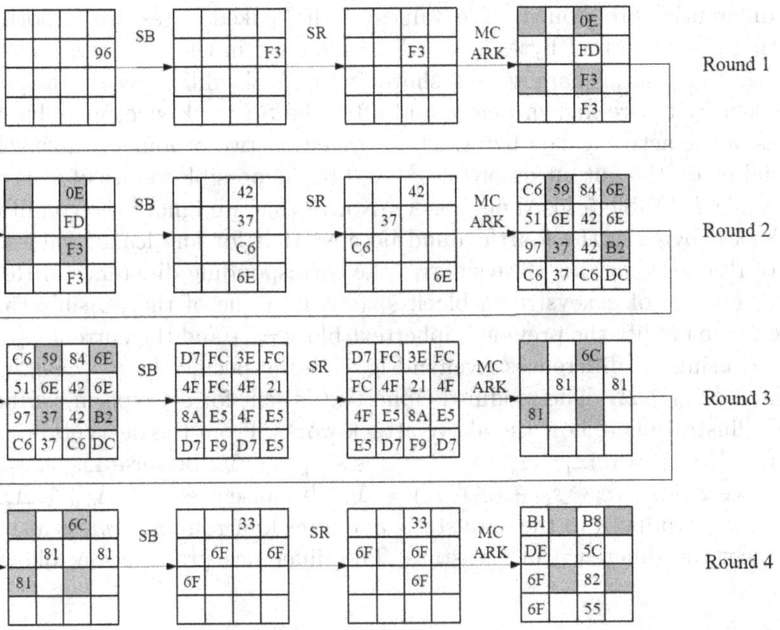

Fig. 5. A differential characteristic of type "1–4–16–4". The hexadecimal numbers indicate the difference values. The empty squares indicate no difference. The squares of leaked bytes are marked with gray color.

differential probability 2^{-6} after testing all the possible positions of the type "1–4–16–4". In order to find a differential characteristic, we need to allow some active S-box with differential probability 2^{-7}. We managed to find a number of differential characteristics. One of them is given in Fig. 5, and we will use this differential characteristic to demonstrate our basic attack. The differential probability of this differential characteristic is given as $2^{-6\times 16+(-7)\times 9} = 2^{-159}$ (differential probability 2^{-6} for 16 active S-boxes, 2^{-7} for 9 active S-boxes).

Three differences in Fig. 5 will be used in our attack: the input difference Δ_{in}, the output difference Δ_{out} and the keystream difference Δ_s:

$$\Delta_{\text{in}} = (0,0,0,0;\ 0,0,0,0;\ 0,0,0,0;\ 0,96,0,0);$$
$$\Delta_{\text{out}} = (\text{B1,DE,6F,6F};\ 0,0,0,0;\ \text{B8,5C,82,55};\ 0,0,0,0);$$
$$\Delta_s = (0,0,\text{E,F3};\ 59,37,\text{6E,F2};\ 0,81,\text{6C},0;\ 0,0,0,0);$$

Note that the values in Δ_s are obtained by simply concatenating the bytes extracted from the states. The order of those bytes has no effect on the attack, as long as this order is fixed.

3.3 Launching the Forgery Attack

After finding a four-round AES differential characteristic, we need to determine the values of the leaked bytes on the differential characteristic so as to improve the differential probability. The values of the leaked bytes are important for locating the ciphertext bytes that will be modified in the forgery attack.

In the differential characteristic shown in Fig. 5, the differences at the positions of leaked bytes are known before and after the S-box. Hence, we solve for the values of the active leaked bytes. There are either two or four possible solutions depending on the output difference. We store the possible values of leaked bytes in a table T (Table 5 in Appendix A). Notice that we ignore the conditions on the leaked bytes in the fourth round because that for any leaked values at the end of Round 3, we can always derive the corresponding difference in Round 4.

If the value of a keystream block s_i falls into one of the possible values of table T, we modify the previous ciphertext block c_{i-1} and the current ciphertext block c_i using the differences given in Fig. 5. More specifically, $c'_{i-1} = c_{i-1} \oplus \Delta_{\text{in}}$; $c'_i = c_i \oplus \Delta_{\text{out}} \oplus \Delta_s$. The modified ciphertext is sent for decryption/verification.

We illustrate here how the above attack works. From the decryption, the difference $\Delta m_{i-1} = (c_{i-1} \oplus s_{i-1}) \oplus (c'_{i-1} \oplus s'_{i-1}) = \Delta_{\text{in}}$ because $\Delta s_{i-1} = 0$; the difference $\Delta m_i = (c_i \oplus s_i) \oplus (c'_i \oplus s'_i) = \Delta_{\text{out}}$ because $c'_i \oplus c_i = \Delta_{\text{out}} \oplus \Delta_s$. Then Δm_{i-1} is introduced to the data state, and after four rounds, Δm_i is introduced to cancel the difference in the state. The difference propagation follows that in Fig. 5.

Complexity of the Attack. In the attack above, the differential probability of the differential characteristic is 2^{-159} before considering the leaked bytes. There are eight leaked bytes being involved in the differential characteristic, with 5 of them being introduced to the active S-boxes with probability 2^{-7}, and another 3 of them being introduced to the active S-boxes with probability 2^{-6}. According to Property 1, the differential probabilities of those eight active boxes involving the leaked bytes become 1. The overall differential probability becomes $2^{-159} \times 2^{7\times5} \times 2^{6\times3} = 2^{-106}$. The success rate of the above attack is thus 2^{-106}.

In this attack, eight leaked keystream bytes are considered, and the values of 6 leaked bytes (from the first two rounds) should be one of the 128 entries in Table T (as explained above). A random keystream block satisfies the requirement with probability $128/2^{6\times8} = 2^{-41}$. We thus need 2^{41} known plaintext blocks in this attack.

4 Optimizing the Leaked-State-Forgery Attack against ALE

In this section, we optimize the LSFA against ALE. In Sect. 4.1, we improve the success rate of the forgery attack. The optimal success rate of a forgery attack can reach 2^{-97}, while 2^{56} known plaintext blocks are needed. In Sect. 4.2, the number of known plaintext blocks can be reduced to $2^{8.4}$ for achieving a success rate 2^{-102}. Note that the known plaintext blocks can be related to different keys or different nonces.

4.1 Improving the Differential Probability

From the attack presented in Sect. 3, we notice that the success rate of forgery attack is determined by the probability of the differential characteristic after taking into account of the leaked bytes. To evaluate the success rate of the forgery attack against ALE, we use the term *effective active S-boxes* to represent the active S-boxes which cannot be bypassed by exploiting the leaked bytes. In the following, we will analyze different cases to find the smallest number of effective active S-boxes.

We start with recalling some properties of the AES round function. The function `MixColumns` has a property that if it is active, the total number of active bytes in the input and output will be at least five (the property of the maximum distance separable code). By referring to the Lemma 9.4.1 from [9], we have the following lemma.

Lemma 1. *The number of active S-boxes of any two-round AES differential characteristic is lower bounded by* $5N$*, where* N *is the number of active columns in the first round.*

In the four AES rounds in ALE, there are 16 leaked bytes. But the leaked bytes from the fourth round cannot be exploited in the attack as they do not pass through S-boxes directly. Therefore only the leaked bytes in the first three rounds can be exploited, and there are at most 12 active leaked bytes. We use $[l_1,l_2,l_3]$ to indicate the number of active leaked bytes in the first three rounds respectively. For instance, the number of active leaked bytes in the differential characteristic in Fig. 4 is $[2,4,2]$. And we use n_i^A $(i = 1,2,3,4)$ to denote the number of active S-boxes at each S-box layer, which will be used in later analysis.

In the following, we will analyze differential characteristics with the smallest number of effective active S-boxes, using the techniques of solving Mixed-Integer Linear Programming (MILP) problems [25, 32]. MILP is a useful technique for proving security bounds against differential cryptanalysis, by evaluating the minimum number of active S-boxes in several rounds of encryption. Designers and cryptanalysts only require to write out simple (in)equations that are input into an MILP solver, then an optimal solution will be returned.

We denote by X_i the input state of round i, then we have $X_{i+1} = ARK \circ MC \circ SR \circ SB(X_i)$, where $i \in \{1,2,3,4\}$. Let $X_{i,j}$ be the j-th byte of X_i, where $0 \le j \le 15$. For a further step, suppose $Y_i = SB(X_i)$, $Z_i = SR(Y_i)$ and $W_i = MC(Z_i)$. We introduce a function χ to catch whether a byte is nonzero, that is , $\chi(x) = 1$ if $x \neq 0$ and $\chi(x) = 0$ if $x = 0$. Here, the value of $\chi(x)$ is a real number. Then, according to the techniques given in [25, 32], the problem of evaluating the minimum number of effective active S-boxes is translated to an MILP problem as follows.

The Objective Function. The objective function is to minimize the value of

$$\sum_{i=1}^{4}\sum_{j=0}^{15}\chi(\Delta X_{i,j}) - \sum_{k=0,2,8,10}(\chi(\Delta X_{2,k}) + \chi(\Delta X_{4,k})) - \sum_{l=4,6,12,14}\chi(\Delta X_{3,l}), \quad (1)$$

since we would like to evaluate the minimum number of effective active S-boxes. In (1), the number of effective active S-boxes is obtained by first counting the number of active S-boxes in four consecutive rounds of AES and then minus the number of active leaked bytes.

Constraints. According to the property of `MixColumns`, we have $\sum_{j=4k}^{4k+3}(\chi(\Delta Z_{i,j}) + \chi(\Delta W_{i,j})) = 0$ or ≥ 5, where $1 \leq i \leq 4$ and $0 \leq k \leq 3$. On the other hand, we have $\chi(\Delta Y_{i,j}) = \chi(\Delta X_{i,j})$, $\chi(\Delta Z_{i,j}) = \chi(\Delta Y_{i,5j \bmod 16})$ and $\chi(\Delta X_{i+1,j}) = \chi(\Delta W_{i,j})$ $(0 \leq j \leq 15)$. Thus, two consecutive rounds of AES provide us four constraints:

$$5d_{i,1} \leq \sum_{j=0}^{3}(\chi(\Delta X_{i,5j \bmod 16}) + \chi(\Delta X_{i+1,j})) \leq 8d_{i,1}, \tag{2}$$

$$5d_{i,2} \leq \sum_{j=4}^{7}(\chi(\Delta X_{i,5j \bmod 16}) + \chi(\Delta X_{i+1,j})) \leq 8d_{i,2}, \tag{3}$$

$$5d_{i,3} \leq \sum_{j=8}^{11}(\chi(\Delta X_{i,5j \bmod 16}) + \chi(\Delta X_{i+1,j})) \leq 8d_{i,3}, \tag{4}$$

$$5d_{i,4} \leq \sum_{j=12}^{15}(\chi(\Delta X_{i,5j \bmod 16}) + \chi(\Delta X_{i+1,j})) \leq 8d_{i,4}, \tag{5}$$

where $i \in \{1, 2, 3\}$ and $d_{i,j} \in \{0, 1\}$ $(1 \leq j \leq 4)$. Notice that $d_{i,j} = 0$ if and only if all eight differences before and after `MixColumns` are zero and $d_{i,j} = 1$ otherwise. Here, we do not consider the case of $i = 4$ since linear transformations in Round 4 does not influence the probability of a differential characteristic.

Additional Constraints. To avoid trivial solution where the minimum number of active S-boxes is zero, the following constraint

$$\sum_{j=0}^{15}\chi(\Delta X_{1,j}) \geq 1 \tag{6}$$

is added to ensure that at least one S-box is active. For a further step, the constraint

$$\sum_{k=0,2,8,10}(\chi(\Delta X_{2,k}) + \chi(\Delta X_{4,k})) + \sum_{l=4,6,12,14}\chi(\Delta X_{3,l}) = n \text{ (or } \leq n) \tag{7}$$

is added to the system. That is, all differential characteristics are classified by the number of active leaked bytes. Constraint (7) help us quickly locate the pattern of differential characteristics with minimum effective active S-boxes.

Since a four-round differential characteristic has at least 25 active S-boxes, the number of effective active S-boxes is at least $25 - n$ if n active leaked bytes are involved. Experimental results confirm this but bring us more knowledge. We solve 11 MILP problems by setting n to be different values, that is, $n \leq 2, 3, \ldots, 8$ and $n = 9, 10, 11, 12$. Here, we choose Maple software [1] to solve them. The minimum number of effective active S-boxes, denoted by m, classified by the number of active leaked bytes is given in Table 1. Each MILP problem cost few seconds to return the optimal solution by running the code in Appendix B.

Table 1. Minimum number m of effective active S-boxes, if $(\leq)n$ active leaked bytes are included in a differential characteristic

n	≤ 2	≤ 3	≤ 4	≤ 5	≤ 6	≤ 7	≤ 8	9	10	11	12
m	23	22	21	20	19	18	17	16	16	19	18

From Table 1, we conclude that the best probability of a differential characteristic is at most 2^{-96}, since a differential characteristic has at least 16 effective active S-boxes. What is more, exactly 9 or 10 active leaked bytes are involved if a differential characteristic has 16 effective active S-boxes. An interesting observation is that the minimum number of active S-boxes (i.e., $n + m$) may be greater than 25 if too many active leaked bytes are included in a differential characteristic, because it has to cover too many specific positions in these cases.

Now, we demonstrate that only 4 kinds of differential characteristics may have exactly 16 effective active S-boxes by analyzing the distribution of 9 or 10 active leaked bytes in a four-round differential characteristic. This is done by adding more concrete constraints to the MILP step by step. We choose the case $l_1 + l_2 + l_3 = 10$ to show the way of determining the distribution of the 10 active leaked bytes in each round. Similar process is applied to $l_1 + l_2 + l_3 = 9$. The procedure is summarized in Table 2.

Since $l_1 + l_2 + l_3 = 10$, we have $l_2 = 2, 3$ or 4. The minimum number of effective active S-boxes is 17, 20 and 16 if $l_2 = 2, 3$ and 4, respectively. Thus, to find differential characteristics with exactly 16 effective active S-boxes, we only need to consider $l_2 = 4$, which implies $l_1 + l_3 = 6$. For a further step, we have $l_1 = 2, 3$ or 4. The minimum number of effective active S-boxes is 17, 20 and 16 if $[l_1, l_2] = [2, 4]$, $[l_1, l_2] = [3, 4]$ and $[l_1, l_2] = [4, 4]$, respectively. Therefore, differential characteristics with exactly 10 active leaked bytes and 16 effective active S-boxes exist only if $[l_1, l_2, l_3] = [4, 4, 2]$. Combined with Lemma 1, $l_1 = 4$ implies $n_1 \geq 2$ and $n_1^A + n_2^A \geq 10$ since at least two columns are active in the first `MixColumns` layer; $[l_1, l_2] = [4, 4]$ implies $n_2^A + n_3^A \geq 20$; $[l_2, l_3] = [4, 2]$ implies $n_3^A + n_4^A \geq 15$ and $n_4^A \geq 4$, where $n_4^A \geq 4$ since two active leaked bytes appear in round 4 and at least two active bytes will appear in two non-leaking columns. Thus, for case $[l_1, l_2, l_3] = [4, 4, 2]$, only one possible type of differential characteristic 2-8-12-4 can be appeared.

Summary of the Analysis. From the above discussion, we conclude that the number of effective active S-boxes is at least 16 in a differential characteristic. And there are four types of differential characteristics, "2–3–12–8", "2–8–12–4", "2–8–12–3" and "4–6–9–6", which can reach this lower bound.

Table 2. Minimum number m of effective active S-boxes with more constraints, the distribution of 9 or 10 active leaked bytes in these rounds, and the type of possible differential characteristic

n	additional constraints	m	$[l_1, l_2, l_3]$	Type of differential characteristic
10	$l_2 = 2$	17	discard	
	$l_2 = 3$	20	discard	
	$l_2 = 4,\ l_1 = 2$	17	discard	
	$l_2 = 4,\ l_1 = 3$	20	discard	
	$l_2 = 4,\ l_1 = 4$	16	[4,4,2]	2-8-12-4
9	$l_2 = 1$	16	[4,1,4]	4-6-9-6
	$l_2 = 2$	17	discard	
	$l_2 = 3$	21	discard	
	$l_2 = 4,\ l_1 = 1$	16	[1,4,4]	2-3-12-8
	$l_2 = 4,\ l_1 = 2$	17	discard	
	$l_2 = 4,\ l_1 = 3$	21	discard	
	$l_2 = 4,\ l_1 = 4$	16	[4,4,1]	2-8-12-3

After testing these four types of differential characteristics, we conclude that there is no differential characteristic in which each of the effective active S-box reaches the maximum differential probability 2^{-6}. The differential characteristic with best probability is of the type "2–8–12–4", and the details are given in Fig. 6. In this differential characteristic, the probability of one effective active S-box is 2^{-7}. So the overall probability of the differential characteristic is $2^{-6\times 15+(-7)} = 2^{-97}$. This is the best success rate of the forgery attack against ALE. For this differential characteristic, the values of 8 leaked bytes (from the first two rounds) should be one of the 2^8 values given in Table 6 in Appendix A. And the probability of random keystream block satisfying the requirement is $2^8/2^{8\times 8} = 2^{-56}$. If each key is restricted to protect 2^{48} message bits (2^{41} message blocks), we need to observe 2^{15} keys to find a weak keystream block to launch the attack. The experimental results of this attack on a small version of ALE are given in Sect. 6.1.

4.2 Reducing the Number of Known Plaintext Blocks

There are two approaches to reduce the number of known plaintext blocks required in the attack. One approach is to allow differential probability of 2^{-7} for some effective active S-boxes; another approach is to reduce the number of active leaked bytes in a differential characteristic. In these two approaches, with the reduced success rate, we are able to reduce the number of known plaintext blocks drastically.

Relaxing Conditions on Effective Active S-boxes. When we try to find the best probability for the differential characteristics, it is important to restrict as many as effective active S-boxes to have probability 2^{-6} for the input and output differences. However, if we are not satisfied with the large number of

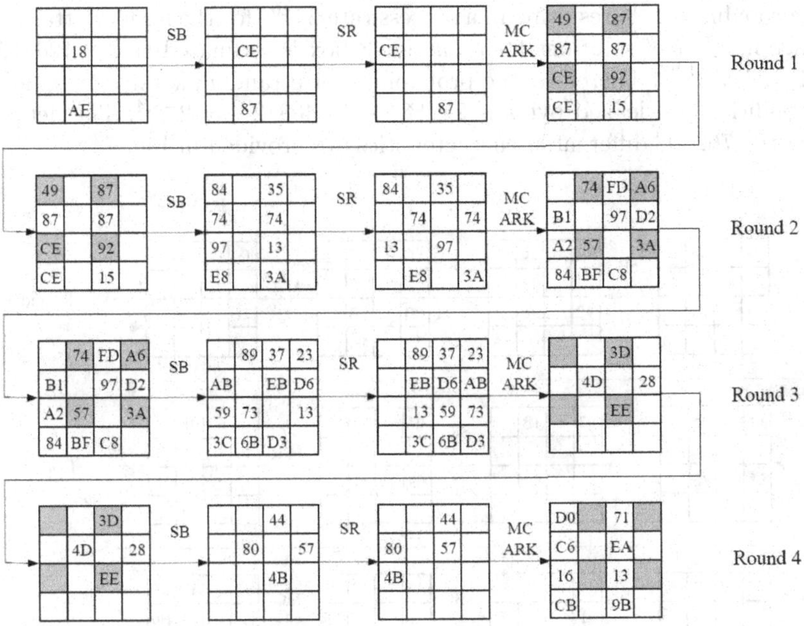

Fig. 6. Differential Path of type "2–8–12–4". The hexadecimal numbers indicate the difference values. The empty squares indicate no difference. The squares of leaked bytes are marked with gray color.

plaintext blocks required to launch the attack, we can relax the condition on some active S-boxes to have probability 2^{-7}. For instance, the probability of random keystream satisfying the requirements for leaked bytes in the differential characteristic presented in Sect. 4.1 is 2^{-56}. However, if we relax the probabilities on two effective active S-boxes to 2^{-7}, this probability increases to at least than 2^{-50} because the increased number of differential characteristics is at least 2^6 by our test. It can be increased further if more conditions on effective active S-boxes are relaxed.

Reducing the Number of Active Leaked Bytes in the First Two Rounds. Another way to reduce the number of known plaintext blocks is to reduce the active leaked bytes in the first two rounds. The reason is that only the active leaked bytes in first two rounds are related to the conditions on leaked bytes. No matter what values the active leaked bytes are taken in Round 3, we can determine the corresponding differences after the S-box layer according to the leaked values. The only cost is an additional pre-computed look-up table. One good choice is let the number of active leaked bytes to be $[4, 0, 4]$, and the type of differential characteristic is "6-4-6-9". In this case, we only need to check conditions on the four active leaked bytes in the first round, yet we can still have a relatively good differential probability. There are 762408 possible differential characteristics in the first two rounds when all the 17 effective active S-boxes are with probability 2^{-6}, resulting in a success rate 2^{-102} for the forgery attack. The average number of solutions for an active S-box is estimated as $2 \times 126/127 + 4 \times 1/127 = 2^{1.01}$. Therefore, the probability for a random keystream satisfying the conditions on leaked bytes is $2^{1.01 \times 4} \times 762408/2^{32} = 2^{-8.4}$. The details of one of the 762408 differential characteristics are provided in Fig. 7.

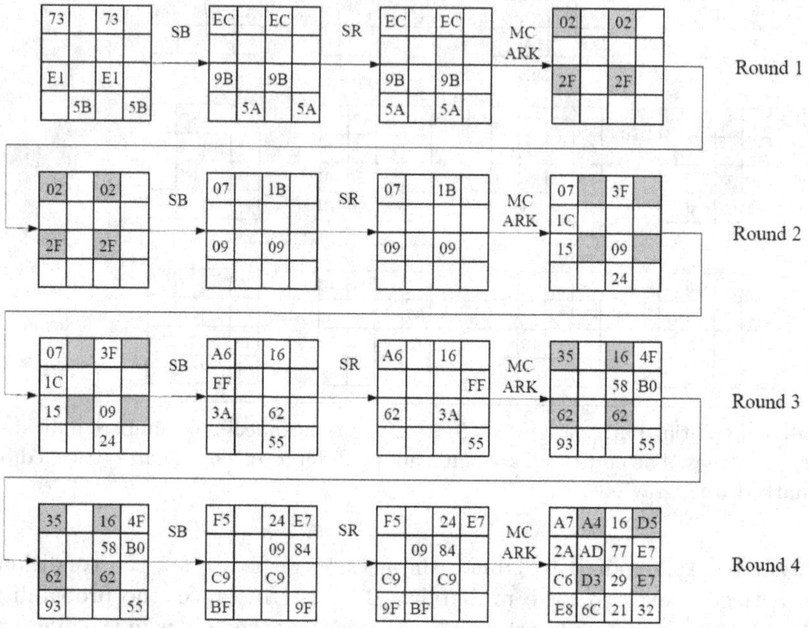

Fig. 7. Differential Path of type "6–4–6–9". The hexadecimal numbers indicate the difference values. The empty squares indicate no difference. The squares of leaked bytes are marked with gray color.

5 Effect of Removing the Whitening Key Layer

In this section, we show that the results may be further improved if the whitening key layer is removed. The success rate of a forgery attack can reach around $2^{-93.1}$, and only one or two plaintext blocks are needed to launch the attack.

Once the whitening key layer is removed, additional four bytes before the first S-box layer are known to an attacker, i.e., byte $X_{1,4}, X_{1,6}, X_{1,12}$ and $X_{1,14}$. They are obtained by xoring the previous message block and the last four leaked bytes of processing the previous message block. Thus, at most 16 leaked bytes can be exploited. In the following discussions, we denote by l_0 the number of active leaked byte before the first S-box layer, while l_1, l_2 and l_3 still indicate the number of active leaked bytes in the first three rounds respectively.

First, we analyze the smallest number of effective active S-boxes in a differential characteristic. The objective function is adjusted to minimize the value of

$$\sum_{i=1}^{4}\sum_{j=0}^{15}\chi(\Delta X_{i,j}) - \sum_{k=4,6,12,14}(\chi(\Delta X_{1,k})+\chi(\Delta X_{3,k})) - \sum_{l=0,2,8,10}(\chi(\Delta X_{2,l})+\chi(\Delta X_{4,l})), \quad (8)$$

since now additional four bytes are leaked before the first S-box layer. Similarly, (7) is adjusted to the following constraint

$$\sum_{k=4,6,12,14}(\chi(\Delta X_{1,k})+\chi(\Delta X_{3,k})) + \sum_{l=0,2,8,10}(\chi(\Delta X_{2,l})+\chi(\Delta Y_{4,l})) = n. \quad (9)$$

Notice that $n = l_0 + l_1 + l_2 + l_3$.

Table 3. Minimum number m of effective active S-boxes, if n active leaked bytes are included in a differential characteristic

n	0	1	2	3	4	5	6	7	8	9	10	11	12	13	14	15	16
m	30	24	23	22	21	20	19	18	17	16	15	19	18	22	21	25	24

The minimum number of effective active S-boxes classified by the number of active leaked bytes is given in Table 3. We conclude that a differential characteristic involves at least 15 effective active S-boxes. Thus, the best probability of a differential characteristic is at most 2^{-90}. For a further step, exactly 10 active leaked bytes are included in a differential characteristic with 15 effective active S-boxes, that is, $l_0 + l_1 + l_2 + l_3 = 10$. Similar to the process of Table 2, the distribution of the 10 active leaked bytes in these four rounds is studied by adding more and more constraints to the MILP problems. This is done by first studying the sum of $l_1 + l_2$, which may be $2, \ldots, 7$ or 8, and then investigating the values of l_i $(0 \le i \le 3)$. The results are given in Table 4.

From Table 4, we conclude that a differential characteristic with 15 effective active S-boxes exists only if the concrete distribution of the 10 active leaked bytes satisfies

1) $[l_0, l_1, l_2, l_3] = [4, 0, 2, 4], [4, 2, 0, 4], [2, 0, 4, 4], [4, 0, 4, 2], [2, 4, 0, 4]$ or $[4, 4, 0, 2]$, and
2) $\chi(\Delta X_{i,4}) = \chi(\Delta X_{i,14})$ if $n_i = 2$ and $i \in \{1, 3\}$; $\chi(\Delta X_{i,0}) = \chi(\Delta X_{i,2})$ if $n_i = 2$ and $i \in \{2, 4\}$.

Then, we analyze all the 12 cases of differential characteristics with 15 effective active S-boxes. For each of the 12 cases listed in Table 4, different types of differential characteristics may satisfy it. In this situation, we maximize the number of effective active S-boxes in Round 1 and Round 4, as the differential probability of effective active S-boxes in these two rounds can always reach the maximum value 2^{-6} once the differential characteristic is constructed using the start-from-the-middle technique, which is also employed by the authors in [22]. The best differential characteristics we found are given as follows.

Table 4. Minimum number m of effective S-boxes with more constraints, and the distribution of 10 active leaked bytes in these rounds

$l_1 + l_2$	additional constraints	m	$[l_0, l_1, l_2, l_3]$	Case number
2	$l_1 = 0,\ \chi(\Delta X_{3,4}) + \chi(\Delta X_{3,14}) = 0$	15	[4,0,2,4]	#1
	$l_1 = 0,\ \chi(\Delta X_{3,4}) + \chi(\Delta X_{3,14}) = 1$	20	discard	
	$l_1 = 0,\ \chi(\Delta X_{3,4}) + \chi(\Delta X_{3,14}) = 2$	15	[4,0,2,4]	#2
	$l_1 = 1$	20	discard	
	$l_1 = 2,\ \chi(\Delta X_{2,0}) + \chi(\Delta X_{2,2}) = 0$	15	[4,2,0,4]	#3
	$l_1 = 2,\ \chi(\Delta X_{2,0}) + \chi(\Delta X_{2,2}) = 1$	20	discard	
	$l_1 = 2,\ \chi(\Delta X_{2,0}) + \chi(\Delta X_{2,2}) = 2$	15	[4,2,0,4]	#4
3		20	discard	
4	$l_1 = 0,\ l_0 = 2,\ \chi(\Delta X_{1,4}) + \chi(\Delta X_{1,14}) = 0$	15	[2,0,4,4]	#5
	$l_1 = 0,\ l_0 = 2,\ \chi(\Delta X_{1,4}) + \chi(\Delta X_{1,14}) = 1$	20	discard	
	$l_1 = 0,\ l_0 = 2,\ \chi(\Delta X_{1,4}) + \chi(\Delta X_{1,14}) = 2$	15	[2,0,4,4]	#6
	$l_1 = 0,\ l_0 = 3$	20	discard	
	$l_1 = 0,\ l_0 = 4,\ \chi(\Delta X_{4,0}) + \chi(\Delta X_{4,2}) = 0$	15	[4,0,4,2]	#7
	$l_1 = 0,\ l_0 = 4,\ \chi(\Delta X_{4,0}) + \chi(\Delta X_{4,2}) = 1$	20	discard	
	$l_1 = 0,\ l_0 = 4,\ \chi(\Delta X_{4,0}) + \chi(\Delta X_{4,2}) = 2$	15	[4,0,4,2]	#8
	$l_1 = 1$	20	discard	
	$l_1 = 2$	18	discard	
	$l_1 = 3$	20	discard	
	$l_1 = 4,\ l_0 = 2,\ \chi(\Delta X_{1,4}) + \chi(\Delta X_{1,14}) = 0$	15	[2,4,0,4]	#9
	$l_1 = 4,\ l_0 = 2,\ \chi(\Delta X_{1,4}) + \chi(\Delta X_{1,14}) = 1$	20	discard	
	$l_1 = 4,\ l_0 = 2,\ \chi(\Delta X_{1,4}) + \chi(\Delta X_{1,14}) = 2$	15	[2,4,0,4]	#10
	$l_1 = 4,\ l_0 = 3$	20	discard	
	$l_1 = 4,\ l_0 = 4,\ \chi(\Delta X_{4,0}) + \chi(\Delta X_{4,2}) = 0$	15	[4,4,0,2]	#11
	$l_1 = 4,\ l_0 = 4,\ \chi(\Delta X_{4,0}) + \chi(\Delta X_{4,2}) = 1$	20	discard	
	$l_1 = 4,\ l_0 = 4,\ \chi(\Delta X_{4,0}) + \chi(\Delta X_{4,2}) = 2$	15	[4,4,0,2]	#12
5		20	discard	
6		17	discard	
7		20	discard	
8		16	discard	

- For each of the 8 cases with $l_1 + l_2 = 4$, that is, case #5 to #12, a differential characteristic with probability of about $2^{-93.1}$ can be construct for almost all of the leaked information. Experimental results show that we can not obtain a differential characteristic for 412, 443, 402 and 373 out of 2^{32} leaked information in case #5 and #6, case #7 and #8, case #9 and #10 and case #11 and #12, respectively. Thus, in average, two plaintext blocks are enough to launch a forgery attack. The differential characteristic of case #10 is given in Appendix C.
- For each of the four cases with $l_1 + l_2 = 2$, that is, case #1 to #4, a class of 1020 differential characteristics with average probability of $2^{-94.1}$ always can be constructed, whatever the leaked information is. Thus, the forgery attack can be launched for any plaintext block. Differential paths of the case #4 are given in Appendix D.

Summary of the Analysis. From the above discussion, the whitening key layer is important for ALE. Once it is removed, more internal information will be leaked to an attacker, resulting in forgery attacks with higher success rates and less required plaintext blocks. The success rate of a forgery attack now is about $2^{-93.1}$ to $2^{-94.1}$, and at most 2 plaintext blocks are needed.

6 Experiments on a Reduced Version of ALE

As a proof of concept, we would apply our attacks to ALE (with the whitening key). However, it is impossible to directly attack the original ALE as the complexity is too high. Instead, we choose to attack a reduced ALE construction based on an AES-like light-weight block cipher, LED [16].

The LED block cipher has similar round function as AES except that the operation `AddConstants` is used before the S-box layer in each round, and the round keys are added every four rounds. The S-box in LED has difference propagation probability at most 2^{-2}. Unlike the AES S-box, the output difference may not be unique to attain the best difference propagation probability. And for input difference 14, the probability 2^{-2} can never be obtained. So we need to take care of these differences in the attack.

In our experiments, we modified the LED round function so that it has the same ordered operations: `SubCells`, `ShiftRows`, `MixColumns`, `AddRoundKeys` as AES. Since the differential characteristic is not related to the key schedule, we use random round keys rather than deriving them from the key schedule. In addition, we simplified the input message to the two-block case without considering the initialization, padding and the associated data. The initial state is randomly generated.

6.1 The "2–8–12–4" Differential Characteristic

In the optimized forgery attacks presented in Sect. 4.1, the differential characteristic of type "2–8–12–4" is one of those have the highest success rate. We will experimentally verify the results on this type of differential characteristics.

Estimations. Using the above reduced ALE, we searched the differential characteristics of type "2–8–12–4". Like the case discussed in original ALE, we need to relax the difference propagation probability of one effective active S-box to find a valid differential characteristic. Fig. 10 in Appendix E illustrates one of the differential characteristics we found.

To estimate the probability that a random keystream block is vulnerable to the attack, we analyze the number of solutions for the values of active leaked bytes in first two rounds. In each of the first two rounds, there are 2^6 possible solutions for the values of the four active leaked bytes. Therefore, the probability of a random keystream block satisfies the conditions on leaked bytes is estimated as $2^6 \times 2^6 \times 2^{(-4)\times 8} = 2^{-20}$. The average number of plaintext blocks needed to get a vulnerable keystream block is thus $1 + 1/2^{-20} = 1 + 2^{20}$. Notice that we need an extra plaintext block to introduce the initial differences.

There are 16 effective active S-boxes in the chosen differential characteristic: 15 of the active effect S-boxes with differential probability 2^{-2}, and one with probability 2^{-3}. So the estimated probability of the differential characteristic is $2^{(-2)\times 15+(-3)\times 1} = 2^{-33}$ which is also the success rate of the forgery attack.

Experimental Results. First, we check the probability of the vulnerable keystream blocks. After encrypting $2^{27.1}$ random plaintext blocks, we found 2^7 vulnerable keystram blocks. Hence, the average plaintext blocks needed to find a vulnerable keystream block is $2^{27.1-7} = 2^{20.1}$ which matches the estimated value.

Then, we verify the success rate of the forgery attack. For a vulnerable keystream block, the value of final state is xored with the second message block and stored as t_1. The differences in the final state (thus the leaked bytes) in Round 4 are determined by the values of leaked bytes in Round 3. Then we compute two forged ciphertext blocks similar to the attack procedure in Sect. 3 (but using the difference shown in Fig. 10 in Appendix E). We decrypt the forged ciphertext blocks and xor the second plaintext block from decryption with the final state to get t_2. If the two internal states t_1 and t_2 collide, we get a successful forgery. After examining $2^{36.36}$ vulnerable keystream blocks, we managed to get 10 collisions at the internal states after two blocks. So the average probability for one successful forgery is $2^{-33.04}$. One of the successful forgeries is given in Appendix E.

6.2 The "6–4–6–9" Differential Characteristic

In Sect. 4.2, the differential characteristics of type "6–4–6–9" (Fig. 7) are observed to require a small number of known plaintext blocks yet have good success rate. We experimentally tested this case on the reduced version of ALE.

Estimations. For this type, we found 1400 differential characteristics for the first two rounds, resulting in 21311 different values for the leaked bytes in the first round. Details of one of the differential characteristics are given in Fig. 11 in Appendix F. It is interesting to notice that certain leaked values may be used in more than one differential characteristic. If we take this into consideration, there are 28657 different leaked values related to the 1400 differential characteristics. Since there are only four active leaked bytes in the first two rounds, the probability that a random keystream is vulnerable is $28657/2^{4\times 4} = 2^{-1.12}$. Thus, the estimated number of plaintext blocks needed to find a vulnerable keystream block is $1 + 1/2^{-1.12} = 2^{1.7}$.

There are 17 effective active S-boxes in the differential characteristic. All of them attain the maximum differential probability 2^{-2}. So the estimated probability of the differential characteristic is $2^{(-2)\times 17} = 2^{-34}$, which is also the success rate of the forgery attack.

Experimental Results. In our experiments, $2^{20.7}$ vulnerable keystream blocks are generated from the encryption of $2^{21.6}$ random 2-block plaintexts. So the average number of blocks needed to find one vulnerable keystream block is $2 \times 2^{21.6}/2^{20.7} = 2^{1.9}$, which is close to the estimated value.

After querying $2^{37.7}$ forged ciphertexts, we found 10 collisions in the internal states. So the average probability of successful forgery is around $2^{-34.4}$ which is close to the estimated 2^{-34}. One of the successful forgeries is given in Appendix F.

7 Conclusion

The ALE authenticated encryption algorithm is claimed with a forgery success rate of 2^{-128}. In this paper, we show that the success rate is significantly higher than the claimed rate. By applying the proposed leaked-state-forgery attack, the success rate can reach 2^{-97}. For a success rate 2^{-102}, every one out of $2^{8.4}$ plaintext blocks is vulnerable to the forgery attack. We also show that the whitening key layer is important for ALE, as the complexity of forgery attack can be improved with probabilities from $2^{-93.1}$ to $2^{-94.1}$, and at most two plaintext blocks are needed if the whitening key layer is removed. Our attacks are well-supported by the experimental results on a reduced version of ALE. Our attack confirms again that "it is very easy to accidentally combine secure encryption schemes with secure MACs and still get insecure authenticated encryption schemes" [23]. Hence, in the design of authenticated encryption algorithms, we should be very cautious in analyzing the interaction between encryption and authentication.

Acknowledgements. The authors would like to thank the anonymous reviewers of ASIACRYPT 2013 for their insightful and helpful comments on this paper. Shengbao Wu, Mingsheng Wang and Wenling Wu were supported by the National Basic Research Program of China (Grant No. 2013CB834203 and Grant No. 2013CB338002) and the National Natural Science Foundation of China (Grant No. 61272476, 61232009 and 11171323). Hongjun Wu and Tao Huang were supported by the National Research Foundation Singapore under its Competitive Research Programme (CRP Award No. NRF-CRP2-2007-03) and Nanyang Technological University NAP startup grant (M4080529.110).

References

1. Maple. Maple Software, `http://www.maplesoft.com/products/maple/`
2. 3GPP. Specification of the 3GPP Confidentiality and Integrity Algorithms 128-EEA3 & 128-EIA3, Document 1, 128-EEA3 and 128-EIA3 specification. The 3rd Generation Partnership Project (3GPP) (2010)
3. Agren, M., Hell, M., Johansson, T., Meier, W.: Grain-128a: A New Version of Grain-128 with Optional Authentication. International Journal of Wireless and Mobile Computing 5(1), 48–59 (2011)
4. Bellare, M., Rogaway, P., Wagner, D.: The EAX mode of operation. In: Roy, B., Meier, W. (eds.) FSE 2004. LNCS, vol. 3017, pp. 389–407. Springer, Heidelberg (2004)
5. Biryukov, A.: A new 128-bit key stream cipher LEX. eSTREAM, ECRYPT Stream Cipher Project, Report, 13:2005 (2005)

6. Bogdanov, A., Mendel, F., Regazzoni, F., Rijmen, V., Tischhauser, E.: ALE: AES-Based Lightweight Authenticated Encryption. In: Fast Software Encryption (2013)
7. CAESAR. Competition for Authenticated Encryption: Security, Applicability, and Robustness, `http://competitions.cr.yp.to/caesar.html`
8. Daemen, J., Rijmen, V.: The Wide Trail Design Strategy. In: Honary, B. (ed.) Cryptography and Coding 2001. LNCS, vol. 2260, pp. 222–238. Springer, Heidelberg (2001)
9. Daemen, J., Rijmen, V.: The Design of Rijndael: AES–the Advanced Encryption Standard. Springer (2002)
10. Daemen, J., Rijmen, V.: The Pelican MAC Function. IACR ePrint Archive, Report 2005/212 (2005)
11. Dunkelman, O., Keller, N.: A New Attack on the LEX Stream Cipher. In: Pieprzyk, J. (ed.) ASIACRYPT 2008. LNCS, vol. 5350, pp. 539–556. Springer, Heidelberg (2008)
12. Dunkelman, O., Keller, N.: Cryptanalysis of the Stream Cipher LEX. In: Des. Codes Cryptogr., vol. 67, pp. 357–373. Springer (2013)
13. Engels, D., Saarinen, M.-J.O., Schweitzer, P., Smith, E.M.: The Hummingbird-2 Lightweight Authenticated Encryption Algorithm. In: Juels, A., Paar, C. (eds.) RFIDSec 2011. LNCS, vol. 7055, pp. 19–31. Springer, Heidelberg (2012)
14. Ferguson, N., Whiting, D., Schneier, B., Kelsey, J., Lucks, S., Kohno, T.: Helix: Fast Encryption and Authentication in a Single Cryptographic Primitive. In: Johansson, T. (ed.) FSE 2003. LNCS, vol. 2887, pp. 330–346. Springer, Heidelberg (2003)
15. Fleischmann, E., Forler, C., Lucks, S.: McOE: A Family of Almost Foolproof On-Line Authenticated Encryption Schemes. In: Canteaut, A. (ed.) FSE 2012. LNCS, vol. 7549, pp. 196–215. Springer, Heidelberg (2012)
16. Guo, J., Peyrin, T., Poschmann, A., Robshaw, M.: The LED Block Cipher. In: Preneel, B., Takagi, T. (eds.) CHES 2011. LNCS, vol. 6917, pp. 326–341. Springer, Heidelberg (2011)
17. ISO/IEC 19772:2009. Information technology – Security techniques – Authenticated encryption. ISO, Geneva, Switzerland (2009)
18. Iwata, T., Yasuda, K.: BTM: A Single-Key, Inverse-Cipher-Free Mode for Deterministic Authenticated Encryption. In: Jacobson Jr., M.J., Rijmen, V., Safavi-Naini, R. (eds.) SAC 2009. LNCS, vol. 5867, pp. 313–330. Springer, Heidelberg (2009)
19. Iwata, T., Yasuda, K.: HBS: A Single-Key Mode of Operation for Deterministic Authenticated Encryption. In: Dunkelman, O. (ed.) FSE 2009. LNCS, vol. 5665, pp. 394–415. Springer, Heidelberg (2009)
20. Jakimoski, G., Khajuria, S.: ASC-1: An Authenticated Encryption Stream Cipher. In: Miri, A., Vaudenay, S. (eds.) SAC 2011. LNCS, vol. 7118, pp. 356–372. Springer, Heidelberg (2012)
21. Jutla, C.S.: Encryption Modes with Almost Free Message Integrity. In: Pfitzmann, B. (ed.) EUROCRYPT 2001. LNCS, vol. 2045, pp. 529–544. Springer, Heidelberg (2001)
22. Khovratovich, D., Rechberger, C.: The LOCAL attack: Cryptanalysis of the authenticated encryption scheme ALE. In: Selected Areas in Cryptography – SAC 2013. Springer, Heidelberg (2013)
23. Kohno, T., Viega, J., Whiting, D.: CWC: A High-Performance Conventional Authenticated Encryption Mode. In: Roy, B., Meier, W. (eds.) FSE 2004. LNCS, vol. 3017, pp. 408–426. Springer, Heidelberg (2004)
24. McGrew, D., Viega, J.: The Galois/Counter Mode of Operation (GCM), `http://csrc.nist.gov/CryptoToolkit/modes/proposedmodes/gcm/gcm-spec.pdf`
25. Mouha, N., Wang, Q., Gu, D., Preneel, B.: Differential and Linear Cryptanalysis Using Mixed-Integer Linear Programming. In: Wu, C.-K., Yung, M., Lin, D. (eds.) Inscrypt 2011. LNCS, vol. 7537, pp. 57–76. Springer, Heidelberg (2012)

26. NIST. Recommendation for Block Cipher Modes of Operation. NIST special publication 800–38A, 2001 edn. (2001)
27. NIST. The Keyed-Hash Message Authentication Code (HMAC). FIPS PUB 198
28. Rogaway, P., Bellare, M., Black, J., Krovetz, T.: OCB: A block-cipher mode of operation for efficient authenticated encryption. In: Proceedings of the 8th ACM conference on Computer and Communications Security, pp. 196–205. ACM (2001)
29. Whiting, D., Housley, R., Ferguson, N.: Counter with CBC-MAC (CCM) (2003), csrc.nist.gov/encryption/modes/proposedmodes/ccm/ccm.pdf
30. Whiting, D., Schneier, B., Lucks, S., Muller, F.: Phelix: Fast Encryption and Authentication in a Single Cryptographic Primitive. eSTREAM, ECRYPT Stream Cipher Project Report 2005/027
31. Wu, H., Preneel, B.: Resynchronization Attacks on WG and LEX. In: Robshaw, M. (ed.) FSE 2006. LNCS, vol. 4047, pp. 422–432. Springer, Heidelberg (2006)
32. Wu, S., Wang, M.: Security Evaluation against Differential Cryptanalysis for Block Cipher Structures. Cryptology ePrint Archive: Report 2011/551 (2011), http://eprint.iacr.org/
33. Yuan, Z., Wang, W., Jia, K., Xu, G., Wang, X.: New Birthday Attacks on Some MACs Based on Block Ciphers. In: Halevi, S. (ed.) CRYPTO 2009. LNCS, vol. 5677, pp. 209–230. Springer, Heidelberg (2009)

A Values of Leaked-Bytes

The values of leaked bytes for the differential characteristic used in the basic LSFA in Sect. 3 are given in Table 5. The index is the byte position in the keystream block. δ_{in} and δ_{out} are the input and output differences for the S-box. α and β can be arbitrary values extracted from the leaked bytes in Round 3. From the table, the total number of possible values at the active leaked bytes in first two rounds is $2 \times 2 \times 2 \times 2 \times 4 \times 2 = 128$.

Table 5. Possible values of leaked bytes in hexadecimal for the basic LSFA. "-" indicates no difference. "$\star$" indicates arbitrary values. α and β are values from the leaked bytes.

Index	δ_{in}	δ_{out}	Value(s)
0 – 1	-	-	$\star$
2	E	42	11 or 1F
3	F3	C6	F, FC
4	59	FC	23, 7A
5	37	E5	19, 2E
6	6E	FC	0, 6E, 8C, E2
7	B2	E5	46, F4
8	-	-	$\star$
9	81	$S(\alpha) \oplus S(81 \oplus \alpha)$	α
10	6C	$S(\beta) \oplus S(6C \oplus \beta)$	β
11 – 15	-	-	$\star$

The values of leaked byes for the differential characteristic used in the optimized LSFA in Sect. 4.2 are given in Table 6. The total number of possible values at the active leaked bytes in first two rounds is 2^8.

Table 6. Possible values of leaked bytes in hexadecimal for the optimized LSFA in Sect. 4.2. "-" indicates no difference. "⋆" indicates arbitrary values. α and β are values from the leaked bytes.

Index	δ_{in}	δ_{out}	Value(s)
0	49	84	1D or 54
1	CE	97	33, FD
2	87	35	44, C3
3	92	13	5E, CC
4	74	89	10, 64
5	57	73	B0, E7
6	A6	23	6D, CB
7	3A	13	08, 32
8 – 9	-	-	$\star$
10	3D	$S(\alpha) \oplus S(3D \oplus \alpha)$	α
11	EE	$S(\beta) \oplus S(EE \oplus \beta)$	β
12 – 15	-	-	$\star$

B Maple Program for Solving MILP Problems

We employ the function "LPSolve" included in the "Optimization" package of Maple software to solve MILP Problems. To simplify the variables in the MILP problems given in Sect. 4.1, we compress $\chi(\Delta X_{i,j})$ and $d_{i,j}$ to xij and dij here. Then, results in Table 1 are obtained by running the following program.

```
with(Optimization);
%if n<=8, the last constraint x20+x22+...+x48+x410>=n will be removed.
n:=9;
LPSolve(x10+x11+x12+x13+x14+x15+x16+x17+x18+x19+x110+x111+x112+x113
+x114+x115+x21+x23+x24+x25+x26+x27+x29+x211+x212+x213+x214
+x215+x30+x31+x32+x33+x35+x37+x38+x39+x310+x311+x313+x315
+x41+x43+x44+x45+x46+x47+x49+x411+x412+x413+x414+x415,
{x10+x15+x110+x115+x20+x21+x22+x23>=5*d11,
 x10+x15+x110+x115+x20+x21+x22+x23<=8*d11,
 x14+x19+x114+x13+x24+x25+x26+x27>=5*d12,
 x14+x19+x114+x13+x24+x25+x26+x27<=8*d12,
 x18+x113+x12+x17+x28+x29+x210+x211>=5*d13,
 x18+x113+x12+x17+x28+x29+x210+x211<=8*d13,
 x112+x11+x16+x111+x212+x213+x214+x215>=5*d14,
 x112+x11+x16+x111+x212+x213+x214+x215<=8*d14,
 x20+x25+x210+x215+x30+x31+x32+x33>=5*d21,
 x20+x25+x210+x215+x30+x31+x32+x33<=8*d21,
  x24+x29+x214+x23+x34+x35+x36+x37>=5*d22,
 x24+x29+x214+x23+x34+x35+x36+x37<=8*d22,
 x28+x213+x22+x27+x38+x39+x310+x311>=5*d23,
 x28+x213+x22+x27+x38+x39+x310+x311<=8*d23,
 x212+x21+x26+x211+x312+x313+x314+x315>=5*d24,
 x212+x21+x26+x211+x312+x313+x314+x315<=8*d24,
 x30+x35+x310+x315+x40+x41+x42+x43>=5*d31,
 x30+x35+x310+x315+x40+x41+x42+x43<=8*d31,
 x34+x39+x314+x33+x44+x45+x46+x47>=5*d32,
 x34+x39+x314+x33+x44+x45+x46+x47<=8*d32,
 x38+x313+x32+x37+x48+x49+x410+x411>=5*d33,
 x38+x313+x32+x37+x48+x49+x410+x411<=8*d33,
 x312+x31+x36+x311+x412+x413+x414+x415>=5*d34,
 x312+x31+x36+x311+x412+x413+x414+x415<=8*d34,
```

```
x14+x16+x112+x114+x10+x11+x12+x13+x111+x110+x15+x17+x18+x19+x113+x115>=1,
x20+x22+x28+x210+x34+x36+x312+x314+x40+x42+x48+x410<=n,
x20+x22+x28+x210+x34+x36+x312+x314+x40+x42+x48+x410>=n
},assume=binary);
```

C Case #10: $[l_0, l_1, l_2, l_3] = [2, 4, 0, 4]$ with $\chi(\Delta X_{1,4}) = \chi(\Delta X_{1,14}) = 1$

The type of a differential characteristic is proposed in Fig. 8. The distribution of active S-boxes in these rounds is $9 \to 6 \to 4 \to 6$, totally 25 active S-boxes. In Fig. 8, from ΔX_1 to ΔZ_4, squares marked with broken line are active, squares marked with backslash should be chosen to satisfy some conditions, and empty squares have no difference.

We denote by MC the matrix used in the `MixColumns` layer. Based on the MDS property of matrix MC, once any four out of the eight differences before and after the matrix MC are given, then another four differences are uniquely determined and can be calculated efficiently.

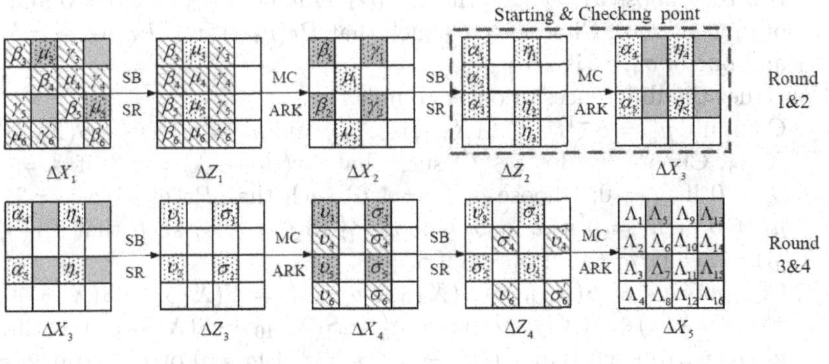

Fig. 8. A differential characteristic with $[l_0, l_1, l_2, l_3] = [2, 4, 0, 4]$ and $\chi(\Delta X_{1,4}) = \chi(\Delta X_{1,14}) = 1$. Gray squares denote leaked bytes. Squares marked with broken line are active, squares marked with backslash should be chosen to satisfy some conditions, and empty squares have no difference.

Now, we specify the differential characteristic following the type of Fig. 8. From ΔX_1 to ΔZ_4, bytes without a Greek alphabet have difference zero, and the difference of a byte with a Greek alphabet (i.e., α, β, γ, η, μ, ν and σ) will be determined in the subsequent discussions. Since $\Delta X_5 = MC(\Delta Z_4)$, we obtain the values of Λ_j $(1 \le j \le 16)$ once ν_i's and σ_i's $(3 \le i \le 6)$ are determined. The procedure of constructing this differential characteristic is given as follows.

1. Construct a differential characteristic from ΔX_2 to ΔZ_3.
   1-1. We start at the `MixColumns` layer of round 2, and match the differences $(\alpha_1, \alpha_2, \ldots, \alpha_5)$ first (see the starting point of Fig. 8). That is, we have to choose nonzero $\alpha_1, \alpha_2, \alpha_3, \alpha_4$ and α_5 such that $(\alpha_4, 0, \alpha_5, 0) = (\alpha_1, \alpha_2, \alpha_3, 0) \cdot MC^t$. This is done by choosing an arbitrary difference $\alpha_1 \neq 0$ and computing $(\alpha_2, \alpha_3, \alpha_4, \alpha_5) = (4\alpha_1, 7\alpha_1, 9\alpha_1, B\alpha_1)$.
   1-2. Compute $\beta_1 = S^{-1}(\alpha_1 \oplus S(X_{2,0})) \oplus X_{2,0}$ and $\gamma_2 = S^{-1}(\alpha_3 \oplus S(X_{2,10})) \oplus X_{2,10}$.
   1-3. Choose β_2 such that one of $\beta_3, \ldots, \beta_6$ is zero, where $(\beta_3, \beta_4, \beta_5, \beta_6)^t = MC^{-1} \cdot (\beta_1, 0, \beta_2, 0)^t$. Thus, $\beta_2 \in \{D^{-1}E\beta_1, B^{-1}9\beta_1, E^{-1}D\beta_1, 9^{-1}B\beta_1\}$. Similarly, choose γ_1 such that one of $\gamma_3, \ldots, \gamma_6$ is zero. Thus, $\gamma_1 \in \{E^{-1}D\gamma_2, 9^{-1}B\gamma_2, D^{-1}E\gamma_2, B^{-1}9\gamma_2\}$.
   1-4. Compute $\eta_1 = S(X_{2,8}) \oplus S(X_{2,8} \oplus \gamma_1)$ and $\eta_2 = S(X_{2,2}) \oplus S(X_{2,2} \oplus \beta_2)$. Now, we have to check whether there are nonzero η_3, η_4 and η_5 such that $(\eta_4, 0, \eta_5, 0) = (\eta_1, 0, \eta_2, \eta_3) \cdot MC^t$. It is equivalent to check whether $\eta_1 = 7\eta_2$ (see the checking point of Fig. 8).
   1-5. If there is a $(\alpha_1, \beta_2, \gamma_1)$ such that $\eta_1 = 7\eta_2$, compute $(\eta_3, \eta_4, \eta_5) = (4\eta_2, B\eta_2, 9\eta_2)$ and go on. Else, return "construction failure" and abort.
   1-6. Choose μ_1, μ_2 such that $Pr(\mu_1 \to \alpha_2) \cdot Pr(\mu_2 \to \eta_3) \neq 0$ and one of μ_4, μ_6 is zero; Choose ν_1, ν_2 such that $Pr(\alpha_4 \to \nu_1) \cdot Pr(\eta_5 \to \nu_2) \neq 0$ and one of ν_4, ν_6 is zero; Choose σ_1, σ_2 such that $Pr(\eta_4 \to \sigma_1) \cdot Pr(\alpha_5 \to \sigma_2) \neq 0$ and one of σ_4, σ_6 is zero.
2. Construct the differences of outer rounds.
   2-1. Compute $\mu'_3 = S^{-1}(\mu_3 \oplus S(X_{1,4})) \oplus X_{1,4}$ and $\mu'_5 = S^{-1}(\mu_5 \oplus S(X_{1,14})) \oplus X_{1,14}$. Choose β'_i $(3 \leq i \leq 6)$ such that $Pr(\beta'_i \to \beta_i) = 2^{-6}$ if $\beta_i \neq 0$ or $\beta'_i = 0$ if $\beta_i = 0$; Choose μ'_i $(i = 4, 6)$ such that $Pr(\mu'_i \to \mu_i) = 2^{-6}$ if $\mu_i \neq 0$ or $\mu'_i = 0$ if $\mu_i = 0$; Choose γ'_i $(3 \leq i \leq 6)$ such that $Pr(\gamma'_i \to \gamma_i) = 2^{-6}$ if $\gamma_i \neq 0$ or $\gamma'_i = 0$ if $\gamma_i = 0$.
   2-2. Compute $\nu'_3 = S(X_{4,0}) \oplus S(X_{4,0} \oplus \nu_3)$, $\nu'_5 = S(X_{4,2}) \oplus S(X_{4,2} \oplus \nu_5)$, $\sigma'_3 = S(X_{4,8}) \oplus S(X_{4,8} \oplus \sigma_3)$ and $\sigma'_5 = S(X_{4,10}) \oplus S(X_{4,10} \oplus \sigma_5)$. Choose ν'_i $(i = 4, 6)$ such that $Pr(\nu'_i \to \nu_i) = 2^{-6}$ if $\nu_i \neq 0$ or $\nu'_i = 0$ if $\nu_i = 0$; Choose σ'_i $(i = 4, 6)$ such that $Pr(\sigma'_i \to \sigma_i) = 2^{-6}$ if $\sigma_i \neq 0$ or $\sigma'_i = 0$ if $\sigma_i = 0$.
3. Compute $\Delta X_5 = MC(\Delta Z_4)$.

Notice that 9 effective active S-boxes in Round 1 and 4 can always reach the maximum differential probability 2^{-6}. Thus, the probability of this differential characteristic is between $2^{-7 \cdot 6 - 9 \cdot 6} = 2^{-96}$ and $2^{-15 \cdot 6} = 2^{-90}$ if it exists. The existence of this differential characteristic is only related to the existence of a differential characteristic in Round 2 and 3. Two questions **Q1** and **Q2** are experimentally verified to ensure the existence of a differential characteristic from ΔX_2 to ΔZ_3:

Q1: For each $X = (X_{2,0}, X_{2,2}, X_{2,8}, X_{2,10})$, can we find a triple $(\alpha_1, \beta_2, \gamma_1)$ in step 1-1 and step 1-3 such that the condition $\eta_1 = 7\eta_2$ in step 1-4 is satisfied?

For each X, it's very likely to find such a triple, because the choices of $(\alpha_1, \beta_2, \gamma_1)$ are about 2^{12} and the probability of $\eta_1 = 7\eta_2$ is about 2^{-8}. We enumerate all 2^{32} values of X and find that the number of "construction failure" is 402, that is, there is at least one $(\alpha_1, \beta_2, \gamma_1)$ such that $\eta_1 = 7\eta_2$ for $2^{32} - 402$ out of 2^{32} X. For each of these X, we may store a candidate of $(\alpha_1, \beta_2, \gamma_1)$ in a table, which is indexed by the value of X (A redundant triple pair $(0, 0, 0)$ may be included for failure cases). The size of this table is 3×2^{32} bytes. The time complexity of this step is at most 2^{44}.

Q2: For any nonzero (α_2, η_3) (resp. (α_4, η_5) and (η_4, α_5)), can we find a pair of (μ_1, μ_2) (resp. (ν_1, ν_2) and (σ_1, σ_2)) which satisfies the conditions given in step 1-6?

Notice that (α_2, η_3) has 255^2 choices, μ_1 and μ_2 have 127 choices once (α_2, η_3) is given. Thus, **Q2** can be verified in time complexity of about 2^{30}. For a given (α_2, η_3), more than one pair of (μ_1, μ_2) may be found to satisfy the condition given in step 1-6. In this case, we choose the pair (μ_1, μ_2) such that $Pr(\mu_1 \to \alpha_2) \cdot Pr(\mu_2 \to \eta_3)$ is maximum. Experimental results show that the condition given in step 1-6 can be satisfied for each pair of (α_2, η_3), and the maximum probability of $Pr(\mu_1 \to \alpha_2) \cdot Pr(\mu_2 \to \eta_3)$ is 2^{-14}, 2^{-13} and 2^{-12} for 3825, 60690 and 510 pairs of (α_2, η_3), respectively. The average probability of $Pr(\mu_1 \to \alpha_2) \cdot Pr(\mu_2 \to \eta_3)$ is $2^{-13.03}$. Similarly, the condition given in step 1-6 can be satisfied for each pair of (α_4, η_5) (resp. (η_4, α_5)), and the maximum probability of $Pr(\alpha_4 \to \nu_1) \cdot Pr(\eta_5 \to \nu_2)$ (resp. $Pr(\eta_4 \to \sigma_1) \cdot Pr(\alpha_5 \to \sigma_2)$) is 2^{-14}, 2^{-13} and 2^{-12} for 4312, 60203 and 510 pairs of (α_4, η_5) (resp. (η_4, α_5)), respectively. The average probability of $Pr(\alpha_4 \to \nu_1) \cdot Pr(\eta_5 \to \nu_2)$ (resp. $Pr(\eta_4 \to \sigma_1) \cdot Pr(\alpha_5 \to \sigma_2)$) is $2^{-13.04}$. The best choices of (μ_1, μ_2) and (ν_1, ν_2) (resp. (σ_1, σ_2)) can be stored in two tables.

Thus, the probability of a four-round differential characteristic proposed in this subsection is $2^{-6 \cdot 9 - 13.03 - 2 \cdot 13.04} \approx 2^{-93.1}$ on average. Notice that it always exists and can be easily rebuilt by looking up several tables.

Similar process is done to case #5 to #12 except case #10. Two questions similar to **Q1** and **Q2** are also experimentally verified to check the existence of these differential characteristics. To answer question **Q1**, 2^{32} values of $X = (X_{3,4}, X_{3,6}, X_{3,12}, X_{3,14})$ are enumerated for case #5 to case #8, and 2^{32} values of $X = (X_{2,0}, X_{2,2}, X_{2,8}, X_{2,10})$ are enumerated for case #9, #11 and #12. The number of "construction failure" is 412 for case #5 and #6, 443 for case #7 and #8, 402 for case #9, and 373 for case #11 and #12, respectively. Experimental results show that question **Q2** always can be satisfied. Therefore, we can construct these differential characteristics for almost all cases of the leaked X. The probabilities of these 7 differential characteristics are around $2^{-93.1}$ with a small deviation.

D Case #4: $[l_0, l_1, l_2, l_3] = [4, 2, 0, 4]$ with $\chi(\Delta X_{2,0}) = \chi(\Delta X_{2,2}) = 1$

The type of a differential characteristic is illustrated in Fig. 9. The distribution of active S-boxes in these rounds is $9 \to 6 \to 4 \to 6$, totally 25 active S-boxes. In Fig. 9, from ΔX_1 to ΔZ_4, squares marked with broken line are active, squares marked with backslash should be chosen to satisfy some conditions, and empty squares have no difference.

From ΔX_1 to ΔZ_4, bytes without a Greek alphabet have difference zero, and the difference of a byte with a Greek alphabet (i.e., α, β, γ, η, μ, ν and σ) will be determined in the subsequent discussions. Since $\Delta X_5 = MC(\Delta Z_4)$, we obtain the value of Λ_j $(1 \leq j \leq 16)$ once ν_i' and σ_i' $(3 \leq i \leq 6)$ are determined. The procedure of constructing this differential characteristic is briefly described as follows.

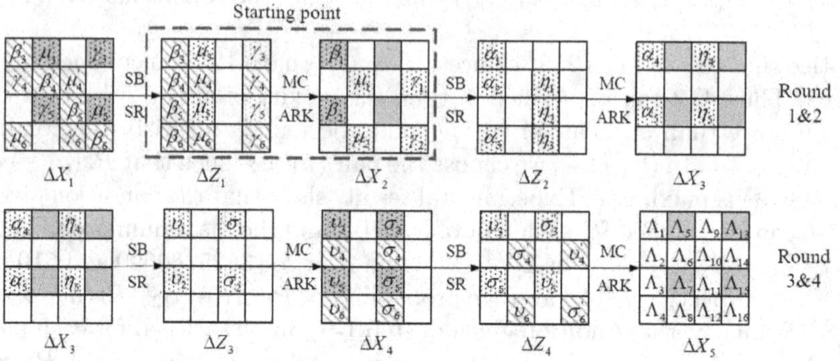

Fig. 9. Differential characteristics with $[l_0, l_1, l_2, l_3] = [4, 2, 0, 4]$ and $\chi(\Delta X_{2,0}) = \chi(\Delta X_{2,2}) = 1$. Gray squares denote leaked bytes. Squares marked with broken line are active, squares marked with backslash should be chosen to satisfy some conditions, and empty squares have no difference.

1. We start at the MC step of Round 1 here, and choose nonzero β_1 and β_2 such that one of $\beta_3, \ldots, \beta_6$ is zero, where $(\beta_3, \beta_4, \beta_5, \beta_6)^t = MC^{-1} \cdot (\beta_1, 0, \beta_2, 0)^t$. Thus, for arbitrary $\beta_1 \neq 0$, we can choose $\beta_2 \in \{D^{-1}E\beta_1, B^{-1}9\beta_1, E^{-1}D\beta_1, 9^{-1}B\beta_1\}$. $\beta_3, \ldots, \beta_6$ are obtained once β_1 and β_2 are determined. Notice that we have 4 choices of β_2 for each $\beta_1 \neq 0$.
2. Compute α_1 and η_2 using the pair $(X_{2,0}, \beta_1)$ and $(X_{2,2}, \beta_2)$, respectively.
3. Compute $\alpha_2, \ldots, \alpha_5$ by solving $(\alpha_4, 0, \alpha_5, 0) = (\alpha_1, \alpha_2, 0, \alpha_3) \cdot MC^t$; Compute η_1, η_3, η_4 and η_5 by solving $(\eta_4, 0, \eta_5, 0) = (0, \eta_1, \eta_2, \eta_3) \cdot MC^t$.
4. Choose (μ_1, μ_2) (resp. (γ_1, γ_2)) such that $Pr(\mu_1 \to \alpha_2) \cdot Pr(\mu_2 \to \eta_3) \neq 0$ (resp. $Pr(\gamma_1 \to \eta_1) \cdot Pr(\gamma_2 \to \alpha_3) \neq 0$) and one of μ_4 and μ_6 (resp. γ_4 and γ_6) is zero. Choose (ν_1, ν_2) (resp. (σ_1, σ_2)) such that $Pr(\alpha_4 \to \nu_1) \cdot Pr(\eta_5 \to \nu_2) \neq 0$ (resp. $Pr(\eta_4 \to \sigma_1) \cdot Pr(\alpha_5 \to \delta_2) \neq 0$) and one of ν_4 and ν_6 (resp. δ_4 and δ_6) is zero.
5. Compute μ_3', μ_5', γ_3' and γ_5' using the pair $(X_{1,4}, \mu_3)$, $(X_{1,14}, \mu_5)$, $(X_{1,12}, \gamma_3)$ and $(X_{1,6}, \gamma_5)$, respectively. Choose β_i' $(3 \leq i \leq 6)$ such that $Pr(\beta_i' \to \beta_i) = 2^{-6}$ if $\beta_i \neq 0$ or $\beta_i' = 0$ if $\beta_i = 0$; Choose μ_i' $(i = 4, 6)$ such that $Pr(\mu_i' \to \mu_i) = 2^{-6}$ if $\mu_i \neq 0$ or $\mu_i' = 0$ if $\mu_i = 0$; Choose γ_i' $(i = 4, 6)$ such that $Pr(\gamma_i' \to \gamma_i) = 2^{-6}$ if $\gamma_i \neq 0$ or $\gamma_i' = 0$ if $\gamma_i = 0$.
6. Compute ν_3', ν_5', σ_3' and σ_5' using the pair $(X_{4,0}, \nu_3)$, $(X_{4,2}, \nu_5)$, $(X_{4,8}, \sigma_3)$ and $(X_{4,10}, \sigma_5)$ respectively. Choose ν_i' $(i = 4, 6)$ such that $Pr(\nu_i' \to \nu_i) = 2^{-6}$ if

$\nu_i \neq 0$ or $\nu'_i = 0$ if $\nu_i = 0$; Choose σ'_i $(i = 4, 6)$ such that $Pr(\sigma'_i \to \sigma_i) = 2^{-6}$ if $\sigma_i \neq 0$ or $\sigma'_i = 0$ if $\sigma_i = 0$.

7. Compute $\Delta X_{r+4} = MC(\Delta Z_4)$.

The existence of these differential characteristics is only related to the existence of pairs (μ_1, μ_2), (γ_1, γ_2), (ν_1, ν_2) and (σ_1, σ_2) in step 4. Based on the experimental results given in the construction of Fig. 8, they always exist. Thus, we have $255 \times 4 = 1020$ differential characteristics here because β_1 has 255 choices and β_2 has four choices for each β_1. The average probability of them is $2^{-6\cdot 7-13.03\cdot 2-13.04\cdot 2} = 2^{-94.1}$.

E Details of One Forgery in the "2–8–12–4" Experiment

The initial state is: $0x7745fe4fa948da9$.

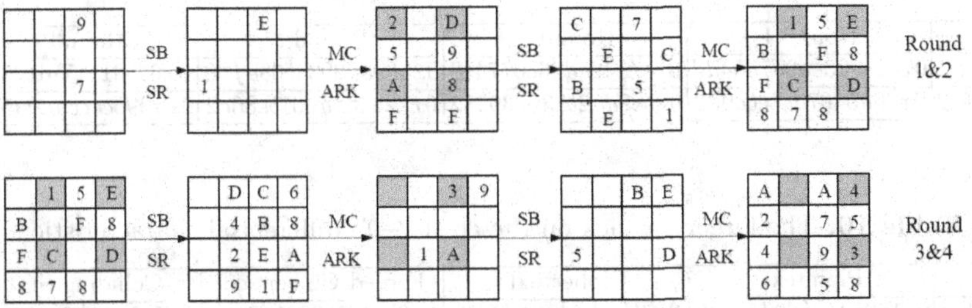

Fig. 10. Differential Path of type "2–8–12–4". The hexadecimal numbers indicate the difference values. The empty squares indicate there is no difference. The squares of leaked bytes are marked with gray color.

Table 7. The values of round keys

	Round 1	Round 2	Round 3	Round 4
Block 1	$0x27de69bc8bbc6a71$	$0x0eda00f69a70d28f$	$0xcaa2cab4fb3cf8a8$	$0x8034f88c57ed2766$
Block 2	$0xb9cacf23fb387dd8$	$0xe9d293e0d9550016$	$0x7537baeca8ed970e$	$0xe1c9150ac5564aad$

F Details of one Forgery in the "6–4–6–9" Experiment

The initial state is: $0x92304e6d9b7c7373$.

Table 8. The forgery attack on the "2–8–12–4" differential characteristic

	Plaintext	Ciphertext	Forged Ciphertext	Colliding State
Block 1	$0x37dc069161450099$	$0x6c2b36071e45d85d$	$0x6cbb36071e35d85d$	$0xb23d4f8eeb91a13e$
Block 2	$0xb1469433d739a810$	$0x39d7ac987dd694a8$	$0x53ba102c0d1b4435$	

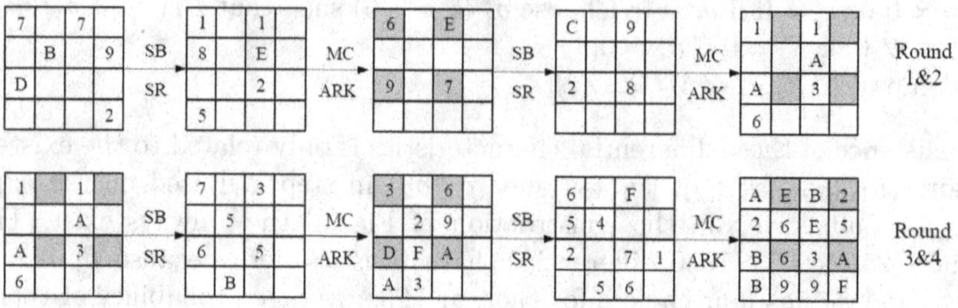

Fig. 11. Differential Path of type "6–4–6–9". The hexadecimal numbers indicate the difference values. The empty squares indicate there is no difference. The squares of leaked bytes are marked with gray color.

Table 9. The values of round keys

	Round 1	Round 2	Round 3	Round 4
Block 1	0x60ee23ea2d7054dd	0xcf849ed86e6774c0	0x569d49934b68af00	0x64b01cb5561255c8
Block 2	0x36a5467dc8ebe9d2	0xbe9da2b83ae39382	0x724461aa61be86e2	0xa396ceccaa9d57f6

Table 10. The forgery attack on the "6–4–6–9" differential characteristic

	Plaintext	Ciphertext	Forged Ciphertext	Colliding State
Block 1	0x182841a869f5e890	0x7bb0dce1e61d0d43	0x0bc0d7e8361d0d41	0xf134343fa5b20472
Block 2	0x35bdb2a519a0818f	0xa3398abfcd7fcd1d	0x646cac5a462f92a8	

A Modular Framework for Building Variable-Input-Length Tweakable Ciphers

Thomas Shrimpton and R. Seth Terashima

Dept. of Computer Science, Portland State University
{teshrim,seth}@cs.pdx.edu

Abstract. We present the Protected-IV construction (PIV) a simple, modular method for building variable-input-length tweakable ciphers. At our level of abstraction, many interesting design opportunities surface. For example, an obvious pathway to building beyond birthday-bound secure tweakable ciphers with performance competitive with existing birthday-bound-limited constructions. As part of our design space exploration, we give two fully instantiated PIV constructions, TCT_1 and TCT_2; the latter is fast and has beyond birthday-bound security, the former is faster and has birthday-bound security. Finally, we consider a generic method for turning a VIL tweakable cipher (like PIV) into an authenticated encryption scheme that admits associated data, can withstand nonce-misuse, and allows for multiple decryption error messages. Thus, the method offers robustness even in the face of certain sidechannels, and common implementation mistakes.

Keywords: tweakable blockciphers, beyond-birthday-bound security, authenticated encryption, associated data, full-disk encryption.

1 Introduction

The main contribution of this paper is the Protected-IV construction (PIV), see Figure 1. PIV offers a simple, modular method for building length-preserving, tweakable ciphers that:

(1) may take plaintext inputs of essentially any length;
(2) provably achieves the strongest possible security property for this type of primitive, that of being a strong, tweakable-PRP (STPRP);
(3) admit instantiations from n-bit primitives that are STPRP-secure well beyond the birthday-bound of $2^{n/2}$ invocations.

Moreover, by some measures of efficiency, beyond-birthday secure instantiations of PIV are competitive with existing constructions that are only secure to the birthday bound. (See Table 1.) We will give a concrete instantiation of PIV that has beyond birthday-bound security and, when compared to EME [16], the overhead is a few extra modular arithmetic operations for each n-bit block of input.

K. Sako and P. Sarkar (Eds.) ASIACRYPT 2013 Part I, LNCS 8269, pp. 405–423, 2013.

Tweakable ciphers with beyond birthday-bound security may have important implications for cryptographic practice. For example, in large-scale data-at-rest settings, where the amount of data that must be protected by a single key is typically greater than in settings where keys can be easily renegotiated.

At least two important applications have already made tweakable ciphers their tool-of-choice, namely full-disk encryption (FDE) and format-preserving encryption (FPE). Our work provides interesting new results for both FDE and FPE.

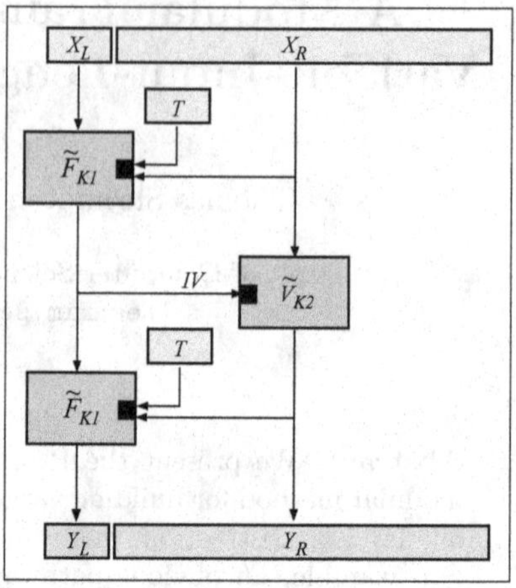

Fig. 1. The $\mathsf{PIV}[\widetilde{F}, \widetilde{V}]$ tweakable cipher. Input T is the tweak, and $X = X_L \parallel X_R$ is a bit string, where $|X_L| = N$ and X_R is any length accepted by $\widetilde{V}$. The filled-in box is the tweak input.

We also show that tweakable ciphers enable a simple mechanism for building authenticated encryption schemes with associated data (AEAD), via an extension of the encode-then-encipher approach of Bellare and Rogaway [4]. This approach has some practical benefits, for example, it securely handles the reporting of multiple types of decryption errors. It can also eliminate ciphertext expansion by exploiting any existing nonces, randomness, or redundancies appearing in either the plaintext or associated data inputs. Combined with our other results, encode-then-encipher over PIV gives a new way to build AEAD schemes with beyond birthday-bound security.

Background. Tweakable blockciphers (TBCs) were introduced and formalized by Liskov, Rivest and Wagner [20]. An n-bit TBC $\widetilde{E}$ is a family of permutations over $\{0,1\}^n$, each permutation named by specifying a key and a *tweak*. In typical usage, the key is secret and fixed across many calls, while the tweak is not secret, and may change from call to call; this allows variability in the behavior of the primitive, even though the key is fixed. A tweakable cipher[1] is the natural extension of a tweakable blockcipher to the variable-input-length (VIL) setting, forming a family of length-preserving permutations.

Since the initial work of Liskov, Rivest and Wagner, there has been substantial work on building tweakable ciphers. Examples capable of handling long inputs (required for FDE) include CMC [15], EME [16], HEH [30], HCH [10], and HCTR [33]. Loosely speaking, the common approach has been to build up the VIL primitive from an underlying n-bit blockcipher, sometimes in concert with one or more hashing operations. The security guaranteed by each of these constructions become vacuous after about $2^{n/2}$ bits have been enciphered. One of

[1] Sometimes called a "tweakable enciphering scheme", or even a "large-block cipher".

our main goals is to break through this birthday bound, i.e., to build a tweakable cipher that remains secure long after $2^{n/2}$ bits have been enciphered.

The PIV *construction.* To this end, we begin by adopting a top-down, compositional viewpoint on the design of tweakable ciphers, our PIV construction. It is a type of three-round, unbalanced Feistel network, where the left "half" of the input is of a fixed bit length N, and the right "half" has variable length. The first and third round-functions are an N-bit tweakable blockcipher ($\widetilde{F}$), where N is a parameter of the construction, e.g. $N = 128$ or $N = 256$. The middle round-function ($\widetilde{V}$) is itself a VIL tweakable cipher, whose tweak is the output of first round.

It may seem as though little has been accomplished, since we need a VIL tweakable cipher $\widetilde{V}$ in order to build our VIL tweakable cipher $\mathsf{PIV}[\widetilde{F}, \widetilde{V}]$. However, we require substantially less of $\widetilde{V}$ than we do of $\mathsf{PIV}[\widetilde{F}, \widetilde{V}]$. In particular, the target security property for PIV is that of being a strong tweakable pseudorandom permutation. Informally, being STPRP-secure means withstanding chosen-ciphertext attacks in which the attacker also has full control over all inputs. The attacker can, for example, repeat a tweak an arbitrary number of times. Our PIV security theorem (Theorem 1) says the following: given (1) a TBC $\widetilde{F}$ that is STPRP-secure over a domain of N-bit strings, and (2) a tweakable cipher $\widetilde{V}$ that is secure against attacks *that never repeat a tweak*, then the tweakable cipher $\mathsf{PIV}[\widetilde{F}, \widetilde{V}]$ is STPRP-secure. Thus, qualitatively, the PIV construction promotes security (over a large domain) against a restricted kind of attacker, into security against arbitrary chosen-ciphertext attacks.

Quantitatively, the PIV security bound contains an additive term $q^2/2^N$, where q is the number of times PIV is queried. Now, N might be the blocksize n of some underlying blockcipher; in this case the PIV composition delivers a bound comparable to those achieved by existing constructions. But $N = 2n$ presents the possibility of using an n-bit primitive to instantiate $\widetilde{F}$ and $\widetilde{V}$, and yet deliver a tweakable cipher with security well beyond beyond-birthday of $2^{n/2}$ queries.

As a small, additional benefit, the PIV proof of STPRP-security is short and easy to verify.

Impacts of modularity on instantiations. Adopting this modular viewpoint allows us to explore constructions of $\widetilde{F}$ and $\widetilde{V}$ independently. This is particularly beneficial, since building efficient and secure instantiations of VIL tweakable ciphers ($\widetilde{V}$) is relatively easy, when tweaks can be assumed not to repeat. The more difficult design task, of building a tweakable blockcipher ($\widetilde{F}$) that remains secure when tweaks may be repeated, is also made easier, by restricting to plaintext inputs of a fixed bit length N. In practice, when (say) $N = 128$ or 256, inefficiencies incurred by $\widetilde{F}$ can be offset by efficiency gains in $\widetilde{V}$.

To make thing concrete, we give two fully-specified PIV tweakable ciphers, each underlain by n-bit blockciphers. The first, TCT_1, provides birthday-bound security. It requires only one blockcipher invocation and some arithmetic, modulo

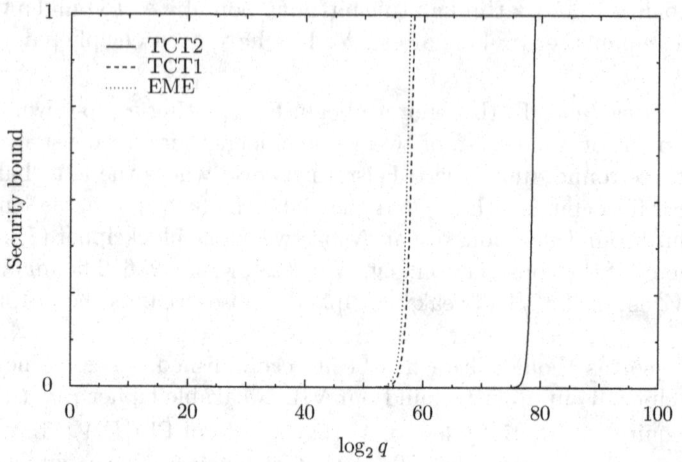

Fig. 2. Security bounds for TCT_1, EME and TCT_2, all using an underlying 128-bit primitive and 4096-byte inputs, typical for FDE. The EME curve is representative of other prior constructions.

a power of two, per n-bit block of input. In contrast, previous modes either require two blockcipher invocations per n-bit block, or require per-block finite field operations.

The second, TCT_2, delivers security beyond the birthday-bound. When compared to existing VIL tweakable ciphers with only birthday-bound security, like EME* construction, TCT_2 incurs only some additional, simple arithmetic operations per n bit block of input. Again, this arithmetic is performed modulo powers of two, rather than in a finite field.

In both TCT_1 and TCT_2, the VIL component is instantiated using counter-mode encryption, but over a TBC instead of a blockcipher. The additional tweak input of the TBC allows us to consider various 'tweak-scheduling' approaches, e.g. fixing a single per-message tweak across all blocks, or changing the tweak each message block.[2] We will see that the latter approach of re-tweaking on a block-by-block basis leads to a beyond birthday-bound secure PIV construction that admits strings of any length at least N.

AEAD via encode-then-(tweakable)encipher. The ability to construct beyond birthday-bound secure tweakable ciphers with large and flexible domains motivates us to consider their use for traditional encryption. Specifically, we build upon the "encode-then-encipher" results of Bellare and Rogaway [4]. They show that messages endowed with randomness (or nonces) and redundancy do not need to be processed by a authenticated encryption (AE) scheme in order to

[2] There is a natural connection between changing the tweak of a TBC, and changing the key of a blockcipher. Both can be used to boost security, but the former is cleaner because tweaks do not need to be secret.

enjoy privacy and authenticity guarantees; a VIL strong-PRP suffices. This is valuable when typical messages are short, as there is no need to waste bandwidth upon transmitting an AE scheme's IV and a dedicated authenticity tag.

We find that the tweakable setting gives additional advantages to the encode-then-encipher approach. An obvious one is that the tweak empowers support for associated data. More interesting, one can explore the effects of randomness, state or redundancy present in the message *and* tweak inputs. We find that randomness and state can be shifted to from the message to the tweak without loss of security, potentially reducing the number of bits that must be processed cryptographically.

We also find that AEAD schemes are built this way, via encode-then-encipher over a tweakable cipher, can accommodate multiple decryption error messages. Multiple, descriptive error messages can be quite useful in practice, but have often empowered damaging attacks (e.g. padding-oracle attacks [32,7,27,1,12]). These attacks don't work against our AEAD schemes because, loosely, changing any bit of a ciphertext will randomize every bit of the decrypted string.

Our work in this direction suggests useful implications for FPE [3,5], and for layered-encryption schemes, for example the onion-encryption scheme used by Tor [23].

Due to space limitations, we refer the reader to the full version of this paper for our results on AEAD, and a discussion of their potential impacts.

Related work. Here we give a much abbreviated discussion of other related work. Please refer to Table 1 for a summary comparison of $\mathsf{TCT}_1, \mathsf{TCT}_2$ with other constructions. A more complete discussion will appear in the full version.

Researchers have developed three general approach for constructing tweakable ciphers from n-bit blockciphers. Each approach has yielded a series of increasingly refined algorithms. The first, Encrypt-Mask-Encrypt, places a light-weight "masking" layer between two encryption layers; examples include CMC [15] and EME* [13]. The second, Hash-ECB-Hash, sandwiches ECB-mode encryption between two invertible hashes. PEP [9], TET [14], and HEH [30,31] are examples. Finally, Hash-CTR-Hash uses non-invertible hashes with CTR-mode encryption. Both HCH [10] and HCTR [33] use this approach. Mancillas-Lópeze et al. [22] report on the hardware performance of most of these modes. Chakraborty et al. [8] discuss implementations of the more recent HEH [30] construction and its refinement [31], which halves the number of finite field multiplications.

We contribute a new, top-down approach that leads us to the first beyond-birthday-bound secure tweakable cipher suitable for encrypting long inputs (i.e., longer than the blocksize of an underlying blockcipher). Table 1 and Figure 2 compare some of these algorithms with our new TCT_1 and TCT_2 constructions in terms of computational cost and security, respectively. Note that the finite field operations counted in Table 1 take hundreds of cycles in software [21,2], whereas their cost relative to an AES blockcipher invocation is much lower in hardware [22]. TCT_1 is the first tweakable cipher to require only a single blockcipher invocation and no extra finite field multiplications for each additional n

Table 1. Tweakable ciphers and their computational costs for ℓn-bit inputs. Costs measured in n-bit blockcipher calls [BC], finite field multiplications [$\mathbb{F}_{2^n}\times$], and ring operations [$\mathbb{Z}_w+$] and [$\mathbb{Z}_{2^w}$], for some word size w. Typically, $\ell = 32$ for FDE, and we anticipate $n = 128$, $w = 64$.

	Cost				
Cipher	[BC]	[$\mathbb{F}_{2^n}\times$]	[$\mathbb{Z}_w+$]	[$\mathbb{Z}_{2^w}$]	Ref.
HCTR	ℓ	$2\ell+2$	–	–	[33]
CMC	$2\ell+1$	–	–	–	[15]
EME	$2\ell+1$	–	–	–	[16]
EME*	$2\ell+3$	–	–	–	[13]
PEP	$\ell+5$	$4\ell-6$	–	–	[9]
HCH	$\ell+3$	$2\ell-2$	–	–	[10]
TET	ℓ	2ℓ	–	–	[14]
HEH	$\ell+1$	$\ell+2$	–	–	[30,31]
TCT_1	$\ell+1$	5	$2\ell\left(\frac{n}{w}\right)^2$	$2\ell\left(\frac{n}{w}\right)^2$	–
TCT_2	$2\ell+8$	32	$4\ell\left(\frac{n}{w}\right)^2$	$4\ell\left(\frac{n}{w}\right)^2$	–

bits of input, while TCT_2 is the first to provide beyond-birthday-bound security (and still gets away with a fixed number of finite field multiplications).

We mention the LargeBlock constructions due to Minematsu and Iwata [25], since they provide ciphers with beyond-birthday-bound security. These do not support tweaking, but it seems plausible that they could without significant degradation of performance or security. These constructions overcome the birthday bound by using $2n$-bit blockciphers as primitives, which are in turn constructed from an n-bit TBC. To our knowledge, CLRW2 [19] is the most efficient n-bit TBC with beyond-birthday-bound security that supports the necessary tweakspace (Minematsu's TBC [24] limits tweak lengths to fewer than $n/2$ bits). Compared to TCT_2, instantiating the LargeBlock constructions with this primitive ultimately requires an extra six finite field multiplications for each n bits of input. Thus, we suspect the LargeBlock designs would be impractical even if adding tweak support proves feasible.

A construction due to Coron, et al. [11], which we refer to as CDMS (after the authors), builds a $2n$-bit TBC from an n-bit TBC, providing beyond-birthday-bound security in n. Like PIV, CDMS uses three rounds of a Feistel-like structure. However, our middle round uses a VIL tweakable cipher, and we require a weaker security property from the round. This allows PIV to efficiently process long inputs. That said, CDMS provides an excellent way to implement a highly-secure $2n$-bit TBC, and we will use it for this purpose inside of TCT_2 to build $\widetilde{F}$. (Nesting CDMS constructions could create $(2^m n)$-bit tweakable blockciphers for any $m > 1$, but again, this would not be practical). We note that Coron, et al. were primarily concerned with constructions indifferentiable from an ideal cipher, a goal quite different from ours.

The Thorp shuffle [26] and its successor, swap-or-not [17], are highly-secure ciphers targeting very small domains (e.g., $\{0,1\}^n$ for $n \le 64$). Swap-or-not could almost certainly become a VIL tweakable cipher, without changing the security bounds, by using domain separation for each input length and tweak in the underlying PRF. Essentially, one would make an input-length parameterized family of (tweakable) swap-or-not ciphers, with independent round-keys for each length. While still offering reasonable performance and unmatched security for very small inputs, the result would be wildly impractical for the large domains we are considering: swap-or-not's PRF needs to be invoked at least $6b$ times to securely encipher a b-bit input (below that, the bound becomes vacuous against even $q = 1$ query), and disk sectors are often 4096 bytes. Also, to match TCT_2's security, the PRF itself would need to be secure beyond the birthday bound (with respect to n).

Finally, we note that Rogaway and Shrimpton [29] considered some forms of tweakable encode-then-encipher in the context of deterministic AE ("keywrapping"), and our work generalizes theirs.

2 Tweakable Primitives

Preliminary notation. Let $\mathbb{N} = \{0, 1, 2, \ldots\}$ be the set of non-negative integers. For $n \in \mathbb{N}$, $\{0,1\}^n$ denotes the set of all n-bit binary strings, and $\{0,1\}^*$ denotes the set of all (finite) binary strings. We write ε for the empty string. Let $s, t \in \{0,1\}^*$. Then $|s|$ is the length of s in bits, and $|(s,t)| = |s \,\|\, t|$, where $s \| t$ denotes the string formed by concatenating s and t. If $s \in \{0,1\}^{nm}$ for some $m \in \mathbb{N}$, $s_1 s_2 \cdots s_m \xleftarrow{n} s$ indicates that each s_i should be defined so that $|s_i| = n$ and $s = s_1 s_2 \cdots s_m$. When n is implicit from context, it will be omitted from the notation. If $s = b_1 b_2 \cdots b_n$ is an n-bit string (each $b_i \in \{0,1\}$), then $s[i..j] = b_i b_{i+1} \cdots b_j$, $s[i..] = s[i..n]$, and $s[..j] = s[1..j]$. The string $s \oplus t$ is the bitwise xor of s and t; if, for example, $|s| < |t|$, then $s \oplus t$ is the bitwise xor of s and $t[..|s|]$. Given $R \subseteq \mathbb{N}$ and $n \in \mathbb{N}$ with $n \leq \min(R)$, $\{0,1\}^R = \bigcup_{i \in R} \{0,1\}^i$, and by abuse of notation, $\{0,1\}^{R-n} = \bigcup_{i \in R} \{0,1\}^{i-n}$. Given a finite set $\mathcal{X}$, we write $X \xleftarrow{\$} \mathcal{X}$ to indicate that the random variable X is sampled uniformly at random from $\mathcal{X}$. Throughout, the distinguished symbol $\perp$ is assumed not to be part of any set except $\{\perp\}$. Given an integer n known to be in some range, $\langle n \rangle$ denotes some fixed-length (e.g., 64-bit) encoding of n.

Let $H : \mathcal{K} \times \mathcal{D} \to \mathcal{R} \subseteq \{0,1\}^*$ be a function. Writing its first argument as a subscripted key, H is ϵ-almost universal (ϵ-AU) if for all distinct $X, Y \in \mathcal{D}$, $\Pr[\, H_K(X) = H_K(Y) \,] \leq \epsilon$ (where the probability is over $K \xleftarrow{\$} \mathcal{K}$). Similarly, H is ϵ-almost 2-XOR universal if for all distinct $X, Y \in \mathcal{D}$ and $C \in \mathcal{R}$, $\Pr[\, H_K(X) \oplus H_K(Y) = C \,] \leq \epsilon$.

An adversary is an algorithm taking zero or more oracles as inputs, which it queries in a black-box manner before returning some output. Adversaries may be random. The notation $A^f \Rightarrow b$ denotes the event that an adversary A outputs b after running with oracle f as its input.

Syntax. Let $\mathcal{K}$ be a non-empty set, and let $\mathcal{T}, \mathcal{X} \subseteq \{0,1\}^*$. A *tweakable cipher* is a mapping $\widetilde{E} : \mathcal{K} \times \mathcal{T} \times \mathcal{X} \to \mathcal{X}$ with the property that, for all $(K, T) \in \mathcal{K} \times \mathcal{T}$,

$\widetilde{E}(K,T,\cdot)$ is a permutation on $\mathcal{X}$. We typically write the first argument (the key) as a subscript, so that $\widetilde{E}_K(T,X) = \widetilde{E}(K,T,X)$. As $\widetilde{E}_K(T,\cdot)$ is invertible, we let $\widetilde{E}_K^{-1}(T,\cdot)$ denote this mapping. We refer to $\mathcal{K}$ as the *key space*, $\mathcal{T}$ as the *tweak space*, and $\mathcal{X}$ as the *message space*. We say that a tweakable cipher $\widetilde{E}$ is *length preserving* if $|\widetilde{E}_K(T,X)| = |X|$ for all $X \in \mathcal{X}$, $T \in \mathcal{T}$, and $K \in \mathcal{K}$. All tweakable ciphers in this paper will be length preserving. Restricting the tweak or message spaces of a tweakable cipher gives rise to other objects. When $\mathcal{X} = \{0,1\}^n$ for some $n > 0$, then $\widetilde{E}$ is a *tweakable blockcipher* with blocksize n. When $|\mathcal{T}| = 1$, we make the tweak implicit, giving a *cipher* $E\colon \mathcal{K} \times \mathcal{X} \to \mathcal{X}$, where $E_K(\cdot)$ is a (length-preserving) permutation over $\mathcal{X}$ and E_K^{-1} is its inverse. Finally, when $\mathcal{X} = \{0,1\}^n$ and $|\mathcal{T}| = 1$, we have a conventional *blockcipher* $E : \mathcal{K} \times \{0,1\}^n \to \{0,1\}^n$.

Security notions. Let Perm$(\mathcal{X})$ denote the set of all permutations on $\mathcal{X}$. Similarly, we define BC$(\mathcal{K},\mathcal{X})$ be the set of all ciphers with keyspace $\mathcal{K}$ and message space $\mathcal{X}$. When $\mathcal{X},\mathcal{X}'$ are sets, we define Func$(\mathcal{X},\mathcal{X}')$ to be the set of all functions $f\colon \mathcal{X} \to \mathcal{X}'$.

Fix a tweakable cipher $\widetilde{E}\colon \mathcal{K} \times \mathcal{T} \times \mathcal{X} \to \mathcal{X}$. We define the strong, tweakable pseudorandom-permutation (STPRP) advantage measure as $\mathbf{Adv}^{\mathrm{sprp}}_{\widetilde{E}}(A) = \Pr\left[K \xleftarrow{\$} \mathcal{K} :\, A^{\widetilde{E}_K(\cdot,\cdot),\widetilde{E}_K^{-1}(\cdot,\cdot)} \Rightarrow 1\right] - \Pr\left[\Pi \xleftarrow{\$} \mathrm{BC}(\mathcal{T},\mathcal{X}) :\, A^{\Pi(\cdot,\cdot),\Pi^{-1}(\cdot,\cdot)} \Rightarrow 1\right]$. The TPRP advantage measure is defined analogously, by dropping the $\widetilde{E}_K^{-1}$ oracle from the first probability, and the Π^{-1} oracle from the second. We assume that A never makes *pointless* queries. By this we mean that for the (S)TPRPexperiments, the adversary never repeats a query to an oracle. For the STPRP advantage measure, this also means that if A queries (T,X) to its leftmost oracle and receives Y in return, then it never queries (T,Y) to its rightmost oracle, and vice versa. These assumptions are without loss of generality.

The strong, indistinguishable-from-random-bits (SRND) advtantage is defined as $\mathbf{Adv}^{\mathrm{srnd}}_{\widetilde{E}}(A) = \Pr\left[K \xleftarrow{\$} \mathcal{K} :\, A^{\widetilde{E}_K(\cdot,\cdot),\widetilde{E}_K^{-1}(\cdot,\cdot)} \Rightarrow 1\right] - \Pr\left[A^{\$(\cdot,\cdot),\$(\cdot,\cdot)} \Rightarrow 1\right]$, where the $\$(\cdot,\cdot)$ oracle always outputs a random string equal in length to its second input: $|\$(T,X)| = |X|$ for all T and X. As before, we assume that A never makes a pointless query. Here, these assumptions are not without loss of generality, but instead prevent trivial wins. Adversaries for the (S)TPRP and SRND advantages are *nonce-respecting* if the transcript of their oracle queries $(T_1,X_1),\ldots,(T_q,X_q)$ does not include $T_i = T_j$ for any $i \neq j$.

For a cipher $E\colon \mathcal{K} \times \mathcal{X} \to \mathcal{X}$, we define the strong, pseudorandom permutation (SPRP) advantage as $\mathbf{Adv}^{\mathrm{sprp}}_{E}(A) = \Pr\left[K \xleftarrow{\$} \mathcal{K} :\, A^{E_K(\cdot),E_K^{-1}(\cdot)} \Rightarrow 1\right] - \Pr\left[\pi \xleftarrow{\$} \mathrm{Perm}(\mathcal{X}) :\, A^{\pi(\cdot),\pi^{-1}(\cdot)} \Rightarrow 1\right]$. As above, the PRP advantage is defined analogously, by dropping the E_K^{-1} oracle from the first probability, and the π^{-1} oracle from the second. We again assume (without loss of generality) that the adversary does not make pointless queries.

For all security notions in this paper, we track three adversarial resources: the time complexity t, the number of oracle queries q, and the total length of these queries μ. The time complexity of A is defined to include the complexity of its enveloping probability experiment (including sampling of keys, oracle computations, etc.), and we define the parameter t to be the maximum time complexity of A, taken over both experiments in the advantage measure.[3]

3 The PIV Construction

We begin by introducing our high-level abstraction, PIV, shown in Figure 1. Let $\mathcal{T} = \{0,1\}^t$ for some $t \geq 0$, and let $\mathcal{Y} \subseteq \{0,1\}^*$ be such that if $Y \in \mathcal{Y}$, then $\{0,1\}^{|Y|} \subseteq \mathcal{Y}$. Define $\mathcal{T}' = \mathcal{T} \times \mathcal{Y}$. Fix an integer $N > 0$. Let $\widetilde{F}\colon \mathcal{K}' \times \mathcal{T}' \times \{0,1\}^N \to \{0,1\}^N$ be a tweakable blockcipher and let $\widetilde{V}\colon \mathcal{K} \times \{0,1\}^N \times \mathcal{Y} \to \mathcal{Y}$ be a tweakable cipher. From these, we produce a new tweakable cipher $\mathsf{PIV}[\widetilde{F},\widetilde{V}]\colon (\mathcal{K}' \times \mathcal{K}) \times \mathcal{T} \times \mathcal{X} \to \mathcal{X}$, where $\mathcal{X} = \{0,1\}^N \times \mathcal{Y}$. As shown in Figure 1, the PIV composition of $\widetilde{F},\widetilde{V}$ is a three-round Feistel construction, working as follows. On input (T, X), let $X = X_L \parallel X_R$ where $|X_L| = N$. First, create an N-bit string $IV = \widetilde{F}_{K'}(T \parallel X_R, X_L)$. Next, use this IV to encipher X_R, creating a string $Y_R = \widetilde{V}_K(IV, X_R)$. Now create an N-bit string $Y_L = \widetilde{F}_{K'}(T \parallel Y_R, IV)$, and return $Y_L \parallel Y_R$ as the value of $\mathsf{PIV}[\widetilde{F},\widetilde{V}]_{K',K}(T, X)$. The inverse $\mathsf{PIV}[\widetilde{F},\widetilde{V}]^{-1}_{K',K}(T, Y)$ is computed in the obvious manner.

At first glance, it seems that nothing interesting has been accomplished: we took an N-bit TBC and a tweakable cipher, and produced a tweakable cipher with a slightly larger domain. However, the following theorem statement begins to surface what our abstraction delivers.

Theorem 1. *Let sets $\mathcal{T}, \mathcal{Y}, \mathcal{T}', \mathcal{X}$ and integer N be as above. Let $\widetilde{F}\colon \mathcal{K}' \times \mathcal{T}' \times \{0,1\}^N \to \{0,1\}^N$ be a tweakable blockcipher, and let $\widetilde{V}\colon \mathcal{K} \times \{0,1\}^N \times \mathcal{Y} \to \mathcal{Y}$ be a tweakable cipher. Let $\mathsf{PIV}[\widetilde{F},\widetilde{V}]$ be as just described. Let A be an adversary making $q < 2^N/4$ queries totaling μ bits and running in time t. Then there exist adversaries B and C, making q and $2q$ queries, respectively, and both running in $O(t)$ time such that $\mathbf{Adv}^{\mathrm{sprp}}_{\mathsf{PIV}[\widetilde{F},\widetilde{V}]}(A) \leq \mathbf{Adv}^{\mathrm{srnd}}_{\widetilde{V}}(B) + \mathbf{Adv}^{\mathrm{sprp}}_{\widetilde{F}}(C) + \frac{4q^2}{2^N}$, where B is* nonce-respecting *and whose queries total $\mu - qN$ bits in length.*

The first thing to notice is that the VIL portion of the PIV composition, $\widetilde{V}$, need be SRND-secure against *nonce-respecting* adversaries only. As we will see in the next section, it is easy to build efficient schemes meeting this requirement. Only the FIL portion, $\widetilde{F}$, needs to be secure against STPRP adversaries that can use arbitrary querying strategies. Thus the PIV composition promotes nonce-respecting security over a large domain into full STPRP security over a slightly larger domain.

The intuition for why this should work is made clear by the picture. Namely, if $\widetilde{F}$ is a good STPRP, then if any part of T or X is "fresh", then the string

[3] We do this simply to make our theorem statements easier to read. A more explicit accounting of time resources in reductions, e.g. separating the running time of A from the time to run cryptographic objects "locally", would not significantly alter any of our results.

IV should be random. Hence it is unlikely that an IV value is repeated, and so nonce-respecting security of the VIL component is enough. Likewise when deciphering, if any part of T, Y is "fresh".

The term $4q^2/2^N$ accounts for collisions in IV and the difference between $\widetilde{F}$ and a random function. This is a birthday-bound term in N, the blocksize of $\widetilde{F}$. Since most TBC designs employ (one or more) underlying blockciphers, we have deliberately chosen the notation N, rather than n, to stress that the blocksize of $\widetilde{F}$ can be larger than that of some underlying blockcipher upon which it might be built. Indeed, we'll see in the next section that, given an n-bit blockcipher (and a hash function), we can build $\widetilde{F}$ with $N = 2n$. This gives us hope of building beyond birthday-bound secure VIL STPRPs in a modular fashion; we will do so, and with relatively efficient constructions, too.

It will come as no surprise that, if one does away with the lower $\widetilde{F}$ invocation and returns $IV \parallel Y_R$, the resulting composition does not generically deliver a secure STPRP. On the other hand, it *is* secure as a TPRP (just not a *strong* TPRP). This can be seen through a straight-forward modification of the PIV security proof.

4 Concrete Instantiations of PIV

Instantiating a PIV composition requires two objects, a (fixed-input-length) tweakable blockcipher $\widetilde{F}$ with an N-bit blocksize, and a variable-input-length tweakable cipher $\widetilde{V}$. In this section we explore various ways to instantiate these two objects, under the guidance of Theorem 1 and practical concerns.

Theorem 1 suggests setting N to be as large as possible, so that the final term is vanishingly small for any realistic number of queries. But for this to be useful, one must already know how to build a TBC $\widetilde{F}$ with domain $\{0,1\}^N$ for a large N, and for which $\mathbf{Adv}^{\widetilde{\mathrm{sprp}}}_{\widetilde{F}}(C)$ approaches $q^2/2^N$. To our knowledge, there are no efficient constructions that permit $\mathbf{Adv}^{\widetilde{\mathrm{sprp}}}_{\widetilde{F}}(C)$ to be smaller than $\mathcal{O}(q^3/2^{2n})$ when using an n-bit blockcipher as a starting point. (A recent result by Lampe and Seurin [18] shows how to beat this security bound, but at a substantial performance cost.) A construction by Coron, et al., which will be discussed in more detail shortly, does meet this bound[4] while providing $N = 2n$.

So we restrict our attention to building TBC $\widetilde{F}$ with small N. In particular, we follow the common approach of building TBCs out of blockciphers. Letting n be the blockcipher blocksize, we will consider $N = n$, and $N = 2n$. In the former case, Theorem 1 only promises us security up to roughly $q = 2^{n/2}$, which is the birthday bound with respect to the blockcipher. With this security bound in mind, we can use simple and efficient constructions of both $\widetilde{F}$ and the VIL tweakable cipher $\widetilde{V}$. On the other hand, when $N = 2n$, Theorem 1 lets us hope for security to roughly $q = 2^n$ queries. To realize this hope we will need a bit

[4] However, nesting this construction to provide a VIL tweakable cipher is prohibitively inefficient.

more from both $\widetilde{F}$ and $\widetilde{V}$, but we will still find reasonably efficient constructions delivering beyond birthday bound security.

In what follows, we will sometimes refer to objects constructed in other works. These are summarized for convenience in Figure 5, found in Appendix A.

An efficient VIL tweakable cipher. We will start by considering general methods for constructing the VIL tweakable cipher, $\widetilde{V}$. Recall that $\widetilde{V}$ need only be secure against adversaries that never repeat a tweak. In Figure 3, we see an analogue of conventional counter-mode encryption, but over an n-bit TBC $\widetilde{E}$ instead of a blockcipher. Within a call (T, X) to TCTR, each n-bit block X_i of the input X is

procedure $\mathrm{TCTR}[\widetilde{E}]_K(T,X)$:	**procedure** $\mathrm{TCTR}[\widetilde{E}]_K^{-1}(T,Y)$:
$X_1, X_2, \ldots, X_\nu \xleftarrow{n} X$	$Y_1, Y_2 \ldots, Y_\nu \xleftarrow{n} Y$
for $i = 1$ to ν	for $i = 1$ to ν
$\quad T_i \leftarrow g(T,i)$; $Z_i \leftarrow \langle i \rangle$	$\quad T_i \leftarrow g(T,i)$; $Z_i \leftarrow \langle i \rangle$
$\quad Y_i \leftarrow \widetilde{E}_K(T_i, Z_i) \oplus X_i$	$\quad X_i \leftarrow Y_i \oplus \widetilde{E}_K(T_i, Z_i)$
Return $Y_1, Y_2, \ldots, Y_\nu$	Return $X_1, \ldots, X_\nu$

Fig. 3. The TCTR VIL tweakable cipher

processed using a per-block tweak T_i, this being determined by a function $g\colon \mathcal{T}' \times \mathbb{N} \to \mathcal{T}$ of the input tweak T and the block index i.

Consider the behavior of TCTR when $g(T, i) = T$. The following result is easily obtained using standard techniques.

Theorem 2. *Let* $\widetilde{E}\colon \{0,1\}^k \times \mathcal{T} \times \{0,1\}^n \to \{0,1\}^n$ *be a tweakable blockcipher, and let* $\mathrm{TCTR}[\widetilde{E}]_K$ *and* $\mathrm{TCTR}[\widetilde{E}]_K^{-1}$ *be defined as above, with* $g(T,i) = T \in \mathcal{T}$. *Let A be a nonce-respecting adversary that runs in time t, and asks q queries, each of length at most ℓn bits (so, $\mu \le q\ell n$). Then for some adversary B making at most $q\ell$ queries and running in time $\mathcal{O}(t)$,* $\mathbf{Adv}^{\widetilde{\mathrm{srnd}}}_{\mathrm{TCTR}[\widetilde{E}]}(A) \le \mathbf{Adv}^{\widetilde{\mathrm{prp}}}_{\widetilde{E}}(B) + 0.5q\ell^2/2^n$.

We note that the bound displays birthday-type behavior when $\ell = o(\sqrt{q})$, and is tightest when ℓ is a small constant. An important application with small, constant ℓ is full-disk encryption. Here plaintexts X would typically be 4096 bytes long, so if the underlying TBC has blocksize $n = 128$, we get $\ell = 256$ blocks.[5]

Extending tweakspaces. In PIV, the TBC $\widetilde{F}$ will need to handle long tweaks. Fortunately, a result by Coron, et al. [11] shows that one can compress tweaks

[5] Actually, slightly less than this when used in the PIV composition, since the first N bits are enciphered by $\widetilde{F}$.

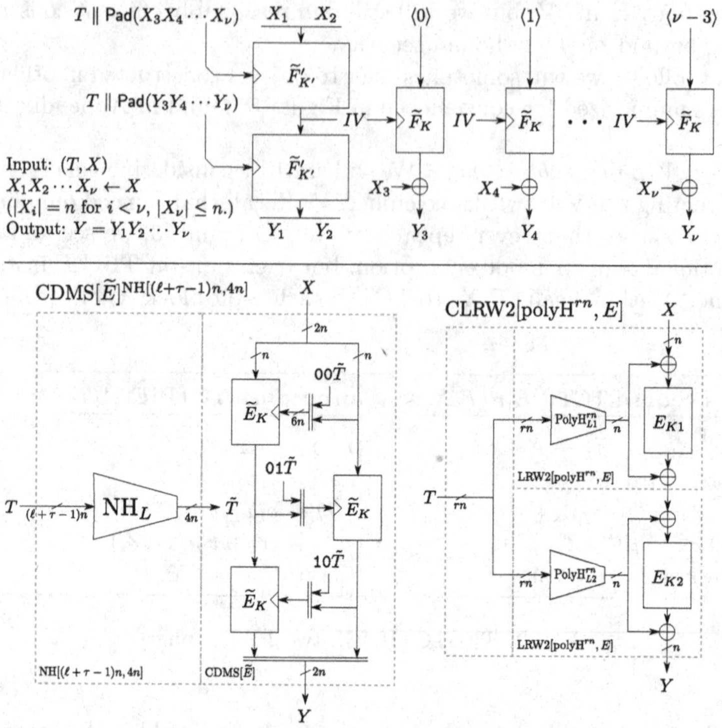

Fig. 4. The TCT_2 construction (top). TCT_2 takes τn-bit tweaks, and the input length is between $2n$ and ℓn bits, inclusive. Here, $\widetilde{F}$ is implemented using the $2n$-bit CDMS construction coupled with the NH hash function (bottom left). Both $\widetilde{V}$ and the TBC $\widetilde{E}$ used inside of CDMS are implemented using CLRW2[polyH^{rn}, E] (bottom right), with $r = 6$ and $r = 2$, respectively. The function Pad maps s to $s \parallel 10^{(\ell+1)n-1-|s|}$. In the diagram for CDMS, the strings $00\widetilde{T}$, $01\widetilde{T}$, and $10\widetilde{T}$ are padded with 0s to length $5n$ before being used.

using an ϵ-AU hash function at the cost of adding a $q^2\epsilon$ term to the tweakable cipher's TPRP security bound. In particular, we will use (a slight specialization of) the NH hash, defined by Black, et al. [6]; $\mathrm{NH}[r,s]_L$ takes r-bit keys ($|L| = r$), maps r-bit strings to s-bit strings, and is $2^{s/2}$-AU. Please see Table 5 for the description. Given a TBC $\widetilde{E}$, $\widetilde{E}^{\mathrm{NH}}$ denotes the resulting TBC, whose tweakspace is now the domain of NH, rather than its range.

4.1 Targeting Efficiency at Birthday-Type Security: TCT_1

Let us begin with the case of $N = n$. To instantiate the n-bit TBC $\widetilde{F}$ in PIV we refer to the pioneering TBC work of Liskov, Rivest and Wagner [20], from which we draw the LRW2 TBC; please refer to Figure 5 for a description.

Before we give the TCT_1 construction, a few notes. In Figure 5 we see that in addition to a blockcipher E, LRW2[H, E] uses an ϵ-AXU$_2$ hash function, H, and so, in theory, it could natively accommodate large tweaks. But for practical purposes, it will be more efficient to implement LRW2 with a small tweakspace, and then extend this using a fast ϵ-AU hash function.[6] For the ϵ-AXU$_2$ hash function itself, we use the polynomial hash polyH (also described in Table 5).

Now are ready to give our TCT_1 construction, which is birthday-bound secure for applications with small plaintext messages (e.g. FDE).

The TCT_1 Construction. Fix $k, n > 0$, and let $N = n$. Let $E\colon \{0,1\}^k \times \{0,1\}^n \to \{0,1\}^n$ be a blockcipher, and let polyH^{mn}, and NH be as defined in Table 5. Then define $\mathsf{TCT}_1 = \mathsf{PIV}[\widetilde{F}, \widetilde{V}]$, where to obtain a τn-bit tweakspace and domain $\{0,1\}^{\{n,n+1,\ldots,\ell n\}}$ we set:

1. n-bit TBC $\widetilde{F} = \text{LRW2}[\text{polyH}^{2n}, E]^{\text{NH}[(\ell+\tau)n,2n]}$, i.e. LRW2 with its tweakspace extended using NH. The keyspace for $\widetilde{F}$ is $\{0,1\}^k \times \{0,1\}^{2n} \times \{0,1\}^{(\ell+\tau)n}$, with key K' partitioning into keys for E, polyH^{2n}, and $\text{NH}[(\ell+\tau)n, 2n]$. (Since NH supports only fixed length inputs, we implicitly pad NH inputs with a 1 and then as many 0s as are required to reach a total length of $(\ell+\tau)n$ bits.) The tweakspace for $\widetilde{F}$ is $\{0,1\}^{\{0,1,2,\ldots,(\ell+\tau-1)n\}}$.
2. VIL tweakable cipher $\widetilde{V} = \text{TCTR}[\text{LRW2}[\text{polyH}^n, E]]$ with the TCTR function $g\colon \{0,1\}^n \times \mathbb{N} \to \{0,1\}^n$ as $g(T,i) = T$. The keyspace for $\widetilde{V}$ is $\{0,1\}^k \times \{0,1\}^n$, with key K partitioning into keys for E and polyH^n. The tweakspace for $\widetilde{V}$ is $\{0,1\}^n$, and its domain is $\{0,1\}^{\{0,1,\ldots,(\ell-1)n\}}$.

Putting together Theorems 1,2, and results from previous works [6,20], we have the following security bound.

Theorem 3 (STPRP-security of TCT_1). *Define TCT_1 as above, and let A be an adversary making $q < 2^n/4$ queries and running in time t. Then there exist adversaries B and C, both running in time $\mathcal{O}(t)$ and making $(\ell-1)q$ and $2q$ queries, respectively, such that* $\mathbf{Adv}^{\widetilde{\text{sprp}}}_{\mathsf{TCT}_1[E]}(A) \le \mathbf{Adv}^{\text{prp}}_E(B) + \mathbf{Adv}^{\text{sprp}}_E(C) + \frac{32q^2}{2^n} + \frac{4q^2(\ell-1)^2}{2^n}$.

The proof appears in the full version. This algorithm requires $2k + (3+\tau+\ell)n$ bits of key material, including two keys for $\widetilde{E}$. As we show at the end of this section, we can get away with a single key for E with no significant damage to our security bound, although this improvement is motivated primarily by performance concerns.

Thus TCT_1 retains the security of previous constructions (see Figure 2 for a visual comparison), uses arithmetic in rings with powers-of-two moduli, rather than in a finite field. This may potentially improve performance in some architectures.

6 Indeed, one can show composing an ϵ-AU hash function with an ϵ'-AXU$_2$ hash function yields an $(\epsilon+\epsilon')$-AXU$_2$ hash function; however, we prefer to work on a higher level of abstraction.

4.2 Aiming for beyond Birthday-Bound Security: TCT_2

Now let us consider the PIV composition with $N = 2n$. For the FIL component, we can use Coron et al.'s [11] CDMS construction to get a $2n$-bit TBC from an n-bit TBC, and implement the latter using the CLRW2, a recent beyond-birthday-bound secure construction by Landecker, Shrimpton, and Terashima [19]. Table 5 describes both constructions.[7] We again extend the tweakspace using NH. (To stay above the birthday bound, we set the range of NH to $\{0,1\}^{2n}$). Ultimately, setting $\widetilde{F} = \text{CDMS}[\text{CLRW2}]^{\text{NH}}$ is secure against up to around $2^{2n/3}$ queries.

CLRW2 also gives us a way to realize a beyond birthday-bound secure VIL component, namely $\widetilde{V} = \text{TCTR}[\text{CLRW2}[E, H]$, at least for $\ell = o(q^{1/4})$. (We'll see how to avoid this restriction, if desired, in a moment.)

We are now ready to give our second fully concrete PIV composition, TCT_2, targeted at applications that would benefit from beyond birthday-bound security. This algorithm requires us to nest four layers of other constructions, so we provide an illustration in Figure 4. Again we emphasize that the (admittedly significant) cost of $\widetilde{F}$ can be amortized.

TCT_2 supports τn-bit tweaks and has domain $\{0,1\}^{\{2n,2n+1,\ldots,\ell n\}}$.

The TCT_2 Construction. Fix $k, \ell, n, \tau > 0$, and let $N = 2n$. Let $E\colon \{0,1\}^k \times \{0,1\}^n \to \{0,1\}^n$ be a blockcipher, and let $\text{polyH}^{\ell n}$, and NH be as defined in Table 5. Then define $\mathsf{TCT}_2 = \mathsf{PIV}[\widetilde{F}, \widetilde{V}]$, where:

1. $\widetilde{F} = \text{CDMS}\left[\text{CLRW2}[\text{polyH}^{6n}, E]\right]^{\text{NH}[(\ell+\tau-1)n,4n]}$, that is, the $2n$-bit TBC $\text{CDMS}\left[\text{CLRW2}[\text{polyH}^{6n}, E]\right]$ with its tweakspace extended using NH. The keyspace for $\widetilde{F}$ is $\{0,1\}^{2k} \times \{0,1\}^{12n} \times \{0,1\}^{(\ell+\tau-1)n}$, with key K' partitioning into two keys for E, two keys for polyH^{6n}, and a key for $\text{NH}[\ell n, 4n]$. The tweakspace for $\widetilde{F}$ is $\{0,1\}^{\tau n}$.
2. $\widetilde{V} = \text{TCTR}\left[\text{CLRW2}[\text{polyH}^{2n}, E]\right]$, with the TCTR function $g\colon \{0,1\}^n \times \mathbb{N} \to \{0,1\}^n$ as $g(T, i) = T$. The keyspace for $\widetilde{V}$ is $\{0,1\}^{2k} \times \{0,1\}^{4n}$ with key K partitioning into two keys for E and two keys for polyH^{2n}. The tweakspace for $\widetilde{V}$ is $\{0,1\}^{2n}$, and its domain is $\{0,1\}^{\{0,1,2,\ldots,(\ell-2)n\}}$.

TCT_2 requires $4k+(\ell+\tau+15)n$ bits of key material. Putting together Theorems 1, 5, and results from previous works [6,11,19], we have the following security result.

Theorem 4 (STPRP-security of TCT_2). *Define TCT_2 as above, and let A be an adversary making q queries and running in time t, where $6q, \ell q < 2^{2n}/4$. Then there exist adversaries B and C, both running in $\mathcal{O}(t)$ time and making $(\ell-1)q$ and $6q$ queries, respectively, such that* $\mathbf{Adv}^{\widetilde{\text{sprp}}}_{\mathsf{TCT}_2}(A) \leq 2\mathbf{Adv}^{\text{prp}}_E(B) + 2\mathbf{Adv}^{\text{sprp}}_E(C) + \frac{12q^2}{2^{2n}} + \frac{q(\ell-1)^2}{2^n} + \frac{6\ell^3 q^3}{2^{2n-2} - \ell^3 q^3} + \frac{6^4 q^3}{2^{2n-2} - 6^3 q^3}$.

[7] We note that for $\text{CDMS}[\widetilde{E}]$, we enforce domain separation via $\widetilde{E}$'s tweak, whereas the authors of [11] use multiple keys for $\widetilde{E}$. The proof of our construction follows easily from that of the original.

Again, the proof appears in the full version. Some of the constants in this bound are rather significant. However, as Figure 2 shows, TCT_2 nevertheless provides substantially better security bounds than TCT_1 and previous constructions.

4.3 Additional Practical Considerations

Several variations and optimizations on TCT_1 and TCT_2 are possible. We highlight a few of them here. None of these changes significantly impact the above security bounds, unless otherwise noted.

Reducing the number of blockcipher keys. In the case of TCT_1, we can use a single key for both LRW2 instances provided we enforce domain separation through the tweak. This allows us to use a single key for the underlying blockcipher, which in some situations may allow for significant implementation benefits (for example, by allowing a single AES pipeline). One method that accomplishes this is to replace $\mathrm{LRW2}[\mathrm{polyH}^{2n}, E]^{\mathrm{NH}[(\ell+1)n,2n]}$ with $\mathrm{LRW2}[\mathrm{polyH}^{3n}, E]^{f(\varepsilon,\cdot)}$ and $\mathrm{LRW2}[\mathrm{polyH}^{n}, E]$ with $\mathrm{LRW2}[\mathrm{polyH}^{3n}, E]^{f(\cdot,\varepsilon)}$. Here, f is a 2^{-n}-AU function with keyspace $\{0,1\}^{3n} \times \{0,1\}^{\ell n}$, taking inputs of the form (X, ε) (for some $X \in \{0,1\}^n$) or (ε, Y) (for some $Y \in \{0,1\}^{\{0,1,\ldots,\ell n\}}$), and outputting a $3n$-bit string. Let $f_L(X,\varepsilon) = 0^{2n} \,\|\, X$ and $f_L(\varepsilon, Y) = 1^n \,\|\, \mathrm{NH}[(\ell+1)n, 2n]_L(Y)$. The function f described here is a mathematical convenience to unify the signatures of the two LRW2 instances, thereby bringing tweak-based domain separation into scope; in practice, we imagine the two instances would be implemented independently, save for a shared blockcipher key. We note that TCT_2 can be modified in a similar manner to require only two blockcipher keys.

Performance optimizations. If we need only a tweakable (FIL) blockcipher, we can use $\mathrm{NH}[\ell n, 2n]$ in place of $\mathrm{NH}[(\ell+1)n, 2n]$ by adjusting our padding scheme appropriately. We emphasize that in the TCTR portion, the polyH functions only need to be computed once, since each LRW2 invocation uses the same tweak. The corresponding optimizations apply to TCT_2, as well.

A naïve implementation of TCT_2 would make a total 72 finite field multiplications during the two FIL phases (a result of evaluating polyH^{6n} twelve times). We can cache an intermediate value of the polyH^{6n} hash used inside of CDMS (four n-bit tweak blocks are constant per invocation), and this saves 32 finite field multiplications. Precomputing the terms of the polynomial hash corresponding to the domain-separation constants eliminates 12 more multiplications, leaving 28 in total. Four more are required during the VIL phase, giving the count of 32 reported in Table 1.

Handling large message spaces. Both TCT_1 and TCT_2 are designed with FDE applications in mind. In particular, they require ℓ to be fixed ahead of time, and require more than ℓn bits of key material.

These limitations are a consequence of using the NH hash function; however, a simple extension to NH (described by the original authors [6]) accommodates arbitrarily long strings. Fix a positive integer r and define $\mathrm{NH}^*_L(M_1 M_2 \cdots M_\nu) =$

$\mathrm{NH}_L(M_1) \parallel \mathrm{NH}_L(M_2) \parallel \cdots \parallel \mathrm{NH}_L(M_\nu) \parallel \langle |M| \bmod rn \rangle$, where $|M_i| = rn$ for $i < \nu$, $|M_\nu| \leq rn$, and NH_L abbreviates $\mathrm{NH}_L[rn, 2N]$. Thus defined, NH^* is 2^{-N}-almost universal, has domain $\{0,1\}^*$, and requires rn bits of key material. This modification shifts some of the weight to the polyH hash; we now require eight extra finite field multiplications for each additional rn bits of input. As long as $r > 4$, however, we require fewer of these multiplications when compared to previous hash-ECB-hash or hash-CTR-hash constructions.

With these modifications, the final two terms in TCT_1's security bound (Theorem 3) would become $8q^2/2^n + 600q^2\ell^2/r^2 2^n + 4q^2(\ell-1)^2/2^n$, where ℓn is now the length of the adversary's longest query, $\ell > 2.5r$, and the remaining terms measure the (S)PRP security of the underlying blockcipher. We also assume $2^n \geq rn$, so that $|M| \bmod rn$ can be encoded within a single n-bit block. Although the constant of 600 is large, we note that setting $r = 16$, for example, reduces it to a more comfortable size — in this case to less than three. The bound for TCT_2 changes in a similar manner. (Note that if $2^{n-2} \geq rn$, we can use a single n-bit block for both the tweak domain-separation constants and $\langle |M| \bmod rn \rangle$.)

Beyond birthday-bound security for long messages. When ℓ is not bounded to some small or moderate value, TCT_2 no longer provides beyond-birthday-bound security. The problematic term in the security bound is $q(\ell-1)^2/2^n$. To address this, we return to TCTR (Figure 3) and consider a different per-block tweak function.

In particular, $g(T, i) = T \parallel \langle i \rangle$. In the nonce-respecting case, the underlying TBC $\widetilde{E}$ is then retweaked with a never-before-seen value on each message block. Again, think about what happens when $\widetilde{E}$ is replaced by an ideal cipher Π: in the nonce-respecting case, every *block* of plaintext is masked by the output of a fresh random permutation.[8] In other words, every block returned will be uniformly random. Thus we expect a tight bound, in this case. Formalizing this logic yields the following theorem.

Theorem 5. *Let* $\widetilde{E}\colon \{0,1\}^k \times \mathcal{T} \times \{0,1\}^n \to \{0,1\}^n$ *be a tweakable blockcipher, and let* $\mathrm{TCTR}[\widetilde{E}]_K$ *and* $\mathrm{TCTR}[\widetilde{E}]_K^{-1}$ *be defined as above, with* $g\colon \mathcal{T}' \times \mathbb{N} \to \mathcal{T}$ *an arbitrary injective mapping. Let A be a nonce-respecting adversary that runs in time t, and asks q queries of total length at most* $\mu = \sigma n$ *bits. Then there exists some adversary B making at most* σ *queries and running in time* $\mathcal{O}(t)$ *such that* $\mathbf{Adv}^{\mathrm{srnd}}_{\mathrm{TCTR}[\widetilde{E}]}(A) \leq \mathbf{Adv}^{\widetilde{\mathrm{prp}}}_{\widetilde{E}}(B)$.

Consequently, using this variation of TCTR in Theorems 3 and 4 would remove the $q(\ell-1)^2$ term from the bounds, thereby lifting message length concerns. Note that if this change is made, $g(T, i)$ needs to be computed up to ℓ times per invocation, rather than just once. This problem may be mitigated by using the XEX [28] TBC in place of LRW2, which makes incrementing the tweak extremely fast without significantly changing our security bound.

[8] Notice that one could use (say) $Z_i \leftarrow 0^n$ and the same would be true. We present it as $Z_i \leftarrow \langle i \rangle$ for expositional purposes.

When the above change are made, TCT_1 and TCT_2 offer efficient tweakable ciphers on an unbounded domain, losing security guarantees only after $\mathcal{O}(2^{n/2})$ (resp., $\mathcal{O}(2^{2n/3})$) bits have been enciphered. Finally, we note that one can use a conventional blockcipher mode of operation to build the VIL component. We report on this in the full version.

Acknowledgements. Portions of this work were carried out while Terashima was visiting Voltage Security. We thank them, especially Terence Spies, for their support. We also thank the CRYPTO and Asiacrypt 2013 reviewers for their diligence and useful feedback. Both Terashima and Shrimpton were supported by NSF grants CNS-0845610 and CNS-1319061.

References

1. AlFardan, N., Paterson, K.G.: Lucky 13: Breaking the TLS and DTLS record protocols. In: IEEE Symposium on Security and Privacy (2013)
2. Aranha, D.F., López, J., Hankerson, D.: Efficient software implementation of binary field arithmetic using vector instruction sets. In: Abdalla, M., Barreto, P.S.L.M. (eds.) LATINCRYPT 2010. LNCS, vol. 6212, pp. 144–161. Springer, Heidelberg (2010)
3. Bellare, M., Ristenpart, T., Rogaway, P., Stegers, T.: Format-preserving encryption. In: Jacobson Jr., M.J., Rijmen, V., Safavi-Naini, R. (eds.) SAC 2009. LNCS, vol. 5867, pp. 295–312. Springer, Heidelberg (2009)
4. Bellare, M., Rogaway, P.: Encode-then-encipher encryption: How to exploit nonces or redundancy in plaintexts for efficient cryptography. In: Okamoto, T. (ed.) ASIACRYPT 2000. LNCS, vol. 1976, pp. 317–330. Springer, Heidelberg (2000)
5. Bellare, M., Rogaway, P., Spies, T.: The FFX mode of operation for format-preserving encryption. Unpublished NIST proposal (2010)
6. Black, J., Halevi, S., Krawczyk, H., Krovetz, T., Rogaway, P.: UMAC: Fast and secure message authentication. In: Wiener, M. (ed.) CRYPTO 1999. LNCS, vol. 1666, pp. 216–233. Springer, Heidelberg (1999)
7. Canvel, B., Hiltgen, A.P., Vaudenay, S., Vuagnoux, M.: Password interception in a SSL/TLS channel. In: Boneh, D. (ed.) CRYPTO 2003. LNCS, vol. 2729, pp. 583–599. Springer, Heidelberg (2003)
8. Chakraborty, D., Mancillas-López, C., Rodríguez-Henríquez, F., Sarkar, P.: Efficient hardware implementations of brw polynomials and tweakable enciphering schemes. IEEE Transactions on Computers 62(2), 279–294 (2013)
9. Chakraborty, D., Sarkar, P.: A new mode of encryption providing a tweakable strong pseudo-random permutation. In: Robshaw, M. (ed.) FSE 2006. LNCS, vol. 4047, pp. 293–309. Springer, Heidelberg (2006)
10. Chakraborty, D., Sarkar, P.: HCH: A new tweakable enciphering scheme using the hash-counter-hash approach. IACR Cryptology ePrint Archive, 2007:28 (2007)
11. Coron, J.-S., Dodis, Y., Mandal, A., Seurin, Y.: A domain extender for the ideal cipher. In: Micciancio, D. (ed.) TCC 2010. LNCS, vol. 5978, pp. 273–289. Springer, Heidelberg (2010)
12. Degabriele, J.P., Paterson, K.G.: On the (in)security of IPsec in MAC-then-encrypt configurations. In: Proceedings of the 17th ACM Conference on Computer and Communications Security, pp. 493–504. ACM (2010)

13. Halevi, S.: EME*: Extending EME to handle arbitrary-length messages with associated data. In: Canteaut, A., Viswanathan, K. (eds.) INDOCRYPT 2004. LNCS, vol. 3348, pp. 315–327. Springer, Heidelberg (2004)
14. Halevi, S.: Invertible universal hashing and the TET encryption mode. In: Menezes, A. (ed.) CRYPTO 2007. LNCS, vol. 4622, pp. 412–429. Springer, Heidelberg (2007)
15. Halevi, S., Rogaway, P.: A tweakable enciphering mode. In: Boneh, D. (ed.) CRYPTO 2003. LNCS, vol. 2729, pp. 482–499. Springer, Heidelberg (2003)
16. Halevi, S., Rogaway, P.: A parallelizable enciphering mode. In: Okamoto, T. (ed.) CT-RSA 2004. LNCS, vol. 2964, pp. 292–304. Springer, Heidelberg (2004)
17. Hoang, V.T., Morris, B., Rogaway, P.: An enciphering scheme based on a card shuffle. In: Safavi-Naini, R., Canetti, R. (eds.) CRYPTO 2012. LNCS, vol. 7417, pp. 1–13. Springer, Heidelberg (2012)
18. Lampe, R., Seurin, Y.: Tweakable blockciphers with asymptotically optimal security. In: FSE (2013)
19. Landecker, W., Shrimpton, T., Terashima, R.S.: Tweakable blockciphers with beyond birthday-bound security. In: Safavi-Naini, R., Canetti, R. (eds.) CRYPTO 2012. LNCS, vol. 7417, pp. 14–30. Springer, Heidelberg (2012)
20. Liskov, M., Rivest, R.L., Wagner, D.: Tweakable block ciphers. In: Yung, M. (ed.) CRYPTO 2002. LNCS, vol. 2442, pp. 31–46. Springer, Heidelberg (2002)
21. Luo, J., Bowers, K.D., Oprea, A., Xu, L.: Efficient software implementations of large finite fields GF(2^n) for secure storage applications. ACM Transactions on Storage (TOS) 8(1), 2 (2012)
22. Mancillas-Lopez, C., Chakraborty, D., Rodriguez-Henriquez, F.: Reconfigurable hardware impementations of tweakable enciphering schemes. IEEE Transactions on Computers 59, 1547–1561 (2010)
23. Mathewson, N.: Cryptographic challenges in and around Tor. Talk Given at the Workshop on Real-World Cryptography (January 2013)
24. Minematsu, K.: Beyond-birthday-bound security based on tweakable block cipher. In: Dunkelman, O. (ed.) FSE 2009. LNCS, vol. 5665, pp. 308–326. Springer, Heidelberg (2009)
25. Minematsu, K., Iwata, T.: Building blockcipher from tweakable blockcipher: Extending FSE 2009 proposal. In: Chen, L. (ed.) IMACC 2011. LNCS, vol. 7089, pp. 391–412. Springer, Heidelberg (2011)
26. Morris, B., Rogaway, P., Stegers, T.: How to encipher messages on a small domain. In: Halevi, S. (ed.) CRYPTO 2009. LNCS, vol. 5677, pp. 286–302. Springer, Heidelberg (2009)
27. Paterson, K.G., Yau, A.K.L.: Padding oracle attacks on the ISO CBC mode encryption standard. In: Okamoto, T. (ed.) CT-RSA 2004. LNCS, vol. 2964, pp. 305–323. Springer, Heidelberg (2004)
28. Rogaway, P.: Efficient instantiations of tweakable blockciphers and refinements to modes OCB and PMAC. In: Lee, P.J. (ed.) ASIACRYPT 2004. LNCS, vol. 3329, pp. 16–31. Springer, Heidelberg (2004)
29. Rogaway, P., Shrimpton, T.: A provable-security treatment of the key-wrap problem. In: Vaudenay, S. (ed.) EUROCRYPT 2006. LNCS, vol. 4004, pp. 373–390. Springer, Heidelberg (2006)
30. Sarkar, P.: Improving upon the TET mode of operation. In: Nam, K.-H., Rhee, G. (eds.) ICISC 2007. LNCS, vol. 4817, pp. 180–192. Springer, Heidelberg (2007)
31. Sarkar, P.: Efficient tweakable enciphering schemes from (block-wise) universal hash functions. IEEE Trans. Inf. Theor. 55(10), 4749–4760 (2009)

32. Vaudenay, S.: Security flaws induced by CBC padding - applications to SSL, IPSEC, WTLS... In: Knudsen, L.R. (ed.) EUROCRYPT 2002. LNCS, vol. 2332, pp. 534–545. Springer, Heidelberg (2002)
33. Wang, P., Feng, D., Wu, W.: HCTR: A variable-input-length enciphering mode. In: Feng, D., Lin, D., Yung, M. (eds.) CISC 2005. LNCS, vol. 3822, pp. 175–188. Springer, Heidelberg (2005)
34. Wegman, M.N., Carter, J.L.: New hash functions and their use in authentication and set equality. Journal of Computer and System Sciences 22(3), 265–279 (1981)

A Components for TCT1 and TCT2

LRW2 [20]: Birthday-bound TBC. Needs blockcipher E, ϵ-AXU$_2$ function H.

$$\mathrm{LRW2}[H,E]_{(K,L)}(T,X) = E_K(X \oplus H_L(T)) \oplus H_L(T)$$

CLRW2[19]: TBC with beyond-birthday-bound security. Requires blockcipher E and ϵ-AXU$_2$ function H.

$$\mathrm{CLRW2}[H,E]_{(K_1,K_2,L_1,L_2)}(T,X) = \mathrm{LRW2}[H,E]_{(K_2,L_2)}(T, \mathrm{LRW2}[H,E]_{(K_1,L_1)}(T,X))$$

polyH^{mn} [34]: ϵ-AXU$_2$ function with domain $(\{0,1\}^n)^m$ and $\epsilon = m/2^n$. All operations in $\mathbb{F}_{2^n}$.

$$\mathrm{polyH}_L^{mn}(T_1 T_2 \cdots T_m) = \bigoplus_{i=1}^{m} T_i \otimes L^i,$$

NH($\nu w, 2tw$) [6]: ϵ-AU hash function with $\epsilon = 1/2^{tw}$. Inputs are νw bits, where ν is even and $w > 0$ is fixed.

$$\mathrm{NH}[\nu,t]_{K_1 \,\|\, \cdots \,\|\, K_{\nu+2(t-1)}}(M) = H_{K_1 \cdots K_\nu}(M) \,\|\, H_{K_3 \cdots K_{\nu+2}}(M) \,\|\, \cdots \,\|\, H_{K_{2t-1} \cdots K_{\nu+2t-2}}(M)$$

where $H_{K_1 \,\|\, \cdots \,\|\, K_\nu}(X_1 \cdots X_\nu) = \sum_{i=1}^{\nu/2} (K_{2i-1} +_w X_{2i-1}) \cdot (K_{2i} +_w X_{2i}) \bmod 2^{2w}$.

CDMS [11]: Feistel-like domain extender for TBC $\widetilde{E}$.

$$\mathrm{CDMS}[\widetilde{E}]_K(T, L \,\|\, R) = \widetilde{E}_K(10 \,\|\, T \,\|\, R', L') \,\|\, R'$$

where $R' = \widetilde{E}_K(01 \,\|\, T \,\|\, L', R)$ and $L' = \widetilde{E}_K(00 \,\|\, T \,\|\, R, L)$.

Fig. 5. TCT$_1$ and TCT$_2$ use these constructions as components

Parallelizable and Authenticated Online Ciphers

Elena Andreeva[1,2], Andrey Bogdanov[3], Atul Luykx[1,2], Bart Mennink[1,2], Elmar Tischhauser[1,2], and Kan Yasuda[1,4]

[1] Department of Electrical Engineering, ESAT/COSIC, KU Leuven, Belgium
[2] iMinds, Belgium
[3] Department of Mathematics, Technical University of Denmark, Denmark
[4] NTT Secure Platform Laboratories, Japan

Abstract. Online ciphers encrypt an arbitrary number of plaintext blocks and output ciphertext blocks which only depend on the preceding plaintext blocks. All online ciphers proposed so far are essentially serial, which significantly limits their performance on parallel architectures such as modern general-purpose CPUs or dedicated hardware. We propose the first parallelizable online cipher, COPE. It performs two calls to the underlying block cipher per plaintext block and is fully parallelizable in both encryption and decryption. COPE is proven secure against chosen-plaintext attacks assuming the underlying block cipher is a strong PRP. We then extend COPE to create COPA, the first parallelizable, online authenticated cipher with nonce-misuse resistance. COPA only requires two extra block cipher calls to provide integrity. The privacy and integrity of the scheme is proven secure assuming the underlying block cipher is a strong PRP. Our implementation with Intel AES-NI on a Sandy Bridge CPU architecture shows that both COPE and COPA are about *5 times faster* than their closest competition: TC1, TC3, and McOE-G. This high factor of advantage emphasizes the paramount role of parallelizability on up-to-date computing platforms.

Keywords: Block cipher, tweakable cipher, online cipher, authenticated encryption, nonce-misuse resistance, parallelizability, AES.

1 Introduction

Online Ciphers. A cipher which takes input of arbitrary length is said to be an *online cipher* if it can output ciphertext blocks as it is receiving the plaintext blocks. Specifically, the ith ciphertext block should only depend on the key and the first i plaintext blocks. This desirable functionality known more generally as online data processing is characteristic for other cryptographic primitives such as standard encryption schemes like CTR, CBC, OFB, and CFB.

The first theoretical treatment of online ciphers was put forward by Bellare, Boldyreva, Knudsen, and Namprempre [4]. They introduce the online ciphers HCBC1 and HCBC2, both of which require the use of two keys, one for the underlying block cipher and the other for the almost-xor-universal hash family [24]. Subsequently, Nandi [21] proposed two more efficient online ciphers MHCBC and

K. Sako and P. Sarkar (Eds.) ASIACRYPT 2013 Part I, LNCS 8269, pp. 424–443, 2013.

MCBC. MHCBC improves upon HCBC2 by using a smaller hashing key with a finite field multiplication as universal hash function, whereas MCBC does not even require a universal hash function, thus needing only one block cipher key and calling the block cipher twice per plaintext block. Rogaway and Zhang in [27] recast the formalism of Bellare et al. [4] in terms of tweakable block ciphers [17] and provide three generalizations of the previous online ciphers: TC1, TC2, and TC3.

Authenticated Encryption from Online Ciphers. While online authenticated encryption (AE) schemes are not a novelty,[1] presently we are aware of only one family of online *and* misuse-resistant AE schemes, McOE [11]. McOE makes use of the online cipher TC3 [27] to build its general structure and adds two calls to the tweakable cipher to achieve authenticity. To process messages of arbitrary lengths, McOE applies a tag splitting method, similar to the ciphertext stealing method [9].

Bellare et al. [4] give a few generic transformations to turn an online cipher into a secure authenticated encryption scheme.

Problem Statement. All existing online ciphers are highly sequential and none of them offer any possibility for parallelizing the computation between distinct block cipher calls. The only exception can be seen in TC1, which allows parallelization only in decryption but not in encryption. As a consequence, the McOE AE schemes are not parallelizable either, due to the fact that they are based on existing online ciphers.

At the same time, in the overwhelming majority of cases in practice, the underlying cipher is AES which is very well parallelizable on many platforms. Parallelization is a rather inherent feature of hardware implementations, both in ASIC and FPGA. Also in general-purpose software, parallelizable encryption algorithms have profited in terms of performance due to the bitslice approach for a long time already [6, 14, 18]. However, with the introduction of the hardware supported AES by Intel in general-purpose x86 CPUs as an instruction set AES-NI in Intel Westmere, Sandy Bridge, and Ivy Bridge — followed by the AMD adoption of AES-NI in AMD Bulldozer and Piledriver — the parallelizability of the AES modes of operation has become of truly paramount importance. With AES-NI, using a parallelizable mode of operation enables performance advantages of a factor 3 and higher — see, for instance, the case of the (serial) CBC encryption vs (parallel) CBC decryption [1].

Our Contributions. We present the first parallelizable online cipher, COPE, and the first parallelizable online authenticated encryption scheme with nonce-misuse resistance, COPA.

COPE: Our novel design is illustrated in Fig. 1. To process a single plaintext block two block cipher calls are required. A secret mask (*tweak*) is applied

[1] Examples of online AE schemes include EAX [5], GCM [19], and OCB1-3 [16,25,26].

to the plaintext block and used as input to the first block cipher call. Then the output of the second block cipher call is masked again to produce the ciphertext block.

By introducing dummy masks, each block cipher call can be viewed as an instance of the XEX construction [25], which uses the so-called "doubling" mask generation. Our basic design only deals with message lengths that are a multiple of the block length. In order to handle messages of arbitrary lengths we use the technique prescribed in the XLS domain extender by Ristenpart et al. in [23]. In contrast with previous designs, our scheme only uses a *single key* and a *single cryptographic primitive*, namely a block cipher.

COPE is proven IND-CPA up to the birthday bound of $n/2$-bit security, where n denotes the block size of the underlying block cipher.

COPA: We transform COPE to support authentication, while maintaining parallelizability. The modifications are limited to computing an XOR sum of the plaintext data and using two extra block cipher calls; these can be seen in Fig. 2. The scheme also supports associated data in a way similar to how PMAC1 [25] operates. The privacy and integrity of COPA are proven up to the birthday bound.

To illustrate the impact of the parallelizability of our online schemes, we implement them with AES-NI on an Intel Sandy Brigde processor. We systematically compare the performance we attain with the online ciphers TC1, TC3, and MCBC as well as the online AE scheme McOE-G when instantiated with the AES. When compared to these closest online competitors, which are all explicitly not parallelizable, our modes provide performance improvements between a factor of 4.5 and 5, being below 2 cycles per byte on a single core. We expect almost a linear speed-up when several cores are available.

Organization of the Paper. The remainder of the paper is organized as follows. We recall the necessary background on block ciphers in Section 2. Section 3 provides the specification of our new parallel modes. Sections 4 and 5 deal with the security proofs. Section 6 gives AES-NI implementations of our modes along with a systematic comparison to the state-of-the-art schemes. We conclude in Section 7.

2 Preliminaries

2.1 Block Ciphers

A block cipher $E : \mathcal{K} \times \{0,1\}^n \to \{0,1\}^n$ is a function that takes as input a key $k \in \mathcal{K}$ and a plaintext $M \in \{0,1\}^n$, and produces a ciphertext $C = E(k, M)$. We sometimes write $E_k(\cdot) = E(k, \cdot)$. For a fixed key k, a block cipher is a permutation on n bits, and we denote the inverse permutation (decryption function) by E_k^{-1}.

Let $\mathrm{Perm}(n)$ be the set of all permutations on n bits. When writing $x \xleftarrow{\$} X$ for some finite set X we mean that x is sampled uniformly from X. We write $\Pr[\mathsf{A} \mid \mathsf{B}]$ to denote the probability of event A given B.

Definition 1. *Let E be a block cipher. The* prp±1 *advantage of a distinguisher $\mathcal{D}$ is defined as*

$$\mathbf{Adv}_E^{\mathrm{prp}\pm 1}(\mathcal{D}) = \left| \Pr_k\left[\mathcal{D}^{E_k, E_k^{-1}} = 1\right] - \Pr_\pi\left[\mathcal{D}^{\pi,\pi^{-1}} = 1\right] \right| .$$

Here, $\mathcal{D}$ is a distinguisher with oracle access to either (E_k, E_k^{-1}) or (π, π^{-1}). The probabilities are taken over $k \xleftarrow{\$} \mathcal{K}$, $\pi \xleftarrow{\$} \mathrm{Perm}(n)$ and random coins of $\mathcal{D}$, if any. By $\mathbf{Adv}_E^{\mathrm{prp}\pm 1}(t, q)$ we denote the maximum advantage taken over all distinguishers that run in time t and make q queries.

We shall also write $E_k^{\pm 1}$ for (E_k, E_k^{-1}). Similarly, $\pi^{\pm 1}$ means (π, π^{-1}), and so on.

2.2 Binary Fields

The set $\{0,1\}^n$ of bit strings can be considered as the finite field $\mathrm{GF}(2^n)$ consisting of 2^n elements. To do this, we represent an element of $\mathrm{GF}(2^n)$ as a polynomial over the field GF(2) of degree less than n. A string $a_{n-1}a_{n-2}\cdots a_1 a_0 \in \{0,1\}^n$ corresponds to the polynomial $a_{n-1}\mathrm{x}^{n-1} + a_{n-2}\mathrm{x}^{n-2} + \cdots + a_1\mathrm{x} + a_0 \in \mathrm{GF}(2^n)$. The addition in the field is just the addition of polynomials over GF(2) (that is, bitwise XOR, denoted by $\oplus$). To define multiplication in the field, we fix an irreducible polynomial $f(\mathrm{x})$ of degree n over the field GF(2). Given two elements $a(\mathrm{x}), b(\mathrm{x}) \in \mathrm{GF}(2^n)$, their product is defined as $a(\mathrm{x})b(\mathrm{x}) \bmod f(\mathrm{x})$—polynomial multiplication over the field GF(2) reduced modulo $f(\mathrm{x})$. We simply write $a(\mathrm{x})b(\mathrm{x})$ and $a(\mathrm{x}) \cdot b(\mathrm{x})$ to mean the product in the field $\mathrm{GF}(2^n)$.

The set $\{0,1\}^n$ can be also regarded as a set of integers ranging from 0 through $2^n - 1$. A string $a_{n-1}a_{n-2}\cdots a_1 a_0 \in \{0,1\}^n$ corresponds to the integer $a_{n-1}2^{n-1} + a_{n-2}2^{n-2} + \cdots + a_1 2 + a_0 \in [0, 2^n - 1]$. We often write elements of $\mathrm{GF}(2^n)$ as integers, based on these conversions. So, for example, "2" means x, "3" means $\mathrm{x} + 1$, and "7" means $\mathrm{x}^2 + \mathrm{x} + 1$. When we write multiplications such as $2 \cdot 3$ and 7^2, we mean those in the field $\mathrm{GF}(2^n)$.

2.3 XE and XEX Constructions of Tweakable Ciphers

Given a block cipher $E : \mathcal{K} \times \{0,1\}^n \to \{0,1\}^n$ and a secret mask $\Delta \in \{0,1\}^n$, the ciphers

$$E'_{k,\Delta}(x) \stackrel{\mathrm{def}}{=} E_k(x \oplus \Delta) \text{ or } \boldsymbol{E}'_{k,\Delta}(x) \stackrel{\mathrm{def}}{=} E_k(x \oplus \Delta) \oplus \Delta$$

behave like another block cipher independent of E_k (up to some bound). In the case of $E'_{k,\Delta}$, adversaries are allowed to make only forward queries, whereas $\boldsymbol{E}'_{k,\Delta}$ accepts both encryption and decryption queries. Now consider a set of secret masks $\{\Delta_i\}_{i \in \mathcal{T}}$, where Δ_i and Δ_j may not be necessarily independent. An index $i \in \mathcal{T}$ is called a tweak, which is not secret. We obtain a tweakable cipher $\widetilde{E} : \mathcal{K} \times \mathcal{T} \times \{0,1\}^n \to \{0,1\}^n$ by defining $\widetilde{E}_{k,i} \stackrel{\mathrm{def}}{=} E'_{k,\Delta_i}$, and similarly $\widetilde{\boldsymbol{E}}_{k,i}$. We consider $\widetilde{E}_{k,i}$ and $\widetilde{\boldsymbol{E}}_{k,j}$ together, where $i \in \mathcal{T}_0$, $j \in \mathcal{T}_1$ and $\mathcal{T}_0 \cap \mathcal{T}_1 = \varnothing$.

Definition 2. *Let $\widetilde{E}, \widetilde{\boldsymbol{E}}$ be tweakable ciphers. The* twk *advantage of a distinguisher $\mathcal{D}$ is defined as*

$$\mathbf{Adv}^{\mathrm{twk}}_{\widetilde{E},\widetilde{\boldsymbol{E}}}(\mathcal{D}) = \left| \Pr_k\left[\mathcal{D}^{\widetilde{E}_{k,i},\widetilde{\boldsymbol{E}}^{\pm 1}_{k,j}} = 1\right] - \Pr_{\pi_i,\boldsymbol{\pi}_j}\left[\mathcal{D}^{\pi_i,\boldsymbol{\pi}^{\pm 1}_j} = 1\right] \right|.$$

Here, $\mathcal{D}$ is a distinguisher with oracle access to a series of permutations. The tweaks run over $i \in \mathcal{T}_0$ and $j \in \mathcal{T}_1$ where $\mathcal{T}_0 \cap \mathcal{T}_1 = \varnothing$. By $\mathbf{Adv}^{\mathrm{twk}}_{\widetilde{E},\widetilde{\boldsymbol{E}}}(t,q)$ we denote the maximum advantage taken over all distinguishers that run in time t and make q queries in total.

The doubling method [25] enables us to produce many different values of the mask Δ from just one secret value $L \stackrel{\mathrm{def}}{=} E_k(0)$. Namely, the masks are produced as $\Delta = 2^\alpha 3^\beta 7^\gamma L$ for varying indices of α, β and γ. To do this, we need to choose our irreducible polynomial $f(\mathrm{x})$ carefully. First, $f(\mathrm{x})$ needs to be primitive, meaning that 2 generates the whole multiplicative group. Second, we make sure that $\log_2 3$ and $\log_2 7$ are both "huge." Third, we check if $\log_2 3$ and $\log_2 7$ are "apart enough" (modulo $2^n - 1$). We impose these conditions to ensure that values $2^\alpha 3^\beta 7^\gamma$ do not collide or become equal to 1. For example, when $n = 128$, the irreducible polynomial $f(\mathrm{x}) = \mathrm{x}^{128} + \mathrm{x}^7 + \mathrm{x}^2 + \mathrm{x} + 1$ satisfies these requirements, making values $2^\alpha 3^\beta 7^\gamma$ all distinct and not equal to 1 for $\alpha \in [-2^{108}, 2^{108}]$ and $\beta, \gamma \in [-2^7, 2^7]$ [25], except for $(\alpha, \beta, \gamma) = (0, 0, 0)$. So we obtain tweakable ciphers $\widetilde{E}_{k,\alpha\beta\gamma}$ and $\widetilde{\boldsymbol{E}}_{k,\alpha\beta\gamma}$.

Lemma 1 (XE and XEX [25]). *Let $\mathcal{T}_0, \mathcal{T}_1 = \{(\alpha, \beta, \gamma)\}$ be two sets of integer triples such that $2^\alpha 3^\beta 7^\gamma$ are all distinct and not equal to 1, in particular $\mathcal{T}_0 \cap \mathcal{T}_1 = \varnothing$. Then the permutations $\{\widetilde{E}_{k,\alpha\beta\gamma}\}_{\mathcal{T}_0} \cup \{\widetilde{\boldsymbol{E}}^{\pm 1}_{k,\alpha\beta\gamma}\}_{\mathcal{T}_1}$ are indistinguishable from independently random permutations $\{\pi_{\alpha\beta\gamma}\}_{\mathcal{T}_0} \cup \{\boldsymbol{\pi}^{\pm 1}_{\alpha\beta\gamma}\}_{\mathcal{T}_1}$. Specifically, for given t, q, there exists a $t' \approx t$ such that*

$$\mathbf{Adv}^{\mathrm{twk}}_{\widetilde{E},\widetilde{\boldsymbol{E}}}(t,q) \leq \frac{9.5q^2}{2^n} + \mathbf{Adv}^{\mathrm{prp}\pm 1}_{E}(t', 2q).$$

3 COPE and COPA: Design and Specification

We define COPE and COPA. COPE is an online cipher secure against chosen plaintext attacks. COPE makes two calls to the underlying block cipher per message block. COPA is an authenticated online cipher that builds on COPE. The additional cost of producing a tag is kept minimal—a message checksum and two extra block cipher calls. COPA accepts associated data input.

In this section we assume that the message length is a positive multiple of n. The length of associated data can be fractional. In App. A we show how to handle fractional messages with COPE and COPA. At the end of this section we give the design rationale for our constructions, explaining our choice of operations.

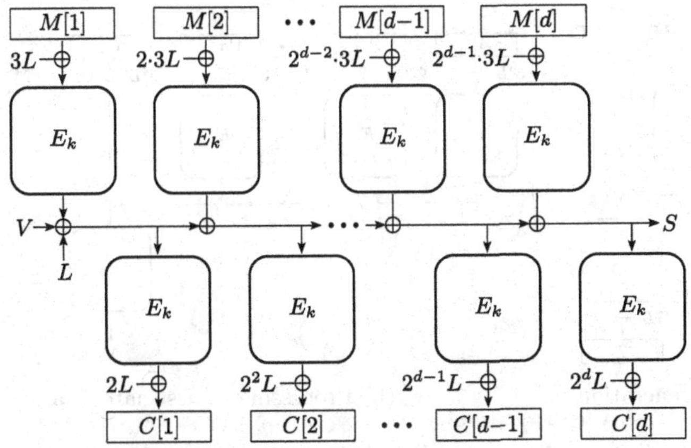

Fig. 1. Online cipher COPE. Set $V \stackrel{\text{def}}{=} 0$ for COPE. Variable S will be used later by COPA.

3.1 COPE Definition

Let $E : \mathcal{K} \times \{0,1\}^n \rightarrow \{0,1\}^n$ be an n-bit block cipher, and denote $L \stackrel{\text{def}}{=} E_k(0)$. The encryption and decryption procedures of the COPE online cipher on a message $M[1]M[2]\cdots M[d]$ of d n-bit blocks and on a ciphertext $C[1]C[2]\cdots C[d]$ are then defined as:

COPE-Encrypt $\mathcal{E}[E]$:

$V[0] \leftarrow L, \Delta_0 \leftarrow 3L, \Delta_1 \leftarrow 2L$
for $i = 1, \ldots, d$ **do**
  $V[i] \leftarrow E_k(M[i] \oplus \Delta_0) \oplus V[i-1]$
  $C[i] \leftarrow E_k(V[i]) \oplus \Delta_1$
  $\Delta_0 \leftarrow 2\Delta_0, \Delta_1 \leftarrow 2\Delta_1$
end for

COPE-Decrypt $\mathcal{E}^{-1}[E]$:

$V[0] \leftarrow L, \Delta_0 \leftarrow 3L, \Delta_1 \leftarrow 2L$
for $i = 1, \ldots, d$ **do**
  $V[i] \leftarrow E_k^{-1}(C[i] \oplus \Delta_1)$
  $M[i] \leftarrow E_k^{-1}(V[i] \oplus V[i-1]) \oplus \Delta_0$
  $\Delta_0 \leftarrow 2\Delta_0, \Delta_1 \leftarrow 2\Delta_1$
end for

The encryption operation is illustrated in Fig. 1.

3.2 COPA Definition

The core of the authenticated online cipher COPA is identical to COPE. The only differences are that first, an authentication tag T is generated after the COPE cipher invocation, and second, that associated data (if any) is processed before the cipher iteration to produce a value V that is XOR-ed into the first intermediate block chaining (see Fig. 1): $V[0] \leftarrow V \oplus L$. If there is no associated data, then we set $V \stackrel{\text{def}}{=} 0$.

The tag T is computed by keeping a XOR checksum of the message blocks $\Sigma \stackrel{\text{def}}{=} M[1] \oplus \cdots \oplus M[d]$ and computing

$$T \leftarrow E_k\big(E_k(\Sigma \oplus 2^{d-1}3^2L) \oplus S\big) \oplus 2^{d-1}7L,$$

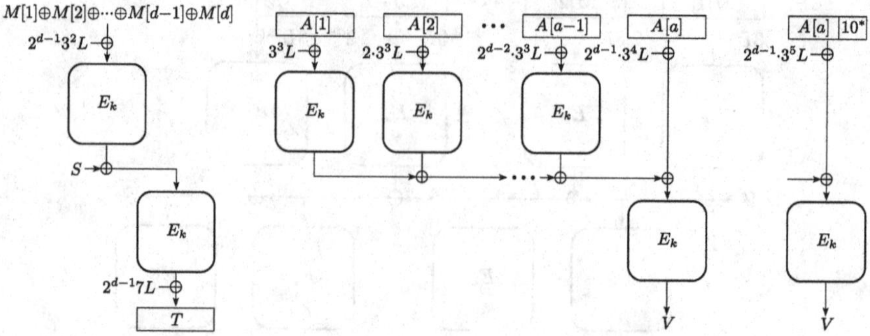

(a) Tag generation (b) Processing of associated data

Fig. 2. Authenticated online cipher COPA: tag generation and processing of associated data

with $S \stackrel{\text{def}}{=} V[d]$ denoting the last intermediate value in COPE's block chaining, as in Fig. 1. The tag computation is illustrated in Fig. 2a. The value V is generated as follows: any associated data $A[1], \ldots, A[a]$ is padded (if not a multiple of n bits) by a one and as many zeroes as necessary to obtain a multiple of the block size n. These blocks are then processed by a PMAC1-like [25] iteration as illustrated in Fig. 2b. Here, the block "$A[a]10^*$" replaces the block "$A[a]$" if $A[a]$ itself is not n bits. Tag verification occurs by checking if

$$S \oplus E_k(\Sigma \oplus 2^{d-1}3^2L) = E_k^{-1}(T \oplus 2^{d-1}7L),$$

where the tag is rejected if the equality is not true.

3.3 COPE and COPA for Arbitrary-Length Messages

We explain how to extend our schemes to accept "fractional" messages M in App. A. Here the length $|M|$ is not necessarily a positive multiple of the block size n. Note that simply using 10^* padding to M would result in ciphertext expansion. The methods described in App. A avoid such expansion with minimal loss of efficiency.

3.4 Design Rationale

One could combine universal hashing with a block cipher to design an AE scheme. Indeed, McOE-G [11] follows this approach. However, we decided to avoid the use of universal hashing, for three reasons. First, the use of universal hashing would result in additional implementation cost, in particular with hardware. Second, recent study shows that there is an issue of weak keys with polynomial-based hashing [22]. Third, on the latest Intel CPUs, one call of AES is faster than one multiplication over the finite field GF(2^{128}), which is an operation used for polynomial-based hashing.

There has been discussion of whether one should use the doubling method or Gray code to produce tweak masks. We decided to use doubling, for three reasons. First, doubling provides us with the framework of XE and XEX constructions, which makes our constructions and proofs simple and easy to analyze. Neither our constructions nor our proofs can be directly translated into a Gray code version, as it is not immediately clear which masks we should use for the construction to make the proof work. Second, although it was reported that Gray code performs better than doubling on Intel CPUs [16], recent study shows that the doubling method can be implemented equally efficiently [3]. Third, the speedup of Gray code mask generation requires precomputation and memory, whereas doubling does not.

4 Privacy of COPE

4.1 Security Definition of Online Ciphers

We use the security definition of online ciphers from Rogaway and Zhang [27]. Let $(\{0,1\}^n)^+$ denote the set of strings whose length is a positive multiple of n bits and is at most $2^n \cdot n$ bits. An online cipher $\mathcal{E} : \mathcal{K} \times (\{0,1\}^n)^+ \to (\{0,1\}^n)^+$ is a function such that it is a permutation on every block of n bits, having the additional feature that the outputs are the same for a common prefix. In other words, the first $|M|$ bits of $\mathcal{E}_k(M\|N)$ and $\mathcal{E}_k(M\|N')$ are the same for any $M, N, N' \in (\{0,1\}^n)^+$. So an online cipher $\mathcal{E}_k$ yields a permutation of i-th blocks, where the permutation is determined by the prefix (i.e. the first $i-1$ blocks). Let OPerm(n) be the set of all such permutations $\pi : (\{0,1\}^n)^+ \to (\{0,1\}^n)^+$.

Definition 3. *Let $\mathcal{E}$ be an online cipher. The IND-CPA advantage of a distinguisher $\mathcal{D}$ is defined as*

$$\mathbf{Adv}_{\mathcal{E}}^{\mathrm{cpa}}(\mathcal{D}) = \left| \Pr_k[\mathcal{D}^{\mathcal{E}_k} = 1] - \Pr_\pi[\mathcal{D}^{\pi} = 1] \right| .$$

Here, $\mathcal{D}$ is a distinguisher with oracle access to either $\mathcal{E}_k$ or π. The probabilities are taken over $k \xleftarrow{\$} \mathcal{K}$, $\pi \xleftarrow{\$}$ OPerm(n) and random coins of $\mathcal{D}$, if any. By $\mathbf{Adv}_{\mathcal{E}}^{\mathrm{cpa}}(t, q, \sigma, \ell)$ we denote the maximum advantage taken over all distinguishers that run in time t and make q queries, each of length at most ℓ blocks, and of total length at most σ blocks.

4.2 IND-CPA Proof Sketch

This section gives a sketch of the proof showing that COPE is secure against chosen-plaintext attacks with respect to privacy (IND-CPA). The details of the proof can be found in the full paper [2].

Theorem 1. *Let $\mathcal{E}[E]$ denote COPE, where E is the underlying block cipher. We have*

$$\mathbf{Adv}_{\mathcal{E}[E]}^{\mathrm{cpa}}(t, q, \sigma, \ell) \le \frac{38\sigma^2}{2^n} + \mathbf{Adv}_E^{\mathrm{prp}\pm 1}(t', 4\sigma) + \frac{(\ell+1)(q-1)^2}{2^n},$$

where $t' \approx t$.

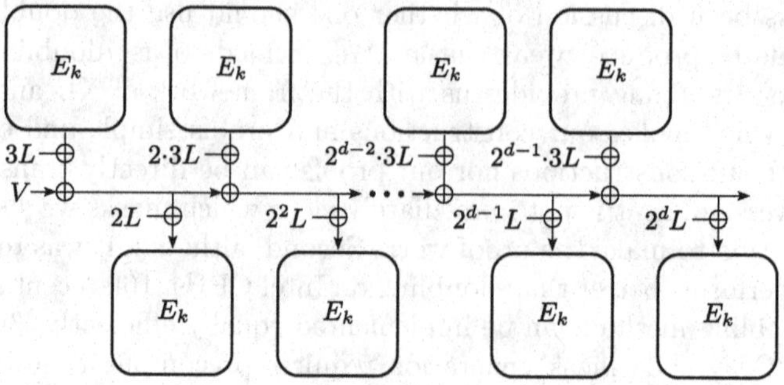

Fig. 3. IND-CPA proofs of COPE: introducing dummy masks rewriting the scheme in terms of XEX

The proof consists of two steps. First, we rewrite COPE in terms of XEX constructions.[2] We introduce dummy masks to the state values, as shown in Fig. 3. The block cipher calls in the upper layer are now $\widetilde{\boldsymbol{E}}_{k,\alpha-1,1,0}$ and those in the lower layer $\widetilde{\boldsymbol{E}}_{k,\alpha,0,0}$. Note that the "$L$" initially XORed to the state now disappears. We use Lem. 1 to replace the block cipher calls in the upper layer with random permutations $\boldsymbol{\pi}_{\alpha-1,1,0}$ and those in the lower layer with $\boldsymbol{\pi}_{\alpha,0,0}$ (for $\alpha = 1, 2, \ldots$). Such a replacement costs us

$$\frac{9.5 \cdot (2\sigma)^2}{2^n} + \mathbf{Adv}_E^{\mathrm{prp}\pm1}(t', 2 \cdot 2\sigma) = \frac{38\sigma^2}{2^n} + \mathbf{Adv}_E^{\mathrm{prp}\pm1}(t', 4\sigma).$$

We write $\mathcal{E}[\boldsymbol{\pi}]$ to denote the COPE scheme making calls to independent random permutations $\boldsymbol{\pi}_{\alpha\beta\gamma}$ rather than to a block cipher.

Second, we show that $\mathcal{E}[\boldsymbol{\pi}]$ behaves exactly the same as the ideal functionality, as long as collisions of state values do not occur. Define variables $V[\alpha]$ of state values as $V[\alpha] \stackrel{\mathrm{def}}{=} \bigoplus_{i=1}^{\alpha} \boldsymbol{\pi}_{i-1,1,0}(M[i])$ which is also equal to $\boldsymbol{\pi}_{\alpha,0,0}^{-1}(C[\alpha])$.

We look for *collisions* of these variables. Here by a "collision" roughly we mean the same value of $V[\alpha]$ coming from different prefixes $M[1]M[2]\cdots M[\alpha]$ and $M'[1]M'[2]\cdots M'[\alpha]$, for some α. More precisely, we have a collision of $V[\alpha] = V'[\alpha]$ if we have $V[\alpha-1] \neq V'[\alpha-1]$ and $V[\alpha] = V'[\alpha]$, which implies we must have $M[\alpha] \neq M'[\alpha]$ and also $M[i] \neq M'[i]$ for some $i < \alpha$. Let C denote the event that a collision of $V[\alpha]$ occurs for some α.

Claim. Unless C occurs, $\mathcal{E}[\boldsymbol{\pi}]$ is indistinguishable from the ideal functionality. Furthermore, we have $\Pr\left[\mathcal{D}^{\mathcal{E}[\boldsymbol{\pi}]} \text{ sets } \mathsf{C}\right] \leq (\ell + 1)/2^n$.

[2] The reason why our IND-CPA COPE is based on XEX constructions, and not on XEs, is because our COPA, which gives decryption oracle access to adversaries, builds upon COPE.

5 Privacy and Integrity of COPA

5.1 Security Definition of Authenticated Online Ciphers

Also for authenticated online ciphers, we use the IND-CPA security advantage of Def. 3, except that the ideal encryption oracle now has an additional random function that maps $\{0,1\}^* \times (\{0,1\}^n)^+$ to $\{0,1\}^n$, corresponding to $(A, M) \mapsto T$.

We use the notion of integrity of authenticated encryption schemes from Fleischmann et al. [11]. By $\perp$, we denote a function that returns $\perp$ on every input.

Definition 4. *Let $\mathcal{E}$ be an online cipher. The integrity advantage of a distinguisher $\mathcal{D}$ is defined as*

$$\mathbf{Adv}_{\mathcal{E}}^{\mathrm{int}}(\mathcal{D}) = \left| \Pr_k\left[\mathcal{D}^{\mathcal{E}_k^{\pm 1}} = 1\right] - \Pr_k\left[\mathcal{D}^{\mathcal{E}_k, \perp} = 1\right] \right| .$$

Here, $\mathcal{D}$ is a distinguisher with oracle access to either $(\mathcal{E}_k, \mathcal{E}_k^{-1})$ or $(\mathcal{E}_k, \perp)$. To avoid a trivial win, we assume that the distinguisher does not make a query (A, C, T) if it has made a query (A, M) to the encryption oracle and obtained (C, T) from the oracle. By $\mathbf{Adv}_{\mathcal{E}}^{\mathrm{int}}(t, q, \sigma, \ell)$ we denote the maximum advantage taken over all distinguishers that run in time t and make q queries, each of length at most ℓ blocks, and of total length at most σ blocks.

5.2 Privacy of COPA

We now give a proof sketch of the IND-CPA security of COPA. The details can be found in the full paper [2].

Theorem 2. *Let $\mathcal{E}[E]$ denote COPA, where E is the underlying block cipher. We have*

$$\mathbf{Adv}_{\mathcal{E}[E]}^{\mathrm{cpa}}(t, q, \sigma, \ell) \leq \frac{39(\sigma + q)^2}{2^n} + \mathbf{Adv}_E^{\mathrm{prp}\pm 1}(t', 4(\sigma + q)) + \frac{(\ell + 2)(q-1)^2}{2^n},$$

where $t' \approx t$.

The IND-CPA security analysis of COPE carries over, with only minor modifications. First, we introduce dummy masks in a similar way (to the encryption part), and replace all XE (in the associated-data part) and XEX constructions by random permutations. This replacement costs us

$$\frac{9.5 \cdot (2\sigma + 2q)^2}{2^n} + \mathbf{Adv}_E^{\mathrm{prp}\pm 1}(t', 2 \cdot 2(\sigma + q)) = \frac{38(\sigma + q)^2}{2^n} + \mathbf{Adv}_E^{\mathrm{prp}\pm 1}(t', 4(\sigma + q)).$$

Write $\mathcal{E}[\pi, \boldsymbol{\pi}]$ to denote the COPA scheme calling random permutations instead of a block cipher.

Next, we again use the collision event **C**, but introduce two more events. One is **A**, which is the event that we have a collision of V for two different associated data. Recall that for $A = \varnothing$, we have $V = 0$. The other is T, which is the event that we have a collision of tag values for messages of the same length (or more precisely, a collision of input values to a random permutation that produces tags).

Claim. Unless $\mathsf{A} \vee \mathsf{C} \vee \mathsf{T}$ occurs, $\mathcal{E}[\pi, \boldsymbol{\pi}]$ is indistinguishable from the ideal functionality.

Lemma 2 (PMAC1, [25]). *The function $H[\pi] : \{0,1\}^* \to \{0,1\}^n$ $(A \mapsto V)$ is indistinguishable from a random function $\Phi : \{0,1\}^* \to \{0,1\}^n$. Specifically, the distinguishing advantage (defined accordingly, only forward queries) is at most $\sigma^2/2^n$. Here, $\{0,1\}^*$ denotes the set of strings whose length is at most $2^n \cdot n$ bits.*

So now we replace XE and XEX constructions with random permutations and $H[\pi]$ with a random function Φ. Denote the scheme by $\mathcal{E}[\Phi, \boldsymbol{\pi}]$. Then we have the following.

Claim. We have $\Pr[\mathcal{D}^{\mathcal{E}[\Phi,\boldsymbol{\pi}]} \text{ sets } \mathsf{A}] \le q^2/2^n$ and $\Pr[\mathcal{D}^{\mathcal{E}[\Phi,\boldsymbol{\pi}]} \text{ sets } \mathsf{C} \vee \mathsf{T} \mid \neg\mathsf{A}] \le (\ell + 2)(q-1)^2/2^n$.

5.3 Integrity of COPA

The integrity proof of COPA is more involved than the privacy proofs and we include the full proof in this paper. We prove the following theorem:

Theorem 3. *Let $\mathcal{E}[E]$ denote COPA, where E is the underlying block cipher. We have*

$$\mathbf{Adv}^{\mathrm{int}}_{\mathcal{E}[E]}(t, q, \sigma, \ell) \le \frac{39(\sigma+q)^2}{2^n} + \mathbf{Adv}^{\mathrm{prp}\pm 1}_{E}(t', 4(\sigma+q)) + \frac{(\ell+2)(q-1)^2}{2^n} + \frac{2q}{2^n},$$

where $t' \approx t$.

Let F denote the event that the decryption oracle $\mathcal{E}_k^{-1}$ returns something other than $\perp$. Clearly the two games are the same as long as the event F does not occur, so we have

$$\Pr[\mathcal{D}^{\mathcal{E}_k^{\pm 1}} = 1] - \Pr[\mathcal{D}^{\mathcal{E}_k, \perp} = 1] \le \Pr[\mathcal{D}^{\mathcal{E}_k^{\pm 1}} \text{ sets } \mathsf{F}].$$

In the rest of this section we shall bound this probability. First, as usual, we replace block cipher calls with random permutations $\pi, \boldsymbol{\pi}$. Then we replace the PMAC1 part of processing associated data with a random function Φ. These all together cost us (cf. the proof of Thm. 2)

$$\frac{38(\sigma+q)^2}{2^n} + \frac{\sigma^2}{2^n} + \mathbf{Adv}^{\mathrm{prp}\pm 1}_{E}(t', 4(\sigma+q)).$$

Removing "Privacy Part". Define events A, C and T as we have done in the privacy proof of Thm. 2. Note that these events are defined in terms of variables $V[\alpha]$, where we also define $V[0] \stackrel{\text{def}}{=} V$ and $V'[\alpha+1]$ the input value to the block cipher that produces a tag. We define these values as being set only by the queries to the encryption oracle $\mathcal{E}$. We do not let queries to the decryption oracle $\mathcal{E}^{-1}$ affect variables $V[\cdot], V'[\cdot]$, whether or not it returns a message or $\perp$. Set $\mathsf{E} \stackrel{\text{def}}{=} \mathsf{A} \vee \mathsf{C} \vee \mathsf{T}$.

Next we define similar events A', C' and T'. These are exactly the same as the previous ones, except that now we consider only those events (i.e. collisions of $V[\cdot]$ or $V'[\cdot]$) that occur prior to a forgery (that is, under the condition $\neg\mathsf{F}$). Again, set $\mathsf{E}' \stackrel{\text{def}}{=} \mathsf{A}' \vee \mathsf{C}' \vee \mathsf{T}'$.

When we consider event F, we would like to assume that we are under the condition $\neg\mathsf{E}'$, meaning that the encryption oracle $\mathcal{E}$ has behaved ideally so far (till forgery). To do this, we use the inequality

$$\Pr\big[\mathcal{D}^{\mathcal{E}^{\pm 1}[\Phi,\pi]} \text{ sets } \mathsf{F}\big] \le \Pr\big[\mathcal{D}^{\mathcal{E}^{\pm 1}[\Phi,\pi]} \text{ sets } \mathsf{F} \bigm| \neg\mathsf{E}'\big] + \Pr\big[\mathcal{D}^{\mathcal{E}^{\pm 1}[\Phi,\pi]} \text{ sets } \mathsf{E}'\big].$$

We shall construct a distinguisher $\mathcal{D}'$ that breaks the privacy of the encryption oracle $\mathcal{E}$. The distinguisher $\mathcal{D}'$ uses $\mathcal{D}$, and the query complexity of $\mathcal{D}'$ is at most that of $\mathcal{D}$. Specifically, $\mathcal{D}'$ starts running $\mathcal{D}$, answering $\mathcal{E}$-queries using its $\mathcal{E}$ oracle, and whenever $\mathcal{D}$ makes a query to the decryption oracle $\mathcal{E}^{-1}$, $\mathcal{D}'$ replies with a $\perp$.

Claim. We have $\Pr\big[\mathcal{D}^{\mathcal{E}^{\pm 1}[\Phi,\pi]} \text{ sets } \mathsf{E}'\big] \le q^2/2^n + (\ell+2)(q-1)^2/2^n$.

Proof. Note that if $\mathcal{D}^{\mathcal{E}^{\pm 1}}$ sets E', then till this event $\mathcal{D}'$ simulates the environment of $\mathcal{D}$ correctly. Hence we get $\Pr\big[\mathcal{D}^{\mathcal{E}^{\pm 1}} \text{ sets } \mathsf{E}'\big] \le \Pr\big[\mathcal{D}'^{\mathcal{E}} \text{ sets } \mathsf{E}\big]$, which is less than $q^2/2^n + (\ell+2)(q-1)^2/2^n$ as shown in the privacy proof. □

Passing to a Single-Query Adversary. So it remains to evaluate the probability that $\mathcal{D}$ sets F under the condition $\neg\mathsf{E}'$. We shall construct a forger $\mathcal{D}_1$ from $\mathcal{D}$. The forger $\mathcal{D}_1$ makes multiple queries to the encryption oracle $\mathcal{E}$ but makes only one query to the decryption oracle $\mathcal{E}^{-1}$ at the end of its run. We define $\mathcal{D}_1$ as follows: it chooses a random index $i \in [1,q]$. It then runs $\mathcal{D}$, answering its $\mathcal{E}$-queries using the $\mathcal{E}$ oracle of $\mathcal{D}_1$ and answering the queries to the decryption oracle $\mathcal{E}^{-1}$ with $\perp$. When $\mathcal{D}$ makes the i-th query $(A^\star, C^\star, T^\star)$ to the decryption oracle, $\mathcal{D}_1$ outputs the query $(A^\star, C^\star, T^\star)$ and stops (or more precisely, makes that query to the decryption oracle $\mathcal{E}^{-1}$ and stops.)

Claim. We have $\Pr\big[\mathcal{D}^{\mathcal{E}^{\pm 1}[\Phi,\pi]} \text{ sets } \mathsf{F} \bigm| \neg\mathsf{E}'\big] \le q \Pr\big[\mathcal{D}_1^{\mathcal{E}^{\pm 1}[\Phi,\pi]} \text{ sets } \mathsf{F} \bigm| \neg\mathsf{E}'\big]$.

Proof. Let F_h denote the event that at the h-th query the decryption oracle $\mathcal{E}^{-1}$ returns something other than $\perp$ for the first time; that is, the oracle has returned only $\perp$ so far. Clearly these are disjoint events, and we have $\mathsf{F} = \bigvee_{h=1}^{q} \mathsf{F}_h$. Then, under the events $\neg\mathsf{E}'$ and $i = h$, the forger $\mathcal{D}_1$ correctly simulates the game of $\mathcal{D}$. Therefore, we get $\Pr\big[\mathcal{D}_1^{\mathcal{E}^{\pm 1}} \text{ sets } \mathsf{F} \bigm| \neg\mathsf{E}'\big] \ge \Pr\big[(i = h) \wedge \mathcal{D}^{\mathcal{E}^{\pm 1}} \text{ sets } \mathsf{F} \bigm| \neg\mathsf{E}'\big] \ge (1/q) \Pr\big[\mathcal{D}^{\mathcal{E}^{\pm 1}} \text{ sets } \mathsf{F} \bigm| \neg\mathsf{E}'\big]$. □

Evaluating Forgery Probabilities. Let $(A^\star, C^\star, T^\star)$ denote the (non-trivial) query made by $\mathcal{D}_1$ to the decryption oracle $\mathcal{E}^{-1}[\Phi, \pi]$. We shall evaluate the probability that this would make $\mathcal{E}^{-1}$ return something other than $\perp$. To evaluate the probability, we shall consider the following cases.

Lemma 3 (Case 1). *If $A^\star$ or $T^\star$ is new, or $C^\star$ contains a new block, then the probability of a forgery is at most $2/2^n$.*

Proof. If $A^\star$ is new, then it means that it triggers the random function Φ and yields a fresh random value $V \leftarrow \Phi(A^\star)$. This value is XORed to the value that is input to the block cipher to produce the tag, which must be equal to $T^\star$. All other values XORed to the value are independent of V. Hence, regardless of the values $C^\star, T^\star$, the probability of such an event is at most $1/2^n$.

Say that $A^\star$ is not new, but that $C^\star$ contains a new block. Let $C^\star[\alpha]$ be one of the new blocks. The decryption invokes $\pi^{-1}_{\alpha,0,0}(C^\star[\alpha])$, which is sampled from the set of at least $2^n - q$ points. Therefore, the probability of a forgery is at most $1/(2^n - q) \leq 2/2^n$, assuming $q \leq 2^{n-1}$.

Say that $A^\star$ is not new, $C^\star$ does not contains a new block, but $T^\star$ is new. This is similar to the previous case. This would trigger a fresh point of $\pi^{-1}_{d^\star-1,0,1}(T^\star)$, where $d^\star$ denotes the number of blocks in the message $M^\star$. The point is sampled from the set of at least $2^n - q$ points. Therefore, the probability of a forgery is at most $1/(2^n - q) \leq 2/2^n$. □

Lemma 4 (Case 2). *If $A^\star$ and $T^\star$ are old, and $C^\star$ consists of old blocks, then the probability of a forgery is at most $2/2^n$.*

Proof. To handle this case, we introduce some notation. For the query $(A^\star, C^\star, T^\star)$ in question, divide $C^\star$ into blocks as $C^\star[1]C^\star[2]\cdots C^\star[d^\star] \leftarrow C^\star$ and define $C^\star[0] \stackrel{\text{def}}{=} A^\star$ and $C^\star[d^\star+1] \stackrel{\text{def}}{=} T$. We then focus on a pair of adjacent "blocks" $(C^\star[\alpha-1], C^\star[\alpha])$ for $\alpha = 1, 2, \ldots, d^\star + 1$. We call a pair *old* if it (as a pair) has already appeared in some previous query made to the encryption oracle $\mathcal{E}$ and in the corresponding value returned by the oracle. That is, if $\mathcal{D}$ has made a query (A', M') to the oracle and got (C', T') back, then we check if the pair in question $(C^\star[\alpha-1], C^\star[\alpha])$ is contained in (A', C', T')—that is, we check if $(C^\star[\alpha-1], C^\star[\alpha]) = (C'[\alpha-1], C'[\alpha])$ holds, where $C'[0]$ and $C'[d'+1]$ are defined similarly. We do this for all previous queries. We call the pair $(C^\star[\alpha-1], C^\star[\alpha])$ *new* otherwise.

Note that the query $(A^\star, C^\star, T^\star)$ always contains a new pair. If $(A^\star, C^\star, T^\star)$ contains no new pairs, then, given the non-triviality of the query, we must have observed a collision, contradicting the assumption $\neg\mathsf{E}'$.

We now make a distinction among new pairs $(C^\star[\alpha-1], C^\star[\alpha])$ based on the decrypted message block $M^\star[\alpha]$ from the two adjacent ciphertext blocks. We say that a pair is *collapsing* if there exists a previous query (A', M') made by $\mathcal{D}$ to the encryption oracle $\mathcal{E}$ such that $M'[\alpha] = M^\star[\alpha]$.

There exists a new pair $(C^\star[\alpha-1], C^\star[\alpha])$ that is not collapsing. This case means that we trigger a random sampling of $\pi^{-1}_{\alpha,1,0}$ to compute $M^\star[\alpha]$. Then, note that the value $\Sigma^\star = M^\star[1] \oplus M^\star[2] \oplus \cdots \oplus M^\star[d^\star]$ is already uniquely determined by the values $C^\star[d^\star]$ and $T^\star$ (via Fig. 2a). There are at least $2^n - q$ possible values for $M^\star[\alpha]$, and the message blocks must sum up to this particular value $\Sigma^\star$, which happens with a probability at most $1/(2^n - q) \le 2/2^n$.

All new pairs in $(A^\star, C^\star, T^\star)$ *are collapsing.* This final case is quite different from the previous ones above, as we do not have any fresh sampling of permutations $\pi^{\pm 1}_{\alpha,\beta,\gamma}$ or the random function Φ in evaluating $\mathcal{E}^{-1}[\Phi, \pi](A^\star, C^\star, T^\star)$. To tackle this case, we shall convert this forgery game into one where the adversary $\mathcal{D}^\circ$ tries to find multiple collisions by outputting the following set of values, without making any query to the oracles:

1. $r \in [1, \ell]$,
2. $1 \le \alpha_1 < \alpha_2 < \cdots < \alpha_r \le \ell + 1$,
3. (A_1, M_1), (A_2, M_2), ..., (A_r, M_r), and
4. (A'_1, M'_1), (A'_2, M'_2), ..., (A'_r, M'_r).

The adversary $\mathcal{D}^\circ$ "wins" if the submitted values form a multi-collision in the following sense: (A_i, M_i) and $(A'_i; M'_i)$ collides at the α_i-th block, for all $i \in [1, r]$. The adversary $\mathcal{D}^\circ$ runs $\mathcal{D}_1$, simulating the $\mathcal{E}$ oracle with an ideal functionality. Note that this simulation is correct under the condition $\neg\mathsf{E}'$. When $\mathcal{D}_1$ outputs $(A^\star, C^\star, T^\star)$, $\mathcal{D}^\circ$ first checks for new pairs contained in it. Let $1 \le \alpha_1 < \alpha_2 < \cdots < \alpha_r \le \ell + 1$ be the positions of new pairs. Then $\mathcal{D}^\circ$ checks the history of values (C, T) that it returned. Note that under $\neg\mathsf{E}'$, a block $C^\star[\alpha]$ determines a unique prefix AM. We choose (A_i, M_i) to be the prefix determined by $C^\star[\alpha_i]$. To choose (A'_i, M'_i), let $A'_i M''$ be the prefix determined by $C^\star[\alpha_i - 1]$. Then $\mathcal{D}^\circ$ chooses randomly, from the previously queried values, a message block $M[\alpha] \neq M_i[\alpha]$. Set $M'_i \overset{\text{def}}{=} M''M[\alpha]$. The adversary $\mathcal{D}^\circ$ does this for $i = 1, 2, \ldots$ except for the last block.

- If $\alpha_r < d^\star + 1$, then we know the message checksum $\Sigma^\star = M^\star[1] \oplus \cdots \oplus M^\star[d^\star]$, so $\mathcal{D}^\circ$ does not have to guess the value of $M'_{\alpha_r}[\alpha_r]$.
- If $\alpha_r = d^\star + 1$, then we simply set the last input value to be the checksum of all previous (guessed) message blocks.

Now we observe that as long as all the guesses of the message blocks are correct, $\mathcal{D}^\circ$ wins if $\mathcal{D}_1$ succeeds in forgery of this type. It should be noted that the values returned by $\mathcal{E}$ are independent of the success probabilities in question, under the event $\neg\mathsf{E}'$. Therefore, for a fixed r,

$$\Pr[\mathcal{D}^\circ \text{ wins} \mid r] \ge \frac{1}{q-1} \cdot \cdots \cdot \frac{1}{q-1} \Pr[\mathcal{D}_1 \text{ forges} \mid r] = \frac{1}{(q-1)^{r-1}} \Pr[\mathcal{D}_1 \text{ forges} \mid r].$$

We then calculate $\Pr[\mathcal{D}^\circ \text{ wins} \mid r]$. We do this by lazy sampling of the permutations, and we see that, for a fixed r,

$$\Pr[\mathcal{D}^\circ \text{ wins} \mid r] \le \frac{1}{2^n - 1} \cdot \frac{1}{2^n - 1} \cdot \cdots \cdot \frac{1}{2^n - 1} = \frac{1}{(2^n - 1)^r}.$$

Hence by varying r we get in total

$$\Pr[\mathcal{D}_1 \text{ forges}] \le \sum_{i=1}^{\ell} \frac{(q-1)^{i-1}}{(2^n - 1)^i} \Pr[i = r] \le \frac{1}{(2^n - 1)} \sum_{i=1}^{\ell} \Pr[i = r] = \frac{1}{(2^n - 1)}.$$

6 Efficient Parallel Implementation

6.1 The Setting

We compare our schemes to some prominent existing online ciphers: TC1 and TC3 [27] being the most efficient previous schemes; and MCBC [21] as a representative for a scheme relying only on block cipher invocations (as opposed to tweakable block ciphers or universal hashing). The modes HCBC1 and MHCBC are implicitly covered by TC1 and TC3, and HCBC2 has a performance inferior to TC3.

For the case of authenticated online ciphers, we exclude modes of operation and dedicated designs that are based on a nonce and rely on its non-reuse (e.g., GCM [19], OCB [16], ALE [7], and AEGIS [28]). Therefore, we compare our COPA design to the McOE family of authenticated encryption algorithms [11], which, to the best of our knowledge, is the only other online scheme not relying on the non-reuse of a nonce. We focus on the McOE-G instance, since McOE-X itself is not secure [20], featuring a key recovery with birthday complexity.

For the concrete instantiation of all schemes, we use the AES-128 [10] as the underlying block cipher, and multiplication in $GF(2^{128})$ as an almost XOR-universal hash function [15]. As target platform for the implementations, we chose the recent generation of Intel microprocessors (Westmere or later) which support the AES-NI instruction set [12] and carryless multiplication [13].

6.2 Implementation Characteristics of COPE and COPA

The online modes proposed in this paper can utilize parallelized execution of block cipher calls in two ways: for messages longer than one block, the encryptions of subsequent message blocks can be carried out independently of each other once the respective masks have been XORed. The same holds for the second series of block cipher calls, once the chaining XORs have been executed.

This parallelism can be exploited in a single-core scenario by pipelining the block cipher rounds for several consecutive block cipher invocations. Similarly, these invocations can be processed independently by multiple threads, with the recombination being the computation of the chaining. Note that both scenarios can be combined when multiple cores with pipelined block cipher calls are available, which is typically the case for Intel's AES-NI architecture.

On the recent Sandy and Ivy Bridge platforms, the AES round function can be computed at a latency of 8 cycles with a throughput of 1 cycle. Consequently, to fully utilize the pipeline, our implementation issues 8 AES round function evaluations on the next 8 consecutive blocks (independent data and same key). The tweak masks are computed using dedicated multiplication routines for 2^α, 3^β and $7^\gamma \in GF(2^{128})$. By contrast, the general $GF(2^{128})$ multiplication needed for TC1, TC3, and McOE-G is implemented using the `PCLMULQDQ` carryless multiplication instruction followed by modular reduction.

Table 1. Software performance of (authenticated) online ciphers based on the AES on the Intel Sandy Bridge platform (AES-NI). All numbers are given in cycles per byte (cpb).

	message length (bytes)						
Algorithm	128	256	512	1024	2048	4096	8192
CTR	1.74	1.27	1.05	0.93	0.86	0.83	0.82
TC1	9.00	8.75	8.65	8.60	8.56	8.56	8.56
TC3	9.08	8.82	8.72	8.67	8.63	8.63	8.62
MCBC	11.66	11.00	10.68	10.52	10.44	10.40	10.38
COPE	2.56	2.08	1.89	1.78	1.72	1.70	1.69
McOE-G	10.85	9.73	9.14	8.90	8.74	8.69	8.66
COPA	3.78	2.85	2.31	2.06	1.94	1.88	1.85

6.3 Performance Measurements

We provide performance data for the (authenticated) encryption of messages of length $16 \cdot 2^b$ bytes, with $3 \leq b \leq 10$. The performance of AES-CTR is provided as a reference point. All measurements were taken on a single core of an Intel Core i5-2520M CPU at 2500 MHz, averaged over $5 \cdot 10^5$ repetitions, processing one message at a time. Our findings are summarized in Table 1. All numbers are given in cycles per byte (cpb).

One can observe that for all message lengths, the parallelizability of the proposed schemes results in speed-ups of factor $4.5 - 5$ in comparison to the existing modes, at least for somewhat longer messages. By fully utilizing the pipeline, our schemes are only marginally slower than two times AES-CTR, which implies that the overhead imposed by the computation of the masks and the chaining is kept at a minimum. The authenticated mode COPA carries the additional overhead of two more AES calls plus field arithmetic for finalization, but this quickly becomes insignificant as the message length increases. Note, however, that some constant overhead in comparison to the unauthenticated mode remains even for very long messages: this can be attributed to the fact that the computation of the checksum does not allow overwriting the message blocks, leading to increased register pressure. We also note that with the availability of carryless multiplication, TC1 and TC3 can be implemented more efficiently than the purely block cipher-based MCBC which was created with the goal to improve performance by avoiding field arithmetic.

The performance of our parallelizable schemes COPE and COPA can be further improved by utilizing multiple cores. Our implementation of multithreaded encryption confirms the intuition that one can expect a nearly linear speedup when using multiple cores for computing our schemes (i.e., the cost is < 1 cpb for two cores and so on).

7 Conclusion

By presenting COPE, our work provides the first solution for a parallelizable online cipher. Building on COPE, we go on to construct COPA, the first parallelizable and nonce-misuse resistant online authenticated encryption scheme. Our implementations of COPE and COPA with Intel AES-NI on a Sandy Bridge processor architecture benefit strongly from the parallelism, which gives us speed-ups of about factor 5 in comparison to existing (serial) online ciphers TC1, TC3, MCBC and the online AE scheme McOE-G.

Our designs additionally employ only a single key and use only a block cipher as a building block—as opposed to tweakable block ciphers or universal hash functions. We prove that our cipher COPE is an IND-CPA secure online permutation. The privacy result is also carried over to COPA. The integrity proof of COPA uses a technique of converting a forgery to a set of multiple collisions. It seems that the technique has not been used before by security proofs of parallelizable authenticated encryption mode or message authentication code. The technique may be applicable to other new types of parallelizable modes of operation. We leave it as an interesting open problem to construct a scheme with less primitive calls but with comparable security guarantees.

Acknowledgments. This work has been funded in part by the IAP Program P6/26 BCRYPT of the Belgian State (Belgian Science Policy), in part by the European Commission through the ICT program under contract ICT-2007-216676 ECRYPT II, in part by the Research Council KU Leuven: GOA TENSE, and in part by the Research Fund KU Leuven, OT/08/027. Elena Andreeva is supported by a Postdoctoral Fellowship from the Flemish Research Foundation (FWO-Vlaanderen). Bart Mennink is supported by a Ph.D. Fellowship from the Institute for the Promotion of Innovation through Science and Technology in Flanders (IWT-Vlaanderen).

References

1. Akdemir, K., Dixon, M., Feghali, W., Fay, P., Gopal, V., Guilford, J., Erdinc Ozturk, G.W., Zohar, R.: Breakthrough AES Performance with Intel AES New Instructions. Intel white paper (January 2010)
2. Andreeva, E., Bogdanov, A., Luykx, A., Mennink, B., Tischhauser, E., Yasuda, K.: Parallelizable and authenticated online ciphers. Cryptology ePrint Archive (2013), full version of this paper
3. Aoki, K., Iwata, T., Yasuda, K.: How Fast Can a Two-Pass Mode Go? A Parallel Deterministic Authenticated Encryption Mode for AES-NI (Extended Abstract of Work in Progress). In: Directions in Authenticated Ciphers (DIAC) (July 2012)
4. Bellare, M., Boldyreva, A., Knudsen, L.R., Namprempre, C.: Online Ciphers and the Hash-CBC Construction. In: Kilian, J. (ed.) CRYPTO 2001. LNCS, vol. 2139, pp. 292–309. Springer, Heidelberg (2001)
5. Bellare, M., Rogaway, P., Wagner, D.: The EAX Mode of Operation. In: Roy, B., Meier, W. (eds.) FSE 2004. LNCS, vol. 3017, pp. 389–407. Springer, Heidelberg (2004)

6. Bernstein, D.J., Schwabe, P.: New AES Software Speed Records. In: Chowdhury, et al. [8], pp. 322–336
7. Bogdanov, A., Mendel, F., Regazzoni, F., Rijmen, V., Tischhauser, E.: ALE: AES-Based Lightweight Authenticated Encryption. In: FSE 2013. LNCS. Springer (to appear, 2013)
8. Chowdhury, D.R., Rijmen, V., Das, A. (eds.): INDOCRYPT 2008. LNCS, vol. 5365. Springer, Heidelberg (2008)
9. Daemen, J.: Hash Function and Cipher Design: Strategies Based on Linear and Differential Cryptanalysis. Ph.D. thesis, Katholieke Universiteit Leuven, Leuven, Belgium (1995)
10. Daemen, J., Rijmen, V.: The Design of Rijndael: AES - The Advanced Encryption Standard. Springer (2002)
11. Fleischmann, E., Forler, C., Lucks, S.: McOE: A Family of Almost Foolproof On-Line Authenticated Encryption Schemes. In: Canteaut, A. (ed.) FSE 2012. LNCS, vol. 7549, pp. 196–215. Springer, Heidelberg (2012)
12. Gueron, S.: Intel Advanced Encryption Standard (AES) Instructions Set. Intel white paper (September 2012)
13. Gueron, S., Kounavis, M.: Intel Carry-Less Multiplication Instruction and its Usage for Computing the GCM mode. Intel white paper (September 2012)
14. Käsper, E., Schwabe, P.: Faster and Timing-Attack Resistant AES-GCM. In: Clavier, C., Gaj, K. (eds.) CHES 2009. LNCS, vol. 5747, pp. 1–17. Springer, Heidelberg (2009)
15. Krawczyk, H.: LFSR-Based Hashing and Authentication. In: Desmedt, Y.G. (ed.) CRYPTO 1994. LNCS, vol. 839, pp. 129–139. Springer, Heidelberg (1994)
16. Krovetz, T., Rogaway, P.: The Software Performance of Authenticated-Encryption Modes. In: Joux, A. (ed.) FSE 2011. LNCS, vol. 6733, pp. 306–327. Springer, Heidelberg (2011)
17. Liskov, M., Rivest, R.L., Wagner, D.: Tweakable Block Ciphers. In: Yung, M. (ed.) CRYPTO 2002. LNCS, vol. 2442, pp. 31–46. Springer, Heidelberg (2002)
18. Matsui, M., Nakajima, J.: On the Power of Bitslice Implementation on Intel Core2 Processor. In: Paillier, P., Verbauwhede, I. (eds.) CHES 2007. LNCS, vol. 4727, pp. 121–134. Springer, Heidelberg (2007)
19. McGrew, D.A., Viega, J.: The Security and Performance of the Galois/Counter Mode (GCM) of Operation. In: Canteaut, A., Viswanathan, K. (eds.) INDOCRYPT 2004. LNCS, vol. 3348, pp. 343–355. Springer, Heidelberg (2004)
20. Mendel, F., Mennink, B., Rijmen, V., Tischhauser, E.: A Simple Key-Recovery Attack on McOE-X. In: Pieprzyk, J., Sadeghi, A.-R., Manulis, M. (eds.) CANS 2012. LNCS, vol. 7712, pp. 23–31. Springer, Heidelberg (2012)
21. Nandi, M.: Two New Efficient CCA-Secure Online Ciphers: MHCBC and MCBC. In: Chowdhury, et al. [8], pp. 350–362
22. Procter, G., Cid, C.: On Weak Keys and Forgery Attacks against Polynomial-based MAC Schemes. In: FSE 2013. LNCS. Springer (to appear, 2013)
23. Ristenpart, T., Rogaway, P.: How to Enrich the Message Space of a Cipher. In: Biryukov, A. (ed.) FSE 2007. LNCS, vol. 4593, pp. 101–118. Springer, Heidelberg (2007)
24. Rogaway, P.: Bucket Hashing and Its Application to Fast Message Authentication. In: Coppersmith, D. (ed.) CRYPTO 1995. LNCS, vol. 963, pp. 29–42. Springer, Heidelberg (1995)
25. Rogaway, P.: Efficient Instantiations of Tweakable Blockciphers and Refinements to Modes OCB and PMAC. In: Lee, P.J. (ed.) ASIACRYPT 2004. LNCS, vol. 3329, pp. 16–31. Springer, Heidelberg (2004)

26. Rogaway, P., Bellare, M., Black, J., Krovetz, T.: OCB: a block-cipher mode of operation for efficient authenticated encryption. In: Reiter, M.K., Samarati, P. (eds.) ACM Conference on Computer and Communications Security, pp. 196–205. ACM (2001)
27. Rogaway, P., Zhang, H.: Online Ciphers from Tweakable Blockciphers. In: Kiayias, A. (ed.) CT-RSA 2011. LNCS, vol. 6558, pp. 237–249. Springer, Heidelberg (2011)
28. Wu, H., Preneel, B.: AEGIS: A Fast Authenticated Encryption Algorithm. Directions in Authenticated Ciphers. In: SAC 2013. LNCS. Springer (to appear, 2013)

A Handling Arbitrary-Length Messages

A.1 COPE for Arbitrary-Length Messages

Our solution relies on the XLS construction [23] of VIL tweakable ciphers. XLS makes only three block-cipher calls and requires only simple bit operations outside block-cipher calls.

Let $\widetilde{\boldsymbol{E}} : \mathcal{K}\times\mathcal{T}\times\{0,1\}^n \to \{0,1\}^n$ be a tweakable cipher and $E : \mathcal{K}'\times\{0,1\}^n \to \{0,1\}^n$ a block cipher. Then $\mathrm{XLS}[\widetilde{\boldsymbol{E}}, E]$ yields a VIL permutation on $\{0,1\}^{n+*}$, the set of string whose length is between n bits and $2n-1$ bits. Specifically, we get $\mathrm{XLS}[\widetilde{\boldsymbol{E}}, E] : \mathcal{K}\times\mathcal{K}'\times\mathcal{T}\times\{0,1\}^{n+*} \to \{0,1\}^{n+*}$. Using appropriate choice of (α,β,γ), we can realize the ciphers used in XLS by the underlying block cipher in COPE encryption scheme $\mathcal{E}$, dependent on the message length d. So we write $\mathrm{XLS}_{k,d}$ to denote the XLS invocation in COPE.

Let M be a message of at least n bits. Divide it into blocks as $M[1]M[2]\cdots M[d-1]M[d] \leftarrow M$, and assume that we have $1 \le |M[d]| \le n-1$. Then we can define $C \leftarrow \mathcal{E}_k(M)$ as

$$\begin{aligned} C[1]C[2]\cdots C[d-2], S &\leftarrow \mathcal{E}_k\big(M[1]M[2]\cdots M[d-2]\big) \text{ (let } \mathcal{E}_k \text{ output } S \text{ for now)} \\ C[d-1]C[d] &\leftarrow \mathrm{XLS}_{k,d}\big((M[d-1]\oplus S)\|M[d]\big) \\ C &\leftarrow C[1]C[2]\cdots C[d]. \end{aligned}$$

The IND-CPA proof of COPE carries over with minor modifications. Note that we have to "wait" the processing of $M[d-1]$ till receiving $M[d]$ (or "redo" after receiving), making the scheme less online. Yet, we make only three calls to the block cipher to process these two blocks.

We require $|M| \ge n$. As pointed out by [27], it seems a challenging problem to handle the case $|M| < n$ with encryption-only online ciphers in a secure manner.

A.2 COPA for Arbitrary-Length Messages

There are solutions of arbitrary-length messages for COPA also. This time we can take the advantage of the tag to handle even the case $|M| < n$. The solution for the case $|M| > n$ also becomes more efficient owing to the presence of tags.

Tag Splitting for $|M| < n$. We can do a trick similar to tag splitting [11] if $|M| < n$. We first choose appropriate parameters (α, β, γ) to make it independent of the ordinary COPA encryption algorithm $\mathcal{E}$. Write it $\mathcal{E}_k^*$ (which will be used only for fractional one-block messages). Given M such that $|M| = s < n$, we can define $(C, T) \leftarrow \mathcal{E}_k(M)$ as

$$\begin{aligned}(C', T') &\leftarrow \mathcal{E}_k^*(M10^*)\\ C &\leftarrow \lceil C' \rceil_s \text{ (leftmost } s \text{ bits)}\\ T &\leftarrow \lfloor C' \rfloor_{n-s} \lceil T' \rceil_s.\end{aligned}$$

One can directly verify the security of this extension. Note that the integrity relies on the 10^* padding as well as on the "partial" tag $\lceil T' \rceil_s$.

XLS for $|M| > n$. Our solution for this case is similar to that of COPE but is more efficient, in that COPA still remains fully online. Again, let M be a message whose length is more than n bits. Divide it into blocks as $M[1]M[2] \cdots M[d-1]M[d] \leftarrow M$, and assume that we have $1 \leq |M[d]| \leq n-1$. Then we can define $(C, T) \leftarrow \mathcal{E}_k(M)$ as

$$\begin{aligned}(C', T') &\leftarrow \mathcal{E}_k(M[1]M[2] \cdots M[d-1])\\ C[d]T &\leftarrow \mathrm{XLS}_{k,d}(M[d]T')\\ C &\leftarrow C'C[d],\end{aligned}$$

where $\mathrm{XLS}_{k,d}$ is defined similarly to the case of COPE. Given the security of COPA and XLS, it is straightforward to verify that this extension is also secure.

How to Construct an Ideal Cipher from a Small Set of Public Permutations

Rodolphe Lampe[1,*] and Yannick Seurin[2,**]

[1] University of Versailles, France
[2] ANSSI, Paris, France
rodolphe.lampe@gmail.com, yannick.seurin@m4x.org

Abstract. We show how to construct an ideal cipher with n-bit blocks and n-bit keys (*i.e.* a set of 2^n public n-bit permutations) from a small constant number of n-bit random public permutations. The construction that we consider is the *single-key iterated Even-Mansour cipher*, which encrypts a plaintext $x \in \{0,1\}^n$ under a key $k \in \{0,1\}^n$ by alternatively xoring the key k and applying independent random public n-bit permutations $P_1, \ldots, P_r$ (this construction is also named a *key-alternating cipher*). We analyze this construction in the plain indifferentiability framework of Maurer, Renner, and Holenstein (TCC 2004), and show that twelve rounds are sufficient to achieve indifferentiability from an ideal cipher. We also show that four rounds are necessary by exhibiting attacks for three rounds or less.

Keywords: block cipher, ideal cipher, iterated Even-Mansour cipher, key-alternating cipher, indifferentiability.

1 Introduction

Block Ciphers. Block ciphers are one of the most important classes of primitives in cryptography. They are mainly used to provide confidentiality and authenticity to communication channels or local data storage means, but also to construct hash functions and in other more advanced cryptographic tasks. Syntactically, a block cipher E with message space $\{0,1\}^n$ and key space $\{0,1\}^m$ is a mapping from $\{0,1\}^m \times \{0,1\}^n$ to $\{0,1\}^n$ such that for each key $k \in \{0,1\}^m$, $E(k,\cdot)$ is an (efficiently invertible) permutation. Block cipher designs (virtually all of which rely on the iteration of some key-dependent round function) can be roughly split into two families (with some rare exceptions such as IDEA):

1) Feistel networks [23] and their generalizations, where the round function is given by $(x,y) \mapsto (y, x \oplus F(k_i, y))$, where x and y are the left and right $n/2$ bits of the state, and k_i is the round key; prominent examples include DES, Blowfish, KASUMI, and Camellia for "classical" Feistel networks, and CAST-256 and MARS for generalized Feistel networks;

* This author is partially supported by the French Direction Générale de l'Armement.
** This author is partially supported by the French National Agency of Research: ANR-11-INS-011.

K. Sako and P. Sarkar (Eds.) ASIACRYPT 2013 Part I, LNCS 8269, pp. 444–463, 2013.

2) substitution-permutation networks (SPNs), where one round generally consists of the composition of a round-key addition, a non-linear mixing layer, and a linear diffusion layer; notable examples include AES, SAFER, CRYPTON, SERPENT, PRESENT, and LED.

At an even higher design level, SPNs can be described (by collapsing the non-linear mixing layer and the linear diffusion layer at i-th round into a single n-bit permutation P_i) as successive applications of round-key additions and permutations P_i. Such a structure was named a *key-alternating cipher* by the designers of AES [17,18].

The traditional security notion for a block cipher is pseudorandomness, *i.e.* indistinguishability from a random permutation [41]: namely, no distinguisher with reasonable resources and having black-box access to a permutation (and also to its inverse in a more stringent variant of the security notion) should be able to distinguish whether it is interacting with the block cipher $E(k, \cdot)$ for a randomly chosen key k, or with a truly random permutation. In an asymptotic and more theoretical language, a family of block ciphers indexed by a security parameter meeting this security notion is called a pseudorandom permutation (PRP), or a strong pseudorandom permutation (SPRP) when the distinguisher has also access to the inverse permutation. The classical example of a construction for which we have some provable security results with respect to indistinguishability is the Feistel network. Starting from the seminal Luby-Rackoff paper [42] which showed that the Feistel construction with three rounds yields a PRP when its round functions are pseudorandom [28], and followed by a paper by Patarin [49] showing that four rounds yield a SPRP (which was stated in [42] without proof), a long series of works established refined results in the same vein, such as [43,44,59,50] to name a few.

THE IDEAL CIPHER MODEL. Though there are numerous examples where the standard pseudorandomness assumption is sufficient to prove (in a reductionist sense) the security of a cryptographic scheme (*e.g.* for building a symmetric encryption scheme [3] or a MAC scheme [4]), there are also some settings where it might not be strong enough to derive a security proof. Indeed, in some situations, the adversary has more abilities than merely querying in a black-box way an encryption/decryption oracle. For example, there are some cases where the attacker might have access to a more powerful "related-key" oracle [9,5,1], *i.e.* it can ask encryption and decryption queries for keys that are related (in some limited and attack-dependent way) to the main key of the system.

Ideally, the ultimate security goal for a block cipher would be that it "behaves" as a random and independent permutation for each possible key. This naturally leads to the so-called *ideal cipher model* (ICM), the origin of which can be traced back to Shannon [56]. In the ICM, a block cipher E with n-bit blocks and m-bit keys is drawn at random from the set of $(2^n!)^{2^m}$ possible block ciphers of this form, and made available through oracle queries (for both encryption and decryption) to all parties (including the adversary). This is very similar in spirit to the random oracle model (ROM) [24,8] used to model a perfect hash function.

To the best of our knowledge, this model was first formally used in a security proof by Winternitz [60] and later by Merkle [47] to show respectively the preimage and collision resistance of the Davies-Meyer compression function. The ICM became increasingly popular after Black *et al.* [12] used it to extensively analyze the security of the PGV block cipher-based compression functions [51]. Since then, the ICM has been used to prove the security of a variety of other block cipher-based hash functions [30,31,58,40,46], of key length extension methods for block ciphers [35,21,7,25,26], of symmetric encryption schemes [33], and even of some public-key protocols such as signature schemes [29], ring signature schemes [53], public-key encryption [34], and key exchange protocols [6]. Despite these numerous successful applications, one must not lose from sight that the ICM only gives heuristic insurance just as the ROM [14]. In particular, Black [11] exhibited an (arguably artificial) block cipher-based hash function which is provably collision resistant in the ICM, but becomes insecure when the ideal cipher is instantiated with any concrete block cipher.

With the ICM at hand, the question now becomes: is it possible to argue that a given block cipher design is as close as possible to an ideal cipher? In the standard model, one immediately faces the problem that, unlike for pseudorandomness, it even seems hard to come with a satisfactory definition of what this formally means, without running into impossibility results (similarly to [14] and [11]) following from the fact that a concrete block cipher has a short description, whereas an ideal cipher does not. This unfortunate state of affairs has not prevented cryptanalysts from *disproving* that a concrete block cipher behaves as an ideal cipher by exhibiting some non-random behavior, *i.e.* some non-trivial[1] relation between inputs and outputs of the block cipher that can be found faster than for an ideal cipher, in a setting where the key is random and given to the attacker (known-key attacks), or when the attacker can freely choose the key(s) (chosen-key attacks). A classical example is the complementation property of DES which, despite being often viewed as a "benign" undesirable property, implies that DES does not behave as an ideal cipher. For AES, no such non-random properties were known until Biryukov *et al.* [10] showed that so-called q-multicollisions can be found faster for AES-256 than for an ideal cipher. Known-key and chosen-key attacks were first put forward as an important cryptanalysis goal by Knudsen ans Rijmen [36], and have since then become an active area of research [48,27,54].

Indifferentiability. Though we cannot hope to formalize (not to say prove) that a concrete block cipher behaves as an ideal cipher in any reasonable sense in the standard model, it is possible to obtain positive results in idealized models, *i.e.* by viewing some subcomponent of the block cipher as perfectly random. This perfect subcomponent is made available to all parties as a public oracle, which makes this setting formally distinct from classical indistinguishability. In order to assess whether a cryptographic construction based on an ideal subcomponent

[1] We stress that because of the lack of a rigorous definition, the meaning of non-trivial here is somehow subjective.

is secure, one has to employ the formalism of *indifferentiability*, introduced by Maurer *et al.* [45]. A construction $\mathcal{C}$ (*e.g.* a block cipher), based on some ideal primitive $\boldsymbol{F}$ (*e.g.* a random permutation), is said to be indifferentiable from some target ideal primitive $\boldsymbol{G}$ (*e.g.* an ideal cipher) if there exists an efficient simulator $\mathcal{S}$ (with black-box access to the primitive $\boldsymbol{G}$) such that the two systems $(\mathcal{C}^{\boldsymbol{F}}, \boldsymbol{F})$ and $(\boldsymbol{G}, \mathcal{S}^{\boldsymbol{G}})$ are indistinguishable. Informally, the goal of the simulator is to provide answers which are consistent with what a distinguisher can obtain from $\boldsymbol{G}$, without deviating too much from the distribution of answers of $\boldsymbol{F}$. An indifferentiability result can be interpreted as a way to make sure that the high-level design of the construction $\mathcal{C}$ has no structural defect. More importantly, a composition theorem [45] asserts that if $\mathcal{C}^{\boldsymbol{F}}$ is indifferentiable from $\boldsymbol{G}$, then any cryptosystem proved secure when used with $\boldsymbol{G}$ remains secure when used with $\mathcal{C}^{\boldsymbol{F}}$, therefore allowing modular proofs of security in idealized models.[2]

Soon after its introduction, Coron *et al.* [15] used the indifferentiability framework to revisit the design of a hash function from an ideal cipher: namely they showed that a number of variants of the Merkle-Damgård domain extension method [19,47], used with an ideal cipher in Davies-Meyer mode, are indifferentiable from a random oracle. The converse direction, *i.e.* proving that it is possible to construct an ideal cipher from a random oracle, turned out to be harder to achieve. A first attempt to prove that the Feistel construction with public random round functions is indifferentiable from a random permutation (and hence from an ideal cipher by prepending the key to each input to the random round functions) was made by Coron *et al.* for six rounds [16], and later by Seurin for ten rounds [55], but serious flaws were found in both proofs [37,32]. The situation was corrected with a proof by Holenstein *et al.* [32] that the 14-round Feistel construction with public random round functions is indifferentiable from a random permutation. This must be contrasted with the classical Luby-Rackoff result stating that the 4-round Feistel construction with pseudorandom round functions yield a SPRP.

OUR CONTRIBUTION. The indifferentiability result for the Feistel construction mentioned above is fundamentally about how to obtain a random permutation from a random (function) oracle. The step to obtain an ideal cipher (*i.e.* an exponential number of independent permutations) is trivially achieved through domain separation of the underlying primitive (namely by prepending the key to each call to the random function oracles). However, it does not tell us anything about how the key should be concretely mixed into the state. In a departure from this approach, we ask the following question: given a small number of objects with n-bit inputs (*e.g.* n-bit permutations $P_1, \ldots, P_r$), is there a way to "combine" them together with an m-bit key in order to obtain a construction which is close to an n-bit block and m-bit key ideal cipher, *i.e.* a set of 2^m independent permutations, without appealing to a trivial domain separation argument? This

[2] Care has to be taken with this composition result when the security definition for the cryptosystem puts some limitations on the adversary, such as an upper bound on its memory [52,20]

naturally prompts us to turn our attention towards the second class of designs, namely key-alternating ciphers.[3] More formally, we consider the construction of a block cipher with n-bit blocks and m-bit keys from r public n-bit permutations $P_1, \ldots, P_r$ defined as follows: derive $(r+1)$ n-bit round keys $(k_0, \ldots, k_r)$ from a master key K through some key derivation function, and encrypt the plaintext $x \in \{0,1\}^n$ by computing the ciphertext y defined as:

$$y = k_r \oplus P_r(k_{r-1} \oplus P_{r-1}(\cdots P_2(k_1 \oplus P_1(k_0 \oplus x)) \cdots)) .$$

When $r = 1$ and two independent n-bit keys (k_0, k_1) are used, so that the ciphertext is simply $y = k_1 \oplus P_1(k_0 \oplus x)$, one obtains the so-called Even-Mansour cipher [22]. When P_1 is modeled as a public random permutation (that the adversary can query in a black-box way), Even and Mansour [22] showed that the resulting block cipher is a SPRP, with security ensured up to $\mathcal{O}(2^{n/2})$ distinguisher queries. The indistinguishability of the general construction for $r > 1$ with independent keys $(k_0, \ldots, k_r)$ was later studied for two rounds by Bogdanov *et al.* [13], for three rounds by Steinberger [57], and for any number r of rounds (with non-tight security bounds) by Lampe *et al.* [38]. Unsurprisingly, the number of adversarial queries up to which the key-alternating cipher is indistinguishable from a random permutation increases with the number of rounds. Following [38], and to emphasize that we work in the random permutation model for $P_1, \ldots, P_r$, we will use the naming *r-round iterated Even-Mansour cipher* to designate the idealized key-alternating cipher where the permutations $P_1, \ldots, P_r$ are public and perfectly random permutations oracles.

In this paper, we consider the iterated Even-Mansour cipher from the point of view of indifferentiability, and ask whether this construction is indifferentiable from an ideal cipher for a sufficient number of rounds when the permutations $P_1, \ldots, P_r$ are public and random. A first simple observation is that the construction with $r+1$ independent n-bit keys $(k_0, \ldots, k_r)$ (resulting in a total key space $\{0,1\}^m = \{0,1\}^{(r+1)n}$) is never indifferentiable (for any r) from an ideal cipher with n-bit blocks and $(r+1)n$-bit keys (this had already been informally observed by [13]). In a sense, independent keys offer too much freedom to the attacker, enabling to easily find related-key relations. There are two possible approaches to solve this problem. The first one is to derive the round keys $(k_0, \ldots, k_r)$ from the master key using some cryptographic function (modeled as a random oracle for the indifferentiability proof). This was considered in an earlier and independent work by Andreeva *et al.* [2] (see below for a discussion of their result). The second possibility (not relying on any cryptographic assumption about the key derivation function) is to "correlate" the round keys. This is the approach we adopt: namely, we consider the iterated Even-Mansour cipher where the n-bit round keys $(k_0, \ldots, k_r)$ are obtained by applying efficiently invertible n-bit permutations $(\gamma_0, \ldots \gamma_r)$ to the n-bit master key k (see Figure 1 on page 453). As will appear clearly in view of its proof, the fact that the master key length is equal to the block length is crucial for our result. To insist on this

[3] One could certainly undertake the same study for Feistel-based block ciphers, but this seems more complicated.

particular point, we call this construction the *single-key* iterated Even-Mansour cipher. Our main result is the following one.

Theorem. *The 12-round single-key iterated Even-Mansour cipher with twelve independent random public n-bit permutations* $(P_1, \ldots, P_{12})$ *and any efficiently invertible (public) n-bit permutations* $(\gamma_0, \ldots, \gamma_{12})$ *for the key schedule is indifferentiable from an ideal cipher with n-bit blocks and n-bit keys.*

In fact, the key derivation permutations γ_i will not play any role in the proof, so that we will focus on the simple case where they are all equal to the identity. Additionally, we show that at least four rounds are necessary by describing attacks (using only a constant number of queries) for three rounds or less (see the full version of the paper [39]).

Together with the result of [2] discussed below, our main theorem validates the design strategy underlying SPNs and more generally key-alternating ciphers as a sound way to ensure security beyond pseudorandomness: it (theoretically) enables to achieve resistance against related-key, known-key and chosen-key attacks (that an ideal cipher can withstand). We stress that our result cannot be used as is to take *concrete* design decisions: first, our bounds (as is often the case with indifferentiability results) are extremely loose.[4] More importantly, the permutations P_i used in concrete block ciphers such as AES are often too simple to be deemed close to random permutations (not to say independent: they are often the same).

Our Techniques. The techniques used to prove our main theorem are very similar to the ones introduced in [16,55,32] for the Feistel construction (while the formalism we adopt is very close to [32]). We simply give a very cursory overview of the main ideas here (assuming all γ_i's are the identity). The simulator works by detecting and completing "partial chains" created by the queries of the distinguisher. Define the computation path for a plaintext x and a key k as the sequence of pairs (x_1, y_1), ..., (x_{12}, y_{12}) of corresponding input and output values for the simulated permutations $P_1, \ldots, P_{12}$. It must hold that the value y obtained through this computation path matches the value $\boldsymbol{E}(k, x)$ obtained from the ideal cipher, otherwise one could straightforwardly distinguish the "simulated" world from the "real" world. Hence, simply answering the distinguisher queries randomly will not work: the simulator must somehow "adapt" the computation path to match the ideal cipher $\boldsymbol{E}$. Observe now the following important property of the single-key iterated Even-Mansour cipher: given only two consecutive values y_i and x_{i+1} of the computation path (*i.e.* the output value of permutation P_i and the input value to permutation P_{i+1}), it is possible to deduce the corresponding key $k = y_i \oplus x_{i+1}$, and hence to move forward and backward along the path. Note that this property essentially relies on the fact that the master key length is equal to the block length of the permutations (would the master key be larger, then it could not be uniquely determined by

[4] Since the proof is already quite involved, we favored simplicity rather than tightness, but the bounds can probably be improved at some places.

y_i and x_{i+1}). Note also that this is the exact analogue of the property of the Feistel network that the input and output values to two consecutive round functions enable to uniquely move forward and backward inside the construction. With this in mind, the strategy of the simulator will be to detect *partial chains* in computation paths created by queries of the distinguisher to two consecutive permutations, and "complete" them by moving forward and backward inside the iterated Even-Mansour construction (randomly setting undefined permutation values encountered along the way, and making a call to the ideal cipher to "wrap around") until the input x_ℓ and the output y_ℓ for one particular permutation P_ℓ are obtained (but still undefined inside P_ℓ history). This permutation is then "adapted" by setting $P_\ell(x_\ell) := y_\ell$ so that the corresponding input and output for the simulated Even-Mansour cipher and for the ideal cipher match. A moment of thinking should make clear that the simulator cannot complete each and every partial chain created in its history, since this would create a "chain reaction" leading to an exponential running time and an exponential number of ideal cipher queries from the simulator. Hence, one must make a careful and parsimonious choice of "detection zones" for deciding which partial chains to complete. In addition, one must ensure that the simulator never overwrites an entry when adapting permutation P_ℓ, thereby rendering a previously completed chain inconsistent. How exactly this is done is very similar to the case of the Feistel construction [55,32], and we refer to Section 3.1 for a more detailed overview.

As a retrospective afterthought, we note that the Feistel and the iterated Even-Mansour indifferentiability results are not that far apart: they both tell how to construct a "big object" (which in both cases has some specific syntactic constraints which are relevant only from a cryptographic perspective) taking $2n$ bits of input (the left and right n-bit halves of the input in the case of the Feistel network, and the key and the plaintext in the case of the iterated Even-Mansour cipher) from smaller objects with only n bits of input (fourteen n-bit to n-bit functions for the Feistel network, and twelve n-bit permutations for the iterated Even-Mansour cipher).

Related Work. In a prior and independent work [2], Andreeva *et al.* proved a result which is close and complementary to ours: they showed that the iterated Even-Mansour construction with *five* rounds and a key derivation function *modeled as a random oracle* is indifferentiable from an ideal cipher. Though significantly reducing the number of rounds required for the proof to go through, and lifting the restriction that the master key length be equal to the block length of the permutations, their technique puts a strong burden on the key derivation function, which can hardly be seen as close to a random oracle in most concrete block ciphers. In fact, most key schedules, such as the one of AES, are "lightweight" and invertible, which makes our result (where the key derivation function has no cryptographic role) more relevant to practice. On the other hand, the bounds obtained by [2] are better: the number of queries, the running time, and the indistinguishability bound achieved by their simulator are respectively $\mathcal{O}(q^2)$, $\mathcal{O}(q^3)$, and $\mathcal{O}(q^{10}/2^n)$, while for our simulator they are respectively $\mathcal{O}(q^4)$, $\mathcal{O}(q^6)$, and $\mathcal{O}(q^{12}/2^n)$.

Taken together, the two results indicate, not too surprisingly, that using a cryptographically strong key schedule, though not being necessary, enables to lower the number of rounds needed to obtain an ideal cipher (however this interpretation must be taken cautiously: it may well be that, say, the iterated Even-Mansour cipher with four rounds is indifferentiable from an ideal cipher, independently of the cryptographic strength of the key schedule).

Regarding the purely theoretical question of the minimal number of n-bit permutations needed to construct an n-bit block and n-bit key ideal cipher, it was additionally showed in [2] that six independent permutations is sufficient, by using a 5-round key-alternating cipher and an independent random permutation P_0 to build a key derivation function $k \mapsto P_0(k) \oplus k$.

2 Preliminaries

2.1 Notation and Definitions

Given a finite non-empty set S, we write $s \leftarrow_{\$} S$ to mean that a value is sampled uniformly at random from S and assigned to s. The security parameter will be denoted n and will be identified with the block length of permutations in the Even-Mansour construction. We will write $f \in \mathtt{poly}(n)$ to denote a polynomially bounded function and $f \in \mathtt{negl}(n)$ to denote a negligible function. For $\delta \in \{+,-\}$, we denote $\bar{\delta}$ the opposite of δ.

In the following, we will use calligraphic fonts $(\mathcal{A}, \mathcal{B}, \ldots)$ to denote interactive Turing machines, and typewriter fonts to denote `Procedures` attached to these machines. A distinguisher is an oracle Turing Machine $\mathcal{D}$ which takes as input a security parameter 1^n, has access to a set of oracles $O_1, \ldots, O_m$, and outputs a bit b, an experiment we denote $\mathcal{D}^{O_1,\ldots,O_m} = b$. We will always consider distinguishers that are deterministic and computationally unbounded, and restricted only with respect to the number of oracle queries they make.

An ideal primitive is a probability distribution on some set of functions, and will be denoted with bold fonts. In the corresponding *model*, a function is drawn at random from the corresponding distribution (say $\boldsymbol{F}$) and all parties (say $\mathcal{M}$) involved in the security experiment are given oracle access to the corresponding function, which we simply denote $\mathcal{M}^{\boldsymbol{F}}$. In the following we will consider the following two ideal primitives:

- a random permutation $\boldsymbol{P}_i$ on $\{0,1\}^n$, which is a permutation drawn at random from the set of all permutations on $\{0,1\}^n$, and which can be accessed in the two directions $\boldsymbol{P}_i(x)$ and $\boldsymbol{P}_i^{-1}(y)$; we will use the notation $\boldsymbol{P} = (\boldsymbol{P}_1, \ldots, \boldsymbol{P}_r)$ to denote a tuple of independent random permutations;
- an ideal cipher $\boldsymbol{E}$ with message space and key space $\{0,1\}^n$, which is drawn at random from the set of all block ciphers of this form, and which can be accessed in encryption, denoted $\boldsymbol{E}(k,x)$, and decryption, denoted $\boldsymbol{E}^{-1}(k,y)$.

2.2 Indifferentiability

We recall the usual definition of indifferentiability.

Definition 1. *Let $q, \sigma, t : \mathbb{N} \to \mathbb{N}$ and $\varepsilon : \mathbb{N} \to \mathbb{R}$ be four functions of the security parameter n. A Turing machine $\mathcal{C}$ with oracle access to an ideal primitive $\mathbf{F}$ is said to be statistically and strongly $(q, \sigma, t, \varepsilon)$-indifferentiable from an ideal primitive $\mathbf{G}$ if there exists an interactive Turing machine $\mathcal{S}$ with oracle access to $\mathbf{G}$ such that for any distinguisher $\mathcal{D}$ making at most q queries, $\mathcal{S}$ makes at most σ oracle queries, runs in time at most t, and the following holds:*

$$\left| \Pr\left[\mathcal{D}^{\mathbf{G}, \mathcal{S}^{\mathbf{G}}} = 1\right] - \Pr\left[\mathcal{D}^{\mathcal{C}^{\mathbf{F}}, \mathbf{F}} = 1\right] \right| \leq \varepsilon \ .$$

$\mathcal{C}^{\mathbf{F}}$ is simply said to be statistically and strongly indifferentiable from $\mathbf{G}$ if for any $q \in \mathtt{poly}(n)$, the above definition is fulfilled with $\sigma, t \in \mathtt{poly}(n)$ and $\varepsilon \in \mathtt{negl}(n)$.

This definition does not refer to the running time of $\mathcal{D}$. When only polynomial-time distinguishers are considered, indifferentiability is said to be *computational.* Weak indifferentiability is defined as above, but the order of quantifiers for the distinguisher and the simulator are switched (for all distinguisher, there is a simulator...).

In this paper, and similarly to [32], we will slightly tweak the definition of strong indifferentiability as follows: we will describe a simulator which, for any distinguisher $\mathcal{D}$ making a polynomial number q of queries, runs in time at most t and makes at most σ queries *with overwhelming probability* (rather than probability one) in system $\mathcal{D}^{\mathbf{G}, \mathcal{S}^{\mathbf{G}}}$. This is not a big concern since any such simulator $\mathcal{S}$ can be transformed into a simulator $\mathcal{S}'$ for weak indifferentiability (which is sufficient for the composition theorem of [45] to hold) which takes the maximal number of queries q of $\mathcal{D}$ as input, and aborts when its number of queries becomes larger than σ (computed as a function of q), hence making at most σ queries with probability one.

2.3 The Iterated Even-Mansour Cipher

Fix an integer $r \geq 1$. Let $P = (P_1, \ldots, P_r)$ be a tuple of permutations on $\{0,1\}^n$. The r-round iterated Even-Mansour construction associated with P, denoted $\bar{\mathcal{C}}_r^P$, is the block cipher with message space $\{0,1\}^n$ and key space $(\{0,1\}^n)^{r+1}$ which maps a message x and a key $(k_0, \ldots, k_r)$ to the ciphertext defined by:

$$\bar{\mathcal{C}}_r^P((k_0, \ldots, k_r), x) = k_r \oplus P_r(k_{r-1} \oplus P_{r-1}(\cdots P_2(k_1 \oplus P_1(k_0 \oplus x)) \cdots)) \ .$$

Let $\gamma = (\gamma_0, \ldots, \gamma_r)$ be a tuple of efficiently invertible permutations on $\{0,1\}^n$. The *single-key* r-round iterated Even-Mansour construction associated with P and γ, denoted $\mathcal{C}_r^{P,\gamma}$, is the block cipher with message space $\{0,1\}^n$ and key space $\{0,1\}^n$ which maps a message x and a key k to the ciphertext defined by (see Figure 1):

$$\mathcal{C}_r^{P,\gamma}(k, x) = \gamma_r(k) \oplus P_r(\gamma_{r-1}(k) \oplus P_{r-1}(\cdots P_2(\gamma_1(k) \oplus P_1(\gamma_0(k) \oplus x)) \cdots)) \ .$$

In all the following, we will focus on the case where all permutations γ_i are the identity, and simply denote $\mathcal{C}_r^P$ the resulting cipher, namely:

$$\mathcal{C}_r^P(k, x) = k \oplus P_r(k \oplus P_{r-1}(\cdots P_2(k \oplus P_1(k \oplus x)) \cdots)) \ .$$

We stress that our main result (Theorem 1) holds for arbitrary permutations γ_i as long as they are efficiently invertible.

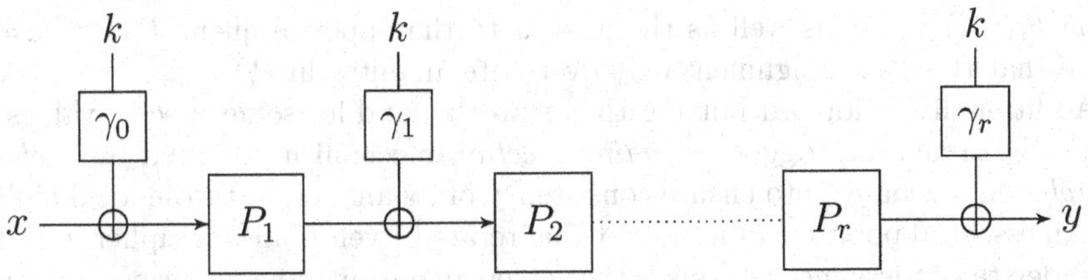

Fig. 1. The single-key iterated Even-Mansour cipher with r rounds $\mathcal{C}_r^{P,\gamma}$. We focus in this paper on the special case $\gamma_i = \text{Id}$ for $i = 0, \ldots, r$.

3 Indifferentiability for Twelve Rounds

In this section we prove the main result of this paper, which is the following theorem.

Theorem 1. *For any q, the 12-round single-key iterated Even-Mansour cipher $\mathcal{C}_{12}^{\boldsymbol{P},\gamma}$ with twelve independent random n-bit permutations $\boldsymbol{P} = (\boldsymbol{P}_1, \ldots, \boldsymbol{P}_{12})$, and fixed, efficiently invertible n-bit permutations $\gamma = (\gamma_0, \ldots, \gamma_{12})$ for the key schedule, is strongly and statistically $(q, \sigma, t, \varepsilon)$-indifferentiable from an ideal cipher $\boldsymbol{E}$ with n-bit blocks and n-bit keys, where:*

$$\sigma = 2^7 \times q^4, \quad t = \mathcal{O}(q^6), \quad \text{and} \quad \varepsilon = \frac{2^{91} \times q^{12}}{2^n}.$$

To prove this, we will describe an efficient simulator $\mathcal{S}$, and show that the two systems $(\mathcal{C}_{12}^{\boldsymbol{P},\gamma}, \boldsymbol{P})$ and $(\boldsymbol{E}, \mathcal{S}^{\boldsymbol{E}})$ are indistinguishable. For simplicity we focus on the case where all γ_i's are the identity, but the generalization is straightforward.

Notational Convention. In all this section, we will use the following useful notational convention: we will interchangeably denote the input to the ideal cipher or the iterated Even-Mansour cipher x or y_0, and the output y or x_{13}.

3.1 Informal Description of the Simulator

We start with a high-level view of the simulator (see also Figure 2). It offers an interface to the distinguisher for querying the simulated permutations, which formally takes the form of a public procedure $\texttt{Query}(i, \delta, z)$, where $i \in \{1, \ldots, 12\}$ names the permutation, $\delta \in \{+, -\}$ tells whether this is a direct or indirect query, and $z \in \{0,1\}^n$ is the actual value queried. The simulator maintains an history for the simulated permutations under the form of hash tables $P_1, \ldots, P_{12}$. Each such table maps entries $(\delta, z) \in \{+, -\} \times \{0,1\}^n$ to values $z' \in \{0,1\}^n$. We denote P_i^+, resp. P_i^-, the (time-dependent) sets of strings $z \in \{0,1\}^n$ such that

$P_i(+,z)$, resp. $P_i(-,z)$, is defined. When the simulator receives a query (i,δ,z), it looks in hash table P_i to see whether the corresponding answer $P_i(\delta,z)$ is already defined. When this is the case, it outputs the answer and waits for the next query. Otherwise, it draws a uniformly random answer z' and defines in hash table $P_i(\delta,z) := z'$, as well as the answer to the opposite query $P_i(\bar{\delta},z') := z$ (note that this last assignment may overwrite an entry in P_i).

Additionally, before outputting the answer z', and for some specific values of (i,δ), the simulator triggers a *chain detection* mechanism followed by a *chain completion* mechanism to ensure consistency of its answers with the ideal cipher $\boldsymbol{E}$. An essential point to notice about the iterated Even-Mansour cipher in order to understand these mechanisms is that given an output value y_i for permutation P_i and an input value x_{i+1} for permutation P_{i+1}, it is possible to compute the corresponding key $k = y_i \oplus x_{i+1}$, and therefore to move forward and backward in the construction up and down to the corresponding input x and output y to the cipher. Hence, any tuple (y_i, x_{i+1}, i) (a so-called *partial chain* later in the reasoning) defines a unique computation path inside the whole construction. This is the exact analogue of the property of the Feistel construction that the input values to two consecutive round functions uniquely define the computation path inside the Feistel network.

There are exactly six such values of (i,δ) for which the simulator performs additional steps: $(2,+)$, $(6,+)$, $(6,-)$, $(7,+)$, $(7,-)$, and $(11,-)$. The cases $(2,+)$ and $(11,-)$ are similar. When receiving a query $(2,+,x_2)$ for which the answer is still undefined, the simulator, after having drawn a random answer y_2 to this query, considers all values $y_1 \in P_1^-$, computes the corresponding key $k := y_1 \oplus x_2$, and moves backward in the iterated Even-Mansour cipher by computing $x_1 := P_1(-,y_1)$, $y_0 := x_1 \oplus k$, $x_{13} := \boldsymbol{E}(k,y_0)$ (hence making a query to the ideal cipher), and $y_{12} := x_{13} \oplus k$, and checks whether $y_{12} \in P_{12}^-$. When this is the case, it enqueues in a queue QUEUE the tuple $(y_0, x_1, 0, 4)$. The first three elements $(y_0, x_1, 0)$ specify the partial chain that must be completed, while the last element $\ell = 4$ specifies which permutation will be adapted during completion of the chain to ensure consistency with $\boldsymbol{E}$. The behavior of the simulator when receiving a query $(11,-,y_{11})$ is symmetric: after having drawn a random answer x_{11}, for all $x_{12} \in P_{12}^+$, it moves forward in the iterated Even-Mansour cipher to check whether the corresponding value x_1 is in P_1^+, and if so enqueues the corresponding tuple $(y_0, x_1, 0, 9)$ (note that in this case adaptation will take place at permutation P_9).

The four remaining cases $(i,\delta) = (6,+)$, $(6,-)$, $(7,+)$, and $(7,-)$ are similar, except that there is no check: the simulator enqueues a tuple $(y_6, x_7, 6, \ell)$ for each newly generated pair $(y_6, x_7) \in P_6^- \times P_7^+$. If this was a query with $i = 6$, then adaptation will take place at $\ell = 4$, while if this was a query with $i = 7$, adaptation will take place at $\ell = 9$. Assume for a concrete example that the simulator receives a query $(6,+,x_6)$ whose answer is undefined yet. Then it draws a random answer $y_6 \leftarrow_\$ \{0,1\}^n$, and enqueues $(y_6, x_7, 6, 4)$ for all $x_7 \in P_7^+$.

Immediately after having enqueued newly created chains (y_i, x_{i+1}, i, ℓ), the simulator starts completing the partial chains, by dequeuing tuples from QUEUE.

For this, when dequeuing (y_i, x_{i+1}, i, ℓ), it computes the key $k := y_i \oplus x_{i+1}$, and moves forward and backward in the iterated Even-Mansour cipher, possibly defining missing permutations values $P_i(+, \cdot)$ or $P_i(-, \cdot)$, and making a query to $\boldsymbol{E}(k, \cdot)$ to "wrap around", until it reaches the input value x_ℓ for P_ℓ (when moving forward) and the corresponding output y_ℓ (when moving backward). It finally "adapts" permutation P_ℓ by setting $P_\ell(+, x_\ell) := y_\ell$ and $P_\ell(-, y_\ell) := x_\ell$ in order to ensure consistency of the entire chain with $\boldsymbol{E}$. It also adds chains that have been completed in a set COMPLETED in order to avoid completing them twice. While completing a chain and adding possibly missing permutation values, the simulator uses the same chain detection mechanism as when receiving a direct query from the distinguisher. Hence new tuples may be enqueued while dequeuing and completing a chain, and the simulator keeps dequeuing tuples until the queue is empty. When this is the case, it returns the answer to the original query of the distinguisher.

As in the indifferentiability proof of the Feistel construction, there will be two crucial points to show: first, that the recursive chain completion mechanism terminates in polynomial time (except maybe with negligible probability); second, that the simulator can always adapt, *i.e.* that it never has (or only with negligible probability) to overwrite previously defined entries when adapting a chain, which would render previously completed chains inconsistent with the ideal cipher $\boldsymbol{E}$. Permutations P_3, P_5, P_8, and P_{10} (*i.e.* the permutations surrounding the two adaptation rounds P_4 and P_9) will play a key role while proving this last point: they will ensure that no bad collisions occur at the input or output of the two permutations used for adapting chains.

We defer the formal definition of the simulator to the full version of the paper [39].

3.2 Sketch of the Proof of Theorem 1

We sketch the main ideas of the proof of Theorem 1. The detailed proof is deferred to the full version of the paper [39].

We use intermediate systems that are depicted on Figure 3. System Σ_1 is the simulated world $(\boldsymbol{E}, \mathcal{S}^{\boldsymbol{E}})$, while Σ_4 is the real world $(\mathcal{C}_{12}^{\boldsymbol{P}}, \boldsymbol{P})$. In system Σ_2, the ideal cipher $\boldsymbol{E}$ is replaced with a so-called keyed two-sided random function $\mathcal{F}(\eta)$ which offers the same interface for encryption and decryption as the ideal cipher. However, when asked for an encryption query (k, x) or a decryption query (k, y), $\mathcal{F}$ first checks (by maintaining a hash table denoted E) whether this value appeared in a previous query, and if so answers consistently. Otherwise it draws a uniformly random answer (the randomness is made explicit through a uniformly random table η) and updates E. Besides, $\mathcal{F}$ has an additional interface $\mathcal{F}.\texttt{Check}(k, x, y)$ (only used by the simulator) which returns true if and only if $E(+, k, x) = y$ or $E(-, k, y) = x$ (in particular, if neither (k, x) was queried for encryption nor (k, y) for decryption, $\texttt{Check}(k, x, y)$ returns false). In Σ_2, the simulator $\mathcal{S}$ is slightly modified into a new simulator $\mathcal{T}$ which queries $\texttt{Check}$ rather than the encryption or decryption interface when deciding whether a tuple $(y_0, x_1, 0, \ell)$ must be enqueued. Moreover the randomness of $\mathcal{T}$ is made

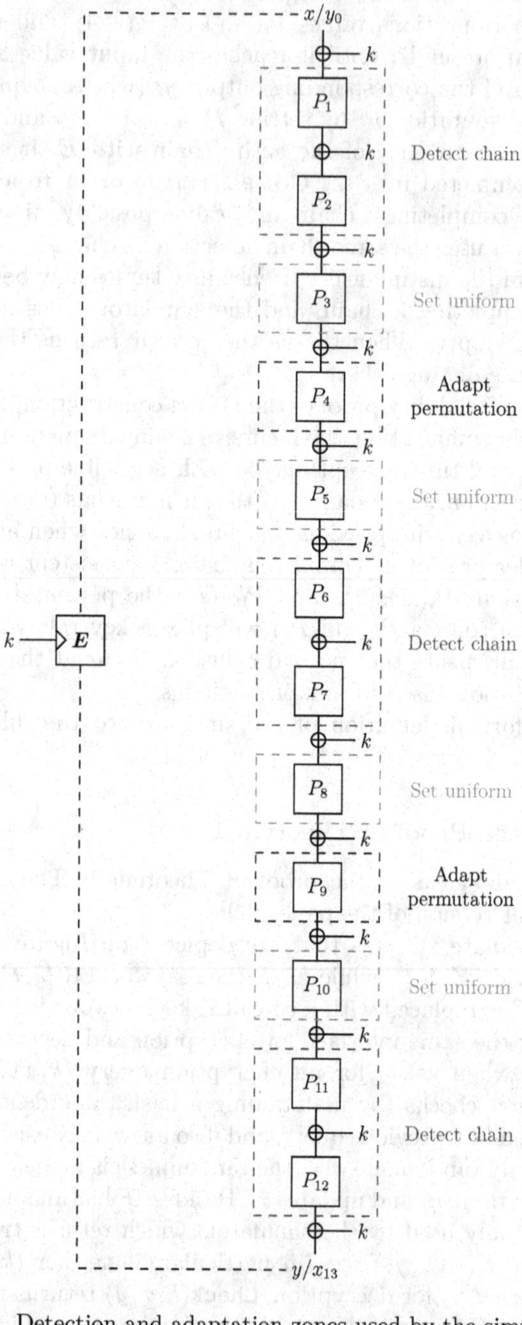

Fig. 2. Detection and adaptation zones used by the simulator

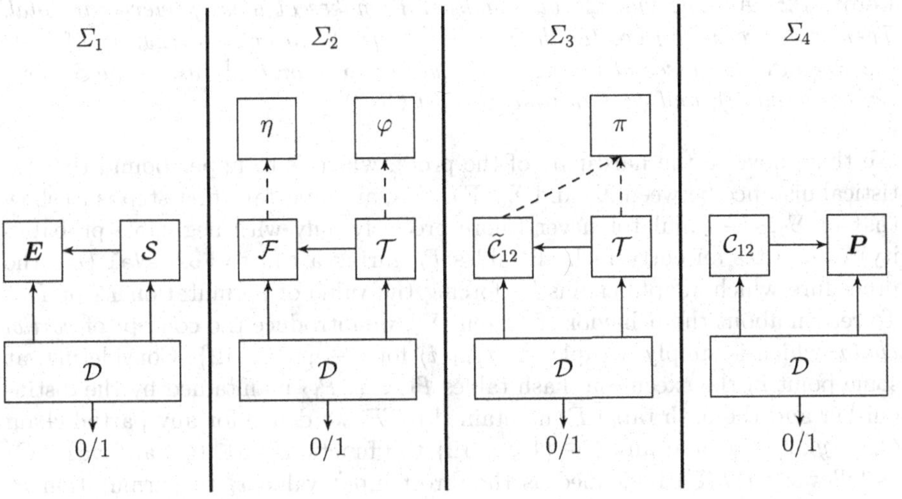

Fig. 3. Systems used in the indifferentiability proof

explicit with uniformly random tables $\varphi = (\varphi_1, \ldots, \varphi_{12})$. In system Σ_3, the keyed two-sided random function is replaced with an iterated Even-Mansour cipher using uniformly random permutations $\pi = (\pi_1, \ldots, \pi_{12})$, enhanced with a `Check` procedure similarly to $\mathcal{F}$. The simulator $\mathcal{T}$ now uses tables π as well for its random draws.

To prove Theorem 1, we will upper bound the statistical distance between successive worlds Σ_i. Additionally, we must show that $\mathcal{S}$ makes a polynomial number of oracle queries and runs in polynomial time in Σ_1 with overwhelming probability. We start the analysis in Σ_2: namely we show that in this system, $\mathcal{T}$ will always complete at most q chains of the form $(y_0, x_1, 0, \ell)$. The reason for this is quite simple: since $\mathcal{T}$ uses interface $\mathcal{F}$.`Check` to decide whether such a tuple must be enqueued, such a chain can be detected and enqueued only if (k, y_0) with $k = y_0 \oplus x_1$ appeared in the queries (or the answers) of the distinguisher to $\mathcal{F}$. Since by assumption the distinguisher makes at most q queries, this implies the result. Starting from this observation, one can then upper bound the size of the hash tables P_i maintained by the simulator as well as the number of queries of $\mathcal{T}$ to $\mathcal{F}$.

We then upper bound the statistical distance between Σ_1 and Σ_2. For this, we appeal to a previous result from [32] to obtain the following lemma.

Lemma 1. *For any distinguisher $\mathcal{D}$ which makes at most q queries in total, we have:*

$$\left| \Pr\left[\mathcal{D}^{\Sigma_1} = 1\right] - \Pr\left[\mathcal{D}^{\Sigma_2(\eta,\varphi)} = 1\right] \right| \leq \frac{2^{22} \times q^{12}}{2^n} .$$

As a side result, this directly implies that with overwhelming probability, $\mathcal{S}$ runs in polynomial time and makes a polynomial number of queries to $\boldsymbol{E}$ in system Σ_1, as captured by the following lemma.

Lemma 2. *Assume that the distinguisher $\mathcal{D}$ makes at most q queries in total. Then with probability greater than $1 - 2^{21} \times q^{12}/2^n$ over an execution of $\mathcal{D}^{\Sigma_1}$, the simulator $\mathcal{S}$ makes at most $2^7 \times q^4$ queries to $\mathbf{E}$ or $\mathbf{E}^{-1}$ (assuming $\mathcal{S}$ never repeats a query), and runs in time at most $\mathcal{O}(q^6)$.*

We then move to the hard part of the proof, which is to upper bound the statistical distance between Σ_2 and Σ_3. For this, an important first step is to show that in Σ_2, the simulator never (more precisely only with negligible probability) overwrites an entry in hash tables P_i during a call to `ForceVal` (*i.e.* the procedure which adapts chains by forcing the value of permutation P_4 or P_9). To reason about the behavior of system Σ_2, we introduce the concept of *partial chain*, which is simply a tuple (y_i, x_{i+1}, i) for $i \in \{1, \ldots, 12\}$. Considering, at some point in the execution, hash tables $P_1, \ldots, P_{12}$ maintained by the distinguisher and the hash table E maintained by $\mathcal{F}$, we define for any partial chain $C = (y_i, x_{i+1}, i)$ and any $\ell \in \{1, \ldots, 12\}$ the functions $\mathtt{val}^+_\ell(C)$ and $\mathtt{val}^-_\ell(C)$ as follows: $\mathtt{val}^+_\ell(C)$ is defined as the direct input value x_ℓ to permutation P_ℓ obtained when moving forward in the Even-Mansour construction (possibly looking in hash table E to wrap around), or $\perp$ is at some point the computation stops because the necessary value was missing in some hash table (including E). Similarly $\mathtt{val}^-_\ell(C)$ is defined as the indirect input value y_ℓ to permutation P_ℓ obtained when moving backward in the Even-Mansour construction, or $\perp$ if the computation stops at some point.

As a preliminary step, we need to exclude some bad events that lead to a pathological behavior of Σ_2. These bad events correspond to the draw of bad values when the simulator randomly defines the value of some permutation P_i or when $\mathcal{F}$ draws a random answer. More precisely, the bad values are exactly those that can be written as the bitwise xor of up to five values in the history, where the history includes all n-bit strings appearing in hash tables P_i and E at the moment where the random answer is drawn. Since the size of the history remains polynomial, the probability of these bad events is negligible.

Then, the proof that the simulator never overwrites an entry in hash tables P_i during a call to `ForceVal` roughly consists of two steps. First, we show that just before the query which leads to some partial chain C being enqueued to be adapted at position ℓ, one has $\mathtt{val}^+_{\ell-1}(C) = \perp$ and $\mathtt{val}^-_{\ell+1}(C) = \perp$, unless an equivalent chain B (where equivalent means that one can obtain B from C by moving forward or backward in the Even-Mansour construction) has been previously enqueued. This crucially relies on fact that the two chain detection zones ("border" and "center") are "protecting" each other. For example, consider the case where some chain $C = (y_0, x_1, 0)$ is enqueued to be adapted at position $\ell = 4$ due to a query for $P_2(x_2)$. Then clearly, before $P_2(x_2)$ is defined, one has $\mathtt{val}^+_3(C) = \perp$. On the other side, if $\mathtt{val}^-_5(C) \neq \perp$, then this means that C is equivalent to some partial chain $B = (y_6, x_7, 6)$ with $y_6 \in P^-_6$ and $x_7 \in P^+_7$, so that D would have been enqueued previously due to some query to P_6 or P_7.

The second step is to show that between the moment where C is enqueued, and the moment where C is dequeued, the completion of other chains (possibly) in the queue will not lead to $\mathtt{val}^+_{\ell-1}(C) \in P^+_{\ell-1}$ or $\mathtt{val}^-_{\ell+1}(C) \in P^-_{\ell+1}$. In particular this requires to show that C cannot collision with another, previously enqueued chain D at round $\ell - 1$ or $\ell + 1$. This is carried out via a careful analysis of all the ways this could happen, which would all imply the occurrence of the bad event previously discussed. Once this is done, it is easy to show that no entry is overwritten during the call to `ForceVal` when adapting C. To finalize the reasoning, we use a randomness mapping argument similar to the one that was introduced in [32], and obtain the following lemma.

Lemma 3. *For any distinguisher $\mathcal{D}$ which makes at most q queries in total, we have:*

$$\left|\Pr\left[\mathcal{D}^{\Sigma_2(\eta,\varphi)} = 1\right] - \Pr\left[\mathcal{D}^{\Sigma_3(\pi)} = 1\right]\right| \leq \frac{2^{89} \times q^{12}}{2^n} .$$

Finally, upper bounding the statistical distance between Σ_3 and Σ_4 is easily handled, and yields the following lemma.

Lemma 4. *For any distinguisher $\mathcal{D}$ which makes at most q queries in total, we have:*

$$\left|\Pr\left[\mathcal{D}^{\Sigma_3(\pi)} = 1\right] - \Pr\left[\mathcal{D}^{\Sigma_4} = 1\right]\right| \leq \frac{2^{89} \times q^{12}}{2^n} .$$

Combining Lemmas 1, 2, 3, and 4 finally enables to prove Theorem 1.

Remark 1. Our choice to use a keyed two-sided random *function* and a simulator $\mathcal{T}$ accessing random *function* tables φ in system Σ_2 allows to handle uniformly random values, which slightly simplifies the computation of various bounds in the proof. It is however possible (and conceptually more satisfying) to use an ideal cipher enhanced with a `Check` procedure rather than a keyed two-sided random function, and random permutation tables rather than random function tables. This would have some nice effects in the analysis of system Σ_2, in particular this would exclude some bad events such as potential overwrites in the hash table E when $\mathcal{F}$ defines an answer by reading table η or in hash tables P_i when $\mathcal{T}$ defines an answer by reading tables φ_i. This kind of approach was taken in [2].

Remark 2. If one contents oneself with weak indifferentiability (where the simulator is allowed to depend on the distinguisher), one can slightly simplify the simulator by having it abort when it is about to complete more than q chains of the form $(y_0, x_1, 0)$; this allows to get rid of the intermediate system Σ_2 where the `Check` procedure is added to the keyed two-sided random function (or the ideal cipher) in order to ensure that the simulator makes a polynomial number of queries and runs in polynomial time with probability 1. Such a simplification does not seem to be possible if one wants to define a universal simulator which does not depend on q.

References

1. Albrecht, M.R., Farshim, P., Paterson, K.G., Watson, G.J.: On Cipher-Dependent Related-Key Attacks in the Ideal-Cipher Model. In: Joux, A. (ed.) FSE 2011. LNCS, vol. 6733, pp. 128–145. Springer, Heidelberg (2011)

2. Andreeva, E., Bogdanov, A., Dodis, Y., Mennink, B., Steinberger, J.P.: On the Indifferentiability of Key-Alternating Ciphers. In: Canetti, R., Garay, J.A. (eds.) CRYPTO 2013, Part I. LNCS, vol. 8042, pp. 531–550. Springer, Heidelberg (2013), Full version available at `http://eprint.iacr.org/2013/061`
3. Bellare, M., Desai, A., Jokipii, E., Rogaway, P.: A Concrete Security Treatment of Symmetric Encryption. In: Symposium on Foundations of Computer Science - FOCS 1997, pp. 394–403. IEEE Computer Society (1997)
4. Bellare, M., Kilian, J., Rogaway, P.: The Security of the Cipher Block Chaining Message Authentication Code. Journal of Computer and System Sciences 61(3), 362–399 (2000)
5. Bellare, M., Kohno, T.: A Theoretical Treatment of Related-Key Attacks: RKA-PRPs, RKA-PRFs, and Applications. In: Biham, E. (ed.) EUROCRYPT 2003. LNCS, vol. 2656, pp. 491–506. Springer, Heidelberg (2003)
6. Bellare, M., Pointcheval, D., Rogaway, P.: Authenticated Key Exchange Secure against Dictionary Attacks. In: Preneel, B. (ed.) EUROCRYPT 2000. LNCS, vol. 1807, pp. 139–155. Springer, Heidelberg (2000)
7. Bellare, M., Ristenpart, T.: Multi-Property-Preserving Hash Domain Extension and the EMD Transform. In: Lai, X., Chen, K. (eds.) ASIACRYPT 2006. LNCS, vol. 4284, pp. 299–314. Springer, Heidelberg (2006)
8. Bellare, M., Rogaway, P.: Random Oracles are Practical: A Paradigm for Designing Efficient Protocols. In: ACM Conference on Computer and Communications Security, pp. 62–73 (1993)
9. Biham, E.: New Types of Cryptanalytic Attacks Using Related Keys. Journal of Cryptology 7(4), 229–246 (1994)
10. Biryukov, A., Khovratovich, D., Nikolić, I.: Distinguisher and Related-Key Attack on the Full AES-256. In: Halevi, S. (ed.) CRYPTO 2009. LNCS, vol. 5677, pp. 231–249. Springer, Heidelberg (2009)
11. Black, J.A.: The Ideal-Cipher Model, Revisited: An Uninstantiable Blockcipher-Based Hash Function. In: Robshaw, M. (ed.) FSE 2006. LNCS, vol. 4047, pp. 328–340. Springer, Heidelberg (2006)
12. Black, J.A., Rogaway, P., Shrimpton, T.: Black-Box Analysis of the Block-Cipher-Based Hash-Function Constructions from PGV. In: Yung, M. (ed.) CRYPTO 2002. LNCS, vol. 2442, pp. 320–335. Springer, Heidelberg (2002)
13. Bogdanov, A., Knudsen, L.R., Leander, G., Standaert, F.-X., Steinberger, J., Tischhauser, E.: Key-Alternating Ciphers in a Provable Setting: Encryption Using a Small Number of Public Permutations - (Extended Abstract). In: Pointcheval, D., Johansson, T. (eds.) EUROCRYPT 2012. LNCS, vol. 7237, pp. 45–62. Springer, Heidelberg (2012)
14. Canetti, R., Goldreich, O., Halevi, S.: The Random Oracle Methodology, Revisited (Preliminary Version). In: Symposium on Theory of Computing - STOC 1998, pp. 209–218. ACM (1998), Full version available at `http://arxiv.org/abs/cs.CR/0010019`
15. Coron, J.-S., Dodis, Y., Malinaud, C., Puniya, P.: Merkle-Damgård Revisited: How to Construct a Hash Function. In: Shoup, V. (ed.) CRYPTO 2005. LNCS, vol. 3621, pp. 430–448. Springer, Heidelberg (2005)
16. Coron, J.-S., Patarin, J., Seurin, Y.: The Random Oracle Model and the Ideal Cipher Model are Equivalent. In: Wagner, D. (ed.) CRYPTO 2008. LNCS, vol. 5157, pp. 1–20. Springer, Heidelberg (2008)
17. Daemen, J., Rijmen, V.: The Wide Trail Design Strategy. In: Honary, B. (ed.) Cryptography and Coding 2001. LNCS, vol. 2260, pp. 222–238. Springer, Heidelberg (2001)

18. Daemen, J., Rijmen, V.: The Design of Rijndael: AES - The Advanced Encryption Standard. Springer (2002)
19. Damgård, I.B.: A Design Principle for Hash Functions. In: Brassard, G. (ed.) CRYPTO 1989. LNCS, vol. 435, pp. 416–427. Springer, Heidelberg (1990)
20. Demay, G., Gaži, P., Hirt, M., Maurer, U.: Resource-Restricted Indifferentiability. In: Johansson, T., Nguyen, P.Q. (eds.) EUROCRYPT 2013. LNCS, vol. 7881, pp. 664–683. Springer, Heidelberg (2013), Full version available at `http://eprint.iacr.org/2012/613`
21. Desai, A.: The Security of All-or-Nothing Encryption: Protecting against Exhaustive Key Search. In: Bellare, M. (ed.) CRYPTO 2000. LNCS, vol. 1880, pp. 359–375. Springer, Heidelberg (2000)
22. Even, S., Mansour, Y.: A Construction of a Cipher from a Single Pseudorandom Permutation. Journal of Cryptology 10(3), 151–162 (1997)
23. Feistel, H.: Cryptography and computer privacy. Scientific American 228(5), 15–23 (1973)
24. Fiat, A., Shamir, A.: How to Prove Yourself: Practical Solutions to Identification and Signature Problems. In: Odlyzko, A.M. (ed.) CRYPTO 1986. LNCS, vol. 263, pp. 186–194. Springer, Heidelberg (1987)
25. Gaži, P., Maurer, U.: Cascade Encryption Revisited. In: Matsui, M. (ed.) ASIACRYPT 2009. LNCS, vol. 5912, pp. 37–51. Springer, Heidelberg (2009)
26. Gaži, P., Tessaro, S.: Efficient and Optimally Secure Key-Length Extension for Block Ciphers via Randomized Cascading. In: Pointcheval, D., Johansson, T. (eds.) EUROCRYPT 2012. LNCS, vol. 7237, pp. 63–80. Springer, Heidelberg (2012)
27. Gilbert, H., Peyrin, T.: Super-Sbox Cryptanalysis: Improved Attacks for AES-Like Permutations. In: Hong, S., Iwata, T. (eds.) FSE 2010. LNCS, vol. 6147, pp. 365–383. Springer, Heidelberg (2010)
28. Goldreich, O., Goldwasser, S., Micali, S.: How to construct random functions. J. ACM 33(4), 792–807 (1986)
29. Granboulan, L.: Short Signatures in the Random Oracle Model. In: Zheng, Y. (ed.) ASIACRYPT 2002. LNCS, vol. 2501, pp. 364–378. Springer, Heidelberg (2002)
30. Hirose, S.: Provably Secure Double-Block-Length Hash Functions in a Black-Box Model. In: Park, C.-s., Chee, S. (eds.) ICISC 2004. LNCS, vol. 3506, pp. 330–342. Springer, Heidelberg (2005)
31. Hirose, S.: Some Plausible Constructions of Double-Block-Length Hash Functions. In: Robshaw, M. (ed.) FSE 2006. LNCS, vol. 4047, pp. 210–225. Springer, Heidelberg (2006)
32. Holenstein, T., Künzler, R., Tessaro, S.: The Equivalence of the Random Oracle Model and the Ideal Cipher Model, Revisited. In: Fortnow, L., Vadhan, S.P. (eds.) Symposium on Theory of Computing - STOC 2011, pp. 89–98. ACM (2011), Full version available at `http://arxiv.org/abs/1011.1264`
33. Jaulmes, É., Joux, A., Valette, F.: On the Security of Randomized CBC-MAC Beyond the Birthday Paradox Limit: A New Construction. In: Daemen, J., Rijmen, V. (eds.) FSE 2002. LNCS, vol. 2365, pp. 237–251. Springer, Heidelberg (2002)
34. Jonsson, J.: An OAEP Variant With a Tight Security Proof. IACR Cryptology ePrint Archive Report 2002/034 (2002), `http://eprint.iacr.org/2002/034`
35. Kilian, J., Rogaway, P.: How to Protect DES against Exhaustive Key Search. In: Koblitz, N. (ed.) CRYPTO 1996. LNCS, vol. 1109, pp. 252–267. Springer, Heidelberg (1996)
36. Knudsen, L.R., Rijmen, V.: Known-Key Distinguishers for Some Block Ciphers. In: Kurosawa, K. (ed.) ASIACRYPT 2007. LNCS, vol. 4833, pp. 315–324. Springer, Heidelberg (2007)

37. Künzler, R.: Are the random oracle and the ideal cipher models equivalent? Master's thesis, ETH Zurich, Switzerland (2009)
38. Lampe, R., Patarin, J., Seurin, Y.: An Asymptotically Tight Security Analysis of the Iterated Even-Mansour Cipher. In: Wang, X., Sako, K. (eds.) ASIACRYPT 2012. LNCS, vol. 7658, pp. 278–295. Springer, Heidelberg (2012)
39. Lampe, R., Seurin, Y.: How to Construct an Ideal Cipher from a Small Set of Public Permutations, Full version of this paper http://eprint.iacr.org/2013/255
40. Lee, J., Stam, M., Steinberger, J.: The Collision Security of Tandem-DM in the Ideal Cipher Model. In: Rogaway, P. (ed.) CRYPTO 2011. LNCS, vol. 6841, pp. 561–577. Springer, Heidelberg (2011)
41. Luby, M., Rackoff, C.: Pseudo-random Permutation Generators and Cryptographic Composition. In: Symposium on Theory of Computing - STOC 1986, pp. 356–363. ACM (1986)
42. Luby, M., Rackoff, C.: How to Construct Pseudorandom Permutations from Pseudorandom Functions. SIAM Journal on Computing 17(2), 373–386 (1988)
43. Maurer, U.M.: A Simplified and Generalized Treatment of Luby-Rackoff Pseudorandom Permutation Generators. In: Rueppel, R.A. (ed.) EUROCRYPT 1992. LNCS, vol. 658, pp. 239–255. Springer, Heidelberg (1993)
44. Maurer, U.M., Pietrzak, K.: The Security of Many-Round Luby-Rackoff Pseudo-Random Permutations. In: Biham, E. (ed.) EUROCRYPT 2003. LNCS, vol. 2656, pp. 544–561. Springer, Heidelberg (2003)
45. Maurer, U.M., Renner, R.S., Holenstein, C.: Indifferentiability, Impossibility Results on Reductions, and Applications to the Random Oracle Methodology. In: Naor, M. (ed.) TCC 2004. LNCS, vol. 2951, pp. 21–39. Springer, Heidelberg (2004)
46. Mennink, B.: Optimal Collision Security in Double Block Length Hashing with Single Length Key. In: Wang, X., Sako, K. (eds.) ASIACRYPT 2012. LNCS, vol. 7658, pp. 526–543. Springer, Heidelberg (2012)
47. Merkle, R.C.: One Way Hash Functions and DES. In: Brassard, G. (ed.) CRYPTO 1989. LNCS, vol. 435, pp. 428–446. Springer, Heidelberg (1990)
48. Minier, M., Phan, R.C.-W., Pousse, B.: Distinguishers for Ciphers and Known Key Attack against Rijndael with Large Blocks. In: Preneel, B. (ed.) AFRICACRYPT 2009. LNCS, vol. 5580, pp. 60–76. Springer, Heidelberg (2009)
49. Patarin, J.: Pseudorandom Permutations Based on the DES Scheme. In: Charpin, P., Cohen, G. (eds.) EUROCODE 1990. LNCS, vol. 514, pp. 193–204. Springer, Heidelberg (1991)
50. Patarin, J.: Security of Random Feistel Schemes with 5 or More Rounds. In: Franklin, M. (ed.) CRYPTO 2004. LNCS, vol. 3152, pp. 106–122. Springer, Heidelberg (2004)
51. Preneel, B., Govaerts, R., Vandewalle, J.: Hash Functions Based on Block Ciphers: A Synthetic Approach. In: Stinson, D.R. (ed.) CRYPTO 1993. LNCS, vol. 773, pp. 368–378. Springer, Heidelberg (1994)
52. Ristenpart, T., Shacham, H., Shrimpton, T.: Careful with Composition: Limitations of the Indifferentiability Framework. In: Paterson, K.G. (ed.) EUROCRYPT 2011. LNCS, vol. 6632, pp. 487–506. Springer, Heidelberg (2011)
53. Rivest, R.L., Shamir, A., Tauman, Y.: How to Leak a Secret. In: Boyd, C. (ed.) ASIACRYPT 2001. LNCS, vol. 2248, pp. 552–565. Springer, Heidelberg (2001)
54. Sasaki, Y., Yasuda, K.: Known-Key Distinguishers on 11-Round Feistel and Collision Attacks on Its Hashing Modes. In: Joux, A. (ed.) FSE 2011. LNCS, vol. 6733, pp. 397–415. Springer, Heidelberg (2011)
55. Seurin, Y.: Primitives et protocoles cryptographiques à sécurité prouvée. PhD thesis, Université de Versailles Saint-Quentin-en-Yvelines, France (2009)

56. Shannon, C.: Communication Theory of Secrecy Systems. Bell System Technical Journal 28(4), 656–715 (1949)
57. Steinberger, J.: Improved Security Bounds for Key-Alternating Ciphers via Hellinger Distance. IACR Cryptology ePrint Archive Report 2012/481 (2012), `http://eprint.iacr.org/2012/481`
58. Steinberger, J.P.: The Collision Intractability of MDC-2 in the Ideal-Cipher Model. In: Naor, M. (ed.) EUROCRYPT 2007. LNCS, vol. 4515, pp. 34–51. Springer, Heidelberg (2007)
59. Vaudenay, S.: Decorrelation: A Theory for Block Cipher Security. Journal of Cryptology 16(4), 249–286 (2003)
60. Winternitz, R.S.: A Secure One-Way Hash Function Built from DES. In: IEEE Symposium on Security and Privacy, pp. 88–90 (1984)

Generic Key Recovery Attack on Feistel Scheme

Takanori Isobe and Kyoji Shibutani

Sony Corporation
1-7-1 Konan, Minato-ku, Tokyo 108-0075, Japan
{Takanori.Isobe,Kyoji.Shibutani}@jp.sony.com

Abstract. We propose new generic key recovery attacks on Feistel-type block ciphers. The proposed attack is based on the all subkeys recovery approach presented in SAC 2012, which determines all subkeys instead of the master key. This enables us to construct a key recovery attack without taking into account a key scheduling function. With our advanced techniques, we apply several key recovery attacks to Feistel-type block ciphers. For instance, we show 8-, 9- and 11-round key recovery attacks on n-bit Feistel ciphers with $2n$-bit key employing random keyed F-functions, random F-functions, and SP-type F-functions, respectively. Moreover, thanks to the meet-in-the-middle approach, our attack leads to *low-data complexity*. To demonstrate the usefulness of our approach, we show a key recovery attack on the 8-round reduced CAST-128, which is the best attack with respect to the number of attacked rounds. Since our approach derives the lower bounds on the numbers of rounds to be secure under the single secret key setting, it can be considered that we unveil the limitation of designing an efficient block cipher by a Feistel scheme such as a low-latency cipher.

Keywords: block cipher, key scheduling function, all-subkeys-recovery attack, meet-in-the-middle attack, key recovery attack, low-data complexity attack.

1 Introduction

A block cipher is considered as an essential technology on modern cryptography, since it is one of the most widely used primitives. Moreover, studies on designing a secure and efficient block cipher are useful also for designing other symmetric primitives such as hash functions and stream ciphers. Since DES was developed in 1977 [19], a lot of progress has taken place in this area. Recently, with the large deployment of network devices requiring security, block ciphers satisfying new demands such as lightweight and low-latency have received a lot of attention. In fact, several block ciphers designed for a lightweight hardware implementation have been proposed such as PRESENT [9], KATAN/KTANTAN [16], LED [20] and Piccolo [34]. The concept of a low-latency encryption, which is used for an application requiring an instant response, was discussed in [24]. Since a low-latency encryption requires a quick response, the number of rounds must be reduced as much as possible compared to a general-purpose block cipher such

K. Sako and P. Sarkar (Eds.) ASIACRYPT 2013 Part I, LNCS 8269, pp. 464–485, 2013.

as AES. In 2012, PRINCE was proposed as an instantiation of a low-latency cipher [12]. Note that PRINCE is not only a low-latency cipher, but also a lightweight block cipher even after supporting both encryption and decryption. Those features are considered to be important in practical use of the cipher, since its lightweightness directly leads to low power and energy consumption and supporting decryption function without much cost leading to this cipher being used more widely.

In general, an SPN cipher requires an inverse function when supporting decryption, and thus an SPN cipher with a decryption function needs additional gate areas. In spite of the fact that PRINCE is an SPN cipher, it is efficiently implemented even when implementing a decryption function due to its novel property called α-reflection. However, as pointed out by the designers, it has been known that α-reflection reduces the security of the cipher [12,23,35] and thus the cipher having α-reflection does not have optimal security. Meanwhile, it has been known that a Feistel cipher, another traditional structure of block cipher, is suitable for a lightweight block cipher especially when supporting both encryption and decryption, since it does not require an inverse function. Thus, a Feistel cipher is considered as a possible candidate of a low-latency cipher, if it has sufficiently small number of rounds. However, it has been still unknown how many rounds are sufficient for a Feistel cipher to be secure. Note that, for low-latency encryption, since the key scheduling function can be precomputed, it can be a heavy function. Thus, its performance with respect to low-latency is considered to mainly depend on the data processing part, namely its number of rounds. Hence, our question is "how many rounds can be reduced without loss of security requirements for Feistel schemes".

In this paper, we tackle the security evaluations of several Feistel schemes, assuming that the key scheduling function is an ideal function. We deal with key recovery attacks under the single secret key setting by extending the all subkeys recovery approach [22]. Since our approach derives the lower bounds on the numbers of rounds to be secure against a key recovery attack even if the underlying key scheduling function is an ideal function, our results show the limitation of designing a low-latency encryption by a Feistel scheme. We introduce several advanced techniques including *function reduction* and *key linearization*. Using those advanced techniques and with the help of the meet-in-the-middle approach [10,21], we show several key recovery attacks on various Feistel ciphers. Table 1 summarizes the number of attacked rounds for Feistel schemes by both distinguishers and key recovery attacks under the single secret key and known-key settings. Compared to the previous results, some of our attacks are the first generic key recovery attacks and also the best for several Feistel schemes with respect to the number of attacked rounds, even if the attacker is allowed to use the known secret key. Moreover, our attack does not restrict the underlying F-function to a permutation, which is a limitation of some of the previous attacks. Furthermore, one of the advantages of our approach is its low data requirement thanks to the meet-in-the-middle approach, in contrast to the classical statistical attacks such as an impossible differential attack [6]. As an example for the

Table 1. Numbers of Attacked Rounds by Generic Attacks on Feistel Schemes

Single Secret Key Setting				
Attack Type	Feistel-1		Feistel-2	Feistel-3
Distinguisher	5 [30]		5 [30]	5 [30]
	5* [25]		5* [25]	5* [25]
	5* [11]		5* [11]	5* [11]
Key Recovery Attack ($k = 2n$)	7 [22]	**8** (Ours)	**9** (Ours)	**11** (Ours)
Key Recovery Attack ($k = 3n/2$)	5 [22]	**6** (Ours)	**7** (Ours)	**9** (Ours)
Key Recovery Attack ($k = n$)	3 [22]	**4** (Ours)	**5** (Ours)	**7** (Ours)

Known Key Setting			
Distinguisher	not given	7 [26]	11* [33]

* : Each F function is restricted to a permutation

practical impact of our work, we show the best attack on the reduced CAST-128 [1] even when its key scheduling function is ideal. Also, we show extremely low-data attacks on the reduced Camellia [5] with less than 60 data sets.

This paper is organized as follows: Section 2 gives notations and definitions used throughout this paper, and gives a brief review of the all subkeys recovery approach. We review the related work and show its improvement in Section 3. Our key recovery attacks on two types of Feistel ciphers and those applications to CAST-128 and Camellia are described in Sections 4 and 5. Section 6 discusses the usefulness of our attack. Finally, we conclude in Section 7.

2 Preliminary

In this section, we give notations used throughout this paper, then define our target Feistel ciphers. Finally, we briefly review the all subkeys recovery approach presented in [22].

2.1 Notation

The following notation will be used throughout this paper:

n : block size.
k : the size of the master key.
L_i, R_i : left or right half of the i-th round input.
K_i : the i-th round subkey ($n/2$ bits).
ℓ : the size of an S-box.
m : the number of S-boxes in an S-box layer.
X_i : the i-th round state.
$X_{i,j}$: the j-th S-box word (ℓ-bit data) of X_i.
X_{iL}, X_{iR} : left or right half bits of X_i.
$a|b$ or $(a|b)$: Concatenation.

2.2 Feistel Cipher

In this paper, we focus on balanced Feistel networks as illustrated in Fig. 1. An n-bit plaintext P is divided into two sub-blocks as $P = (L_1, R_1)$, where

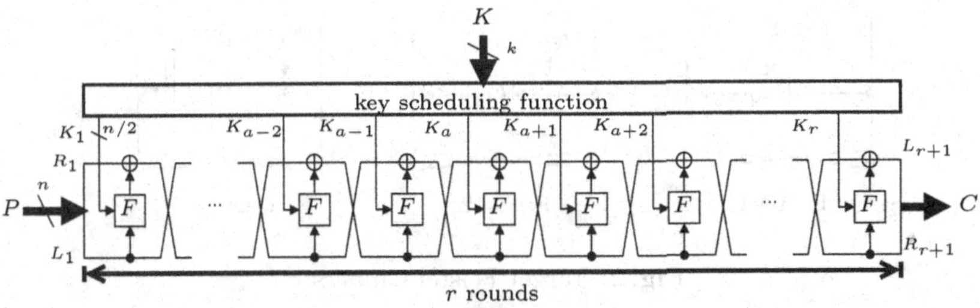

Fig. 1. Balanced Feistel Network (Feistel-1)

$L_i, R_i \in \{0,1\}^{n/2}$. Then the $(i+1)$-th round input state is calculated as follows:

$$(L_{i+1}, R_{i+1}) \leftarrow (R_i \oplus \mathcal{F}_i^{K_i}(L_i), L_i),$$

where $\mathcal{F}_i^{K_i} : \{0,1\}^{n/2} \rightarrow \{0,1\}^{n/2}$ is a keyed function in the i-th round using the i-th round $(n/2)$-bit subkey K_i. An n-bit ciphertext C for the r-round encryption function is derived as $C = (R_{r+1}, L_{r+1})$. Note that the last round of the Feistel cipher does not have a swap operation. Hereafter, the size of each subkey used in one round is assumed to be half of the block size (i.e., $K_i \in \{0,1\}^{n/2}$).

In this work, we deal with three types of Feistel block ciphers illustrated in Fig. 2. Feistel-1 denotes the Feistel cipher with random keyed F-functions. Each subkey is assumed to be randomly independent. Thus each keyed F-function is also independent from each other. In concrete ciphers, each subkey is usually XORed before an F-function. Feistel-2 reflects such ciphers. In other words, the output of the F-function $Y_i = \mathcal{F}_i^{K_i}(X_i)$ is represented as $Y_i = F_i(X_i \oplus K_i)$, where F_i is a fixed function in the i-th round (not limited to a permutation). Similarly, Feistel-3 is the Feistel-2 cipher whose F_i is limited to an SP-type F-function, where each F-function consists of a bijective S-box layer (S-layer) and a linear diffusion layer (P-layer), and an $n/2$-bit subkey is XORed before the S-box layer. Each S-box layer consists of m ℓ-bit S-boxes (i.e., $m \cdot \ell = n/2$), and each P-layer consists of an $m \times m$ linear matrix represented as M_i. Note that Feistel-1 includes Feistel-2 and Feistel-3, also Feistel-3 is a subset of Feistel-2. The size of the master key is denoted as Feistel-$[k]$. For example, Feistel-2$[n]$ is the Feistel cipher with fixed F-functions XORed by a subkey before the function whose master key size is the same as the block size (e.g., a 128-bit block cipher taking a 128-bit key).

2.3 All Subkeys Recovery Approach [22]

The all subkeys recovery (ASR) attack was proposed by Isobe and Shibutani at SAC 2012 [22]. The ASR attack is considered as an extension of the meet-in-the-middle (MITM) attack, which mainly exploits a low key-dependency in the key scheduling function. The basic concept of the ASR attack is guessing all subkeys instead of the master key so that the attack can be constructed independently from the structure of the key scheduling function, by regarding all subkeys as

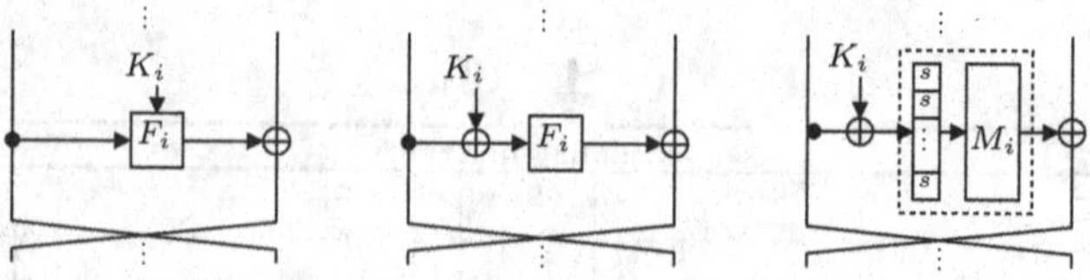

Fig. 2. Target Feistel Ciphers

independent variables. Thus the attack can also be applied to a block cipher having a complex key scheduling function.

Let us briefly review the procedure of the ASR attack. In the ASR attack, an attacker first determines a t-bit matching state X, where $X \in \{0,1\}^t$. In the forward direction, the matching state derived from a plaintext P and a set of subkeys $\mathcal{K}_{(1)}$ by a function $\mathcal{F}_{(1)}$ is represented as $X = \mathcal{F}_{(1)}(P, \mathcal{K}_{(1)})$. Similarly, the state computed from a ciphertext C and another set of subkeys $\mathcal{K}_{(2)}$ by a function $\mathcal{F}_{(2)}$ in the backward direction is denoted as $X = \mathcal{F}_{(2)}^{-1}(C, \mathcal{K}_{(2)})$. $\mathcal{K}_{(3)}$ denotes a set of the remaining subkeys not required for computing X, i.e., $|\mathcal{K}_{(1)}| + |\mathcal{K}_{(2)}| + |\mathcal{K}_{(3)}| = r \cdot n/2$. The attacker guesses $\mathcal{K}_{(1)}$ and $\mathcal{K}_{(2)}$ in parallel, then checks if the equation $\mathcal{F}_{(1)}(P, \mathcal{K}_{(1)}) = \mathcal{F}_{(2)}^{-1}(C, \mathcal{K}_{(2)})$ holds. Note that the equation holds when the guessed subkey bits are correct. After this process, it is expected that there will be $2^{r \cdot n/2 - t}$ key candidates. Finally, the attacker exhaustively searches the correct key from the surviving key candidates. The required computations of the attack in total C_{comp} using N plaintext/ciphertext pairs is estimated as

$$C_{comp} = \max(2^{|\mathcal{K}_{(1)}|}, 2^{|\mathcal{K}_{(2)}|}) \times N + 2^{r \cdot n/2 - N \cdot t}. \quad (1)$$

The number of required plaintext/ciphertext pairs is $\max(N, \lceil (r \cdot n/2 - N \cdot t)/n \rceil)$. The required memory is about $\min(2^{|\mathcal{K}_{(1)}|}, 2^{|\mathcal{K}_{(2)}|}) \times N$ blocks, which is the cost of the table used for the matching. Clearly, the ASR attack works faster than the brute force attack when Eq.(1) is less than 2^k, which is the required computations for the brute force attack.

3 Generic Key Recovery Attack on Feistel-1

In this section, we first review key recovery attacks on balanced Feistel networks presented in [22] and generalize it to Feistel-1$[n]$, -1$[\frac{3}{2}n]$ and -1$[2n]$. After that, we show that the basic attack can be improved by using *splice and cut* [3] and *key linearization* techniques. By the improved attack, the numbers of attacked rounds for the Feistel-1 are increased by one round.

For a Feistel-1 cipher, an $(n/2)$-bit matching state X is computed from a plaintext P and a set of subkeys $\mathcal{K}_{(1)} \in \{K^{(1)}, K^{(2)}, ..., K^{(a-1)}\}$ as shown in Fig. 1 (i.e., $X = \mathcal{F}_{(1)}(P, \mathcal{K}_{(1)})$). Similarly, the matching state is obtained from

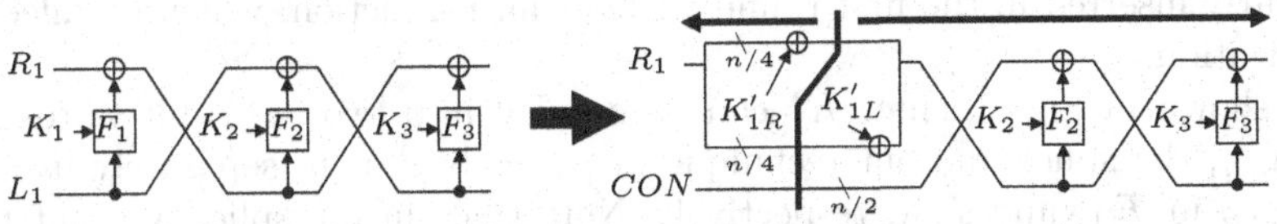

Fig. 3. Splice and Cut Technique for Feistel-1

a ciphertext C and another set of subkeys $\mathcal{K}_{(2)} \in \{K^{(a+1)}, K^{(a+2)}, ..., K^{(r)}\}$ as $X = \mathcal{F}_{(2)}^{-1}(C, \mathcal{K}_{(2)})$. Also, X is computed independently from an $(n/2)$-bit subkey $K^{(a)}$, i.e., $\mathcal{K}_{(3)} \in \{K^{(a)}\}$.

3.1 Basic Attack on Feistel-1 [22]

For Feistel-1$[2n]$ (e.g., a 128-bit block cipher accepting a 256-bit key), 7 rounds of the cipher can be attacked in a straightforward manner, since $\mathcal{F}_{(1)}$ and $\mathcal{F}_{(2)}$ are composed of 3 rounds of the cipher and thus the sizes of $\mathcal{K}_{(1)}$ and $\mathcal{K}_{(2)}$ are both $3 \cdot n/2$ bits. In this attack, the total time complexity C_{comp} using four plaintext/ciphertext pairs is estimated as

$$C_{comp} = \max(2^{3n/2}, 2^{3n/2}) \times 4 + 2^{7 \cdot n/2 - 4 \cdot n/2} \approx 2^{3n/2+2} \quad (= 2^{3k/4+2})$$

The required memory is about $4 \times 2^{3n/2}$ blocks. Since C_{comp} is less than $2^{2n} (= 2^k)$ when $(4 < n)$, the attack works faster than the exhaustive key search.

Similarly to this, for Feistel-1$[\frac{3}{2}n]$ and Feistel-1$[n]$ (e.g., a 128-bit block cipher accepting a 192-bit key or a 128-bit key), key recovery attacks of at least 5 and 3 rounds of the cipher are constructed, respectively. For Feistel-1$[\frac{3}{2}n]$, $\mathcal{F}_{(1)}$ and $\mathcal{F}_{(2)}$ consist of 2 rounds of the cipher, and thus the sizes of $\mathcal{K}_{(1)}$ and $\mathcal{K}_{(2)}$ are both n bits. Therefore, the required time complexity using 3 plaintext/ciphertext pairs is estimated as $C_{comp} = \max(2^n, 2^n) \times 3 + 2^{5n/2-3n/2} \approx 2^{n+2}$, and the required memory is about 2^{n+2} blocks. For Feistel-1$[n]$, a similar attack on 3 rounds requiring $2^{n/2+1}$ ($\approx 2^{n/2} \times 2 + 2^{n/2}$) computations and $(2 \times 2^{n/2})$ blocks memory is mounted by using 1 round of $\mathcal{F}_{(1)}$ and $\mathcal{F}_{(2)}$. Roughly speaking, when Eq.(1) is less than 2^k, the ASR attack works faster than the brute force attack. Therefore, the necessary condition for the basic ASR attack is that each size of all subkeys in $\mathcal{F}_{(1)}$ and $\mathcal{F}_{(2)}$ is less than the size of the master key.

3.2 Improved Attack on Feistel-1

We demonstrate that the basic attack on Feistel-1 presented in [22] is improved by controlling the value of plaintexts. It allows us to attack one more round on Feistel-1, e.g., an 8-round attack on Feistel-1$[2n]$.

Suppose that an input $L_1 (= R_2)$ is fixed to an arbitrary $(n/2)$-bit constant CON, then L_2 is expressed as $L_2 = R_1 \oplus K_1'$, where $K_1' = F_1(K_1 \oplus CON)$. Since K_1' depends only on K_1, it is regarded that a new $(n/2)$-bit subkey K_1'

is linearly inserted in the first round without an F-function, which is called key linearization.

As shown in Fig. 3, since K'_1 can be divided into two $(n/4)$-bit words K'_{1L} and K'_{1R}, the splice and cut technique in [4] enables us to separately use K'_{1L} and K'_{1R} in $\mathcal{F}_{(1)}$ and $\mathcal{F}_{(2)}$, respectively. Note that, in the splice and cut technique, the MITM attack starts from multiple values of start states for parallel guesses of $\mathcal{K}_{(1)}$ and $\mathcal{K}_{(2)}$, while the basic MITM attack starts from multiple plaintext/ciphertext pairs.

For Feistel-1$[2n]$, an 8-round generic key recovery attack is mounted thanks to the splice and cut technique, while each cost (namely time, memory and data) for the attack is increased by $\mathcal{O}(2^{n/4})$ compared to the basic attack. The size of each key set $\mathcal{K}_{(1)}$ and $\mathcal{K}_{(2)}$ is increased by $(n/4)$ bits due to the splice and cut, and thus the size of each set $\mathcal{K}_{(1)}$ and $\mathcal{K}_{(2)}$ is $7n/4(= 3 \cdot n/2 + n/4)$ bits long. In this attack, the total time complexity C_{comp} using five start states is estimated as

$$C_{comp} = \max(2^{7n/4}, 2^{7n/4}) \times 5 + 2^{8 \cdot n/2 - 5 \cdot n/2} \approx 2^{7n/4+3} \quad (= 2^{7k/8+3}).$$

The required memory is about $5 \times 2^{7n/4}$ blocks. Since $(n/4)$ bits of plaintexts are varied depending on $\mathcal{K}_{(2)}$ and the start states, the required data is $2^{n/4}$ chosen plaintexts when the other $3n/4$ bits of the start state are fixed.

For Feistel-1$[\frac{3}{2}n]$ and Feistel-1$[n]$, by using the splice and cut technique, key recovery attacks of at least 6 and 4 rounds of the cipher are constructed, respectively. For Feistel-1$[\frac{3}{2}n]$, the sizes of $\mathcal{K}_{(1)}$ and $\mathcal{K}_{(2)}$ are $5n/4$ bits each. Therefore, the required time complexity with four start states is estimated as $C_{comp} = \max(2^{5n/4}, 2^{5n/4}) \times 4 + 2^{6n/2-4n/2} \approx 2^{5n/4+2}$, and the required memory is about $2^{5n/4+2}$ blocks. For Feistel-1$[n]$, a similar attack requiring $2^{3n/4+2}$ $(\approx 2^{3n/4} \times 3 + 2^{n/2})$ computations and $(3 \times 2^{3n/4})$ blocks memory is mounted. These attacks also require $2^{n/4}$ chosen plaintexts. Those results are summarized in Table 2.

4 Key Recovery Attack on Feistel-2

This section shows generic key recovery attacks on Feistel-2 ciphers. In contrast to Feistel-1 ciphers, key injections of Feistel-2 ciphers are restricted to XOR operations. This allows an attacker to equivalently transform subkeys, then more rounds can be attacked. To begin with, we introduce an advanced technique called *function reduction*, which enables us to reduce the number of involved subkey bits by exploiting degrees of freedom of a plaintext/ciphertext pair. Combining it with a (multi-)collision technique, 5, 7 and 9 rounds attacks on Feistel-2$[n]$, -2$[\frac{3}{2}n]$ and -2$[2n]$ are demonstrated, respectively. The overview of the function reduction is depicted in Fig. 4. The required complexities for those attacks are summarized in Table 2, and the overview of the attacks are illustrated in Fig. 5. Note that the key additions of Feistel-2 are limited to XOR operations, however, similar idea may be applied to other key additions such as modular additions. Moreover, as an application of our approach on Feistel-2,

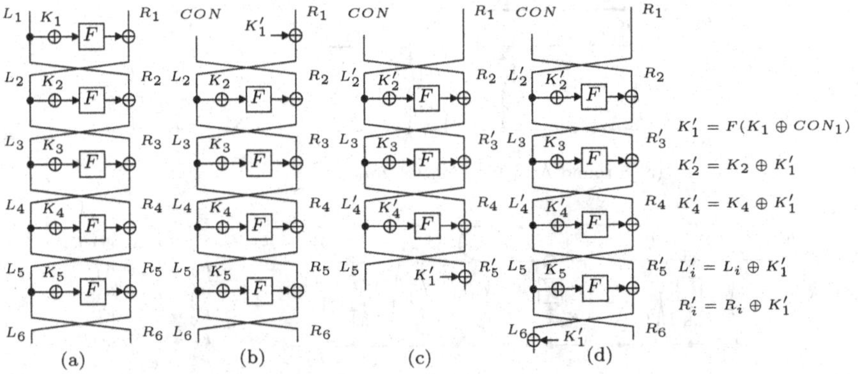

Fig. 4. Function Reduction Technique

we show a key recovery attack on the reduced CAST-128 [1,2]. The structure of CAST-128 is similar to Feistel-2, however, the size of each round key of CAST-128 is larger than that of Feistel-2 and the key additions are not only XOR operations but also modular additions and subtractions. Since the larger round key generally requires more computations to guess, it seems to be hard to directly mount an attack on CAST-128. We use the improved function reduction technique to make an attack feasible, then show a key recovery attack on the 8-round reduced CAST-128, which is the best attack known in literature.

4.1 Function Reduction Technique

Suppose that the half outputs of the r-round Feistel-2 cipher L_{r+1} and R_{r+1} are represented by functions $\mathcal{F}_{L,r}$ and $\mathcal{F}_{R,r}$ as $L_{r+1} = \mathcal{F}_{L,r}(\mathcal{K}_L, L_1|R_1)$ and $R_{r+1} = \mathcal{F}_{R,r}(\mathcal{K}_R, L_1|R_1)$, where $\mathcal{K}_L$ and $\mathcal{K}_R$ denote sets of subkeys used in $\mathcal{F}_{L,r}$ and $\mathcal{F}_{R,r}$, respectively. In general, after sufficient number of round operations, all subkeys are required to compute L_{r+1}, i.e., $|\mathcal{K}_L| = n/2 \cdot r$, while R_{r+1} is derived independently from the last subkey K_r, i.e., $|\mathcal{K}_R| = n/2 \cdot (r-1)$. For the Feistel-2 cipher, fixing half bits of inputs, one more round of subkey data can be reduced as follows:

Theorem 1 (Function Reduction). For the Feistel-2 cipher, if L_1 is fixed, $\mathcal{K}_L$ and $\mathcal{K}_R$ used in $\mathcal{F}_{L,r}$ and $\mathcal{F}_{R,r}$ contain at most $(n/2 \cdot r)$ and $(n/2 \cdot (r-2))$ subkey bits when r is odd, and contain at most $(n/2 \cdot (r-1))$ and $(n/2 \cdot (r-1))$ subkey bits when r is even, respectively.

Proof. By using the key linearization, L_2 is considered to be linearly affected by the subkey K'_1 as follows. Assuming that L_1 is an arbitrary $(n/2)$-bit constant CON, L_2 and R_2 are expressed as $L_2 = R_1 \oplus K'_1$ and $R_2 = CON$, where $K'_1 = F(K_1 \oplus CON)$[1]. Since K'_1 depends only on K_1, it can be regarded as a

[1] For simplicity, we assume that all F-functions are identical. However, our attack works even if each F-function is distinct from each other.

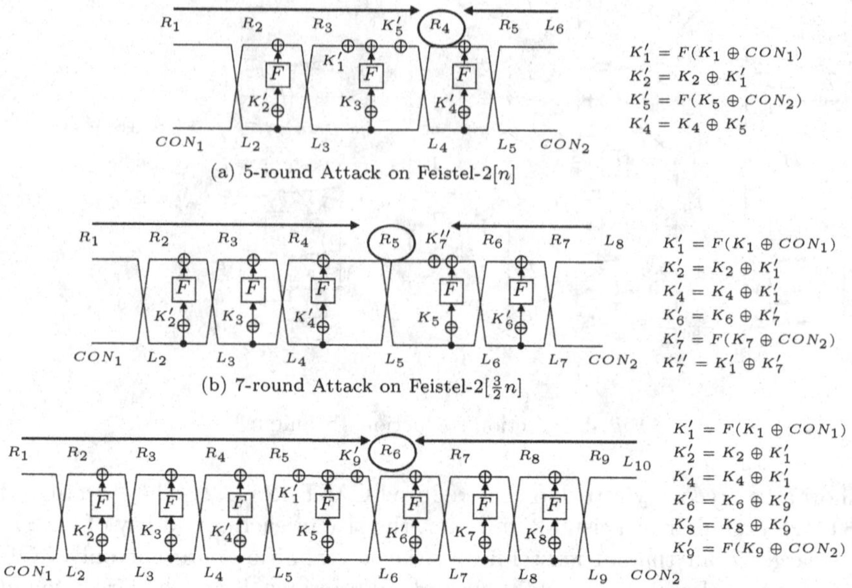

(a) 5-round Attack on Feistel-2[n]

$$K_1' = F(K_1 \oplus CON_1)$$
$$K_2' = K_2 \oplus K_1'$$
$$K_5' = F(K_5 \oplus CON_2)$$
$$K_4' = K_4 \oplus K_5'$$

(b) 7-round Attack on Feistel-2[$\frac{3}{2}n$]

$$K_1' = F(K_1 \oplus CON_1)$$
$$K_2' = K_2 \oplus K_1'$$
$$K_4' = K_4 \oplus K_1'$$
$$K_6' = K_6 \oplus K_7'$$
$$K_7' = F(K_7 \oplus CON_2)$$
$$K_7'' = K_1' \oplus K_7'$$

(c) 9-round Attack on Feistel-2[$2n$]

$$K_1' = F(K_1 \oplus CON_1)$$
$$K_2' = K_2 \oplus K_1'$$
$$K_4' = K_4 \oplus K_1'$$
$$K_6' = K_6 \oplus K_9'$$
$$K_8' = K_8 \oplus K_9'$$
$$K_9' = F(K_9 \oplus CON_2)$$

Fig. 5. Key Recovery Attacks on Feistel-2 Ciphers

new subkey instead of K_1 (see Fig. 4-(b)). By using an equivalent transform, K_1' is moved to the end of the cipher as shown in Figs. 4-(c) and (d). After the transform, each subkey introduced in even round is XORed with K_1', and thus it can be redefined as $K_p' = K_p \oplus K_1'$ (p is even). When r is even, K_1' is linearly affecting to R_{r+1} in the last as shown in Fig. 4-(c). Therefore, both L_{r+1} and R_{r+1} contain at most $(n/2 \cdot (r-1))$ bits of subkeys. When r is odd, K_1' is linearly affecting to L_{r+1} in the last as shown in Fig. 4-(d). Consequently, R_{r+1} contains at most $(n/2 \cdot (r-2))$ bits of subkeys, while the amount of subkey bits required for computing L_{r+1} is not reduced (i.e., $|\mathcal{K}_L| = n/2 \cdot r$). □

The function reduction technique, which consists of equivalent transforms of round keys and the key linearization, is related to the complementation properties of Feistel networks in which the round keys of even (or odd) rounds are complemented by some fixed values. It essentially exploits the property of Feistel network that an input of a keyed F-function in the i-th round (L_i) linearly affects an input of a keyed F-function in the $(i + 2)$-th round (L_{i+2}). In other words, the relation of L_i and L_{i+2} is expressed as $L_{i+2} = L_i \oplus X_{i+1}$, where X_{i+1} is an output of an F-function of (L_{i+1}). We exploit it in the line of a MITM attack to reduce the subkey data for the computation of the intermediate values, while the previous attacks are used for differential attacks [15,8] and speeding up keysearches using equivalent keys [7,18].

4.2 Key Recovery Attack on 5-Round Feistel-2[n]

In order to apply the function reduction to both the forward and backward computations, we prepare plaintext/ciphertext pairs in the form of $L_1 = CON_1$ and $R_6 = CON_2$, where CON_1 and CON_2 denote arbitrary $(n/2)$-bit constants.

Let R_4 be an $(n/2)$-bit matching state. From Theorem 1, in the forward computation, R_4 can be computed by an $(n/2)$-bit subkey $K'_2(= K_2 \oplus K'_1)$, where $K'_1 = F(K_1 \oplus CON_1)$. In the backward computation, since R_4 can be regarded as an output of the even round ($r = 2$), R_4 can also be computed by an $(n/2)$-bit subkey $K'_4(= K_4 \oplus K'_5)$, where $K'_5 = F(K_5 \oplus CON_2)$, i.e., $\mathcal{K}_{(1)} \in K'_2$ and $\mathcal{K}_{(2)} \in K'_4$. Since $|\mathcal{K}_{(1)}| = |\mathcal{K}_{(2)}| = n/2$ and the size of the matching state is also $n/2$, two plaintext/ciphertext pairs are sufficient to determine $\mathcal{K}_{(1)}$ and $\mathcal{K}_{(2)}$. In order to obtain such two pairs that have the form of $L_1 = CON_1$ and $R_6 = CON_2$, we use $2^{n/4}$ chosen plaintexts by randomly changing R_1 as $P = (CON_1|R_1)$. After this process, we have $2^{n/4}$ corresponding ciphertexts, and thus there will exist $(n/2)$ bits colliding R_6 with high probability due to the birthday paradox.

The time complexity of determining K'_2 and K'_4 by the MITM approach is estimated as $C_{comp} = \max(2^{n/2}, 2^{n/2}) \times 2 = 2^{n/2+1}$. In order to determine all subkeys, we use the following equation: $F(R_4 \oplus K_3) = R_1 \oplus K'_1 \oplus L_6 \oplus K_5 = R_1 \oplus L_6 \oplus K''_1$, where $K''_1 = K'_1 \oplus K_5$. Since R_4 can be computed from K'_2 or K'_4, we can recursively mount the MITM approach to determine K_3 and K''_1 with complexity of $2^{n/2+1}(= \max(2^{n/2}, 2^{n/2}) \times 2)$. After exhaustively guessing K_1 with a time complexity of $2^{n/2}$, all subkeys K_i $(1 \leq i \leq 5)$ are determined from the previously obtained subkeys K'_2, K'_4 and K''_1. Therefore, the whole time complexity is estimated as $2^{n/2+2}(\approx 2^{n/2+1} + 2^{n/2+1} + 2^{n/2})$. Due to $k = n$, the time complexity $2^{n/2+2} = 2^{k/2+2}$ is less than 2^k which is required computations for the brute force attack. The required data is $2^{n/4}$ chosen plaintext, and the required memory is about $2^{n/2+1}$ words. If the function reduction technique is used only in the forward computation, a 4-round attack is constructed with less data (see Fig. 5-(a) and Table 2).

4.3 Key Recovery Attack on 9-Round Feistel-2[$2n$]

A key recovery attack on a 9-round Feistel-2[$2n$] is constructed in a similar way to the 5-round attack on Feistel-2[n]. In this attack, we can add 2 more rounds in each direction, and a 6-multicollision is required to obtain desired plaintext/ciphertext pairs unlike the attack on Feistel-2[n]. It has been known that an n-bit t-multicollision is found in $t! \cdot 2^{n \cdot (t-1)/t}$ random data with high probability [36]. Thus, the six plaintext/ciphertext pairs whose form are $P = (CON_1|R_1)$ and $C = (CON_2|L_{10})$ could be found from $6!^{1/6} \cdot (2^{n/2})^{5/6} \approx 3 \cdot (2^{n/2})^{5/6}$ chosen plaintexts. More precisely, after querying $3 \cdot (2^{n/2})^{5/6}$ chosen plaintexts with distinct R_1, there will exist a 6-collision of R_{10} in corresponding ciphertexts with high probability (see Fig. 5-(c) and Table 2).

Table 2. Details of Our Attacks

Target	key size	Round	Time	Memory	Data	Reference
Feistel-1	n	4	$2^{3n/4+2}$	$2^{3n/4+2}$	$2^{n/4}$	Sect. 3
	$\frac{3}{2}n$	6	$2^{5n/4+2}$	$2^{5n/4+2}$	$2^{n/4}$	Sect. 3
	$2n$	8	$2^{7n/4+3}$	$2^{7n/4+3}$	$2^{n/4}$	Sect. 3
Feistel-2	n	4	$2^{n/2+2}$	$2^{n/2+1}$	2	Sect. 4.2
		5	$2^{n/2+2}$	$2^{n/2+1}$	$2^{n/4}$	Sect. 4.2
	$\frac{3}{2}n$	6	$2^{5n/4+4}$	2^{n+4}	9	Sect. 4.4
		7	$2^{5n/4+4}$	2^{n+4}	$9!^{1/9} \cdot (2^{n/2})^{8/9}$	Sect. 4.4
	$2n$	8	$2^{3n/2+3}$	$2^{3n/2+3}$	6	Sect. 4.3
		9	$2^{3n/2+3}$	$2^{3n/2+3}$	$6!^{1/6} \cdot (2^{n/2})^{5/6}$	Sect. 4.3
Feistel-3	n	7	$2^{3n/4+\ell} \cdot N_1$	$2^{3n/4+\ell} \cdot N_1$	N_1	Sect. 5.3
	$\frac{3}{2}n$	8	$2^{n+\ell} \cdot N_2$	$2^{n+\ell} \cdot N_2$	N_2	Sect. 5.5
		9	$2^{n+\ell} \cdot N_2$	$2^{n+\ell} \cdot N_2$	$N!^{1/N_2} \cdot (2^{n/2})^{(N_2-1)/N_2}$	Sect. 5.5
	$2n$	11	$2^{7n/4+\ell} \cdot N_3$	$2^{7n/4+\ell} \cdot N_3$	N_3	Sect. 5.4

$N_1 = (3n/2 + 2\ell)/\ell$, $N_2 = (2n + 2\ell)/\ell$, $N_3 = (7n/2 + 2\ell)/\ell$

4.4 Key Recovery Attack on 7-Round Feistel-2[$\frac{3}{2}n$]

In this attack, R_5 is used as the matching state. From Theorem 1, in the forward computation, R_5 can be computed from $3 \cdot n/2$ bits subkeys K_2', K_3 and K_1', where $K_2' = K_2 \oplus K_1'$ and $K_1' = F(K_1 \oplus CON_1)$. In the backward computation, R_5 can be computed from $3 \cdot n/2$ bits subkeys K_6', K_5 and K_7', where $K_6' = K_6 \oplus K_7'$ and $K_7' = F(K_7 \oplus CON_2)$. Since R_5 is expressed as $K_1' \oplus L_4$ and $K_7' \oplus (F(R_6 \oplus K_5) \oplus L_6)$, if only $(n/4)$ bits of $K_1' \oplus K_7'$ are guessed, $(n/4)$-bit matching is feasible. It is regarded that $K_7''(= K_1' \oplus K_7')$ is included in the backward computation (see Fig. 5-(b)).

Then, since $|\mathcal{K}_{(1)}| = n/4$, $|\mathcal{K}_{(2)}| = 5n/4$, and the size of the matching state is $n/4$, nine plaintext/ciphertext pairs are required to determine $\mathcal{K}_{(1)}$ and $\mathcal{K}_{(2)}$ due to the relation $(n + 5n/4)/(n/4) = 9$. Such nine plaintext/ciphertext pairs whose form are $P = (CON_1|R_1)$ and $C = (CON_2|L_8)$ can be found from $9!^{1/9} \cdot (2^{n/2})^{8/9} \approx 4.2 \cdot (2^{n/2})^{8/9}$ chosen plaintexts. The other complexities required for this attack and the low data attack on 6-round Feistel-2[$\frac{3}{2}n$] are described in Table 2.

4.5 Application to 8-Round Reduced CAST-128

In order to demonstrate the practical impact of our work on Feistel-2, we apply it to CAST-128 block cipher. Using the improved function reduction techniques, we show an attack on the 8-round reduced CAST-128 having more than 118 bits key, which is the best attack with respect to the number of attacked rounds in literature even when its key scheduling is an ideal function.

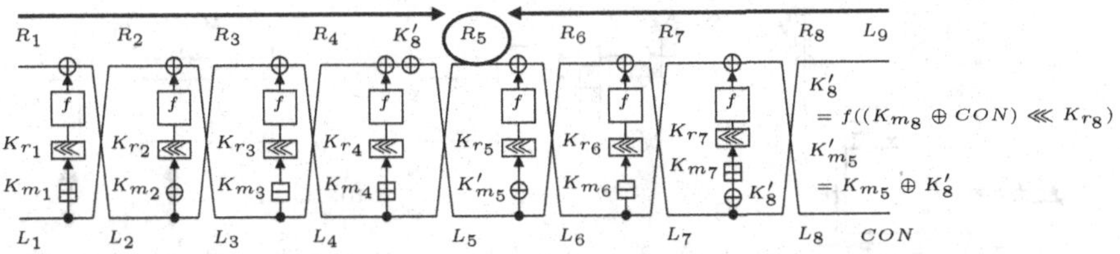

Fig. 6. Key Recovery Attack on 8-Round CAST-128

Description of CAST-128. CAST-128 [1,2] is a 64-bit Feistel block cipher accepting a variable key size from 40 up to 128 bits (but only in 8-bit increments). The number of rounds is 16 when the key size is longer than 80 bits. First, the algorithm divides the 64-bit plaintext into two 32-bit words L_0 and R_0, then the i-th round function outputs two 32-bit data L_{i+1} and R_{i+1} as follows:

$$L_{i+1} = R_i \oplus F_i(L_i, K_i^{rnd}), R_{i+1} = L_i,$$

where F_i denotes the i-th round function and K_i^{rnd} is the i-th round key consisting of a 32-bit masking key K_{m_i} and a 5-bit rotation key K_{r_i}. The detail of F_i is expressed as

$$F_i = f((L_i \bigcirc_i K_{m_i}) \lll K_{r_i}),$$

where f consists of four 8 to 32-bit S-boxes, $\lll K_{r_i}$ denotes a K_{r_i}-bit left rotation, and $\bigcirc_i$ denotes addition, XOR or subtraction depending on the round number i, i.e., $\bigcirc_i$ denotes addition for $i \in \{1, 4, 7, 10, 13\}$, XOR for $i \in \{2, 5, 8, 11, 14\}$ and subtraction for $i \in \{3, 6, 9, 12, 15\}$. We omit the details of f, since, in our analysis, it is regarded as the random function that outputs a 32-bit random value from a 32-bit input.

Key Recovery Attack on 8-Round CAST-128. The structure and the parameter of CAST-128 having sufficiently large key are similar to Feistel-2[$2n$]. However, for CAST-128, a $37(= 32 + 5)$-bit subkey is inserted into each F_i, i.e., a 32-bit subkey is used in $\bigcirc_i$ and the remaining 5-bit subkey is used in a key dependent rotation, while a 32-bit subkey is inserted in each round for Feistel-2[$2n$] with $n = 32$. Thus, the 9-round attack on Feistel-2[$2n$] is not directly applicable to CAST-128. However, the improved function reduction technique allows us to construct an 8-round attack on CAST-128.

Let R_5 be an $(n/2)$-bit matching state. In the backward computation, R_9 is fixed as CON, and $K'_8 = f((CON \oplus K_{m_8}) \lll K_{r_8})$ is moved to L_5 and an input of the 7-th round function, by converting K_{m_5} into $K'_{m_5} = K'_8 \oplus K_{m_5}$, as shown in Fig. 6. Then, the input of f in the 7-th round is expressed as $(L_9 \oplus K'_8) + K_{m_7}$. If the lower b bits of L_9, which are controllable by the ciphertext, are fixed to 0, the lower b bits of this computation are expressed as $K'_8 + K_{m_7}$. Thus, $(K'_8 + K_{m_7})$ is regarded as a new b-bit subkey $K'_{m_7} = (K'_8 + K_{m_7})$, while the upper $(n/2 - b)$ bits remain $(L_9 \oplus K'_8) + K_{m_7}$. In the backward computation of R_5, $|\mathcal{K}_{(2)}| = 37 \times 2 + (b + (n/2 - b) \times 2 + 5)$ bits of the key are involved.

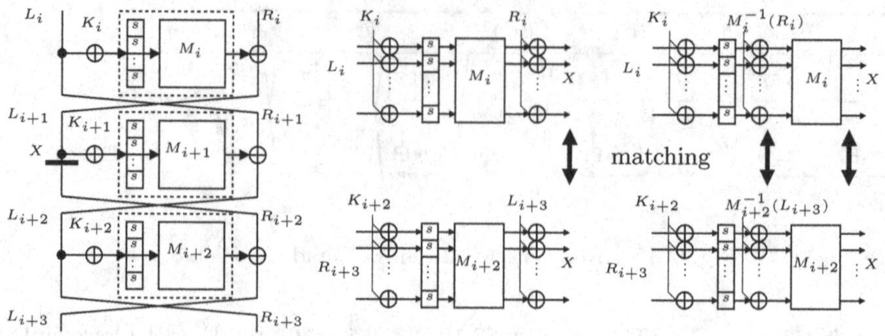

Fig. 7. Matching without Matrix

Evaluation. Since $|\mathcal{K}_{(1)}| = 111$, $|\mathcal{K}_{(2)}| = 114$ ($b = 29$) and the size of the matching state is 32 bits, eight plaintext/ciphertext pairs are required to determine $\mathcal{K}_{(1)}$ and $\mathcal{K}_{(2)}$ due to the relation $(111 + 114)/32 < 8 (= 2^{32-29})$. The required time complexity to determine subkeys K_1^{rnd}, K_2^{rnd}, K_3^{rnd}, K_6^{rnd}, K'_{m_5}, K_{r_5}, the lower 29 bits of K'_{m_7}, the upper 3 bits of K'_8, and K_{m_7} is estimated as $C_{comp} = \max(2^{111}, 2^{114}) \times 10 \approx 2^{118}$. The remaining K_{r_4} and K'_8 are exhaustively searched with the time complexity of 2^{64}. Then, all subkeys are obtained by using the relations of $K'_{m_7} = K'_8 + K_{m_7}$, $K'_{m_5} = K'_8 \oplus K'_{m_5}$ and $K'_8 = f((CON \oplus K_{m_8}) \lll K_{r_8})$. The required data is eight chosen ciphertexts, and the required memory is 2^{111} words. Therefore, when the key size is more than 118 bits long, our attack works faster than the brute force attack.

5 Key Recovery Attack on Feistel-3

This section presents generic key recovery attacks on Feistel-3 ciphers. Feistel-3 ciphers are the Feistel-2 ciphers whose F-functions are restricted to be SP-type F-functions, which consist of an S-box layer followed by a linear matrix operation. This allows an attacker to exploit a linearity of a matrix computation, and thus the number of attacked rounds can be increased. To begin with, we review two techniques which exploit a linearity of a matrix computation. We refer those two techniques as *matching without matrix* and *matrix separation* to make our explanation simple. However, those techniques have already been introduced, for example, in [29,32]. Combining them with a (multi-)collision technique and *function reduction*, 7, 9 and 11 rounds attacks on Feistel-3$[n]$, -3$[\frac{3}{2}n]$ and -3$[2n]$ are demonstrated, respectively. Furthermore, as an application of our approach on Feistel-3, we show several key recovery attacks on the reduced Camellia [5]. Since Camellia is a Feistel cipher with SP-type F-functions, our attack on Feistel-3 can be directly applied to it even if its key scheduling function is ideal. Besides, the number of attacked rounds by our attack is further increased by one round for Camellia due to its non-MDS matrix. Consequently, we present generic key recovery attacks requiring extremely low data on the 8-, 10-, 12-round reduced Camellia without FL/FL^{-1} functions and key whitenings.

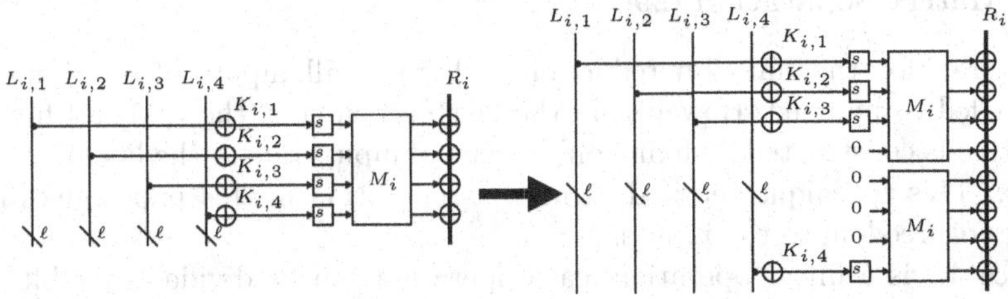

Fig. 8. Matrix Separation

5.1 Matching without Matrix [32]

Let us consider three consecutive rounds of the Feistel-3 cipher whose input and output are represented as $(L_i|R_i)$ and $(L_{i+3}|R_{i+3})$ as shown in Fig. 7. Assuming that an attacker knows those input and output variables, the following equation holds:

$$F_i(L_i \oplus K_i) \oplus R_i = F_{i+2}(R_{i+3} \oplus K_{i+2}) \oplus L_{i+3}. \tag{2}$$

In order to check if the equation holds, we need to guess $2 \cdot n/2$ bits subkeys K_i and K_{i+2}, while K_{i+1} is not needed to be guessed. However, if F-functions are SP-type F-functions (i.e., $F_i = M_i \circ S_i$, where M_i and S_i denote an $m \times m$ matrix and an S-box layer consisting of m ℓ-bit S-boxes, respectively), the size of guessing subkey bits can be reduced by exploiting the linearity of the matrix operation. Since M_i is a linear function, Eq.(2) is redescribed as:

$$M_i(S_i(L_i \oplus K_i)) \oplus R_i = M_{i+2}(S_{i+2}(R_{i+3} \oplus K_{i+2})) \oplus L_{i+3},$$
$$M_i(S_i(L_i \oplus K_i) \oplus M_i^{-1}(R_i)) = M_{i+2}(S_{i+2}(R_{i+3} \oplus K_{i+2}) \oplus M_{i+2}^{-1}(L_{i+3})).$$

When $M_i = M_{i+2}$, we have

$$S_i(L_i \oplus K_i) \oplus M_i^{-1}(R_i) = S_{i+2}(R_{i+3} \oplus K_{i+2}) \oplus M_{i+2}^{-1}(L_{i+3}). \tag{3}$$

Unlike Eq.(2), we can separately check if Eq.(3) holds by the size of the S-box ℓ. Therefore, this technique enables us to reduce the number of subkey bits to be guessed for the 3-round matching from $2^{2 \cdot n/2}$ to $2^{2\ell}$. When $M_i \neq M_{i+2}$, the matching technique called matching through matrix presented in [31] is utilized. In this case, more than $m \cdot \ell$ bits subkeys are required to be guessed. For simplicity, from now on, we assume that $M_i = M_{i+2}$.

In the function reduction, the modified subkey K_1' affects L_{2t} and R_{2t-1} ($t = 1, 2, ...$). Also, in the matching without matrix, we utilize the relation of L_{i+1} as the matching state. This implies that if $(i+1)$ is even (i.e., $(i+1) = 2t$), L_{i+1} is affected by K_1' and it cannot be used as the matching state. Therefore, if the matching without matrix is used with the function reduction, the starting round of the matching i must be even (i.e., $(i+1)$ must be odd).

5.2 Matrix Separation [29]

In general, for the function reduction technique, all inputs of an F-function are needed to be fixed. However, in the Feistel-3 ciphers, the (partial) function reduction is constructed by fixing only a part of inputs due to the linearity of the matrix. This technique referred as matrix separation in this paper gives more degrees of freedom to the inputs.

Since M_i is a linear operation, each operation can be divided by ℓ bits. For instance, we show the case of $m = 4$ as an example (see Fig. 8). Suppose that $K_i = (K_{i,1}|K_{i,2}|K_{i,3}|K_{i,4})$, $K_{i,j} \in \{0,1\}^\ell$ and $L_i = (L_{i,1}|L_{i,2}|L_{i,3}|L_{i,4})$, $L_{i,j} \in \{0,1\}^\ell$. If three input words $L_{i,1}$, $L_{i,2}$ and $L_{i,3}$ are fixed, only $3/4 \times n/2$ bits of K_i are linearly inserted into the $(i+1)$-round by regarding $T = M(S'((L_{i,1} \oplus K_{i,1})|(L_{i,2} \oplus K_{i,2})|(L_{i,3} \oplus K_{i,3}))|\mathbf{0}^\ell)$ as new subkey bits, where S' consists of three S-boxes and $\mathbf{0}^\ell$ denotes ℓ bits of 0. Note that T is an $(n/2)$-bit data, however, it is determined by $(3/4 \cdot n/2)$ bits subkeys $K_{i,1}$, $K_{i,2}$ and $K_{i,3}$. Since $L_{i,4}$ is not fixed, $M(\mathbf{0}^{3/4 \cdot n/2}|s(L_{i,4} \oplus K_{i,4}))$ is non-linearly inserted into the $(i+1)$-th round.

5.3 Key Recovery Attack on 7-Round Feistel-3$[n]$

For the 7-round Feistel-3$[n]$, it seems that the function reduction is applied to both directions and the matching without matrix is used in the rounds 3 to 5. However, this approach does not work due to the restriction of the combination of the matching without matrix and the function reduction. To overcome this problem, we utilize the partial function reduction in conjunction with the matching without matrix.

At first, L_1 is fixed as CON_1, and $K'_1 = F(K_1 \oplus CON_1)$ is moved to R_5 by converting K_2 and K_4 into $K_2 \oplus K'_1$ and $K_4 \oplus K'_1$, respectively. In addition, R_{1L}, which is the left half of R_1 ($n/4$ bits), is also fixed as an $n/4$-bit constant CON_L. Using the matrix separation technique, the partial function reduction technique is applicable to the left half of K'_2 represented as K'_{2L}. Specifically, let an $n/2$-bit variable K''_2 be $K''_2 = M(S'(K'_{2L} \oplus CON_L)|\mathbf{0}^{n/4})$, where S' consisting of $m/2$ S-boxes and $\mathbf{0}^{n/4}$ denotes $n/4$ bits of 0. Since K''_2 is linearly inserted in round 2 by the matrix separation, it is possible to move to L_7 (see Fig. 9-(a)).

The matching without matrix technique is applied to the three consecutive rounds from rounds 4 to 6. In the forward and backward computations, $(L_4|R_4)$ and $(L_7|R_7)$ are computable from (K'_{2R}, K'_3) and (K'_{2L}, K'_7), respectively. Then, if ℓ bits of K'_4 and K_6 are guessed, an ℓ-bit matching is feasible, i.e., $\mathcal{K}_{(1)} \in \{K'_{2R}, K'_3, K'_{4,a}\}$ and $\mathcal{K}_{(2)} \in \{K'_{2L}, K'_7, K_{6,a}\}$, where $(1 \le a \le m)$, and $K'_{4,a}$ and $K_{6,a}$ denote arbitrary ℓ bits data of K'_4 and K_6, respectively.

Since $|\mathcal{K}_{(1)}| = |\mathcal{K}_{(2)}| = 3/2 \cdot n/2 + \ell$ and the matching size is ℓ bits, $N_1 = (3n/2 + 2\ell)/\ell$ plaintext/ciphertext pairs are required to determine $\mathcal{K}_{(1)}$ and $\mathcal{K}_{(2)}$. The complexity of determining $\mathcal{K}_{(1)}$ and $\mathcal{K}_{(2)}$ is estimated as $C_{comp} = \max(2^{3n/4+\ell}, 2^{3n/4+\ell}) \times N_1$. After that, we are able to determine the other bits for finding all subkey bits by using a simple MITM attack on the remaining K'_4 and K_6, and K'_1 and K_5, respectively.

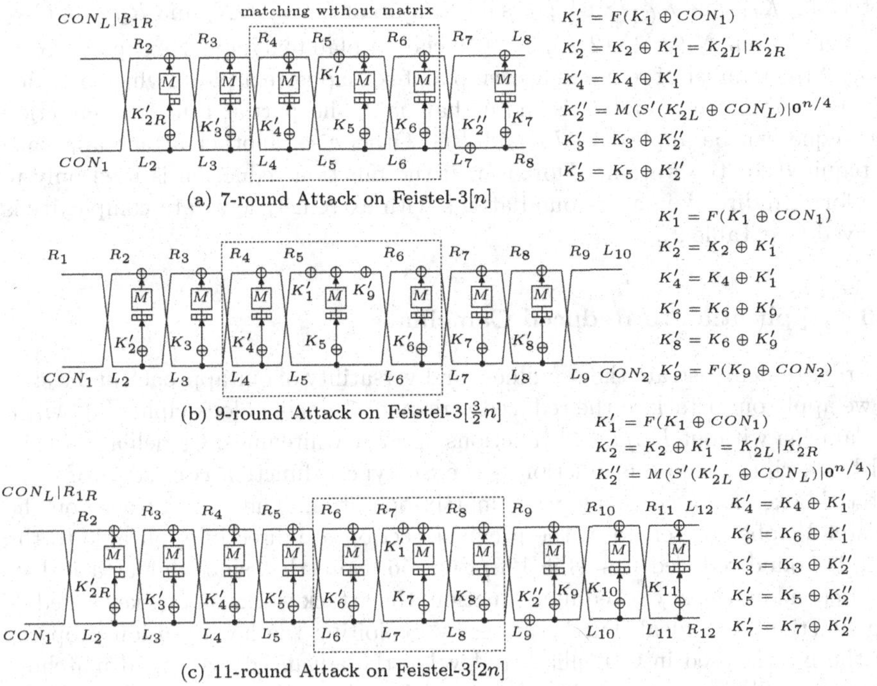
matching without matrix
$K'_1 = F(K_1 \oplus CON_1)$
$K'_2 = K_2 \oplus K'_1 = K'_{2L}|K'_{2R}$
$K'_4 = K_4 \oplus K'_1$
$K''_2 = M(S'(K'_{2L} \oplus CON_L)|0^{n/4}$
$K'_3 = K_3 \oplus K''_2$
$K'_5 = K_5 \oplus K''_2$
(a) 7-round Attack on Feistel-3[n]
$K'_1 = F(K_1 \oplus CON_1)$
$K'_2 = K_2 \oplus K'_1$
$K'_4 = K_4 \oplus K'_1$
$K'_6 = K_6 \oplus K'_9$
$K'_8 = K_6 \oplus K'_9$
$K'_9 = F(K_9 \oplus CON_2)$
(b) 9-round Attack on Feistel-3[$\frac{3}{2}n$]
$K'_1 = F(K_1 \oplus CON_1)$
$K'_2 = K_2 \oplus K'_1 = K'_{2L}|K'_{2R}$
$K''_2 = M(S'(K'_{2L} \oplus CON_L)|0^{n/4})$
$K'_4 = K_4 \oplus K'_1$
$K'_6 = K_6 \oplus K'_1$
$K'_3 = K_3 \oplus K''_2$
$K'_5 = K_5 \oplus K''_2$
$K'_7 = K_7 \oplus K''_2$
(c) 11-round Attack on Feistel-3[$2n$]

Fig. 9. Key Recovery Attacks on Feistel-3 Ciphers

Therefore, the whole time complexity is estimated as $2^{3n/4+\ell} \times N_1$. Due to $k = n$, the required complexity $2^{3k/4+\ell} \cdot N_1$ is less than 2^k. The required data is $N_1 = (3n/2 + 2\ell)/\ell$ chosen plaintexts, and the memory is $2^{3n/4+\ell} \cdot N_1$ words.

5.4 Key Recovery Attack on 11-Round Feistel-3[$2n$]

Similarly to the attack on the 7-round Feistel-3[n], chosen plaintexts in the form of $P = (L_1|R_{1L}|R_{1R}) = (CON|CON_L|R_{1R})$ are used. Then two more rounds can be added to both forward and backward directions due to increasing the master key size. Thus, an 11-round attack is constructed. For the detailed parameters, see Table 2 and Fig. 9.

5.5 Key Recovery Attack on 9-Round Feistel-3[$\frac{3}{2}n$]

As shown in Fig. 9-(b), for the 9-round Feistel-3[$\frac{3}{2}n$], the function reduction is applied to both directions combined with the matching without matrix to the rounds 4 to 6, since the middle of the matching is odd indexed round. Thus, a key recovery attack is constructed in a straightforward way, unlike the attacks on Feistel-3[n] and -3[$2n$].

In this attack, since $|\mathcal{K}_{(1)}| = |\mathcal{K}_{(2)}| = 2n/2 + \ell$ and the matching size is ℓ bits from Fig. 9-(b), $N_2 = (2n + 2\ell)/\ell$ plaintext/ciphertext pairs are required to determine $\mathcal{K}_{(1)}$ and $\mathcal{K}_{(2)}$. Such pairs in the form of $L_1 = CON_1$ and $R_{10} = CON_2$ are found from $N_2!^{1/N_2} \cdot (2^{n/2})^{N_2-1/N_2}$ chosen plaintext/ciphertext pairs. Note that, if the number of required chosen plaintext/ciphertext pairs, which depends on the parameter n and ℓ, is more than $n/2$, the partial function reduction technique can be applied to L_1 and R_{10}. Otherwise, another attack approach is required for this variant. Moreover, if the function reduction is used only in the forward direction, an 8-round attack with extremely low data complexity is derived (see Table 2).

5.6 Application to Reduced Camellia

In order to demonstrate the usefulness and versatility of our approach on Feistel-3, we apply our attack to the reduced version of Camellia block cipher [5], which is Camellia without FL/FL^{-1} functions and key whitenings. Camellia is a Feistel block cipher whose F-function is the SP-type F-function consisting of eight 8-bit S-boxes followed by an 8×8 matrix operation. Thus, our attacks on the Feistel-3 cipher presented in the previous section are directly applicable to the 7/9/11-round reduced Camellia-128/192/256. Note that since our attack does not depend on the key scheduling function, the attack works on any key scheduling function even ideal. Furthermore, by exploiting the low diffusion property on the matrix used in Camellia, we develop the advanced five round matching technique. Then we present low-data complexity attacks requiring less than 60 plaintext/ciphertext pairs on the 8/10/12-round reduced Camellia-128/192/256 without FL/FL^{-1} and key whitenings.

Five Round Matching for Non-MDS Matrix. Let us consider five consecutive rounds of the Camellia whose input and output are represented as $(L_i|R_i)$ and $(L_{i+5}|R_{i+5})$, respectively. By using the three-round matching without matrix technique in the middle, the following equation holds.

$$S(L_{i+1} \oplus K_{i+1}) \oplus M^{-1}(L_i) = S(R_{i+4} \oplus K_{i+3}) \oplus M^{-1}(R_{i+5}).$$

Since the S-box layer consists of eight 8-bit S-boxes, by guessing two bytes of subkeys with the same byte position $K_{i+1,j}$ and $K_{i+3,j}$, the 8-bit matching is possible if the same indexed 8 bits data $L_{i+1,j}$ and $R_{i+4,j}$ are also known. Since $L_{i+1} = M(S(L_i \oplus K_i)) \oplus R_i$ and $R_{i+4} = M(S(L_{i+5} \oplus K_{i+4})) \oplus R_{i+5}$, all bits of K_i and K_{i+4} are required to be guessed to obtain any byte of L_{i+1} and R_{i+4} if the underlying matrix M is optimal (i.e., MDS matrix). However, for Camellia, the 8 bits data $L_{i+1,j}$ and $R_{i+4,j}$ are derived by guessing corresponding $40 (= 8 \times 5)$ bits of K_i and K_{i+4} when $(5 \le j \le 8)$, since Camellia utilizes non-MDS matrix (See [5] for the details of the matrix used in Camellia). For example, $L_{i+1,5}$ and $R_{i+4,5}$ are derived from $K_{i,p} (p \in \{1, 2, 6, 7, 8\})$ and $K_{i+4,q} (q \in \{1, 2, 6, 7, 8\}$, respectively. Therefore, the number of key bits to be guessed for the 5-round matching in each direction is reduced from 128 bits $(= 64 \times 2)$ to 48 bits $(= 8 + 40)$.

$$K_1' = F(K_1 \oplus CON_1)$$
$$K_2' = K_2 \oplus K_1'$$
$$K_4' = K_4 \oplus K_1'$$
$$K_6' = K_6 \oplus K_1'$$
$$K_2'' = M(S'(K_{2,1-7}' \oplus CON_L)|0^8)$$
$$K_3' = K_3 \oplus K_2''$$
$$K_5' = K_5 \oplus K_2''$$
$$K_7' = K_7 \oplus K_2''$$
$$K_9' = K_9 \oplus K_2''$$

(a) 8-round Attack on Camellia-128

(b) 10-round Attack on Camellia-192

(c) 12-round Attack on Camellia-256

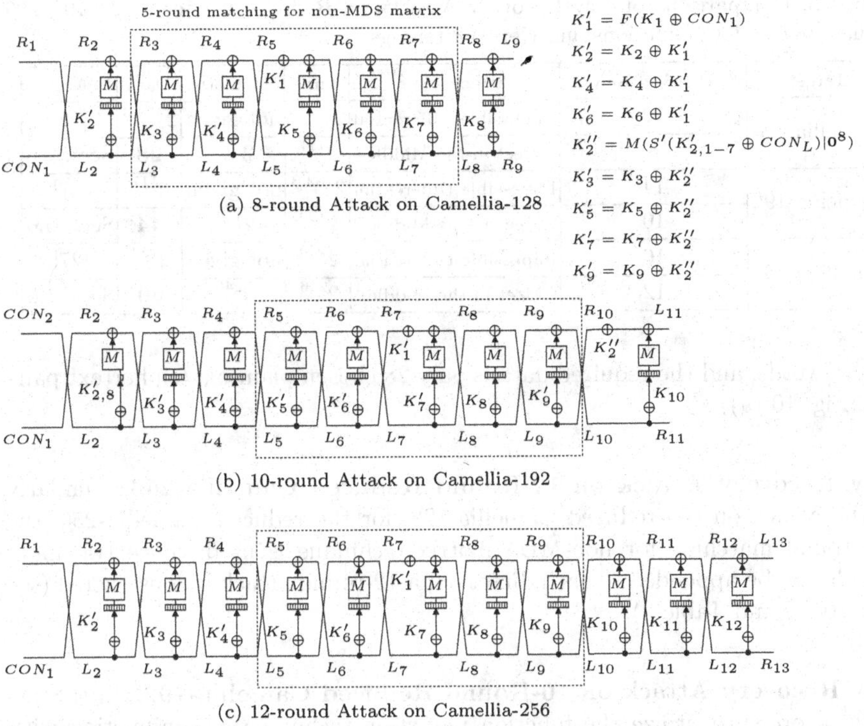

Fig. 10. Key Recovery Attacks on Reduced Camellia-128/192/256

Key Recovery Attack on 8-Round Reduced Camellia-128.

Let us consider the 8-round reduced Camellia-128. In order to use the function reduction technique in the forward process, we collect chosen plaintexts in the form of $L_1 = CON_1$.

The five round matching for non-MDS matrix technique is used from rounds 3 to 7. In the forward and backward computations, $(L_3|R_3)$ and $(L_8|R_8)$ are computable by using $K_2'(= K_2 \oplus K_1')$ and K_8, respectively. Then, for the 8-bit matching, 8 bits subkey $K_{4,a}'$ and the corresponding 40 bits of subkey K_3 in the forward computation are required to be guessed, where $K_4' = K_4 \oplus K_1'$. Similarly, we need to guess 8 bits subkey $K_{6,a}$ and the corresponding 40 bits of subkey K_7 in the backward computation. In other words, $\mathcal{K}_{(1)} \in \{K_2', K_{4,a}', 40 \text{ bits of } K_3\}$ and $\mathcal{K}_{(2)} \in \{K_8, K_{6,a}, 40 \text{ bits of } K_7\}$, where $5 \le a \le 8$.

Since $|\mathcal{K}_{(1)}| = |\mathcal{K}_{(2)}| = 112$ and the matching size is 8 bits, $28(= (112+112)/8)$ plaintext/ciphertexts are sufficient to determine $\mathcal{K}_{(1)}$ and $\mathcal{K}_{(2)}$. The complexity of determining $\mathcal{K}_{(1)}$ and $\mathcal{K}_{(2)}$ is estimated as $C_{comp} = \max(2^{112}, 2^{112}) \times 28 \approx 2^{117}$. After that, we are able to determine the other bits for finding all subkeys by using the simple MITM attack on the remaining 24 bits of K_3 and K_7, and 56 bits of K_4' and K_6 in the forward and backward computations, respectively. Therefore, the whole complexity is estimated as $2^{117}(\approx 2^{117} + 2^{80})$. The required memory

Table 3. Comparisons of Key Recovery Attacks on Reduced Camellia-128/192/256 without FL/FL^{-1} Functions and Key Whitenings

Target	# Attacked Rounds	Attack Type	Time	Memory	Data	Reference
Camellia-128	12	Impossible Differential	$2^{116.6}$	not given	$2^{116.3}$	[28]
	8	Meet-in-the-Middle	2^{117}	2^{117}	**28**	Sect. 5.6
Camellia-192	14	Impossible Differential	$2^{182.2}$	not given	2^{117}	[27]
	10	Meet-in-the-Middle	2^{190}	2^{174}	**44**	Sect. 5.6
Camellia-256	16	Impossible Differential	2^{249}	not given	2^{123}	[27]
	12	Meet-in-the-Middle	2^{246}	2^{246}	**60**	Sect. 5.6

is 2^{117} words, and the required data is only 28 chosen plaintext/ciphertext pairs (see Fig. 10-(a)).

Key Recovery Attack on 12-Round Reduced Camellia-256. Similarly to the attack on the reduced Camellia-128, for the reduced Camellia-256, the five round matching for non-MDS matrix technique is used. Since two more rounds can be appended to each direction, a 12-round attack is constructed (see Fig. 10-(c) and Table 3).

Key Recovery Attack on 10-Round Reduced Camellia-192. In this attack, in order to utilize the function reduction technique in conjunction with the matrix separation technique, we collect chosen plaintexts in the form of $L_1 = CON_1$ and $R_{1,1-7} = CON_L$, where $R_{1,1-7}$ denotes the left 56 bits of R_1 and CON_L is a 56-bit constant. Then $K_1' = F(CON_1 \oplus K_1)$ is moved to R_7 by redefining $K_p' = K_p \oplus K_1'(p = 2, 4, 6)$. In addition, the left 56 bits of K_2' defined as $K_{2,1-7}'$ is also moved to R_{10} by using the partial function reduction technique. Namely, we assume that $K_2'' = M(S'(K_{2,1-7}' \oplus CON_L)|0^8)$ is linearly inserted in round 2 and the remaining 8-bit subkey $K_{2,8}'$ is non-linearly inserted in round 2, where S' consists of seven 8-bit S-boxes.

The five round matching for non-MDS matrix technique is used from rounds 5 to 9. Here, $(L_5|R_5)$ and (L_{10}, R_{10}) are computed from $(K_{2,8}', K_3'(= K_3 \oplus K_2''), K_4'(= K_4 \oplus K_{1L}'))$ and $(K_{2,1-7}', K_9)$, respectively. For the 8-bit matching, $K_{6,8}'$ and the corresponding 40 bits of $K_5'(= K_5 \oplus K_2'')$ are required to be guessed in the forward computation, where $K_6' = K_6 \oplus K_1'$. Similarly, in the backward computation, $K_{8,8}$ and the corresponding 40 bits of $K_9'(= K_9 \oplus K_2'')$ are required to be guessed. Namely, $\mathcal{K}_{(1)} \in \{K_{2,8}', K_3', K_4', K_{6,8}',$ 40 bits of $K_5'\}$ and $\mathcal{K}_{(2)} \in \{K_{2,1-7}', K_9, K_{8,8},$ 40 bits of $K_9'\}$. In this attack, the whole complexity to determine all subkey bits is estimated as $2^{190} (\approx 2^{190} + 2^{80})$. The required memory is $2^{174} (\approx 2^{168} \times 44)$ words, and the required data is only 44 chosen plaintext/ciphertext pairs (see Fig. 10-(b) and Table 3).

6 Discussion

In order to compare the numbers of attacked rounds by our attacks with the previous results, we consider key recovery attacks from a 5-round impossible differential distinguisher or a 5-round zero-correlation linear distinguisher on the Feistel ciphers employing bijective F-functions [6,11]. Note that those distinguishers depend only on the structure of the cipher unlike the other distinguishers such as a differential and a linear distinguisher. When $k = n$, guessing $n/2$ bits subkey involved in the 6-th round, it is possible to construct a 6-round key recovery attack from the 5-round distinguishers. Similarly, for $k = 3n/2$ and $k = 2n$, a 7 and an 8-round key recovery attacks are constructed by additionally guessing $n/2$ and n bits subkeys, respectively. Compared to those results, our attacks are the best attacks with respect to the number of attacked rounds for Feistel-2[$2n$], -3[n], -3[$\frac{3}{2}n$] and -3[$2n$] as described in Table 1. Also, for Feistel-1[$2n$] and Feistel-2[$\frac{3}{2}n$], the same numbers of rounds are attacked by our approach. Especially, the attack on the 11-round Feistel-3[$2n$] greatly exceeds the number of attacked rounds given by the distinguisher based attacks. More importantly, Feistel-3[$2n$] structure is well used in concrete block ciphers such as a 128-bit block cipher taking a 256-bit key, e.g., Camellia-256.

In addition, thanks to the MITM approach, most of our attacks require an extremely small data complexity, in contrast to the classical statistical attacks such as the impossible differential and zero correlation linear attacks that generally require huge amount of data. This implies that our attacks may work even if the number of queries to the encryption oracle is restricted. In fact, the similar approach, which is the low-data complexity attacks on AES, has already been studied in [13,14]. Thus, our work is also regarded as the first evaluation results on the low-data complexity attacks on the Feistel schemes.

7 Conclusion

This paper has shown the improved generic key recovery attacks on Feistel schemes independent of the key scheduling function. The proposed approach is based on the all subkeys recovery attack. With several advanced techniques such as function reduction and key linearization, which basically reduce the number of involved subkey bits, we presented several new key recovery attacks on the Feistel schemes.

To demonstrate the usefulness and the versatility of our approach, we showed several attacks on the concrete block ciphers including CAST-128 and Camellia. Among them, we would like to stress that the presented attack on the 8-round reduced CAST-128 having more than 118 bits key is the best attack with respect to the number of attacked rounds. Since our approach is generic, it is expected to be applied to other Feistel-type block ciphers. We believe that our results are useful not only for a deeper understanding the security of the Feistel schemes, but also for designing an efficient block cipher such as a low-latency cipher. Moreover, we expect that our attacks could be improved by combining with the recent attack called sieve-in-the-middle attack [17].

References

1. Adams, C.: The CAST-128 encryption algorithm. RFC-2144 (May 1997)
2. Adams, C.: Constructing symmetric ciphers using the CAST design procedure. Des. Codes Cryptography 12(3), 283–316 (1997)
3. Aoki, K., Sasaki, Y.: Preimage attacks on one-block MD4, 63-step MD5 and more. In: Avanzi, R.M., Keliher, L., Sica, F. (eds.) SAC 2008. LNCS, vol. 5381, pp. 103–119. Springer, Heidelberg (2009)
4. Aoki, K., Guo, J., Matusiewicz, K., Sasaki, Y., Wang, L.: Preimages for step-reduced SHA-2. In: Matsui, M. (ed.) ASIACRYPT 2009. LNCS, vol. 5912, pp. 578–597. Springer, Heidelberg (2009)
5. Aoki, K., Ichikawa, T., Kanda, M., Matsui, M., Moriai, S., Nakajima, J., Tokita, T.: Camellia: A 128-bit block cipher suitable for multiple platforms - design and analysis. In: Stinson, D.R., Tavares, S. (eds.) SAC 2000. LNCS, vol. 2012, pp. 39–56. Springer, Heidelberg (2001)
6. Biham, E., Biryukov, A., Shamir, A.: Cryptanalysis of Skipjack reduced to 31 rounds using impossible differentials. In: Stern, J. (ed.) EUROCRYPT 1999. LNCS, vol. 1592, pp. 12–23. Springer, Heidelberg (1999)
7. Biham, E., Shamir, A.: Differential cryptanalysis of Snefru, Khafre, REDOC-II, LOKI and Lucifer. In: Feigenbaum, J. (ed.) CRYPTO 1991. LNCS, vol. 576, pp. 156–171. Springer, Heidelberg (1992)
8. Biryukov, A., Nikolić, I.: Complementing Feistel ciphers. In: FSE 2013. LNCS. Springer (2013)
9. Bogdanov, A.A., Knudsen, L.R., Leander, G., Paar, C., Poschmann, A., Robshaw, M., Seurin, Y., Vikkelsoe, C.: PRESENT: An ultra-lightweight block cipher. In: Paillier, P., Verbauwhede, I. (eds.) CHES 2007. LNCS, vol. 4727, pp. 450–466. Springer, Heidelberg (2007)
10. Bogdanov, A., Rechberger, C.: A 3-Subset meet-in-the-middle attack: Cryptanalysis of the lightweight block cipher KTANTAN. In: Biryukov, A., Gong, G., Stinson, D.R. (eds.) SAC 2010. LNCS, vol. 6544, pp. 229–240. Springer, Heidelberg (2011)
11. Bogdanov, A., Rijmen, V.: Linear hulls with correlation zero and linear cryptanalysis of block ciphers. IACR Cryptology ePrint Archive, vol. 2011, p. 123 (2011)
12. Borghoff, J., Canteaut, A., Güneysu, T., Kavun, E.B., Knezevic, M., Knudsen, L.R., Leander, G., Nikov, V., Paar, C., Rechberger, C., Rombouts, P., Thomsen, S.S., Yalçın, T.: PRINCE - A low-latency block cipher for pervasive computing applications - extended abstract. In: Wang, X., Sako, K. (eds.) ASIACRYPT 2012. LNCS, vol. 7658, pp. 208–225. Springer, Heidelberg (2012)
13. Bouillaguet, C., Derbez, P., Dunkelman, O., Keller, N., Rijmen, V., Fouque, P.-A.: Low data complexity attacks on AES. IEEE Transactions on Information Theory 58(11), 7002–7017 (2012)
14. Bouillaguet, C., Derbez, P., Fouque, P.-A.: Automatic search of attacks on round-reduced AES and applications. In: Rogaway, P. (ed.) CRYPTO 2011. LNCS, vol. 6841, pp. 169–187. Springer, Heidelberg (2011)
15. Bouillaguet, C., Dunkelman, O., Leurent, G., Fouque, P.-A.: Another look at complementation properties. In: Hong, S., Iwata, T. (eds.) FSE 2010. LNCS, vol. 6147, pp. 347–364. Springer, Heidelberg (2010)
16. De Cannière, C., Dunkelman, O., Knežević, M.: KATAN and KTANTAN — A family of small and efficient hardware-oriented block ciphers. In: Clavier, C., Gaj, K. (eds.) CHES 2009. LNCS, vol. 5747, pp. 272–288. Springer, Heidelberg (2009)
17. Canteaut, A., Naya-Plasencia, M., Vayssière, B.: Sieve-in-the-middle: Improved MITM attacks. In: Canetti, R., Garay, J.A. (eds.) CRYPTO 2013, Part I. LNCS, vol. 8042, pp. 222–240. Springer, Heidelberg (2013)

18. Dinur, I., Dunkelman, O., Shamir, A.: Improved attacks on full GOST. In: Canteaut, A. (ed.) FSE 2012. LNCS, vol. 7549, pp. 9–28. Springer, Heidelberg (2012)
19. FIPS, Data Encryption Standard. Federal Information Processing Standards Publication 46
20. Guo, J., Peyrin, T., Poschmann, A., Robshaw, M.: The LED block cipher. In: Preneel, B., Takagi, T. (eds.) CHES 2011. LNCS, vol. 6917, pp. 326–341. Springer, Heidelberg (2011)
21. Isobe, T.: A single-key attack on the full GOST block cipher. J. Cryptology 26(1), 172–189 (2013)
22. Isobe, T., Shibutani, K.: All subkeys recovery attack on block ciphers: Extending meet-in-the-middle approach. In: Knudsen, L.R., Wu, H. (eds.) SAC 2012. LNCS, vol. 7707, pp. 202–221. Springer, Heidelberg (2013)
23. Jean, J., Nikolić, I., Peyrin, T., Wang, L., Wu, S.: Security analysis of PRINCE. In: Pre-proceeding of FSE 2013. LNCS. Springer (2013)
24. Knežević, M., Nikov, V., Rombouts, P.: Low-latency encryption – is "Lightweight = light + wait"? In: Prouff, E., Schaumont, P. (eds.) CHES 2012. LNCS, vol. 7428, pp. 426–446. Springer, Heidelberg (2012)
25. Knudsen, L.R.: DEAL - a 128-bit block cipher. Technical Report 151, University of Bergen, Department of Informatics, Norway (February 1998)
26. Knudsen, L.R., Rijmen, V.: Known-key distinguishers for some block ciphers. In: Kurosawa, K. (ed.) ASIACRYPT 2007. LNCS, vol. 4833, pp. 315–324. Springer, Heidelberg (2007)
27. Lu, J., Wei, Y., Kim, J., Fouque, P.-A.: Cryptanalysis of reduced versions of the Camellia block cipher. In: Pre-Proceedings of SAC 2011 (2011)
28. Mala, H., Shakiba, M., Dakhilalian, M., Bagherikaram, G.: New results on impossible differential cryptanalysis of reduced-round Camellia-128. In: Jacobson Jr., M.J., Rijmen, V., Safavi-Naini, R. (eds.) SAC 2009. LNCS, vol. 5867, pp. 281–294. Springer, Heidelberg (2009)
29. Ohtahara, C., Okada, K., Sasaki, Y., Shimoyama, T.: Preimage attacks on full-ARIRANG: Analysis of DM-mode with middle feed-forward. In: Jung, S., Yung, M. (eds.) WISA 2011. LNCS, vol. 7115, pp. 40–54. Springer, Heidelberg (2012)
30. Patarin, J.: Security of random Feistel schemes with 5 or more rounds. In: Franklin, M. (ed.) CRYPTO 2004. LNCS, vol. 3152, pp. 106–122. Springer, Heidelberg (2004)
31. Sasaki, Y.: Meet-in-the-middle preimage attacks on AES hashing modes and an application to Whirlpool. In: Joux, A. (ed.) FSE 2011. LNCS, vol. 6733, pp. 378–396. Springer, Heidelberg (2011)
32. Sasaki, Y.: Preimage attacks on Feistel-SP functions: Impact of omitting the last network twist. In: Jacobson, M., Locasto, M., Mohassel, P., Safavi-Naini, R. (eds.) ACNS 2013. LNCS, vol. 7954, pp. 170–185. Springer, Heidelberg (2013)
33. Sasaki, Y., Yasuda, K.: Known-key distinguishers on 11-round Feistel and collision attacks on its hashing modes. In: Joux, A. (ed.) FSE 2011. LNCS, vol. 6733, pp. 397–415. Springer, Heidelberg (2011)
34. Shibutani, K., Isobe, T., Hiwatari, H., Mitsuda, A., Akishita, T., Shirai, T.: Piccolo: An ultra-lightweight blockcipher. In: Preneel, B., Takagi, T. (eds.) CHES 2011. LNCS, vol. 6917, pp. 342–357. Springer, Heidelberg (2011)
35. Soleimany, H., Blondeau, C., Yu, X., Wu, W., Nyberg, K., Zhang, H., Zhang, L., Wang, Y.: Reflection cryptanalysis of PRINCE-like ciphers. In: Pre-proceeding of FSE 2013. LNCS. Springer (2013)
36. Suzuki, K., Tonien, D., Kurosawa, K., Toyota, K.: Birthday paradox for multi-collisions. In: Rhee, M.S., Lee, B. (eds.) ICISC 2006. LNCS, vol. 4296, pp. 29–40. Springer, Heidelberg (2006)

Does My Device Leak Information? An *a priori* Statistical Power Analysis of Leakage Detection Tests

Luke Mather, Elisabeth Oswald, Joe Bandenburg, and Marcin Wójcik

University of Bristol, Department of Computer Science,
Merchant Venturers Building, Woodland Road, BS8 1UB, Bristol, UK
{Luke.Mather,Elisabeth.Oswald,Marcin.Wojcik}@bris.ac.uk,
joe@bandenburg.com

Abstract. The development of a leakage detection testing methodology for the side-channel resistance of cryptographic devices is an issue that has received recent focus from standardisation bodies such as NIST. Statistical techniques such as hypothesis and significance testing appear to be ideally suited for this purpose. In this work we evaluate the candidacy of three such detection tests: a t-test proposed by Cryptography Research Inc., and two mutual information-based tests, one in which data is treated as continuous and one as discrete. Our evaluation investigates three particular areas: statistical power, the effectiveness of multiplicity corrections, and computational complexity. To facilitate a fair comparison we conduct a novel *a priori* statistical power analysis of the three tests in the context of side-channel analysis, finding surprisingly that the continuous mutual information and t-tests exhibit similar levels of power. We also show how the inherently parallel nature of the continuous mutual information test can be leveraged to reduce a large computational cost to insignificant levels. To complement the *a priori* statistical power analysis we include two real-world case studies of the tests applied to software and hardware implementations of the AES.

1 Introduction

The evaluation of the resilience of cryptographic devices against side-channel adversaries is an issue of increasing importance. The potential of side-channel analysis (SCA) as an attack vector is driving the need for standards organisations and governing bodies to establish an acceptance-testing methodology capable of robustly assessing the vulnerability of devices; the National Institute of Standards and Technology (NIST) held a workshop in 2011 driving the requirements [4] and recent papers have been published on this topic by industry [13,16].

Current evaluation methodologies such as Common Criteria [2], used by bodies such as ANSSI [1] and BSI [3], consist of executing a battery of known side-channel attacks on a device and considering whether the attack succeeds and, if so, the quantity of resources expended by an adversary to break the device. This methodology is likely to prove unsustainable in the long-term: the number and

K. Sako and P. Sarkar (Eds.) ASIACRYPT 2013 Part I, LNCS 8269, pp. 486–505, 2013.

type of Simple Power Analysis (SPA), and particularly Differential Power Analysis (DPA) attacks is steadily increasing year-on-year, lengthening the testing process and forcing evaluation bodies to keep up-to-date with an increasingly large, technically complex and diverse number of researched strategies.

A desirable complement or alternative to an attack-focused evaluation strategy is to take a 'black-box' approach; rather than attempting to assess security by trying to find the data or computational complexity of an optimal adversary against a specific device, we can attempt to quantify whether *any* side-channel information is contained in power consumption data about underlying secrets without having to precisely characterise and exploit leakage distributions. We describe this as a *detection* strategy; the question any detection test answers is whether *any* side-channel information is present, and *not* to precisely quantify the exact amount or how much of it is exploitable. Detection-based strategies can be used to support 'pass or fail' type decisions about the security of a device [13], or can be used to identify time points that warrant further investigation.

In practice we *estimate* information leakage, and so any reasonable detection strategy should ideally incorporate a degree of statistical rigour. In this paper we provide a comprehensive evaluation of three leakage detection hypothesis tests in the context of power analysis attacks: a t-test proposed by [13], and two tests for detecting the presence of zero mutual information (MI)—one in which power traces are treated as continuous data (hereafter the CMI test) [10], and one as discrete (hereafter the DMI test) [9].

Our contribution. Previous work in the context of side-channel analysis has assessed detection tests through practical experimentation only [13]. This approach creates flawed comparisons of tests for reasons similar to those encountered in the practical analysis of distinguishers in DPA [28]; the effects of sample size and estimation error on detection test performance cannot be quantified in a practical experiment and consequently it becomes difficult to draw fair comparisons that apply in a general context. To ensure a fair comparison in this work we perform an *a priori* statistical power analysis[1] of the three detection tests using a variety of practically relevant side-channel analysis scenarios. The analysis allows us to study the effects that sample size, leakage functions, noise and other hypothesis testing criteria have on the performance of the detection tests in a fair manner. In addition to statistical power, we also investigate the computational complexity of the tests and the effectiveness of multiplicity corrections.

Related work. An alternative to the black-box strategy is the 'white-box' leakage evaluation methodology proposed by Standaert et al. [26]. Their methodology requires an estimation of the conditional entropy of a device's leakage distribution using an estimated leakage model. This allows for a tighter bound on the amount

[1] The overlap in terminology of the *statistical* power analysis of hypothesis tests with the entirely different differential or simple power analysis technique is unfortunate. To establish a reasonable separation of terminology we will use 'DPA' or 'SPA' to address the latter technique, and 'statistical power' when referencing the former topic.

of information available to an adversary, but requires additional computational expense and the ability to profile a device, and bounding estimation error in the results is non-trivial. The black-box detection approach outlined in this work does not require any device profiling, trading-off the ability to estimate the *exploitable* information leakage contained within the device for efficiency gains and the ability to increase robustness through statistical hypothesis testing. A detection strategy may be used as a complement to the approach of Standaert et al. by identifying a subset of time points that are known to leak information and can be further explored in a white-box analysis.

There is no previous *a priori* power analysis study of these three tests in the context of SCA. A generic analysis of the CMI test and additional non-parametric hypothesis tests was conducted in [10], but does not consider the influence of variables such as noise and leakage function in the context of side-channel analysis, and cannot be used in comparison with the DMI or t-tests.

Organisation In Section 4 of this work we present the results of the first *a priori* statistical power analysis of the three detection tests in the context of side-channel analysis. To support the *a priori* analysis we also provide a case study illustrating an example application of the tests to real-world traces acquired from a software and a hardware implementation of the AES in Section 5. Section 6 discusses the computational complexity of the three tests.

2 Introduction to Selected Hypothesis Tests

2.1 Side-Channel Analysis

We will consider a 'standard' SCA scenario whereby the power consumption T of a device is dependent on the value of some internal function $f_k(x)$ of plaintexts and secret keys evaluated by the device. Using the random variable $X \in \mathcal{X}$ to represent a plaintext and the random variable $K \in \mathcal{K}$ to represent a sub-key, the power consumption T of the device can be modelled using $T = L \circ f_k(x) + \varepsilon$, where L is a function that describes the data-dependent component of the power consumption and ε represents the remaining component of the power consumption modelled as additive random noise.

2.2 Candidate Tests

There are many hypothesis tests that may be used to detect information leakage: one can test for differences between particular moments (such as the mean) of leakage distributions, or one can test for any general differences between leakage distributions. In this work we consider three tests, one from the former category and two from the latter. In the former category, the Welch t-test [27], used to assess the difference between the means of two distributions, has been proposed by Cryptography Research Inc. [13]. One can also analyse higher moments using tests such as the F-test [20]. Information leakage solely occurring in a particular higher moment is rare—to our knowledge, one example of this is in [20]—and so a natural progression is to use a generic non-parametric test instead. Chatzikokolakis et al. and Chothia et al. present hypothesis tests capable of detecting the presence of discrete and continuous mutual information [9,10].

Whilst alternative non-parametric tests are available, mutual information-based methods provide an intuitive measure and are frequently used in other contexts [23,26]. There is a generic *a priori* power analysis comparing the CMI test and additional non-parametric hypothesis tests in [10], finding that the CMI test compared favourably. The analysis does not discuss any of the side-channel specific variables described in Section 2.1 and cannot be used in comparison with the t-test, but does suggests that an MI-based test is a natural choice for a generic test candidate. As such, we focus on the t-test and the two MI-based methods, and note that our evaluation strategy can be easily applied to other detection tests in the future.

The null hypothesis for any hypothesis testing procedure used in a detection context is that there is no information leakage: using the t-test, any statistically significant difference of means is evidence for an information leak, and using MI-based tests, any significant non-zero mutual information is evidence.

The generic strategy followed by each test is to systematically evaluate each individual time point in a set of traces in turn. This is a 'univariate' approach, and in many cases is likely to be sufficient; vulnerabilities arising from sub-optimal security measures are likely to manifest themselves as leakage detectable within a single time point. To detect leakage exploitable by n-th order attacks would necessitate the joint comparison of n time points. This results in a considerable increase on the the amount of computation required—the brute force strategy would be to analyse the joint distribution of every possible n-tuple of points—and additionally can substantially increase the complexity of the test statistics, with multivariate mutual information in particular becoming costly. Whilst an efficient multivariate strategy would be desirable, it is beyond the scope of this initial work.

2.3 Difference-of-means and the t-test

Exploiting the difference-of-means $\overline{T_1} - \overline{T_2}$ between two sets of power traces T_1 and T_2 partitioned on a single bit of a targeted intermediate state was proposed by Kocher et al. and is the canonical example of a generic DPA attack [17]. The same difference-of-means can also be used to detect information leakage, and was proposed as a candidate detection test in [13].

Welch's t-test is a hypothesis test that (in the two-tailed case) tests the null hypothesis that the population means of two variables are equal, where the variables have possibly unequal variances, yielding a p-value that may or may not provide sufficient evidence to reject this hypothesis. The test statistic t is:

$$t = \frac{\overline{T_1} - \overline{T_2}}{\sqrt{\frac{s_1^2}{N_1} + \frac{s_2^2}{N_2}}}, \tag{1}$$

where $\overline{T_i}$, s_i^2 and N_i are the sample means, sample variances and sample size of the i-th set T_i. Using this test statistic and the Welch-Satterthwaite equation[2] to compute the degrees of freedom ν, a p-value can be computed to determine whether there is sufficient evidence to reject the null hypothesis at a particular significance level $1 - \alpha$. Using the quantile function for the t distribution at a significance level α and with ν degrees of freedom, a confidence interval for the difference-of-means can also be computed.

Leveraging the t-test requires a partitioning of the traces based on the value of a particular bit of an intermediate state with the targeted algorithm, and therefore to comprehensively evaluate a device every single bit of every single intermediate state must be tested. To assess the i-th bit of a particular state for leakage (e.g. the output of SubBytes in a particular round), an evaluator must compute the intermediate values for the chosen state, using a set of chosen messages. Having recorded the encryption or decryption of the chosen messages, the resulting traces can be partitioned into two sets T_1 and T_2, depending on the value of the i-th bit of the intermediate state. The test statistic t and corresponding p-values or confidence intervals can then be used to determine whether a difference between the means exists.

The t-test by design can only detect differences between subkeys that are contained within the mean of the leakage samples, and assumes that the populations being compared are normally distributed. In practice univariate leakage from unprotected devices is typically close enough to Gaussian for this condition to not be too restrictive [7,8,17].

2.4 Mutual Information

Given two random variables X and Y, the MI $\mathrm{I}(X;Y)$ computes the *average* information gained about X if we observe Y (and vice-versa). The application of MI to detecting information leaks from a cryptographic device is straightforward: any dependence between subkeys and the power consumed by the device, giving $\mathrm{I}(K;T) > 0$, may be evidence for an exploitable information leak[3].

The rationale for using MI to detect information leaks is that it compares distributions in a general way, incorporating all linear and non-linear dependencies between sub-keys and power values. Unfortunately, the estimation of MI is well-known to be a difficult problem. There are no unbiased estimators, and it has been proven that there is no estimator that does not perform differently depending on the underlying structure of the data [22].

Recent results on the behaviour of zero MI can help to alleviate this problem. Chatzikokolakis et al. find the sampling distribution of MI between two discrete random variables when it is zero, where the distribution of one of the variables is known and the other unknown, and use this to construct a confidence interval

[2] Using Welch-Satterthwaite, the degrees of freedom ν for a t-distribution can be calculated as $\nu = \frac{(s_1^2/N_1+s_2^2/N_2)^2}{(s_1^2/N_1)^2/(N_1-1)+(s_2^2/N_2)^2/(N_2-1)}$.

[3] Under the assumption of the 'equal images under different sub-keys' property [24] we can safely compute $\mathrm{I}(X;T)$, if simpler.

test [9]. A second result from Chothia and Guha establishes a rate of convergence, under reasonable assumptions, for the sampled estimate for zero MI between one discrete random variable with a known distribution and one *continuous* random variable with an unknown distribution [10]. This result is then used to construct a non-parametric hypothesis test to assess whether sampled data provides evidence of an information leak within a system.

Discrete mutual information As side-channel measurements are typically sampled using digital equipment, it may be viable to treat the sampled data as discrete. The most common way to make continuous data discrete is to split the continuous domain into a finite number of bins. Using the standard formula for marginal and conditional entropy, the discrete MI estimate can be computed as

$$\hat{\mathrm{I}}(K;T) = \sum_{k \in \mathcal{K}} \sum_{t \in \mathcal{T}} \hat{p}(k,t) \log_2 \left(\frac{\hat{p}(k,t)}{p(k)\hat{p}(t)} \right). \tag{2}$$

The test of Chatzikokolakis et al. is biased by $(I-1)(J-1)/2n$, where I and J are the sizes of the distribution domains of two random variables in question, and n is the number of samples acquired. In our context, $I = |\mathcal{K}|$, the number of possible sub-keys, and $J = |\mathcal{T}|$, the number of possible power values as a result of discretisation. Consequently, the point estimate e for MI is the estimated value minus this bias: $e = \hat{\mathrm{I}}(K;T) - (I-1)(J-1)/2n$. We can use this to compute $100(1-\alpha)\%$ confidence intervals for zero and non-zero MI (full details can be found in [9]).

As a result of the bias of the test, to be sure of good results it is necessary to ensure that the number of traces sampled is larger than the product of the number of sub-keys and the number of possible power values. The applicability of this discrete test is then dictated by the ability of an evaluator to sample enough traces to meet this condition.

Continuous mutual information. The test of Chothia and Guha requires two assumptions about the data to guarantee a convergence result for zero MI [10]. The first is that the power values are continuous, real-valued random variables with finite support. This may or may not hold theoretically, depending on the distribution of the leakages, but in practice will be true; the sampling resolution used dictates the range of the recorded power consumption. The second is that for $u = \{0,1\}$, the probability $p(u,t)$ must have a continuous bounded second derivative in t. This can be fulfilled with the leakage analysis of a single bit of a key only. However, Chothia and Guha also demonstrate experimentally that the test works well in cases of multiple inputs, often outperforming other two-sample tests [10].

Under the assumption of a continuous leakage distribution, we are estimating a hybrid version of the MI:

$$\hat{\mathrm{I}}(K;T) = \sum_{k \in \mathcal{K}} \int_{T} \hat{p}(k,t) \log_2 \left(\frac{\hat{p}(k,t)}{p(k)\hat{p}(t)} \right) dt. \tag{3}$$

To compute this estimate we are required to estimate a conditional probability density function $\hat{\Pr}\{t|k\}$ using kernel density estimation. The assumptions underlying the test's convergence result dictate the use of a function such as the Epanechnikov kernel[4] as the chosen kernel function, and a bandwidth function such as Silverman's [25] general purpose bandwidth[5].

Using this estimated density function, we can compute an estimate of the MI, $\hat{\text{I}}(K;T)$. The following step of the hypothesis test is a permutation stage requiring s permutations of the sampled data T': for each sampled power value, we randomly assign a new sub-key to the value without replacement. The power values contained in each permuted set should now have no relation with the sub-keys, and so the MI of the s sets can now be computed $\hat{\text{I}}_1(K;T'_1),\ldots,\hat{\text{I}}_s(K;T'_s)$, providing a baseline for zero MI.

An *estimated* p-value can be computed by computing the percentage of the MI estimates $\hat{\text{I}}_1,\ldots,\hat{\text{I}}_s$ that have a value greater than the observed point estimate $\hat{\text{I}}(K;T)$. The suggested number of shuffled estimates to achieve useful baseline results is given to be 100 by Chothia and Guha, but to increase the power of the test and the precision of the estimated p-values a few thousand shuffles may be required.

3 Evaluation Methodology

3.1 Comparing Detection Tests

The most important notion in hypothesis testing is of the quantification and classification of the error involved. The type I error rate α is defined as the probability of incorrectly rejecting a true null hypothesis, usually termed the significance criterion. Tests are also associated with a type II error rate β: the probability of failing to reject a false null hypothesis. The exact valuation assigned to these error rates is an important factor to balance; typically decreasing one error rate will result in an increase in the other, and the only way to reduce both in tandem is to increase the sample size available to the test. The statistical power of a test is defined as the probability of correctly rejecting a false null hypothesis, $\pi = 1 - \beta$. This is the key factor for our detection tests: higher statistical power indicates increased robustness and lessens reliance on large sample sizes.

A common motivation for performing an *a priori* statistical power analysis[6] is to compute or estimate the minimum sample size required to detect an *effect* of a given size, or to determine the minimum effect size a test is likely to detect when supplied with a particular sample size. The determination of sample sizes required to achieve acceptable power has two-fold uses: firstly, data acquisition from a cryptographic device is an expensive and time-consuming operation, and so tests that are less data-hungry are likely to be preferable, and secondly,

[4] Epanechikov's kernel function is defined as $K(u) = 3/4(1-u^2)_{\chi\{|u|\leq 1\}}$.

[5] $h = 1.06 s_T N^{-1/5}$, where s_T is the sample standard deviation of T and N is the number of sampled traces.

[6] For further discussion of statistical power analysis, see [11].

knowledge of the sample sizes required to detect a particular effect can serve as a guideline for evaluators to determine the number of trace acquisitions sufficient for detecting an information leak.

3.2 Multiple Testing

When considering the results of large numbers of simultaneously-computed hypothesis tests, we must take into account that the probability a single test falsely rejects the null hypothesis will increase in proportion with the number of tests computed. A single test computed at significance level $\alpha = 0.05$ has a 5% chance of incorrectly rejecting the null hypothesis; when conducting a large number of simultaneous tests the probability of a false positive increases. The intuitive solution is to control the overall false rejection rate by selecting a smaller significance level for each test. There are two main classes of procedure: controlling the *familywise error rate* (FWER) and controlling the *false discovery rate* (FDR).

Familywise error rate. The FWER is defined as the probability of falsely rejecting one or more true null hypotheses (one or more type I errors) across a family of hypothesis tests. The FWER can be controlled, allowing us to bound the number of false null hypothesis rejections we are willing to make—in our device evaluation context this would allow the evaluator to control the probability a device is falsely rejected. FWER controlling procedures are conservative, and typically trade-off FWER for increasing type II error.

False discovery rate. Proposed by Benjamini and Hochberg in 2005, the FDR is defined as the *expected* proportion of false positives (false discoveries) within the hypothesis tests that are found to be significant (all discoveries). Procedures that control the FDR are typically less stringent than FWER-based methods, and have a strong candidacy for situations where test power is important. The Benjamini-Hochberg (BH) procedure is a 'step-up' method that strongly controls the FDR at a rate α [6]. Given m simultaneous hypothesis tests, the BH procedure sorts the p-values and selects the largest k such that $p_k \leq \frac{k}{m}\alpha$, where all tests with p-values less than or equal to p_k can be rejected. Many additional FWER and FDR controlling methods exist, e.g. [14,15], but are beyond the scope of this paper.

A trade-off with multiplicity corrections that control the FWER is that generally decreasing the FWER results in an increase in type II error. As a consequence the FDR approach may be more suitable if an evaluator is particularly concerned with ensuring that the type II error rate is kept low—that the statistical power remains high. It may also serve a useful purpose by identifying a small candidate set of time points that are *likely* to contain information leakage—the evaluator can then perform further analysis on the set of points, for example by inspecting the effect sizes reported for each of the points, re-sampling additional data and performing new hypothesis tests, or even by trying to attack the points using an appropriate method. We demonstrate an example application of the BH procedure in Section 5.

3.3 Why Perform an *a priori* Power Analysis?

Having established the importance of statistical power to our detection tests, the motivation for performing an *a priori* power analysis for our three candidate tests is that it is not possible to make generally true inferences based on practical experiments alone; given that it is only possible to establish the vulnerability of a time point by successfully attacking it, it becomes impossible to establish whether a reported rejection of the null hypothesis is a false positive—in other words, the type II error rate β cannot be estimated—and hence any *a posteriori* (or post-hoc) power analysis is likely to be misleading.

To be able to perform an *a priori* statistical power analysis, we need to be able to produce or simulate data, ideally with characteristics as close as possible to those observed in practice, for which we are sure of the presence of information leakage. The most straightforward way to do this is to simulate trace data under the 'standard' DPA model commonly used throughout the existing body of literature, detailed in Section 2.1.

4 *A priori* Power Analysis

As all of the variables in the standard SCA model outlined in Section 2.1 have an effect on detection test performance, to perform a useful *a priori* power analysis we defined a variety of leakage scenarios that have relevance to practice, and then estimated the power π of each of the detection tests under many combinations of the different parameters in the SCA model for each scenario. For each leakage scenario, power was estimated under varying sample sizes, noise levels and using two different significance criteria: $\alpha = 0.05$ and $\alpha = 0.00001$. The former provides a general indication of test power with a common level of significance, and the intention with the latter level of significance is to gain an understanding of how much statistical power is degraded by the typical tightening of the significance criteria enforced by multiple testing corrections.

Leakage model. We defined five different practically-relevant leakage models L under which to simulate trace data:

1. Hamming weight—a standard model under which the device leaks the Hamming weight of the intermediate state;
2. Weighted sum—the device leaks an unevenly weighted sum of the bits of the intermediate state, where the least significant bit (LSB) dominates with a relative weight of 10, as motivated by [5];
3. Toggle count—the power consumption of hardware implementations has been shown to depend on the number of *transitions* that occur in the S-Box. The model used here is computed from back-annotated netlists as in [19], and creates non-linear leakage distributions;
4. Zero value—for this model we set the power consumption for every non-zero intermediate value to be 1, and for the value zero we set the power consumption to be 0; this will typically produce small amounts of information leakage and should stress the data efficiency of the tests;

5. VARIANCE—the mean of the power consumption does not leak, and the variance of the power consumption follows the distribution given in Maghrebi et al. [18]. The t-test will not be able to detect any leakage, but the model can be used to evaluate the relative performances of the MI tests.

A statistical power analysis would ideally be performed for each candidate target function; given the limited space available we have focused on the AES. For this comparison we targeted, without loss of generality, the first byte of the key. For each leakage model, we simulated traces under a wide range of signal-to-noise ratios (SNRs), ranging from 2^{-14} to 2^{12}, enabling us to assess the maximum amount of noise a test can overcome when provided with a particular sample size.

Estimation process. The estimated power for the test is computed as the fraction of times the test correctly[7] rejects the null hypothesis for $1,000$ tests run. For the CMI and t-tests we used the significance criterion α to determine rejection or acceptance, and for the DMI test we checked whether the corrected estimate for the MI was inside the $100(1-\alpha)\%$ confidence interval for zero MI.

In the following section we present the results of our *a priori* statistical power analysis on the five leakage models in terms of the number of samples required to achieve 80% power for each combination of model, SNR and sample size. We performed $1,000$ permutations of the simulated traces for each CMI test, and used the Epanechnikov kernel with Silverman's bandwidth for the kernel density estimation. To enable a fair comparison between the bit and byte level tests, we chose to represent the results for the t-test corresponding to the most leaky bit of the state. Graphs illustrating the number of samples required by each test to achieve 80% power for each leakage model and SNR are shown in Figure 1.

HAMMING WEIGHT. We can see that the t-test is the most powerful test in general, as we would expect given the unbiased estimator for the mean values and the Gaussian noise assumption holding true in the model. The CMI test requires slightly more samples to achieve the requisite power in the presence of high noise, and both tests seem to perform equivalently for mid-range and low levels of noise.

The DMI test appears to be significantly less powerful; this is unsurprising given a loss of information from the treatment of continuous data as discrete is to be expected, and we also see that the test struggles to cope with high levels of noise—the lowest SNR for which we could detect an information leak with up to 192,000 samples was 2^{-3}. A closer inspection indicates that this is caused by the bias correction required; the size of the input space for the AES often necessitates a large sample size to minimise the size of the correction to within manageable bounds.

[7] Each of these scenarios contain information leakage; even for the extremely low SNRs, given sufficiently large data an attacker will eventually be able to exploit the leakage, and as a consequence candidate detection tests should, for some level of sample size, be able to consistently detect information leakage.

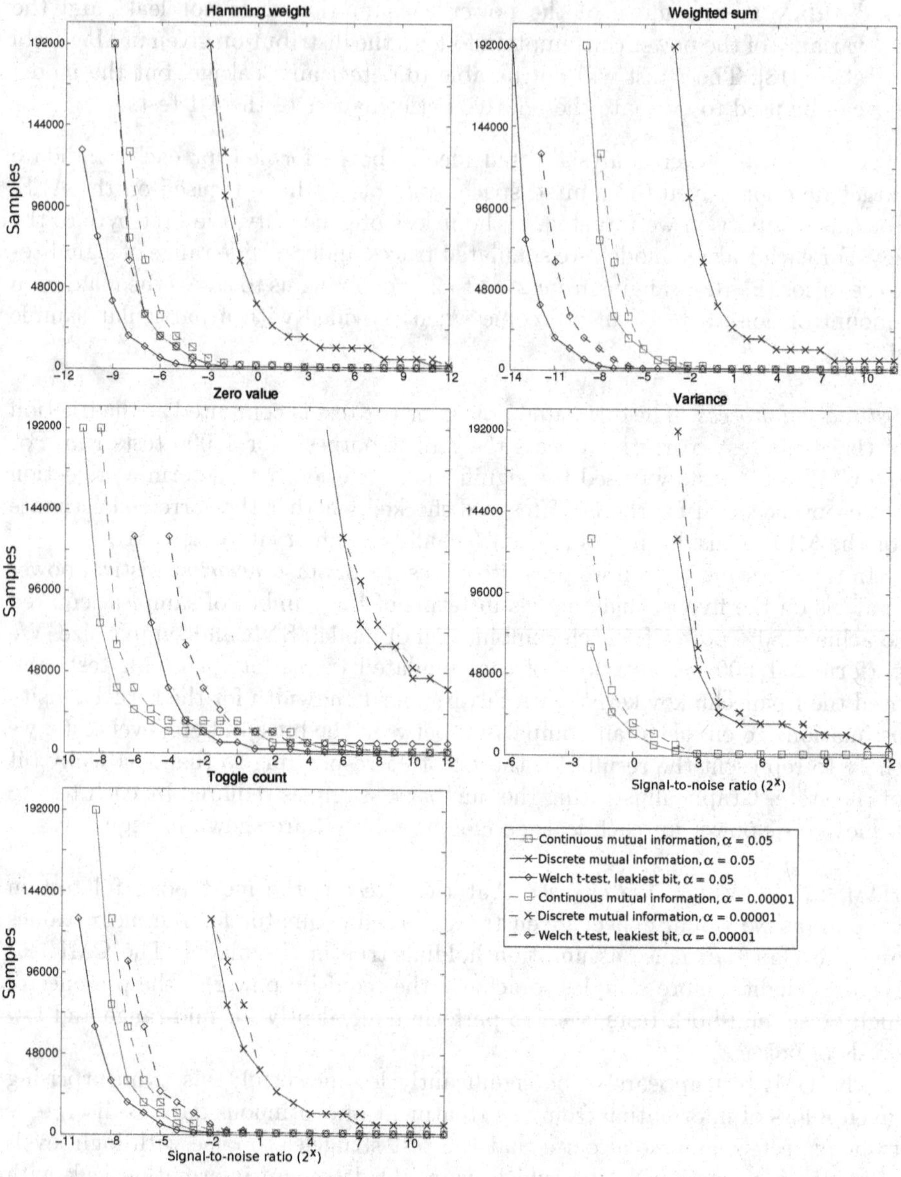

Fig. 1. Number of samples required for the t-test, CMI and DMI tests to achieve estimated 80% power for a variety of leakage models and SNRs.

The stricter significance criterion $\alpha = 0.00001$ seems to have a small but noticeable effect on the test power for the CMI and t-tests. Under the DMI test we see little change in behaviour; the dominant factor influencing power is the bias correction rather than the precise width of the confidence intervals.

WEIGHTED SUM. The relative dominance of the LSB in the leakage provides an additional advantage for the t-test and we found as expected that the test achieved its highest power when evaluating this bit. This results in a relative increase in overall power compared to the CMI test than we observed in the Hamming weight scenario and also allows for detection of leakage at lower SNRs. The CMI test seems to exhibit performance consistent with that under the Hamming weight model, and similarly for the DMI test. The effects of the stricter significance criterion are also similar, with noticeable reductions in power observed for each of the tests under the smaller α values save for the DMI test, where again the bias correction is the predominant factor.

TOGGLE COUNT. An analysis of the underlying true distance of means for the TOGGLE COUNT model indicated that the largest information leakage was contained within the second-least significant bit, which was also twice the leakage in the next most leaky bit. As with the WEIGHTED SUM model, the relative dominance of this bit supplies the t-test with an advantage over the CMI test but in this instance the advantage is by a smaller margin. We can also see that the CMI test appears to be significantly more robust to the stricter significance criterion, outperforming the more sensitive t-test in all of the high noise settings. Here we also see the DMI test exhibiting an increased sensitivity to the significance criterion.

ZERO VALUE. The size of the information leak present in a noise-free setting for the ZERO VALUE model is small relative to those in the other models: the true MI in a noise-free setting is 0.0369 and the true distance-of-means 0.0078. As such it is interesting to note the stronger performance of the CMI test in high noise settings relative to that of the t-test observed in these results—the additional information on the non-linear dependencies contained in the estimated MI values increases the power of the CMI test whereas the quantity of noise has a stronger effect on the difference-in-means estimated by the t-test. The low power estimates for the DMI test are consistent with the small size of the information leak in the model coupled with the loss of information in the conversion process of continuous to discrete data.

VARIANCE. By design the mean of the power consumption for all sub-key values is equivalent in the VARIANCE model, and so the t-test cannot be applied. As a test for the applicability of the CMI and DMI to situations in which only higher-order moments leak, the CMI test appears to be robust, so that small sample sizes suffice to achieve the requisite power at medium and low noise levels. The true information leakage contained within the variances is strongly affected by the amount of noise in the samples, which explains why both tests soon begin to struggle as the SNR drops below 2^0.

Conclusion. The t-test was generally shown by the *a priori* power analysis to be the most powerful. This is not unexpected: the sample mean is a consistent, unbiased estimator for the population mean and converges quickly to the true value. The performance of the CMI test was close to that of the t-test in all scenarios, indicating that it remains a robust, if slightly inferior alternative in the majority of settings. The DMI test was expected to be less powerful due to the loss of information by the conversion of continuous data to discrete, and this was observed in our analysis; the results indicate that the test is a viable choice only when supplied with large amounts of trace data and only when the SNR is high.

Of note was the superior performance of the CMI test when detecting the small leaks produced by our Zero value model, particularly in high-noise settings. This suggests that the CMI test may be a better, or safer, choice when applied to devices with these sorts of characteristics. The results obtained under the Variance model indicate that the CMI test is sufficiently robust to handle 'tough' leakage scenarios in which the leakage is solely contained in higher moments of the power consumption distribution.

5 Case Studies

The *a priori* statistical power analysis is the primary method for comparison of the detection tests. To complement the analysis, and to further explore the effectiveness of multiplicity corrections, in the following section we demonstrate the application of the three detection tests to the evaluation of two cryptographic devices implementing the AES. The first device we analyse is an ARM7 microcontroller implementing the AES in software, with no countermeasures applied. This device would be expected to exhibit significant information leakage in Hamming-weight form, and hence is a good opportunity to analyse the efficacy of multiple testing correction procedures. The second device analysed is a Sasebo-R evaluation board manufactured using a 90nm process implementing AES in hardware with a Positive-Prime Reed-Muller (PPRM) based SubBytes operation using single-stage AND-XOR logic [21]. This second case study is intended to investigate the performance of the detection tests under increasingly complex leakage distributions as well as acting as a further test for the multiplicity corrections.

5.1 ARM7 Microcontroller

Our data set contained 32,000 traces from the device and we chose to evaluate the first key byte for information leakage. For the t-test we analysed the output of the first SubBytes operation. Figure 2 illustrates the estimated MI values and t-test statistics produced by the detection tests ran at a significance level $\alpha = 0.05$ for each of the 200,000 time points in our traces. For the CMI test we performed 1,000 permutations of the traces at each time point, and as we found that all 8 of the bits in the intermediate state produced similar information leakage we elected to display the results for the LSB.

At the initial significance level $\alpha = 0.05$, the CMI test identified 9,360 time points consistent with information leakage, the discrete test 178, and the t-test

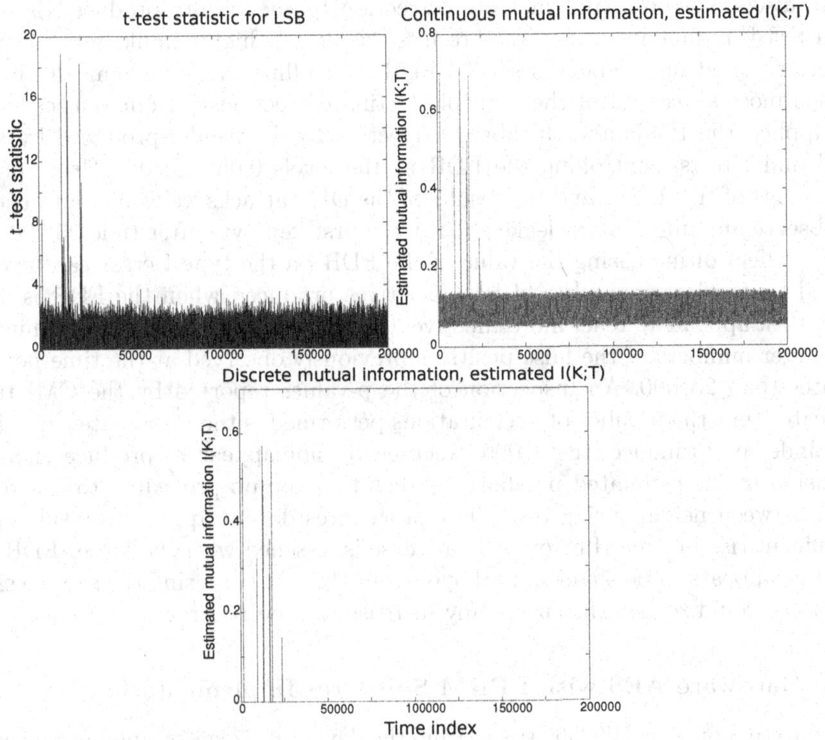

Fig. 2. Estimated CMI and DMI values and t-test statistics produced using 32,000 traces during an evaluation of an ARM7 microcontroller implementing a software version of the AES.

9,713. These occur across the full range of the traces, and account for around 4.8% of the total in the CMI and t-test cases. Using our prior knowledge of the device we could ascertain that many of these points are likely to be false positives.

To gain an *indication* of how many of these time points actually contain exploitable leakage, we conducted a battery of attacks on the output of the SubBytes operation on all of the time points using the same set of traces including Brier et al.'s correlation (CPA) [7], Gierlichs et al.'s mutual information analysis (MIA) [12], both using a Hamming weight power model, and Kocher et al.'s difference of means [17]. Whilst we have argued that practical results should not be used to perform a *post hoc* power analysis, the results of the DPA attacks can be used to quantify under-performances of the three tests—time points that *can* be successfully attacked that are missed by detection tests are indicative of low statistical power given the available sample size. In this regard the only notable false acceptances of time points occurred under the DMI test, with the CMI and t-tests able to spot the vast majority of the vulnerable time points. These results appear to be consistent with those observed under the Hamming-weight scenario in the statistical *a priori* power analysis.

False discovery rate Applying any correction to the results produced by the DMI test is redundant as the 'raw' results are already highly unlikely to contain falsely rejected null hypotheses. The FDR controlling procedures are likely to be the most successful of the multiple testing corrections for our purposes, so we applied the Benjamini-Hochberg correction to the results produced by the CMI and t-tests, controlling the FDR at the levels 0.05 and 0.5. Using prior knowledge of the device and the results of the DPA attacks we would not expect to observe any information leaked about the first key byte after time 25,000.

The effect of increasing the value of the FDR on the type I error can be observed by the larger number of false positives produced when the FDR is 0.5. The t-test appears to react more effectively to the corrective procedure, eliminating larger numbers of the false positives previously observed at the time points greater than $25,000$. An inspection of the p-values reported by the CMI test indicates that the number of permutations performed is the proximate cause for the under-performance: the $1,000$ executed do not appear to produce enough precision in the estimated p-values to allow the step-up procedure to differentiate between neighbouring tests. The procedures do not appear to result in a significant rise in type II error—the increase is lessened with the looser FDR of 0.5, but appears to be slight in both cases. As always, increasing the sample size available would reduce the size of any increase in type II error.

5.2 Hardware AES with PPRM SubBytes Implementation

The dataset contained $79,360$ traces from the device at 5 giga-samples per second and we again chose to evaluate the first key byte for information leakage; for the t-test we analysed the output of the first SubBytes operation. Figure 4 illustrates the estimated MI values and t-test statistics produced by the detection tests run at a significance level $\alpha = 0.05$ for each of the $50,000$ time points in our traces. The first and last $10,000$ points are not displayed as they do not correspond to any part of the full AES operation. For the CMI test we increased the number of permutations to $10,000$ per time point in an attempt to gain additional precision on the estimated p-values. Information leakage was found to occur to a varying degree across all 8 bits of the intermediate state when using the t-test—as such, we have elected to superimpose the results for all of the state bits on a single graph. The DMI test was not able to identify any information leakage.

A visual inspection of the results produced by both the CMI test and t-tests indicate that there are 10 groups of points within the power traces that contain significant amounts of information leakage. As would be expected the shape and scale of the leakages differ: the t-test is only assessing the SubBytes operation *and* the leakage of individual bits. We were able to confirm the vulnerability of the device by successfully executing a reduced Bayesian template attack on the intermediate values of the SubBytes operation at the time points the detection tests indicated would be vulnerable. The hardware device exhibits less, but still significant leaking behaviour when compared to the ARM7 microcontroller implementation, as evidenced by the lower mutual information estimates and the smaller t-test statistic scores.

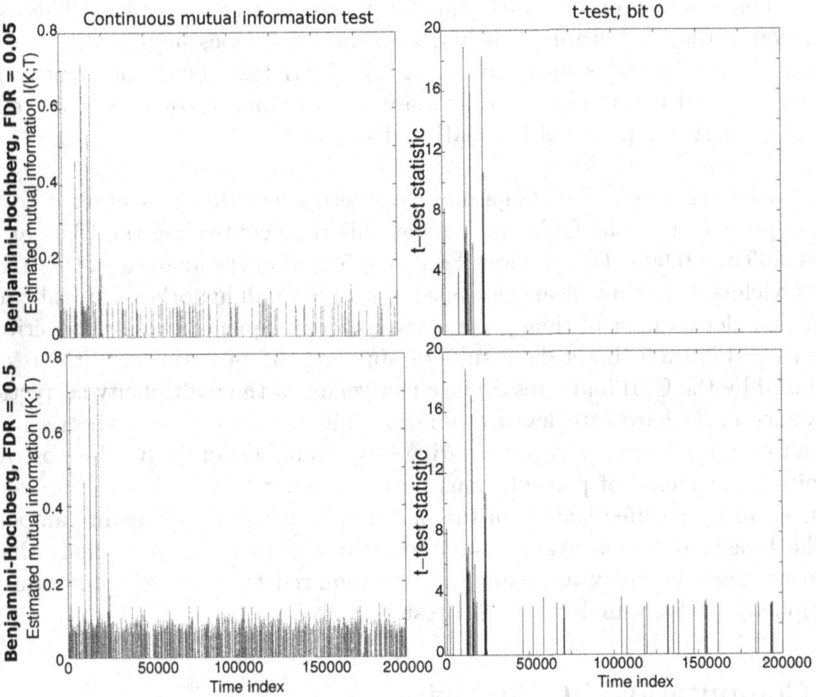

Fig. 3. Plots of the time points consistent with information leakage after applying the Benjamini-Hochberg FDR controlling procedure to the results produced by the t-test and CMI test.

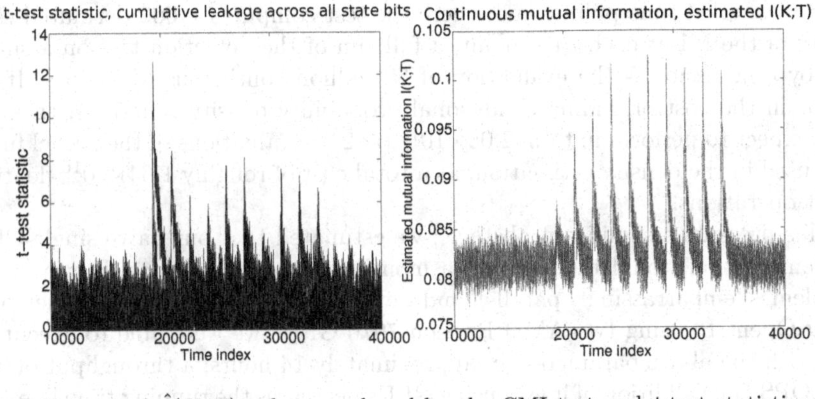

Fig. 4. Estimated $\hat{I}(K;T)$ values produced by the CMI test and t-test statistics produced using 79,360 traces taken from an evaluation of a hardware AES device with the SubBytes operation using Positive-Prime-Reed-Muller (PPRM) logic.

The performance of the CMI and t-tests appears to be similar. The extra definition in the CMI graph is likely due to the t-test assessing leakage from the output of the SubBytes operation only. The DMI test could not identify any information leakage, indicating that many more samples would be required to begin to match the power of the CMI and t-tests.

False discovery rate. The Benjamini-Hochberg correction was applied to the results produced by the CMI and t-tests, this time controlling the FDR at the levels 0.05 and 0.005. The previous FDR of 0.5 used in the analysis of the ARM7 device yielded too many clear false rejections of the null hypothesis, possibly due to the smaller number of time points, and as a consequence two stricter criteria were used. Figure 5 shows the results of applying the two criteria to the results produced by the CMI and t-test. The effectiveness of the multiplicity corrections is lessened in the hardware device evaluation. The t-test again reacts better to the stricter corrective procedure, eliminating larger numbers of likely false positives. Despite the increase of permutations per time point from $1,000$ to $10,000$ for the CMI test, the effectiveness of the multiplicity correction is again dampened by the lack of precision available in the estimated p-values. It is likely that a different, more complex approach may be required to effectively mitigate the multiplicity problem under the CMI test.

6 Computational Complexity

If we consider commercial and logistical pressures on the evaluation process then we must also include the computational complexity of the detection tests as a factor in our evaluation. In this regard, the CMI test is particularly expensive. Under reasonable parameters of a data set of $80,000$ traces each consisting of $50,000$ sampled time points, and where the test computes $1,000$ permuted estimates of the MI at each time point, a full run of the detection test on a single key byte necessitates the evaluation of 50 million continuous MI values. If we factor in the cost of finding conditional probability density functions, then we may expect to perform in total 2.05×10^{15} ($\approx 2^{51}$) evaluations of the kernel function used in the density estimation, at a total cost of roughly 1.64×10^{16} floating point operations.

This presents a significant obstacle; we estimated that our naive single-CPU implementation would take around a month to analyse a device. However the problem is 'embarrassingly parallel' and we implemented the test in parallel form using OpenCL: using two AMD Radeon 7970 GPUs we were able to execute a test with the above parameters in approximately 14 hours; a throughput of 300 GFLOPS. The addition of inexpensive GPUs decreases the running time linearly, ensuring that the CMI test, even with large data set parameters, is feasible to run. By comparison the DMI and t-tests are efficient; a key byte can be fully assessed for leakage in under 30 minutes.

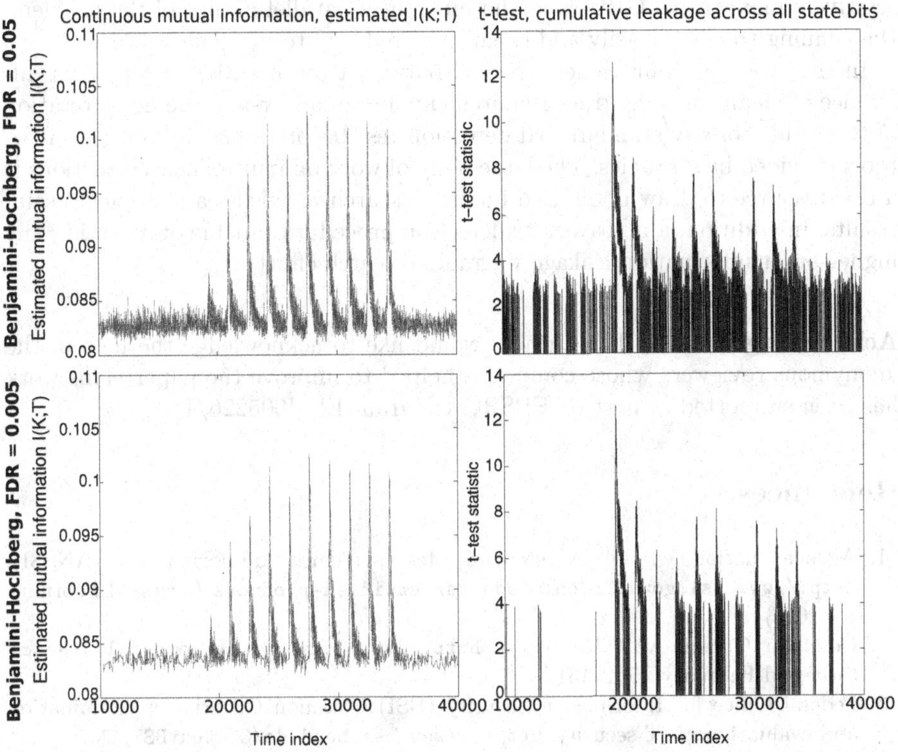

Fig. 5. Plots of the time points consistent with information leakage after applying the Benjamini-Hochberg FDR controlling procedure at levels 0.05 and 0.005 to the results produced by the t-test and CMI test for the hardware AES implementation.

7 Conclusion

Taking the perspective of a 'black-box' evaluation, in which the evaluator may have little knowledge about the leakage characteristics of the device, it would be desirable to select a leakage detection test that is the most generally applicable and that has the best all-round performance. In the majority of our *a priori* analysis this was, by a small margin, the t-test. However we must also take into account the inherent limitations in the t-test's inability to measure leakage in any moment other than the mean. If an evaluator wished to gain the most coverage over *all* possible leakage scenarios, then, given the significant under-performance of the discrete version in the *a priori* analysis, the CMI test is the *only* viable candidate.

The complexity of the tests is an additional factor to consider. The t-test must be re-run for every bit and every intermediate operation within the algorithm implemented on the device, whereas the CMI and DMI tests need only to be run once per bit or byte of key analysed. At first glance the computational cost of the CMI test appears to be prohibitive, but we have demonstrated that using relatively inexpensive GPUs and the inherently parallel nature of the problem, the running time can easily and cheaply be reduced to insignificant levels.

In the absence of any general result that can translate MI, entropy or a difference of means into the trace requirements for an adversary, the interpretation of the results of any standardised detection test becomes heavily reliant on the tools provided by statistics. The large body of work on multiplicity corrections is a rich resource to draw upon, and further research in this area may yield useful results. In addition, a multivariate detection procedure capable of detecting any higher-order information leakage warrants research effort.

Acknowledgements. The authors would like to acknowledge the help of the anonymous reviewers whose comments helped to improve the paper. This work has been supported in part by EPSRC via grant EP/I005226/1.

References

1. Agence nationale de la sécurité des systèmes d'information (ANSSI), `http://www.ssi.gouv.fr/en/products/certified-products` (accessed February 25, 2013)
2. Common Criteria v3.1 Release 4, `http://www.commoncriteriaportal.org/cc/` (accessed February 25, 2013)
3. Federal Office for Information Security (BSI) Common Criteria for examination and evaluation of IT security, `https://www.bsi.bund.de/ContentBSI/EN/Topics/CommonCriteria/commoncriteria.html` (accessed February 25, 2013)
4. National Institute of Standards and Technology: Non-Invasive Attack Testing Workshop (2011), `http://csrc.nist.gov/news_events/non-invasive-attack-testing-workshop` (accessed February 25, 2013)
5. Akkar, M.-L., Bévan, R., Dischamp, P., Moyart, D.: Power Analysis, What Is Now Possible... In: Okamoto, T. (ed.) ASIACRYPT 2000. LNCS, vol. 1976, pp. 489–502. Springer, Heidelberg (2000)
6. Benjamini, Y., Hochberg, Y.: Controlling the False Discovery Rate: A Practical and Powerful Approach to Multiple Testing. Journal of the Royal Statistical Society. Series B (Methodological) 57(1), 289–300 (1995)
7. Brier, E., Clavier, C., Olivier, F.: Correlation Power Analysis with a Leakage Model. In: Joye, M., Quisquater, J.-J. (eds.) CHES 2004. LNCS, vol. 3156, pp. 16–29. Springer, Heidelberg (2004)
8. Chari, S., Rao, J.R., Rohatgi, P.: Template Attacks. In: Kaliski Jr., B.S., Koç, Ç.K., Paar, C. (eds.) CHES 2002. LNCS, vol. 2523, pp. 13–28. Springer, Heidelberg (2003)
9. Chatzikokolakis, K., Chothia, T., Guha, A.: Statistical Measurement of Information Leakage. In: Esparza, J., Majumdar, R. (eds.) TACAS 2010. LNCS, vol. 6015, pp. 390–404. Springer, Heidelberg (2010)
10. Chothia, T., Guha, A.: A Statistical Test for Information Leaks Using Continuous Mutual Information. In: CSF, pp. 177–190. IEEE Computer Society (2011)
11. Ellis, P.D.: The Essential Guide to Effect Sizes: An Introduction to Statistical Power, Meta-Analysis and the Interpretation of Research Results. Cambridge University Press, United Kingdom (2010)

12. Gierlichs, B., Batina, L., Tuyls, P., Preneel, B.: Mutual Information Analysis. In: Oswald, E., Rohatgi, P. (eds.) CHES 2008. LNCS, vol. 5154, pp. 426–442. Springer, Heidelberg (2008)
13. Goodwill, G., Jun, B., Jaffe, J., Rohatgi, P.: A Testing Methodology for Side-Channel Resistance Validation. In: NIST Non-Invasive Attack Testing Workshop (2011)
14. Hochberg, Y., Tamhane, A.C.: Multiple Comparison Procedures. John Wiley & Sons, Inc., New York (1987)
15. Holm, S.: A Simple Sequentially Rejective Multiple Test Procedure. Scandinavian Journal of Statistics 2(6), 65–70 (1979)
16. Jaffe, J., Rohatgi, P., Witteman, M.: Efficient Side-Channel Testing For Public Key Algorithms: RSA Case Study. In: NIST Non-Invasive Attack Testing Workshop (2011)
17. Kocher, P.C., Jaffe, J., Jun, B.: Differential Power Analysis. In: Wiener, M. (ed.) CRYPTO 1999. LNCS, vol. 1666, pp. 388–397. Springer, Heidelberg (1999)
18. Maghrebi, H., Danger, J.-L., Flament, F., Guilley, S.: Evaluation of Countermeasures Implementation Based on Boolean Masking to Thwart First and Second Order Side-Channel Attacks. In: Signals, Circuits and Systems, SCS (2009)
19. Mangard, S., Pramstaller, N., Oswald, E.: Successfully Attacking Masked AES Hardware Implementations. In: Rao, J.R., Sunar, B. (eds.) CHES 2005. LNCS, vol. 3659, pp. 157–171. Springer, Heidelberg (2005)
20. Moradi, A., Mischke, O., Eisenbarth, T.: Correlation-Enhanced Power Analysis Collision Attack. In: Mangard, S., Standaert, F.-X. (eds.) CHES 2010. LNCS, vol. 6225, pp. 125–139. Springer, Heidelberg (2010)
21. Morioka, S., Satoh, A.: An Optimized S-Box Circuit Architecture for Low Power AES Design. In: Kaliski Jr., B.S., Koç, Ç.K., Paar, C. (eds.) CHES 2002. LNCS, vol. 2523, pp. 172–186. Springer, Heidelberg (2003)
22. Paninski, L.: Estimation of Entropy and Mutual Information. Neural Computation 15(6), 1191–1253 (2003)
23. Reparaz, O., Gierlichs, B., Verbauwhede, I.: Selecting Time Samples for Multivariate DPA Attacks. In: Prouff, E., Schaumont, P. (eds.) CHES 2012. LNCS, vol. 7428, pp. 155–174. Springer, Heidelberg (2012)
24. Schindler, W., Lemke, K., Paar, C.: A Stochastic Model for Differential Side Channel Cryptanalysis. In: Rao, J.R., Sunar, B. (eds.) CHES 2005. LNCS, vol. 3659, pp. 30–46. Springer, Heidelberg (2005)
25. Silverman, B.W.: Density Estimation for Statistics and Data Analysis. Chapman and Hall, London (1986)
26. Standaert, F.-X., Malkin, T.G., Yung, M.: A Unified Framework for the Analysis of Side-Channel Key Recovery Attacks. In: Joux, A. (ed.) EUROCRYPT 2009. LNCS, vol. 5479, pp. 443–461. Springer, Heidelberg (2009)
27. Welch, B.L.: The generalization of "Student's" problem when several different population variances are involved. Biometrika 34(1-2), 28–35 (1947)
28. Whitnall, C., Oswald, E.: A fair evaluation framework for comparing side-channel distinguishers. J. Cryptographic Engineering 1(2), 145–160 (2011)

Behind the Scene of Side Channel Attacks

Victor Lomné, Emmanuel Prouff, and Thomas Roche

ANSSI
51, Bd de la Tour-Maubourg, 75700 Paris 07 SP, France
firstname.lastname@ssi.gouv.fr

Abstract. Since the introduction of side channel attacks in the nineties, a large amount of work has been devoted to their effectiveness and efficiency improvements. On the one side, general results and conclusions are drawn in theoretical frameworks, but the latter ones are often set in a too ideal context to capture the full complexity of an attack performed in real conditions. On the other side, practical improvements are proposed for specific contexts but the big picture is often put aside, which makes them difficult to adapt to different contexts. This paper tries to bridge the gap between both worlds. We specifically investigate which kind of issues is faced by a security evaluator when performing a state of the art attack. This analysis leads us to focus on the very common situation where the exact time of the sensitive processing is drown in a large number of leakage points. In this context we propose new ideas to improve the effectiveness and/or efficiency of the three considered attacks. In the particular case of stochastic attacks, we show that the existing literature, essentially developed under the assumption that the exact sensitive time is known, cannot be directly applied when the latter assumption is relaxed. To deal with this issue, we propose an improvement which makes stochastic attack a real alternative to the classical correlation power analysis. Our study is illustrated by various attack experiments performed on several copies of three micro-controllers with different CMOS technologies (respectively 350, 130 and 90 nanometers).

1 Introduction

Since the seminal differential power analysis of Kocher *et al.* [17], various side channel Attacks (SCA) have been proposed and improved (*e.g.* [8,9,11,12,31]). In order to to compare and classify them, **theoretical** frameworks have then been introduced [11,22,35,39]. Their main purpose is to identify the attacks similarities and differences, and to exhibit contexts where one is better than another. They have laid the foundation stones for a general comparison and evaluation framework. In parallel, several **practical** works have addressed issues arising when applying an SCA in the real world (*e.g.* in an industrial context) [2,5,16,24,37]. Those works essentially attempt to fill the gap between the theoretical analysis of the attacks and their application in non-idealized contexts. However, whereas the published theoretical analyses usually tend towards generic and formal statements (sometimes at the cost of too simple models), many of

K. Sako and P. Sarkar (Eds.) ASIACRYPT 2013 Part I, LNCS 8269, pp. 506–525, 2013.

the practical analyses only focus on a particular attack specificity and often put the big picture aside. The latter analyses are indeed usually dedicated to one specific attack running against a specific target device, which makes them hard to generalize. This paper tries to be at the intersection of both worlds: we study practice-driven issues while keeping a generic approach w.r.t. attacks mechanisms and targeted platforms. This approach and our final purpose are close to those in the works of Standaert *et al.* [33] and Renauld *et al.* [29].

The starting observation of our study is that side channel traces are never reduced to one point in practice, even when they rely on the manipulation of a single variable. In contrary, those traces are often composed of a large number of points (typically several thousands). In spite of the evidence of this observation, it is rarely taken into account when analysing the effectiveness of a side channel attack. Such an analysis is indeed frequently done under the assumption, sometimes implicit, that a small number of points of interest (POI) has already been extracted from the traces either by pattern matching [21], or by dimension reduction [1,6,7,32] or thanks to a previous successful attack [10,29]. However, the two first categories of techniques are not yet perfect and, after reduction, the traces are often still composed of several points in practice. And, what is more important, the risk of information loss during the reduction process leads most of attack practitioners to not apply them. The third technique (performing a first attack to identify the POI) allows for interesting analyses, but it does not correspond to a *real attack* context. Moreover, the best POI for one attack type may not be so good for another one. Eventually, we come to a situation where attacks are analysed in a (uni-dimensional) context which does not fit with the (multi-dimensional) reality faced by the attack practitioners.

We argue in this paper that the state-of-the-art uni-dimensional analyses cannot be straightforwardly adapted to multi-dimensional contexts, which raises new interesting issues. The selection of the most likely candidate among the results of several instantaneous attacks is one of them. Indeed when the leakage traces are composed of several points, a side channel attack against the targeted sensitive variable must be performed for each point (*a.k.a* time index) in the traces. Then, the adversary must apply a strategy to select the most likely candidate among the different *instantaneous attacks* results. A classical method is to select the one with the highest score (*e.g.* the highest correlation coefficient in a Correlation Power Analysis – `CPA`– [8]). Nevertheless we argue that this strategy can be ineffective for some attack categories, including the case of the Linear Regression Analysis (`LRA`) [11,30,31]. For the latter one, we propose a new strategy to select the most likely candidate and we demonstrate its effectiveness in practice.

Another interesting issue when dealing with a large number of high dimensional traces is the reduction of the computational complexity. Here again, some works have investigated the use of parallel computing to decrease the data processing time [4,19] but their goal was not to diminish the algorithmic complexity of the attacks. This work studies the `LRA`[1] and the Template Attacks (`TA`) with

[1] In this paper we only consider the unprofiled version of `LRA` [11].

this goal in mind. A common structure in their algorithmic description is exhibited and then used to propose a new general *modus operandi* which enables to significantly reduce the computation time when the number of traces is non-negligible. The strategy can also be applied to other attack (*e.g.* the CPA).

Finally, to make sure that our analysis is consistent with the reality, we completed our investigations by several experiments performed on three micro-controllers based on different CMOS technologies (350, 130 and 90 nanometers process). We report here on these experiments results. We moreover use them to confirm and complete the interesting behaviours observed in [29]: (1) the leakage seems to diverge from the classical Hamming weight model as the CMOS technology tends to the nanometer scale, which makes LRA a promising tool for side channel evaluations of nano-scale devices and (2) TA is effective in practice, even when the *templates* are built on one copy of the device and the attack is done on another copy.

The paper is organized as follows. In Section 2, we introduce the theoretical background for our study and we present the outlines of our proposal. Then, two sections are dedicated to the application of our ideas to the LRA and TA attacks respectively[2].

2 SCA: Practical Issues

In this section, we introduce some basics and we get into the specifics of the problematic focussed in this paper.

2.1 Notations

Throughout this paper, random variables are denoted by large letters. A realization of a random variable, said X, is denoted by the corresponding lowercase letter, said x. A *sample* of several observations of X is denoted by $(x_i)_i$. It will sometimes be viewed as a vector defined over the definition set of X. The notation $(x_i)_i \hookleftarrow X$ denotes the instantiation of the set of observations $(x_i)_i$ from X. The *mean* of X is denoted $\mathbb{E}[X]$, its standard deviation by $\sigma[X]$ and its *variance* by $\mathrm{var}[X]$. The latter equals $\mathbb{E}\left[(X - \mathbb{E}[X])^2\right]$. The *covariance* of two random variables X and Y is denoted by $\mathrm{cov}(X, Y)$ and satisfies $\mathrm{cov}(X, Y) = \mathbb{E}[(X - \mathbb{E}[X])(Y - \mathbb{E}[Y])]$. When we will need to specify the variable on which statistics are computed, we will write the variable in subscript (*e.g.* $\mathbb{E}_X[Y]$ instead of $\mathbb{E}[Y]$).

The notation $\overrightarrow{X}$ will be used to denote column vectors and $\overrightarrow{X}[u]$ will denote its uth coordinate. Calligraphic letters will be used to denote a matrix. The elements of a matrix $\mathcal{M}$ will be denoted by $\mathcal{M}[i][j]$. Classical additions and multiplications (over real values, vectors or matrices) are denoted by $+$ and $\times$ respectively. Scalar-vector operations are denoted by $\cdot$ and $/$ (all the coordinates

[2] This work is completed in the extended version of this paper with a similar study on CPA.

of the vector are multiplied, respectively divided, by the scalar). When applied to vectors or matrices, the symbols $\cdot^2$ and $\sqrt{\cdot}$ denote the operation consisting in computing the square (resp. the square root) of all the vector/matrix coordinates. Eventually, a function from $\mathbb{F}_2^n$ to $\mathbb{F}_2^m$ will be called a (n, m)-function.

2.2 General Attacks Framework

In this paper, the attacks framework is described by considering that the adversary targets the manipulation of a single sensitive variable Z, but the study and results directly extend to contexts where several variables are targeted in parallel. The variable Z is supposed to functionally depend on a public variable X and a secret sub-part k such that $Z = F(X, k)$ where F is a $(n+n, m)$-function (which implies $X, k \in \mathbb{F}_2^n$ and $Z \in \mathbb{F}_2^m$). The bit-lengths n and m depend on the cryptographic algorithm and the device architecture[3].

The attacks are described under the assumption that the adversary owns N side channel traces $\vec{\ell}_0$, ..., $\vec{\ell}_{N-1}$, each of them containing information about Z. Namely, the ith leakage trace $\vec{\ell}_i \hookleftarrow \vec{\mathbf{L}}$ corresponds to the processing of a public value $x_i \hookleftarrow X$ and contains information on the value $z_i \hookleftarrow Z$ such that $z_i = F(x_i, k)$. The dimension of the traces (*i.e.* the number of different instantaneous leakage points) is denoted by d. By definition, we have $d \doteq \dim \vec{\mathbf{L}}$.

When little information is known about the implementation and the device (which is usually the case in practice), the exact manipulation time of z_i cannot be precisely determined *a priori*. Also, precision in the observation often comes at the cost of a high sampling rate[4]. As a consequence, the dimension of the traces is usually high (from several thousand of points up to millions) and the attack must be repeated on all of their coordinates independently (as *e.g.* in LRA) or must consider huge traces chunks globally (as *e.g.* for TA). Although bearing differences, most of side-channel attacks (including LRA and TA) follow a common process flow. Starting from this generic description, this paper studies, in Sections 3 and 4 respectively, the *effectiveness* and *efficiency* of the LRA and TA attacks. The core ideas of those analyses are presented in the two next subsections.

2.3 Effectiveness Discussions

A part of our study is dedicated to the *distinguisher value* definition and, more precisely its relevance when considering side channel traces with a large number of points. This study was motivated by the observation that the classical LRA distinguisher value for one leakage time is not comparable as such to that computed for another leakage time. Figure 1(a) illustrates this claim for an LRA targeting the device B described in Section 2.5: when directly applying the protocol given in [11, 31], the correct key candidate does not maximize the distinguisher value

---

[3] An example of function F is the function that applies a so-called *sbox* transformation to the bitwise addition between k and X.

[4] Especially in the case of Electro-Magnetic side channel measurements

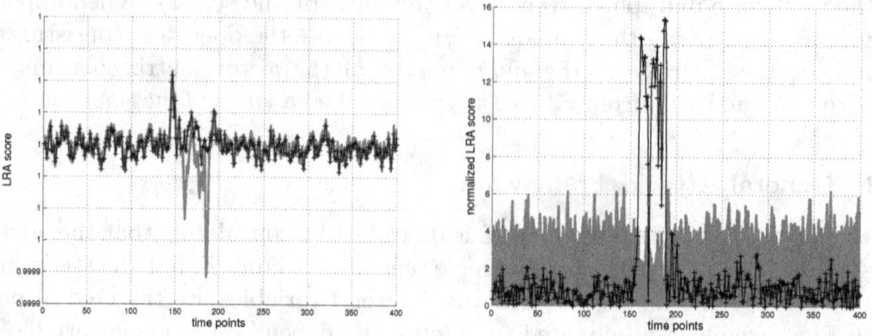

(a) without normlalization (b) with normlalization

Fig. 1. Instantaneous LRA scores computed over 10000 traces (scores for the correct key in black)

globally but only in a local area, which makes the attack unsuccessful unless this area is known by the adversary (which is not assumed here). This observation led us to study the handling of distinguishing values in SCA attacks. We for instance show that by normalizing the LRA distinguishing values, the correct key candidate becomes clearly distinguishable even when considering the full vector of instantaneous attack results (as depicted on Figure 1(b)).

More generally, our study relies on a well studied problem which is the comparison of the results of two different instantaneous attacks [11,20,33,34,36,38]. For the LRA, it will lead to a modification of the candidate selection rule.

2.4 Algorithmic Complexity Improvements Proposals

The other important issue an evaluator faces when performing SCA, is the computational complexity of the attack when the number of traces N and their dimension grows to millions. Indeed, the execution time of naive attack implementations can easily reach several days of processing and this is not compatible with standard evaluation processes[5].

We show in Sections 3 and 4 that the two considered attacks may be re-written in a *partitioning* fashion that can be exploited to significantly decrease the algorithmic complexity. Roughly speaking, the basic idea is to lower the impact of the heavy computations so that its complexity does no longer depend on the traces number N but on the dimension n of the targeted data. To that purpose, we propose to modify the attack first step so that it processes separately the traces with respect to their input value x_i. As a result, the algorithmic complexity of the attacks is divided by $\frac{N}{2^n}$ making the algorithmic improvement interesting when $N \gg 2^n$ (which is often the case in practice).

[5] In Common Criteria evaluations applied on hardware security devices, all penetration tests (including invasive and non-invasive attacks) have usually to be performed in 3 months, leaving only few weeks for the whole side channel evaluation.

Almost every SCA may be rewritten as a combination of tests on statistics estimated on leakage partitions. Some of them (*e.g.* the `DPA` [17] or the multi-bit `DPA` [23]) were actually originally written as such, whereas the other ones were developed in a partitioning way after their introduction (see *e.g.* [18] for the `CPA`, [10] for the `LRA` and [33] for the `MIA`). To the best of our knowledge, this property has however never been exploited to improve the attacks efficiency.

2.5 Experimental Setup

For each studied SCA, practical experiments were performed on three Micro-Controller Units (MCUs for short) with different CMOS technologies (350, 130 and 90 nanometers processes). The observed processing was that of an AES128 encryption handling one byte at a time. Each attack was performed against 4 sbox outputs of the first round. Furthermore, to measure the variability of our experiments, we used three different copies for each MCU (called copy 1, 2 and 3 in the sequel). This choice enabled us to perform the `TA` profiling step on one copy and to use the results to attack other ones. Also, it gives more credit to our experimental results as the templates consistency was checked on three different versions of the same MCU.

The side channel observations were obtained by measuring the electromagnetic (EM) radiations emitted by the device. To this aim, several sensors were used, all made of several coils of copper (the diameters of the coils were respectively of 1mm, 500μm and 250μm for the 350, 130 and 90nm MCUs), and were plugged into a low-noise amplifier. To sample measurements, a digital oscilloscope was used with a sampling rate of 1G samples per second for the 350nm MCU and 10G samples per second for the others, whereas the MCUs were running at few dozen of MHz.

We insist on the fact that the temporal acquisition window was set to record the first round of the AES only. This synchronization has been done thanks to *simple electromagnetic analysis* [26]. As the MCU clocks were not stable, we had to resynchronize the measurements. This process is out of the scope of this work, but we emphasize that it is always needed in a practical context and it impacts the measurements noise.

We sum-up the specificities of the three experimental campaigns hereafter:

- **Device A** (3 copies): 90nm CMOS technology with MCU based on a 8-bit 8051 architecture. EM traces composed of 12800 points each after resynchronization. Highest Signal to Noise Ratio (SNR) over the full traces equals to 0.09.
- **Device B** (3 copies): 130nm CMOS technology with MCU based on a 8-bit 8051 architecture. EM traces composed of 16800 points each after resynchronization. Highest SNR equals to 0.6.
- **Device C** (3 copies): 350nm CMOS technology with MCU based on a 8-bit AVR architecture. EM traces composed of 51600 points each after resynchronization. Highest SNR equals to 0.3.

3 Practical Evaluation of Linear Regression Attacks

Linear regression attacks (a.k.a. stochastic attacks) have been introduced by Schindler *et al.* in 2005 [31]. Initially, they were presented with a profiling step and were viewed as an alternative to the template attacks [13]. In [11], the authors have shown how to express the linear regression attacks such that the profiling stage is no longer required. They also argued that the LRA can be applied in the same context as the CPA, but with weaker assumption on the device behavior. Subsequently, these results of Doget *et al.* have been extended in [10] to apply against masked implementations. In parallel, linear regression attacks have been used to analyse/model the deterministic part of the information leakage for complex circuits [14,15]. As a matter of fact, all those analyses assume that the side-channel traces are composed of a single leakage point: the issue raised in Section 2.3 is thus put aside. Moreover, the question of the efficient processing of the attack, when applied against high dimensional leakage traces, is not tackled. The rest of this section aims at dealing with two issues.

3.1 Attack Description

In LRA, the adversary chooses a so-called *basis* of functions[6] $(\mathtt{m}_p)_{1\leqslant p\leqslant s}$ with the only condition that $\mathtt{m}_1$ is a constant function (usually $\mathtt{m}_1 = 1$). Then, for each x_i and each sub-key hypothesis $\hat{k}$, the prediction $\hat{z}_i = F(x_i, \hat{k})$ is calculated. The basis functions $\mathtt{m}_p$ are then applied to the $\hat{z}_i$ independently, leading to the construction of a $(N \times s)$-matrix $\mathcal{M}_{\hat{k}} \doteq (\mathtt{m}_p(F(x_i, \hat{k})))_{i,p}$. The comparison of this matrix with the set of d-dimensional leakages $(\overrightarrow{\ell}_i)_{i\leqslant N} \hookleftarrow \overrightarrow{\mathbf{L}}$ is done by processing a linear regression of each coordinate of $\overrightarrow{\ell}_i$ in the basis formed by the row elements of $\mathcal{M}_{\hat{k}}$. Namely, a real-valued $(s\times d)$-matrix $\mathcal{B}_{\hat{k}}$ with column vectors $\overrightarrow{\beta}_1, \cdots, \overrightarrow{\beta}_d$ is estimated in order to minimize the error when approximating $\overrightarrow{\ell}_i^{\top}$ by $(\mathtt{m}_1(F(x_i, \hat{k})), \cdots, \mathtt{m}_s(F(x_i, \hat{k}))) \times \mathcal{B}_{\hat{k}}$. The matrix $\mathcal{B}_{\hat{k}}$ is defined such that:

$$\mathcal{B}_{\hat{k}} = \underbrace{\left(\mathcal{M}_{\hat{k}}^{\top} \times \mathcal{M}_{\hat{k}}\right)^{-1} \times \mathcal{M}_{\hat{k}}^{\top}}_{\mathcal{P}_{\hat{k}}} \times \mathcal{L} \ , \tag{1}$$

where $\mathcal{L}$ denotes the $(N \times d)$-matrix with the $\overrightarrow{\ell}_i^{\top}$ as row vectors. In the following, the uth column vector of $\mathcal{L}$ (composed of the uth coordinate of all the $\overrightarrow{\ell}_i$) is denoted by $\overrightarrow{\mathcal{L}}[u]$. Moreover, the $(s \times N)$-matrix $\left(\mathcal{M}_{\hat{k}}^{\top} \times \mathcal{M}_{\hat{k}}\right)^{-1} \times \mathcal{M}_{\hat{k}}^{\top}$, which does not depend on the leakage values, is denoted by $\mathcal{P}_{\hat{k}}$.

To quantify the estimation error, the *goodness of fit model* is used and the *correlation coefficient of determination* $\mathcal{R}^2$ is computed for each u. The latter is defined by $\mathcal{R}^2 = 1 - \text{SSR}/\text{SST}$, where SSR and SST respectively denote the *residual sum of squares* (deduced from $\mathcal{B}_{\hat{k}}$) and the *total sum of squares*[7] (deduced from $\mathcal{L}$). We give in Algorithm 1 the pseudo-code corresponding to a classical LRA attack processing.

---

[6] The basis choice and its impact are not a trivial matter, see [10] for a detailed study.
[7] For their exact definitions, see their construction in Alg. 1

Algorithm 1: LRA - Linear Regression Analysis

Input : a set of d-dimensional leakages $(\overrightarrow{\ell}_i)_{i \leqslant N}$ and the corresponding plaintexts $(x_i)_{i \leqslant N}$, a set of model functions $(\mathtt{m}_p)_{p \leqslant s}$
Output: A candidate sub-key $\hat{k}$

/* Processing of the leakage Total Sum of Squares ($\overrightarrow{\mathrm{SST}}$) */
for $i = 0$ **to** $N-1$ **do**
  $\mu_{\overrightarrow{L}} = \mu_{\overrightarrow{L}} + \overrightarrow{\ell}_i$
  $\sigma_{\overrightarrow{L}} = \sigma_{\overrightarrow{L}} + \overrightarrow{\ell}_i^2$
$\overrightarrow{\mathrm{SST}} = \sigma_{\overrightarrow{L}} - 1/N \cdot \mu_{\overrightarrow{L}}^2$
/* Processing of the 2^n predictions matrices $\mathcal{M}_{\hat{k}}$ and $\mathcal{P}_{\hat{k}}$ */
for $\hat{k} = 0$ **to** $2^n - 1$ **do**
  /* Construct the matrix $\mathcal{M}_{\hat{k}}$ and $\mathcal{P}_{\hat{k}}$ */
  for $p = 1$ **to** s **do**
    for $i = 0$ **to** $N-1$ **do**
      $\mathcal{M}_{\hat{k}}[i][p] \leftarrow \mathtt{m}_p[F(x_i, \hat{k})]$
  $\mathcal{P}_{\hat{k}} = (\mathcal{M}_{\hat{k}}^\top \times \mathcal{M}_{\hat{k}})^{-1} \times \mathcal{M}_{\hat{k}}^\top$
for $\hat{k} = 0$ **to** $2^n - 1$ **do**
  /* Test hyp. $\hat{k}$ for all leakage coordinates */
  for $u = 0$ **to** $d-1$ **do**
    /* *Instantaneous attack (at time u)* */
    $\overrightarrow{\beta} = \mathcal{P}_{\hat{k}} \times \overrightarrow{\mathcal{L}}[u]$
    /* Compute an estimator $\overrightarrow{E}$ of $\overrightarrow{\mathcal{L}}[u] = (\overrightarrow{\ell}_0[u], \cdots, \overrightarrow{\ell}_{N-1}[u])^\top$ */
    $\overrightarrow{E} = \mathcal{M}_{\hat{k}} \times \overrightarrow{\beta}$
    /* *Compute the estimation error (i.e. the SSR)* */
    $\mathrm{SSR} = 0$
    for $i = 0$ **to** $N-1$ **do**
      $\mathrm{SSR} = \mathrm{SSR} + (\overrightarrow{E}[u] - \overrightarrow{\ell}_i[u])^2$
    /* Compute the coefficient of determination */
    $\mathcal{R}[\hat{k}][u] = 1 - \mathrm{SSR}/\overrightarrow{\mathrm{SST}}[u]$
/* Most likely candidate selections */
candidate = $\mathrm{argmax}_{\hat{k}}(\max_u \mathcal{R}[\hat{k}][u])$
return candidate

3.2 On the LRA Effectiveness

Let us focus on the best candidate selection step in a classical LRA. Each sub-key hypothesis $\hat{k}$ is first associated with a score which is the greatest *instantaneous coefficient of determination* when testing it for all temporal coordinates u. It is denoted by $\max_u \mathcal{R}[\hat{k}][u]$ in Alg. 1. The second phase of the selection consists in the processing of the maximum $\mathrm{argmax}_{\hat{k}}(\max_u \mathcal{R}[\hat{k}][u])$. The purpose of the latter step is to identify the candidate that maximises the greatest instantaneous coefficient. Implicitly, such a classical approach by *total maximisation of the distinguisher value* assumes that the most likely candidate corresponds to the greatest value taken by the distinguisher not only over all sub-key hypotheses but also over all the leakage times. This assumption relies on another one, often done in the embedded security community, which states that the value of a distinguisher computed between wrong hypotheses (*i.e.* computed for a wrong sub-key value or a wrong time) and the leakage values tends toward its minimum value (often 0) when the sample size N increases (see *e.g.* [20]). However, as already noticed in several papers (*e.g.* by Messerges in [23], Brier *et al.* in [8] or by Whitnal *et al.* in [40]), both assumptions are often not verified in practice, where the adversary must for instance deal with the *ghost peaks* phenomenon. The situation is even worst for the LRA attacks since the vector of coefficients $\overrightarrow{\beta}$ (and thus the set of predictions) depends not only on $\hat{k}$ but also on the attack time u. The strength of the LRA, namely its ability to adapt to the instantaneous leakage, is also its weakness as it makes it difficult to compare the different instantaneous attacks results.

To illustrate the issue raised in the previous paragraph, we experimented a LRA against an AES sbox processing running on Device B (see Section 2.5). The full leakage traces were composed of 16800 points. We performed the attack on the full trace length and, for each time coordinate, we recorded the scores of all the 256 key-candidates after $N = 1000$ observations. For clarity reasons, we present in Figure 2 the results only for a temporal window of size 250 points where the targeted variable was known to occur. In the top of the figure, the rank of the correct key is plotted and it can yet be observed that it is 0 for few times. In the second trace of Figure 2, the instantaneous maximum scores comprised in $[0.9982, 0.999]$ are plotted[8]: it may be checked that the maximum among those scores corresponds to a time ($t = 238$) when the correct key is not ranked first. This explains why the total maximisation approach fails in returning the correct key candidate in this case.

To build a better rule than the total maximisation test, we respectively plotted in the third and fourth traces of Figure 2 the mean (plain green trace) and the variance (plain red trace) of the instantaneous scores (*i.e.* the values $\mu(u) = 2^{-8}\sum_{\hat{k}} \mathcal{R}[\hat{k}][u]$ and $\sigma(u) = 2^{-8}\sum_{\hat{k}}(\mathcal{R}[\hat{k}][u] - \mu(u))^2$ with u denoting the time coordinate in abscissa). For each time, we also plotted in black dashed line, the maximum score $\max(u) = \max_{\hat{k}}(\mathcal{R}[\hat{k}][u])$. It may be observed that the correct key is ranked first at the time u when the distance $\max(u) - \mu(u)$ is large and $\sigma(u)$ is small. The third (red) trace and the fourth (gray) trace aim at supporting this claim. Eventually, they suggest us the following pre-processing before comparing the instantaneous attack results: for each leakage coordinate, center the maximum of the coefficients of determination and divide it by their standard deviation. The resulting scoring is plotted in the fifth (magenta) trace, where it can been checked that the maximum is indeed achieved for the correct key.

[8] For visibility purpose, we chose to not plot the scores lower than 0.9982.

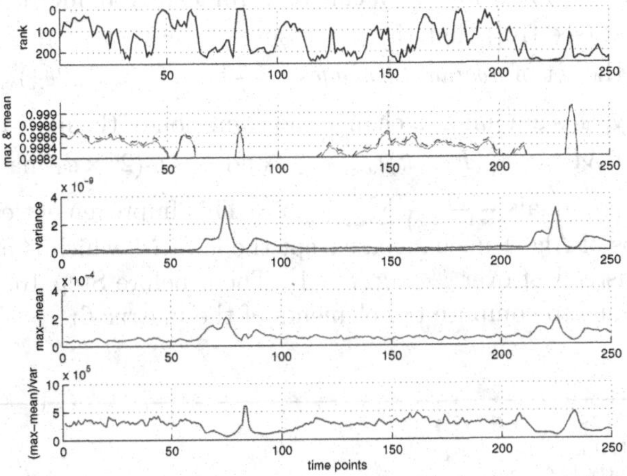

Fig. 2. LRA on Device B (over 1000 traces): Scores Statistics

As a conclusion, and in the light of our analysis, we propose to replace the candidate selection step of the LRA by the following ones[9]:

> **for** $u = 0$ **to** $d - 1$ **do**
>
> $\quad$ attackRes$[u] = \{\text{argmax}_{\hat{k}}(\mathcal{R}[\hat{k}][u]), \frac{\max_{\hat{k}}(\mathcal{R}[\hat{k}][u]) - \mathbb{E}_{\hat{k}}[\mathcal{R}[\hat{k}][u]]}{\sigma_{\hat{k}}[\mathcal{R}[\hat{k}][u]]}\}$
>
> candidate = arg1max2(attackRes)

In Section 3.4, our scores pre-processing technique is applied to attack samples of Device A and Device C in order to test whether our observations, about (1) the ineffectiveness of the classical LRA and (2) the soundness of the new pre-processing, stay valid for other devices than Device B.

3.3 On the LRA Efficiency

The construction of the prediction matrices in Alg. 1 implies, for each $\hat{k}$, the processing of 3 products of matrices with one dimension equal to s (number of basis functions) and the second dimension equal to N (number of leakage traces). The processing of the instantaneous attacks also requires two such matrix products for each pair $(\hat{k}, u)$ with $u \leqslant d$. This makes the application of a linear regression attack as depicted in Alg. 1 difficult to perform (and even impossible) when the number N of leakage traces and/or the number d of attack times are

[9] Where arg1max2 is a function returning the first coordinate of the maximum of an array of 2-dimensional elements, the maximisation being computed with respect to the second coordinate of the array elements.

large. Fortunately this complexity can be significantly reduced. It can indeed be easily shown (see [10]) that the processing of the vectors $\overrightarrow{\beta}$ is unchanged if performed for the set of *averaged leakages* $(\frac{1}{\#\{i:x_i=x\}}\sum_{i,x_i=x}\overrightarrow{\ell}_i)_{x\in\mathbb{F}_2^n}$ instead of $(\overrightarrow{\ell}_i)_i$. Actually, this amounts to change the definition of the matrices $\mathcal{M}_{\hat{k}}$ and $\mathcal{L}$ in (1) such that $\mathcal{M}_{\hat{k}} \doteq (\mathtt{m}_p(F(x,\hat{k}))_{x\in\mathbb{F}_2^n,p\leqslant s}$ and $\mathcal{L}$ is a $(2^n \times d)$-matrix whose x^{th} row vector $\overrightarrow{\mathcal{L}^\top}[x]$ equals $\frac{1}{\#\{x_i=x\}}\sum_{i,x_i=x}\overrightarrow{\ell}_i$. This improvement essentially lets the first 9 steps of Alg. 1 unchanged except the loop 7-8 which is now computed over $x \in \mathbb{F}_2^n$ instead of over $i \in [0; N-1]$. Then, before Step 10, the following processing is done to compute the elements of the matrix $\mathcal{L}$:

```
for i = 0 to N − 1 do
 | L^T[x_i] = L^T[x_i] + ℓ_i
 | count[x_i] = count[x_i] + 1
for x = 0 to 2^n − 1 do
 | L^T[x] = L^T[x]/count[x]
```

Eventually, Steps 10-17 are replaced by the following ones where we recall that $\overrightarrow{\mathcal{L}}[u]$ denotes the uth column vector of $\mathcal{L}$.

```
for k̂ = 0 to 2^n − 1 do
 | /* Test hypothesis k̂ for all leakage coordinates                              */
 | for u = 0 to d − 1 do
 |  | /* Instantaneous attack (at time u)                                        */
 |  | β = P_k̂ × L[u]
 |  | /* Compute an estimator E of L[u] = (L[0][u], ··· , L[2^n − 1][u])^T       */
 |  | E = M_k̂ × β
 |  | /* Compute the estimation error (i.e. the residual sum of squares)         */
 |  | SSR = 0
 |  | for x = 0 to 2^n − 1 do
 |  |  | SSR = SSR + (E[x] − L[x][u])^2
 |  | /* Compute the coefficient of determination                                */
 |  | R[k̂][u] = 1 − SSR/SST[u]
```

The efficiency improvements proposed here for the LRA attack allows for a significant time/memory gain. First, it replaces the $(N \times s)$-matrix products at Step 13 by $(2^n \times s)$-matrix products. More globally, the complexity is reduced from $O(s \times d \times N)$ to $O(s \times d \times 2^n)$. If the $\overrightarrow{\beta}$ values are not needed (*i.e.* the weights of the linear regression is of no interest to the attacker), the matrix products $\mathcal{M}_{\hat{k}} \times \mathcal{P}_{\hat{k}}$ can also be pre-processed. This enables to save one matrices product per loop iteration (over $\hat{k}$ and u).

3.4 Experiments

We experimented the classical and improved LRA against against three copies of Devices A, B and C (see Section 2.5). The attacks target four bytes of the AES state after the first SubBytes operation and they are applied on the full side channel traces. Each attack has been performed 10 times against each of the three copies. The average rank over the four correct sub-keys is plotted in Figure 3 for each device. We recall that the rank of a sub-key k is here defined as the position of $\max_u \mathcal{R}[k][u]$ in the vector $(\max_u \mathcal{R}[\hat{k}][u])_{\hat{k}}$ after sorting (see Section 2.2 for a discussion about this choice). The experiments reported in Figure 3(a)-(c) are done with a *linear basis* (*i.e.* the functions $\mathtt{m}_i$ were chosen such that $\mathtt{m}_0$ is constant equal to 1 and $\mathtt{m}_i$, with $i \leqslant 8$, returns the ith bit of its inputs). It may be observed that the classical attack always failed whereas the improved one succeeded with less than 2500 observations (and even less than 800 for Device B).

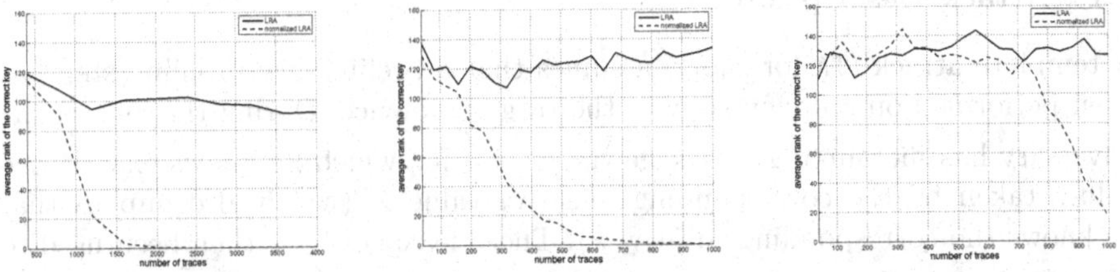

(a) LRA on Device A (b) LRA on Device B (c) LRA on Device C

Fig. 3. LRA campaign – Rank evolution *versus* number of observations

4 Practical Evaluation of Template Attacks

Template attacks have been introduced in 2002 by Chari *et al.* [9]. Subsequent works have then been published which either show how to apply them against particular implementations (*e.g.* AES, RSA or ECDSA) or propose efficiency/effectiveness improvements [1,3,27,29]. In [27], the authors reduce the complexity of template attacks by first applying a pre-processing on the measurements (to go from time domain to frequency domain) and then by applying dimension reduction techniques (*e.g.* PCA). The latter idea is also followed in [1] and [3]. In all those papers, the improvement of the template attacks efficiency is not studied at the algorithmic level. Moreover, the reported template attack experiments involve the same device for the profiling and matching phases of the attacks, which strongly reduces the practical significance of the argumentations. Indeed, as the profiling phase requires a full access to the device (and in particular the ability to chose the secret parameter), the latter experiments do not fit with the large majority of real attack/evaluation contexts where the adversary has no (or very few) control on the target device. In a more realistic attacker model the profiling phase is conducted on a different device. For such a model, we have the following well known question[10] about the efficiency of template attack: how sound/relevant is a profiling done on a device A when attacking another device B ? The first work, and to the best of our knowledge, the single one reporting on template attacks in such context is due to Renauld *et al.* [29]. On the latter article, the two devices used for the experiments are test chips implementing an AES s-box and made in 65-nanometer CMOS technology.

The results presented in the rest of this section improve the state-of-the-art recalled previously on two points. First, the efficiency improvement is done at the algorithmic level. It can hence be combined with the previous improvements which essentially correspond to measurements traces pre-processing. Secondly, the reported experiments concern a full AES implementation running on 3 different samples of 3 different technologies. This allowed us to complete the analyses done in [29] and to draw, for the first time, conclusions about the template attack efficiency for realisitic scenarios.

4.1 Attack Description

A template attack (TA for short) assumes that a preliminary profiling step has been performed on an open copy of the targeted device. During this phase, the adversary has measured N' leakage traces $\vec{\ell}_i'$ for which he knows exactly the values taken by the corresponding sensitive value Z (which also implies that he knows the corresponding sub-key k). Those leakages have then been used to compute estimations $f_z(\cdot)$ of the *probability density function* of $(\vec{\mathbf{L}} \mid Z = z)$ for all possible z (which imposes $N' \gg 2^m$). The pdf estimations $f_z(\cdot)$ will play in a template attack, a similar role as the model functions in a CPA or LRA.

Once the adversary has the set of pdf estimations $(f_z(\cdot))_{z\in\mathbb{F}_2^m}$ in hand, a TA against the set of traces $(\vec{\ell}_i)_{i\leqslant N}$ (for which the secrets are unknown) follows essentially the same outlines as the LRA: the hypothesis $\hat{k}$ is tested by first computing the predictions $\hat{z}_i = F(x_i, \hat{k})$ and then by calculating the product $\prod_{i\leqslant N} f_{\hat{z}_i}(\vec{\ell}_i)$. Usually, the pdf of the variables $(\vec{\mathbf{L}} \mid Z = z)$ is estimated by a multivariate normal law, which implies that f_z can be developed s.t.:

$$f_z(\vec{\ell}_i) = \frac{1}{(2\pi)^d \det(\Sigma_z)} \exp\left(-\frac{1}{2}(\vec{\ell}_i - \vec{\mu}_z)^\top \Sigma_z^{-1} (\vec{\ell}_i - \vec{\mu}_z)\right) , \quad (2)$$

where Σ_z denotes the $(d \times d)$-matrix of covariances of $\vec{\mathbf{L}} \mid Z = z$ and where the (d)-dimensional vector $\vec{\mu}_z$ denotes its mean.

To minimize approximation errors induced by the processing of the product of exponential values, one usually prefers, in practice, a log-maximum likelihood processing to the classical maximum likelihood[11]. Together with (2), this leads to the following computation to test the hypothesis $\hat{k}$:

[10] This question is sometimes also related to the statistical problem of pdf estimations robustness [25].

[11] The two processes discriminate equivalently.

$$\mathcal{ML}[\hat{k}] = -\sum_{i\leqslant N}\left(\vec{\ell}_i - \vec{\mu}_{\hat{z}_i}\right)^{\top}\Sigma_{\hat{z}_i}^{-1}\left(\vec{\ell}_i - \vec{\mu}_{\hat{z}_i}\right) - \sum_{i\leqslant N}\log((2\pi)^{d+1}\det(\Sigma_{\hat{z}_i})) \ . \quad (3)$$

We give in Alg. 2 the pseudo-code corresponding to the TA attack discussed previously.

Algorithm 2: TA - Template Attacks

Input : a set of d-dimensional leakages $(\vec{\ell}_i)_{i\leqslant N}$ and the corresponding plaintexts $(x_i)_{i\leqslant N}$, a set of pdf estimations $(\vec{\mu}_z, \Sigma_z)_{z\in\mathbb{F}_2^m}$

Output: A candidate sub-key $\hat{k}$

/* Pre-Processing of the 2^m log-determinants $\log(2\pi^{d+1}\Sigma_z)$ and inverse-matrices Σ_z^{-1} */

for $z = 0$ **to** $2^m - 1$ **do**
- $\text{logDet}_z = \log(2\pi^{d+1}\Sigma_z)$
- $\text{invCov}_z = \Sigma_z^{-1}$

/* Instantaneous TA attacks Processing */

for $\hat{k} = 0$ **to** $2^n - 1$ **do**
- /* Test hyp. $\hat{k}$ */
- $\mathcal{ML}[\hat{k}] = 0$
- **for** $i = 0$ **to** $N - 1$ **do**
  - $\hat{z} = F(x_i, \hat{k})$
  - $\mathcal{ML}[\hat{k}] = \mathcal{ML}[\hat{k}] - (\vec{\ell}_i - \vec{\mu}_{\hat{z}})^{\top} \times \text{invCov}_{\hat{z}} \times (\vec{\ell}_i - \vec{\mu}_{\hat{z}}) - \text{logDet}_{\hat{z}}$

/* Most likely candidate selections */

candidate = $\text{argmax}_{\hat{k}}(\max \mathcal{ML}[\hat{k}])$

return candidate

4.2 On the TA Effectiveness

The idea developed in previous sections to improve the selection of the best candidate among the results of several instantaneous attacks is not relevant here. Indeed, for both the profiling and attack phases, a template attack exploits, by nature, all the leakage coordinates of the $\vec{\ell}_i$ simultaneously. There is consequently no need to compare the results of several (different) instantaneous attacks.

4.3 On the TA Efficiency

Applying the same idea as for the LRA, we propose hereafter an alternative writing of $\mathcal{ML}[\hat{k}]$ that leads to a much faster attack processing. For such a purpose, we focus on the term $\left(\vec{\ell}_i - \vec{\mu}_{\hat{z}_i}\right)^{\top}\Sigma_{\hat{z}_i}^{-1}\left(\vec{\ell}_i - \vec{\mu}_{\hat{z}_i}\right)$ in (3).

After denoting by $\mathcal{L}_i$ each $(d\times d)$-matrix $(\vec{\ell}_i[u]\vec{\ell}_i[u'])_{u,u'}$, we get the following rewriting of the latter term:

$$\sum_{u,u'}\left(\mathcal{L}_i[u][u'] \times \Sigma_{\hat{z}_i}^{-1}[u][u']\right) - \vec{\mu}_{\hat{z}_i}^{\top} \times \left(\Sigma_{\hat{z}_i}^{-1} + {\Sigma_{\hat{z}_i}^{-1}}^{\top}\right) \times \vec{\ell}_i + \vec{\mu}_{\hat{z}_i}^{\top} \times \Sigma_{\hat{z}_i}^{-1} \times \vec{\mu}_{\hat{z}_i} \ .$$

After recalling that $\hat{z}_i$ equals $F(x_i, \hat{k})$ and after denoting $F(x, \hat{k})$ by $\hat{z}$ and $\#\{i, x_i = x\}$ by N_x, we deduce that the sum $\sum_i \left(\overrightarrow{\ell}_i - \overrightarrow{\mu}_{\hat{z}_i}\right)^{\top} \Sigma_{\hat{z}_i}^{-1} \left(\overrightarrow{\ell}_i - \overrightarrow{\mu}_{\hat{z}_i}\right)$ may be rewritten:

$$\sum_{x \in \mathbb{F}_2^n} \left(\sum_{u,u'} \Big(\sum_{i, x_i = x} \mathcal{L}_i \Big)[u][u'] \times \Sigma_{\hat{z}}^{-1}[u][u'] \right.$$

$$\left. - \overrightarrow{\mu}_{\hat{z}}^{\top} \times \left(\Sigma_{\hat{z}}^{-1} + {\Sigma_{\hat{z}}^{-1}}^{\top}\right) \times \Big(\sum_{i, x_i = x} \overrightarrow{\ell}_i \Big) + N_x \times \overrightarrow{\mu}_z^{\top} \times \Sigma_z^{-1} \times \overrightarrow{\mu}_z \right) .$$

As a consequence, if the 2^n possible sums $\sum_{i, x_i = x} \mathcal{L}_i$ and $\sum_{i, x_i = x} \overrightarrow{\ell}_i$ have been precomputed, then the complexity of evaluating (3) for each $\hat{k}$ goes from $O(Nd^2)$ to $O(2^n d^2)$. Algorithm 3 describes the improved TA attack.

Algorithm 3: TA - Template Attacks (Improved Version)

Input : a set of N leakages $(\overrightarrow{\ell}_i)_i$ and the corresponding plaintexts $(x_i)_i$, a set of pdf estimations $(\overrightarrow{\mu}_z, \Sigma_z)_{z \in \mathbb{F}_2^m}$

Output: A candidate subkey $\hat{k}$

/* Pre-Processing of the predictions data */

for $z = 0$ **to** $2^m - 1$ **do**

  $\text{logDet}_z = \log(2\pi^{d+1}\Sigma_z)$; $\text{invCov}_z = \Sigma_z^{-1}$; $\text{meanCov}_z = \overrightarrow{\mu}_z^{\top} \times \text{invCov}_z \times \overrightarrow{\mu}_z$; $\text{sumMeanCov}_z = \overrightarrow{\mu}_z^{\top} \times (\Sigma_z^{-1} + {\Sigma_z^{-1}}^{\top})$

/* Pre-Processing of the leakage data[12] $\overline{\ell_x} = \sum_{i, x_i = x} \overrightarrow{\ell}_i$, $\overline{\mathcal{L}_x} = \sum_{i, x_i = x} \mathcal{L}_i$ and $N[x] = \#\{i; x_i = x\}$. */

for $i = 0$ **to** $N - 1$ **do**

  $x = x_i$; $N[x] = N[x] + 1$; $\overline{\mathcal{L}_x} = \overline{\mathcal{L}_x} + \overrightarrow{\ell}_i \overrightarrow{\ell}_i^{\top}$; $\overline{\ell_x} = \overline{\ell_x} + \overrightarrow{\ell}_i$

/* Instantaneous TA attacks Processing */

for $\hat{k} = 0$ **to** $2^n - 1$ **do**

  /* Test hyp. $\hat{k}$ */

  $\mathcal{ML}[\hat{k}] = 0$

  for $x = 0$ **to** $2^n - 1$ **do**

    $\hat{z} = F_j(x, \hat{k})$

    for $u = 0$ **to** $d - 1$ **do**

      for $u' = 0$ **to** $d - 1$ **do**

        $\mathcal{ML}[\hat{k}] = \mathcal{ML}[\hat{k}] - \overline{\mathcal{L}_x}[u][u'] \times \text{invCov}_{\hat{z}}[u][u']$

    $\mathcal{ML}[\hat{k}] = \mathcal{ML}[\hat{k}] + \text{sumMeanCov}_z \times \overline{\ell_x} - N[x] \times (\text{meanCov}_{\hat{z}} + \text{logDet}_{\hat{z}})$

/* Most likely candidate selections */

candidate = $\text{argmax}_{\hat{k}}(\max \mathcal{ML}[\hat{k}])$

return candidate

4.4 Experiments

To study the effectiveness of TA attacks in practice (and to confirm the observations reported in [29]) we experimented them against the families of devices A, B and C for three different scenarios. In the first scenario (referred to as "copy 1 → copy 1"), the profiling and the attacks are performed on the same device copy. In the second and third scenarios (respectively referred to as "copy 1 → copy 2" and "copy 1 → copy 3"), the profiling made for copy 1 is used to attack the second and third copies. For each of these 9 attacks frameworks, we plot in Figure 4 the average rank of the correct sub-key (in color) with respect to both the number of traces used for the profiling (in ordinate) and the number of traces used for the attack (in abscissa). The rank averaging has been done over 10 attacks.

In the first scenario, a profiling done on 15000 (resp. 47000) traces on Device B (resp. Device C) allows for a very efficient attack phase (the correct sub-key ranked first with less than ten traces). Moreover, it may be observed that a profiling on 8000 traces for Devices B and C is sufficient to have a successful attack in less than 23 (resp. 90) traces for device B (resp. C). For Device A, the TA attack in Scenario 1 is one order of magnitude less efficient (roughly speaking the values are multiplied per ten w.r.t. the traces for devices B and C).

Attacks on Devices B (resp. C) perform quite similarly in Scenarios 2 and 3. For Device A, a profiling performed on copy 1 for 18000 traces is sufficient to successfully attack copies 2 and 3 with less than 10 traces. Moreover, a profiling on 8000 traces enables successful attacks for less than 30 traces. For Device C, it may be observed that, even for a profiling performed on 50000 traces, the attacks on copies 2 and 3 require at least 80 traces to succeed. However, a profiling on 9000 traces is sufficient to have the TA succeeding in less than 130 traces.

As expected, we may observe a significant variability for the attack results in Scenarios 2 and 3 for Device A: templates done on copy 1 are almost as efficient to attack copy 2 than they were to attack copy 1 itself. They are however much less informative on the copy 3 behaviour since the profiling on copy 1 must be performed on at least 130000 traces to see the attack working on copy 3 with less than 700 traces. This observation is in-line with those done in [28] about the high variability of nano-scale technologies (we recall that Device A is made in a 90nm CMOS technology).

In the full version of this paper, we report on similar experiments results when only the leakage means (and not the covariance matrices) are involved in the templates. This approach indeed seems to be a natural alternative to the attacks described here since the traces contain instantaneous leakages. Our results actually confirm this feeling and it can even be noticed that it leads to improve the TA efficiency for Scenario 3 on Device A[13]. Another general remark on these simplified templates is that they perform much better than the full ones when the number of traces used for the profiling is small (around 4000).

[13] This could be explained by the fact that the technology variability has more impact on the electromagnetic leakage variances than it has on the means.

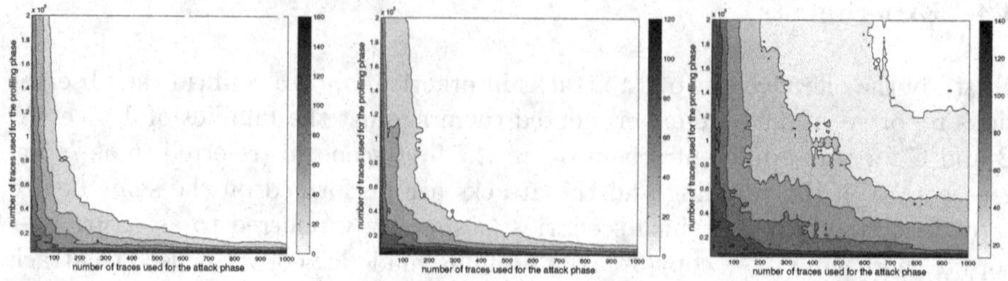

(a) Dev. A: copy 1→copy 1 (b) Dev. A: copy 1→copy 2 (c) Dev. A: copy 1→copy 3

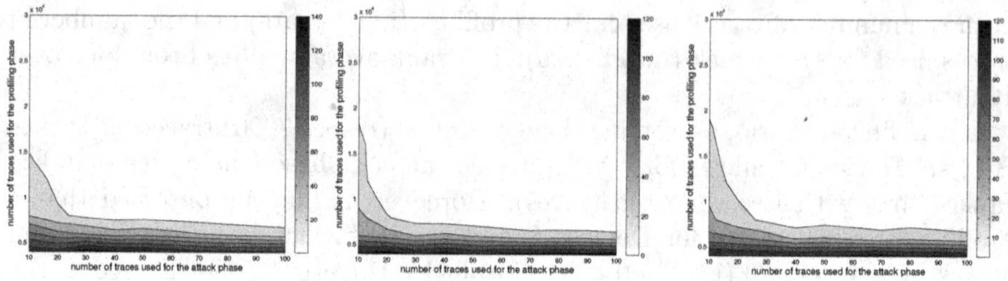

(d) Dev. B: copy 1→copy 1 (e) Dev. B: copy 1→copy 2 (f) Dev. B: copy 1→copy 3

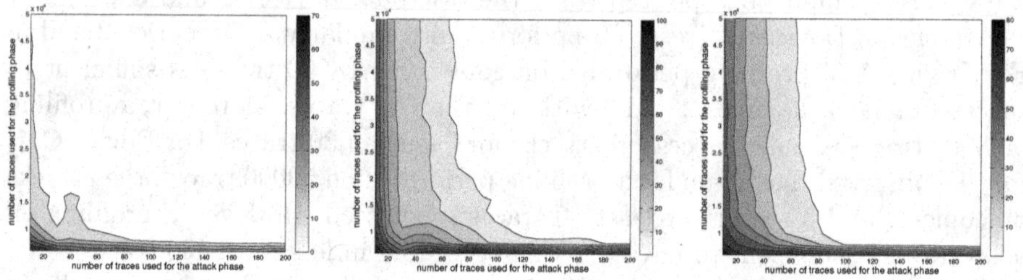

(g) Dev. C: copy 1→copy 1 (h) Dev. C: copy 1→copy 2 (i) Dev. C: copy 1→copy 3

Fig. 4. TA campaigns – Rank evolution *vs.* nb. of traces for the attack phase (x-axis) and the profiling (y-axis)

5 Conclusion

In this paper, we have studied the effectiveness and efficiency of the LRA and the TA attacks when performed in a context where the exact time of the sensitive computations is not known. In this situation, and even after the application of pattern matching or resynchronization techniques, the exploited leakage traces may be composed of several thousands of points and the same attack must be processed for each of those points. We noticed that the study of the side channel attacks effectiveness and efficiency in this multivariate context is an over-estimated problem. Most of the time, it is indeed assumed that the adversary succeeded in significantly reducing the traces size (*e.g.* by priorly processing a SNR, or a test attack, or even a dimension reduction). However, as argued in this paper, those techniques are either unrealistic or may lead to a significant loss of useful information (a dimension reduction technique like the PCA may be sound for one attack – *e.g.* the `CPA` – and not for another one – *e.g.* the `MIA` or the `LRA` –). As a consequence, there was no work discussing about the rule to apply in order to select a candidate among all of those returned by a same attack performed against several time coordinates. To the best of our knowledge, the *de facto* rule was hence to simply choose the key candidate maximising all the attacks scores. In this paper, we have shown that this rule does not work for a `LRA` attack and we have conducted a statistical analysis to deduce a new selection rule that renders it effective in practice, even when the traces are composed of huge number of points. In this paper, we have also tackled out the efficiency problem for the multivariate `LRA` and `TA` attacks. For each of them, we have followed a similar approach which led us to significantly reduce their complexity when the number of traces and their dimension are high. It may be noticed that the approach could also be applied (almost straightforwardly) to improve the efficiency of the correlation power attack and of the mutual information attack (with histogram pdf estimation). Eventually, all our results and analyses have been illustrated by several attack experiments on three different copies of three different technologies. In particular, the latter experiments have enabled us to confirm the practicability of template attacks when the profiling phase and the attack are performed on different copies of the same device.

References

1. Archambeau, C., Peeters, E., Standaert, F.-X., Quisquater, J.-J.: Template Attacks in Principal Subspaces. In: Goubin, L., Matsui, M. (eds.) CHES 2006. LNCS, vol. 4249, pp. 1–14. Springer, Heidelberg (2006)
2. Balasch, J., Gierlichs, B., Verdult, R., Batina, L., Verbauwhede, I.: Power Analysis of Atmel CryptoMemory – Recovering Keys from Secure EEPROMs. In: Dunkelman, O. (ed.) CT-RSA 2012. LNCS, vol. 7178, pp. 19–34. Springer, Heidelberg (2012)
3. Bär, M., Drexler, H., Pulkus, J.: Improved Template Attacks. In: Proceedings of the Constructive Side-Channel Analysis and Secure Design - First International Workshop, COSADE 2010, Darmstadt, Germany, February 4-5 (2010)
4. Bartkewitz, T., Lemke-Rust, K.: A high-performance implementation of differential power analysis on graphics cards. In: Prouff, E. (ed.) CARDIS 2011. LNCS, vol. 7079, pp. 252–265. Springer, Heidelberg (2011)
5. Batina, L., Gierlichs, B., Lemke-Rust, K.: Comparative Evaluation of Rank Correlation Based DPA on an AES Prototype Chip. In: Wu, T.-C., Lei, C.-L., Rijmen, V., Lee, D.-T. (eds.) ISC 2008. LNCS, vol. 5222, pp. 341–354. Springer, Heidelberg (2008)
6. Batina, L., Hogenboom, J., van Woudenberg, J.G.J.: Getting More from PCA: First Results of Using Principal Component Analysis for Extensive Power Analysis. In: Dunkelman, O. (ed.) CT-RSA 2012. LNCS, vol. 7178, pp. 383–397. Springer, Heidelberg (2012)
7. Bohy, L., Neve, M., Samyde, D., Quisquater, J.-J.: Principal and Independent Component Analysis for Crypto-systems With Hardware Unmasked Units. In: Proceedings of E-Smart (2003)

8. Brier, E., Clavier, C., Olivier, F.: Correlation Power Analysis with a Leakage Model. In: Joye, M., Quisquater, J.-J. (eds.) CHES 2004. LNCS, vol. 3156, pp. 16–29. Springer, Heidelberg (2004)
9. Chari, S., Rao, J., Rohatgi, P.: Template attacks. In: Kaliski Jr., B.S., Kocc, cC.K., Paar, C. (eds.) CHES 2002. LNCS, vol. 2523, pp. 13–28. Springer, Heidelberg (2003)
10. Dabosville, G., Doget, J., Prouff, E.: A New Second Order Side Channel Attack Based on Linear Regression. IEEE Transactions on Computers (2012) (to appear)
11. Doget, J., Prouff, E., Rivain, M., Standaert, F.-X.: Univariate Side Channel Attacks and Leakage Modeling. Journal of Cryptographic Engineering 1(2), 123–144 (2011)
12. Gierlichs, B., Batina, L., Tuyls, P., Preneel, B.: Mutual Information Analysis. In: Oswald, E., Rohatgi, P. (eds.) CHES 2008. LNCS, vol. 5154, pp. 426–442. Springer, Heidelberg (2008)
13. Gierlichs, B., Lemke-Rust, K., Paar, C.: Templates vs. Stochastic Methods. In: Goubin, L., Matsui, M. (eds.) CHES 2006. LNCS, vol. 4249, pp. 15–29. Springer, Heidelberg (2006)
14. Heuser, A., Schindler, W., Stöttinger, M.: Revealing Side-Channel Issues of Complex Circuits by Enhanced Leakage Models. In: Rosenstiel, W., Thiele, L. (eds.) DATE, pp. 1179–1184. IEEE (2012)
15. Kasper, M., Schindler, W., Stoettinger, M.: A Stochastic Method for Security Evaluation of Cryptographic FPGA Implementations. In: IEE International Conference on Field-Programmable Technology (FPT 2010). IEEE Press (December 2010)
16. Kizhvatov, I.: Side Channel Analysis of AVR XMEGA Crypto Engine. In: Serpanos, D.N., Wolf, W. (eds.) WESS. ACM (2009)
17. Kocher, P.C., Jaffe, J., Jun, B.: Differential Power Analysis. In: Wiener, M. (ed.) CRYPTO 1999. LNCS, vol. 1666, pp. 388–397. Springer, Heidelberg (1999)
18. Le, T.-H., Clédière, J., Canovas, C., Robisson, B., Servière, C., Lacoume, J.-L.: A Proposition for Correlation Power Analysis Enhancement. In: Goubin, L., Matsui, M. (eds.) CHES 2006. LNCS, vol. 4249, pp. 174–186. Springer, Heidelberg (2006)
19. Lee, S.J., Seo, S.C., Han, D.-G., Hong, S., Lee, S.: Acceleration of Differential Power Analysis through the Parallel Use of GPU and CPU. IEICE Transactions 93-A(9), 1688–1692 (2010)
20. Mangard, S.: Hardware Countermeasures against DPA – A Statistical Analysis of Their Effectiveness. In: Okamoto, T. (ed.) CT-RSA 2004. LNCS, vol. 2964, pp. 222–235. Springer, Heidelberg (2004)
21. Mangard, S., Oswald, E., Popp, T.: Power Analysis Attacks – Revealing the Secrets of Smartcards. Springer (2007)
22. Mangard, S., Oswald, E., Standaert, F.-X.: One for All - All for One: Unifying Standard DPA Attacks. IET Information Security (2011)
23. Messerges, T.S.: Using Second-Order Power Analysis to Attack DPA Resistant Software. In: Paar, C., Kocc, cC.K. (eds.) CHES 2000. LNCS, vol. 1965, pp. 238–251. Springer, Heidelberg (2000)
24. Oswald, D., Paar, C.: Breaking Mifare DESFire MF3ICD40: Power Analysis and Templates in the Real World. In: Preneel, B., Takagi, T. (eds.) CHES 2011. LNCS, vol. 6917, pp. 207–222. Springer, Heidelberg (2011)
25. Press, W.: Numerical Recipes in Fortran 77: The Art of Scientific Computing. Fortran Numerical Recipes. Cambridge University Press (1992)
26. Quisquater, J.-J., Samyde, D.: ElectroMagnetic Analysis (EMA): Measures and Counter-Measures for Smart Cards. In: Attali, S., Jensen, T. (eds.) E-smart 2001. LNCS, vol. 2140, pp. 200–210. Springer, Heidelberg (2001)

27. Rechberger, C., Oswald, E.: Practical Template Attacks. In: Lim, C.H., Yung, M. (eds.) WISA 2004. LNCS, vol. 3325, pp. 440–456. Springer, Heidelberg (2005)
28. Renauld, M., Kamel, D., Standaert, F.-X., Flandre, D.: Information Theoretic and Security Analysis of a 65-Nanometer DDSLL AES S-Box. In: Preneel, B., Takagi, T. (eds.) CHES 2011. LNCS, vol. 6917, pp. 223–239. Springer, Heidelberg (2011)
29. Renauld, M., Standaert, F.-X., Veyrat-Charvillon, N., Kamel, D., Flandre, D.: A Formal Study of Power Variability Issues and Side-Channel Attacks for Nanoscale Devices. In: Paterson, K.G. (ed.) EUROCRYPT 2011. LNCS, vol. 6632, pp. 109–128. Springer, Heidelberg (2011)
30. Schindler, W.: Advanced Stochastic Methods in Side Channel Analysis on Block Ciphers in the Presence of Masking. Journal of Mathematical Cryptology 2, 291–310 (2008)
31. Schindler, W., Lemke, K., Paar, C.: A Stochastic Model for Differential Side Channel Cryptanalysis. In: Rao, J.R., Sunar, B. (eds.) CHES 2005. LNCS, vol. 3659, pp. 30–46. Springer, Heidelberg (2005)
32. Standaert, F.-X., Archambeau, C.: Using Subspace-Based Template Attacks to Compare and Combine Power and Electromagnetic Information Leakages. In: Oswald, E., Rohatgi, P. (eds.) CHES 2008. LNCS, vol. 5154, pp. 411–425. Springer, Heidelberg (2008)
33. Standaert, F.-X., Gierlichs, B., Verbauwhede, I.: Partition *vs.* Comparison Side-Channel Distinguishers: An Empirical Evaluation of Statistical Tests for Univariate Side-Channel Attacks against Two Unprotected CMOS Devices. In: Lee, P.J., Cheon, J.H. (eds.) ICISC 2008. LNCS, vol. 5461, pp. 253–267. Springer, Heidelberg (2009)
34. Standaert, F.-X., Koeune, F., Schindler, W.: How to Compare Profiled Side-Channel Attacks? In: Abdalla, M., Pointcheval, D., Fouque, P.-A., Vergnaud, D. (eds.) ACNS 2009. LNCS, vol. 5536, pp. 485–498. Springer, Heidelberg (2009)
35. Standaert, F.-X., Malkin, T.G., Yung, M.: A Unified Framework for the Analysis of Side-Channel Key Recovery Attacks. In: Joux, A. (ed.) EUROCRYPT 2009. LNCS, vol. 5479, pp. 443–461. Springer, Heidelberg (2009)
36. Standaert, F.-X., Veyrat-Charvillon, N., Oswald, E., Gierlichs, B., Medwed, M., Kasper, M., Mangard, S.: The World is not Enough: Another Look on Second-Order DPA. In: Abe, M. (ed.) ASIACRYPT 2010. LNCS, vol. 6477, pp. 112–129. Springer, Heidelberg (2010)
37. Télécom ParisTech. DPA Contest v1 and v2, `http://www.dpacontest.org/` (retrieved on August 1, 2012)
38. Whitnall, C., Oswald, E.: A Comprehensive Evaluation of Mutual Information Analysis Using a Fair Evaluation Framework. In: Rogaway, P. (ed.) CRYPTO 2011. LNCS, vol. 6841, pp. 316–334. Springer, Heidelberg (2011)
39. Whitnall, C., Oswald, E.: A Fair Evaluation Framework for Comparing Side-Channel Distinguishers. J. Cryptographic Engineering 1(2), 145–160 (2011)
40. Whitnall, C., Oswald, E., Mather, L.: An Exploration of the Kolmogorov-Smirnov Test as Competitor to Mutual Information Analysis. In: Prouff, E. (ed.) CARDIS 2011. LNCS, vol. 7079, pp. 234–251. Springer, Heidelberg (2011)

SCARE of Secret Ciphers with SPN Structures

Matthieu Rivain[1] and Thomas Roche[2]

[1] CryptoExperts, France
matthieu.rivain@cryptoexperts.com
[2] ANSSI, Fance
thomas.roche@ssi.gouv.fr

Abstract. Side-Channel Analysis (SCA) is commonly used to recover secret keys involved in the implementation of publicly known cryptographic algorithms. On the other hand, Side-Channel Analysis for Reverse Engineering (SCARE) considers an adversary who aims at recovering the secret design of some cryptographic algorithm from its implementation. Most of previously published SCARE attacks enable the recovery of some secret parts of a cipher design –*e.g.* the substitution box(es)– assuming that the rest of the cipher is known. Moreover, these attacks are often based on idealized leakage assumption where the adversary recovers noise-free side-channel information. In this paper, we address these limitations and describe a generic SCARE attack that can recover the full secret design of any iterated block cipher with common structure. Specifically we consider the family of Substitution-Permutation Networks with either a classical structure (as the AES) or with a Feistel structure. Based on a simple and usual assumption on the side-channel leakage we show how to recover all parts of the design of such ciphers. We then relax our assumption and describe a practical SCARE attack that deals with noisy side-channel leakages.

1 Introduction

Side-Channel Analysis for Reverse Engineering (SCARE) refers to a set of techniques that exploit side-channel information to recover secret algorithms and/or software/hardware designs. One of the main application of SCARE is the recovery of symmetric ciphering algorithms of private design, as often used in Pay-TV and GSM authentication protocols. The first SCARE attack in this context was introduced by Novak [25], who showed how to recover one out of two s-boxes from a secret instance of A3/A8 algorithm (used in GSM protocol). This work was subsequently improved by Clavier [10] who described how to recover both s-boxes altogether with the secret key used by the cipher. In parallel to these results, Daudigny *et al.* [13] showed that simple secret modifications of the DES cipher could also be recovered from side-channel observations. In a more recent work, Réal *et al.* [27] took a closer look at Feistel schemes in a more general sense. They showed how an adversary that gets the Hamming weight of some intermediate result can interpolate the round function of the cipher. Eventually,

K. Sako and P. Sarkar (Eds.) ASIACRYPT 2013 Part I, LNCS 8269, pp. 526–544, 2013.

a SCARE attack on stream ciphers was proposed by Guilley *et al.* [19]. They showed how to retrieve the overall design when either the linear or the nonlinear part of the cipher is known.

Our Contribution. In this paper, we introduce a SCARE attack that recovers the full secret design of an iterated Substitution-Permutation Network (SPN for short), namely an iterated cipher composed of substitution boxes (or s-boxes), linear layers and key additions. As in [25,10], our attack is based on the simple assumption that the side-channel leakage enables the detection of colliding s-box computations. Specifically, the attacker is able to select strips of side-channel traces where the s-box computations are located and decide on collisions between the processed values from the observation of these traces. This assumption has been the basis of various previously published side-channel key-recovery attacks (see for instance [32,31,3,4,2,6,5,24,11,17]). We first show how a perfect detection of colliding s-box computations enables an efficient recovery of a secret cipher with *classical* SPN structure as the one of the AES [14]. Roughly speaking, the collision detection mechanism allows us to build simple linear equation systems involving the different unknowns of the cipher algorithm (*i.e.* the s-box values, the linear layer coefficients, the secret round key coordinates). In the full version of the paper [29], we further extend our basic attack to relax as much as possible the constraints on the design, allowing several different s-boxes, binary linear layers, and Feistel structures, so that we cover a wide spectrum of usual block cipher designs. In the second part of this paper, we address the practical aspects of our attack and relax the perfect detection assumption. We describe a practical SCARE attack working in the presence of noise in the side-channel leakage and we provide experimental results showing its practicability.

Related Work. In a recent independent work [12], Clavier *et al.* present a SCARE attack against AES-like block ciphers. The authors consider a chosen-plaintext and known-ciphertext scenario with perfect detection of colliding s-boxes. Under these assumptions, they show how to efficiently recover the secret parameters of a modified AES. They further address the case of protected implementations with common software countermeasures against side-channel attacks. In comparison, our attack targets a wider class of SPN ciphers, including modified AES ciphers as a particular case. Moreover, we extend our attack to deal with noisy leakages, hence relaxing the perfect detection assumption. However, we do not deal with the case of protected implementations (though we give a few insights about it in Section 6).

Paper Organization. In the first section we describe the design of target SPN block ciphers. Then we present our generic SCARE attack in Section 3. The practical SCARE attack dealing with noisy leakages is described in Section 4, and experimental results are presented in Section 5. Finally, we give some discussions and perspectives in Section 6.

2 Substitution-Permutation Networks

We consider a block cipher E computing an ℓ-bit ciphertext block c from an ℓ-bit plaintext block p through the repetition of a key-dependent permutation, called *round function* ρ. Each round is parameterized by a different round key k_i derived from the secret key k through a key scheduling process. Let r denote the number of rounds, the ciphertext block is then defined as

$$c = E_k(p) = \rho_{k_r} \circ \rho_{k_{r-1}} \circ \cdots \circ \rho_{k_1}(p) .$$

In an SPN block cipher, the round function is composed of linear permutations and nonlinear substitutions, and the key is introduced by addition. The addition and linearity are considered over the vectorial space $\mathbb{F}_2^\ell$. Namely round keys are introduced by a simple exclusive-or (XOR), and linear permutations are homomorphic with respect to the XOR operation. Non-linear substitutions are applied on small blocks of bits which are replaced by new blocks looked-up from a predefined table usually called *s-box* (for substitution-box). In what we shall call a *classical SPN structure*, the different s-boxes and linear transformations are bijective (*e.g.* the Advanced Encryption Standard [14]). But when they are not, it is common to use a so-called *Feistel scheme* in order to make the round function, and hence the overall cipher, invertible (*e.g.* the Data Encryption Standard [15]). In the following, we only focus on the classical SPN structures. Extension of our work to Feistel schemes is provided in the full version of the paper.

In a classical SPN structure, the plaintext is considered as a n-dimensional vector of m-bit coordinates: $p = (p_1, p_2, \ldots, p_n)$, with $\ell = nm$. The round function is composed of a key addition layer σ_{k_i}, a nonlinear layer γ, and a linear layer λ, that is

$$\rho_{k_i} = \lambda \circ \gamma \circ \sigma_{k_i} .$$

The key addition layer is a simple XOR-ing of the round key:

$$\sigma_k(p) = p \oplus k .$$

The nonlinear layer consists of the parallel application of an $m \times m$ s-box S:

$$\gamma(p) = (S(p_1), S(p_2), \ldots, S(p_n)) ,$$

And the linear layer is a linear transformation over $(\mathbb{F}_{2^m})^n$:

$$\lambda(p) = \begin{pmatrix} a_{1,1} & a_{1,2} & \cdots & a_{1,n} \\ a_{2,1} & a_{2,2} & \cdots & a_{2,n} \\ \vdots & \vdots & \ddots & \vdots \\ a_{n,1} & a_{n,2} & \cdots & a_{n,n} \end{pmatrix} \cdot \begin{pmatrix} p_1 \\ p_2 \\ \vdots \\ p_n \end{pmatrix} \tag{1}$$

where the a_{ij} and the p_j are considered as elements of $\mathbb{F}_{2^m}$.

Remark 1. The final round sometimes skips the linear layer and an additional key addition is often performed after the final nonlinear layer. The attack described in this paper works as well for these variants.

3 Basic SCARE of Classical SPN Structures

3.1 Attacker Model

We present a generic SCARE attack in a known-plaintext scenario, and we show how its complexity can be lowered in a chosen-plaintext scenario. Our attack does not require the knowledge of the ciphertext but only exploits the side-channel leakage of the cipher execution. Moreover, it is assumed that *colliding s-box computations can be detected from the side-channel leakage.* Specifically, we assume that the attacker is able to

(i) identify the s-box computations in the side-channel leakage trace and extract the leakage corresponding to each s-box computation,

(ii) decide whether two s-box computations $y_1 \leftarrow S(x_1)$ and $y_2 \leftarrow S(x_2)$ are such that $x_1 = x_2$ or not from their respective leakages.

Remark 2. This assumption implicitly means that the cipher implementation processes the s-box computations in a sequential way and that two s-box computations of the same input at two different points in the execution produce identical side-channel leakages. These constraints are further discussed in Section 6.

Under the above assumption, the attacker can identify r different groups of n s-box computations, and hence recover the number r of rounds, the number n of s-boxes per round and hence the s-box size $m = \ell/n$, where ℓ is the block size. We will therefore assume these parameters to be known in our attack description.

In what follows, we first show how the above assumption enables the complete recovery of a secret cipher with SPN structure as described in Section 2. In Section 4, we relax this assumption and extend our attack to deal with noisy leakages which can lead to decision errors in the collision detections.

3.2 Equivalent Representations

Several equivalent representations are possible for an SPN cipher such as considered here. For instance one can change the s-box S for the s-box S' defined as

$$S'(x) = S(x \oplus \delta)$$

for some $\delta \in \mathbb{F}_{2^m}$, and replace every round key $k_i = (k_{i,1}, k_{i,2}, \dots, k_{i,n})$ by

$$k_i' = (k_{i,1} \oplus \delta, k_{i,2} \oplus \delta, \dots, k_{i,n} \oplus \delta) .$$

The two representations are clearly equivalent in a functional sense. Moreover, the ability of detecting collisions in s-box computations does not make it possible to distinguish between two different equivalent representations.

Another way to obtain equivalent representations is by changing the s-box S for the s-box S' defined as

$$S'(x) = \alpha \cdot S(x)$$

for some $\alpha \in \mathbb{F}_{2^m}^*$, and by replacing the linear layer λ defined in (1) by the linear layer λ' obtained from the matrix $(a'_{i,j})_{i,j}$ whose coefficients satisfy

$$a'_{i,j} = a_{i,j}/\alpha$$

for every (i, j).

In our attack, we fix the first round key coordinate $k_{1,1}$ to 0 and we fix the coefficient $a_{1,1}$ to 1, which is equivalent to fixing the variables δ and α. Note that $a_{1,1}$ may equal 0 (which is revealed by the attack), in which case we try fixing $a_{1,2}$, then $a_{1,3}$, and so on. We describe hereafter the successive stages of the attack.

3.3 Stage 1: Recovering k_1

Since we have fixed $k_{1,1} = 0$, we aim to recover the $n - 1$ remaining subkeys $k_{1,2}, k_{1,3}, \ldots, k_{1,n}$. Let $\mathcal{I}$ denote the set of indices i for which $k_{1,i}$ is known. At the beginning of the attack $\mathcal{I} = \{1\}$. Then for any collision $[y_i \leftarrow S(p_i \oplus k_{1,i})] \sim [y_j \leftarrow S(p_j \oplus k_{1,j})]$ for some $i \in \mathcal{I}$, one deduces

$$k_{1,j} = p_j \oplus p_i \oplus k_{1,i} \ ,$$

and the index j is added to $\mathcal{I}$. We expect to retrieve all subkeys with less than $2^{m/2}$ encryptions.

3.4 Stage 2: Recovering λ, S and k_2

Once k_1 has been recovered, one knows the inputs of the s-box in the first round. Let us define $x_i = S(i)$ for every $i \in \{0, 1, \ldots, 2^m - 1\}$, so that recovering the s-box means recovering the 2^m unknowns $x_0, x_1, \ldots, x_{2^m-1}$. The attack consists in constructing a set of equations in the x_i's, the $a_{i,j}$'s and the $k_{2,i}$'s. Solving the obtained system hence amounts to recover λ, S and k_2.

The first step of this stage consists in collecting the leakages ℓ_β from s-box computations $\mu \leftarrow S(\beta)$ for every $\beta \in \mathbb{F}_{2^m}$. We shall denote by $\mathcal{B}$ the obtained leakage basis $\{\ell_\beta \mid \beta \in \mathbb{F}_{2^m}\}$. Such a basis can be constructed since k_1 is known from the first stage, hence the inputs of the s-box computations in the first round are known. This basis is then used to detect collisions between s-box computations in the second round and s-box computations $\mu \leftarrow S(\beta)$. Let w_j be the jth s-box input before key addition in the second round (*i.e.* w_j is the jth m-bit output of the first round), in the encryption of some plaintext p. Then w_i satisfies

$$w_i = a_{i,1}\, x_{j_1} \oplus a_{i,2}\, x_{j_2} \oplus \cdots \oplus a_{i,n}\, x_{j_n} \ ,$$

where $j_t = p_t \oplus k_{1,t}$ is a known index. If the corresponding s-box computation $y_i \leftarrow S(w_i \oplus k_{2,i})$ collides with some s-box computation $\mu \leftarrow S(\beta)$ from $\mathcal{B}$, then we get the following quadratic equation

$$a_{i,1}\, x_{j_1} \oplus a_{i,2}\, x_{j_2} \oplus \cdots \oplus a_{i,n}\, x_{j_n} \oplus k_{2,i} = \beta \ .$$

Once several such equations have been collected, one can solve the system and recover all the unknowns (*i.e.* the x_i's, the $a_{i,j}$'s and the $k_{2,i}$'s).

Solving the System. In order to solve the quadratic system obtained from all the collected equations, one can use the linearization method. The monomial $a_{i,j}\, x_u$ is replaced by a new unknown y_t for every triplet $t \equiv (i, j, u)$ where $1 \le i, j \le n$ and $0 \le u \le 2^m - 1$. We get a linear system with $2^m n^2 + n$ unknowns (the y_t and the $k_{2,i}$), which can be solved based on $2^m n^2 + n$ independent equations. Since every encryption provides n new equations, the required number of encryptions is $2^m n + 1$.

However, using linearization is not mandatory and we show hereafter that the system can be directly rewritten as a linear system. To do so, we consider the n equations obtained for the different s-box computations at the same time. Let $\beta_1, \beta_2, \ldots, \beta_n$ be the values such that $y_i \leftarrow S(w_i \oplus k_{2,i})$ collides with $\mu_i \leftarrow S(\beta_i)$. The obtained system for the n equations can be written in matrix form as

$$A \cdot \boldsymbol{x} \oplus \boldsymbol{k_2} = \boldsymbol{\beta} \ ,$$

where $A = (a_{i,j})_{i,j}$, $\boldsymbol{x} = (x_{j_1}, x_{j_2}, \ldots, x_{j_n})^T$, $\boldsymbol{k_2} = (k_{2,1}, k_{2,2}, \ldots, k_{2,n})^T$ and $\boldsymbol{\beta} = (\beta_1, \beta_2, \ldots, \beta_n)^T$. Since λ is invertible, we have

$$\boldsymbol{x} \oplus A^{-1} \cdot \boldsymbol{k_2} = A^{-1} \cdot \boldsymbol{\beta} \ .$$

Let $\boldsymbol{k_2'} = (k'_{2,1}, k'_{2,2}, \ldots, k'_{2,n})$ denote the vector resulting from the product $A^{-1} \cdot \boldsymbol{k_2}$ and let $a'_{i,j}$ denote the coefficients of A^{-1}. We obtained the n following equations:

$$\begin{aligned} x_{j_1} \oplus k'_{2,1} &= a'_{1,1}\, \beta_1 \oplus a'_{1,2}\, \beta_2 \oplus \cdots \oplus a'_{1,n}\, \beta_n \ , \\ x_{j_2} \oplus k'_{2,2} &= a'_{2,1}\, \beta_1 \oplus a'_{2,2}\, \beta_2 \oplus \cdots \oplus a'_{2,n}\, \beta_n \ , \\ &\ \ \vdots \\ x_{j_n} \oplus k'_{2,n} &= a'_{n,1}\, \beta_1 \oplus a'_{n,2}\, \beta_2 \oplus \cdots \oplus a'_{n,n}\, \beta_n \ . \end{aligned}$$

After collecting several such equations, we obtained a linear system with $n^2 + n + 2^m$ unknowns: the x_i's, the $a'_{i,j}$'s and the $k'_{2,i}$'s. This system can hence be solved based on $n^2 + n + 2^m$ independent equations. Since every encryption provides n new equations, the required number of encryptions is at least $n + 1 + 2^m/n$. Once all the $a'_{i,j}$'s and the $k'_{2,i}$'s have been recovered, we can inverse the matrix A^{-1} to get λ and then compute $\boldsymbol{k_2} = A \cdot \boldsymbol{k_2'}$.

As explained in Section 3.2, we must fix $a_{1,1} = 1$ in order to fix a representation among the equivalence class of the cipher. For the above system, this amounts to fixing $a'_{1,1} = 1$. Here again, $a'_{1,1}$ may equal 0 in which case the solving fails and the attacker must try again by fixing $a'_{1,2}$ and so on. Another degree of freedom exists that is not recovered by solving the above system: one can add a fixed offset δ to every s-box output and to every coordinate of $\boldsymbol{k_2'}$ (which amounts to add $A \cdot (\delta, \delta, \ldots, \delta)$ to $\boldsymbol{k_2}$). Clearly, such a modification would not change the collected equations. In order to set this degree of freedom, we can fix one of the s-box output, say x_0 to 0. To summarize, additionally to the collected n-equation groups from each encryption, we add the equations $a'_{1,1} = 1$ and $x_0 = 0$ in order to obtain a full rank system.

Note that fixing $x_0 = 0$ may induce a non-equivalent representation of the cipher. Indeed, the recovered cipher is equivalent to the real cipher but a fixed offset δ is xor-ed to each s-box outputs in the last round. As a consequence the resulting ciphertexts are xor-ed with the constant value $A \cdot (\delta, \delta, \ldots, \delta)$. Note that if a key-addition is performed after the nonlinear layer in the last round then its recovery absorbs this offset as for the other rounds. Otherwise, one must recover the offset δ in order to correct the ciphertext values and get an equivalent representation of the cipher. This can be easily done by comparing a real ciphertext with the one obtained from the recovered cipher.

Chosen Plaintexts Attack. To optimize the attack, one shall select the plaintexts in order to make every unknown of the system appear with the least possible number of requested encryptions. The $a'_{i,j}$'s and the $k'_{2,i}$'s all appear in each group of n equations resulting from a single encryption. On the other hand such a group of equations only involves n out of 2^m unknowns x_i's. The best approach is hence to make n different x_i's appear for each encryption request. To do so, one can simply ask for the encryption of the plaintext

$$(i \cdot n + 0, i \cdot n + 1, i \cdot n + 2, \ldots, (i + 1)n - 1) \oplus k_1 ,$$

for $i = 0, 1, \ldots, \lceil 2^m/n \rceil - 1$. The s-box inputs in the first round of the corresponding encryptions then equal $(0, 1, 2, \ldots, n - 1)$, $(n, n + 1, \ldots, 2n - 1)$, *etc.* Every possible s-box value thus appears in the system after $\lceil 2^m/n \rceil$ encryptions. It just remains to ask for the encryption of $n + 1$ additional plaintexts to get a full rank linear system in the $n^2 + n + 2^m$ unknowns.

3.5 Stage 3: Recovering k_3, k_4, ..., k_r

Once the two first stages have been completed, it only remains to recover the last round keys k_3, k_4, ..., k_r. This is simply done by detecting a collision $[y_i \leftarrow S(p_{j,i} \oplus k_{j,i})] \sim [\mu_{j,i} \leftarrow S(\beta_{j,i})]$ giving $k_{j,i} = p_{j,i} \oplus \beta_{j,i}$ for every round $j \in \{3, 4, \ldots, r\}$ and every s-box index $i \in \{1, 2, \ldots, n\}$.

4 SCARE in the Presence of Noisy Leakage

So far, we have considered an idealized model in which the attacker is able to detect a collision between two s-box computations from their respective leakages with a 100% confidence. As a matter of facts, the proposed SCARE attack do not tolerate any false-positive error in the collision detections. In this section, we relax this assumption and describe a practical SCARE attack in the presence of noise in the side-channel leakage. As for the basic attack, the principle is to exploit equations arising from collisions in s-box computations. We explain hereafter how to collect sound equations with high confidence in the presence of noisy leakage.

4.1 Stage 1: Recovering k_1

In our SCARE attack, the first stage exactly corresponds to the usual scenario of *linear collision attacks* that aim at recovering key bytes differences $k_{1,i} \oplus k_{1,j}$ by detecting collisions between s-box computations in the first round from the side-channel leakage [3,4,24,17].

In a linear collision attack, the attacker is assumed to possess the leakage traces corresponding to the encryption of N random plaintexts $((p_t)_{t \leq N})$. Let $\ell_{t,i}$ denote the leakage associated to ith s-box computation in the encryption of p_t. The principle is to compute the mean leakage $\bar{\ell}_{i,x}$ of the set $\{\ell_{t,i}\ ;\ p_{t,i} = x\}$ for every i and x, in order to average the leakage noise and detect collisions more easily. As explained in Section 3.3, detecting a collision between $\bar{\ell}_{i,x}$ and $\bar{\ell}_{j,y}$ implies the equality of the two s-box inputs $x \oplus k_{1,i}$ and $y \oplus k_{1,j}$ and provides the linear equation $k_{1,i} \oplus k_{1,j} = x \oplus y$. In [3], Bogdanov points out that the equation system arising from the key byte differences is overdetermined and that the redundant information could be used to tolerate some erroneous equations. In [17], Gérard and Standaert further show that solving such an equation system can be written as a LDPC[1] code decoding problem for which an efficient algorithm is known. We suggest to use their method for the first stage of our practical SCARE attack.

4.2 Stage 2: Recovering λ, S and k_2

As for the attack without collision errors, the second stage is the main task. To deal with the leakage noise, we make the well admitted *Gaussian noise assumption.* Namely, we assume that the leakage corresponding to an s-box computation $\mu \leftarrow S(\beta)$ follows a multivariate Gaussian distribution with mean m_β and covariance matrix Σ_β, denoted $\mathcal{N}(m_\beta, \Sigma_\beta)$.

Building Leakage Templates. The first step of the second stage consists in estimating the leakage parameters. Namely, for each $\beta \in \mathbb{F}_{2^m}$ we estimate the mean m_β and the covariance matrix Σ_β of the leakage from the s-box computation $\mu \leftarrow S(\beta)$. The leakage basis of the noise-free attack is then replaced by a leakage template basis $\mathcal{B} = \{(\widehat{m}_\beta, \widehat{\Sigma}_\beta)_\beta \mid \beta \in \mathbb{F}_{2^m}\}$ where $\widehat{m}_\beta$ and $\widehat{\Sigma}_\beta$ denote the estimated values for the leakage parameters. The estimation is obtained from the leakages used in the first stage, and possibly more, until the estimated means converge.

Our convergence criterion is based on the Hotelling T^2-test which is the natural extension of the Student T-test for multinormal distributions (see for instance [21]). Let d denote the dimension of the distribution $\mathcal{N}(m_\beta, \Sigma_\beta)$ *i.e.* the number of points in an s-box leakage trace, and let $F^{-1}_{(d_1,d_2)}$ denote the quantile function of the Fisher's F-distribution with parameters (d_1, d_2) (*i.e.* $F_{(d_1,d_2)}$ is the distribution CDF). For some confidence parameter $\alpha \in [0; 1]$ and some estimation quality parameter $q \in [0; 1]$, our convergence criterion is satisfied when we have:

[1] Low Density Parity Check

$$R_\alpha\Big(\frac{\widehat{\sigma}^2_\beta}{\det(\widehat{\mathbf{S}})}\Big)^{1/d} \leq q \qquad \text{where } R_\alpha := \frac{d}{N-d}\, F^{-1}_{(d,N-d)}(\alpha)\,.$$

The rationale of this definition is detailed in the full version of the paper.

Based on this criterion, the template basis is built iteratively: we first collect N leakage samples for every s-box input value β. Based on these samples, we estimate the distribution parameters $(\widehat{m}_\beta, \widehat{\Sigma}_\beta)$ for every β, as well as the inter-class covariance matrix $\widehat{\mathbf{S}}$. Then if we have $\max_\beta R_\alpha(\widehat{\sigma}^2_\beta / \det(\widehat{\mathbf{S}}))^{1/d} \leq q$ for some chosen confidence α and estimation quality parameter q we stop. Otherwise we continue with twice more samples (namely we collect N more leakage samples and set N to $2N$), and so on until we get a satisfying estimation quality. In practice, we shall use $\alpha = 99\%$ and $q = 0.5$.

Remark 3. A possible variant for building the template basis is to make the identical noise assumption which considers that Σ_β is equal to some constant matrix Σ for every β. This enables a better estimation $\widehat{\Sigma}$ based on all leakage samples.

Collecting Equations. Once the template basis has been built, we collect several groups of n equations of the form $\boldsymbol{x} \oplus \boldsymbol{k}'_2 = A^{-1} \cdot \boldsymbol{\beta}$, as in the basic attack (see Section 3.4). Due to the noise, we cannot determine the value of $\boldsymbol{\beta}$ with a 100% confidence. To deal with this issue we use averaging. Namely, the encryption of the same plaintext p is requested several – say N – times and we compute the average leakage for each s-box computation in the second round. Let ℓ_i denote the average leakage for the ith s-box, and let β^*_i denote the corresponding (unknown) s-box input. The average leakage ℓ_i follows a distribution $\mathcal{N}(m_{\beta^*_i}, \frac{1}{N}\Sigma_{\beta^*_i})$. Then we must recover the n corresponding values β^*_1, β^*_2, ..., β^*_n in order to get a group of equations. The problem is hence to determine to which distribution $\mathcal{N}(m_\beta, \frac{1}{N}\Sigma_\beta)$ belongs each leakage ℓ_i based on the template basis. For such a purpose, we use a maximum likelihood approach, namely we follow the classical approach of template attacks [9]. Given the leakage observation ℓ_i, the probability that the ith s-box input value β^*_i equals some value β satisfies

$$\Pr[\beta^*_i = \beta \mid \ell_i] = \frac{\phi_\beta(\ell_i)}{\sum_{\beta' \in \mathbb{F}_{2^m}} \phi_{\beta'}(\ell_i)}\,,$$

where ϕ_β denotes the pdf of $\mathcal{N}(m_\beta, \frac{1}{N}\Sigma_\beta)$ satisfying

$$\phi_\beta(\ell) \propto \exp\Big(-\frac{N}{2}(\ell - m_\beta)^T \cdot \Sigma^{-1}_\beta \cdot (\ell - m_\beta)\Big)\,.$$

The likelihood of the candidate β for β^*_i based on the estimations $(\widehat{m}_\beta)_\beta$ and $(\widehat{\Sigma}_\beta)_\beta$ is hence defined as

$$\mathsf{L}(\beta \mid \ell_i) := \frac{\exp\big(-\frac{N}{2}(\ell_i - \widehat{m}_\beta)^T \cdot \widehat{\Sigma}^{-1}_\beta \cdot (\ell_i - \widehat{m}_\beta)\big)}{\sum_{\beta' \in \mathbb{F}_{2^m}} \exp\big(-\frac{N}{2}(\ell_i - \widehat{m}_{\beta'})^T \cdot \widehat{\Sigma}^{-1}_{\beta'} \cdot (\ell_i - \widehat{m}_{\beta'})\big)}\,. \tag{2}$$

The corresponding likelihood for a vector $\boldsymbol{\beta} = (\beta_1, \beta_2, \ldots, \beta_n)$ given the average leakage vector $\boldsymbol{\ell} = (\ell_1, \ell_2, \ldots, \ell_n)$ can then be defined as $\mathsf{L}(\boldsymbol{\beta} \mid \boldsymbol{\ell}) := \prod_i \mathsf{L}(\beta_i \mid \ell_i)$. Note that the most likely candidate $\text{argmax}_{\boldsymbol{\beta}}\ \mathsf{L}(\boldsymbol{\beta} \mid \boldsymbol{\ell})$ is also the one whose coordinates are the most likely *i.e.* equal to $\text{argmax}_{\beta_i}\ \mathsf{L}(\beta_i \mid \ell_i)$ for every i.

In practice, we shall select the most likely value of $\boldsymbol{\beta}$ as the good one with a confidence $\mathsf{L}_{\boldsymbol{\beta}}$. However we not only want to select the best candidate, we further want its likelihood to be high (*i.e.* close to 1) in order to have a high confidence in the selected candidate. Getting a vector $\boldsymbol{\beta}$ with high likelihood may however be far more difficult than getting a single coordinate β_i with high likelihood since for the vector one needs all coordinates to have high likelihood. Indeed, the probability of having a high likelihood for the vector $\boldsymbol{\beta}$ is the probability of having a high likelihood for all coordinates β_i which is exponentially smaller in n.

To deal with this issue, our approach is to restrict the number of equations of the form $\boldsymbol{x} \oplus \boldsymbol{k}_2' = A^{-1} \cdot \boldsymbol{\beta}$ that are needed to succeed the attack. For such a purpose, we first solve a subsytem (*i.e.* with less unknowns) for which we require less equations than in the original attack, and then we recover the remaining unknowns based on simpler forms of equations.

Solving a Subsystem. We first solve a subsystem involving the $a'_{i,j}$'s, the $k_{2,i}$'s and a restricted number of x_i's. To do so we select a set of s values β, say $\mathcal{S} = \{0, 1, \ldots, s-1\}$, and we only request the device for the encryption of plaintexts from the set

$$\mathcal{P}_s = \{(p_1, p_2, \ldots, p_n) \mid \forall i : 0 \leq p_i \oplus k_{1,i} \leq s-1\} .$$

These plaintexts are such that all s-box inputs in the first encryption rounds are in $\mathcal{S}$. We hence obtained a linear system as described in Section 3.4 but with $n^2 + n + s - 2$ unknowns: the $a'_{i,j}$'s (but $a'_{1,1}$ which is set to 1), the $k'_{2,i}$'s, and the x_i's for $i \in \mathcal{S}$ (but x_0 which is set 0). Such a system can be solved based on $t = n + 1 + \lceil (s-2)/n \rceil$ good groups of equations. The value of s must hence be selected to ensure that the plaintext subspace $\mathcal{P}_s$ is large enough to get t good groups with high confidence, while making t the smallest possible.

In order to increase our chances to actually come up with t groups of correct equation, one direction would be to select a larger set of say q groups of equations (instead of only taking the t best) and test all combinations of t groups among them. The complexity of the resulting attack will however increase dramatically with q.

So, let us assume that we have a computing power of 2^k, meaning that we can try to solve 2^k linear systems, and that we can get the leakage measurement from T encryptions. Then our approach is to request N times for the encryption of T/N different plaintexts in $\mathcal{P}_s$. For each of the T/N plaintexts, we compute the more likely candidate $\boldsymbol{\beta}$ for the s-box inputs in the second round, based on the N-averaged leakages. We thus obtained T/N groups of n equations with a corresponding confidence (*i.e.* the likelihood of the best candidate $\boldsymbol{\beta}$). Then we select the q groups for which we get the highest confidence in the best candidate $\boldsymbol{\beta}$, where q is such that $\binom{q}{t} \approx 2^k$ (that is $q = c_0 t\, 2^{k/t}$ for some $c_0 \in [\mathrm{e}^{-1}; 1]$), and we try to solve each system arising from t of these q groups. In order to make sure that a found solution is the good one, we make the system over determined. This can be done without increasing the number t of needed equation groups. Namely, we take $s \leq n+2$ in order to get $t = n+1+\lceil (s-2)/n \rceil = n+2$. We thus obtain systems of $n^2 + 2n$ equations with $n^2 + n + s - 2$ unknowns. Obtaining a bad system that has a solution roughly occurs with probability $p_e \approx \left(\frac{1}{2^m}\right)^{n-s+2}$. So we take s to make this probability small, typically $s = n + 2 - 32/m$ giving $p_e \approx 2^{-32}$. For instance, for $n = 16$ and taking $s = 14$, we then have to select $t = 18$ good groups of equations from $|\mathcal{P}_s| \approx 2^{61}$ possible encryptions (which is quite enough). Another direction in increasing our chances of success would be to select the optimal averaging level.

Selecting the Averaging Level. We now explain how to select the averaging level N in order to optimize the success probability of the attack. Increasing the averaging level is good on the one hand to lower the noise and get better confidence in the recovered s-box inputs. On the other hand, the lower N, the greater the number T/N of different equation groups among which we can select the q best ones. To select a good tradeoff, we adopt the approach of [28] which estimates the success probability of an attack based on estimated leakage parameters. Namely we assume that the estimated paremeters $(m_\beta)_\beta$, and $(\Sigma_\beta)_\beta$ are the real leakage parameters and we simulate the attack accordingly. To simulate an attack, we fill two lists Succ and Fail by repeating the following steps:

1: $\boldsymbol{\beta}^* \xleftarrow{\$} (\mathbb{F}_{2^m})^n$
2: **for** $i = 1$ **to** n **do** $\ell_i \xleftarrow{\$} \mathcal{N}(\widehat{m}_{\beta_i^*}, \frac{1}{N}\widehat{\Sigma}_{\beta_i^*})$
3: $\mathsf{L}^{\max} \leftarrow \prod_i \max_\beta \mathsf{L}(\beta \mid \ell_i)$
4: **if** $\mathrm{argmax}_\beta \mathsf{L}(\beta \mid \ell_i) = \beta_i^*$ for every i
5: **then** add $\mathsf{L}^{\max}$ to Succ
6: **else** add $\mathsf{L}^{\max}$ to Fail

After iterating the above steps T/N times, one checks whether the q maximum values of $\mathsf{Succ} \cup \mathsf{Fail}$ include at least t value from Succ or not. In the affirmative, the simulated attack succeeded, otherwise it failed. Once the attack simulation has been performed several times, we obtain an estimation for the success probability of the attack.

Compared to a real attack experiment, the obtained success probability is affected by two differences: the actual leakage distributions $\mathcal{N}(m_{\beta_i^*}, \frac{1}{N}\Sigma_{\beta_i^*})$ are replaced by the estimated distributions $\mathcal{N}(\widehat{m}_{\beta_i^*}, \frac{1}{N}\widehat{\Sigma}_{\beta_i^*})$ and the distribution of the vector $\boldsymbol{\beta}^*$ of s-box inputs in the second round is replaced by the uniform distribution although it is not the case in practice since the plaintexts are randomly drawn from $\mathcal{P}_s$ instead of $\{0,1\}^\ell$. However for good estimations of the leakage parameters, we expect to get a good estimation of the trade-off of choice for averaging.

Recovering Remaining Unknowns. For the remaining unknowns x_s, x_{s+1}, $\ldots$, x_{2^m-1} we will here again use an iterative approach that recovers them one by one. For the sake of clarity, we assume that the linear layer is such that the matrix A has a column j_0 with no zero coefficients. Then our approach is to take a random plaintext p in $\mathcal{P}_s$ and to set its j_0th coordinates to $s \oplus k_{1,j_0}$ so that the j_0th s-box inputs equals s. By definition, the ith s-box input in the second round satisfies

$$\beta_i^* = a_{i,1}\, x_{t_1} \oplus a_{i,2}\, x_{t_2} \oplus \cdots \oplus a_{i,n}\, x_{t_n} \oplus k_{2,i} ,$$

where $t_j \leq s-1$ for every $j \neq j_0$ and $t_{j_0} = s$. This can be rewritten

$$\beta_i^* = a_{i,j}\, x_s \oplus k_{2,i} \oplus \bigoplus_{j \neq j_0} a_{i,j}\, x_{t_j} . \tag{3}$$

Since we know the values of the $a_{i,j}$'s, the $k_{2,i}$'s and the x_{t_j}'s for $t_j \leq s-1$, recovering x_s amounts to recovering β_i^*. And as we cannot recover β_i^* with a 100% success probability, we use a maximum likelihood approach.

Specifically, the likelihood of each candidate value $\omega \in \mathbb{F}_{2^m}$ for x_s is initialized to 0 if $\omega \in \{x_0, x_1, \ldots, x_{s-1}\}$ (indeed $x_s \notin \{x_0, x_1, \cdots, x_{s-1}\}$ as the s-box is bijective) and to $(2^m - s)^{-1}$ otherwise. Then the leakage ℓ_i resulting from each s-box computation is used to update the likelihood of each candidate for x_s. Namely, the likelihood $\mathsf{L}(\omega)$ of the candidate ω is multiplied by the likelihood of the candidate β_i^ω for the ith s-box input, where $\beta_i^\omega = a_{i,j}\,\omega \oplus k_{2,i} \oplus \bigoplus_{j \neq j_0} a_{i,j}\, x_{t_j}$ according to (3). Doing so for every s-box, $\mathsf{L}(\omega)$ is updated by

$$\mathsf{L}(\omega) \leftarrow \mathsf{L}(\omega) \times \prod_{i=1}^{n} \mathsf{L}(\beta_i^\omega \mid \ell_i) ,$$

where $\mathsf{L}(\cdot \mid \ell_i)$ is computed as in (2) with $N = 1$ (since we do not use averaging here). Eventually, the likelihood vector is normalized, that is all the coordinates are divided by $\sum_\omega \mathsf{L}(\omega)$. We iterate this process for several encryptions until one likelihood value $\mathsf{L}(\omega)$ get close enough to 1. Then we deduce $x_s = \omega$, and start again with x_{s+1}, and so on until x_{2^m-1}. Note that we can stop once x_{2^m-2} since a single value remains for x_{2^m-1}.

4.3 Stage 3: Recovering $k_3, k_4, \ldots, k_r$

Eventually, the last round keys can be recovered one by one by performing any classical side-channel key recovery attack (since we now know the design of the cipher). We suggest to use a maximum likelihood approach based on the template basis.[2]

5 Experiments

We report hereafter the results of various simulations of the practical SCARE attack described in the previous section. Each simulated attack aims at recovering a secret cipher with classical SPN structure (such as described in Section 2). We consider two different settings for the cipher dimensions:

[2] Such technique is well known and pretty similar to that used in the previous section so we do not detail it here.

- **the (128,8)-setting:** 128-bit message block and 8-bit s-boxes, as in the AES block cipher [14] (*i.e.* $\ell = 128$, $n = 16$, $m = 8$),
- **the (64,4)-setting:** 64-bit message block and 4-bit s-boxes, as in the LED [20] and PRESENT [7] lightweight block ciphers (*i.e.* $\ell = 64$, $n = 16$, $m = 4$).

For each attack experiment, a random secret cipher is picked up. Namely, we randomly generate a full-rank $n \times n$ matrix over $\mathbb{F}_{2^m}$, a bijective m-bit s-box, and several ℓ-bit round keys. The attack succeed if it recovers an equivalent representation of the generated cipher.

In order to evaluate our attack under a realistic leakage model, we have profiled the leakage of an 8-bit s-box computation on an AVR chip.[3] The side-channel leakage was captured by the means of an electromagnetic probe and a digital oscilloscope with a sampling rate of 1G sample per second. To infer a leakage model from the measurements we made the Gaussian and independent noise assumptions. We therefore estimated the mean leakage for every s-box input value and the mean leakage for every s-box output value based on 100000 leakage traces. We then selected three leakage points for the input and three leakage points for the output. We thus obtained 256 means $(m_{1,\beta}, m_{2,\beta}, m_{3,\beta})_\beta$ for the 256 possible input values $\beta \in \{0, 1, \ldots, 255\}$ and the 256 means $(m_{4,\mu}, m_{5,\mu}, m_{6,\mu})_\mu$ for the 256 possible output values $\mu \in \{0, 1, \ldots, 255\}$. Afterwards we estimated the noise covariance matrix Σ for the selected points (*i.e.* the matrix of covariances between the 6 points after subtracting the means). A preview of the obtained parameters can be found in the full version of the paper. In particular we get a multivariate SNR[4] of 0.033 and univariate SNRs[5] of 0.13, 0.033, 0.099, 0.058, 0.047, and 0.051, for the different leakage points. These inferred parameters provide us with a leakage model for our attack simulations. Namely, for a given cipher with s-box S, the leakage associated to the s-box computation with input β is randomly drawn from the multivariate Gaussian distribution $\mathcal{N}(m_\beta, \Sigma)$ with mean satisfying

$$m_\beta = (m_{1,\beta}, m_{2,\beta}, m_{3,\beta}, m_{4,S(\beta)}, m_{5,S(\beta)}, m_{6,S(\beta)}) .$$

Stage 1. For the recovering of k_1, we implemented the Gérard and Standaert method based on the normalized Euclidean distance. For the (128,8)-setting, we obtained a 100% success rate using a few thousands of leakage traces while for the (64,4)-setting a few hundreds were sufficient. We did not try to optimize this stage of the attack (in particular we did not use the Bayesian extension proposed in [17]) as it requires a very small amount of leakage traces compared to the next stage.

---

[3] ATMega 32A, 8-bit architecture, 8Mz.

[4] The multivariate SNR is defined as the ratio of the interclass generalized variance (*i.e.* the determinant of the leakage means covariance matrix) over the intraclass generalized variance (*i.e.* the determinant of the noise covariance matrix) to the power $1/d$ (where d is the dimension equal to 6 in our case).

[5] The univariate SNR is defined as the variance of the means over the variance of the noise.

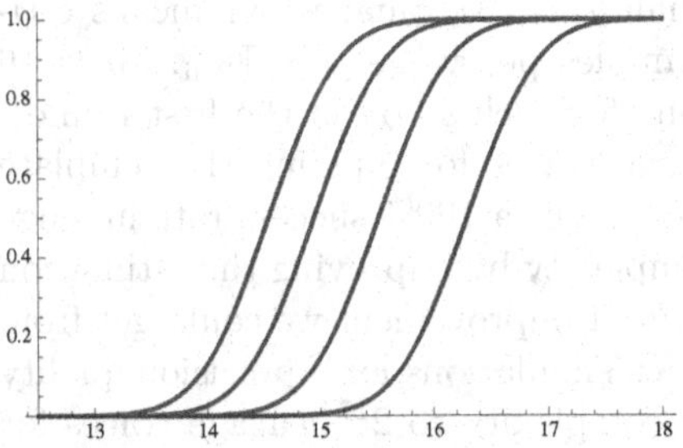

Fig. 1. Stage 2.1 for the (128,8)-setting: success rate over an increasing number of leakage traces (in $\log_2$-scale) for a computing power of 2^k with $k \in \{0, 1, 8, 32\}$

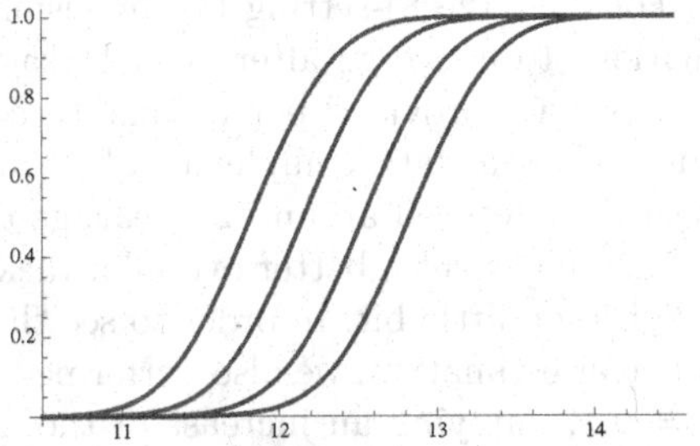

Fig. 2. Stage 2.1 for the (64,4)-setting: success rate of stage 2.1 over an increasing number of leakage traces (in $\log_2$-scale) for a computing power of 2^k with $k \in \{0, 1, 8, 32\}$

Stage 2.1. For this stage (recovery of λ, k_2, $S(0)$, $S(1)$, ..., $S(s-1)$) we fixed the number s of s-box outputs in the system to 14 for the (128,8)-setting and to 10 for the (64,4)-setting (according to the suggested formula $s = n+2-32/m$). For both settings, we chose a precision quality parameter $q = 0.5$ for the building of the template basis and we simulated the attack for a computing power of 2^k with $k \in \{0, 8, 16, 32\}$ (*i.e.* 2^k systems among the likeliest ones are tested). The obtained success rates are plotted in Figure 1 for the (128,8)-setting and in Figure 2 for the (64,4)-setting. Each curve represents a different computing power. Naturally the leftmost curves (*i.e.* the most successful) correspond to the 2^{32} computing power and the rightmost ones to the 2^0 computing power. As one can see, with a reasonable computing power, a 100% success rate is reached with less than 2^{16} leakage traces for the (128,8)-setting, and with less than 2^{13} leakage traces for the (64,4)-setting.

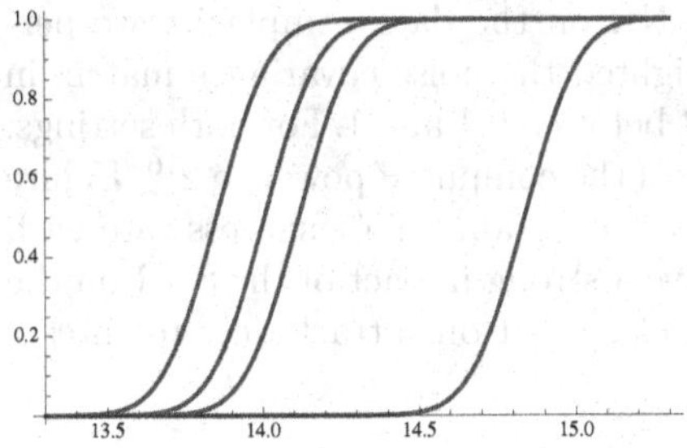

Fig. 3. Stage 2.1 for the (128,8)-setting: success rate over an increasing number of leakage measurements (in $\log_2$-scale) for a estimation quality $q = 0.1$

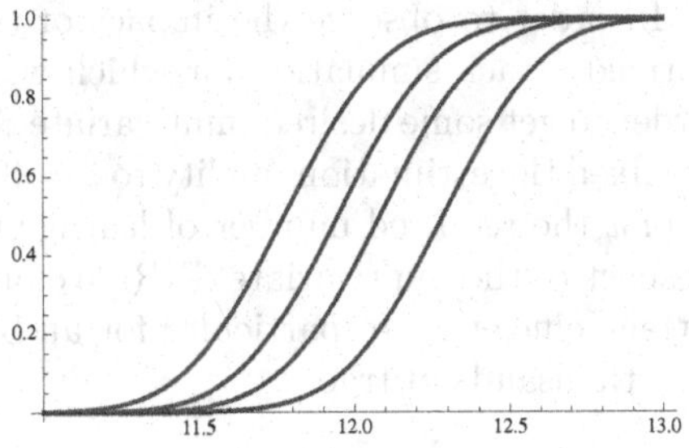

Fig. 4. Stage 2.1 for the (64,4)-setting: success rate over an increasing number of leakage measurements (in $\log_2$-scale) for a estimation quality $q = 0.1$

For the (128,8)-setting the precision quality $q = 0.5$ makes our means estimations to converge after 1024 leakage samples per value $\beta \in \mathbb{F}_{256}$. Since 16 samples are provided per leakage trace (one for each s-box in the first round), this makes a data complexity of 2^{14} leakage traces for building the template basis. As we need around 2^{16} leakage traces to get a 100% success rate in stage 2.1 we might get a better overall attack complexity by improving the estimation precision a little bit. In order to see the kind of improvement we could get from a better estimation, we also performed attack simulations for a precision quality $q = 0.1$, implying an increase of the data complexity to 2^{17} leakage traces for the template basis. The obtained success rates are given in Figure 3. We get a 100% success rate with between 2^{14} and $2^{14.5}$ leakage traces for all computing powers except for $k = 0$ which requires 2^{15} traces.

For the (64,4)-setting, the estimated means converge after 2048 samples per value $\beta \in \mathbb{F}_{16}$, making a data complexity of 2048 for template basis. Here again we also performed attack simulations for a precision quality of $q = 0.1$ (see results in Figure 4). We get a data complexity of 2^{13} leakage traces for the template basis and around $2^{12.5}$ leakage traces for the system solving. This precision therefore seems to give the best tradeoff for the (64,4)-setting.

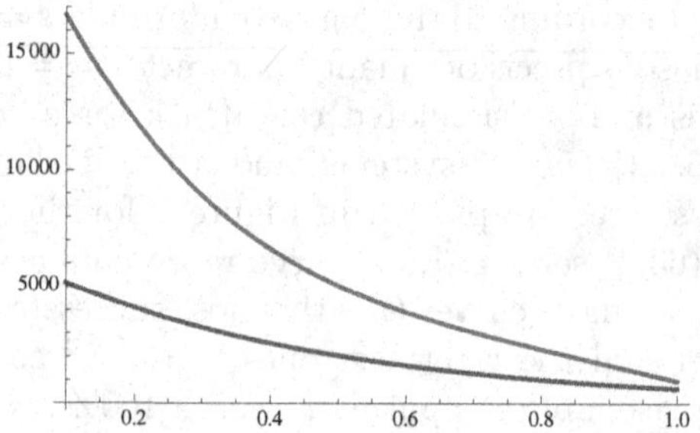

Fig. 5. Number of leakage traces to get a 90% success rate over an increasing SNR in [0.1; 1] for the (128,8)-setting (green curve) and the (64,4)-setting (red curve)

In order to observe the impact of the SNR on the data complexity we performed attack simulation for which we weighted the noise covariance matrix in order to get some desired multivariate SNR between 0.1 and 1. For both settings, we fixed the estimation quality to $q = 0.5$ and the computed power to 2^{16}. Figure 5 plot the required number of leakage traces to obtain a 90% success rate with respect to the multivariate SNR. We observe a strong impact of the SNR on the attack efficiency. In particular for an SNR close to 1 our attack only requires a few thousands of traces.

Stage 2.2 and 3. The recovery of the remaining s-box outputs based on the maximum likelihood approach is very efficient. Taking a lower bound of 0.999 on the likelihood to decide that a candidate is the good one, the attack stops after 640 leakage traces on average and reaches a 97% success rate for the (128,8)-setting (a tighter likelihood bound would yield a 100% success rate). For the

(64,4)-setting, it stops after 10 leakage traces on average and reaches a 100% success rate. The high efficiency of the attack for the (64,4)-setting comes from the fact that it only has to recover 6 remaining s-box outputs. Therefore the likelihoods quickly converge.

We did not implement attack simulation for the third step but we would clearly get comparable figures than for stage 2.2, *i.e.* negligible data requirements compared to stage 2.1 which is clearly the bottleneck of our attack.

6 Discussions and Perspectives

In this paper we have described a generic SCARE attack against a wide class of SPN block ciphers. The attacker model defined in Section 3.1 assumes that colliding s-box computations can be detected from the side-channel leakage. We have first investigated the case of perfect collision detection and then we have extended our attack to deal with noisy leakages.

About the Attacker Model. As mentioned in Section 3.1 (Remark 2), our attacker model implicitly means that the cipher implementation processes the s-box computations in a sequential way, which is therefore more suited for software implementations. This makes sense for secret ciphers which are rarely implemented at the hardware level. Note that it is also common to use a sequential approach for the s-box computations in light-weight hardware implementations of block ciphers, and our attack naturally applies to this context. Our model further implicitly assumes that two s-box computations with the same input at two different points in the execution produce identical side-channel leakages (or identically distributed in the noisy context). Although this assumption seems fair in practice, it might not always be satisfied. It was for instance observed in [17,30] that for some software implementations the side-channel leakage of an s-box computation may vary according to the s-box index and the target register. For such implementations, it might not be possible to detect collisions between two s-box computations at different indices. This issue can be addressed by considering each s-box index independently, which amounts to deal with the *multiple s-boxes setting* studied in the full version of the paper (except that we need to recover a single s-box). In this context, one only detects collisions between s-box computations at the same index. Note that our attack still assumes that s-box computations at a given index leak identically in the successive rounds.

Countermeasures to Our Attack. Our work shows that under a practically relevant assumption, it is possible to retrieve the complete secret design of a block cipher with a common SPN structure. This clearly emphasizes that the secrecy of the design is not sufficient to prevent side-channel attacks, and that one should include countermeasures to the implementation of secret ciphers as well. A typical choice for block cipher implementations in software is to use masking with table recomputation for the s-box (see for instance [23,1]). As studied by Roche and Lomné in [30], such a countermeasure only prevents collision detections between different cipher executions but it still allows the detection of intra-execution collisions. In a variant of their attack against AES-like secret ciphers, Clavier *et al.* take this constraint into account in order to bypass the masking countermeasure with table recomputation [12]. Our attack in the idealized leakage model (perfect collision detection) could also be extended to work with this constraint. It would be more tricky in the presence of noise as averaging would not be an option anymore, but our attack could still be generalized using a similar approach as [30]. In order to thwart our attack, one should therefore favor masking schemes enabling the use of different masks for the different s-box computations (see for instance [26,8]), so that intra-execution collisions would not be detectable anymore. Another common software countermeasure is operation shuffling (see for instance [22]). This countermeasure has a direct impact on our attack as it randomizes the indices of the s-box computations from one execution to another. As shown by Clavier *et al.* [12], such a countermeasure can be simply bypassed in the idealized leakage model. However, it seems more complicated to deal with in a noisy leakage model especially if combined with masking. We therefore suggest to use such a combination of countermeasure against our attack.

Perspectives. Our work opens several interesting issues for further research. First, our attack could probably be improved by using better/optimal approaches to solve the set of noisy equations arising in Stage 2.1 (see Section 4.2). One could for instance follow the approach of [18,16] by rewriting the system as a decoding problem. Our attack could also be improved by considering a known ciphertext scenario (as *e.g.* done in [12]). On the other hand, our attack was only validated by simulations (although from a practically inferred leakage model). It would be interesting to mount the attack against a real implementation of a secret SPN cipher *e.g.* on a smart card, to check how the different steps work in practice. Another interesting direction would be to investigate extensions of our attack against protected implementations in order to determine to what extent an implementation should be protected in practice.

Acknowledgements. This work has been financially supported by the French national FUI12 project MARSHAL+ (Mechanisms Against Reverse-Engineering for Secure Hardware and Algorithms). We would like to thank Victor Lomné for providing the microcontroller side-channel traces and the anonymous reviewers for their useful comments.

References

1. Akkar, M.-L., Giraud, C.: An Implementation of DES and AES, Secure against Some Attacks. In: Koç, Ç.K., Naccache, D., Paar, C. (eds.) CHES 2001. LNCS, vol. 2162, pp. 309–318. Springer, Heidelberg (2001)

2. Biryukov, A., Khovratovich, D.: Two New Techniques of Side-Channel Cryptanalysis. In: Paillier, P., Verbauwhede, I. (eds.) CHES 2007. LNCS, vol. 4727, pp. 195–208. Springer, Heidelberg (2007)
3. Bogdanov, A.: Improved Side-Channel Collision Attacks on AES. In: Adams, C., Miri, A., Wiener, M. (eds.) SAC 2007. LNCS, vol. 4876, pp. 84–95. Springer, Heidelberg (2007)
4. Bogdanov, A.: Multiple-Differential Side-Channel Collision Attacks on AES. In: Oswald, E., Rohatgi, P. (eds.) CHES 2008. LNCS, vol. 5154, pp. 30–44. Springer, Heidelberg (2008)
5. Bogdanov, A., Kizhvatov, I.: Beyond the Limits of DPA: Combined Side-Channel Collision Attacks. IEEE Trans. Computers 61(8), 1153–1164 (2012)
6. Bogdanov, A., Kizhvatov, I., Pyshkin, A.: Algebraic Methods in Side-Channel Collision Attacks and Practical Collision Detection. In: Chowdhury, D.R., Rijmen, V., Das, A. (eds.) INDOCRYPT 2008. LNCS, vol. 5365, pp. 251–265. Springer, Heidelberg (2008)
7. Bogdanov, A.A., Knudsen, L.R., Leander, G., Paar, C., Poschmann, A., Robshaw, M., Seurin, Y., Vikkelsoe, C.: PRESENT: An Ultra-Lightweight Block Cipher. In: Paillier, P., Verbauwhede, I. (eds.) CHES 2007. LNCS, vol. 4727, pp. 450–466. Springer, Heidelberg (2007)
8. Carlet, C., Goubin, L., Prouff, E., Quisquater, M., Rivain, M.: Higher-Order Masking Schemes for S-Boxes. In: Canteaut, A. (ed.) FSE 2012. LNCS, vol. 7549, pp. 366–384. Springer, Heidelberg (2012)
9. Chari, S., Rao, J.R., Rohatgi, P.: Template Attacks. In: Kaliski Jr., B.S., Koç, Ç.K., Paar, C. (eds.) CHES 2002. LNCS, vol. 2523, pp. 13–28. Springer, Heidelberg (2003)
10. Clavier, C.: An Improved SCARE Cryptanalysis Against a Secret A3/A8 GSM Algorithm. In: McDaniel, P., Gupta, S.K. (eds.) ICISS 2007. LNCS, vol. 4812, pp. 143–155. Springer, Heidelberg (2007)
11. Clavier, C., Feix, B., Gagnerot, G., Roussellet, M., Verneuil, V.: Improved Collision-Correlation Power Analysis on First Order Protected AES. In: Preneel, B., Takagi, T. (eds.) CHES 2011. LNCS, vol. 6917, pp. 49–62. Springer, Heidelberg (2011)
12. Clavier, C., Isorez, Q., Wurcker, A.: Complete SCARE of AES-like Block Ciphers by Chosen Plaintext Collision Power Analysis. In: INDOCRYPT 2013 (to Appear, 2013)
13. Daudigny, R., Ledig, H., Muller, F., Valette, F.: SCARE of the DES. In: Ioannidis, J., Keromytis, A.D., Yung, M. (eds.) ACNS 2005. LNCS, vol. 3531, pp. 393–406. Springer, Heidelberg (2005)
14. FIPS PUB 197. Advanced Encryption Standard. National Bureau of Standards (November 2001)
15. FIPS PUB 46. The Data Encryption Standard. National Bureau of Standards (January 1977)
16. Fourquet, R., Loidreau, P., Tavernier, C.: Finding good linear approximations of block ciphers and its application to cryptanalysis of reduced round DES. In: The 6th International Workshop on Coding and Cryptography (WCC 2009), Ullensvang, Norvège (May 2009)
17. Gérard, B., Standaert, F.-X.: Unified and Optimized Linear Collision Attacks and Their Application in a Non-profiled Setting. In: Prouff, E., Schaumont, P. (eds.) CHES 2012. LNCS, vol. 7428, pp. 175–192. Springer, Heidelberg (2012)

18. Gérard, B., Standaert, F.-X.: Unified and optimized linear collision attacks and their application in a non-profiled setting: extended version. J. Cryptographic Engineering 3(1), 45–58 (2013)
19. Guilley, S., Sauvage, L., Micolod, J., Réal, D., Valette, F.: Defeating Any Secret Cryptography with SCARE Attacks. In: Abdalla, M., Barreto, P.S.L.M. (eds.) LATINCRYPT 2010. LNCS, vol. 6212, pp. 273–293. Springer, Heidelberg (2010)
20. Guo, J., Peyrin, T., Poschmann, A., Robshaw, M.: The LED Block Cipher. In: Preneel, B., Takagi, T. (eds.) CHES 2011. LNCS, vol. 6917, pp. 326–341. Springer, Heidelberg (2011)
21. Härdle, W., Simar, L.: Applied Multivariate Statistical Analysis. Springer (2003)
22. Herbst, C., Oswald, E., Mangard, S.: An AES Smart Card Implementation Resistant to Power Analysis Attacks. In: Zhou, J., Yung, M., Bao, F. (eds.) ACNS 2006. LNCS, vol. 3989, pp. 239–252. Springer, Heidelberg (2006)
23. Messerges, T.S.: Securing the AES Finalists Against Power Analysis Attacks. In: Schneier, B. (ed.) FSE 2000. LNCS, vol. 1978, pp. 150–164. Springer, Heidelberg (2001)
24. Moradi, A., Mischke, O., Eisenbarth, T.: Correlation-Enhanced Power Analysis Collision Attack. In: Mangard, S., Standaert, F.-X. (eds.) CHES 2010. LNCS, vol. 6225, pp. 125–139. Springer, Heidelberg (2010)
25. Novak, R.: Side-Channel Attack on Substitution Blocks. In: Zhou, J., Yung, M., Han, Y. (eds.) ACNS 2003. LNCS, vol. 2846, pp. 307–318. Springer, Heidelberg (2003)
26. Prouff, E., Rivain, M.: A Generic Method for Secure SBox Implementation. In: Kim, S., Yung, M., Lee, H.-W. (eds.) WISA 2007. LNCS, vol. 4867, pp. 227–244. Springer, Heidelberg (2008)
27. Réal, D., Dubois, V., Guilloux, A.-M., Valette, F., Drissi, M.: SCARE of an Unknown Hardware Feistel Implementation. In: Grimaud, G., Standaert, F.-X. (eds.) CARDIS 2008. LNCS, vol. 5189, pp. 218–227. Springer, Heidelberg (2008)
28. Rivain, M.: On the Exact Success Rate of Side Channel Analysis in the Gaussian Model. In: Avanzi, R.M., Keliher, L., Sica, F. (eds.) SAC 2008. LNCS, vol. 5381, pp. 165–183. Springer, Heidelberg (2009)
29. Rivain, M., Roche, T.: SCARE of Secret Ciphers with SPN Structures. Cryptology ePrint Archive (2013), `http://eprint.iacr.org/`
30. Roche, T., Lomné, V.: Collision-correlation attack against some 1st-order boolean masking schemes in the context of secure devices. In: Prouff, E. (ed.) COSADE 2013. LNCS, vol. 7864, pp. 114–136. Springer, Heidelberg (2013)
31. Schramm, K., Leander, G., Felke, P., Paar, C.: A Collision-Attack on AES (Combining Side Channel and Differential-Attack). In: Joye, M., Quisquater, J.-J. (eds.) CHES 2004. LNCS, vol. 3156, pp. 163–175. Springer, Heidelberg (2004)
32. Schramm, K., Wollinger, T., Paar, C.: A New Class of Collision Attacks and Its Application to DES. In: Johansson, T. (ed.) FSE 2003. LNCS, vol. 2887, pp. 206–222. Springer, Heidelberg (2003)

Author Index